ANNUAL REVIEW OF ECOLOGY AND SYSTEMATICS

ANNUAL REVIEW OF ECOLOGY AND SYSTEMATICS

VOLUME 15, 1984

RICHARD F. JOHNSTON, *Editor*

University of Kansas

PETER W. FRANK, *Associate Editor*

University of Oregon

CHARLES D. MICHENER, *Associate Editor*

University of Kansas

ANNUAL REVIEWS INC.   4139 EL CAMINO WAY   PALO ALTO, CALIFORNIA 94306 USA

R ANNUAL REVIEWS INC.
Palo Alto, California, USA

International Standard Serial Number: 0066–4162
International Standard Book Number: 0–8243–1415–8
Library of Congress Catalog Card Number 71–135616

Annual Review and publication titles are registered trademarks of Annual Reviews Inc.

Annual Reviews Inc. and the Editors of its publications assume no responsibility for the statements expressed by the contributors to this *Review*.

Typesetting by Kachina Typesetting Inc., Tempe, Arizona; John Olson, President Typesetting coordinator, Janis Hoffman

PRINTED AND BOUND IN THE UNITED STATES OF AMERICA

PREFACE

From its inception, this Review has offered readers an eclectic menu. This volume is no exception. It is intended to provide evolutionary biologists, whether ecologists or systematists, with varied food for thought and for further exploration. Despite the recent advances in optimal foraging theory detailed in Graham H. Pyke's review, this theory offers no clear prescription for readers. We are thus free to reassert our view that the pervasive forces of evolution integrate our disciplines, making it essential for all of us to keep abreast of advances in the study of seemingly disparate phenomena.

At the systematic end of our spectrum is an assessment (by Claude Dupuis) of Hennig's effect on taxonomic thought; in a different vein, Donald S. Buth considers the applicability of electrophoretic data to systematic studies. Two separate views of genetic revolutions and their relation to speciation (by Hampton L. Carson & Alan R. Templeton and by N. H. Barton & B. Charlesworth) serve to document, and perhaps clarify, issues in a controversial area with clear implications for both systematists and ecologists. We include reviews of the ecological determinants of genetic structure in plants (by M. D. Loveless & J. L. Hamrick) and of the constraints on the expression of plant phenotypic plasticity (by Maxine A. Watson & Brenda B. Casper); of associations among protein heterozygosity, growth rate, and developmental homeostasis (by J. B. Mitton & M. C. Grant); of the central-marginal model for evolution of *Drosophila* (by Peter F. Brussard); and of migration and genetic population structure (by E. M. Wijsman & L. L. Cavalli-Sforza). Likewise, several of the more overtly ecological topics have important implications for systematics, e.g. the evolution of eusociality (by Malte Andersson) and life history patterns and the comparative social ecology of carnivores (by Marc Bekoff, Thomas J. Daniels, & John L. Gittleman). Three other reviews deal primarily with problems in evolutionary ecology: A. Dafni's discussion of mimicry and deception in pollination, Mark S. Boyce's view of r- and K-selection as a model of density-dependent selection, and C. C. Smith & O. J. Reichman's treatment of the evolution of food caching.

The remaining reviews are explicitly ecological. One is impressed by the frequent emphasis on the importance of the time scale as a major variable. This group includes reviews by Wayne P. Sousa on the role of disturbance in communities, by A. R. M. Nowell & P. A. Jumars on fluid environments of benthos, by Earl E. Werner & James F. Gilliam on size-specific competition, and by R. S. Loomis on traditional agriculture in America.

We note with deep regret the untimely death of Thomas J. M. Schopf, who, along with his many other interests, consented to serve on our Editorial Committee, beginning his term last year. His incisive and cogent contributions at the one meeting he had the opportunity to attend sharpen our awareness of the scientific community's loss.

THE EDITORS AND THE EDITORIAL COMMITTEE

Annual Review of Ecology and Systematics
Volume 15, 1984

CONTENTS

viii   CONTENTS   *(Continued)*

SOME RELATED ARTICLES IN OTHER *ANNUAL REVIEWS*

From the *Annual Review of Earth and Planetary Sciences*, Volume 12 (1984)

Oceanography from Space, Robert H. Stewart
Rates of Evolution and the Notion of "Living Fossils," Thomas J. M. Schopf
Blancan-Hemphillian Land Mammal Ages and Late Cenozoic Mammal Dispersal Events, Everett H. Lindsay, Neil D. Opdyke, and Noye M. Johnson

From the *Annual Review of Energy,* Volume 9 (1984)

Energy and Acid Rain, Roy R. Gould

From the *Annual Review of Entomology,* Volume 29 (1984)

Frederick Simon Bodenheimer (1897–1959): Idealist, Scholar, Scientist, Isaac Harpaz
Economics of Decision Making in Pest Management, J. D. Mumford and G. A. Norton
The Ecology and Sociobiology of Bumble Bees, R. C. Plowright and T. M. Laverty
Spiders as Biological Control Agents, Susan E. Riechert and Tim Lockley
Assessing and Interpreting the Spatial Distributions of Insect Populations, L. R Taylor.
Insect Molecular Systematics, Stewart H. Berlocher

From the *Annual Review of Genetics,* Volume 18 (1984)

The Evolutionary Implications of Mobile Genetic Elements, Michael Syvanen

From the *Annual Review of Microbiology,* Volume 38 (1984)

Deep-Sea Microbiology, Holger W. Jannasch and Craig D. Taylor

From the *Annual Review of Nutrition,* Volume 4 (1984)

Nutritional Energetics of Animals, R. L. Baldwin and A. C. Bywater

From the *Annual Review of Phytopathology,* Volume 22 (1984)

Acidic Precipitation Effects on Terrestrial Vegetation, Lance S. Evans

(continued)   ix

Willi Hennig (1913–1976) in his fifties.

Ann. Rev. Ecol. Syst. 1984. 15:1-24

WILLI HENNIG'S IMPACT ON TAXONOMIC THOUGHT

C. Dupuis

Muséum National d'Histoire Naturelle et École Pratique des Hautes Études, 75005 Paris, France

The present review is neither bibliographical nor historical. I hope merely to overview the methodological strengths and weaknesses of Hennigian taxonomical thought and of some recent works devoted to theoretical questions that have been revived by an increasing interest in that thought.

A FASCINATING EPISTEMOLOGICAL VENTURE

Willi Hennig (1913–1976) was a German insect taxonomist who was known during his life time among Diptera specialists for numerous monographs assignable at first sight to alpha systematics (100). Although he published a major work on theoretical taxonomy in German as early as 1950 (74), he became famous only after the publication in 1966 of the English translation of a second book (77), which was not published in its original language until 1982 (82). Initially, his work engendered fierce attacks on his "phylogenetic systematics" by members of other taxonomic schools (41). Recently, however, Hennig's ideas have come to form the foundation of the fashionable cladistic school. The epistemological analysis of this venture will be a task for the future, provided that the original works are not forgotten.

Hennig and His Works

On this topic, we only need to supplement a previous review (41). Besides a first theoretical paper based on Diptera (70), in 1936 Hennig published works on the *Rassenkreis* and the biogeography of the lizard genus *Draco* (71, 72). A criticism of the first paper (169) led Hennig to extend his study of Diptera larvae and to refine the notion of larva-imaginal incongruence (73). From this time on,

1

0066-4162/84/1120-0001$02.00

he substituted incongruence for *Rassenkreis* as evidence of the fundamental statement that similarity alone cannot indicate genealogy. This position, a virtual break with what was to become the Mayrian school, was acknowledged as early as 1948 (176). Hennig's 1950 book was a preliminary sketch, eventually improved by introducing the word and concept of *synapomorphy* (75) and by an acquaintance (76) with Woodger's logic. The English "translation" that appeared in 1966 was, then, a new book (77). Nowadays it constitutes the main avenue into Hennigian thought. Yet this is one of the weaknesses of the cladistic school: Having been completed in manuscript form in 1961, this book is not Hennig's *ultima verba*, and its translation is far from satisfactory (see examples of necessary retranslations below and in 55). In the same year (1966), Hennig's work on the Diptera of New Zealand (78) was translated into English; it contains the first hint of Hennig's doubts about what would become biogeographical "cladism" (see 41). An excellent Spanish translation of the 1961 manuscript was published in 1968 (79), but it has been overlooked. In 1969, Hennig's treatise on fossil insects (80) opened with a chapter representing "the most complete, the most balanced, and perhaps the most felicitous expression of his thought" (41, p. 10).

Since Hennig's death, his two fundamental books have been reprinted (74, 77). His treatise of 1969 has been translated [perhaps with too many interpolations and comments (81)], and his son, Professor Wolfgang Hennig, has edited three posthumous works: the most useful authentic German text of the first American book (82), a classification of Chordata (83), and—largely echoing a 1974 polemic with Mayr—a synopsis of the problems of phylogenetic research (84).

Hennig Today, Through Words and Works

Hennig's ideas, commonly considered part of what is called cladism, continue to be discussed in many writings. Some words, such as *cladism*, have to be explained, and a choice made among the various writings.

SOME POLYSEMIC WORDS   When coining the word *clade* from "cladogenesis . . . taken over directly from Rensch," Huxley intended to denote "delimitable monophyletic units" (96, p. 454) or, more precisely, "monophyletic units of whatever magnitude" (97, p. 27). He seems not to have known of an older use of this word (36) referring to a group of great magnitude possessing a particular structural type, i.e. simultaneously applied to a taxon, a higher taxonomic category, and a grade sensu Huxley!

The adjective *cladistic* has been explicitly derived from *clade* sensu Huxley by Cain & Harrison, for whom "closeness of relationship in terms of phyletic lines can be called cladistic . . . [and] similarity due to common ancestry, not to convergence, can be called patristic" (20, p. 3). Both *cladistic* and *patristic*

(20) refer to phyletic affinity and need the clarification afforded by Sokal & Sneath (159, p. 220): *Cladistic* is a relation "through *recency* of common ancestry" and *patristic*, one by common ancestry (not otherwise specified). For these authors, who rely on the relationships of Hennig's opposing types I and I*a* (76, Fig. 6), *cladistic* is "the type I phylogenetic relationship of Hennig, 1957." Thus, they equate a cladistic relation with a Hennigian one.

The noun *cladist* was introduced in 1965 "in line with the terminology developed by Rensch (1960), Huxley (1958) and Cain & Harrison (1960)" (116, p. 167). Although none of these authors referred to Hennig, in coining the word *cladism* Mayr typifies *the cladist* as one who adheres to that "phylogenetic school" of which Hennig is the "most articulate spokesman" (118, p. 167).

The word *cladogram* was introduced in the same year (1965) but has two different meanings. For Mayr (117, p. 81), "the cladogram of the cladist" meant the phyletic diagram of the Hennigian school. For Camin & Sokal (21, p. 312), the term *cladogram* is used "to distinguish a cladistic dendrogram from a phenetic one which might be called a phenogram"; unfortunately, these authors built their "cladograms" by phenetic and not by Hennigian techniques. Consequently, there are two kinds of cladograms: those relying on attributes (Hennig's *Stufenreihe;* Sneath's synapomorphograms, see 125) and those relying on objects (in phenetic constructs); their inadvertent use as synonyms (e.g. in 158, p. 160) is confusing.

In Hennig's works, these words are always set off by quotation marks and appear as mere citations. In works by other authors, *cladist* and *cladism* commonly apply nowadays to the school born of Hennigian thought, whereas *cladistic* and *cladogram* have become equally popular in both the phenetic and the Hennigian schools, and consequently, their meaning remains ambiguous. Even worse, *cladogram* has become a fashionable word that is sometimes used to denote intuitive constructs and no longer refers to a definite procedure.

SOME SIGNIFICANT WORKS   Under such headings as cladism, Hennigism, and phylogenetic systematics, Hennigian thought nowadays underlies many works—by both supporters and opponents—on a wide variety of topics. For example, Hennig's name appears as a key-word in the titles of numerous works: methodological essays (5, 28, 41, 45, 69, 98, 107, 108, 128, 129, 149, 150); reports of meetings (58, 162); or special works on vertebrate paleontology (13, 66, 68, 151, 168), biogeography (33, 37, 143), parasitology (15), entomology (1, 42), and biochemical or molecular taxonomy (95, 139).

Works that do not refer to Hennig—whether in their titles or their lists of references—are becoming much more common. Owing to the increasing volume of this cladistic literature, only a limited number of recent sources can be cited to supplement previous reviews (see 41; and for botany: 62, 85).

Journal papers that offer factual applications of Hennigian methods are too numerous to cite here; those dealing with theoretical questions have also proliferated, particularly in *Systematic Zoology*, and will be considered in the ensuing discussion when necessary. Review articles are scattered in various serials; those recently published in this *Annual Review* and in others bear specifically on Hennigian procedure (161) or document the degree of Hennigian penetration into various fields (biogeography: 99; paleontology: 136; molecular biology: 170) and particular taxa (22, 110, 112).

Symposia and symposium-like volumes with a predominantly Hennigian perspective have appeared in biogeography (133) and paleontology (19), as special journal issues devoted to Hennig (51, 87, 160), and as proceedings of the Hennig Society (60, 142). Others are found in fields where Hennigian thought has only begun to penetrate (88, 126, 163). Most of these volumes, however, reflect a balance between the conflicting taxonomic methods; some are of general or methodological interest (29, 56, 61, 106, 137, 153) and others bear on special fields—e.g. paleontology (31, 115), evolutionary theory (23, 152), biogeography (10, 155), primatology (26, 27), and arthropodology (65). Books or monographs with a Hennigian background include important general treatises (46, 67, 145, 165, 178) or major specialized studies in biogeography (132), paleontology (109, 111), primatology (166, 167), and entomology (2, 12, 35).

Textbooks that discuss Hennigian views from a biological (40, 148) or a methodological (103, 157) standpoint remain scarce. Handbooks, whether historical or epistemological, are somewhat deceiving in their treatment of Hennig. Jahn et al. (101) only give biographical details and an amalgam of Hennig's and Remane's quite different views on homologies (see 69, 98). Mayr (118) reiterates his criticisms of cladism and limits his support to the synapomorphy procedure. [See however, Ross (147a).]

Since Hennig's death, as many as 1000 books and papers illustrating or discussing his thought may have been published. Comparisons of Hennig with Darwin, Mayr, Simpson—even Picasso (33)—and, more recently, with Weismann (43) have also flourished. As impressive as the numbers of works and comparisons may be, an even greater number of taxonomists have never heard of Hennig (25, 138), and some of his peers are among his most determined opponents.

Under these conditions, it is obviously impossible to be content with enumerating numbers of books and papers; the following sections are, therefore, a personal attempt to evaluate the Hennigian impact on present-day taxonomy.

DISCOURSE ON A METHOD

At the beginning of the nineteenth century—besides the phenetic Adansonian and the genealogic Lamarckian concepts, which were both far ahead of their times—the dominant paradigm was the so-called natural method, a Linnean system amended with the ad-hoc teleological principles of subordination and correlation of characters.

Darwin, who had "two distinct objects in view" (38, p. 61), carried out two independent revolutions. His biological revolution proposed a mechanism (natural selection) to account for the amount of difference between organisms. His taxonomic revolution, founded on the inference of descent with modification, advocated genealogical taxonomic arrangements freed from categorical and typological thinking. Darwin repeatedly distinguished *ranking,* which implies amount of difference, from *arrangement,* which implies descent. He recalled (37a, pp. 457–58) that even in the absence of a known mechanism, consideration of taxonomic arrangement alone would have directed him to the concept of descent. Regretably, he carefully elaborated the theory of selection but not the practice of taxonomy. This fact explains why the majority of naturalists—among whom Haeckel holds the foremost position (see 44)—eventually retained pre-Darwinian taxonomic practices.

The genetic revolution has led to two opposing views of taxonomy that correspond to each side of the Darwinian dualism. Those taxonomists devoted to "evolution as a process"—here termed Mayrians—study variation, selection, isolation, etc. When looking at the results of these processes, however, they support the use of categories (first of all, species), as well as the principles of typology (they highly prize "grades") and overall similarity; they discard divisive procedures only at the lower taxonomic levels. *Taxonomically speaking,* the true Darwinians are those naturalists who focus on the products of evolution, whatever its mechanisms. They praise the agglomerative methods and break with typological thinking. They represent two schools that have developed independently of the Mayrians and of one another. The phenetic school rests on similarity and neglects categories; at the beginning, it discarded any concern for genealogy. The Hennigian school rests on genealogy and discards similarity; it initially was haunted by the equivalence of categories.

Such an opposition is highly significant. Any classificatory approach usually considers *objects*—that which is being classified, e.g. biological individuals or taxa—and their *attributes,* often called characters by naturalists. The relations between objects and attributes (in biology, between taxa and characters) may be viewed as *intensional,* as if attributes make objects, or as *extensional,* as if objects make attributes (see below). According to an intensional view, classification rests on the similarity of attributes; according to an extensional view, it depends on possible intrinsic relations between objects. Among biological objects—individuals and taxa—that are linked by a history, these relations make possible an extensional view that is precisely the one expressed by the Hennigian fundamental statement. The next section presents that statement and the corresponding logic and procedure; the following one will compare Hennigism with other schools.

The Hennigian Fundamental Statement

Hennig's approach to taxonomy starts from what I have called the Hennigian fundamental statement (41), which teaches that there is no absolute coincidence between similarity of attributes and genealogy of objects. The proposition "a common origin implies some similarity" is true, while the converse, "some similarity implies a common origin," is false (177). This kind of statement dates back to the fathers of genealogical thinking in taxonomy (Buffon, Darwin, etc.) and was even expressed by some naturalists who, instead of acknowledging actual descent, believed in an ideal affinity between living beings [it was then asserted that "similarities and affinities are two astronomically [*himmelweit*] different things" (122)].

To test for the lack of an absolute linkage between similarity and genealogy, Darwin and Hennig used several kinds of empirical facts.

1. Hennig considered the *Rassenkreis* (which he studied in the genus *Draco;* see 71, 72) as a demonstration of the fundamental statement. In these cases, however, the genealogical arrangements are merely inferred from the *vollständige Übergänge* (i.e. perfect transitions) between varieties (70, p. 171). These *Übergänge* are only similarity relationships, and Hennig eventually turned away from this circularity.

2. From the 5th edition of the *Origin* . . . (1869) on, Darwin adopted Fritz Müller's discovery of incongruences between larval and adult arrangements in Crustacea as a demonstration of the lack of an absolute linkage between similarity and genealogy. Likewise, from 1943 on, Hennig consistently substantiated his fundamental statement with the example of the frequent lack of *Deckung* between larval and adult arrangements in holometabolous insects (73). After 1950 (74), he used the term *Incongruenz,* which Van Emden (171) had revived from Weismann's work.

Incongruence in supraspecific arrangements, although frequently endorsed as such (41, 117, 156, 176), is not a valid demonstration. Sokal & Sneath rightly remark:

> Incongruence between larval and adult classifications shows that the cladistic relationships cannot be exactly proportional to the phenetic relationships, since the cladistic relationships of adults and larvae of the same species must be identical, while the phenetic relationships need not be. This statement is true, but the converse is not necessarily true; that is, congruence does not prove the exact correspondence of phenetic and cladistic relationships because both larval and adult features might have both undergone convergence to an equal degree, although this is unlikely (159, p. 224).

In other words, the Hennigian fundamental statement can be demonstrated using larval-adult incongruences only if the set of organisms being studied is monophyletic. This monophyly, although often probable at lower levels, is precisely the fact to be demonstrated.

3. Besides the two "demonstrations" refuted here because they are founded on merely inferred genealogies, Darwin (37a, pp. 424, 433, 456) and Hennig (70, p. 171) used a third one that validly took into account sets of more probable genealogies. They considered that if one did not know the genealogy, incongruences between characters in different instars, generations, or sexes in a species could lead one to place the variants in unrelated taxa. This demonstration can be freed from the circularity of a categorical approach by considering the various allomorphs in what I call a proved, short lineage (44)—which can be operationally observed—rather than in a "species."

Hennigian Logic and Procedure

As just noted, the Hennigian fundamental statement and its demonstrations are not recent discoveries. What is interesting is the logical consequences that Hennig drew from such a statement and, above all, the corresponding procedure.

HENNIGIAN LOGIC   From an epistemological standpoint, this topic can be most profitably treated by comparing Darwinian and Hennigian approaches.

Major logical approaches   Darwin and Hennig drew the same major logical deductions from the fundamental statement: (*a*) in order to be stable, the arrangement of taxa must reflect their history, i.e. their unique genealogy and (*b*) the hierarchically nested taxa, if genealogical, are monophyletic "individuals." When considering proved, short lineages, it appears indeed that the individuality of the taxa, i.e. their reality, relies on their homogeneity and completeness. Therefore, long lineages can be considered proven only if they simultaneously satisfy the conditions of homogeneity (Darwin's single progenitor) and completeness.

Homogeneity is a classical condition of monophyly. Completeness means that a taxon comprises its ancestor and *all the descendants of that ancestor*. The need for this condition was already foreseen in the third criterion of De Candolle's (39) composite "definition" of species. In the *Origin of Species . . .*, Darwin stressed no less than 20 times the importance of considering the totality of the descendants of a given ancestor; but he used merely the term *all* (listed among the "words suppressed" from a recent Concordance; see 6) to express this condition. He employed *co-descended* for the first time in the 5th edition (1869) and *co-descendants* only in the *Descent of Man* (38, 1871 edition, p. 188). This notion of completeness, which was acknowledged "early in the history of evolutionary theory" (89), occurs commonly in classical studies. It is also explicit in works concerning pure lines (104) or the logic of taxonomy (48, 103, 180). Curiously enough, since 1971 (3) completeness as a criterion for monophyly has been challenged only by criticizing Hennig, and the recent discussions of this criterion (25, 178) continue to ignore its consistent pre-Hennigian use.

Other logical approaches   Darwin and Hennig do not draw the same deductions from the fundamental statement when they consider the treatment of categories and divisive versus agglomerative procedures.

1. Darwin openly denied the reality of categories, including that of species, using them only for convenience. Hennig's constant practice was to contrast species, as a *Wirkungsystem*, and the higher taxa. Beginning in 1936 (70), he searched for many years for the objective equivalence of rank; in 1969 (80) he finally recognized that this quest was premature (see 41). Given his repeated acknowledgments (e.g. in 76) of Woodger's anticategorical thinking and the multiplicity of ranks expressed in his later works (80, 81), it is evident that Hennig's ultima ratio could only be a theoretical rejection of categories. This rejection explicitly included the category "species" (84); thus it follows that Hennig "assumes the association of evolutionary changes with speciation only for the sake of convenience" (86). Today, considering the multiple mechanisms of isolation (already foreseen by Darwin), neo-Hennigians assert that intersterility results from apomorphic acquisitions developed along with the divergence of lineages (14, 178); this implies that interfecundity is not an essential criterion of species but only a plesiomorphic condition of variable categorical significance.

2. Darwin, not having elaborated explicit procedures in taxonomy, did not provide divisive or agglomerative models. By constructing divisive or semi-agglomerative arrangements, his followers seem to have betrayed his anti-typological and anticategorical thought; this is particularly true of Haeckel (see 44). Hennigian logic, on the contrary, is plainly agglomerative. Sneath (156) emphasized that an agglomerative construct exhibits more stability than a divisive one. In fact, a reappraisal at the higher levels in a divisive construct has no more repercussions on the lower levels than a modification at the lower levels in an agglomerative construct has on the higher ones. The best argument in favor of an agglomerative procedure is perhaps that it is emancipated from the teleological principles of subordination and correlation of characters. In this sense, Mayr (118) has rightly underlined the importance of replacing the divisive procedure with the agglomerative one—a "revolution" that lasted two centuries. It is worthwhile to recall, however, that when applied to a few large, scattered taxa, there are operational uncertainties (see below).

HENNIGIAN PROCEDURE   Since the Hennigian procedure is agglomerative as to attributes, it is opposed to the Haeckelian one (44) and implies the rejection of predefined objects (and therefore of categories; see above). It proceeds in two steps: (*a*) the analysis of attributes, called character analysis, and (*b*) the clustering of attributes.

Character analysis   This constitutes the first step of the procedure and requires numerous small, densely distributed taxa (as often occurs in entomology where the procedure was born; see 42). These objects allow a relative evaluation (*Wertung*) of a few states—ideally two: apomorph and plesiomorph—for each attribute. *Apomorphy* refers to an attribute (not object) that is unique among the numerous objects (*autapomorphy*) or shared by only a few of them (*synapomorphy*); *plesiomorphy* denotes the converse condition, found in most objects. These terms are necessarily relative and may vary for each attribute of an object.

Hennig identified the "frequency of occurrence" (*Häufigkeit des Vorkommens*) of an attribute state as the foremost criterion of plesiomorphy (common occurrence) or apomorphy (unique occurrence). Hennig believed, however, that it is possible to determine attribute polarity on the basis of the intrinsic criterion of ontogenetic precedence of states in a single taxon. He also thought that the extrinsic, diachronic criteria of geological precedence and chorological progression might be helpful. He disregarded all subjective criteria on principle (as presuming a functional significance or a success in evolution) but retained logical criteria such as complexity of characters (see an implicit critique in 144) and correlation of transformation series.

From the beginnings of modern descent theory, all of these criteria for evaluating states of attributes have been used in deciding between primitive and derived states. Their critical study has been enhanced by both Hennig (74, 77) and his followers (for bibliography and critiques, see 41). A prime example of such a discussion is offered by Stevens (161), who seems to be the first to have taken into account the data and opinions of both zoologists and botanists. Having carefully scrutinized all the criteria, Stevens concluded that the best assessment of the polarity of attribute states rests on out-group comparison. I would even argue that most of the criteria should be reduced to this ultimate one, which has been carefully worked out (175) and is an *R*-technique (see below).

Two kinds of criteria, however—those linked to chorological or to paleontological documentation—seem impossible to reduce to out-group comparison of attributes. In fact, the distribution, or date, of a taxon is not an attribute but a document because it depends on the extrinsic and never exhaustively explored dimensions of space and time. I use the word *document* for what, despite the Darwinian critique (see 37a, p. 486), has been frequently viewed as "armorial bearings." Although such documents have given rise to considerable discussion, the conditionality of their use in phylogenetic taxonomy has not yet been logically formalized. It is not clear under what circumstances distributions and dates can be viewed as extrinsic but questionable documents or as having the same significance as intrinsic attributes. The biogeographical aspect will be treated below; as for fossils, the reader is referred to Hennig himself and to numerous reviews or symposia (see, for example, 13, 19, 23, 31, 41, 59, 87, 111, 115, 136, 151, 152, 168).

Clustering of oriented attributes This second step constitutes the most original part of Hennigian procedure. Except for a few precursors, phylogenetic taxonomists before Hennig clustered objects according to a Haeckelian view of conservative evolution by searching for the most primitive ancestor and often for plesiomorphic attributes (see 44). In contrast, Hennig's principle (149, 150) requires clustering to be started using the most divergent apomorphies. The smallest objects that share the most diverging apomorphy are considered as sister objects and form a synapomorphic taxon. By continuous chaining, the successive-sister apomorphic taxa of t rank constitute in ascending order the successive taxa of $t-1$ rank. In my opinion, "the word synapomorphy expresses the full spirit of a true genealogical and Hennigian procedure" (44). Logically, synapomorphic clustering is equivalent to the single-link method advocated in phenetics (49, 103). Biologically, it corresponds to the proven fact of descent with modification. Genealogically, it affords the only conceivable way of building strictly monophyletic arrangements—i.e. in which the taxa are homogeneous and complete. Consequently, the monophyletic taxa resulting from the chained clustering of synapomorphies are defined not by attributes (similarity) but by members (genealogy); this is a requisite of the Hennigian fundamental statement and of an R-method (see below).

Obviously, such a chaining of synapomorphies—from those that substantiate the smallest taxa to those that indicate the largest ones—reflects a parsimony procedure. It could be faced with unsolved convergences and parallelisms, especially among attributes of simple pattern, e.g. molecular ones (see examples in 144). Perhaps in this case the alternative procedure of character compatibility would be appropriate (see 113, 164); its para-Hennigian value is a subject for discussion (54, 55a).

The Hennigian clustering procedure results in a multiplicity of categories (which leads to the rejection of that concept). The corresponding nomenclature of the nested taxa—if necessary at each rank—could raise practical difficulties (see 41). Yet, despite the place it is given in some phylogenetic treatises (178), nomenclature is not taxonomy; it is nothing but its servant and must consent to some compromises.

THE CHALLENGE OF METHODS AND OBJECTS

During the last 30 years, all scientific fields have been invaded by numerical taxonomy (see 157, 159). Similarly, Hennigism now seems to have come of age. The mere expansion of these methods, however, does not demonstrate their universal value. During the eighteenth century, all kinds of objects—minerals, illnesses, etc.—were classified according to Linnean procedures, and at the end of the nineteenth century, scholars brought up on natural history and Darwinism saw genealogies in every field. Because of the evident misuses, such past generalizations aroused numerous, useful discussions among naturalists. Likewise, one must acquire a critical understanding of both objects and methods before deciding whether present-day methods can legitimately be applied to each kind of object and, conversely, if the objects are being viewed with the right method.

Various Concepts of Methods and Objects

Despite the psychogenetic relation that ultimately unites the concepts of objects and methods, logically one can only consider them as separate notions.

TAXONOMIC OBJECTS   The inclination to apply a "neutral" method to the objects of taxonomy has been strengthened by the view that the historical relations between these objects can be known only by inference. It is precisely this question—theory now or theory later?—that Hull (89) raised when he recalled some biologists' opinion that "classification should be theoretically neutral; no theoretical considerations should ever intrude during the formation stages of classification although theoretical inferences may be drawn from the classification afterwards" (90). Such a petitio principii reflects the major conflict between Mayrian and Hennigian taxonomists, which rests not on methods (Mayr supports the value of synapomorphy) but on objects. Mayrians emphasize the amount of difference and Hennigians, the pattern of descent.

A poor translation ascribed the following opinion to Hennig:

> This is probably a lingering effect of the ancient concept of the "ladder of organisms", expecting that a developmental series from "lower" to higher, or at least from more "primitive" to derived forms, must be expressed in the phylogenetic system. But the task of the phylogenetic system is not to present the result of evolution [sic], but only to present the phylogenetic relationships of species and species groups on the basis of the temporal sequence of origin of sister groups (77, p. 194).

If critics puzzled by this declaration (25) had read it in Spanish (79, p. 263) or German (82, p. 188), they would have understood it as follows:

> The expectation that the phylogenetic system must also express an evolutionary series from "lower" to "higher" forms—or at least from "more primitive" to "more derived" forms—is probably a lingering effect of the very old concept of the "chain of beings." In fact, the task of the phylogenetic system is only to describe the phylogenetic relatedness of the species and groups of species according to the temporal sequence of the birth of the sister groups; it is not to describe the final success [éxito or Erfolg] of the evolution.

Such a statement demonstrates that, according to Hennig's view of phylogenetic research, the pattern of descent must take precedence over the final success (e.g. fitness, adaptation, progress) of the descendants. The distinction corresponds to the dualism of Darwin, who asserted that independent of any explanation of process, the arrangement of products alone would have convinced him of descent with modification. It also reflects the point of contention between Hennigians and Mayrians. Hennigian taxa are strictly monophyletic lineages, i.e. complete and homogeneous, regardless of the degree of modification of the attributes of their members. Mayrian taxa may be incomplete or heterogeneous, depending upon the gaps introduced among them by the differential success of attributes in nature—or in the minds of naturalists.

This conceptual duality affects research objectives: For biologists who stress processes, models of evolution can generate heuristic taxonomies; for those who focus on the results, models of genealogies can produce heuristic views of evolution. Among the bicentenary profusion of alleged arrangements and mechanisms—whatever may by chance be the predictive value of a particular one—only the research is actually heuristic. In this sense, the analysis of object discontinuities (genetic diversity, reproductive isolation, population ecology) is as heuristic as that of attribute discontinuities (plesiomorphies, apomorphies, and incongruences at various levels of development or organization). Since all of these discontinuities are accessible to observation and experimentation, a theory of objects may be devised now or later, i.e. it may either precede or follow the development of a theory of attributes.

TAXONOMIC METHODS   The use of so-called numerical methods in taxonomy has been favored not only because of the "neutrality" of mathematics and the aura of computers, but also because of the following epistemological constraints:

1. Mathematization is frequently viewed as self-justified (e.g. one refrains from examining the Hennigian principles "since they are primarily non-numerical"; see 127). Such an attitude neglects a more basic distinction than that between quantitative and qualitative thinking, i.e. that between intensional and extensional thinking (see 180, pp. 15, 23, 64), of which only a few taxonomists are aware (14, 16, 18, 105, 144).

2. The intensional methods that antedate the reception of Boolean thought ["based squarely upon the relations of extension" (180, p. 64)] have routinely been applied. Williams avows this when he says:

> Purely logical considerations have played little part in clustering theory. Few problems are amenable to the "X is A or not − A" type of approach; clustering procedures have developed because so many problems involve the many-valued or continuous concept of "X is more like Y than it is like Z." Set theory may have its place in consideration of the nature or purpose of classification . . ., but for the construction of classifications more conventional algebraic techniques are unavoidable (179, p. 304).

3. Clustering has obscured analysis, which necessarily constitutes a prior step.

4. The phenetic intensional technique usually refers to the taxonomic philosophy of Gilmour, for whom "a natural classification is that grouping which endeavours to utilize all the attributes of the individuals under consideration" (63). This postulate is even commonly substantiated by referring to the notion of information content. In fact, "classifications in the narrow sense are incapable of storing much in the way of specific information. Rather than being storage-and-retrieval systems themselves, they serve as indexes to such storage-and-retrieval systems. The information resides in the monograph, not in the classification" (89, p. 28), and perhaps not even in the monograph but in the object. This statement seems to be true for the classifications of all schools (11). Nevertheless, Gilmour's postulate has been accepted by pheneticists for whom *natural classification* means one "whose constituent groups describe the distributions among organisms of as many features as possible" (50, 52).

Obviously, the dispute between pheneticists and Hennigians is not about objects (today, pheneticists admit phylogeny) but about method. Although both advocate an agglomerative procedure, they start from opposite considerations. The functors of the phenetic analysis of objects are the attributes; those of the Hennigian analysis of attributes are the objects. Thus, phenetic taxonomy is intensional, since for appraisal of each object, it requires an intensive study of its various attributes, Q's (*qualitas*). Conversely, Hennigian taxonomy is extensional since, for appraisal of each attribute, it requires an extensive consideration of various objects, R's (*res*). The Q-procedure of the phenetic taxonomy best defines each OTU (class or exemplar) when, at the analysis step, it uses a large number of attributes, whether discrete or not. At the clustering step, three techniques can be applied: complete-linkage, average-linkage, or single-linkage; some authors recommend only the last one (49, 103). The clusters delimit the supraindividual taxa that are polythetic classes (monophyletic or not) defined by a diagnosis (not by content) and nested in continuous phenetic ranks with increasing distances between objects. The R-procedure of Hennigian taxonomy best defines each state of an attribute when, at the analysis step, it uses a large number of objects, whether they are of the same "rank" or not. At the clustering step, three techniques (homologous to the former ones) can be applied: symplesiomorphy, typology, and synapomorphy; Hennig uses only the last one. The clusters delimit the supraindividual taxa that are monophyletic individuals (monothetic or not) defined by content (not by a diagnosis) and nested in discontinuous phyletic ranks of successive divergences of attributes.

The mere operationality of the above methods implies, besides the intentional postulate of the Q-naturalness of objects as a function of attributes, the recognition of the inverse extensional postulate of the R-naturalness of attributes as a function of objects (which seems to have been formulated only implicitly; see 91). Ultimately such "naturalnesses" are only partial; according to De Candolle's (39) criticism of Adanson, total naturalness would require that all the attributes of all the objects be known.

Generalizations as Empirical Tests

Various instances of expanding or interchanging procedures and objects afford the candid taxonomist an opportunity to test empirically the legitimacy of a number of inadvertent generalizations of methods.

GENERALIZATIONS OF R-TECHNIQUE   The extensional R-technique of agglomerating attributes tested against objects encounters difficulties when attributes are under- or overestimated. An underestimation occurs when intrinsic, reproducible (i.e. homologous) attributes, such as those of living beings, are considered as documents independent of the descent of objects. Conversely, an overestimation occurs when documents independent of the descent of objects are viewed as intrinsic, reproducible attributes.

"Transformed cladism"   The disagreement over the question of theory or no theory illustrates an underestimation of true attributes. Discarding some precursory remarks, it seems to have arisen from a polemical paper by Nelson (130) and an expanded version of this paper by Nelson & Platnick (132) expressing strictly formal views about phylogeny. Numerous attack were launched against the authors, labeled pattern cladists (8), and against their alleged sterile enterprise (4), which was called dendronomics (120) or transformed cladism (24, 25). Platnick convincingly assumed the defense but, unfortunately, in an article whose title at first sight credited the school with a "transformation of cladistics" (140); there were numerous other advocates as well (e.g. 14, 59, 134, 138). Apparently these Nelsonians—despite a formal and unattractive treatment—strictly follow the Hennigian principles, both in terms of character analysis and synapomorphic clustering. Without any discontinuity or transformation of method, their "methodological cladistics" remains, as before, grounded on parsimony. This technique is, moreover, thoroughly defended by Farris (54).

The conflict rests ultimately on a philosophical requisite that constructs be independent from evolutionary theory (not merely from a particular one). Patterson (138) reduced the crucial point of the debate to whether homologies and monophyly must be *constructed* by an operative parsimony or *defined* by real descent from a common ancestor. Darwin resolved this problem in terms of probabilities: If there are so many homologies, they cannot be without a cause and the only scientifically conceivable one is descent with modification. Hennig explicitly adopted this solution (75, 77; see also 41), and his position on this point has been contrasted with Remane's (see 69, 98).

The consideration of objects governed by a reproducible syntax (although they are not living organisms) may demonstrate that despite the Nelsonians' reluctance, homologies must have a cause. The significant messages (or texts) appear suitable for copying and in this respect are similar to organisms. For a long time, they have represented the field of application of a classifying *R*-technique, i.e. the Lachmann (ca. 1850) or "stemmatic" method, also called the "method of the common faults" (9, 59a, 114). Two transformed cladists have argued that this old method corresponds to Hennigian synapomorphy (141). It is notably more objective than stylistic judgment in the humanities, which is the homologue of the gradistic one in biology.

It is significant that through stemmatics one can chronologize related texts, the only possible history of which results from copying with modification. If the existence of a syntax (molecular or linguistic) makes the reproduction of organized individuals or texts possible, descent (whether repetitive or with modification) is the prime cause of their history. The descent—that of Darwin's *Descent of Man* and of *Descendenztheorie*—is real for organisms (i.e. *omne vivum e vivo*) and also for texts (i.e. no copy without a model). The modifications, whatever their mechanism (e.g. fault in autoreproduction, clerical error), are equally real. When they arise, descent with modification necessarily follows. This statement does not imply a particular theory about the mechanism involved, and consequently, Nelsonians could acknowledge evolution as an outcome of reproduction.

Use of nonhomologous attributes   This type of overestimation of "attributes" is illustrated by the so-called Hennig-Brundin biogeography (37). For these authors, geographical distributions represent attributes of the object-taxa. In fact, the distributions are acquired and nonreproducible and constitute documents concerning the taxa, rather than inherited, intrinsic attributes of these objects. Such documents can be viewed as attributes characterizing the objects only for those objects whose intrinsic divergence exhibits a coincidence (parallelism) with extrinsic distribution in space. This corresponds to the "progression rule" originally enunciated by Hennig, who eventually expressed increasing doubts about its universality (see 41).

GENERALIZATIONS OF *Q*-TECHNIQUE   The intensional *Q*-technique of agglomerating objects tested against attributes encounters difficulties when objects are under- or overestimated. An underestimation occurs when individual objects linked by descent with modification, such as living taxa, are considered as extrinsic assemblages of attributes. Conversely, an overestimation occurs when extrinsic assemblages of elements are viewed as true individual objects.

Numerical cladistics   This field illustrates the underestimation of living taxa. Since a *Q*-matrix can be converted easily into its inverse *R*-matrix, pheneticists have ad libitum elaborated mathematical techniques for introducing some consideration of attribute phylogeny into their intensional method. All these techniques may ultimately be "phenetic methods in disguise" (158) and generalizations of Mayrian "objects" (delimited by a posteriori gaps). In recent years, one has witnessed the phenomenal growth of an inconclusive literature concerning the comparative value of these techniques (e.g. see 123, 146, 147, 158). Such controversies have fostered a return to standard Hennigian procedures in morphological (64), biochemical (144), and karyological (174) systematics. Two recent events will accelerate this movement.

In 1979 at the 13th Numerical Taxonomy Conference, "J. S. Farris amazed friend and foe alike by rejecting as philosophically defective most previous approaches to the logical basis of phylogenetic inferences, including his own elegant statistical work" (124); the Willi Hennig Society arose from this "speciation event" (121). Later, when speaking of (molecular) distance data in phylogenetic analysis, Farris confirmed that "none of the known measures of genetic distances seems able to provide a logically defensible method" and advocated carrying out "phylogenetic character analysis directly on electromorphs" (53).

Such a return to extensional methodology, which Mickevich had long defended (e.g. 123), has recently culminated in a theoretical work by Farris (54) and an interesting factual study by Patton & Avise (139). In the latter, trees obtained directly from qualitative attribute states and indirectly from distance matrices are tested against organismal "model" (i.e. syncretic) classifications. "In each case, the qualitative cladistic trees provided fits to model phylogenies which were strong and as good or better than those resulting from phenetic clustering of distance-Wagner trees based on manipulation of quantitative values in matrices of genetic distance." Despite some weaknesses inherent to electrophoretic attributes, the Hennigian constructs retain one major strength: Any point of ambiguity in a tree may be specifically identified. In other words, "there is a potential loss of information in first generating a distance matrix," whereas such a loss is avoidable when one can "focus upon analyses of the character states themselves." The treatment of the attributes available in an R-matrix via the detour of a Q-matrix appears, therefore, "an unnecessary and hazardous manipulation because the particular characters ultimately contributing to the tree structure are first submerged in a distance matrix." (All preceding quotations are from 139.)

Vicariance biogeography This "method" illustrates the difficulties arising from an overestimation of "objects." It considers the biogeographic units (usually areas) as objects and their biogeographic elements (the taxa living therein) as attributes. In this approach (30, 132, 133), it is argued that the method allows a relative dating of areas based on the genealogy of their attribute-taxa. Although qualitative (the homologous quantitative one would be Croizat's "generalized tracks"; see 135), this is a true Q-method and, therefore, the inverse of the Brundin R-method; such an opposition is implicit in some discussions (17, 135). The feasibility (17, 47, 154) and value (32, 34) of the method have been strongly questioned by opponents (see also 155); even sympathizers (93) have recalled incongruent examples. In fact, biogeographic units are determined by events extrinsic to their elements (if not merely by geographical assumptions); they appear as assemblages of elements rather than as objects with intrinsic attributes. Such assemblages can be viewed as objects characterized by attributes only for those elements whose intrinsic divergence exhibits a coincidence (synchronism) with the extrinsic patterning of the areas in time. Such elements are termed vicariants and their study constitutes vicariance biogeography. Sometimes an annexation of R-terminology to vicariance biogeography results in labeling areas geological taxa and their trees (without a syntax of reproduction) geological cladograms, as if cladograms of organisms were built by dividing preexisting sets.

The Operational and Logical Relevance of Methods to Objects

Using an agglomerative method, whether R or Q, is not always feasible. When tracing historic relations between higher taxa or between areas, technical problems can arise from gaps in the required information. The value of clusters rests on the number and the *density* of the functors (objects or attributes). When there are important gaps among objects or in the knowledge of their attributes (or both), the higher clusters—whether phyletic or phenetic—cannot be determined from a chained agglomeration of lower clusters of increasing rank. One is compelled to suppose that the few a priori high-ranked objects (i.e. higher taxa, continents) are representative of a set of lower-ranked ones or that the scattered attributes (characters or elements) are "typical" of such objects.

Among extant objects and attributes, the uncertainty depends on the "circumstantial availability of data" (44). For fossils or continents, the uncertainty depends on the same factor plus the losses and transformations of attributes accumulated over time. The higher the rank of taxa (or faunal assemblages), the more ancient their common origin and, consequently, the more insufficient the data available for tracing this origin. It follows that at the higher levels an agglomerative technique is not operationally better than a divisive one. This fact explains the slow reception of Hennigism among "higher taxonomists" (see examples in 88, 153) and the debates on the "taxonomy" of biogeographic areas.

Although such impediments are operational and not logical, they force us to recognize that the limits of the methods depend on the objects. It must be pointed out that when the founders of phenetics and the transformed cladists advocate no theory, they are in fact, advocating no theory of methods and no theory of objects, respectively. Whatever their excuses (e.g. difficulties in circumscribing the prerequisites of the methods or the ultimate properties of objects), such attitudes may lead to misuses or misinterpretations. In intuitive terms, although the numerical cladists reintroduce R-considerations in their Q-method, in overlooking the biases of the mathematical techniques they seem to equate such distinct concepts as extension and intension. In this case, the weakness proceeds from numerical routines that disregard other ways of thinking; this elimination can be equated with a theory. Similarly, although the Nelsonians apply specific methods to objects, in denying the importance of reproduction they seem to equate such distinct objects as taxa and areas. The source of the weakness here is a rigorous philosophy (Popperian?) that is opposed to inductive thinking; this choice in itself also constitutes a theory.

GENERAL CONCLUSIONS

Its sparring with both evolutionary and phenetic approaches to taxonomy has prevented Hennigism from yielding all it might have. Today, as in 1979 (see 41), it still has not been tested either at the infra-specific level or with asexual organisms. Attempts to test conflicting procedures empirically using lineages of "known" pedigree are few and inconclusive. Some rely on simulation models (158; other references in 57) that "produce situations where phenetic methods are superior to phylogenetic methods and conversely" (102); others rely on organisms whose "pedigree" is more alleged than proved, such as cultivars (7) and natural hybrids (119). Nevertheless, hybridization and reticulate evolution remain, as in 1979, a stumbling block for the Hennigian method (94, 131, 172, 173). Likewise, incongruence remains a procedural debate (54, 123, 147) and not a biological one, despite the need for analyzing the feedback between levels of development or organization (see 145).

Such deficiencies may reflect traditional taxonomists' reluctance to enter into critical areas of biological research and may also be due to fundamental difficulties relating to the link between taxonomy and evolutionary theory. These difficulties bear on the following topics: the *three* kinds of "species" (i.e. genealogical individuals, phenetic classes, and biologic systems), homology, objects and assemblages, the value of dates and distributions as documents or as attributes, congruence, parsimony versus compatibility, and the biogenetic law. The revival of these old questions, which could not be discussed sufficiently in a short review, bears witness to the stimulating impact of Hennigian thought.

Perhaps such discussions also could not have progressed further because of the number of redundant philosophical and mathematical considerations that submerge the subject. As to philosophy, the Nelsonians have apparently been locked into one position by their Popperianism, and it is necessary to arrange a way out (92). Doing so, however, requires a more diversified philosophical debate than is currently available in the taxonomic literature. As to mathematics, due to inconclusive disputes about the variety of "dialects" (48) applied to taxonomy, it is not surprising that naturalists share the reluctance expressed by Darwin about what is often "a parade of mathematical accuracy" [37a, 6th edition (1872), p. 168].

Authors seem to choose a philosophical or mathematical method primarily for subjective reasons—e.g. the author's training in that particular field. On the contrary, the debate should focus on the objective conditions for applying methods to a reality of which naturalists have the best knowledge. Although considerations of propositional logic remain scarce in present-day taxonomy (4), this debate can ultimately be resolved only by relying on the common ancestor of philosophy and mathematics—logic. The acceptance of one or another method or theory would be furthered more by a sound, candid examination than by polemics, cleverness, and redundance. This concern for logic was a constant in Hennig's personal thought; it has been the leaven of the present renewal in taxonomy.

ACKNOWLEDGMENTS

I am indebted to Professor Wolfgang Hennig for the gift of the posthumous work of his father, to Miss J. Laurent for patient secretarial assistance, to Mrs. P. Dupuis-Certain for valuable documents and references, to Dr. M. A. Schlee for linguistic elaboration, and to the editors for such an exciting enterprise, achieved despite the lack of a specific grant.

Literature Cited

1. Alberti, B. 1981. Ueber Wesen und Aussagegrenzen der "Phylogenetischen Systematik" von Hennig, untersucht am Beispiel der Zygaenidae (Lepidoptera). *Mitt. Muench. Entomol. Ges.* 71:1–31
2. Andersen, N. M. 1982. *The Semiaquatic Bugs (Hemiptera, Gerromorpha): Phylogeny, Adaptations, Biogeography and Classification.* Klampenborg, Denmark: Scandinavian Science. 455 pp.
3. Ashlock, P. D. 1971. Monophyly and associated terms. *Syst. Zool.* 20:63–69
4. Ball, I. R. 1982. Implication, conditionality and taxonomic statements. *Bijdr. Dierk.* 52:186–90
5. Banarescu, P. 1978. Some critical reflexions on Hennig's phyletical concepts. *Z. Zool. Syst. Evolutionsforsch.* 16:91–101
6. Barrett, P. H., Weinshank, D. J., Gottleber, T. T. 1981. *A Concordance to Darwin's Origin of Species First Edition.* Ithaca, NY: Cornell Univ. Press. 834 pp.
7. Baum, B. R. 1983. Relationships between transformation series and some numerical cladistic methods at the infraspecific level, when genealogies are known. See Ref. 56, pp. 340–45
8. Beatty, J. 1982. Classes and cladists. *Syst. Zool.* 31:25–34
9. Bedier, J. 1928. La tradition manuscrite du Lai de l'Ombre: réflexions sur l'art d'éditer les anciens textes. *Romania* 54:161–96, 321–56. Paris: Champion. 100 pp. Reprinted 1970
10. Bernardi, G., ed. 1981. Tendances actuelles de la biogéographie. *Mém. Soc. Biogéogr. 3e ser.* 2:1–162
11. Bottjer, P. D. 1980. Farris' "Information content" and phylogenetic versus evolutionary classification: The philosophical differences remain. *Syst. Zool.* 29:382–86
12. Boudreaux, H. B. 1979. *Arthropod Phylogeny with Special Reference to Insects.* New York: Wiley. 320 pp.

13. Boy, J. A. 1981. Zur Anwendung der Hennigschen Methode in der Wirbeltierpaläontologie. *Palaeontol. Z.* 55:87–107
14. Brady, R. H. 1983. Parsimony, hierarchy, and biological implications. *Adv. Cladistics* 2:49–60
15. Brooks, D. R. 1981. Hennig's parasitological method: A proposed solution. *Syst. Zool.* 30:229–49
16. Brothers, D. J. 1983. Nomenclature at the ordinal and higher levels. *Syst. Zool.* 32:34–42
17. Brundin, L. Z. 1981. Croizat's panbiogeography versus phylogenetic biogeography. See Ref. 133, pp. 94–158
18. Buck, R. C., Hull, D. L. 1966. The logical structure of the Linnean hierarchy. *Syst. Zool.* 15:97–111
19. Buffetaut, E., Janvier, P., Rage, J. C., Tassy, P., eds. 1982. Phylogénie et Paléobiogéographie. Livre Jubilaire en l'Honneur de Robert Hoffstetter. *Géobios.* 492 pp. (Spec. issue No. 6)
20. Cain, A. J., Harrison, G. A. 1960. Phyletic weighting. *Proc. Zool. Soc. London* 135:1–31
21. Camin, J. H., Sokal, R. R. 1965. A method for deducing branching sequences in phylogeny. *Evolution* 19:311–26
22. Carroll, R. L. 1982. Early evolution of reptiles. *Ann. Rev. Ecol. Syst.* 13:87–109
23. Chaline, J., ed. 1983. Modalités, rythmes, mécanismes de l'évolution biologique. Gradualisme phylétique ou équilibres ponctués. *Colloq. Int. CNRS.* 330:337 pp.
24. Charig, A. 1981. Cladistics: A different point of view. *Biologist* 28:19–20 Reprinted 1982 in *Evolution Now*, ed. J. M. Smith, pp. 121–24. London: Macmillan
25. Charig, A. 1982. Systematics in biology: A fundamental comparison of some major schools of thought. *Syst. Assoc. Spec. Vol.* 21:363–440
26. Ciochon, R. L., Chiarelli, A. B., eds. 1980. *Evolutionary Biology of the New World Monkeys and Continental Drift.* New York: Plenum. 528 pp.
27. Ciochon, R. L., Corruccini, R. S., eds. 1983. *New Interpretations of Ape and Human Ancestry.* New York/London: Plenum. 843 pp.
28. Colless, D. H. 1969. The interpretation of Hennig's "Phylogenetic Systematics"—a reply to Dr. Schlee. *Syst. Zool.* 18:134–44
29. Coombs, E. A. K., Donoghue, M. J., McGinley, R. J. 1981. Characters, computers, and cladograms: A review of the Berkeley cladistics workshop. *Syst. Bot.* 6:359–72
30. Cracraft, J. 1983. Cladistic analysis and vicariance biogeography. *Am. Sci.* 71:273–81
31. Cracraft, J., Eldredge, N., eds. 1979. *Phylogenetic Analysis and Paleontology.* New York: Columbia Univ. Press. 233 pp.
32. Craw, R. C. 1982. Phylogenetics, areas, geology and the biogeography of Croizat: A radical view. *Syst. Zool.* 31:304–16
33. Croizat, L. 1979. Hennig (1966) entre Rosa (1918) y Lovtrup (1977): Medio siglo de sistemática filogenética. *Bol. Acad. Cien. Fis. Mat. Nat. Caracas* 38(116):59–147
34. Croizat, L. 1982. Vicariance/Vicariism, Panbiogeography, "Vicariance biogeography", etc.: A clarification. *Syst. Zool.* 31:291–304
35. Crowson, R. A. 1981. *The Biology of the Coleoptera.* London: Academic 802 pp.
36. Cuénot, L. 1940. Remarques sur un essai d'arbre généalogique du règne animal. *C. R. Acad. Sci.* 210:23–27
37. Darlington, P. J. Jr. 1970. A practical criticism of Hennig-Brundin "Phylogenetic systematics" and antarctic biogeography. *Syst. Zool.* 19:1–18
37a. Darwin, C. 1859. *On the Origin of Species by Means of Natural Selection.* London: Murray. 513 pp. (Facsimile ed., Cambridge, Mass: Harvard Univ. Press, 1964)
38. Darwin, C. 1877. *The Descent of Man, and Selection in Relation to Sex.* London: Murray. 693 pp. 2nd ed. rev. & augmented. 12th thousand [Definitive version].
39. De Candolle, A. P. 1813. *Théorie élémentaire de la Botanique.* Paris: Deterville. 527 pp.
40. Dobzhansky, T., Ayala, F. J., Stebbins, G. L., Valentine, J. W. 1977. *Evolution.* San Francisco: Freeman. 572 pp.
41. Dupuis, C. 1979. Permanence et actualité de la systématique: La "Systématique phylogénétique" de W. Hennig. *Cah. Nat.* (NS) 34:1–69
42. Dupuis, C. 1980. The hennigo-cladism: A taxonomic method born of entomology. *Abstr. 16th Int. Congr. Entomol. Kyoto*, p. 15
43. Dupuis, C. 1983. La volonté d'être entomologiste . . . *Bull. Soc. Entomol. Fr.* 88:18–38
44. Dupuis, C. 1984. Haeckel or Hennig? The Gordian knot of characters, development and procedures in phylogeny. In *Phylogenesis and Ontogenesis, Fond. Arch. Jean Piaget, Geneva, 5th Adv. Course, 1983. Hum. Dev.* In press
45. Eichler, W. 1978. Kritische Einwände

gegen die hennigische kladistische Systematik. *Biol. Rundsch.* 16:175–85
46. Eldredge, N., Cracraft, J. 1980. *Phylogenetic Patterns and the Evolutionary Process: Method and Theory in Comparative Biology.* New York: Columbia Univ. Press. 349 pp.
47. Endler, J. A. 1982. Problems in distinguishing historical from ecological factors in biogeography. *Am. Zool.* 22:441–52
48. Estabrook, G. F. 1972. Cladistic methodology: A discussion of the theoretical basis for the induction of evolutionary history. *Ann. Rev. Ecol. Syst.* 3:427–56
49. Estabrook, G. F. 1972. Theoretical methods in systematic and evolutionary studies. *Prog. Theoret. Biol.* 2:23–86
50. Farris, J. S. 1977. On the phenetic approach to vertebrate classification. In *Major Patterns in Vertebrate Evolution,* ed. M. K. Hecht, P. C. Goody, B. M. Hecht, pp. 823–50. New York: Plenum. 908 pp.
51. Farris, J. S., ed. 1979. The Willi Hennig memorial symposium. *Syst. Zool.* 28:415–519
52. Farris, J. S. 1980. Naturalness, information, invariance, and the consequences of phenetic criteria. *Syst. Zool.* 29:360–81
53. Farris, J. S. 1981. Distance data in phylogenetic analysis. *Adv. Cladistics* 1:3–23
54. Farris, J. S. 1983. The logical basis of phylogenetic analysis. *Adv. Cladistics* 2:7–36
55. Farris, J. S., Kluge, A. G., Eckardt, M. J. 1970. A numerical approach to phylogenetic systematics. *Syst. Zool.* 19:172–89
55a. Felsenstein, J. 1982. Numerical methods for inferring evolutionary trees. *Q. Rev. Biol.* 57:379–404
56. Felsenstein, J., ed. 1983. *Numerical Taxonomy, Proc. NATO Adv. Study Inst. Numer. Taxon., Bad Windsheim, Germany, July 4–16, 1982.* Berlin: Springer-Verlag. 644 pp.
57. Fiala, K. L. 1983. A simulation model for comparing numerical taxonomic methods. See Ref. 56, pp. 87–91
58. Fink, S. V. 1982. Report on the second annual meeting of the Willi Hennig Society. *Syst. Zool.* 31:180–97
59. Forey, P. L. 1982. Neontological analysis versus palaeontological stories. *Syst. Assoc. Spec. Vol.* 21:119–57
59a. Froger, J. 1968. *La critique des textes et son automatisation.* Paris: Dunod. 280 pp.
60. Funk, V. A., Brooks, D. R., eds. 1981. *Advances in Cladistics, Proc. 1st Meet.*

Willi Hennig Soc. Bronx: NY Bot. Gard. 250 pp.
61. Funk, V. A., Brooks, D. R. 1981. National Science Foundation workshop on the theory and application of cladistic methodology. Organized by T. Duncan and T. Stuessy. University of California, Berkeley, 22–28 March 1981. *Syst. Zool.* 30:491–98
62. Funk, V. A., Wagner, W. H. Jr. 1982. A bibliography of botanical cladistics: I. 1981. *Brittonia* 34:118–24
63. Gilmour, J. S. L. 1940. Taxonomy and philosophy. In *The New Systematics,* ed. J. S. Huxley, pp. 461–74. Oxford: Oxford Univ. Press
64. Griffiths, T. A. 1983. On the phylogeny of the *Glossophaginae* and the proper use of outgroup analysis. *Syst. Zool.* 32:283–85
65. Gupta, A. P., ed. 1979. *Arthropod Phylogeny.* New York: Van Nostrand-Reinhold. 762 pp.
66. Gutmann, W. F., Bonik, K. 1981. Hennigs Theorem und die Strategie des stammesgeschichtlichen Rekonstruierens: Die Agnathen-Gnathostomen-Beziehung als Beispiel. *Palaeontol. Z.* 55:51–70
67. Hanson, E. D. 1977. *The Origin and Early Evolution of Animals.* Middletown, Conn: Wesleyan Univ. Press. 670 pp.
68. Hemmer, H. 1981. Die Evolution der Pantherkatzen: Modell zur Ueberpruefung der Brauchbarkeit der hennigschen Prinzipien der phylogenetischen Systematik für wirbeltierpalaeontologische Studien. *Palaeontol. Z.* 55:109–16
69. Hengsbach, R. 1973. Zum Verständnis der phylogenetischen Ansichten von Schindewolf, Hennig und Remane. *Zool. Beitr.* (NF) 19:315–17
70. Hennig, W. 1936. Beziehungen zwischen geographischer Verbreitung und systematischer Gliederung bei einiger Dipterenfamilien: Ein Beitrag zum Problem der Gliederung systematischer Kategorien höherer Ordnung. *Zool. Anz.* 116:161–75
71. Hennig, W. 1936. Revision der Gattung *Draco* (Agamidae). *Temminckia* 1:153–220
72. Hennig, W. 1936. Ueber einige Gesetzmässigkeiten der geographischen Variation in der Reptiliengattung *Draco* L.: "Parallele" und "konvergente" Rassenbildung. *Biol. Zentralbl.* 56:549–59
73. Hennig, W. 1943. Ein Beitrag zum Problem der "Beziehungen zwischen Larven und Imaginalsystematik". *Arb. Morphol. Taxon. Entomol.* 10:138–44

74. Hennig, W. 1950. *Grundzüge einer Theorie der Phylogenetischen Systematik*. Berlin: Deutscher Zentralverlag. 370 pp. Reprinted 1981. Koenigstein, West Germany: Koeltz

75. Hennig, W. 1953. Kritische Bemerkungen zum phylogenetischen System der Insekten. *Beitr. Entomol.* 3 (Sonderheft: Festschr. Sachtleben): 1–85

76. Hennig, W. 1957. Systematik und Phylogenese. In *Bericht über die Hundertjahrfeier d. Dtsch. Entom. Gesells. Berlin, 30 Sept.–5 Oct. 1956*, pp. 50–71. Berlin: Akad

77. Hennig, W. 1966. *Phylogenetic Systematics*. Urbana: Univ. Ill. Press. 263 pp. Reprinted 1979

78. Hennig, W. 1966. The Diptera fauna of New Zealand as a problem in systematics and zoogeography. Transl. P. Wygodzinsky. *Pac. Insects Monogr.* 9:1–81 (From German)

79. Hennig, W. 1968. *Elementos de una sistemática filogenética*. Buenos Aires: Ed. Univ. 353 pp.

80. Hennig, W. 1969. *Die Stammesgeschichte der Insekten*. Frankfurt: Kramer. 436 pp.

81. Hennig, W. 1981. *Insect Phylogeny*. Transl. A. C. Pont. Rev. notes, D. Schlee. Chichester: Wiley. 514 pp. (From German)

82. Hennig, W. 1982. *Phylogenetische Systematik*. Berlin/Hamburg: Parey. 246 pp.

83. Hennig, W. 1983. Stammesgeschichte der Chordaten. *Fortschr. Zool. Syst. Evolutionsforsch.* 2:1–208

84. Hennig, W. 1984. *Aufgaben und Probleme Stammesgeschichtlicher Forschung*. Berlin/Hamburg: Parey. 64 pp.

85. Hill, C. R., Crane, P. R. 1982. Evolutionary cladistics and the origin of Angiosperms. *Syst. Assoc. Spec. Vol.* 21:269–361

86. Hoffman, A. 1982. Punctuated versus gradual mode of evolution. A reconsideration. *Evol. Biol.* 15:411–36

87. Hölder, H., ed. 1981. Hennig's kladistische Methode aus paläontologischer Sicht. *Palaeontol. Z.* 55:9–131

88. House, M. R., ed. 1979. The origin of major invertebrate groups. *Syst. Assoc. Spec. Vol.* 12:1–515

89. Hull, D. L. 1970. Contemporary systematic philosophies. *Ann. Rev. Ecol. Syst.* 1:19–54

90. Hull, D. L. 1976. Are species really individuals? *Syst. Zool.* 25:174–91

91. Hull, D. L. 1979. The limits of cladism. *Syst. Zool.* 28:416–40

92. Hull, D. L. 1983. Karl Popper and Plato's metaphor. *Adv. Cladistics* 2:177–89

93. Humphries, C. J. 1981. Biogeographical methods and the southern beeches (Fagaceae: Nothofagus). *Adv. Cladistics* 1:177–207. Abridged reprint 1983. *Syst. Assoc. Spec. Vol.* 23:335–65

94. Humphries, C. J. 1983. Primary data in hybrid analysis. *Adv. Cladistics* 2:89–103

95. Humphries, C. J., Richardson, P. M. 1980. Hennig's methods and phytochemistry. *Syst. Assoc. Spec. Vol.* 16:353–78

96. Huxley, J. S. 1957. The three types of evolutionary process. *Nature* 180:454–55

97. Huxley, J. S. 1958. Evolutionary processes and taxonomy with special references to grades. *Uppsala Univ. Årsskr.* 1958 (6):21–38

98. Illies, J. 1967. Zur modernen Systematik: Ein Vergleich der Methoden von Hennig und Remane. *Zool. Beitr.* (NF) 13:521–28

99. Illies, J. 1983. Changing concepts in biogeography. *Ann. Rev. Entomol.* 28:391–406

100. In memoriam. 1978. Willi Hennig (20.4.1913 † 5.11.1976). *Beitr. Entomol.* 28:169–77

101. Jahn, I., Löther, R., Senglaub, K., eds. 1982. *Geschichte der Biologie: Theorien, Methoden, Institutionen, Kurzbiographien*. Jena: Fischer. 859 pp.

102. Janowitz, M. F. 1980. Similarity measures on binary data. *Syst. Zool.* 29:342–59

103. Jardine, N., Sibson, R. 1971. *Mathematical Taxonomy*. London: Wiley. 286 pp. Reprinted 1977

104. Johannsen, W. 1909. *Elemente der Exakten Erblichkeitslehre*. Jena: Fischer. 516 pp.

105. Johnson, L. A. S. 1968. Rainbow's end: The quest for an optimal taxonomy. *Proc. Linn. Soc. N. S. Wales* 93:8–45

106. Joysey, K. A., Friday, A. E., eds. 1982. Problems of phylogenetic reconstruction. *Syst. Assoc. Spec. Vol.* 21:442 pp.

107. Kavanaugh, D. H. 1972. Hennig's principles and methods of phylogenetic systematics. *Biologist* (Denver) 54:115–27

108. Kavanaugh, D. H. 1978. Hennigian phylogenetics in contemporary systematics: Principles, methods and uses. In *Beltsville Symposia in Agriculture Research*. Vol. 2, *Biosystematics in Agriculture*. *Beltsville 1977*, ed. J. A. Romberger, pp. 139–50. Montclair, NJ: Allanheld, Osmun

109. Kemp, T. S. 1982. *Mammal-Like Reptiles and the Origin of Mammals*. New York/London: Academic. 364 pp.

110. Kristensen, N. P. 1981. Phylogeny of insect orders. *Ann. Rev. Entomol.* 26:135–57

111. Kühne, W. G. 1979. *Paläontologie und*

dialektischer Materialismus. Jena: Fischer. 131 pp.

112. Lawrence, J. F., Newton, A. F. Jr. 1982. Evolution and classification of beetles. *Ann. Rev. Ecol. Syst.* 13:261–90

113. Le Quesne, W. J. 1982. Compatibility analysis and its applications. *Zool. J. Linn. Soc. London* 74:267–75

114. Marichal, R. 1961. La critique des textes. In *L'Histoire et ses méthodes,* ed. C. Samaran, pp. 1247–1366. Paris: Gallimard

115. Martinell, J., ed. 1981. *International Symposium on "Concept and Method in Paleontology": Contributed Papers.* Barcelona: Dep. Paleontol., Univ. Barcelona. 329 pp.

116. Mayr, E. 1965. Classification and phylogeny. *Am. Zool.* 5:165–74

117. Mayr, E. 1965. Numerical phenetics and taxonomic theory. *Syst. Zool.* 14:73–97

118. Mayr, E. 1982. *The Growth of Biological Thought: Diversity, Evolution and Inheritance.* Cambridge, Mass: Belknap. 974 pp.

119. McAllister, D. E., Coad, B. W. 1978. A test between relationships based on phenetic and cladistic taxonomic methods. *Can. J. Zool.* 56:2198–2210

120. McGinley, R. J., Michener, C. D. 1980. Dr. Nelson on dendronomics. *Syst. Zool.* 29:91–93

121. McNeill, J. 1982. Report on the fifteenth annual numerical taxonomy conference. *Syst. Zool.* 31:197–201

122. Medicus, F. K. 1789–1791. *Philosophische Botanik, mit kritischen Bemerkungen.* Vols. 1, 2. Mannheim: n. Hof- u. Akad. Buchhandl. 266 pp., 112 pp.

123. Mickevich, M. F. 1980. Taxonomic congruence: Rohlf and Sokal's misunderstanding. *Syst. Zool.* 29:162–76

124. Mitter, C. 1980. The thirteenth annual numerical taxonomy conference. *Syst. Zool.* 29:177–90

125. Moss, W. W. 1983. Report on NATO Advanced Study Institute on Numerical Taxonomy. *Syst. Zool.* 32:76–83

126. Moss, W. W., Brooks, D. R., eds. 1979. Contemporary methods in systematic parasitology. *Am. Zool.* 19:1177–1238

127. Moss, W. W., Hendrickson, J. A. Jr. 1973. Numerical taxonomy. *Ann. Rev. Entomol.* 18:227–258

128. Nelson, G. J. 1972. Comments on Hennig's "Phylogenetic systematics" and its influence on ichthyology. *Syst. Zool.* 21:364–74

129. Nelson, G. J. 1974. Darwin-Hennig classification: A reply to Ernst Mayr. *Syst. Zool.* 23:452–58

130. Nelson, G. J. 1979. Cladistic analysis and synthesis: Principles and definitions with a historical note on Adanson's *Famille des Plantes* 1763–1764. *Syst. Zool.* 28:1–21

131. Nelson, G. J. 1983. Reticulation in cladograms. *Adv. Cladistics* 2:105–11

132. Nelson, G. J., Platnick, N. I. 1981. *Systematics and Biogeography: Cladistics and Vicariance.* New York: Columbia Univ. Press. 567 pp.

133. Nelson, G. J., Rosen, D. E., eds. 1981. *Vicariance Biogeography.* New York: Columbia Univ. Press. 593 pp.

134. Patterson, C. 1980. Cladistics. *Biologist* 27:234–40. See also Ref. 24, pp. 110–20

135. Patterson, C. 1981. Methods of paleobiogeography. See Ref. 133, pp. 446–500

136. Patterson, C. 1981. Significance of fossils in determining evolutionary relationships. *Ann. Rev. Ecol. Syst.* 12:195–223

137. Patterson, C., ed. 1982. Methods of phylogenetic reconstruction. *Zool. J. Linn. Soc. London* 74:197–344

138. Patterson, C. 1982. Morphological characters and homology. *Syst. Assoc. Spec. Vol.* 21:21–74

139. Patton, J. C., Avise, J. C. 1983. An empirical evaluation of qualitative Hennigian analyses of protein electrophoretic data. *J. Mol. Evol.* 19:244–54

140. Platnick, N. I. 1979. Philosophy and the transformation of cladistics. *Syst. Zool.* 28:537–46

141. Platnick, N. I., Cameron, H. D. 1977. Cladistic methods in textual, linguistic, and phylogenetic analysis. *Syst. Zool.* 26:380–85

142. Platnick, N. I., Funk, V. A., eds. 1983. *Advances in Cladistics, Proc. 2nd Meet. Willi Hennig Soc.* New York: Columbia Univ. Press. 218 pp.

143. Renous, S. 1981. Développement de l'aspect historique de la biogéographie par la superposition de deux thèses: Proposition d'une hypothèse phylogénétique bâtie selon les principes hennigiens et théorie de la dérive des continents. *C. R. Séances Soc. Biogéogr.* 57:81–102

144. Richardson, P. M. 1983. Flavonoids and phylogenetic systematics. *Adv. Cladistics* 2:115–23

145. Riedl, R. 1978 [1975]. *Order in Living Organisms.* Transl. R. P. S. Jefferies. Chichester, England: Wiley. 313 pp. (From German)

146. Rohlf, F. J., Colless, D. H., Hart, G. 1983. Taxonomic congruence re-examined. *Syst. Zool.* 32:144–58

147. Rohlf, F. J., Sokal, R. R. 1980. Comments on taxonomic congruence. *Syst. Zool.* 29:97–101

147a. Ross, H. H. 1973. Evolution and phylogeny. In *History of Entomology,* ed.

R. F. Smith, T. E. Mittler, C. N. Smith, pp. 171–84. Palo Alto, Calif: Annual Reviews

148. Ross, H. H. 1974. *Biological Systematics*. Reading, Mass: Addison-Wesley. 345 pp.

149. Schlee, D. 1969. Hennig's principle of phylogenetic systematics: An "Intuitive Statistico-phenetic Taxonomy"? *Syst. Zool.* 18:127–34

150. Schlee, D. 1971. Die Rekonstruktion der Phylogenese mit Hennig's Prinzip. *Aufsätze Reden senckenb. naturforsch. Ges.* 20:1–62

151. Schultze, H. P. 1981. Hennig und der Ursprung der Tetrapoda. *Palaeontol. Z.* 55:71–86

152. Schwartz, J. H., Rollins, H. B., eds. 1979. Models and methodologies in evolutionary theory. *Bull. Carnegie Mus. Nat. Hist.* 13:1–105

153. Siewing, R., ed. 1980. Strukturanalyse und Evolutionsforschung: Das Merkmal (Symp., 5–8 März 1979, Erlangen). *Zool. Jahrb. Abt. Anat. Ontog. Tiere* 103:153–485

154. Simberloff, D., Heck, K. L., McCoy, E. D., Connor, E. F. 1981. There have been no statistical tests of cladistic biogeographical hypotheses. See Ref. 133, pp. 40–93

155. Sims, R. W., Price, J. H., Whalley, P. E. S., eds. 1983. Evolution, time and space: The emergence of the biosphere. *Syst. Assoc. Spec. Vol.* 23:492 pp.

156. Sneath, P. H. A. 1962. The construction of taxonomic groups. *Soc. Gen. Microbiol. Symp.* 12:289–332

157. Sneath, P. H. A., Sokal, R. R. 1973. *Numerical Taxonomy*. San Francisco: Freeman. 573 pp.

158. Sokal, R. R. 1983. A phylogenetic analysis of the Caminalcules. I–IV. *Syst. Zool.* 32:159–84; 185–201; 248–58; 259–75

159. Sokal, R. R., Sneath, P. H. A. 1963. *Principles of Numerical Taxonomy*. San Francisco: Freeman. 359 pp.

160. Steffan, A. W., ed. 1978. Dem Gedenken an Professor Dr. phil. Dr. rer. nat. h.c. Willi Hennig. *Entomol. Ger.* 4:193–396

161. Stevens, P. F. 1980. Evolutionary polarity of character states. *Ann. Rev. Ecol. Syst.* 11:333–58

162. Stevens, P. F. 1983. Report of third annual Willi Hennig Society meeting. *Syst. Zool.* 32:285–91

163. Stone, A. R., Platt, H. M., Khalil, L. F., eds. 1983. Concepts in Nematode systematics. *Syst. Assoc. Spec. Vol.* 22: 388 pp.

164. Strauch, J. G. Jr. 1978. The phylogeny of Charadriiformes (Aves): A new estimate using the method of character compatibility analysis. *Trans. Zool. Soc. London* 34:263–345

165. Sucker, U. 1978. *Philosophische Probleme der Arttheorie*. Jena: Fischer. 119 pp.

166. Szalay, F. S., Delson, E. 1979. *Evolutionary History of the Primates*. New York/London: Academic. 580 pp.

167. Tattersall, I. 1982. *The Primates of Madagascar*. New York: Columbia Univ. Press. 382 pp.

168. Thenius, E. 1979. Hennigs phylogenetische Systematik und paläontologische Befunde. *Neues Jahrb. Geol. Palaeontol. Monatsh.* 7:404–14

169. Thienemann, A., Krüger, F. 1937. "*Orthocladius*" *abiskoensis* Edwards und *rubicundus* (Mg.), zwei "Puppen-Species" der Chironomiden. *Zool. Anz.* 117:257–67

170. Thorpe, J. P. 1982. The molecular clock hypothesis: Biochemical evolution, genetic differentiation and systematics. *Ann. Rev. Ecol. Syst.* 13:139–68

171. Van Emden, F. 1929. Ueber den Speciesbegriff vom Standpunkt der Larvensystematik aus. *Wanderversamml. Deutsch. Entomol.* 3:47–56

172. Wagner, W. H. Jr. 1983. Reticulistics: The recognition of hybrids and their role in cladistics and classification. *Adv. Cladistics* 2:63–79

173. Wanntorp, H. E. 1983. Reticulated cladograms and the identification of hybrid taxa. *Adv. Cladistics* 2:81–88

174. Warner, R. M. 1983. Karyotypic megaevolution and phylogenetic analysis: New World nectar-feeding bats revisited. *Syst. Zool.* 32:279–82

175. Watrous, L. E., Wheeler, Q. D. 1981. The out-group comparison method of character analysis. *Syst. Zool.* 30:1–11

176. Weber, H. 1948. Morphologie, Entwicklungsgeschichte, Systematik und specielle Ökologie der Wirbellosen. *FIAT Rev. Germ. Sc. 1939–46, Biol.* 4:29–76

177. Wigand, A. 1872. Die Genealogie der Urzellen. Braunschweig. p. 47 Cited in de Hartmann, E. 1877. *Le Darwinisme*, p. 23. Paris: Baillière

178. Wiley, E. O. 1981. *Phylogenetics: The Theory and Practice of Phylogenetic Systematics*. New York: Wiley. 439 pp.

179. Williams, W. T. 1971. Principles of clustering. *Ann. Rev. Ecol. Syst.* 2:303–26

180. Woodger, J. H. 1952. *Biology and Language*. Cambridge: Cambridge Univ. Press. 364 pp.

Ann. Rev. Ecol. Syst. 1984. 15:25–64

GEOGRAPHIC PATTERNS AND ENVIRONMENTAL GRADIENTS: THE CENTRAL–MARGINAL MODEL IN *DROSOPHILA* REVISITED

Peter F. Brussard

Section of Ecology and Systematics, Cornell University, Ithaca, New York 14853

INTRODUCTION

The Central–Marginal Model

The familiar central-marginal model in evolutionary biology asserts that populations near the center of a species' range usually are contiguous, are at high density, and display high levels of genetic and phenotypic variation, while populations on the margin of the range are isolated, sparse, and chromosomally monomorphic (88, 94). Because of their supposedly unique genetic properties, marginal populations are considered very important in speciation events (11, 94), and some researchers now assume that speciation on the edge of the range is the predominant mode of speciation in the majority of groups (e.g. 62).

Most of the earlier data on the genetic characteristics of central and marginal populations came from studies of chromosomal polymorphism in various species of *Drosophila*. These studies often revealed a common pattern: Populations from the center of the range were highly polymorphic for inverted sequences, while those near the edge of the range usually showed a marked reduction or absence of this kind of variation. By 1973 such a pattern could be documented for 14 of the 15 species that had been examined extensively enough to permit such a comparison (128). Da Cunha et al (37) and da Cunha & Dobzhansky (38) proposed that this pattern could be explained by an ecological hypothesis: The amount of chromosomal polymorphism in a population is proportional to the number of ecological niches it exploits, and more niches are

25

0066-4162/84/1120-0025$02.00

available to a species in the center of its range than on the edge. Thus, these investigators viewed inversions as devices that "locked up" coadapted blocks of genes that conferred specific adaptations to particular niches or ecological conditions.

Carson (21) proposed a different explanation. He felt that inversions do not relate to any specific component of the environment but rather confer a general vigor, or "heterotic buffering," in the heterozygous state. High levels of structural heterozygosity are favored in the center of the range because this heterotic buffering results in better overall performance in the variety of niches that are available there, and central populations can afford the luxury of producing less fit homozygotes because they are large and outcrossed. On the other hand, marginal populations are too small and inbred to afford adaptation by heterotic buffering. Among these populations, selection favors homozygotes and the attainment of adaptation based on specific, genetically fixed, adaptive features.

The advent of electrophoretic surveys, which have been used to estimate the amounts of genetic variation present in natural populations, seriously weakened certain aspects of Carson's hypothesis. The case for marginal reduction in heterozygosity was not supported by the data on enzyme variation; marginal populations usually proved to have levels of overall heterozygosity similar to those in the center of the range. In order to explain the apparent paradox between the patterns of chromosomal and allozyme variation in *Drosophila,* Soulé (128) proposed a third hypothesis. He argued that any novel environments, such as those likely to be encountered on the margins of a range, will cause a transient reduction in the diversity of coadapted gene complexes. Given enough time, increasing coadaptation to these conditions will favor increased levels of structural heterozygosity, but marginal populations rarely persist long enough for this to happen. The advantages of high levels of genic heterozygosity, however, would be as great in marginal populations as in central ones due to heterosis, and for this reason, allozyme variability would be as high on the edge of the range as elsewhere.

Wallace (140) recently proposed a fourth explanation for the central-marginal decline in inversion polymorphism. Assuming that structural heterozygotes have distinctly higher competitive abilities than homozygotes, heterozygotes would be favored in the more dense, crowded populations near the range center. Moving toward the margin of the species distribution, however, the probability that a gravid female can find a suitable site for oviposition declines, as does the probability that her offspring will survive to reproduce in these less than ideal physical surroundings. Under these circumstances, the production of whole broods of reasonably fit individuals has a greater probability of leaving surviving offspring than does the production of a smaller number of competitively superior, inversion polymorphic ones.

The purpose of this review is to examine the existing data on central-marginal contrasts in *Drosophila* and evaluate the four above hypotheses in light of these data. In addition, I will point out that generalist species with large ranges and continental distributions tend to have a predictable, declining level of abundance from the center of the range to the periphery. This decline is usually highly correlated with the reduction in inversion heterozygosity often observed in *Drosophila,* which in turn suggests that the latter may be causally related to the former.

Semantic Difficulties

One of the hardest things to sort out in the central-marginal model is the relationship between peripheral, or "range-edge," effects and ecological marginality. It is usually tacitly assumed that ecologically marginal conditions obtain in peripheral areas. It is possible, however, that certain characteristics of peripheral populations may result from historical events, rather than from a selective response to ecologically marginal conditions. For example, chromosomal polymorphism may be lower in peripheral populations because structural rearrangements arising in central populations have not yet had time to spread throughout the distribution area. Since conditions on the periphery of a range can at least theoretically result from historical factors, ecological conditions, or both, I will hereafter use words such as *optimal, favorable,* and *marginal* when referring to environmental rigor and *central, intermediate,* and *peripheral* when referring to positions within a distribution area.

A second semantic difficulty concerns the words *habitat* and *niche*. Whittaker et al (148, 149) provided an extensive discussion of the concepts associated with each term, and I will follow their usage here. *Habitat* is conceived as the range of environments or communities over which a species occurs, while *niche* applies to the role of a species in a given community or habitat. For *Drosophila,* niche variables would include the type of breeding and feeding substrates, utilization of space (e.g. characteristically flying in the canopy or at ground level, seasonal time, or diurnal time), among others. A species' relationship to its environment can be defined in relation to both habitat and niche variables, which together represent the full range of ecological circumstances to which it is exposed. This relationship to both habitat and niche is referred to as a species' *ecotope*.

PATTERNS OF DISTRIBUTION AND ABUNDANCE

Geographical Patterns and Range Maps

The ranges of most species of *Drosophila* are known in only a most superficial way, but the patterns described for other, better known groups are probably generally applicable. One pattern consists of a central, essentially continuously inhabited area surrounded by peripheral areas where habitation is much patchier. This pattern seems to be fairly typical of widely distributed, common species. An examination of about 200 carefully constructed range maps of North American forest trees (91) indicates that it may hold for about one third of the species shown. Another third have more than one distributional center (each of which may have peripheral outliers), while the rest are distributed in dendritic patterns, as islands or archipelagos, or as highly disjunct isolates.

Range maps generally depict only the limits of distribution areas, providing no indication of relative abundance within the range. When such information is available, however, the densities are usually greatest in the central regions and decline toward the periphery (e.g. 12, 53, 122), but again this pattern is by no means universal. Rabinowitz (119) provided a useful framework for understanding the relationships between distribution and abundance in plants based upon considerations of geographic range, habitat specificity, and local population size. She concluded that only those species with a wide range and broad habitat specificity are likely to show a central-peripheral decline in abundance. Many other species, even some with wide ranges, are not common anywhere but occur sparsely in a variety of habitats; still others, although sometimes widespread, are associated only with particular habitats that are themselves uncommon. Carson (14) supplied some evidence that a similar diversity of patterns exists for *Drosophila* species as well.

Local and Regional Patterns

For any species, most natural environments consist of a mosaic of habitat patches, some of which are more suitable than others. A habitat may be classified as favorable or marginal depending on the relative numbers of individuals of a species that it can maintain. A favorable habitat patch will support consistently higher densities than a marginal one, although eruptions may occur occasionally in the latter. Habitat patches differ in size and frequency (i.e. the probability of finding one in a given area); both parameters must be assessed in relation to the average dispersal tendencies of the species in question. The assessment of patch size for most species of *Drosophila* undoubtedly presents special problems due to their generally small size and inconspicuousness.

Whittaker (144–146) convincingly showed that if the habitat patches occupied by many species of plants and insects are ordered with respect to various underlying environmental gradients—primarily temperature and moisture—it is usually possible to discern a relationship between patterns of species abundance and these gradients. Typically, such an ordering results in bell-shaped response curves for various population measurements. These curves are apparently Gaussian in form, tapering on each side of the mode or population optimum toward asymptotic tails. When measures of species abundance are plotted against both gradients on two other axes, the abundance measures can be bounded with contour lines, and the resulting figures can be visualized in three dimensions as population "hills." The simplest hills are Gaussian surfaces or binomial solids, any transection of which will produce a bell-shaped curve of abundance measures. It is assumed that this Gaussian response is a product of the physiological tolerances of a species, modified by the availability of resources subject to competitive utilization by associated species (143). In the absence of competition, a species should reach its maximum abundance at the center of its environmental range and its abundance should decline toward the extremes of its tolerance. In the presence of closely related species, however, densities may be affected by competition, and it is likely that a species' occupancy of a gradient will be determined both by physical stress and by the outcome of various interspecific interactions. Even under competitive situations, though, the population response curve is expected to be Gaussian (147).

A *Drosophila* population's ultimate size in any habitat patch will depend primarily on two factors. The first is the availability of suitable resources for feeding and breeding, both in an absolute sense and relative to the flies' ability to locate them. The second is the prevailing annual temperature regime within the patch. This strong dependence on environmental temperature results from the fact that the average size a population of any r-strategist insect species, such as most *Drosophila*, can attain every year depends upon two major factors—the rate of per capita population growth (r) and the length of time r remains positive (2). Longevity, fecundity, and speed of development in *Drosophila* are all strongly influenced by temperature (101), and these life history components are clearly highly important elements of r. Thus, not only will population growth rates be highest under optimal temperature conditions, but flies will also select and move into habitats where these conditions are most favorable (96, 102, 108, 118). Other physical factors, particularly humidity, may also exert some influence, but as long as resources for feeding and breeding do not become limiting, environmental temperature conditions should be the major determinant of the population level attained in any particular patch. Furthermore, on a regional basis, temperature fluctuates rather predictably from season to season and from year to year (136). This predictability means that patterns of regional abundance should also be highly correlated with prevailing annual temperature regimes.

The foregoing considerations strongly suggest the following: In most wide-ranging, generalist species of *Drosophila*, the center of the range at any given time will coincide with the geographic region that is the most ecologically favorable for it in terms of its physiological responses to the physical environment *and* its interactions with other species. Here its populations will, on average, be largest; and areas of increasing ecological marginality can be defined by decreasing average levels of abundance. It also follows that the periphery of a range will usually correspond to an area of extreme marginality for the species. It cannot be assumed, however, that all peripheries are in fact marginal until average levels of abundance have been determined empirically at various points within the species' range.

CENTRAL–PERIPHERAL COMPARISONS IN *DROSOPHILA*

Data sets sufficient to make some types of central-peripheral comparisons exist for a score or more *Drosophila* species, but the bulk of the available information concerns four species or species groups, most of which are wide-ranging generalists. They include the Nearctic *D. robusta,* studied primarily by Carson and his collaborators; the Neotropical *D. willistoni* group and the Nearctic sibling species *D. pseudoobscura* and *D. persimilis,* studied by Dobzhansky and his students and colleagues; and the Palearctic *D. subobscura,* the domain of several European investigators. Each of these will be discussed in turn, and I will attempt to relate patterns of distribution and abundance to those of genetic variation whenever possible.

Drosophila robusta

DISTRIBUTION AND ABUNDANCE   This species is distributed throughout the eastern deciduous forests of North America, extending from central Florida northward to southern Ontario and westward to Texas, Oklahoma, Kansas, Nebraska, and South Dakota. It breeds primarily on the sap exudations of various trees (at the time of these investigations, most commonly the American elm, *Ulmus americana*) (27). The flies appear in late spring, produce an unknown number of generations during the summer, and enter reproductive diapause in late summer or early autumn. Females captured in the fall have undeveloped ovaries and a large proportion of body fat; diapause apparently is not obligatory, however. Under laboratory conditions, breeding can occur after 10 days, and continuous laboratory generations are possible thereafter (26).

Carson (16) attempted to estimate the relative abundance of this species in different parts of its range by calculating the percentage of *D. robusta* in the total sample of *D. obscura*–group and *D. robusta*–group flies captured at banana baits. These species seem to be quite similar in their ecological requirements, and Carson considered this procedure valid, although somewhat rough. As can be seen in Figure 1, the relative proportion of *D. robusta* in collections varies considerably from location to location, but the pattern strongly suggests a Gaussian surface with central-peripheral decline in abundance. The central "core" of high relative frequency (15–20%) extends from the southern Appalachians through the Middle West in a northwesterly direction. The proportion of *D. robusta* falls off in areas that surround this central core, and diminishes

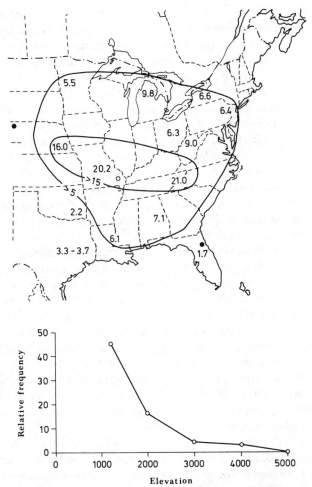

Figure 1   Frequency of *D. robusta* relative to the *affinis* and *robusta* groups in various parts of eastern North America. Data from 3 nearby locations in eastern Tennessee-North Carolina have been averaged; peripheral populations from Nebraska and Florida are shown in solid circles. Insert shows the relative frequency of *D. robusta* along an altitudinal transect in the Great Smoky Mountain National Park and vicinity. Data from Carson (16, 17).

further toward the periphery of the range in Oklahoma, Texas, Nebraska, and Florida. The percentage of *D. robusta* in the Florida collection was estimated to be 1.7 (24). *Drosophila robusta* also shows a regular decline in relative frequency along an elevational transect in the Great Smoky Mountains (16). These data are also shown in Figure 1 and support the hypothesis that there is a symmetry between the geographical and microgeographical patterns.

Although it is clear that the relative proportion of *D. robusta* in collections declines from the center to the periphery, can these data be used to infer a decline in population numbers as well? There are two kinds of estimators of population size: absolute and relative (129). Absolute estimates permit a reasonably accurate assessment of the numbers of individuals per unit area, but in *Drosophila* research they almost always involve highly labor-intensive capture-recapture techniques and are not often attempted, especially in large-scale surveys. Most wild *Drosophila*, however, are collected in the field by trapping, and trapping results can provide assessments of relative abundance provided that the traps themselves are of some standard size and configuration and that an accurate record is kept of the trapping effort involved. The data can then be expressed as numbers of individuals taken per trapping period, and the relative abundance levels of a species can be compared among different regions, habitats, or seasons. Unfortunately, methods of trapping *Drosophila* have rarely been standardized, and estimates of relative abundance among different localities are virtually nonexistent.

However, there are a few data sets available that allow us to calculate the proportional representation of a species of interest relative to other species (or of a subset of other species as Carson has done) taken on various sampling occasions. This provides an estimate of the *relative frequency* of a species in a locality, which may or may not be correlated with its relative abundance. This problem can be seen in the following example. Suppose that a standardized trapping effort produces 50 individuals of species A and 50 individuals of species B in one locality and 100 individuals of A and 900 of B in another. While the relative frequency of species A declines from 0.5 to 0.1 between the two localities, its relative abundance increases twofold. Thus, relative frequency data—without accompanying information on relative abundance—can be highly misleading indicators of population trends.

Returning to *D. robusta*, indirect evidence and anecdotal accounts suggest that relative frequency and relative abundance are actually reasonably well correlated. For example, the species is known to be rare in Florida and common in Missouri and Tennessee. Thus, the data on relative frequency shown in Figure 1 probably reflect a central-peripheral decline in relative abundance as well, as Carson suggests.

CHROMOSOMAL INVERSIONS   *Drosophila robusta* has 14 known inversions on chromosomes X, 2, and 3; some are rare, others widely distributed (25). The widely distributed inversions either were found to be clinal or displayed no recognizable pattern over large geographic areas, and significant frequency differences were observed between populations only a few kilometers apart (13). Some of the clinal changes over broad geographic areas were paralleled at least to some degree among populations along an altitudinal transect in Tennessee (86, 133). On a seasonal basis, gene arrangements were stable from year to year in the populations in New Jersey and Missouri (19, 80), despite some drastic alterations to the habitat in the latter. In contrast, seasonal changes were observed in two arrangements in a population in Virginia, and the extent of these changes differed between males and females (81, 82). Nonrandom associations among certain gene arrangements have been observed in natural populations (13), and rapid, directional changes in inversion frequencies have occurred in laboratory populations (80). Heterosis for mating speed, position effects, and recombination restriction have all been proposed as factors involved in the selective maintenance of these gene arrangements (15, 83–85, 109–111).

Paralleling the previously described pattern of relative frequency in collections, the percentage of individuals heterozygous for relatively long inversions was found to be highest in populations in the range center. Carson (17) devised an "index of free recombination" (IFR)—the percentage of the total haploid euchromatic chromosome length in which free crossing-over can occur—to quantify this trait. He found that the IFR values measured in several different populations fell into 3 statistically significant groups on the basis of their position within the species' range. Populations he referred to as central had an average IFR value of 70.5 ± 1.0, intermediate populations averaged 77.6 ± 1.0, while marginal populations averaged 84.5 ± 1.3. Subsequently, an even more marginal population was discovered at Chadron in northwestern Nebraska, which had an IFR value of 99.7 ± 0.2 (18, 20). The relationships between relative frequency, IFR values, and the percentage of females having 80–100% free crossing-over are shown in Figure 2. Both correlations are statistically significant, although their interpretation is somewhat less clear. Taken at face value, these data certainly imply that a central to peripheral decline in relative frequency is accompanied by an increase in the potential for crossing-over.

The striking contrast in the extent of chromosomal heterozygosity and potential for recombination between central and peripheral populations led Carson (21) to develop his previously mentioned theory of contrasting types of genetic adaptation in different areas of a species' range. Recall that he hypothesized that "heteroselection" prevails in central populations because heterotic buffering produces better general performance in the variety of niches available in the center of the range, while selection favors homozygotes and the attainment of adaptation based on specific, genetically fixed, adaptive features in peripheral populations. Nevertheless, whatever genetic variation is present in peripheral populations should be free to recombine, so flies from these populations should respond more readily to directional selection. This hypothesis was tested by selecting flies derived from the peripheral Chadron, Nebraska, population, and the central Steeleville, Missouri, population for mobility

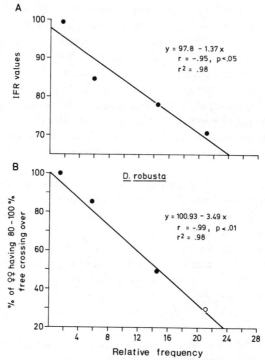

Figure 2 A. The relationship between relative frequency of *D. robusta* in collections and IFR values. B. The relationship between relative frequency and percent of females having 80–100% free crossing over. Data from Carson (16–18, 20).

toward light. From the peripheral population, 8 out of 10 responded positively, while 7 of 15 lines from the central group showed no response. The 8 lines from Steeleville that did respond to selection did so more sluggishly than the Chadron flies (20). In further experiments, Carson (22) also found that laboratory populations derived from the center of the range (Steeleville again, plus one from Michigan) tended to be both larger and more productive than those from the periphery (i.e. Chadron).

GENETIC LOAD   The genetic load, estimated by comparing egg-to-adult mortality in inbred and outbred crosses of F_1 lines derived from wild flies collected during consecutive years, was found to be more or less uniform in central populations from year to year (98). Peripheral populations (this time in Florida), however, showed a considerable difference in genetic loads between years. In 1964 there was little difference in adult survival between inbred and outbred crosses, but in 1965 the load seemed to be as high as it was in central populations. It was hypothesized that a severe frost during the winter of

1962–1963 had reduced the population to extremely low levels, creating a relatively homogeneous pool with a lower genetic load. By the following year, however, the population had returned to its normal size and had recovered most of its genetic variability.

ALLOZYME VARIABILITY   Prakash (113) investigated allozyme variability at 40 loci in 8 populations of *D. robusta* from central, peripheral, and intermediate areas. The average proportion of polymorphic loci per population was 39%, and the average heterozygosity per individual was 11%. There was no central-peripheral decline in these measures of genetic variability, however. Curiously enough, Carson anticipated this result in 1958. The results of his laboratory selection experiments clearly suggested that peripheral populations harbored considerable genetic variability, and he had fully abandoned his notions on marginal homoselection by 1975 (20).

SUMMARY   Table 1 provides an overall comparison of the differences between central and peripheral populations of *D. robusta*. It is clear that such populations differ in several respects. Peripheral populations are almost certainly smaller but not necessarily small enough to be strongly inbred and have less genetic variation. They also appear to respond more readily to some forms of directional selection in the laboratory, probably as a result of their increased recombination potential. This flexibility, however, may involve a certain cost, i.e. lowered productivity, again provided that the results of laboratory experiments are actually relevant to natural situations.

Table 1   Comparison of central and peripheral populations of *Drosophila robusta*

Variable	Central	Peripheral	References
Abundance relative to *robusta* and *affinis* groups	High	Low	16
Extent of chromosomal polymorphism	High	Low or absent	17, 18, 24
Recombination potential	Low	High	15, 17
Polygenic variation available for selection	High	High	21
Response to directional selection	Variable, generally low	Higher	19, 20
Size and productivity of laboratory populations	Higher	Lower	22
Genetic load	Higher; uniform between years	Variable; high one year, low another	98
Allozyme variability	High	High	113

Drosophila willistoni

DISTRIBUTION AND ABUNDANCE   The four sibling species of the *Drosophila willistoni* group are distributed throughout tropical America. The most common and thoroughly studied is *D. willistoni,* which is found from Argentina to Mexico and Florida, while *D. paulistorum, D. tropicalis,* and *D. equinoxialis* are somewhat less wide ranging (42). Although morphologically very similar, the species can be identified by slight differences in morphology, different inversions in the salivary gland chromosomes, and electrophoretic differences. Anecdotal accounts indicate that *D. willistoni* is an exceedingly common fly in many areas of Brazil and that it has the broadest ecological amplitude of the group, inhabiting a variety of ecotopes, including man-made ones. *Drosophila paulistorum* is generally confined to superhumid tropical forests and consists of five partially reproductively isolated "semispecies" plus some transitional forms. *Drosophila tropicalis* and *D. equinoxialis* are most common in drier forests alternating with savannas (130). All species breed in decaying fruits of various species (48).

Even though data exist on the relative frequencies of the four sibling species in various collections (130), they do not provide any useful insights into population trends. It is not possible to tell from these data whether the flies are common or rare in any particular area, since no information is available on the trapping effort that went into each of these collections. To make interpretation even more difficult, demographic patterns are more complex in this group than in *D. robusta,* since these flies have cyclic population buildups and declines at different times of the year and in different parts of their ranges, often out of phase with other sympatric species (48, 65). About the only reliable conclusion is that *D. willistoni* is genuinely uncommon at its northern periphery. Carson & Heed (24) took no *D. willistoni* in collections numbering over 55,000 flies in Florida, although Townsend (139) had taken some there previously.

CHROMOSOMAL INVERSIONS   *Drosophila willistoni* has about 50 different gene arrangements that tend to be short and difficult to distinguish in the homozygous state (37). Thus, the data base for geographic comparisons consists of frequencies of individuals that are heterozygous for a given inversion in each population. These gene arrangements are seasonally rigid for the most part; some are widespread, while others are confined to different geographic regions. Two seem to have an amphitropical distribution, being dominant in Florida and common at the southern periphery as well (128). Although none of these inversions is distributed randomly throughout the species' range, perhaps the only regularity in their overall distribution is their marked central-peripheral decline. The highest frequencies of inversion heterozygotes occur in central and northern Brazil; they decline as one moves away from this central region and reach minimal values in the deserts of northeastern Brazil and in southern

Brazil, Costa Rica, and the West Indies (37, 42). This central-peripheral decline in inversion heterozygosity seems unrelated to a parallel decline in the relative frequency of *D. willistoni* compared to its siblings. However, inversion heterozygosity in *D. willistoni* apparently *is* lower in areas where several of the other sibling species are present, especially if they are in high proportions, than in ecologically similar areas where they are absent or in low proportions (39). In the Caribbean island populations there is also a correlation between island size and the amount of chromosomal diversity present; small island populations are virtually monomorphic (42).

On the basis of these data, da Cunha & Dobzhansky (38) advanced their previously mentioned thesis that genetic variability is smaller in island and peripheral populations compared to continental and central ones because the amount of chromosomal polymorphism in a species is proportional to the number of ecological niches the population exploits. Niche utilization was judged to be a function of the number of niches potentially available for a species, which in turn depends on the length of time it has inhabited a given region and the number and abundance of competing species (37). In order to test this idea, da Cunha & Dobzhansky (38) constructed an environmental "index" for various localities within the distribution area of *D. willistoni* from which inversion data had been obtained. The environmental variables used included climate (whether temperate or tropical, seasonal or unseasonal, wet or dry), biotic provinces (an assessment of potential niche breadth based on the diversity of local vegetation), the number of competing sibling species, and spatial and temporal habitat breadth. This index correlated reasonably well with the mean number of heterozygous inversions per individual; there was a slightly better fit for females than for males. Three years later this hypothesis was elevated to a "rule" (42): The amount of adaptive polymorphism carried by a population is a function of environmental opportunities—or ecological niches—that the population exploits. Unfortunately, da Cunha & Dobzhansky's environmental index does not provide much of an assessment of the number of ecological niches actually exploited by the populations in question. It may provide a measure of potential ecotope breadth, although quantified data on habitat occupancy and resource utilization are not available, nor is a useful measure of relative population size. What the index probably does supply is a first-order assessment of environmental suitability for these flies. If so, it can be concluded that highly heterozygous populations usually occupy highly favorable environments. Such a relationship further suggests that the association between levels of chromosomal polymorphism and proportional frequency in collections seen in *D. robusta* may hold true for this species as well.

ALLOZYME VARIABILITY   The *Drosophila willistoni* group has been surveyed extensively for enzyme polymorphism (for a recent review of this literature see 54). The average levels of heterozygosity (about 20% over 25 loci) are not significantly different in *D. willistoni, D. tropicalis,* or *D. equinoxialis,* nor probably in *D. paulistorum* (21% for 17 loci) (3, 120). In *D. willistoni* allozyme heterozygosity varies somewhat from locality to locality, but there is no clear-cut central-peripheral decline among the continental populations (5). Average levels of genic heterozygosity are slightly lower on Caribbean islands (16.2% for island populations vs 18.4% for continental), although the decline is not nearly as marked as it is for chromosomal inversions. On the average there are 4.03 ± 0.47 heterozygous inversions per individual in continental populations vs 1.06 ± 0.26 in island populations (4). There is no evident relationship between island size and genic heterozygosity, and there are no unique alleles on the islands, suggesting strongly that dispersal is the source of whatever allelic diversity exists.

GENETIC LOAD   Townsend (139) studied the genetics of peripheral populations of *D. willistoni.* He concluded that populations at both the northern and southern boundaries of the range had somewhat lower frequencies of autosomes producing recessive adverse effects on viability and fertility and possibly fewer recessive, visible mutants than populations near the center. Hoenigsberg et al (68) found that flies collected during more stressed ecological circumstances (e.g. during the dry season or flies flying at noon rather than in the early morning) had fewer lethal and subvital loads than those collected under more favorable circumstances. Finally, Tabachnick & Powell (134) demonstrated dramatic differences in the ability of flies from peripheral areas to respond selectively to additions of various toxic chemicals to the culture medium. Populations derived from Caribbean localities, essentially monomorphic for chromosomal inversions ($\overline{H} = 0.07 \pm 0.05$), always survived for more generations than those derived from Venezuelan populations ($\overline{H} = 3.67 \pm 0.22$).

SUMMARY   The characteristics of central and peripheral populations of *D. willistoni* are summarized in Table 2. Although the data are not nearly as extensive as they are for *D. robusta,* many similarities are evident. Chromo-

Table 2 Comparison of central and peripheral populations of *Drosophila willistoni*

Variable	Central	Peripheral	References
Extent of chromosomal poly-morphism	High	Low	37
Response to directional selection	Low	High	134
Allozyme variability	High	Slightly reduced on islands, not in con-tinental populations	4

somal polymorphism shows a distinct central-peripheral decline both in continental populations and in those on outlying islands, and inversion-poor populations seem to respond more effectively to novel selection pressures. There is no regular central-peripheral decline in allozyme heterozygosity in continental areas, although it does decline slightly in island populations. In addition, flies from geographically central localities may have reduced genetic loads (i.e. numbers of detrimental and quasi-normal genes) in ecologically marginal, stressful situations.

Drosophila subobscura

DISTRIBUTION AND ABUNDANCE   *Drosophila subobscura*, one of the most common drosophilids in Europe, is found from North Africa, Spain, and the British Isles to southern Scandinavia, extending into Russia and the Middle East. Although no comparative studies of population levels throughout *D. subobscura's* range have been done, the species is reasonably well studied demographically. Begon (7–9) used capture-mark-release-recapture methods to estimate density and dispersal in British populations. Although the densities per 100 m² ranged between 0.3 and 31, the estimated effective size (N_e) of these populations was judged to be considerably smaller than the census size. This discrepancy was due to the effects of an annual winter bottleneck and to a strikingly nonrandom distribution of progeny sizes.

One can infer from anecdotal accounts that there are large differences in population levels between central and peripheral areas. Loukas et al (92) stated that *D. subobscura* populations in Crete (near the range center) are "essentially infinite." (This statement refers to census size, not effective size.) Lakovaara & Saura (77) provided some information on peripheral populations in southern Finland, on the northern edge of the species' range. In 1968 they found that "several hundred flies could be trapped during a day," but that in 1969 and 1970 the species was rare "even in late summer, though extensive trapping was done by several people" (77, p. 77). Only a few flies could be caught with the baits during any one day, and these workers concluded that these populations, located on the northern periphery of the range, are usually sparse and exhibit wide fluctuations in numbers from year to year.

CHROMOSOMAL INVERSIONS   Krimbas & Loukas (76) have recently reviewed the extensive research on inversion polymorphism in *D. subobscura*. Researchers have detected 58 different arrangements in its 5 acrocentric chromosomes; they are generally distributed in north-south clines, with the standard arrangements prevailing in the north and the derived ones in the south. Krimbas & Loukas (76) then calculated the IFR in 71 natural populations of *D. subobscura*, and as in *D. robusta*, peripheral populations display the highest IFR values and are the least polymorphic for inversions. The most polymorphic populations (those with lower IFR values), however, are in Asia Minor, somewhat displaced from the center of the range. They felt that this eccentricity probably reflects the recent northerly movement of the southern and eastern frontiers of the species' range. Furthermore, they believed that the current distribution results from the colonization of the territories covered by the last glaciation and from the retreat of the southern populations because of the desertification of the Sahara, parts of Asia Minor, and vast territories in the Middle East.

After reviewing the considerable literature on the subject, Krimbas & Loukas (76) concluded that the evidence that inversion polymorphisms of *D. subobscura* are subject to natural selection and thus represent adaptive devices is slim and somewhat contradictory. They argue that there is much stronger evidence that historical processes have had a major impact in determining the patterns of geographical distribution of the polymorphisms. In accord with this view, McFarquhar & Robertson (95) found no evidence supporting genetic coadaptation in local populations in *D. subobscura*. Crosses among Scottish populations and between Scottish and Israeli populations failed to reveal any evidence of F_1 heterosis or F_2 breakdown. These results would be expected if *D. subobscura* has expanded into its current distribution area quite recently and thus extensive adaptation to local conditions has yet to occur. The historical hypothesis is also consistent with Loukas et al's (92) scant evidence of linkage disequilibria between allozymes and gene arrangements.

ALLOZYME VARIABILITY   *Drosophila subobscura* has been studied extensively for allozyme variation. Saura et al (125) found that levels of average heterozygosity ($\overline{H}$) were variable, but that no trends were evident among central ($\overline{H}$ = 0.182–0.234), intermediate ($\overline{H}$ = 0.196), and peripheral ($\overline{H}$ = 0.219) populations. Once again, there is no correlation between patterns of genic and chromosomal heterozygosity. Certain allele frequencies did differ significantly from locality to locality, however. The major heterogeneities were found among geographically adjacent groups of populations separated by major barriers to gene flow such as the Alps or the Mediterranean (105). Prevosti et al (117) described the same pattern for chromosomal polymorphisms.

Peripheral populations also show some striking heterogeneities in allelic frequencies. Lakovaara & Saura (77), studying Finnish populations, found significant differences at two loci among localities and between years at the same locality. They also found that some peripheral island populations were fixed at alleles that were extremely rare ($f < 0.03$) in nearby mainland populations. The data on both chromosomal and allozyme frequencies seem to indicate that the effects of isolation are more important than selection in determining geographic patterns of genetic variation in this species.

GENETIC LOAD   Sperlich et al (131) examined the relative viabilities of homozygotes and random heterozygotes for wild "O" chromosomes (the name applied to one of the four autosomes in this species) derived from peripheral (Norwegian) and central (Greek) populations. Approximately 20% of the chromosomes tested from the Norwegian population proved to be lethal in the homozygous condition, while over 30% of the chromosomes carried by the Greek flies had one or more lethals. The mean viability was generally higher for random heterozygotes than for homozygotes in both populations, but this heterotic effect was more pronounced in the Norwegian flies. There was also considerably more variation in the viabilities of all genotypes collected from the peripheral population than from the central group. These authors attributed this homeostatic action of inversions to heterosis. They hypothesized that the decrease in inversion polymorphism toward the periphery of the range is associated with an increase in phenotypic plasticity. Therefore, the major effect of inversions is to increase canalization via higher order effects on gene regulation or interaction.

SUMMARY   Table 3 compares the characteristics of central and peripheral populations of *D. subobscura*. Population size appears to be high in central localities, low and variable on the periphery. The now familiar pattern of declining chromosomal polymorphism and genetic load from center to periphery is seen in this species as well, accompanied by the usual retention of high levels of allozyme variability. Heterotic effects of gene arrangements seem to be more pronounced in peripheral populations of this species, but the prediction of "homoselection" at the range edge is not supported by the data (151).

Drosophila pseudoobscura *and* Drosophila persimilis

DISTRIBUTION AND ABUNDANCE   *Drosophila pseudoobscura* is widely distributed from Guatemala to British Columbia and from the Pacific Coast into the Great Plains and the Rio Grande valley region of Texas (46). In addition, an isolated "island" population was discovered in the 1950s in the vicinity of Bogotá, Colombia, 2400 km from the nearest known population in Guatemala (47). Ecologically, little is known about the Bogotá population other than that the flies are locally widespread but uncommon and that they can be collected year-round. It is also unknown whether or not this completely isolated population is a relict of former range expansion or a recently established pioneering population, perhaps introduced by human activity, although the former possibility now seems more likely due to the discovery of several alleles unique to this population (32, 127). The distributional area of *D. persimilis* includes parts of Washington, Oregon, and California, and is almost entirely contained within

Table 3 Comparison of central and peripheral populations of *Drosophila subobscura*

Variable	Central	Peripheral	References
Extent of chromosomal poly-morphism	High	Lower	76
Recombination potential (index of free recombination)	Low	Higher	76
Allozyme variability (proportion of loci polymorphic)	High	High	125
Population size	High	Low, variable	77, 78, 131
Effects of chromosomal heterosis on viability	Low	High	131
Percentage of O chromosome lethals	32–44	20	76, 131

the range of *D. pseudoobscura*. Recent evidence indicates that this species is now spreading into parts of southern California and Nevada (45).

Patterson & Wagner (104) calculated the percentages of *D. pseudoobscura* relative to all *Drosophila* captured in various collections made in the Southwest during the 1930s and 1940s. I have averaged these figures for various localities from which inversion data were collected at approximately the same time (reported in 46). The results are shown in Figure 3. The relative frequency of *D. pseudoobscura* clearly declined from Utah and northern Arizona, where it comprises 90% or more of the drosophilids taken, to central Texas, where it comprises only a small fraction of the flies collected. Unfortunately, adequate data are not available for the other localities. Although numerous collections were made in California and elsewhere, proportions of *D. pseudoobscura* relative to other *obscura*-group flies or to all drosophilids taken were rarely recorded (97). Dobzhansky (40) made numerous collections of *D. pseudoobscura* and *D. persimilis* along a transect in the Sierra Nevada of California from 800 to over 10,000 ft, but he did not record the relative numbers. Elevational trends were evident, however; he reported that the former species is more common at lower elevations and that *D. persimilis* is the predominant species at 7,000 ft and above. Cooper & Dobzhansky (29) did record the occurrence of various *Drosophila* species in collections made in the Yosemite Park region, but the relative abundance of neither *D. pseudoobscura* nor *D. persimilis* can be determined accurately from their data.

Several attempts have been made to estimate absolute population densities of *D. pseudoobscura* (35, 51, 52); fortunately, these have been done in different localities. Although the data are not strictly comparable because of differences in data collection and analysis, it is still possible to make some inferences about population densities in both favorable and marginal habitats.

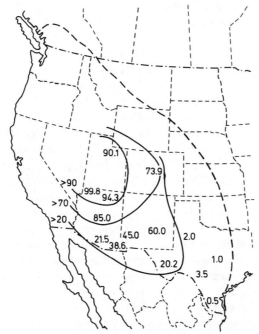

Figure 3 Frequency of *D. pseudoobscura* relative to all *Drosophila* captured in various collections made in the Southwest in the 1930s and 1940s. The dotted line shows the approximate range limits of this species north of Mexico; no comparable data are available from other parts of the range.

Crumpacker & Williams (35) used capture-mark-release-recapture techniques to study population density and dispersal rate in *D. pseudoobscura* from a region called the Black Forest in Colorado. This locality is near the eastern periphery of the species' range and is considered by these authors to be ecologically marginal. They found that midsummer densities in the Black Forest averaged 0.38 flies/100 m². They also reanalyzed density data obtained by Dobzhansky & Wright (51, 52) at two localities in California—Mather and Mount San Jacinto. Although different experimental designs and methods of data analysis had been utilized in the earlier studies, Crumpacker & Williams felt that their reanalysis made the density estimates comparable. They concluded that densities on Mount San Jacinto, an ecologically favorable location, ranged from 3.50 to 6.34 flies/100 m², depending on season and location, with a midsummer estimate of 6.29. The Mather locality, judged to be of moderate ecological favorability, supported an average density of 0.61 flies/100 m² at midsummer. Hence, midsummer density at the peripheral and marginal Colorado locality was three fifths that at Mather and one seventeenth that at Mount San Jacinto. These are the only data on *Drosophila* of which I am aware that provide numerical estimates of the extent of decline in absolute density from favorable to marginal areas.

CHROMOSOMAL INVERSIONS   Most of the chromosomal inversions in *D. pseudoobscura* and *D. persimilis* are located on the third chromosome, and these have been the most intensively investigated. A total of 35 different third-chromosome gene arrangements belonging to 4 major "phylads," or families, have now been described in *D. pseudoobscura*, although only about a third of these arrangements are common and/or widespread (100). Each phylad has a different geographical distribution, and the species has been divided into 4 major subdivisions on the basis of these distributions. The subdivisions are: (*a*) the Pacific Coast of the United States and the Sierra Nevada–Cascade Mountain chains, (*b*) the Intermontane Plateau, (*c*) the Rocky Mountains and Texas, and (*d*) Mexico and Guatemala (46). In this species there appears to be no central-peripheral decline in third-chromosome inversion heterozygosity. In fact, the center of the range more or less coincides with the Intermontane Plateau subdivision, which has the least heterozygosity of the 4 (41).

The third-chromosome gene arrangements in *D. persimilis* belong to 2 closely related phylads, one of which is common in the southern parts of its range and rare or absent in the north, while the situation is reversed in the other (46). There is a marked decline in inversion heterozygosity toward the northern periphery of the range but not toward the south or east.

The right arm of the X chromosome of both species carries an arrangement called SR (sex ratio). Males carrying SR produce offspring that are all—or almost all—females, and the presence of the arrangement can be determined either by progeny testing or cytological examination. In *D. pseudoobscura*, SR is absent in the northern part of the range, reaches frequencies of 25–30% in the southwestern United States and Mexico, and declines to about 1.5% in Guatemala (46). On an east-west axis, however, there appears to be no central-peripheral decline in the frequency of SR. In *D. persimilis* the situation is similar. The arrangement is absent in the north, but reaches moderately high frequencies (25–30%) elsewhere within the range, with no regular pattern of peripheral decline to the south and east (46).

The behavior of the third-chromosome gene arrangements in *D. pseudoobscura* varies considerably from population to population. Altitudinal clines in arrangement frequencies have been shown to occur over short distances (40). In some places, the polymorphisms are "rigid," showing no regular seasonal fluctuations; in others, they are "flexible" and oscillate fairly regularly (44). Longer-term temporal changes in gene arrangement frequencies have also been documented. The "Pikes Peak" arrangement (PP) was virtually absent from California and much of the Southwest in the 1930s, rose to fairly high frequencies in the late 1940s and 1950s, and subsequently declined (1, 43, 44).

Many attempts have been made to correlate seasonal, altitudinal, and long-term changes in third-chromosome arrangements in both species with various environmental parameters. For example, droughts and pesticide usage were thought to be correlated with shifts in inversion frequencies in *D. pseudoobscura*, but these relationships did not hold up in the long run (44). It was also suggested that flexible polymorphisms may be associated with marginal environments (36), but further work indicated that the relationship was not that straightforward and that flexible polymorphisms may be associated with either optimal or marginal habitats, depending on geographic locality (34, 142).

In his previously mentioned study of both species along an altitudinal transect in the Sierra Nevada, Dobzhansky (40) recorded the frequencies of third-chromosome arrangements from several sites. I have converted these frequencies to percentages of heterozygosity (on the assumption of a Hardy-Weinberg equilibrium for the arrangements) and plotted them against elevation in Figure 4. The results provide an interesting contrast: Heterozygosity declines significantly with increasing elevation in *D. persimilis* but remains relatively constant in *D. pseudoobscura*. It is very difficult, however, to determine whether or not trends in third-chromosome heterozygosity are connected with population trends in either species. *Drosophila pseudoobscura* is reported to be more common at lower elevations, being replaced by *D. persimilis* as the predominant species above 7000 ft. Thus, reduced heterozygosity does not seem to be associated with increasing ecological marginality in either species, although quantitative data on population sizes would make a more robust interpretation possible.

Numerous population cage experiments have also been conducted with both species that have been aimed at investigating the possible selective values associated with the different arrangements. Although one polymorphic laboratory population of *D. pseudoobscura* declined to monomorphism over the

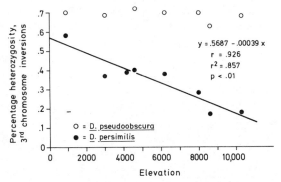

Figure 4   The relationship between third chromosome inversion heterozygosity and elevation in *D. pseudoobscura* (open circles) and *D. persimilis* (closed circles). Data from Dobzhansky (40).

course of an experiment (87), the polymorphisms tend to be maintained under most laboratory conditions (49). At 25°C—near the upper limit of temperature tolerance for *D. pseudoobscura*—certain heterokaryotypes are clearly superior to others, while under less stressful conditions, all karyotype fitnesses are uniform (44). Working with *D. persimilis* collected from high elevations, Spiess (132) found no fitness differentials associated with different gene arrangements in population cages maintained at 25°C, but selective differences could be demonstrated at 16°C. Thus, the third-chromosome arrangements show "adaptive effects" at 25°C in *D. pseudoobscura* and at 16°C for *D. persimilis*. The significance of this result in relation to the altitudinal distribution of the flies remains obscure but tantalizing.

Dobzhansky's work on *D. pseudoobscura* led him to the belief that heterosis was primarily responsible for maintaining these variants in nature (44). A recent analysis of Dobzhansky's data from five natural populations (64) showed, however, that the hypothesis of selective neutrality could be rejected in only two of them. Nevertheless, some form of balancing selection *was* implicated as the primary mode of selection in these two cases. A different view was held by Epling and his collaborators, who also did extensive work on *D. pseudoobscura* populations in southern California. These workers favored the thesis that the adaptive function of the inversion system was to restrict gene recombination in heterozygotes and that structural heterozygotes generally were not adaptively superior (55–57). But if recombination within inverted regions is deleterious, it has to mean that the inverted regions confer superiority in most of the individuals that carry them. This superiority must be the consequence of either heterosis or epistasis. Thus, it is difficult to see how there could be an advantage in the reduction of recombination per se, in the absence of heterosis or epistasis in the inverted sections. Since the data are so complex and contradictory, it seems safe to conclude only that there is a complex interaction among prevailing environmental conditions, the recombination rate, and the genetic contents of the various gene arrangements. Furthermore, whatever factors are responsible for the characteristic central-peripheral decline in chromosomal heterozygosity observed in most other *Drosophila* species seem to work differently in *D. pseudoobscura* and *D. persimilis*.

ALLOZYME VARIABILITY   Investigations involving *D. pseudoobscura* were the first to demonstrate the high levels of protein polymorphisms present in natural populations. Using stocks of flies derived from 5 separate populations, Lewontin & Hubby (90) investigated 18 loci, 7 of which (i.e. 39%) proved to be polymorphic over all samples. They concluded that the average *D. pseudoobscura* population was polymorphic at 30% of its loci, while the average proportion of heterozygous loci in an individual genome was 8 to 15%. Prakash et al (116) extended these investigations to include additional populations sampled from throughout the species' range. They found that the average heterozygosity per individual over 24 loci was 0.14 in Strawberry Canyon, California; 0.11 in Mesa Verde, Colorado; 0.12 in Austin, Texas; and 0.04 in the isolated outlying population in Bogotá, which will be discussed in more detail shortly. The California population was "dense" and highly polymorphic for third-chromosome inversions; the Mesa Verde was "medium sized" and monomorphic for the AR (Arrowhead) gene arrangement; the Austin population was "small," located on the extreme southeastern periphery of the range, and moderately polymorphic for inversions. Any apparently declining relationship between levels of enzyme and chromosomal polymorphism within the main body of the species' range disappears, however, when two loci—Pt-10, α-amylase—located within the third-chromosome inversions are removed. Average heterozygosities then become: Strawberry Canyon, 0.11; Mesa Verde, 0.10; Austin, 0.11; and the Bogotá isolate increases slightly to 0.05. Lewontin (88) provided additional data from more populations. Those from the Great Basin, Arizona, and Colorado had heterozygosities nearly as high as those from Strawberry Canyon; only Guatemala showed slightly reduced heterozygosity. When 22 additional loci were added, bringing the total to 43, the absolute values of the heterozygosity estimates for Strawberry Canyon, Austin, Mesa Verde, and Bogotá changed little, and the rank orders stayed the same (114). The situation with allozyme variability in $D.$ $persimilis$ is much the same as it is in $D.$ $pseudoobscura$. The average levels of heterozygosity are essentially equivalent in both species, and all populations of $D.$ $persimilis$ studied have about the same average level of polymorphism (115).

These data were obtained by using standard electrophoretic techniques. Recent refinements involving sequential electrophoresis have revealed considerable "hidden" variability that is not evident under standard conditions. The discovery of numerous "hidden" alleles at several loci does not seem to change the picture of within-range variation in heterozygosity very much, however. Some loci revealed no additional variability under varied electrophoretic conditions (e.g. 10), and loci that appear to be essentially monomorphic under standard conditions remain so under varied conditions (126). Furthermore, any increases in heterozygosity disclosed by sequential electrophoresis are generally proportional to the levels previously reported (127). Thus, the final picture that emerges is that levels of allozyme heterozygosity differ little within the main range of the species but are sharply reduced in the Bogotá isolate.

THE BOGOTA POPULATION   This population appears to be a depauperate version of the Guatemalan group in terms of its third-chromosome inversions, but it differs in several important respects. Although the Bogotá population is dimorphic for the Santa Cruz (SC) and Tree Line (TL) gene arrangements, the genetic constitution of the former is substantially different from that of SC in

Guatemala or southern California. The Bogotá population has the lowest frequency of lethals and semilethals yet reported in *D. pseudoobscura,* and the frequency of allelism in lethals is much greater there than in Guatemala (47, 93). Furthermore, it has developed some degree of reproductive isolation from the rest of the species (112). Crosses involving females from Bogotá and males from elsewhere produce sterile males and normal females, but the reciprocal cross results in an F_1 with fully fertile males.

Although Prakash (112) originally reported that no unique allozymes occurred in the Bogotá population, subsequent studies involving sequential electrophoresis under a variety of conditions have revealed that substantial differences were present. For example, Bogotá initially appeared to be homozygous for the Xdh^{100} allele, which was also the most common variant in the main distribution area. The Xdh^{100} "allele," however, was later found to consist of several variants that had the same electrophoretic mobility under standard conditions. Under sequential electrophoresis, it was possible to resolve 5 separate alleles in the Bogotá population, 3 of which were unique (127). Likewise 8 of 12 lines examined from Bogotá at the *Adh-6* locus proved to have 4 alleles found nowhere else in the species, and no lines have the allele most commonly found elsewhere (32).

SUMMARY   Superficially, it appears that *D. pseudoobscura* is somewhat of a maverick with respect to population trends and patterns of chromosomal heterozygosity. The western edge of its range, however, is not determined by declining ecological favorability; rather, populations with "central" characteristics simply stop at the Pacific Ocean. On the eastern edge, both relative frequency and abundance seem to decline, although moderate levels of heterozygosity are retained. The geographic center of its range corresponds to the Colorado Plateau area, where it is virtually the only *Drosophila* species trapped. Thus, its relative frequency in this region is high, but anecdotal accounts assess relative population abundance only as "medium-sized" (116) or "not uncommon" (88). It is not unreasonable to assume, therefore, that ecological conditions are somewhat marginal for *D. pseudoobscura* here, and the essential monomorphism for the AR chromosome arrangement should not be completely unexpected.

The only truly familiar aspect of the distribution of genetic variation in *D. pseudoobscura* is its allozyme variability; it is essentially similar throughout the species' range, with the exception of Bogotá—a population that is moderately depleted in chromosomal heterozygosity as well.

Other Species

Soulé (128) lists 9 additional *Drosophila* species that show a central-peripheral decline in inversion heterozygosity, so the pattern is certainly common enough.

The lack of reduced allelic diversity in peripheral situations also seems to be a general trend, although not a hard and fast rule. On the one hand, Lakovaara & Saura (78) studied allozyme variation of *D. obscura* in the northern part of its range and found no reduction in average heterozygosity over 13 polymorphic loci in an isolated outlier in Lapland. Nor were there any differences in average heterozygosity between peripheral populations in central Finland and intermediate populations in southern Finland. On the other hand, peripheral populations of *Drosophila buzzatii* on the southern edge of their range in Australia showed a slight reduction in average heterozygosity, while those toward the arid interior appeared to have slightly higher than average levels (6).

Drosophila pachea is endemic to the Sonoran Desert, where it is obligately associated with necroses in senita cactus *(Lophocereus schottii)* (66). The one chromosomal inversion in this species shows the familiar central-peripheral decline in heterozygosity with the 7^+ arrangement fixed in the southern portions of the range and the 7A arrangement fixed in the north. The polymorphic center in mainland Mexico corresponds to a major change in biogeographic regions and in the morphology of the host plant (141). Apparently the zone of high chromosomal polymorphism extends further south in Baja California, following the major vegetation zones of the peninsula (66). Rockwood-Sluss et al (123) studied allozyme variability in *D. pachea* and found no clear trends in heterozygosity, although it was slightly lower in the three northernmost populations analyzed. Inversion heterozygosity in *Drosophila mojavensis,* another cactophilic species, also shows a significant decline in the northern part of its range. This decline is also associated with a shift in the host plant from agria cactus to organ pipe; the latter species is a less suitable host in many ways and generally provides a marginal environment for these flies (67).

Jaenike (69) studied populations of *Drosophila affinis* and *D. athabasca* on the mainland and on small, nearby islands off the coast of Maine. The abundance of the two species was negatively correlated, with *D. athabasca* more common on the mainland and large islands, and *D. affinis* generally more prevalent on small islands. This study was conducted on the northern periphery of the range of *D. affinis,* where its overall abundance is generally low. Jaenike attributed the greater abundance of *D. affinis* relative to *D. athabasca* on small islands to a "preadaptation" to island life resulting from its small effective population size. He postulated that repeated bouts of inbreeding should have weeded out most of the deleterious alleles in the populations, enabling the species to persist at low densities. On the other hand, this locality is more central on a north-south axis for *D. athabasca,* and its populations are always large there. Thus, the *D. athabasca* populations rarely if ever become so small that they experience inbreeding, and in island situations where population numbers are low, inbreeding affects them deleteriously.

In summary, data from a number of other species clearly support the hypothesis that there is a general tendency toward a central-peripheral decline in chromosomal polymorphism and the lack of a similar attenuation in allozyme heterozygosity, although there are a few exceptions. The author of one study suggested that inbreeding tolerance is increased in island populations near the periphery of a range, perhaps by the elimination of deleterious mutations as a by-product of having small populations.

DISPERSAL RATES IN FAVORABLE AND MARGINAL HABITATS

There is some evidence now accumulating that dispersal rates in *Drosophila* differ significantly between favorable and marginal habitats. Crumpacker & Williams (35) noted that the flies they studied in Colorado dispersed at a rate about 50% greater than did the Mount San Jacinto and Mather (California) flies Dobzhansky & Wright (51, 52) studied. They felt that some of the discrepancy between the two estimates might be explained by differences in experimental design, since the Colorado experiments utilized wild captured flies whereas Dobzhansky & Wright used F_1 hybrids of laboratory strains that carried a visible mutation (orange eyes). They also considered it possible that the dispersal capabilities of both types of flies were essentially equivalent and that *D. pseudoobscura* exhibits more dispersal activity in low-density, ecologically marginal habitats.

Powell et al (107) attempted to reestimate dispersal rates at Mather. Although the experimental procedures and analysis they used were considerably different from Crumpacker & Williams's, their estimates of dispersal distances were remarkably close. Therefore, Powell et al (107) concluded that the earlier experiments underestimated dispersal because the laboratory-reared mutants were less mobile than wild ones.

In further studies, however, Dobzhansky et al (50) were able to show that density and dispersal rates at Mather were quite habitat dependent. In dense woods, the flies were abundant and sedentary; they were less common and dispersed more rapidly in dry woods; and they were rarely caught and moved extensively in meadows. Their results suggest that density and dispersal rates are negatively correlated. This effect probably results from habitat selection and not from any relationship between density and dispersal rates per se.

Moore et al (97) observed *D. pseudoobscura* move rapidly into a recently burned area in southern California; within a year, the species was as abundant there as in the adjacent, unburned habitat. No heterogeneity in chromosomal frequencies was observed among samples taken from the burned area, but considerable genetic structuring in chromosomal arrangements has been recorded from other, nearby localities.

Some of the best evidence for high rates of dispersal in *D. pseudoobscura* comes from the Death Valley region of southeastern California and nearby Nevada. Oases in Death Valley are densely populated by this species during the spring, but the flies totally disappear during the heat of summer. These spring populations apparently result from the dispersal of flies from neighboring mountain habitats (30, 74, 75). Experiments with marked flies show that *D. pseudoobscura* will venture into the surrounding desert and travel from one oasis to another; movements of up to 10 km in 24 hr over hostile terrain have been reported. Additional evidence of high rates of dispersal comes from electrophoretic data: Oasis flies closely resemble those collected from nearby mountains in terms of allele frequencies at the *Est-5* locus. If each oasis had an isolated population of flies, allelic frequencies should be quite heterogeneous and different from those in neighboring populations due to the effects of drift, which would occur during the severe summer bottleneck in population size.

An experimental introduction of flies that artificially increased the frequency of the *Est-5*$^{0.85}$ allele also suggested that dispersal and annual recolonization, rather than survival, accounted for the population trends. Large numbers of flies homozygous for this allele were introduced into an oasis population. The frequency of the allele rose dramatically, with large numbers of heterozygous combinations indicating that the introduced flies were breeding with the natives. The following year, however, the frequency of *Est-5*$^{0.85}$ was back to normal levels, and the experimenters concluded that the decline resulted from recolonization following essentially complete mortality of the resident population (74).

Other desert *Drosophila* have also been shown to have high rates of dispersal (70–72). *Drosophila pachea* individuals are highly mobile, having 8% to 14% migration rates between cacti, and *D. nigrospiracula* is so mobile that the entire species may act as one, effectively panmictic unit. Furthermore, adverse conditions increased dispersal rates.

Johnson & Templeton (73, 138), studying natural populations of *D. mercatorum* on the island of Hawaii, elegantly demonstrated the latter point. These populations also inhabit patches of a large species of cactus, in this case *Opuntia megacantha*, which are distributed on the leeward side of a cinder cone and range in elevation from about 670 m to 1000 m above sea level. The mountain slope forms a moisture gradient as well; the low elevation sites are warmer and less humid than the higher ones. These particular populations had a relatively high frequency of the trait "abnormal abdomen" (*aa*), which has a number of pleiotropic effects, some of which affect life history characteristics of the flies. Adults homozygous or heterozygous for *aa* show increased fecundity and decreased longevity.

During the period 1976–1980, "normal" weather conditions prevailed, and each cactus hosted an isolated subpopulation of *D. mercatorum*, which persisted for several generations with no gene flow in or out. The flies were numerically abundant, and a cline in the frequency of *aa* remained unchanged, with higher frequencies found at the lower, drier site. *Drosophila mercatorum* shared this habitat with populations of *D. hydei,* and the relative abundance of the two species was also clinal. The highest proportions of *D. hydei* occurred at the higher, moister end of the gradient. Thus, under favorable conditions, microgeographic ecological differences determined the average size and genetic composition of each *D. mercatorum* subpopulation, which was reflected in parallel clines of abundance and allele frequencies along the underlying environmental gradient.

In 1981 a drought occurred that had an enormous impact on the *D. mercatorum* populations. Population size was drastically reduced, and population subdivision increased; only a few cactus patches could support flies under these conditions. The cline in the frequency of *aa* disappeared, and values throughout the entire study site converged to those previously observed only at the dry site. Dispersal measurements were made during the drought period. Although the absolute distance moved by the flies was small, it was adequate to homogenize gene frequencies throughout the study site.

Thus, under ecologically favorable conditions, at least some *Drosophila* species are probably sedentary due to habitat selection, and a certain amount of genetic adaptation to local conditions may occur. This differential adaptation may be reflected by clines or other less regular heterogeneities in chromosomal or allelic frequencies over relatively short distances. Under increasingly marginal conditions, however, the flies seem to increase their dispersal activity, even though the subpopulations may become less dense, more patchily distributed, or both. This increase in dispersal activity should obliterate interpopulation genetic differences, unless local selection pressures are incredibly strong.

CONCLUSIONS

Population Trends

Underlying environmental gradients—primarily temperature and perhaps to a lesser extent moisture—are the major determinants of the distribution and abundance of most widely distributed, generalist *Drosophila* species. In the absence of resource limitation and competition, the interaction between the life history attributes of each species and the underlying temperature gradient should produce a bell-shaped curve of average abundances, tapering on each side of the mode (or physiological optimum) to asymptotic tails. Topographic irregularities, nonrandom distribution of appropriate resources for feeding and breeding, and the presence of competing species may distort the Gaussian form of this distribution pattern somewhat. Nevertheless, once the temperature-dependent life history characteristics are known for a species, it should be possible to predict its average levels of abundance at different points along this gradient, on both a regional and a local basis.

Because of generally favorable physical conditions, populations within the central part of the gradient curve should be able to inhabit a variety of ecotopes. The densities reached in these different ecotopes will reflect their relative favorableness at any one time. Less favorable habitats will be colonized regularly, and the resultant populations may sometimes reach high densities. These populations are essentially transient, however, and local extinction is likely to occur before many genetic adjustments can be made. The temporary nature of ecologically marginal populations in geographically central localities can easily explain their lack of local adaptation (e.g. 97).

Effective population sizes in the center of the range will also be high, but perhaps not as high as often assumed. Habitat selection (108, 135) and the tendency for *Drosophila* to remain in ecologically favorable areas may reduce levels of dispersal and gene flow somewhat. Such factors probably account for the microspatial heterogeneities in chromosomal or allozyme frequencies reported in the literature from time to time (e.g. 121).

To whatever extent *Drosophila* populations are governed by density-dependent factors, these should be most important in central areas. Intraspecific competition may be intense, and populations in this part of the gradient will probably be subject to what Whittaker & Goodman (147) termed saturation selection. Like Wallace (140), these authors suggested that individual adaptations to these habitats should reflect specialization for competitive ability.

Moving away from the optimum physical conditions in the central part of the gradient, there will be intermediate areas where the average population size is lower and fluctuates more. Although permanent populations are generally maintained in this region, fluctuations in numbers related to variation in the physical environment will be common and pronounced. Whittaker & Goodman (147) called the selective regime prevalent here exploitation selection, characterized by a flexibility of demographic strategies. There are frequent episodes of selection for fast increase, followed by selection favoring those traits that dampen population decline through resistance to unfavorable conditions. Populations occupying this part of the gradient should be generalists with respect to adaptation to various densities and to different levels of environmental stress.

Populations inhabiting the ecologically marginal areas of the gradient will be even smaller and fluctuate more than those in intermediate areas. Demographic peaks and troughs will be produced by changes in the physical environment, since even relatively small changes in some critical parameter can have very important demographic consequences on populations living close to their tolerance limits. Even short periods of particularly favorable environmental conditions can result in population explosions, but numbers will decline rapidly when normal conditions return. The greatest selective pressures under these conditions would be toward evolving means of surviving unfavorable conditions; Whittaker & Goodman (147) called this adversity selection. They also suggested that the infrequency of opportunities for rapid increase would generally select against "*r*" characteristics, and most of the individuals resulting from the infrequent population eruptions would probably be expended in such relatively reckless gambits as long-distance dispersal. I suspect, however, that *Drosophila* populations are unlikely to persist for long in a truly marginal environment in which Whittaker & Goodman's (147) concept of adversity selection would apply. The ecological strategies of *Drosophila* just do not seem to include adaptations for long-term survival under adverse conditions. Rather, as the edges of the range are approached, an extreme form of exploitation selection favoring maximum flexibility is more likely to become the predominant selective regime.

Although adequate data on even relative population sizes from different localities within the range of any *Drosophila* species are lacking, anecdotal accounts, a few absolute population estimates, and information on the relative frequency of various species in collections suggest that the above model is a reasonable one. The only apparent exceptions to the pattern appear to be found in *D. pseudoobscura*. When more is known about the ecological requirements, population sizes, and community relationships of this species, however, most of these problems will probably be resolved.

Genetic Patterns

ALLOZYMES   None of the available studies suggest that there is a reduction in levels of allozyme heterozygosity in peripheral populations. Soulé (128) theorized that this could be explained by heterosis, the advantages of which would be as large in ecologically marginal environments as elsewhere. Although there is ample data indicating that overall heterozygosity is advantageous in most natural environments (58), the evidence for single-gene heterosis at allozyme loci is weak and rather contradictory, and the kinetic parameters of enzymes from heterozygotes appear to lie between those of the related homozygotes (61). Furthermore, heterotic maintenance of high levels of allozyme variability now seems to be unlikely on theoretical grounds (89).

Since it is now generally agreed that electrophoretically detectable variants are either neutral or have very small selective differentials associated with them, it seems most likely that migration satisfactorily accounts for the lack of geographic differentiation in allele frequencies and the absence of a central-peripheral decline in heterozygosity in *Drosophila* species (60). Even if peripheral populations tend to be two or three orders of magnitude smaller, on average, than those in central areas, this reduction is not likely to lower heterozygosity much. It will not do so in part because the high dispersal rates observed in marginal environments tend to decrease the isolation of individual subpopulations and increase overall effective population size. Furthermore, the effects of population bottlenecks on overall heterozygosity depend not only on bottleneck size but also on the length of time the population remains small (59, 99). If population size increases rapidly after experiencing a bottleneck, the reduction in average heterozygosity is rather small, even if the bottleneck was severe. Thus, periodic recoveries during favorable periods will retard the wholesale erosion of heterozygosity. These effects, plus sporadic gene flow from more central areas, are probably sufficient to maintain the high levels of overall heterozygosity commonly encountered in most peripheral *Drosophila* populations.

CHROMOSOMAL POLYMORPHISMS   As discussed earlier, several explanations have been advanced to explain the characteristic central-peripheral reduction in inversion heterozygosity observed in most *Drosophila* species. Da Cunha and his colleagues (37, 38) felt that high levels of polymorphism were related to ecological amplitude, and they implied that specific arrangements conferred higher fitness to their bearers in some environmental situations than in others. While it is no doubt true that ecotope space is filled out more completely in the center of the range, there is a conspicuous lack of data demonstrating that certain genotypes have a higher fitness in specific niches or habitats (128). Thus a 1:1 correspondence between ecotope breadth and chromosomal heterozygosity seems highly unlikely. Carson (21) proposed the idea that the higher overall vigor of flies heterozygous for chromosomal arrangements would be favored in central areas, but that the specific adaptive features of homozygous arrangements would be selectively advantageous in marginal, peripheral areas. While it is true that some population cage experiments have shown that heterosis is indeed associated with certain inversions under laboratory conditions (e.g. 33), many other experiments do not; and invoking heterosis as a universal explanation for chromosomal polymorphism is not satisfactory. Perhaps the idea that certain inversions do confer higher fitness in the homozygous state in marginal areas will raise fewer objections. Evidence consistent with this explanation includes the apparently nonrandom distribution of certain inversions on the edges of the range, such as the amphitropical arrangements III-A and III-J in *D. willistoni* (128) and the fixation of arrangements XL–1 in northern peripheral areas and XL in Florida in *D. robusta* (21). Likewise, the fixation or near fixation of the AR arrangement in *D. pseudoobscura* populations inhabiting the Colorado Plateau and surrounding areas may argue for its superiority in marginal situations. Although *D. pseudoobscura* is virtually the only *Drosophila* trapped in this region, it only reaches moderate levels of abundance there, strongly suggesting the effects of a rigorous physical environment.

The role of historical factors in the nonrandom distribution of inversions is difficult to factor out completely. Krimbas & Loukas (76) felt that historical factors play the major role in determining the patterns of distribution of inversions in *D. subobscura;* how much of a role history plays in the distribution of these variants in other species is hard to determine. I suspect that the generally high vagility of *Drosophila* would tend to minimize the importance of historical accident in the nonrandom distribution of genetic variation. We do know that events during the Pleistocene had profound effects on distributional patterns in other insects, such as Coleoptera, that leave a respectable fossil record (28). The evidence shows that beetles inhabiting temperate regions tracked favorable environments across the continents, shifting their range centers in response to changing ecological conditions. *Drosophila* are certainly as vagile as beetles. Major range dislocations should have kept gene pools sufficiently well stirred to minimize the possibility that many nonrandom distributions of chromosomal variants can be explained satisfactorily by historical accident, at least in temperate-zone species. Even in tropical regions, the alternating arid and humid periods during the Pleistocene would certainly have resulted in major range contractions and expansions, providing ample opportunities for gene flow.

If not niche breadth, heterosis, or historical accident, what accounts for the almost universally observed pattern of central-peripheral reduction in inversion heterozygosity? I suggest, as Carson (21) did years ago, that as environmental favorableness and predictability for a species decrease gradually from center to periphery, the average size a *Drosophila* population can reach decreases slowly, and the selective mode changes from one emphasizing high average fitness to one favoring maximum genetic flexibility. I envision a somewhat different selective mechanism from that suggested by Carson, however.

Let us consider the organism—not just particular chromosomes or genotypes—as the basic unit of selection. The traits on which selection is most likely to act are quantitative ones, largely determined by polygenic inheritance (103). Selection for these traits will tend to be stabilizing near the center of the range; individuals near the mean of the distribution for all characters will be the ones best adapted to the range of conditions likely to be encountered in this region.

In order to minimize the production of individuals with extreme phenotypes, linkage arrangements that combine alleles of opposite effects will be favored (79). The linkage disequilibrium resulting from these epistatic interactions will tend to increase by the accumulation of modifiers; an inversion will eventually occur that "congeals" the evolving supergene (63). As long as the predominant selective mode is stabilizing, as it will be in the optimal environment of the center of the range, inversions that preserve the linkage disequilibrium among highly epistatic, plus-minus allelic combinations will be favored. Populations farther and farther from the center of the range, however, will encounter increasingly stressful and unpredictable environmental conditions, and average phenotypes are less likely to be favored than extreme ones. For example, tolerance to extreme cold or desiccation stress may be irrelevant in optimal, central environments but may become increasingly important as the periphery is approached. Thus, free recombination—so that these phenotypes can be produced—will also become increasingly important, and any devices such as inversions that tend to retard or prevent recombination will be selected against. There is empirical support for this idea: Expressed variation among phenotypes has been shown to be highest under conditions of environmental stress and to increase with the level of stress, provided that the populations have not previously experienced such stresses (103).

Wallace's (140) suggestion that structural heterozygotes would be favored among the more crowded populations in the range center owing to their higher competitive abilities is also perfectly compatible with this hypothesis. Individuals with a modal phenotype are most likely to be the best overall competitors among individuals of their own species. Furthermore, if Wallace's idea is correct, it may partially explain the lack of peripheral reduction in chromosomal heterozygosity in *D. pseudoobscura* on the eastern and northern edges of its range. He pointed out that the high levels of intraspecific competition in central areas may be replaced by high levels of interspecific competition from other *obscura*-group flies in these regions (*D. athabasca* to the north; *D. affinis* to the east). This effect may also explain the moderate to high levels of chromosomal polymorphism maintained by *D. pseudoobscura* in the ecologically marginal Black Forest region. Crumpacker et al (34) found that during part of the active season the closely related *Drosophila lowei* comprised as much as 35% of the total catch of *obscura*-group flies. If suitable sites for breeding and oviposition are scarce in the Black Forest, *D. lowei* and *D. pseudoobscura* may well compete directly and the levels of chromosomal polymorphism may be related to competitive interactions.

Given enough time, will populations near range edges also accumulate their own epistatic combinations and preserve them in inversions, as Soulé (128) suggested? They are unlikely to do so because the inherent unpredictability associated with increasing marginality will continually favor a genome that is free to recombine, rather than one that has preserved epistatic combinations that are likely only to be transiently favorable. The nonrandom occurrence of a few inversions near the edges of the range does suggest, however, that some sequences are sufficiently valuable to be protected from recombination even under marginal conditions.

Where Do We Go from Here?

The foregoing is not presented as yet another central-marginal model, but rather as a working hypothesis, subject to considerable modification and falsification. However, testing various aspects of this hypothesis will require paying much more attention to the ecology, dynamics, and spatial structure of

Drosophila populations than is currently done. Reliable data on catch per unit effort for the species of interest plus its possible competitors in central, intermediate, and peripheral areas throughout the active season are an absolute prerequisite for any future studies. They will confirm or refute the expected demographic patterns.

Gradient analysis (143–146) should also be a useful tool for relating estimates of relative abundance to environmental parameters. Transects from central to peripheral areas should be ordered with respect to various environmental gradients, especially the annual temperature regime. These data, along with more information on breeding sites and the seasonal dynamics of species with similar ecological requirements, should provide considerable insight into the types of selective factors that might be operating in various localities.

Finally, much more attention should be paid to different kinds of "ecological phenotypes" (103). Coyne et al (31) have made a welcome start in this direction. These authors tested lines of *D. pseudoobscura* collected from several localities in the western United States for resistance to extreme temperatures and desiccation. Although they failed to find differences in tolerance to these stresses in the directions predicted from gross environmental differences among collecting localities, this result is not entirely unexpected if the hypothesis I am proposing here has general validity. Rather, based on this hypothesis I would predict that flies collected from any areas that support large populations of flies with high levels of chromosome heterozygosity—even if these areas have substantially different ecological conditions—should have about the same levels of resistance to environmental extremes. Flies collected from low-density marginal areas, in contrast, should be chromosomally monomorphic, have higher resistances to environmental stresses in the expected directions, and display higher levels of expressed variation in most additive traits as well.

Once we understand the relationships between distribution, levels of abundance, and various kinds of genetic and phenotypic variation better, we should be able to make substantial progress in comprehending the role of peripheral populations in the speciation process. Carson (23) and Templeton (137) have created a theoretical framework emphasizing the interaction among demographic and genetic events that would most likely be found in marginal situations and that would lead to the creation of new gene pools, and Powell (106) has subjected some of these ideas to an experimental test. Exciting as some of these ideas and results may be, I would hate to see research proceed along these lines much further without considerably more detailed knowledge about relevant events in naturally occurring marginal populations.

ACKNOWLEDGMENTS

I thank Janet Wright for reading and criticizing numerous drafts of this manuscript and Evelyn Cook for patiently typing them. Richard Baker, William L.

Brown, Jr., Hampton L. Carson, James Fogleman, William Heed, Robert Lacy, Ross MacIntyre, Deborah Rabinowitz, Michael Soulé, and Bruce Wallace provided many useful insights and critical comments. Monica Howland drew the figures. Funding was provided by the Office of Research and Development, US Environmental Protection Agency, under Cooperative Agreement No. CR 807856 and by Cornell University. The work and conclusions published herein represent the views of the author and do not necessarily represent the opinions, policies, or recommendations of the Environmental Protection Agency or of Cornell University.

Literature Cited

1. Anderson, W., Dobzhansky, T., Pavlovsky, D., Powell, J., Yardley, D. 1975. Genetics of natural populations. XLII. Three decades of genetic change in *Drosophila pseudoobscura*. *Evolution* 29:24–36
2. Andrewartha, H. G., Birch, L. C. 1954. *The Distribution and Abundance of Animals.* Chicago: Univ. Chicago Press
3. Ayala, F. J., Powell, J. R. 1972. Enzyme variability in the *Drosophila willistoni* group. VI. Levels of polymorphism and the physiological function of enzymes. *Biochem. Genet.* 7:331–45
4. Ayala, F. J., Powell, J. R., Dobzhansky, T. 1971. Enzyme variability in the *Drosophila willistoni* group. II. Polymorphisms in continental and island populations of *Drosophila willistoni*. *Proc. Natl. Acad. Sci. USA* 68:2480–83
5. Ayala, F. J., Powell, J. R., Tracey, M. L., Mourão, C. A., Perez-Salas, S. 1972. Enzyme variability in the *Drosophila willistoni* group. IV. Genic variation in natural populations of *Drosophila willistoni*. *Genetics* 70:113–39
6. Barker, J. S. F., Mulley, J. C. 1976. Isozyme variation in natural populations of *Drosophila buzzatii*. *Evolution* 30:213–33
7. Begon, M. 1976. Dispersal, density and microdistribution in *Drosophila subobscura* Collin. *J. Anim. Ecol.* 45:441–56
8. Begon, M. 1977. The effective size of a natural *Drosophila subobscura* population. *Heredity* 38:13–18
9. Begon, M. 1978. Population densities in *Drosophila obscura* Fallér and *D. subobscura* Collin. *Ecol. Entomol.* 3:1–12
10. Bekenbach, A. T., Prakash, S. 1977. Examination of allelic variation at the hexokinase loci of *Drosophila pseudoobscura* and *D. persimilis* by different methods. *Genetics* 87:743–61
11. Bush, G. L. 1975. Modes of animal speciation. *Ann. Rev. Ecol. Syst.* 6:339–64
12. Bystrak, D. 1979. The breeding bird survey. *Sialia* 1:74–79
13. Carson, H. L. 1949. Seasonal variation in gene arrangement frequencies over a three-year period in *Drosophila robusta* Sturtevant. *Evolution* 3:322–29
14. Carson, H. L. 1952. Contrasting types of population structure in *Drosophila. Am. Nat.* 87:239–48
15. Carson, H. L. 1953. The effects of inversions on crossing over in *Drosophila robusta. Genetics* 38:168–86
16. Carson, H. L. 1955. The genetic characteristics of marginal populations of *Drosophila. Cold Spring Harbor Symp. Quant. Biol.* 20:276–87
17. Carson, H. L. 1955. Variation in genetic recombination in natural populations. *J. Cell. Comp. Physiol.* 45:221–36 (Suppl. 2)
18. Carson, H. L. 1956. Marginal homozygosity for gene arrangement in *Drosophila robusta. Science* 123:630–31
19. Carson, H. L. 1958. The population genetics of *Drosophila robusta. Adv. Genet.* 9:1–40
20. Carson, H. L. 1958. Response to selection under different conditions of recombination in *Drosophila. Cold Spring Harbor Symp. Quant. Biol.* 23:291–305
21. Carson, H. L. 1959. Genetic conditions that promote or retard the formation of species. *Cold Spring Harbor Symp. Quant. Biol.* 24:87–103
22. Carson, H. L. 1961. Relative fitness of genetically open and closed experimental populations of *Drosophila robusta. Genetics* 46:553–67
23. Carson, H. L. 1975. The genetics of speciation at the diploid level. *Am. Nat.* 109:83–92
24. Carson, H. L., Heed, W. B. 1964. Structural homozygosity in marginal

populations of Nearctic and Neotropical species of *Drosophila* in Florida. *Proc. Natl. Acad. Sci. USA* 52:427–30

25. Carson, H. L., Stalker, H. D. 1947. Gene arrangements in natural populations of *Drosophila robusta*. *Evolution* 1:113–33

26. Carson, H. L., Stalker, H. D. 1948. Reproductive diapause in *Drosophila robusta*. *Proc. Natl. Acad. Sci. USA* 34: 124–29

27. Carson, H. L., Stalker, H. D. 1951. Natural breeding sites for some wild species of *Drosophila* in the eastern United States. *Ecology* 32:317–30

28. Coope, G. R. 1979. Late Cenozoic fossil Coleoptera: Evolution, biogeography, and ecology. *Ann. Rev. Ecol. Syst.* 10:247–67

29. Cooper, D. M., Dobzhansky, T. 1956. Studies on the ecology of *Drosophila* in the Yosemite region of California. I. The occurrence of species of *Drosophila* in different life zones and at different seasons. *Ecology* 37:526–33

30. Coyne, J. A., Boussy, I. A., Prout, T., Bryant, S. H., Jones, J. S., Moore, J. A. 1982. Long-distance migration of *Drosophila*. *Am. Nat.* 119:589–95

31. Coyne, J. A., Bundgaard, J., Prout, T. 1983. Geographic variation of tolerance to environmental stress in *Drosophila pseudoobscura*. *Am. Nat.* 122:474–88

32. Coyne, J. A., Felton, A. A. 1977. Genic heterogeneity at two alcohol dehydrogenase loci in *Drosophila pseudoobscura* and *Drosophila persimilis*. *Genetics* 87:285–304

33. Crumpacker, D. W. 1968. Uniform heterokaryotypic superiority for viability in a Colorado population of *Drosophila pseudoobscura*. *Evolution* 22:256–61

34. Crumpacker, D. W., Pyati, J., Ehrman, L. 1977. Ecological genetics and chromosomal polymorphism in Colorado populations of *Drosophila pseudoobscura*. *Evol. Biol.* 10:437–69

35. Crumpacker, D. W., Williams, J. S. 1973. Density, dispersion, and population structure in *Drosophila pseudoobscura*. *Ecol. Monogr.* 43:499–538

36. Crumpacker, D. W., Williams, J. S. 1974. Rigid and flexible chromosomal polymorphisms in neighboring populations of *Drosophila pseudoobscura*. *Evolution* 128:57-66

37. da Cunha, A. B., Burla, H., Dobzhansky, T. 1950. Adaptive chromosomal polymorphism in *Drosophila willistoni*. *Evolution* 4:212–35

38. da Cunha, A. B., Dobzhansky, T. 1954. A further study of chromosomal polymorphism in *Drosophila willistoni* in relation to environment. *Evolution* 8:119–34

39. da Cunha, A. B., Dobzhansky, T., Pavlovsky, O., Spassky, B. 1959. Genetics of natural populations. XXVIII. Supplementary data on the chromosomal polymorphism in *Drosophila willistoni* in relation to the environment. *Evolution* 13:389–404

40. Dobzhansky, T. 1948. Genetics of natural populations. XVI. Altitudinal and seasonal changes produced by natural selection in certain populations of *Drosophila pseudoobscura* and *Drosophila persimilis*. *Genetics* 33:158–76

41. Dobzhansky, T. 1951. *Genetics and the Origin of Species*. New York: Columbia Univ. Press. 2nd ed.

42. Dobzhansky, T. 1957. Genetics of natural populations. XXVI. Chromosomal variability in island and continental populations of *Drosophila willistoni* from Central America and the West Indies. *Evolution* 11:280–93

43. Dobzhansky, T. 1958. Genetics of natural populations. XXVII. The genetic changes in populations of *Drosophila pseudoobscura* in the American Southwest. *Evolution* 12:385–401

44. Dobzhansky, T. 1971. Evolutionary oscillations in *Drosophila pseudoobscura*. In *Ecological Genetics and Evolution*, ed. R. Creed, pp. 109–33. Oxford: Blackwell

45. Dobzhansky, T. 1973. Active dispersal and passive transport in *Drosophila*. *Evolution* 27:565–75

46. Dobzhansky, T., Epling, C. 1944. Contribution to the genetics, taxonomy, and ecology of *Drosophila pseudoobscura* and its relatives. *Carnegie Inst. Washington Publ. 554*. 183 pp.

47. Dobzhansky, T., Hunter, A. S., Pavlovsky, O., Spassky, B., Wallace, B. 1963. Genetics of natural populations. XXXI. Genetics of an isolated marginal population of *Drosophila pseudoobscura*. *Genetics* 48:91–103

48. Dobzhansky, T., Pavan, C. 1950. Local and seasonal variations in relative frequencies of species of *Drosophila* in Brazil. *J. Anim. Ecol.* 19:1–14

49. Dobzhansky, T., Pavlovsky, O. 1960. How stable is balanced polymorphism? *Proc. Natl. Acad. Sci. USA* 46:41–47

50. Dobzhansky, T., Powell, J. R., Taylor, C. E., Andregg, M. 1979. Ecological variables affecting the dispersal behavior of *Drosophila pseudoobscura* and its relatives. *Am. Nat.* 114:325–34

51. Dobzhansky, T., Wright, S. 1943. Ge-

netics of natural populations. X. Dispersion rates of *Drosophila pseudoobscura*. *Genetics* 28:304–40

52. Dobzhansky, T., Wright, S. 1947. Genetics of natural populations. XV. Rate of diffusion of a mutant gene through a population of *Drosophila pseudoobscura*. *Genetics* 32:303–24

53. Dow, D. D. 1969. Home range and habitat of the cardinal in peripheral and central populations. *Can. J. Zool.* 47:103–14

54. Ehrman, L., Powell, J. R. 1982. The *Drosophila willistoni* species group. In *The Genetics and Biology of Drosophila*, ed. M. Ashburner, H. L. Carson, J. N. Thompson, Jr., 3b:193–225. New York: Academic

55. Epling, C., Lower, W. R. 1957. Changes in an inversion system during a hundred generations. *Evolution* 11:248–58

56. Epling, C., Mitchell, D. F., Mattoni, R. H. T. 1953. On the role of inversions in wild populations of *Drosophila pseudoobscura*. *Evolution* 7:342–65

57. Epling, C., Mitchell, D. F., Mattoni, R. H. T. 1957. The relation of an inversion system to recombination in wild populations. *Evolution* 11:225–47

58. Frankel, O. H., Soulé, M. E. 1981. *Conservation and Evolution*. Cambridge: Cambridge Univ. Press. 327 pp.

59. Franklin, I. A. 1980. Evolutionary change in small populations. In *Conservation Biology: An Evolutionary-Ecological Perspective*, ed. M. E. Soulé, B. A. Wilcox, pp. 135–50. Sunderland, Mass: Sinauer

60. Gillespie, J. H. 1975. The role of migration in the genetic structure of populations in temporarily and spatially varying environments. I. Conditions for polymorphism. *Am. Nat.* 109:127–35

61. Gillespie, J. H. 1977. A general model to account for enzyme variation in natural populations. IV. The quantitative genetics of viability mutants. In *Measuring Selection in Natural Populations, Lect. Notes in Biomath.*, ed. F. B. Christiansen, T. M. Fenchel, 19:301–14. New York: Springer-Verlag

62. Gould, S. J., Eldredge, N. 1977. Punctuated equilibria: The tempo and mode of evolution reconsidered. *Paleobiology* 3:115–51

63. Hartl, D. L. 1977. How does the genome congeal? See Ref. 61, pp. 65–82

64. Haymer, D. S., Hartl, D. L. 1981. Using frequency distributions to detect selection: Inversion polymorphisms in *Drosophila pseudoobscura*. *Evolution* 35:1243–46

65. Deleted in proof

66. Heed, W. B. 1978. Ecology and genetics of Sonoran Desert *Drosophila*. In *Ecological Genetics: The Interface*, ed. P. F. Brussard, pp. 109–26. New York: Springer-Verlag

67. Heed, W. B. 1981. Central and marginal populations revisited. *Drosophila Inf. Serv.* 56:60–61

68. Hoenigsberg, H. F., Palomino, J. J., Hayes, M. J., Zandstra, I. Z., Rojas, G. G. 1977. Population genetics in the American tropics. X. Genetic load differences in *Drosophila willistoni* from Colombia. *Evolution* 31:805–11

69. Jaenike, J. 1978. Ecological genetics in *Drosophila athabasca*: Its effect on local abundance. *Am. Nat.* 112:287–99

70. Johnston, J. S. 1974. Dispersal in natural populations of the cactiphilic *Drosophila pachea* and *D. mojavensis*. *Genetics* 77:32–33 (Suppl.)

71. Johnston, J. S., Heed, W. B. 1975. Dispersal of *Drosophila*: The effect of baiting on the behavior and distribution of natural populations. *Am. Nat.* 109:207–16

72. Johnston, J. S., Heed, W. B. 1976. Dispersal of desert-adapted *Drosophila*: The saguaro-breeding *D. nigrospiracula*. *Am. Nat.* 110:629–51

73. Johnston, J. S., Templeton, A. R. 1982. Dispersal and clines in *Opuntia* breeding *Drosophila mercatorum* and *D. hydei* at Kamvela, Hawaii. In *Ecological Genetics and Evolution*, ed. J. S. F. Barker, W. T. Starmer, pp. 241–56. Sydney: Academic

74. Jones, J. S., Bryant, S. H., Lewontin, R. C., Moore, J. A., Prout, T. 1981. Gene flow and the geographical distribution of a molecular polymorphism in *Drosophila pseudoobscura*. *Genetics* 98:157–78

75. Jones, J. S., Parkin, D. T. 1977. Attempts to measure natural selection by altering gene frequencies in natural populations. See Ref. 61, pp. 83–96

76. Krimbas, C. B., Loukas, M. 1980. The inversion polymorphism of *Drosophila subobscura*. *Evol. Biol.* 12:163–234

77. Lakovaara, S., Saura, A. 1971. Genic variation in marginal populations of *Drosophila subobscura*. *Hereditas* 69:77–82

78. Lakovaara, S., Saura, A. 1971. Genetic variation in natural populations of *Drosophila obscura*. *Genetics* 69:377–84

79. Lande, R. 1975. The maintenance of genetic variability by mutation in a polygenic character with linked loci. *Genet. Res.* 26:221–35

80. Levitan, M. 1951. Experiments on chro-

mosomal variability in *Drosophila robusta*. *Genetics* 36:285–305

81. Levitan, M. 1951. Response of the chromosomal variability in *Drosophila robusta* to seasonal factors in a southwest Virginia woods. *Genetics* 36:561–62 (Suppl.)

82. Levitan, M. 1951. Selective differences between males and females in *Drosophila robusta*. *Am. Nat.* 85:385–88

83. Levitan, M. 1955. Studies of linkage in populations. I. Associations of second chromosome inversions in *Drosophila robusta*. *Evolution* 9:62–74

84. Levitan, M. 1958. Non-random associations of inversions. *Cold Spring Harbor Symp. Quant. Biol.* 23:251–68

85. Levitan, M. 1973. Studies of linkage in populations. VII. Temporal variation and X-chromosome linkage disequilibriums. *Evolution* 27:476–85

86. Levitan, M. 1978. Studies of linkage in populations. IV. The effect of altitude on X-chromosomal arrangement combinations in *Drosophila robusta*. *Genetics* 89:751–63

87. Lewontin, R. C. 1958. Studies on heterozygosity and homeostasis. II. Loss of heterosis in a constant environment. *Evolution* 12:494–503

88. Lewontin, R. C. 1974. *The Genetic Basis of Evolutionary Change*. New York: Columbia Univ. Press

89. Lewontin, R. C., Ginzburg, L. R., Tuljapurkar, S. D. 1978. Heterosis as an explanation for large amounts of genic polymorphism. *Genetics* 88:149–70

90. Lewontin, R. C., Hubby, J. L. 1966. A molecular approach to the study of genic heterozygosity in natural populations. II. Amount of variation and degree of heterozygosity in natural populations of *Drosophila pseudoobscura*. *Genetics* 54:595–609

91. Little, E. L. Jr. 1971. *Atlas of United States Trees, US Dep. Agric. Misc. Publ. No. 1146*. Vol. 1, *Conifers and Important Hardwoods*. Washington, DC: USGPO

92. Loukas, M., Krimbas, C. B., Morgan, K. 1980. The genetics of *Drosophila subobscura* populations. XIV. Further data on linkage disequilibria. *Genetics* 95:757–68

93. Mayhew, S. H., Kato, S. K., Ball, F. M., Epling, C. 1966. Comparative studies of arrangements within and between populations of *Drosophila pseudoobscura*. *Evolution* 20:646–62

94. Mayr, E. 1963. *Animal Species and Evolution*. Cambridge, Mass: Belknap

95. McFarquhar, A. M., Robertson, F. W.

1963. The lack of evidence for coadaptation in crosses between geographical races of *Drosophila subobscura* Coll. *Genet. Res.* 4:104–31

96. McKenzie, J. A., Parsons, P. A. 1974. Numerical changes and environmental utilization in natural populations of *Drosophila*. *Aust. J. Zool.* 22:175–87

97. Moore, J. A., Taylor, C. E., Moore, B. C. 1979. The *Drosophila* of southern California. I. Colonization after a fire. *Evolution* 33:156–71

98. Nair, P. S. 1969. Genetic load in *Drosophila robusta*. *Genetics* 63:221–28

99. Nei, M., Maruyama, T., Chakraborty, R. 1975. The bottleneck effect and genetic variability in populations. *Evolution* 29:1–10

100. Olvera, O., Powell, J. R., de la Rosa, M. E., Salceda, V. M., Gaso, M. I., Guzman, J., Anderson, W. W., Levine, L. 1979. Population genetics of Mexican *Drosophila*. VI. Cytogenetic aspects of the inversion polymorphism in *Drosophila pseudoobscura*. *Evolution* 33:381–95

101. Parsons, P. A. 1978. Boundary conditions for *Drosophila* resource utilization in temperate regions, especially at low temperatures. *Am. Nat.* 112:1063–74

102. Parsons, P. A. 1978. Habitat selection and evolutionary strategies in *Drosophila*: An invited address. *Behav. Genet.* 8:511–26

103. Parsons, P. A. 1983. *The Evolutionary Biology of Colonizing Species*. Cambridge: Cambridge Univ. Press. 262 pp.

104. Patterson, J. T., Wagner, R. P. 1943. II. Geographical distribution of species of the genus *Drosophila* in the United States and Mexico. *Univ. Tex. Publ.* 4313:217–81

105. Pinsker, W., Sperlich, D. 1979. Allozyme variation in natural populations of *Drosophila subobscura* along a north-south gradient. *Genetica* 50:207–19

106. Powell, J. R. 1978. The founder-flush speciation theory: An experimental approach. *Evolution* 32:465–74

107. Powell, J. R., Dobzhansky, T., Hook, J. E., Wistrand, H. E. 1976. Genetics of natural populations. XLIII. Further studies on rates of dispersal of *Drosophila pseudoobscura* and its relatives. *Genetics* 82:493–506

108. Powell, J. R., Taylor, C. E. 1979. Genetic variation in ecologically diverse environments. *Am. Sci.* 67:590–96

109. Prakash, S. 1967. Association between mating speed and fertility in *Drosophila robusta*. *Genetics* 57:655–63

110. Prakash, S. 1967. Chromosome interac-

tions in *Drosophila robusta. Genetics* 57:385–400

111. Prakash, S. 1968. Chromosome interactions affecting mating speed in *Drosophila robusta. Genetics* 60:589–600

112. Prakash, S. 1972. Origin of reproductive isolation in the absence of apparent genic differentiation in a geographic isolate of *Drosophila pseudoobscura. Genetics* 72: 143–55

113. Prakash, S. 1973. Patterns of gene variation in central and marginal populations of *Drosophila robusta. Genetics* 75:347–69

114. Prakash, S. 1977. Further studies on gene polymorphism in the main body and geographically isolated populations of *Drosophila pseudoobscura. Genetics* 85: 713–19

115. Prakash, S. 1977. Gene polymorphism in natural populations of *Drosophila persimilis. Genetics* 85:513–20

116. Prakash, S., Lewontin, R. C., Hubby, J. L. 1969. A molecular approach to the study of genic heterozygosity in natural populations. IV. Patterns of genic variation in central, marginal and isolated populations of *Drosophila pseudoobscura. Genetics* 61:841–58

117. Prevosti, A., Ocaña, J., Alonso, G. 1975. Distances between populations of *Drosophila subobscura*, based on chromosome arrangements frequencies. *Theor. Appl. Genet.* 45:231–41

118. Prince, G. J., Parsons, P. A. 1977. Adaptive behavior of *Drosophila* adults in relation to temperature and humidity. *Aust. J. Zool.* 25:285–90

119. Rabinowitz, D. 1981. Seven forms of rarity. In *The Biological Aspects of Rare Plant Conservation*, ed. H. Synge, pp. 205–17. New York: Wiley

120. Richmond, R. C. 1972. Enzyme variability in the *Drosophila willistoni* group. III. Amounts of variability in the superspecies, *D. paulistorum. Genetics* 70:87–112

121. Richmond, R. C. 1978. Microspatial genetic differentiation in natural populations of *Drosophila*. See Ref. 66, pp. 127–42

122. Robbins, C. S., Van Velzen, W. T. 1969. *The Breeding Bird Survey 1967 and 1968, Bur. Sport Fish. Wildl. Spec. Sci. Rep.—Wildl. 124*. Washington DC: Bur. Sport Fish. Wildl.

123. Rockwood-Sluss, E. S., Johnston, J. S., Heed, W. B. 1973. Allozyme genotype-environment relationships. I. Variation in natural populations of *D. pachea. Genetics* 73:135–46

124. Saura, A. 1974. Genic variation in Scan-dinavian populations of *Drosophila bifasciata. Hereditas* 76:161–72

125. Saura, A., Lakovaara, S., Lokki, J., Lankinen, P. 1973. Genic variation in central and marginal populations of *Drosophila subobscura. Hereditas* 75:33–46

126. Singh, R. S. 1979. Genic heterogeneity within electrophoretic "alleles" and the pattern of variation among loci in *Drosophila pseudoobscura. Genetics* 93: 997–1018

127. Singh, R. S., Lewontin, R. C., Felton, A. A. 1976. Genetic heterogeneity within electrophoretic "alleles" of xanthine dehydrogenase in *Drosophila pseudoobscura. Genetics* 84:609–29

128. Soulé, M. 1973. The epistasis cycle: A theory of marginal populations. *Ann. Rev. Ecol. Syst.* 4:165–87

129. Southwood, T. R. E. 1978. *Ecological Methods*. New York: Chapman & Hall. 2nd ed.

130. Spassky, B., Richmond, R. C., Pérez-Salas, S., Pavlovsky, O., Mourão, C. A., Hunter, A. S., Hoenigsberg, H., Dobzhansky, T., Ayala, F. J. 1971. Geography of the sibling species related to *Drosophila willistoni* and of the semi-species of the *Drosophila paulistorum* complex. *Evolution* 25:129–43

131. Sperlich, D., Feuerbach-Mravlag, H., Lange, P., Michaelidis, A., Pentzos-Daponte, A. 1977. Genetic load and viability distribution in central and marginal populations of *Drosophila subobscura. Genetics* 86:835–48

132. Spiess, E. B. 1950. Experimental populations of *Drosophila persimilis* from an altitudinal transect of the Sierra Nevada. *Evolution* 4:14–33

133. Stalker, H. D., Carson, H. L. 1948. An altitudinal transect of *Drosophila robusta* Sturtevant. *Evolution* 2:295–305

134. Tabachnick, W. J., Powell, J. R. 1977. Adaptive flexibility of "marginal" versus "central" populations of *Drosophila willistoni. Evolution* 31:692–94

135. Taylor, C. E., Powell, J. R. 1977. Microgeographic differentiation of chromosomal and enzyme polymorphisms in *Drosophila persimilis. Genetics* 85:681–95

136. Taylor, F. 1981. Ecology and evolution of physiological time in insects. *Am. Nat.* 117:1–23

137. Templeton, A. R. 1980. The theory of speciation *via* the founder principle. *Genetics* 94:1011–38

138. Templeton, A. R., Johnston, J. S. 1982. Life history evolution under pleiotropy and *K*-selection in a natural population of

Drosophila mercatorum. See Ref. 73, pp. 225–39

139. Townsend, J. I. 1952. Genetics of marginal populations of *Drosophila willistoni. Evolution* 6:428–42

140. Wallace, B. 1984. A possible explanation for observed differences in the geographical distributions of chromosomal arrangements of plants and *Drosophila. Egypt. J. Genet. Cytol.* In press

141. Ward, O. L., Starmer, W. T., Russell, J. S., Heed, W. B. 1974. The correlation of climate and host plant morphology with a geographic gradient of an inversion polymorphism in *Drosophila pachea. Evolution* 28:565–75

142. Wasserman, M., Koepfer, H. R. 1975. Fitness of karyotypes in *Drosophila pseudoobscura. Genetics* 79:113–26

143. Westman, W. E. 1980. Gaussian analysis: Identifying environmental factors influencing bell-shaped species distributions. *Ecology* 61:733–39

144. Whittaker, R. H. 1952. A study of summer foliage insect communities in the Great Smoky Mountains. *Ecol. Monogr.* 22:1–44

145. Whittaker, R. H. 1967. Gradient analysis of vegetation. *Biol. Rev. Cambridge Philos. Soc.* 42:207–64

146. Whittaker, R. H. 1973. Direct gradient analysis: Results. In *Handbook of Vegetation Science.* Vol. 5, *Ordination and Classification of Communities,* ed. R. H. Whittaker, pp. 33–51. The Hague: Junk

147. Whittaker, R. H., Goodman, D. 1979. Classifying species according to their demographic strategy. I. Population fluctuations and environmental heterogeneity. *Am. Nat.* 113:185–200

148. Whittaker, R. H., Levin, S. A., Root, R. B. 1973. Niche, habitat, and ecotope. *Am. Nat.* 107:321–38

149. Whittaker, R. H., Levin, S. A., Root, R. B. 1975. On the reasons for distinguishing "niche, habitat, and ecotope." *Am. Nat.* 109:479–82

150. Yamaguchi, O., Moriwaki, D. 1971. Chromosomal variation in natural populations of *Drosophila bifasciata. Jpn. J. Genet.* 46:383–91

151. Zouros, E., Krimbas, C. B., Tsakas, S., Loukas, M. 1974. Genic versus chromosomal variation in natural populations of *Drosophila subobscura. Genetics* 78: 1223–44

Ann. Rev. Ecol. Syst. 1984. 15:65–95

ECOLOGICAL DETERMINANTS OF GENETIC STRUCTURE IN PLANT POPULATIONS

M. D. Loveless and J. L. Hamrick

Departments of Systematics & Ecology and Botany, University of Kansas, Lawrence, Kansas 66045

INTRODUCTION

Plant populations are not randomly arranged assemblages of genotypes but are structured in space and time (2, 29, 49, 58, 84, 112). This structure may be manifested among geographically distinct populations, within a local group of plants, or even in the progeny of individuals. Genetic structure results from the joint action of mutation, migration, selection, and drift, which in turn must operate within the historical and biological context of each plant species. Ecological factors affecting reproduction and dispersal are likely to be particularly important in determining genetic structure (2, 31, 58). Reproduction is the process that translates the current genotypic array into that of subsequent generations, while the dispersal of pollen and seeds determines the postreproductive patterns of gene dispersion within and among populations. Although the concept of genetic structure has been used in various ways (58, 130, 154), we limit our definition to the nonrandom distribution of alleles or genotypes in space or time and disregard genome organization and meiotic processes that can also affect allele and genotype frequencies.

Because of the limited mobility of plants, their genetic structure implies spatial structure, or the actual physical distribution of individuals. While spatial patterns often have genetic implications, nonrandom genetic patterns can exist without a nonrandom distribution of individuals. Conversely, a population may have a nonrandom spatial distribution without any accompanying genetic structure. Spatial and genetic patterns are often assumed to result from environmental heterogeneity and differential selection pressures (22, 53, 131, 132). Selection is a ubiquitous feature of natural populations; it alters gene and

65

0066-4162/84/1120-0065$02.00

genotype frequencies and acts in concert with migration, dispersion, and other processes to generate genetic structure (36, 37, 65, 124, 149, 152). In this review, however, we limit our discussion to the deterministic effects of various ecological traits on genetic structure. Selective factors vary in time and space in ways that cannot be directly anticipated by a plant population. As a result, selection is local and idiosyncratic in how it alters genetic structure, and it usually cannot be used to predict the distribution of genetic variation within and among populations.

Comprehensive studies of population genetic structure in plants are biased toward temperate species, usually annuals, short-lived perennials, or conifers. Long-lived herbaceous perennials, angiosperm trees, and monocots (except grasses) have not been studied extensively, and the population genetics of tropical or alpine species is virtually unknown. Lack of information on the genetic structure of plant populations is a serious problem, since any understanding of speciation, adaptation, or genetic change must take into account genetic patterns and the processes by which they are modified (6, 22). The importance of genetic structure in constraining and directing evolutionary change is clearly highlighted in current efforts toward genetic conservation. In order to make reasoned decisions about sampling procedures that preserve maximum levels of genetic diversity, biologists must know how genetic variation is distributed throughout a species; they must know (or guess) its genetic structure (1, 24, 28, 47, 88).

In this paper, we consider a variety of ecological characteristics, summarizing their predicted effects on genetic structure within and among populations. We then analyze available data describing genetic differentiation in plants to determine which ecological variables are consistent predictors of genetic organization. Finally, we suggest directions for future research that will improve our understanding of the factors determining genetic structure in natural populations.

ECOLOGICAL AND LIFE HISTORY TRAITS AFFECTING GENETIC STRUCTURE

Procedures

Table 1 presents a summary, drawn largely from the literature, of life history traits and their genetic implications for plant populations. Within each trait, the character states are ordered beginning with those most likely to enhance genetic structure. Since genetic structure is dependent in part on the amount of genetic variation in a population, we also include a summary of how these traits affect genetic variation within populations (45, 50, 51). Table 1 cannot exhaustively chronicle the factors that alter or influence genetic structure in plants, and in some cases, the categories within each factor are arbitrary. In many species, these traits vary among populations or over time (42, 52, 115), and populations may thus differ in the degree to which they demonstrate predicted genetic effects. Different combinations of characters may result in similar genetic structure in different species (11, 26, 60), or similar suites of characters may interact in different ways. We have tried, however, to consider those ecological variables that should have real and predictable effects on genetic structure.

The major conceptual model for how subdivision occurs within large, randomly organized populations is that of isolation by distance, which Wright (151–154) developed. This model is based on the biologically reasonable expectation that the probability of mating depends on the distance between individuals or the variance in the dispersion of their propagules. The effective population size, N_e, represents the size of a population, or neighborhood, that undergoes the same decay of genetic variance by inbreeding or drift as that in an idealized, panmictic reference population of size N. As Wright and others have demonstrated (66, 78, 79, 149, 150, 152, 153), N_e is usually less than the actual number of individuals in the population; processes that decrease N_e alter genetic structure by increasing differentiation among neighborhoods. The neighborhood concept is thus a useful method of conceptualizing the genetic implications of ecological processes.

Ecological Variables

BREEDING SYSTEMS   Plant breeding systems have been identified as a major factor influencing genetic structure (4, 10, 23, 24, 45, 50, 58, 59, 86, 151, 152). Inbreeding, whether from autogamy or restricted gene flow, increases correlations between uniting gametes, reduces recombination, and maintains gametic phase disequilibrium. As a result, it homogenizes genotypes produced within a family lineage and increases the potential for genetic differentiation among families. Where other factors, such as limited seed dispersal, produce aggregation among related progeny, inbreeding can generate intrapopulation genetic subdivision.

Outcrossing generally decreases correlations between uniting gametes, thus increasing N_e and reducing population subdivision. Where family structure is set up by restricted seed or pollen movement, outcrossed matings may represent inbreeding owing to the consanguinity of adjacent individuals (14, 35, 79, 110). But if self-incompatibility is combined with family structure, pollen dispersal and N_e become larger, and genetic subdivision is less likely. Outcrossing also enforces pollen movement, increasing the probability of long-distance gene flow. Theoretical studies have shown that only a small amount of long-distance gene flow is needed to prevent population differentiation for neutral alleles (125, 129, 153, 154).

In mixed-mating species, the proportions of autogamy and allogamy will determine genetic structure (58, 86, 78, 79). Since only a small amount of

Table 1 Summary of ecological factors that can affect the genetic structure of populations and their predicted effects

Ecological factor	Genetic variation within populations	Genetic structure among populations	Genetic structure within populations	References
Breeding system				
Autogamous or primarily inbreeding	Lower; low heterozygosity	Increased divergence due to drift and reduced gene flow	Reduced heterozygosity and within-family genetic diversity, increased between-family genetic variation; low N[a]; increased gametic phase disequilibrium; restricted gene migration and high population subdivision	2, 4, 7, 10, 14, 23–25, 31, 45, 50, 51, 58–60, 62, 72, 73, 78, 79, 86, 110
Mixed mating	More variability	Potential for differentiation; depends on level of selfing and may vary in time	Potentially subdivided; depends on balance between selfing and outcrossing	
Predominately outcrossed	Higher; high heterozygosity	Reduced divergence due to increased pollen flow	Increased N_e and N_A and reduced subdivision	
Floral morphology				
Hermaphroditic	Moderate levels, if mixed mating; lower if selfing	Depends on breeding system; selfing promotes divergence	Potential for subdivision; depends on mating system and pollen movement; floral morphology affects pollination and pollen carryover, altering N_e and N_A up or down	10, 14, 15, 18, 24, 45, 50, 71–73, 78, 79, 105, 144, 145, 150, 152
Monoecious or dichogamous	Potentially high, if predominantly outcrossed	Increased outbreeding and pollen flow reduce differentiation	Depends on mating system and pollinators; likely to have reduced subdivision and increased homogeneity	
Dioecious or heterostylous	High	Enforced outcrossing and pollen movement reduce differentiation	Enforced outbreeding reduces subdivision; however, assortative mating and unequal sex ratios can reduce N_e and generate differentiation	

Mode of reproduction				
Obligate apomixis	Low, but depends on the number of genets	Founder effects and drift promote divergence; loss of genotypic variability is increased by lack of recombination	Homogeneous clones; population highly subdivided or monomorphic, depending on number of genets	10, 50, 72, 78–80
Facultative apomixis	Moderate; depends on breeding system and other factors	Founder effect may limit number of genets and thus enhance differentiation	Potentially subdivided; depends on breeding system and amount of sexual reproduction	
Sexual reproduction	Potentially high	Depends on other factors	Depends on other factors	

Pollination mechanism				
Small bee General entomophily	Insect-pollinated species have reduced amounts of variability	Limited pollen movement and local foraging, especially by small insects, increase differentiation	Limited, leptokurtic, or nearest neighbor pollen movement reduces N_e, promotes subdivision, family structure, and inbreeding	4, 7, 8, 13, 14–16, 19, 33, 39, 40, 42, 43, 50, 54, 55, 61, 64, 71, 73–75, 77, 79–81, 106–109, 111, 113, 115–117, 133, 136, 139–141, 143–145, 150–153, 157, 158
Large bee Butterfly/moth Bird/bat		Rare long-distance pollen dispersal, long-distance traplining, or low background pollen levels (wind) prevent divergence	Animal vectors with high variance in pollen carryover and delivery will increase N_e; Large, vagile vectors will visit more plants, reduce subdivision, give moderate to large N_e and large N_A	
Wind	High		Wind pollination gives large N_e, N_A; reduces subdivision	

Table 1 (*continued*)

Ecological factor	Genetic variation within populations	Genetic structure among populations	Genetic structure within populations	References
Seed dispersal				
Gravity	Intermediate	Limited dispersal promotes differentiation	Limited seed movement reduces N_e, promotes family structure, inbreeding, increased homozygosity, and subdivision	4, 5, 7, 19, 26, 30, 50, 57, 79, 92, 111, 120, 139
Explosive/capsule	Intermediate	Small amounts of long-distance migration can prevent divergence	Large variance in dispersal distance increases N_e, decreases subdivision	
Winged/plumose (wind)	High			
Animal-ingested	Intermediate	Regular long distance transport promotes homogeneity	Dispersal by wind and animals may reduce clumping and family structure	
Animal-attached	Low			
Seed dormancy				
Absent	Determined by other factors	Determined by other factors	Determined by other factors	59, 72, 82, 135
Present	Increases potential genetic variation	Reduces divergence; retards loss of alleles by drift and isolation	Retards loss of alleles; increases generation time of genotypes, increases N_e, and inhibits subdivision; may be countered by differential fecundities or other factors	

Phenology				
Populations asynchronous	No prediction	Prevents gene exchange, promotes divergence	Restricts mating, reduces N_e, and promotes subdivision	4, 7–9, 15, 17, 24, 41, 44, 116, 145, 153
Populations seasonal and synchronous	No prediction	Potential for extensive gene flow reduces probability of divergence	Large potential N_e; may be restricted by pollinator behavior or family structure, but potentially homogeneous	
Extended, low-level flowering	No prediction	Long-distance pollinator movement prevents divergence	Reduces selfing, increases pollen flow, increases N_A, and prevents subdivision	
Life cycle				
Annual Short-lived	Reduced; less heterozygosity	Increases chances of subdivision	Increases susceptibility to drift due to bottleneck effects and to variable fecundities; smaller N_e promotes local subdivision	5, 10, 24, 72, 82
Long-lived	Increased	Reduces effects of drift, increases chances of migration, and thus hinders divergence	Retards loss of variation; increases N_e, increases mating opportunities, and retards subdivision	
Timing of reproduction				
Monocarpic	No prediction	Promotes drift and between-population divergence	Restricts mating possibilities, shortens effective generation time; reduces N_e, which promotes differentiation in time and space but reduces flowering density, which may increase N_e	72, 82, 147, 150
Polycarpic	No prediction	May inhibit divergence; depends on other factors	Increases N_e by increasing mating pool and generation time, reducing probability of subdivision	

Table 1 *(continued)*

Ecological factor	Genetic variation within populations	Genetic structure among populations	Genetic structure within populations	References
Successional stage				
Early	Reduced variability	Founder and drift effects, short population lifespan promote differentiation	Depends on other factors; generation time, breeding system, and dispersal may have conflicting effects on N_e	10, 11, 26, 50, 58
Late	Increased variability	Stable, long-lived population structure promotes migration, reduces drift, and reduces differentiation	Depends on other factors; longer generation time will reduce population subdivision	
Geographic range				
Endemic	Genetically depauperate	Small, local populations will show more divergence due to drift and isolation	Possibly homogenous, due to size fluctuations, lack of variability	1, 5, 6, 11, 23, 50, 58, 59, 88
Narrow Regional	Moderate levels Maximum variation	Patterns in more widespread species determined by other factors	Patterns influenced by other factors	
Widespread	Less variability			
Population size				
Large and stable	High	Trade-off in populations of all sizes between drift and migration effects: small populations promote divergence due to drift, but are more heavily influenced by small numbers of migrant propagules;	Potentially subdivided, depending on pollinator behavior	1, 4, 6, 60, 61, 72, 73, 75, 79, 86, 88, 96, 101, 111, 123, 126, 150, 152
Small and stable	Lower, due to drift		More likely homogeneous, depending on scale of gene flow and magnitude of drift	

Fluctuating size	Low, due to drift	structure will depend on amount of migration	Homogeneous, due to loss of variability and inbreeding during periods of small size; net N_e is weighted toward length of time spent at small population sizes	5–7, 9, 15, 19, 35, 61, 73–82, 86, 116, 117, 152
Population density				
High	No prediction	Trade-offs analogous to those for population size	Animal-dispersed pollen movement is more susceptible to density; high densities restrict pollen flow and increase subdivision	
Low	No prediction	Low density may promote long-distance pollen flow, increasing homogeneity	Low densities may increase pollen movement (increase N_A) or may reduce pollinator visits (decrease N_A and N_e)	
			With wind pollination, lower density will decrease N_e but not N_A, leading to increased subdivision	
Population spatial distribution				
Patchy	No prediction	Increasing isolation reduces gene flow and enhances differentiation	Patchiness may affect pollinator behavior in complex ways; in general, spatial patchiness increases inbreeding, reduces gene flow and N_e, and enhances genetic patchiness and subdivision	7, 19, 67–69, 72–75, 78–80, 82, 90, 97, 111, 127, 150, 152, 154
Uniform	No prediction	Promotes migration and homogeneity	Promotes gene flow and reduces subdivision	
Population shape	No prediction	Divergence enhanced in linear arrays of populations	Subdivision is increased in linear habitats	

[a] N_e = effective population size; N_A = neighborhood area.

outcrossing can retard differentiation among population subdivisions (4, 59), the genetic structure of plants with mixed mating systems should resemble that of outbreeders. However, since outcrossing rates can vary in time owing to factors such as population density or pollinator behavior (35, 95), generations or age classes of mixed-mating species may be genetically differentiated.

FLORAL MORPHOLOGY   This factor is closely related to the breeding system, since autogamy only occurs in hermaphroditic or monoecious plants. Dioecious and heterostylous species must be outcrossed and thus may have larger neighborhood areas and less pronounced population differentation. If the sex ratios of dioecious populations deviate from 1:1, however, N_e decreases and mating patterns may become restricted (16, 72, 105, 150, 152). Under these conditions, even relatively large dioecious populations could diverge by drift. Family structure due to vegetative reproduction or limited seed dispersal would further increase differentiation.

Floral biology is often correlated with the pollination mechanism: Nectar and pollen rewards, the temporal separation of male and female phases, and the arrangement of flower parts may influence pollen deposition and carry-over (71, 136, 144). Phenotypic plasticity in sexual expression or intermediate morphological conditions such as androdioecy could also affect genetic structure by altering the distribution of sexes in space and time.

MODE OF REPRODUCTION   Under obligate apomixis, all progeny are genotypically identical to their mothers, and each lineage is genetically distinct. Isolated, clonally reproducing populations may diverge by losing alleles differentially through drift and by accumulating unique mutations, increasing differentiation and subdividing populations into clonal patches (89). Facultative apomixis affects genetic structure by slowing the decay of genetic variance (78), producing clumped genotype distributions (121), and altering genotypes' fecundity schedules (72). By increasing generation time, apomixis increases N_e and inhibits population subdivision. Clonal patchiness also increases N_e if the species is self-incompatible (78). With self-compatibility, however, family structure and inbreeding will reduce N_e and promote local subdivision. Varying degrees of asexual reproduction by different genotypes may produce a non-Poisson distribution of offspring that will reduce N_e (150, 152), perhaps completely reversing the effects of the longer generation time (72). Apomixis also prevents recombination, thus perpetuating gametic phase disequilibrium, increasing population divergence, and slowing the population's approach to equilibrium (89). In species that reproduce only sexually, genetic structure will be determined by other aspects of their life history; all things being equal, they will be less likely than apomictics to show either intra- or interpopulation subdivision.

POLLINATION MECHANISMS   Only pollen grains that actually effect fertilization influence genetic structure. Although pollen movement has been measured by several methods (79, 107, 113, 143), realized pollen flow is known for only a few natural populations (52, 113). This distinction between pollen movement and realized pollen flow is especially important when long distances are involved, since migration of ineffective pollen grains leads to overestimates of actual gene flow (36).

The distribution of pollen movement is generally leptokurtic (13, 79), permitting isolation by distance on a scale commensurate with the realized pollen distribution. In wind-pollinated species, background pollen levels are sufficient to prevent differentiation over fairly large geographic areas (5, 61, 122). The amount of effective pollen flow among populations is influenced by spatial separation, relative sizes and densities, phenology, and intervening vegetation (79). Prevailing winds may produce directional movement but will not increase N_e or net dispersion distance and should reduce N_e relative to symmetrical dispersal (149, 152).

In a dense, homogeneous population, animal pollen vectors may conform to optimal foraging models (73, 109, 140, 141). Because of energetic constraints, small insects generally have restricted movement patterns (55). Large insects may fly longer distances and visit more plants, but this depends on the insect's physiology and behavior and the reward structure of the population (42, 52, 54, 73, 74, 106, 108, 109, 115–117, 140, 141, 145, 157, 158). There is increasing evidence that pollen transport is highly stochastic and that pollen deposition is variable even under a regular foraging regime (73, 136). In general, pollinators that routinely effect long-distance pollen movement will increase N_e and neighborhood area and decrease the probability of geographical differentiation, while restricted movements will have opposite effects (8, 43, 75). In simulation studies (61, 80, 111, 139), subdivision increases with limited pollen flow, and this patchiness persists over many generations. These results are also consistent with theoretical studies (67–69, 70, 87, 152, 153); as migration increases, the population approaches panmixia (97, 98, 131).

SEED DISPERSAL   Seed dispersal is analogous to pollen dispersal, since its effect on N_e is a function of the variance in dispersal distance. Long-distance dispersal and seedling establishment prevent population divergence. Dispersal by gravity or explosive capsules deposits seeds near the parents (57), increasing family structure, reducing N_e, and promoting divergence within and among populations. This effect could, however, be modified by secondary seed dispersal (19, 57, 60). The movement of wind-dispersed fruits depends on wind speed, seed weight, pappus or wing characteristics, and height of release (79, 120). Dispersion is leptokurtic, but relatively few propagules are carried long distances from the source (79). Thus wind dispersal will inhibit local differentiation, but long-distance migration may still be limited.

Although frugivore behavior can influence seed dispersal and establishment (92), there is no good evidence documenting the effects of different frugivores on genetic structure. Clumps of related seedlings could arise from seed caches or from fruits with multiple seeds, but the implications for genetic structure will depend on seedling mortality prior to reproduction. Bullock & Primack's (30) experiments suggest that animal-attached fruits may have long and variable seed movement. In general, animal-mediated dispersal probably produces longer, more variable seed dispersal and, all things being equal, increases N_e. On the other hand, if open sites are colonized by only a few animal-dispersed propagules, founder effects could produce genetic heterogeneity among the populations.

Few studies consider the relative effectiveness of pollen vs seed dispersal for increasing realized gene flow (5, 134). Antonovics (5) found that, with low selective pressures, seeds were more effective than pollen in retarding population differentiation. If pollen is abundant, an immigrant pollen grain may have little likelihood of effecting fertilization. If selection is ignored, however, an incoming seed is as likely to reach maturity as a locally produced seed. Since seeds carry two alleles at every locus, effective gene flow is twice that of a pollen grain. If one allows for selection, then the differential effectiveness of pollen and seeds depends on fitness and dominance relationships.

SEED DORMANCY   Seed dormancy increases effective generation time, since a plant's progeny continue to enter the population after its death. Dormancy thus increases N_e, reduces the decay of genetic variance, and retards local subdivision. In the absence of selection, the effect of seed dormancy on N_e depends on the average time spent in the seed pool (135). Differential fecundity, however, may bias the contributions of genotypes to the seed pool, reducing N_e and increasing subdivision. In early successional habitats, seed pools may decrease genetic structure by providing a genetically diverse founding population, in contrast to colonization by only a few propagules after long-distance dispersal. A population derived from the seed pool can be instantaneously subdivided, though, if it reflects genetic structure in the extinct source population.

PHENOLOGY   All other things being equal, synchronous flowering increases both the genetically effective density and N_e and thus retards differentiation (8, 9, 15), but the actual effects will depend on vector movement between plants. With limited pollen and seed dispersal, isolation by distance is likely despite the existence of potentially large flowering populations. Mass flowering of large individuals may result in within-plant pollen transfer (14, 17, 39, 42–44,

64, 133), but several mechanisms have been proposed that might encourage interplant movements (42, 43, 133), reducing inbreeding and increasing N_e. Extended, low-level flowering increases N_e by enforcing long-distance pollinator movement and increasing matings between unrelated individuals (41, 44). If pollinators follow stereotyped "trapline" visitation patterns, however, actual matings may be spatially restricted, reducing neighborhood size and area. In addition, if flowering time is heritable, phenology will enforce assortative mating, reducing N_e and promoting divergence between phenology morphs (17). Asynchrony in flowering severely restricts mating (9), and temporally isolated individuals may undergo complete selfing, which will subdivide the population into isolated lineages. Nevertheless, some pollen vectors could effect regular, long-distance pollen flow between widely separated plants. While N_e would remain small during each flowering period, it could vary among years, and neighborhood area could be large, limiting population divergence (7, 17).

THE LIFE CYCLE   Effective population size is a direct function of the life cycle; shorter generation times reduce N_e, promoting population divergence (82). The effects of generation time are similar to those resulting from asexual reproduction and seed dormancy: Genetic variance will decay more slowly in long-lived species, and populations will be less susceptible to drift.

TIMING OF REPRODUCTION   Monocarpy in a perennial acts, in part, like asynchronous phenology, reducing flowering population density and limiting mating possibilities. Monocarpic species also have shorter generation times than polycarps that begin flowering at the same age (82), but the precise effects depend on age-specific reproductive rates. In general, monocarpy—because of its effects on density and generation time—reduces N_e and promotes population differentiation. In monocarpic populations with highly skewed fecundity distributions, N_e is further reduced. If, in addition, variance in generation time is low, so that cohorts are likely to flower together, a temporal genetic structure could result (147).

SUCCESSIONAL STAGE   Successional stage is not strictly an ecological characteristic of plant species but rather an aggregate category based on other ecological traits. Early successional, weedy species are characterized as short-lived, often annual plants with wind or animal-attached dispersal, seed dormancy, hermaphroditic flowers that are selfed or pollinated by wind or various small insects, and vegetative reproduction. These traits permit colonizing species to exploit open, early successional habitats (11, 26). The life cycles, breeding systems, and pollination syndromes of weedy species should reduce N_e, while seed dispersal mechanisms, seed dormancy, and vegetative reproduction should produce a larger N_e. Founder effects during colonization can lead to pronounced divergence among populations. Because the population is transient, long-distance dispersal is less effective in homogenizing populations, and strong within-population subdivision is less likely. However, selfing, clonal spread, or seed bank effects could generate some population subdivision.

Late successional traits all suggest large N_e and reduced differentiation. Different configurations of these ecological traits will interact differently in generating genetic structure; however, if species of particular successional seres share similar ecological characteristics, they should also share predictable patterns of genetic structure.

GEOGRAPHIC RANGE   A species' range will be influenced, at least in part, by its dispersal ability. Widely distributed species are more likely to have moderate rates of gene flow and less interpopulation divergence. Small, localized populations of endemic species will be more susceptible to drift and limited gene flow, increasing interpopulation differentiation but reducing intrapopulation genetic differentiation. Historical factors and habitat heterogeneity will also be closely associated with species distributions, however. As a result, geographic range may not be a good predictor of genetic structure.

POPULATION SIZE   When size fluctuates between generations, N_e is limited by the harmonic mean of the population numbers, which tends to favor smaller sizes (150, 152, 153). Transient, small population sizes or bottlenecks reduce allelic diversity (96, 101, 123), but reductions in heterozygosity depend on how quickly the original population size is restored. Thus, populations subjected to persistent bottlenecks may diverge owing to drift (96, 123).

Where populations are large, differentiation depends on pollen and seed dispersal. Immigration is less effective in homogenizing gene frequencies among large populations because migrants are swamped by locally produced propagules. In small populations, immigration is more effective in altering gene frequencies because migrant propagules make up a larger proportion of the seed or pollen array. Where populations differ in size, migration proceeds directionally from large to small populations, increasing their genetic similarity. Differences among large populations are likely to result from historical or selective factors. Small populations are more susceptible to drift and fixation, but gene flow, especially in long-lived species, can prevent differentiation. The general effect of frequent extinction and recolonization is to increase N_e and reduce divergence among populations (91, 119); however, the way in which colonization and migration take place can lead to variable results (126).

POPULATION DENSITY   Like large population size, high density minimizes the effectiveness of gene flow. The influence of density on animal-mediated pollen and seed movements is dependent on the vector's behavior. For bees, dense, resource-rich populations reduce mean flight distance and flight directionality (78, 79) but increase turning behavior (54, 75), selectivity with respect to plant size or height (117), and floral handling time, which may in turn enhance pollen deposition and reduce pollen carry-over (136). These behavioral responses decrease pollen migration distance, reduce N_e, and favor population subdivision and divergence. As the density or reward per unit area decreases, bee flight distance and directionality increase, with a corresponding increase in neighborhood area (81). However, over at least a moderate range of population densities, N_e remains constant (73, 74), since bees must visit roughly the same number of flowers to realize an energetic gain (55). At energetically unprofitable densities, pollinators will abandon the resource or show increased inconstancy, and N_e will drop precipitously (73, 74).

With wind pollination, neighborhood area remains constant as density decreases, but N_e declines, increasing the probability of self-pollination. This increase in self-pollination, in combination with reduced pollen density, may impair seed set and generate skewed fecundity distributions among plants, reducing N_e still further and enhancing divergence. There will be less long-distance pollen movement, but it may be more effective if the recipient populations are also sparsely distributed.

If dispersal is passive, gene flow by seeds should demonstrate a similar density effect. For both pollen and seeds, migration into low-density populations may increase the probability of fertilization or establishment and thus enhance realized gene flow.

POPULATION SPATIAL STRUCTURE   Spatial structure, along with density and size, are unique and sometimes short-lived features of specific populations. While they can strongly influence pollen flow, their effects on genetic structure are transient, and their influence on population differentiation depends on other, more generalizable species traits. Wright's models of isolation by distance (149, 150, 152, 153) showed that, given equal migration rates, linear population arrays lead to more extensive subdivision than two-dimensional patterns. Simulation studies and analytical models (82, 90, 111, 127) verify these results; increasing linearity promotes differentiation. Spatial patchiness can have a variety of effects on pollinator behavior, either restricting movements within a single patch or encouraging flights among patches (73, 74). With limited dispersal, gaps between patches may be barriers to gene flow, promoting differentiation.

GENETIC STRUCTURE WITHIN AND AMONG POPULATIONS

Procedures

We reviewed isozyme data from 163 studies representing 124 plant taxa to examine associations between ecological factors and the distribution of allozyme variation. Using original papers and pertinent floras, we classified each taxon for 10 ecological traits (Table 2). Seed dormancy, population size, population density, and spatial patterning were excluded because of insufficient information. Using Nei's statistics of gene diversity (100), we calculated H_T, the total genetic diversity of the polymorphic loci; H_S, the mean genetic diversity within populations at polymorphic loci; and D_{ST}, the genetic diversity among populations. These parameters are related as follows: $H_T = H_S + D_{ST}$. The proportion of the total genetic diversity found among populations, G_{ST} (which equals D_{ST}/H_T), was also calculated, and the arithmetic means of H_T, H_S, and G_{ST} over all loci were obtained for each study. We also calculated the weighted means of each genetic parameter for each character-state, weighting them by the product of the number of polymorphic loci and the number of populations. The weighted means were then compared within each ecological trait using a one-way ANOVA, and multivariate principal components analyses (PC) were performed on G_{ST} and the 10 ecological traits.

Results

GENETIC DIVERSITY WITHIN SPECIES   The magnitude of H_T is influenced by the proportion of polymorphic loci within a species, the number of alleles per locus, and the evenness of mean allele frequencies summed for the species. Since we only used polymorphic loci, differences in H_T among character-states reflect the maintenance of different numbers of alleles at polymorphic loci or the unevenness of allele frequency distributions or both.

There are significant differences (Table 2) among character-states for mode of reproduction, life cycle, successional stage (all $P<.05$), and geographic range ($P<.01$). Short-lived perennials, species that are widespread, plants in late successional habitats, or those that reproduce both sexually and vegetatively have the highest H_T values. Low H_T values are found in sexually reproducing species, long-lived perennials, regionally distributed species, and species in mid-successional stages. Compared to H_S and G_{ST}, H_T varied little among character-states, indicating that these species differ mainly in the way in which variation at polymorphic loci is partitioned among populations. This finding also suggests that differences in H_T result primarily from variation in the proportion of polymorphic loci and not from the levels of variability maintained at polymorphic loci.

GENETIC DIVERSITY WITHIN POPULATIONS H_S, often referred to as the expected heterozygosity or the polymorphic index, measures the mean percentage of loci heterozygous per individual. If only polymorphic loci are analyzed, H_S is a function of the number and evenness of allele frequencies within populations. We found highly significant differences ($P < .01$) among character-states for breeding system, life cycle, and successional stage, while floral morphology, mode of reproduction, pollinator mechanism, timing of reproduction, and geographic range were significant at the 5% level (Table 2). High levels of genetic diversity were maintained within populations of species from later successional stages, species reproducing by both sexual and vegetative means, polycarpic perennials, monoecious species, and plants pollinated by large bees. Species with low H_S values include selfing plants, annual species with wind or butterfly pollination, early successional species, plants with perfect flowers, and regionally or narrowly distributed species.

These results indicate that annual, selfing species have fewer alleles per locus and more skewed allele frequencies, a pattern that is indicative of reduced gene flow (128). Examination of individual species demonstrated that populations of annual selfing plants are often monomorphic at loci polymorphic in the species as a whole (47). *Avena barbata* (32) and *Capsella bursa-pastoris* (21) exemplify this pattern. Conversely, outcrossing species such as *Pinus contorta* (148) and *Populus trichocarpa* (146) maintain most of their alleles within individual populations.

Previous reviews of studies using both monomorphic and polymorphic loci (45, 50) have demonstrated that outcrossing species with large ranges, high fecundities, wind pollination, long generation times, and late successional habitats have large H_S values. Our results are in general agreement with these conclusions. Together, these studies indicate that the percentage of polymorphic loci and the number and frequency of alleles within populations are positively associated and that they are influenced by similar combinations of life history traits.

GENETIC DIVERSITY AMONG POPULATIONS G_{ST} measures the proportion of variation among populations relative to the total species' diversity (H_T). High G_{ST} values were characteristic of selfing species, annuals, early successional species, and those with gravity- or animal-dispersed seed (Table 2). Interpopulation variation was also high for widespread, hermaphroditic species and for those with wind or butterfly pollination. *Hordeum spontaneum* ($G_{ST} = 0.360$) (29), *Capsella bursa-pastoris* ($G_{ST} = 0.814$) (21), and *Chenopodium alba* ($G_{ST} = 0.326$) (142) characterize this group. Low G_{ST} values were found in monoecious or dioecious species, predominantly outcrossing plants, long-lived taxa, polycarps, and late successional species. Wind- or bee-pollinated species also had low G_{ST} values, as did species with endemic to regional distributions.

Table 2 The relationship between the distribution of allozyme variation within and among populations and 10 ecological factors

Ecological factor	No. of studies	Mean diversity within species H_T^a (s.e.)	Mean diversity within populations H_S (s.e.)	Mean diversity among populations G_{ST} (s.e.)
Breeding system		n.s.[b]	**	**
Autogamous	39	.291(.086)	.128(.030)	.523(.211)
Mixed mating	48	.242(.056)	.174(.044)	.243(.059)
Predominantly outcrossed	76	.251(.045)	.214(.034)	.118(.036)
Floral morphology		n.s.	*	**
Hermaphroditic	113	.284(.038)	.161(.026)	.389(.056)
Monoecious	47	.224(.083)	.201(.033)	.092(.046)
Dioecious	3	.155(.066)	.121(.012)	.109(.054)
Mode of reproduction		*	*	n.s.
Obligate apomixis	1	.172	.159	.080
Facultative apomixis	16	.356(.107)	.282(.083)	.205(.078)
Sexual	146	.261(.041)	.170(.021)	.300(.087)
Pollination mechanism		n.s.	n.s.	**
Small bee	31	.305(.133)	.202(.081)	.262(.121)
General entomophily	32	.321(.074)	.244(.066)	.227(.066)
Large bee	8	.393(.163)	.304(.127)	.224(.073)
Butterfly/moth	9	.211(.066)	.141(.050)	.303(.074)
Bird/bat	2	.246(.163)	.181(.123)	.158(.063)
Wind	81	.250(.055)	.154(.024)	.322(.128)
Seed dispersal		n.s.	n.s.	**
Gravity	59	.298(.091)	.150(.029)	.446(.220)
Explosive/capsule	24	.272(.062)	.199(.052)	.262(.066)
Winged/plumose	48	.216(.041)	.196(.037)	.079(.026)
Animal-ingested	14	.348(.221)	.213(.136)	.332(.196)
Animal-attached	18	.243(.063)	.141(.042)	.398(.129)

	n			
Phenology		n.s.	n.s.	n.s.
Asynchronous	3	.512(.283)	.371(.183)	.276(.199)
Seasonal and synchronous	153	.258(.040)	.170(.020)	.295(.089)
Extended	7	.336(.236)	.203(.145)	.332(.204)
		*	**	**
Life cycle				
Annual	62	.264(.066)	.136(.025)	.430(.157)
Short-lived	53	.321(.081)	.228(.056)	.262(.081)
Long-lived	48	.221(.042)	.202(.038)	.077(.027)
		n.s.	*	**
Timing of reproduction				
Monocarpic	78	.273(.058)	.153(.025)	.396(.134)
Polycarpic	85	.250(.047)	.205(.035)	.143(.044)
		*	**	**
Successional stage				
Early	81	.281(.066)	.151(.027)	.411(.153)
Middle	47	.227(.050)	.178(.036)	.184(.049)
Late	35	.299(.057)	.264(.053)	.105(.036)
		**	*	**
Geographic range				
Endemic	11	.272(.085)	.202(.077)	.227(.057)
Narrow	43	.255(.061)	.177(.040)	.249(.058)
Regional	57	.218(.041)	.159(.032)	.227(.069)
Widespread	52	.316(.093)	.183(.036)	.407(.219)

[a] Symbols explained in the text.
[b] n.s. = not significant; * = P<.05; ** = P<.01.

Pseudotsuga menziesii (G_{ST} = 0.026) (156), *Sequoiadendron giganteum* (G_{ST} = 0.097) (38), and *Desmodium nudiflorum* (G_{ST} = 0.105) (114) exemplify this group.

Geographic range, seed dispersal, and pollination mechanism showed patterns of G_{ST} values that were not consistent with our predictions. We argued that widespread species will have lower G_{ST} values than endemic or narrowly distributed species. Several of our widespread species, however, were also selfing, colonizing weeds; these traits are associated with high G_{ST} values, apparently overriding predictions based on geographic range. At the other extreme, some endemic species have large populations, resisting drift effects and preventing population divergence. Thus, as we noted earlier, geographic range is a poor predictor of genetic structure. Similar conflicting ecological factors may explain the unexpectedly high G_{ST} values for animal-dispersed seeds and the lower values for wind-dispersed plants. Many of the species with winged seeds that we reviewed are wind-pollinated, monoecious, outcrossing conifers, while several of the animal-dispersed plants are selfing annuals. Although animal dispersal gives plants the capacity for long-distance movement, many fruits may be locally dispersed. One of us (JLH) has preliminary evidence that the bird-dispersed *Juniperus virginiana* has a mean G_{ST} twice that of the wind-dispersed conifers included here. In a similar way, the relatively high G_{ST} value obtained for wind-pollinated species may reflect conflicts among ecological traits. Several species categorized in this group are predominantly selfing, although they wind-pollinate when they outcross. If the 44 wind-pollinated, outcrossed species are analyzed separately, the mean G_{ST} is 0.068, indicating that the G_{ST} for the original group was inflated by predominantly selfing species.

Because annual plants are often autogamous, we examined the separate effects of longevity and breeding system using a two-way comparison of G_{ST}. For outcrossing species, 43 long-lived perennials, 18 short-lived perennials, and 15 annuals had mean G_{ST} values of 0.068, 0.239, and 0.161 respectively. In the mixed-mating group, 5 long-lived species, 27 short-lived perennials, and 16 annuals had mean G_{ST} values of 0.169, 0.230, and 0.269 respectively. There were no long-lived selfing species, but the 8 short-lived, perennial selfing species had a mean G_{ST} of 0.329, while G_{ST} for the 31 selfing annuals was 0.560, the highest mean G_{ST} for any character-state. Thus, although differences among life-cycle classes are influenced by the breeding system, G_{ST} values tend to increase as the life span shortens, which is consistent with our predictions.

Associations among mating system, longevity, and timing of reproduction also exist. The 5 selfing polycarps had a mean G_{ST} of 0.394; the low mean G_{ST} for the polycarpic group is thus influenced by its predominantly outcrossed breeding system. Similarly, the 15 short-lived perennial monocarps had a mean

G_{ST} of 0.224, indicating that selfing, annual species are responsible for the high G_{ST} values in this group.

By comparing genetic structure among congeneric species that share some ecological characteristics, we can obtain further insights into how particular ecological variables affect genetic structure. Comparisons within 6 genera whose species have different breeding systems demonstrated that in four (*Gilia, Lolium, Oenothera,* and *Phlox*), selfing species had higher G_{ST} values. In *Helianthus* and *Solanum*, however, G_{ST} values did not differ among species. In an especially instructive study of *Gilia achilleifolia*, Schoen (118) showed that the G_{ST} among selfing populations was 0.390, while the G_{ST} for outcrossing populations was 0.182.

MULTIVARIATE ANALYSES   Approximately half of the correlations between G_{ST} and the 10 ecological factors were statistically significant. G_{ST} was significantly correlated ($P < .05$) with breeding system (-0.560), life cycle (-0.440), timing of reproduction (-0.409), stage of succession (-0.294), and floral morphology (-0.272). Among the ecological characteristics, the highest correlations were between life cycle and timing of reproduction (0.816), life cycle and successional stage (0.652), life cycle and floral morphology (0.646), floral morphology and successional stage (0.624), life cycle and breeding system (0.577), and timing of reproduction and successional stage (0.525). In some instances, these correlations result from biases in the data set, while in others they reflect biological interactions—e.g. between life cycle and breeding system—that will jointly influence genetic structure.

To explore interactions between genetic structure and ecological variables, three principal axes describing 61.5% of the variation were extracted from the correlation matrix. The first principal axis explained 34.1% of the variation and had high loadings from G_{ST}, life cycle, timing of reproduction, floral morphology, successional stage, breeding system, and pollination mechanism. Long-lived polycarps common to the later stages of succession had low G_{ST} values on this axis. They are predominantly outcrossed, monoecious or dioecious, and wind-pollinated. Annual monocarps in the early successional stages had high G_{ST} values. They are predominantly self-fertilized, pollinated by small insects, and have perfect flowers. Thus the first principal component (PC) describes an axis with conifer trees such as *Pinus ponderosa* at one extreme and annual, weedy selfers such as *Capsella bursa-pastoris* at the other. Since conifers made up 39 of the 163 studies and could bias these results, we performed a second PC analysis that excluded them. The first axis of this analysis showed a positive association between low G_{ST} values, long-lived polycarps in later successional stages, and species reproducing sexually by outcrossing. The main effect of excluding the conifers was a reduction in the importance of this axis, since it then explained only 25.7% of the variation.

The second PC axis for the complete data set accounted for an additional 15.1% of the variation. It had high loadings for breeding system, pollination mechanism, geographic range, seed dispersal, and phenology and a significant but rather weak loading for G_{ST}. High G_{ST} values were associated with widespread, selfing, animal-dispersed species with wind pollination (e.g. *Avena barbata*). The third PC explained an additional 12.3% of the variation and had significant loadings for seed dispersal, phenology, and mode of reproduction and a moderate loading for G_{ST}. Species represented at one end of this axis had high G_{ST} values, occurred in late successional habitats, reproduced sexually, flowered over extended periods, and had animal-dispersed seeds (e.g. *Lycopersicon* and *Solanum*).

An appreciable proportion of the variation in G_{ST} (48%) is left unexplained by these three suites of characters. The three axes accounted for more of the variation in ecological traits. There are at least two possible explanations. First, because many studies analyzed only a few loci, they may not accurately document the genetic structure of a species; G_{ST} values among loci within a study were often divergent. Second, G_{ST} values are clearly influenced by historical factors, such as extinction, recolonization, and population spread. As a result, while ecological variables should have predictive power for genetic structure, they cannot and do not completely describe the genetic organization of plant populations.

GENETIC STRUCTURE WITHIN POPULATIONS   Ecological factors should also have important effects on the distribution of genetic variation within plant populations. In Table 3, we summarize 29 studies of 19 plant species for which intrapopulation genetic data were available. A population was defined as the individuals in an area experiencing at least 1% pollen or seed movement across its width. We include interpopulation G_{ST} values for comparison. Because of the limited data set, we grouped studies by breeding system, since this factor was highly correlated with interpopulation genetic structure (24, 45, 47). Table 3 shows that breeding systems also have significant effects on the distribution of genetic variation among population subdivisions. Inbreeding species had relatively high mean G_{ST} values (0.220), while outbreeders had much lower values (0.041). These results are consistent with Brown's (24) earlier, more limited analysis. For both selfing and outcrossing species, G_{ST} values among populations were approximately twice those among population subdivisions, suggesting that genetic diversity varies in magnitude at different levels within species.

CONCLUSIONS

Measures of intrapopulation diversity (H_S) based on the breeding system, life cycle, timing of reproduction, successional stage, and geographic range were consistent with our expectations; character-states predicted to increase N_e and enhance gene flow were associated with higher H_S values. Interpopulation genetic differentiation was consistent with predictions based on breeding systems, floral morphology, life cycle, timing of reproduction, and successional stage; again, ecological characters that promote gene flow and increase N_e led to decreased population differentiation. Our results agree with those in other studies (15, 23–25, 45, 47, 50, 51, 58, 60, 73, 78, 88, 153) indicating that the breeding system is a principal factor in the organization of genetic structure. Breeding systems that promote pollen movement between individuals also permit alleles to be shared widely among populations and reduce differentiation within species.

Predictions of genetic structure based on the characteristics of seed and pollen movement were unsatisfying. Although there were significant differences among G_{ST} values in different subcategories for both of these traits, the G_{ST} values did not decline consistently with the hypothesized increases in gene dispersal. This discrepancy between predicted and observed results may be an artifact of the small and relatively biased data set; the studies may not adequately represent species with particular pollen and seed dispersal modes. Another possible explanation is that the effects of pollen and seed dispersal are overshadowed by other ecological factors, such as breeding system or life cycle, and thus are confounded in their ultimate effects on genetic structure. A third possibility hinges on our correct interpretation of the genetic consequences of gene movement by pollen and seed vectors. If realized gene flow patterns, which have seldom been documented in natural populations, differ markedly from patterns of vector movement, the predictions themselves may be in error.

While we have identified several ecological factors that may be important determinants of plant genetic structure, we still lack descriptive and experimental data from natural populations that would permit us to sort out the multiple effects of these ecological traits. Several avenues of future research would contribute significantly to our understanding of genetic structure in plants. First, we need comprehensive studies that sample at different geographical scales. Such hierarchical sampling designs would allow us to better assess the distances over which differentiation can take place in plant species, and they would permit us to make more rigorous distinctions between patterns of genetic structure at regional, local, or subpopulational levels (1, 2, 15, 23, 88). Second, our understanding of temporal genetic structure is extremely limited. We know little about the temporal stability of genetic structure or about how this structure is related to population growth or to demographic changes over time. Finally, we need more comprehensive, comparative studies following Schoen's example (118), in which attempts are made to isolate and examine the effects of particular life history features within a single species or group of related taxa. Only by conducting such carefully designed studies can we

Table 3 Among populations and levels of genetic diversity among population subdivisions

Species[a]	Among populations[b]			Among population subdivisions			References[c]
	No. populations	No. polymorphic loci	G_{ST}	No. subpopulations	No. polymorphic loci	G_{ST}	
Autogamous							
Avena barbata	16	5	.736	7	5	.259	32, 48
				17	5	.333	3
				23	5	.357	49
Avena fatua				16	24	.372	63
Bromus mollis	5	6	.018	10	6	.029	27
Hordeum spontaneum	28	25	.360	4	11	.090	29, 103
	12	19	.491	4	2	.100	28, 102
Mean			.401			.220	
Mixed mating							
Lolium multiflorum				4	4	.010	Mitton et al, unpubl.
Silene maritima	4	6	.060	6	4	.038	12
Mean			.060			.020	
Predominantly outcrossing							
Abies balsamea				5	7	.012	99
Abies lasiocarpa				3	1	.015	46
Anthoxanthum odoratum				3	3	.045	155
Chrysanthemum leucanthemum	6	7	.045	12	7	.050	Griswold & Hamrick, unpubl.

							Reference
Desmodium nudiflorum	5	6	.105	2	3	.015	114
Helianthus annuus	15	9	.055	5	1	.032	Lan & Hamrick, unpubl., 138
Liatris cylindracea	18	1	.144	2	1	.006	137
Picea abies	28	37	.093	66	14	.069	112
				4	1	.148	20
				3	1	.100	46
Picea engelmannii	5	11	.038	18	7	.016	148, Hamrick & Smith, unpubl.
Pinus contorta	11	12	.015	9	11	.023	56
				90	12	.043	Hamrick et al, unpubl.
				11	1	.029	93
Pinus longaeva	11	1	.065	6	7	.042	104, 85
Pinus ponderosa	10	12	.123	3	7	.027	94
				2	3	.003	85
				2	1	.051	83
Pseudotsuga menziesii	11	18	.026	4	18	.060	156, 34
				2	1	.026	84
Mean			.071			.041	

[a] Species are grouped by their breeding systems.
[b] Among-population genetic diversities shown for comparison.
[c] When two citations are given, the first refers to the among-population data.

explore the specific effects of individual ecological variables, as well as their interactions with other ecological, historical, and selective factors that shape genetic structure. In addition, future studies should encompass larger numbers of individuals within populations, more populations, and more genetic loci if their results are to be generalizable.

Perhaps the most striking finding in our analysis is the degree to which genetic diversity in mixed-mating and outcrossing species is distributed among populations. Outcrossing not only allows gene exchange among individuals within a local population, but it seemingly permits significant amounts of gene flow among populations. Outbreeding, long life cycles, and repeated reproduction encourage gene flow and appear to be important in distributing alleles widely within a species (47). Fifteen years ago, Ehrlich & Raven (33) concluded that gene flow in natural populations is strictly limited and is thus an insignificant factor in countering local selective pressures and maintaining species-wide genetic homeostasis. In a recent review, Levin (75) revised upward earlier estimates of gene flow in plant populations (33, 79), but he still regarded gene flow as (a) small enough to permit substantial local differentiation and (b) too small to provide genetic cohesion to a species gene pool. While we agree with his first conclusion, our data suggest that, at least in outcrossing species, gene flow is sufficient to introduce novel alleles into widely separated populations continually. This interpopulation migration repeatedly generates new genetic combinations, although the fate of these genotypes will be determined by local selective pressures and stochastic events. As a result, gene flow among plant populations may, in fact, be an important species-wide cohesive force linking natural plant populations.

ACKNOWLEDGMENTS

Linda Vescio and Catherine Gorman assisted in compiling the data. Robert Kingsolver offered helpful comments on the manuscript. Jan Elder and Coletta Spencer prepared the manuscript and tables with characteristic speed and accuracy. Our thanks to all these persons. Portions of this work were supported by NSF Grants BSR–8206946 and DEB–8213260 to JLH.

Literature Cited

1. Allard, R. W. 1970. Population structure and sampling methods. In *Genetic Resources in Plants—Their Exploration and Conservation, IBP Handb.*, ed. O. H. Frankel, E. Bennett, 11:97–107. Philadelphia: Davis

2. Allard, R. W. 1975. The mating system and microevolution. *Genetics* 79:115–26 (Suppl.)

3. Allard, R. W., Babbel, G. R., Clegg, M. T., Kahler, A. L. 1972. Evidence for coadaptation in *Avena barbata*. *Proc. Natl. Acad. Sci. USA* 69:3043–48

4. Allard, R. W., Jain, S. K., Workman, P. L. 1968. The genetics of inbreeding populations. In *Advances in Genetics*, ed. E. W. Caspari, 14:55–131. New York: Academic

5. Antonovics, J. 1968. Evolution in closely adjacent plant populations. VI. Manifold effects of gene flow. *Heredity* 23: 507–24

6. Antonovics, J. 1976. The nature of limits to natural selection. *Ann. Mo. Bot. Gard.* 63:224–47

7. Ashton, P. S. 1969. Speciation among

tropical forest trees: Some deductions in the light of recent evidence. *Biol. J. Linn. Soc.* 1:155–96

8. Augspurger, C. K. 1980. Mass-flowering of a tropical shrub (*Hybanthus prunifolius*): Influence on pollinator attraction and movement. *Evolution* 34: 475–88

9. Augspurger, C. K. 1981. Reproductive synchrony of a tropical shrub: Experimental studies on effects of pollinators and seed predators on *Hybanthus prunifolius* (Violaceae). *Ecology* 62: 775–88

10. Baker, H. G. 1959. Reproductive methods as factors in speciation in flowering plants. *Cold Spring Harbor Symp. Quant. Biol.* 24:177–99

11. Baker, H. G., Stebbins, G. L., eds. 1965. *The Genetics of Colonizing Species.* New York: Academic

12. Baker, J., Maynard Smith, J., Strobeck, C. 1975. Genetic polymorphism in the bladder campion, *Silene maritima. Biochem. Genet.* 13:393–410

13. Bateman, A. J. 1947. Contamination of seed crops. I. Insect pollination. *J. Genet.* 48:257–75

14. Bawa, K. S. 1974. Breeding systems of tree species of a lowland tropical community. *Evolution* 28:85–92

15. Bawa, K. S. 1976. Breeding of tropical hardwoods: An evaluation of underlying bases, current status and future prospects. In *Tropical Trees: Variation, Breeding, and Conservation,* ed. J. Burley, B. T. Styles, pp. 43–59. London: Academic

16. Bawa, K. S. 1980. Evolution of dioecy in flowering plants. *Ann. Rev. Ecol. Syst.* 11:15–39

17. Bawa, K. S. 1983. Patterns of flowering in tropical plants. In *Handbook of Experimental Pollination Biology,* ed. C. E. Jones, R. J. Little, pp. 394–410. New York: Van Nostrand-Reinhold

18. Bawa, K. S., Opler, P. A. 1975. Dioecism in tropical forest trees. *Evolution* 29:167–79

19. Beattie, A. 1978. Plant-animal interactions affecting gene flow in *Viola.* In *The Pollination of Flowers by Insects,* ed. A. J. Richards, pp. 151–64. London: Academic

20. Bergmann, F. 1978. The allelic distribution at an acid phosphatase locus in Norway spruce (*Picea abies*) along similar climatic gradients. *Theor. Appl. Genet.* 52:57–64

21. Bosbach, K., Hurka, H. 1981. Biosystematic studies on *Capsella bursa-pastoris* (Brassicaceae): Enzyme polymorphism in natural populations. *Plant Syst. Evol.* 137:73–94

22. Bradshaw, A. D. 1972. Some evolutionary consequences of being a plant. In *Evolutionary Biology,* ed. T. Dobzhansky, M. K. Hecht, W. C. Steere, 5: 25–47. New York: Appleton-Century-Crofts

23. Brown, A. H. D. 1978. Isozymes, plant population genetic structure and genetic conservation. *Theor. Appl. Genet.* 52: 145–57

24. Brown, A. H. D. 1979. Enzyme polymorphism in plant populations. *Theor. Popul. Biol.* 15:1–42

25. Brown, A. H. D., Albrecht, L. 1980. Variable outcrossing and the genetic structure of predominantly self-pollinated species. *J. Theor. Biol.* 82:591–606

26. Brown, A. H. D., Marshall, D. R. 1981. Evolutionary changes accompanying colonization in plants. In *Evolution Today: Proc. 2nd Int. Congr. Syst. Evol. Biol.,* ed. G. G. E. Scudder, J. L. Reveal, pp. 351–63. Pittsburgh, Pa: Hunt Inst., Carnegie-Mellon Univ.

27. Brown, A. H. D., Marshall, D. R., Albrecht, L. 1974. The maintenance of alcohol dehydrogenase polymorphism in *Bromus mollis* L. *Aust. J. Biol. Sci.* 27:545–59

28. Brown, A. H. D., Munday, J. 1982. Population-genetic structure and optimal sampling of land races of barley from Iran. *Genetica* 58:85–96

29. Brown, A. H. D., Zohary, D., Nevo, E. 1978. Outcrossing rates and heterozygosity in natural populations of *Hordeum spontaneum* Koch in Israel. *Heredity* 41:49–62

30. Bullock, S. H., Primack, R. B. 1977. Comparative experimental study of seed dispersal on animals. *Ecology* 58:681–86

31. Clegg, M. T. 1980. Measuring plant mating systems. *BioScience* 30:814–18

32. Clegg, M. T., Allard, R. W. 1972. Patterns of genetic differentiation in the slender wild oat species *Avena barbata. Proc. Natl. Acad. Sci. USA* 69:1820–24

33. Ehrlich, P. R., Raven, P. H. 1969. Differentiation of populations. *Science* 165: 1228–32

34. El-Kasaby, Y. A., Sziklai, O. 1982. Genetic variation of allozyme and quantitative traits in a selected Douglas-Fir (*Pseudotsuga menziesii* var. *menziesii* (Mirb.) Franco) population. *For. Ecol. Manage.* 4:115–26

35. Ellstrand, N. C., Torres, A. M., Levin, D. A. 1978. Density and the rate of apparent outcrossing in *Helianthus annuus* (Asteraceae). *Syst. Bot.* 3:403–7

36. Endler, J. A. 1979. Gene flow and life history patterns. *Genetics* 93:263–84
37. Felsenstein, J. 1976. The theoretical population genetics of variable selection and migration. *Ann. Rev. Genet.* 10:253–80
38. Fins, L., Libby, W. J. 1982. Population variation in *Sequoiadendron:* Seed and seedling studies, vegetative propagation, and isozyme variation. *Silvae Genet.* 31:102–10
39. Frankie, G. W. 1975. Tropical forest phenology and pollinator plant coevolution. In *Coevolution of Animals and Plants,* ed. L. E. Gilbert, P. H. Raven, pp. 192–209. Austin: Univ. Tex. Press
40. Frankie, G. W., Baker, H. G. 1974. The importance of pollinator behavior in the reproductive biology of tropical trees. *An. Inst. Biol. Univ. Nac. Auton. Méx. Ser. Bot.* 45(1):1–10
41. Frankie, G. W., Baker, H. G., Opler, P. A. 1974. Comparative phenological studies in tropical wet and dry forests in the lowlands of Costa Rica. *J. Ecol.* 62:881–919
42. Frankie, G. W., Haber, W. A. 1983. Why bees move among mass-flowering neotropical trees. See Ref. 17, pp. 360–72
43. Frankie, G. W., Opler, P. A., Bawa, K. S. 1976. Foraging behavior of solitary bees: Implications for outcrossing of a neotropical forest tree species. *J. Ecol.* 64:1049–57
44. Gentry, A. H. 1974. Flowering phenology and diversity in tropical Bignoniaceae. *Biotropica* 6:64–68
45. Gottlieb, L. D. 1981. Electrophoretic evidence and plant populations. In *Progress in Phytochemistry,* ed. L. Reinhold, J. B. Harborne, T. Swain, 7:1–46. Oxford: Pergamon
46. Grant, M. C., Mitton, J. B. 1977. Genetic differentiation among growth forms of Englemann spruce and subalpine fir at tree line. *Arct. Alp. Res.* 9(3):259–63
47. Hamrick, J. L. 1983. The distribution of genetic variation within and among natural plant populations. In *Genetics and Conservation,* ed. C. M. Schonewald-Cox, S. M. Chambers, B. MacBryde, L. Thomas, pp. 335–48. London: Benjamin-Cummings
48. Hamrick, J. L., Allard, R. W. 1972. Microgeographical variation in allozyme frequencies in *Avena barbata. Proc. Natl. Acad. Sci. USA* 69:2100–4
49. Hamrick, J. L., Holden, L. R. 1979. Influence of microhabitat heterogeneity on gene frequency distribution and gametic phase disequilibrium in *Avena barbata. Evolution* 33:521–33
50. Hamrick, J. L., Linhart, Y. B., Mitton, J. B. 1979. Relationships between life history characteristics and electrophoretically detectable genetic variation in plants. *Ann. Rev. Ecol. Syst.* 10:173–200
51. Hamrick, J. L., Mitton, J. B., Linhart, Y. B. 1981. Levels of genetic variation in trees: Influence of life history characteristics. In *Isozymes of North American Forest Trees and Forest Insects, Gen. Tech. Rep.,* ed. M. T. Conkle, 48:35–41. Berkeley, Calif: Pac. Southwest Forest & Range Exp. Stn.
52. Handel, S. N. 1983. Contrasting gene flow patterns and genetic subdivision in adjacent populations of *Cucumis sativus* (Cucurbitaceae). *Evolution* 37:760–71
53. Hedrick, P. W., Ginevan, M. E., Ewing, E. P. 1976. Genetic polymorphism in heterogeneous environments. *Ann. Rev. Ecol. Syst.* 7:1–32
54. Heinrich, B. 1979. Resource heterogeneity and patterns of movement in foraging bumblebees. *Oecologia* 40:235–45
55. Heinrich, B., Raven, P. H. 1972. Energetics and pollination ecology. *Science* 176:597–602
56. Hiebert, R. D., Hamrick, J. L. 1983. Patterns and levels of genetic variation in Great Basin bristlecone pine, *Pinus longaeva. Evolution* 37:302–10
57. Howe, H. F., Smallwood, J. 1982. Ecology of seed dispersal. *Ann. Rev. Ecol. Syst.* 13:201–28
58. Jain, S. K. 1975. Population structure and the effects of breeding system. In *Crop Genetic Resources for Today and Tomorrow,* ed. O. H. Frankel, J. G. Hawkes, pp. 15–36. London: Cambridge Univ. Press
59. Jain, S. K. 1976. The evolution of inbreeding in plants. *Ann. Rev. Ecol. Syst.* 7:469–95
60. Jain, S. K. 1976. Patterns of survival and microevolution in plant populations. In *Population Genetics and Ecology,* ed. S. Karlin, E. Nevo, pp. 49–90. New York: Academic
61. Jain, S. K., Bradshaw, A. D. 1966. Evolutionary divergence among plant populations. I. The evidence and its theoretical analysis. *Heredity* 21:407–41
62. Jain, S. K., Marshall, D. R. 1968. Simulation of models involving mixed selfing and random mating. I. Stochastic variation in outcrossing and selection parameters. *Heredity* 23:411–32
63. Jain, S. K., Rai, K. N. 1974. Population biology of *Avena.* IV. Polymorphism in small populations of *Avena fatua. Theor. Appl. Genet.* 44:7–11
64. Kalin Arroyo, M. T. 1976. Geitonogamy

in animal pollinated tropical angio-sperms: A stimulus for the evolution of self-incompatibility. *Taxon* 25:543–48

65. Karlin, S. 1976. Population subdivision and selection-migration interaction. See Ref. 60, pp. 617–57

66. Kimura, M., Crow, J. F. 1963. The measurement of effective population number. *Evolution* 17:279–88

67. Kimura, M., Maruyama, T. 1971. Pattern of neutral polymorphism in a geographically structured population. *Genet. Res., Cambridge* 18:125–30

68. Kimura, M., Ohta, T. 1971. *Theoretical Aspects of Population Genetics*. Princeton: Princeton Univ. Press

69. Kimura, M., Weiss, G. H. 1964. The stepping stone model of population structure and the decrease of genetic correlation with distance. *Genetics* 49:561–76

70. Latter, B. D. H. 1973. The island model of population differentiation: A general solution. *Genetics* 73:147–57

71. Lertzman, K. P., Gass, C. L. 1983. Alternative models of pollen transfer. See Ref. 17, pp. 474–89

72. Levin, D. A. 1978. Some genetic consequences of being a plant. In *Ecological Genetics: The Interface*, ed. P. Brussard, pp. 189–212. New York: Springer-Verlag

73. Levin, D. A. 1978. Pollinator behavior and the breeding structure of plant populations. See Ref. 19, pp. 133–50

74. Levin, D. A. 1979. Pollinator foraging behavior: Genetic implications for plants. In *The Population Biology of Plants*, ed. O. T. Solbrig, S. K. Jain, G. B. Johnson, P. H. Raven, pp. 131–53. New York: Columbia Univ. Press

75. Levin, D. A. 1981. Dispersal versus gene flow in plants. *Ann. Mo. Bot. Gard.* 68:233–53

76. Levin, D. A., Kerster, H. W. 1968. Density-dependent gene dispersal in *Liatris*. *Am. Nat.* 103:61–73

77. Levin, D. A., Kerster, H. W. 1969. The dependence of bee-mediated pollen and gene dispersal upon plant density. *Evolution* 23:560–71

78. Levin, D. A., Kerster, H. W. 1971. Neighborhood structure in plants under diverse reproductive methods. *Am. Nat.* 105:345–54

79. Levin, D. A., Kerster, H. W. 1974. Gene flow in seed plants. In *Evolutionary Biology*, ed. T. Dobzhansky, M. K. Hecht, W. D. Steere, 7:139–220. New York: Plenum

80. Levin, D. A., Kerster, H. W. 1975. The effect of gene dispersal on the dynamics and statics of gene substitution in plants. *Heredity* 35:317–36

81. Levin, D. A., Kerster, H. W., Niedzlek, M. 1971. Pollinator flight directionality and its effect on pollen flow. *Evolution* 25:113–18

82. Levin, D. A., Wilson, J. B. 1978. The genetic implications of ecological adaptations in plants. *Verh. K. Ned. Akad. Wet. Afd. Natuurkd. Reeks 2* 70:75–100

83. Linhart, Y. B., Davis, M. L., Mitton, J. B. 1981. Genetic control of allozymes of shikimate dehydrogenase in ponderosa pine. *Biochem. Genet.* 19:641–46

84. Linhart, Y. B., Mitton, J. B., Sturgeon, K. B., Davis, M. L. 1981. Genetic variation in space and time in a population of ponderosa pine. *Heredity* 46:407–26

85. Linhart, Y. B., Mitton, J. B., Sturgeon, K. B., Davis, M. L. 1981. An analysis of genetic architecture in populations of ponderosa pine. See Ref. 51, 48:53–59

86. Lloyd, D. G. 1980. Demographic factors and mating patterns in angiosperms. In *Demography and Evolution in Plant Populations*, ed. O. T. Solbrig, pp. 67–88. Berkeley: Univ. Calif. Press

87. Malécot, G. 1975. Heterozygosity and relationship in regularly subdivided populations. *Theor. Popul. Biol.* 8:212–41

88. Marshall, D. R., Brown, A. H. D. 1975. Optimum sampling strategies in genetic conservation. See Ref. 58, pp. 53–80

89. Marshall, D. R., Weir, B. S. 1979. Maintenance of genetic variation in apomictic plant populations. I. Single locus models. *Heredity* 42:159–72

90. Maruyama, T. 1970. Effective number of alleles in a subdivided population. *Theor. Popul. Biol.* 1:273–306

91. Maruyama, T., Kimura, M. 1980. Genetic variability and effective population size when local extinction and recolonization of subpopulations are frequent. *Proc. Natl. Acad. Sci. USA* 77:6710–14

92. McKey, D. 1975. The ecology of coevolved seed dispersal systems. See Ref. 39, pp. 159–91

93. Mitton, J. B., Linhart, Y. B., Hamrick, J. L., Beckman, J. S. 1977. Observations on the genetic structure and mating system of ponderosa pine in the Colorado front range. *Theor. Appl. Genet.* 51:5–13

94. Mitton, J. B., Sturgeon, K. B., Davis, M. L. 1980. Genetic differentiation in ponderosa pine along a steep elevational transect. *Silvae Genet.* 29:100–3

95. Moran, G. F., Brown, A. H. D. 1980. Temporal heterogeneity of outcrossing rates in alpine ash (*Eucalyptus delegatensis* R. T. Bak.) *Theor. Appl. Genet.* 57:101–5

96. Motro, U., Thomson, G. 1982. On

heterozygosity and the effective size of populations subject to size changes. *Evolution* 36:1059–66

97. Nagylaki, T. 1977. Decay of genetic variability in geographically structured populations. *Proc. Natl. Acad. Sci. USA* 74:2523–25

98. Nagylaki, T. 1980. The strong-migration limit in geographically structured populations. *J. Math. Biol.* 9:101–14

99. Neale, D. 1978. *Allozyme studies in balsam fir.* MS thesis, Univ. NH, Durham

100. Nei, M. 1973. Analysis of gene diversity in subdivided populations. *Proc. Natl. Acad. Sci. USA* 70:3321–23

101. Nei, M., Maruyama, T., Chakraborty, R. 1975. The bottleneck effect and genetic variability in populations. *Evolution* 29:1–10

102. Nevo, E., Beiles, A., Storch, N., Doll, H., Andersen, B. 1983. Microgeographic edaphic differentiation in hordein polymorphisms of wild barley. *Theor. Appl. Genet.* 64:123–32

103. Nevo, E., Brown, A. H. D., Zohary, D., Storch, N., Beiles, A. 1981. Microgeographic edaphic differentiation in allozyme polymorphisms in wild barley (*Hordeum spontaneum*, Poaceae). *Plant Syst. Evol.* 138:287–92

104. O'Malley, D. M., Allendorf, F. W., Blake, G. M. 1979. Inheritance of isozyme variation and heterozygosity in *Pinus ponderosa*. *Biochem. Genet.* 17:233–50

105. Opler, P. A., Bawa, K. S. 1978. Sex ratios in tropical forest trees. *Evolution* 32:812–21

106. Pleasants, J. M., Zimmerman, M. 1979. Patchiness in the dispersion of nectar resources: Evidence for hot and cold spots. *Oecologia* 41:283–88

107. Price, M. V., Waser, N. M. 1982. Experimental studies of pollen carryover: Hummingbirds and *Ipomopsis aggregata*. *Oecologia* 54:353–58

108. Primack, R. B., Silander, J. A. 1975. Measuring the relative importance of different pollinators to plants. *Nature* 255:143–44

109. Pyke, G. H. 1978. Optimal foraging: Movement patterns of bumblebees between inflorescences. *Theor. Popul. Biol.* 13:72–98

110. Ritland, K. 1984. The effective proportion of self-fertilization with consanguineous matings in inbred populations. *Genetics* 106:139–52

111. Rohlf, F. J., Schnell, G. D. 1971. An investigation of the isolation by distance model. *Am. Nat.* 105:295–324

112. Schaal, B. A. 1975. Population structure and local differentiation in *Liatris cylindracea*. *Am. Nat.* 109:511–28

113. Schaal, B. A. 1980. Measurement of gene flow in *Lupinus texensis*. *Nature* 284:450–51

114. Schaal, B. A., Smith, W. G. 1980. The apportionment of genetic variation within and among populations of *Desmodium nudiflorum*. *Evolution* 34:214–21

115. Schmitt, J. 1980. Pollinator foraging behavior and gene dispersal in *Senecio* (Compositae). *Evolution* 34:934–43

116. Schmitt, J. 1983. Density-dependent pollinator foraging, flowering phenology, and temporal pollen dispersal patterns in *Linanthus bicolor*. *Evolution* 37:1247–57

117. Schmitt, J. 1983. Flowering plant density and pollinator visitation in *Senecio*. *Oecologia* 60:97–102

118. Schoen, D. J. 1982. Genetic variation and the breeding system of *Gilia achilleifolia*. *Evolution* 36:361–70

119. Selander, R. K. 1975. Stochastic factors in the genetic structure of populations. In *Proc. 8th Int. Conf. Numer. Taxon.*, ed. G. F. Estabrook, pp. 284–322. San Francisco: Freeman

120. Sheldon, J. C., Burrows, F. M. 1973. The dispersal effectiveness of the achene-pappus units of selected Compositae in steady winds with convection. *New Phytol.* 72:665–75

121. Silander, J. A. 1979. Microevolution and clone structure in *Spartina patens*. *Science* 203:658–60

122. Silen, R. R. 1962. Pollen dispersal consideration for Douglas-fir. *J. For.* 60:790–95

123. Sirkkomaa, S. 1983. Calculations on the decrease of genetic variation due to the founder effect. *Hereditas* 99:11–20

124. Slatkin, M. 1973. Gene flow and selection in a cline. *Genetics* 75:733–56

125. Slatkin, M. 1976. The rate of spread of an advantageous allele in a subdivided population. See Ref. 60, pp. 767–80

126. Slatkin, M. 1977. Gene flow and genetic drift in a species subject to frequent local extinction. *Theor. Popul. Biol.* 12:253–62

127. Slatkin, M. 1981. Fixation probabilities and fixation times in a subdivided population. *Evolution* 35:477–88

128. Slatkin, M., 1981. Estimating levels of gene flow in natural populations. *Genetics* 99:323–25

129. Slatkin, M., Maruyama, T. 1975. The influence of gene flow on genetic distance. *Am. Nat.* 109:597–601

130. Solbrig, O. T. 1980. Genetic structure of plant populations. See Ref. 86, pp. 49–65

131. Spieth, P. T. 1974. Gene flow and genetic differentiation. *Genetics* 78:961–65

132. Spieth, P. T. 1979. Environmental heterogeneity: A problem of contradictory selection pressures, gene flow, and local polymorphism. *Am. Nat.* 113:247–60

133. Stephenson, A. G. 1982. When does outcrossing occur in a mass-flowering plant? *Evolution* 36:762–67

134. Sved, J. A., Latter, B. D. H. 1977. Migration and mutation in stochastic models of gene frequency change. *J. Math Biol.* 5:61–73

135. Templeton, A. R., Levin, D. A. 1979. Evolutionary consequences of seed pools. *Am. Nat.* 114:232–49

136. Thomson, J. D., Plowright, R. C. 1980. Pollen carryover, nectar rewards, and pollinator behavior with special reference to *Diervilla lonicera*. *Oecologia* 46:68–74

137. Torres, A. M., Diedenhofen, U. 1979. Baker sunflower populations revisited. *J. Hered.* 70:275–76

138. Torres, A. M., Diedenhofen, U., Johnstone, I. M. 1977. The early allele of alcohol dehydrogenase in sunflower populations. *J. Hered.* 68:11–16

139. Turner, M. E., Stephens, J. C., Anderson, W. W. 1982. Homozygosity and patch structure in plant populations as a result of nearest-neighbor pollination. *Proc. Natl. Acad. Sci. USA* 79:203–7

140. Waddington, K. D., Heinrich, B. 1981. Patterns of movement and floral choice by foraging bees. In *Foraging Behavior: Ecological, Ethological, and Psychological Approaches,* ed. A. Kamil, T. Sargent, pp. 215–30. New York: Garland

141. Waddington, K. D., Holden, L. R. 1979. Optimal foraging: On flower selection by bees. *Am. Nat.* 114:179–96

142. Warwick, S. I., Marriage, P. B. 1982. Geographical variation in populations of *Chenopodium album* resistant and susceptible to atrazine. I. Between- and within-population variation in growth and response to atrazine. *Can. J. Bot.* 60:483–93

143. Waser, N. M., Price, M. V. 1982. A comparison of pollen and fluorescent dye carry-over by natural pollinators of *Ipomopsis aggregata* (Polemoniaceae). *Ecology* 63:1168–72

144. Waser, N. M., Price, M. V. 1983. Optimal and actual outcrossing in plants, and the nature of plant-pollinator interaction. See Ref. 17, pp. 341–59

145. Webb, C. J., Bawa, K. S. 1983. Pollen dispersal by hummingbirds and butterflies: A comparative study of two lowland tropical plants. *Evolution* 37:1258–70

146. Weber, J. C., Stettler, R. F. 1981. Isoenzyme variation among ten populations of *Populus trichocarpa* Torr. et Gray in the Pacific Northwest. *Silvae Genet.* 30:32–87

147. Wells, H., Wells, P. H. 1980. Are geographic populations equivalent to genetic populations in biennial species? A study using *Verbascum virgatum* (Scrophulariaceae). *Genet. Res. Cambridge* 36:17–28

148. Wheeler, N. C., Guries, R. P. 1982. Population structure, genic diversity, and morphological variation in *Pinus contorta* Dougl. *Can. J. For. Res.* 12(3):595–606

149. Wright, S. 1931. Evolution in Mendelian populations. *Genetics* 16:97–159

150. Wright, S. 1938. Size of population and breeding structure in relation to evolution. *Science* 87:430–31

151. Wright, S. 1940. Breeding structure of populations in relation to speciation. *Am. Nat.* 74:232–48

152. Wright, S. 1943. Isolation-by-distance. *Genetics* 28:114–38

153. Wright, S. 1946. Isolation by distance under diverse systems of mating. *Genetics* 31:39–59

154. Wright, S. 1951. The genetical structure of populations. *Ann. Eugen.* 15:323–54

155. Wu, L., Jain, S. K. 1980. Self-fertility and seed set in natural populations of *Anthoxanthum odoratum* L. *Bot. Gaz.* 141:300–4

156. Yeh, F. Ch.-H., O'Malley, D. 1980. Enzyme variations in natural populations of Douglas fir, *Pseudotsuga menziesii* (Mirb.) Franco, from British Columbia. 1. Genetic variation patterns in coastal populations. *Silvae Genet.* 29:77–164

157. Zimmerman, M. 1979. Optimal foraging: A case for random movement. *Oecologia* 43:261–67

158. Zimmerman, M. 1982. The effect of nectar production on neighborhood size. *Oecologia* 52:104–8

Ann. Rev. Ecol. Syst. 1984. 15:97–131

GENETIC REVOLUTIONS IN RELATION TO SPECIATION PHENOMENA: THE FOUNDING OF NEW POPULATIONS

Hampton L. Carson

Department of Genetics, University of Hawaii, Honolulu, Hawaii 96822

Alan R. Templeton

Department of Biology, Washington University, St. Louis, Missouri 63130

THE BIOGEOGRAPHICAL BACKGROUND

The Role of Population Genetics

As a dynamic process of genetic change, evolution manifests itself in two ways: It produces adaptive characters, and it produces species. To recognize that evolution has occurred by studying these products does not satisfy interested population geneticists. They see their science as providing an opportunity to contribute information that permits us to describe how the process of descent with genetic change occurs in populations of sexually reproducing, cross-fertilizing organisms.

The Origin of Populations and the Founder Effect

Both experimentally and theoretically, much has been learned about the origin and fate of genetic change within populations. Although the minute details may still elude us, the basic mode by which biological adaptations arise is clear enough. What concerns us in the present discussion is the mode of origin of a population that may be genetically competent to go on and give rise to a new species in immediately succeeding generations. Of the various proposed modes of speciation, we will specifically address the notion that a new species may be

97

0066-4162/84/1120-0097$02.00

formed allopatrically following a very simple type of population subdivision, namely, the establishment of a new (daughter) population from one or a few founder individuals (13, 18, 64).

No one denies that such population founding events actually occur. Indeed, they can be seen naturally and induced experimentally. What has been brought into question in two recent papers by Charlesworth & Smith (23) and Charlesworth et al (24) is the extent to which "genetic revolutions" (Mayr's term, 46) take place in the new population started by the founder. "Instant" or "punctuated" speciation based on the random drift of a few genes (33) has not been proposed by either of us. Nevertheless, some critics of our views (e.g. 23, 85) have mistakenly assumed this. Rather, we contend that, under some circumstances, the founder event may set the stage for speciation by altering genetic conditions in the gene pool. In the generations immediately following the founder event, genetic change that is largely recombinational may be profound enough to result in the emergence of a descendant population, recognizable as a new biological species. This idea bears no relationship to macroevolutionary speciation schemes proposed by certain paleontologists (33, 61). In spelling out the conditions in detail, we will invoke such genetic changes as gametic disequilibrium and recombinational disorganization of the mating system; these events appear to be forced on the system by the founder event. In contrast, Charlesworth & Smith conclude that "an important role of bottlenecks in population size as a cause of rapid speciation remains to be established" (24, p. 235).

Life History: Individual and Population

An individual organism's DNA sets the stage for its interaction with the environment during its lifetime. The DNA of a population, however, is not merely a collection of individual genotypes. Bonds of kinship and sexual reproduction result in the emergence of a supraindividual entity—the local gene pool or deme. This genetic system reacts with the environment in such a way that gradual genetic change may occur over many generations; thus, the deme has the capacity to change genetically over time.

Although individual and deme differ profoundly, it may nonetheless be useful to consider an analogy between them: Just as the individual has a life history, so does the population. Individuals have parents and so do populations. Every population, whether in the sense of a species as a whole or of some smaller populational subdivision, ultimately originates from some preexisting population. A population can sometimes be recognized as undergoing a period of youth, during which it may grow, expand geographically, and rapidly incorporate genetic change through mutation, recombination, and selection. Later, a period of maturity may follow, which is characterized by both biogeographic and genetic equilibria. In the face of an ever-changing environment, a mature population may become saturated and be unable to respond to new conditions. Thus, it may decline to a state of genetic rigidity and risk extinction. This stage is analogous to the death of the individual organism. Researchers have devoted little attention to the genetic basis of such a population senescence. The species may rely on the buildup of a genetic system based on ever more complex polygenic balances or polymorphisms to meet the environmental difficulties. Furthermore, there is the apparent necessity to genetically neutralize spreading transposable elements, viruses, and similar chromosomal contaminants. Such theories are frankly speculative but appear to be no more outrageous than those ascribing the decline of certain species in the fossil record to purely ecological factors, such as worldwide catastrophies (3).

The analogy with the individual life history is merely a guide to thinking; one cannot take it too seriously. Nevertheless, it is tempting to speculate that the possibility of reproduction, that is the production of daughter populations, may be a characteristic of youth or maturity rather than old age (18). Knowledge of the life history and reproduction of populations is rudimentary. Prior to developing genetic theories, we will attempt to review relevant facts about the natural history and ecology of populations, especially as they relate to speciation theories that invoke the founder principle. The present discussion focuses not only on the event of founding itself but also on population structures both before and after founding.

As outlined above, the concept of the life history of a population could be viewed on a macroevolutionary scale involving millions of generations and only capable of being scaled geologically. This need not be the case, however, and one of the fascinations of this area is that populations can be manipulated experimentally at the contemporary time level. Thus, populations of fast-breeding organisms can be founded in various ways and monitored in field plots or in population cages (e.g. 57).

Our discussions deal with those cases whose link with the past is extremely tenuous. One or a very small number of propagules somehow become removed from contact with the parent population and found a daughter population in a biogeographical situation that precludes genetic contact with the parent population. We limit our discussion to those cases where clear allopatric separation is assured. Nevertheless, theoretical model systems based on assortative mating that does not involve biogeographical separation of the incipient population segregants can also be considered under the framework of the founder effect, although certain other assumptions are necessary (7).

Is it realistic to think that an entire population may stem, via sexual reproduction, from a single, fertilized female individual? The answer is affirmative, as attested by voluminous literature on isofemale lines in, for example, *Peromyscus, Drosophila,* and certain plant species. The necessity of sib matings in the progeny of the founder individual poses no impenetrable genetic or other biological barrier to survival and reproduction. The extreme, naturally occurring case, wherein a population is founded by a single such propagule, is not only biologically feasible but also provides a degree of operational simplicity for theoretical purposes.

CONTINENTAL BIOGEOGRAPHY

Colonizing Species

Certain species, usually exotic to the basic biota of an area, become active colonizers on the continents and indeed also on oceanic islands. Such species may arise and evolve on one continent and spread to another, often aided by the activities of man. At first glance, such species appear to be good material for the study of the founder effect. Yet such plant or animal "weeds" frequently have genotypes that seem resistant to changes by the founder effect. Baker (6) has suggested that such species may have a "general purpose genotype" that is rather resistant to the genetic repatterning characteristic of the speciation process. This suggestion has been documented only through incomplete biogeographic and genetic studies; unfortunately, comparative study of the genetic system has not proceeded far enough to identify crucial features of weedy colonizing species. Until more data are available, we must seek clarification by recourse to species that do not have these invasive or colonizing properties.

Peripheral Populations of Endemic Species

In seeking out endemic biotas wherein continental population founding may be studied, researchers have examined populations that are geographically peripheral to the main body of the species (10). Peripheral areas of a species' distribution, especially if ecologically marginal, might be where founder effects could be observed. A peripheral isolate is likely to undergo a more or less continuous ebb and flow of population pressure. It is therefore especially vulnerable to periodic recolonization and genetic swamping from the main body of the species. This ebb and flow, brilliantly proposed and developed by Timofeef-Ressovsky (80), may produce conditions that mimic the salient features of the founder effect. In most cases, the minor differentiations observed are unlikely to have permanent biological effects sufficient to break with the norm of the genetic system of the species and thus set speciation in motion.

Continental Subspecies and Islands

Mayr (46) mounts a strong argument that certain subspecies, e.g. of birds on small islands peripheral to New Guinea, are good candidates for incipient species. He argues that the gene pools of such populations in a sense are unaffected by the ebb and flow phenomenon and show effective long-term isolation from the main bulk of the species. Thus, such populations may be prone to undergo "genetic revolutions" that may ultimately lead to speciation; the nature of the "revolutions" is not spelled out in detail. The concept of genetic revolution will be considered later in this paper, where various detailed genetic models will be presented. Mayr's general concept is a valuable one, given three major constraints: (a) the isolation should be very great, (b) the colonizing population should have arisen from a very small number of sexually reproducing propagules (two or even one fertilized or gravid individual) and (c) the genetic system of the donor population must be able to undergo the rather drastic change required if it is to serve as an incipient species.

We feel that the data on continental and peripheral subspecies do not provide many compelling examples of incipient species, contrary to a view expressed earlier by one of us (10). This is principally because points (a) and (b) do not permit sufficient genetic reorganization (18). For much the same reasons, steps in clines or interfaces between subspecies also do not usually appear to serve as substantial enough breaks with the older populations.

Continental Mountaintops

In certain places where the tectonic plates are more or less stationary and volcanic mountains or lava flows occur, conditions exist for evolution mediated by the founder effect. The high East African mountains, reaching 3500 to 6000 meters altitude, have various ecological belts, culminating in an Afroalpine belt. The various peaks are occupied by an island-like biota as the mountains protrude as isolated temperate "islands" above the surrounding warm plains (35). These peaks provide new areas for colonization, and propagules arrive from long distances, rather than from adjacent, ecologically similar areas.

The high African peaks represent a biogeographical situation quite different from the high Andes of South America or the volcanic areas of western North America. There are two important differences: newness and isolation. These factors must operate together to provide geographical opportunities for the founder effect. Thus, certain continental mountain peaks (nunataks, see 26) that have their biological peculiarities because they escaped glaciation represent a very different situation. Rather than supporting a waif biota based on new colonization, they are areas that have retained a relict community formed by partial extinction of an older biota that was formerly common throughout the region.

Biogeographers often find themselves divided into "dispersalists" and "vicariance biogeographers" (53). The biogeography of the continents appears to provide examples of both tendencies. For the evolutionist concerned with genetic changes in populations via the founder effect, it seems prudent to turn to isolated archipelagos of oceanic islands for examples of the virtually unequivocal dispersal by founder individuals. The "virtually" is inserted above, since critics of this view exist (54).

OCEANIC ISLAND BIOGEOGRAPHY

General Properties

Since Darwin's preoccupation with the Galapagos, the terrestrial life of oceanic islands has played a key role in evolutionists' thinking. The emergence of the concept of continental drift and seafloor spreading has had a secondary effect; it provides an explanation of the origin of oceanic islands and archipelagos without recourse to vicariance biogeography. Oceanic islands such as Hawaii, Tristan da Cunha or the Galapagos serve as examples of new lands pushed above the ocean by volcanic activity remote from any continental mass. Some of this activity has been geologically recent. Propagules reach such islands by long-distance dispersal and with an extreme degree of irregularity and chance.

We do not intend to review here the large and complex subject of the biotas of oceanic islands. Their peculiarities are well documented and reviewed (8). What is important to the present discussion is that chance founder individuals have played a large role in the initial and continuing populating of such islands.

Examples from Hawaii

A few examples will be drawn from Hawaii's extensive biota. Endemism is extraordinarily high in many of the taxa, often running over 90% (87, 51). Some endemics, such as the honeycreepers, certain land snails, and the drosophilids, form phylogenetically close but adaptively diverse assemblages of species. Scrutiny of the whole biota indicates that these groups are by no means exceptional; the seeming paradox—extreme adaptive diversity superimposed on fundamental phyletic similarity—is the rule rather than the exception.

Although some Hawaiian groups are luxuriantly speciose, e.g. Drosophilidae (22, 34) and Microlepidoptera (87), there are certain indigenous or endemic organisms that, although recognized at the specific level, have not proliferated many descendant species. Some of these are clearly recent arrivals to the islands, but others appear to be more ancient. Another category of Hawaiian species includes widespread or cosmopolitan species that have been introduced in historic times, presumably by man. These, of course, form a very diverse assemblage of plant weeds and ornamentals, mammals, birds, and many insects. Some of the latter have been introduced as a means of biological control of the more noxious, introduced plants and insects.

With the exception of the biological control organisms, the number and genetic source of the introduced flora and fauna of Hawaii are unknown, except for an occasional historic reference that usually does not give specific numbers or sources. With a very few exceptions—e.g. *Passer domesticus* (39), *Drosophila immigrans* (55)—little attention has been given to the extent to which neoevolution has occurred in introduced organisms. Changes in *Passer domesticus* are at most minor matters of plumage color. *D. immigrans* displays the same three widespread chromosome inversions found in most of the populations from the rest of the world; the species is morphologically unchanged. Thus, neither of these species has shown signs of genetic shift since introduction. Although anecdotal accounts exist (88), the evidence for significant genetic change, especially in the direction of speciation, is virtually nonexistent.

There are a number of reasons why the newly introduced species in Hawaii do not manifest a notable founder effect leading toward speciation, even though enough generations have apparently elapsed. First is the generalist nature of most of the organisms concerned. They appear to fall into the same category as the colonizing species on continents, which were discussed above. In fact, many of them are the same species that have demonstrated their ability to spread throughout continental ecosystems. There appear to be considerably fewer barriers to their spread in the insular ecosystems. Many of these species have complex, balanced genetic systems of heterosis that seem to underlie their colonizing ability. It is suggestive that, even if forced through a single founder, the entire system may be carried through colonization more or less intact, with little loss of the "general purpose genotype" (6).

Speciating and Nonspeciating Lineages

In the more detailed genetic considerations that follow, we present the genetic systems that can probably undergo alterations following the founder effect and suggest a theoretical background. Suffice it to say here that not all genetic conditions appear to have equal potential for speciation mediated through a founder bottleneck.

Many strand organisms are adapted to long-distance dispersal through devices such as flotation and longevity of seeds. Many such organisms employ either vegetative reproduction or obligatory self-fertilization systems that transport a more or less preadapted and fixed genotype intact to a new location. Such gross alterations of the sexually reproducing, cross-fertilizing system are not the subject of this essay, which concentrates on diploid, cross-fertilizing forms. Only in very unusual circumstances could such an apomictic gene complex undergo major genetic shift following a founder event (63). We conclude that those genetic systems that permit exuberant speciation episodes on oceanic islands have certain properties that distinguish them from colonizing and cosmopolitan species.

The extent of the difference between speciating and nonspeciating lineages will not be appreciated until a true science of comparative genetic systems has emerged. As a detailed consideration of the various founder models is developed below, several possibilities should be kept in mind. A population that is old, i.e. that has maintained a large outbred population in a relatively stable environment for a large number of generations (thousands or hundreds of thousands), may become more and more closed to progressive change. This development may be due to the gradual buildup of coadapted gene complexes into balanced polymorphisms in such a way that freely recombining genetic variability is limited. A young species, in contrast, should have balanced polymorphism that is less complex; more of the genetic variability would be segregating in the open genetic system (14), making shifts in balance easier to achieve if a new population were formed via the founder effect.

Evidence for the Participation of Founders in Speciation on Oceanic Islands

GEOGRAPHICAL AND SYSTEMATIC AFFINITIES WITH SOURCE AREAS
Hawaii is an isolated archipelago consisting principally of eight high islands in the central Pacific Ocean (Figure 1). A series of lower islets and seamounts extends from the high islands to the northwest. Basically volcanic in origin, the islands have apparently been formed through successive eruptive episodes, resulting in a series of perforations in the Pacific tectonic plate. Extremely hot lavas emanating from a geographically fixed plume deep in the earth's mantle have periodically perforated the plate, which is moving northwestward at the rate of about 9 cm a year, carrying the islands, as they are formed, away from the "hot spot."

Although several other Pacific archipelagos have apparently been formed in the same manner, the 8 southeastern Hawaiian islands are the best exemplars, dating from the 2 presently active volcanoes at the southeastern end back 5.6 million years to the extinct volcanoes of Kauai, 600 km to the northwest (45). There are 12 major volcanoes in this series, each successively younger as one moves from northwest to southeast.

The implications of this pattern of island development for the upland fauna and flora of the archipelago are numerous. In the first place, since its inception the island chain has been one of the most isolated on the globe. It is currently approximately 3200 km from the nearest island or continent. Furthermore, from the geological history of Hawaii it seems clear that this situation has persisted at least since Miocene times and probably much longer.

Compared with the high islands, the terrestrial life on the low islands northwest of Kauai is sparse, due to erosion and submersion of what apparently were once high islands. Accordingly, biogeographical interpretation of the terrestrial biotas essentially begins with Kauai, although this does not mean that the biota of this island was derived directly from the continents.

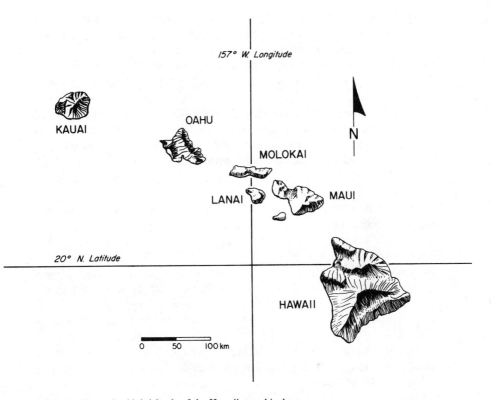

Figure 1   The major high islands of the Hawaiian archipelago.

Whether the biota of modern Hawaii stems largely from former high islands to the northwest or was derived directly from the continents via long-distance dispersal, the latter must be responsible in large measure for the establishment of this insular biota. If a former high island or islands served as the original receptor of colonizers, this merely pushes the original founding events backwards in time but does not change their essential nature. In short, because of the vast ocean distances between the islands and the continents, occupation of the islands clearly must have had a large stochastic element and involved a small number of founding propagules.

At the level of the archipelago, the degree of isolation from continents or other biologic sources is extraordinary. The same phenomenon—implying a small number of founders—has been invoked for other oceanic islands, but their situations are generally less striking than Hawaii's. Based on geographic and systematic considerations, biogeographers have made minimal estimates of the number of potential ancestral introductions into the archipelago (e.g. 27, 87). In addition, they usually advance some hypothesis about the geographic sources of the hypothetical propagules. The Drosophilidae of Hawaii serve as a good example. The endemic Hawaiian species are easily distinguished morphologically from the recently introduced, cosmopolitan species (79) and show extraordinary diversity of size and adaptive characters. Their morphology is diverse, and species diagnosis does not depend on cryptic characters like those sometimes used to distinguish continental sibling species. A study of the conservative characters of the internal anatomy, however, reveals that this enormous assemblage of endemic species may be reduced to at most two lineages represented by the genera *Drosophila* (359 species with approximately 125 additional undescribed forms) and *Scaptomyza* (132 species with approximately 100 additional ones that are undescribed). Certain species of the Hawaiian fauna seem to bridge the gap between the two genera, and it is not unreasonable to suggest that all of them may have descended from a single chance introduction (79).

This situation is repeated in many other Hawaiian endemics (87). Only a few examples will be mentioned here. Compositae of the silversword alliance (9) include the giant herbaceous silverswords *(Argyroxiphium)* and the many ecologically diverse woody shrubs, small trees, and lianas of the genera *Dubautia* and *Wilkesia*. Like *Drosophila,* these forms appear to represent a lineage stemming from a single ancestral stock. Examples of other groups derived from single ancestors include the Drepanididae, consisting of 42 species of honeycreepers; *Pelea* (Rutaceae), 94 species; and *Cyanea* (Campanulaceae), 100 species. Nearly as large as *Drosophila* is the microlepidopteran genus *Hyposmocoma*, a single lineage that has about 350 species (89). As far as we know, polyploidy is rare or absent in these speciose groups; thus the proliferation seems to have occurred at the diploid level.

The numbers of species cited in these accounts may include some local variant populations that might not deserve the status of full species rank, but the numbers exemplify the diversities in a general way.

None of the above can be used to specify in any exact manner whether the ancestral population of one of these lineages stemmed from a single or many individuals. Yet, the inference of single founders is strong, given the limited number of lineages and the systematic disparateness and coherence of each. It has repeatedly been argued (see 8) that the biota of oceanic islands is in large measure descended from waifs or strays that reach the islands by long-distance dispersal. For conspicuous animals like birds, one can roughly estimate the frequency with which such strays reach a remote island archipelago. Not only are these strays highly infrequent and irregular, but there is a second consideration: What proportion of such strays actually become successful colonists in the sense of starting a daughter population? Indeed, there is strong evidence that both events—especially the latter—are rare, rendering colonization by the founder effect (as judged in this manner) a very rare event indeed. This argument also carries the corollary that many lineages must have been established by a single, fortuitous, gravid bisexually produced propagule.

RELATIONSHIPS WITHIN THE ARCHIPELAGO   Evidence that founder individuals are involved in speciation may be gathered on a smaller geographic and systematic scale if an archipelago consists of a series of successively younger islands, as does Hawaii. A striking feature of the drosophilids and *Cyanea*, among others, is the single-island endemism of most species. For example, among the Hawaiian *Drosophila*, the newest island, Hawaii, is separated from the others by an abysmally deep channel 46 km wide. Its 26 species of large "picture-winged" *Drosophila* can be related by chromosomal sequences and morphological attributes to particular existing populations in the fauna of the older islands to the northwest (19).

With one possible exception, the species of the island of Hawaii are unique to the volcanoes of this newest island. These judgments are based on morphological, behavioral, cytological, molecular, and hybridization studies (20). The data confirm that in almost every case each species represents a strongly differentiated gene pool and must be assigned specific rank. The point here is that of the 26 species on the Big Island (i.e. Hawaii), there are 19 lineages with uniquely different inversion-sequence formulae. Thus, each appears to be related to a separate ancestral population on an older island. The major picture of speciation in the archipelago as a whole (i.e. disparate lineages with systematic coherence within each lineage) tends to be recapitulated on the smaller interisland scale.

The situation described above suggests that each new population went through a founder bottleneck at the time it was established. If each had been established by repeated colonizations of individuals across the channel, the likely result would have been the simple spread of the original, undifferentiated species. Systematic interisland distinctions would have been at most at the level of subspecies rather than at the species level. In addition, since much of Hawaii island is less than 400,000 years old (50), these events occurred very recently in geological time.

The basic ecological adaptations of these drosophilids—their breeding substrates, their altitudinal tolerances, their positions in the ecosystem with respect to rainfall, humidity, and temperature—are remarkably similar between the Hawaii forms and the Maui forms from which most were so clearly derived. What has changed most as these insects have moved down the archipelago is the mating system, represented by morphological and behavioral factors pertaining to sexual selection (16). The process brings the sexes together in specific pairs of organisms that produce progeny of high fitness. As especially speciose groups in the islands, *Drosophila* and *Hyposmocoma* (90) are sensitive to such changes.

Carson (16–18) has suggested in several recent papers that it is the genetic basis of this system that is most easily destabilized at the founder event and then restabilized around a new norm as sexual selection is reimposed. Of 103 picture-winged *Drosophila* species, 97% are single-island endemics, suggesting that the act of interisland colonization is overwhelmingly conducive to speciation, even when the environments are similar. Some host shifts, however, do indeed occur as a new island is occupied by migrants from an old. This phenomenon has been observed in both *Drosophila* and a number of other insects. In contrast, the preponderance of cases that do not involve host shifts suggest that the shift is a secondary rather than the primary pivotal change following the founder effect.

Those picture-winged *Drosophila* species found on more than one island are exceptional and thus are worthy of attention. *D. crucigera* (21) is found on both Kauai and Oahu. Chromosomal and behavioral evidence appears to require two exchanges between the islands and a third one between the two major volcanoes on Oahu (29). These 3 exchanges seem to require an increased vagility for this species compared to that of many other picture-winged species. Such a property might lead to the presence of more than one successful interisland colonist and thus colonization without a speciation event, essentially transplanting the major elements of the gene pool of the species from one island to another. The fact that *crucigera* is an ovipositional generalist, breeding on a variety of substrates, fits in with the above, since such generalism should confer a capacity for successful colonization above that of specialists. This generalist tendency recalls the propensities of the recently introduced, nonspeciating forms discussed earlier.

RELATIONSHIPS WITHIN A SINGLE ISLAND   A number of oceanic island groups, like Hawaii, have single islands composed of more than one volcano. When allochronically developed with rather arid lowland saddles between the mountains, there is further opportunity for colonization from an older area to a newer one. For example, the island of Maui is made up of two volcanoes. West Maui is older (1.3 million years) than East Maui (0.8 million years), so the colonization of West Maui presumably occurred first. Certain *Drosophila* species—e.g. *neopicta, planitibia, adiastola, limitata,* and *grimshawi*—have apparently spread from the older to the newer mountain with only minor changes. In other cases, East Maui is populated by a form that is similar to but specifically distinct from that on West Maui; examples include *ingens/melanocephala, hanaulae/cyrtoloma, affinidisjuncta/disjuncta,* and *paenehamifera/hamifera*. The habitats on both volcanoes occupied by these species are similar. Information is sparse on the breeding sites, but the latter two pairs apparently utilize decaying stems of *Freycinetia* and *Cyanea* respectively.

The rather sharp interisland and intervolcano species differences in the drosophilas are not usually paralleled by comparable differences in their host plants; the latter change considerably less as they colonize. Thus, there does not seem to be a direct correlation between the speciation of *Drosophila* and its hosts. This is not to say, however, that certain plant groups do not show exuberant speciation in the islands. Prime examples are *Cyrtandra* and *Bidens;* neither of these is a host for *Drosophila*. The "Drosophila plants" most prone to speciation in the islands are the lobeliads *(Clermontia, Cyanea, Delissea, Lobelia),* but even these do not keep pace with their drosophilid associates. To conclude this section, there is evidence that the founder effect can play a role in speciation even within an island.

LAVA FLOWS ON THE SLOPES OF A SINGLE VOLCANO   Shield volcanoes like those of Hawaii grow only partly from summit flows; a large amount of lava emanates from rifts on the flanks of the shield, resulting in extensive gravity flows down the slopes. These flows frequently cut wide swaths through forest and render large areas, as much as a kilometer or more in width, totally barren of vegetation. Colonization of these areas proceeds from adjacent areas and in time the former forest is replaced with a new one, often having a somewhat different composition.

Although many discussions have centered around the islands of vegetation *(kipukas)* formed by the bifurcation of lava flows, the recolonization effect is more important for the present discussion of the founder effect. Not only are the populations divided by the flows, but many saprophagous insects like *Drosophila* require prior colonization by the host plant before they can become established. If the host species is rare, local, and disseminated by birds, it may take an extended period before recolonization is possible. By that time, lava flows may have overrun part or all of the nearest source area, requiring that recolonization be accomplished by a few founders from some distance.

Figure 2 gives some perspective on the time scales that are involved. In the last 100 years, lava from Mauna Loa has destroyed large areas of forest on its western and southern flanks. The precise age of this forest is not known, but it is probably thousands of years old. If the mountain is 500,000 years old and forest development from bare lava takes about 5,000 years, then such a course of events as seen on southern and western Mauna Loa in the last century must have occurred in the same exposure on the same volcano about 100 times. Even if this estimate is too low or too high, the fact remains that no concept of climax vegetation can be invoked. Rather, the forest and its inhabitants on such very new oceanic islands are undergoing continual replacement, until the eruptive activity of that particular volcano ceases.

The relevance of these patterns for the founder principle is obvious. Figure 2 shows five localities where *Drosophila silvestris* has been studied in recent years (17). The altitudinal range of the species and its principal host plants,

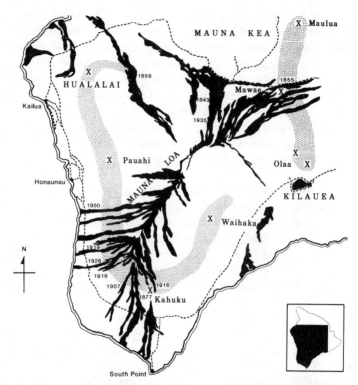

Figure 2 The southern two thirds of the island of Hawaii showing division of *Drosophila silvestris* habitat (stippled rain forest between 1000 and 1700 m altitude) by 100 years of historic lava flows. Sites of collected samples are given by X.

Clermontia coerulea and *hawaiiensis,* are roughly indicated by the stippled area. Although some populations exist far from current volcanic activity, others have been strongly affected (as at Mawae on the map in Figure 2). The inference that those populations unaffected in the last 100 years have been affected in the past is inescapable.

Summary: The Natural History of Populations Likely to Proliferate by Founders

In the preceding sections, we have shown that various ecological and geographical conditions support the notion that a founder effect can be a precondition for genetic changes leading to species formation. The genetic system most relevant to this kind of event is a cross-fertilizing diploid with a high recombination index. The latter point, which has been stressed elsewhere (14), seems necessary since some diploid, cross-fertilizing systems may be rendered inflexible genetically by restraints on segregation and/or crossing-over.

These points lead directly to consideration of the precise genetic nature of the breeding systems in populations from which the founders originate. Accordingly, we will now explore the extent to which certain exuberantly speciating groups, such as the Hawaiian *Drosophila*, provide a clue to the recognition of such a system.

COADAPTED GENE COMPLEXES

Definition and Examples

There are three principal models for speciation induced by founder events. All of these models invoke at one stage or another the idea of coadapted gene complexes, resulting in multiple adaptive peaks. We are using Wallace's definition of coadaptation: "Genes are said to be coadapted if high fitness depends upon specific interactions [i.e. *fitness epistasis*] between them" (83, p. 305). This definition does not necessarily require the existence of linkage disequilibrium, contrary to the assertions by Barton & Charlesworth in the accompanying paper. Charlesworth et al state: "There is no question that epistasis in fitness, which is an essential condition for the existence of multiple adaptive peaks, has been observed in many instances" (23, p. 482). There is no disagreement on this point. Some details are needed, however, and accordingly, we give two examples specifically related to founder-induced speciation.

Our first example is the genetic basis of head shape differences between two Hawaiian picture-winged *Drosophila*, *D. silvestris* and *D. heteroneura*. Templeton (62) and Val (81) have shown that the differences are primarily determined by a major X-linked segregating unit (a locus or linked cluster of loci) interacting strongly with several minor, autosomal genes. Interestingly, hybrids are occasionally found in nature, and recent evidence with mitochondrial DNA (DeSalle, unpublished data) indicates extensive introgression in certain areas. Nevertheless, the head shape phenotypes display remarkably little phenotypic variation (62). Further inter- and intraspecific crosses with these species (2) have revealed extensive intraspecific genetic variation for traits involved with mating success, despite apparent phenotypic uniformity. Thus, the lack of phenotypic variation is *not* based upon a lack of genotypic variation. In light of the introgression and intraspecific genetic variation, the maintenance of this striking contrast between *heteroneura* and *silvestris* is most likely due to strong sexual selection upon the highly epistatic gene complex controlling the head shape.

The second example of a coadapted gene complex comes from an experimental investigation of the founder effect using *Drosophila mercatorum* (63). Parthenogenesis was used to generate the most extreme founder event possible from a natural, outcrossing sexual population: diploidization based on a single haploid genome. By pushing the founder event to its absolute limit, evolutionary responses should be accentuated and hence are amenable to empirical observation. The results were quite extreme, with strong pre- and postmating barriers arising. The genetic basis of one of the postmating barriers, which gave the classical F_2 breakdown pattern associated with coadapted gene complexes, has been worked out in considerable detail (63, 68, 71, 74, 76). This coadapted trait, known as abnormal abdomen *(aa)* because of one of its pleiotropic effects, is determined by a major X-linked segregating unit that engages in very strong epistatic interactions with X-linked, Y-linked, and autosomal genes. Moreover, there are strong epistatic interactions among the autosomal modifiers.

Additional studies (25; unpublished data) have revealed that *aa* is associated with the preferential amplification in polytene tissues of *28S* ribosomal genes bearing large insertions that disrupt normal transcription. The *aa* syndrome displays a wide variety of pleiotropic effects, many of which are consistent with the phenotypes normally associated with ribosomal deficiencies in *Drosophila,* and some of which are associated with the effects *aa* has on juvenile hormone metabolism (76). Among the pleiotropic effects associated with this syndrome are decreased adult longevity and early sexual maturation (68, 71). These pleiotropic effects are associated with extremely strong fitness effects in natural populations (72, 75).

Although *aa* was first discovered in these experiments on the founder effect (63), it was shortly thereafter found in some of the original parthenogenetic strains of *D. mercatorum* that Carson (11) isolated, as well as being polymorphic both for the major locus and the modifiers in natural populations (76). The failure to discover this syndrome earlier was due to the fact that the modifier system is very effective in creating apparent phenotypic uniformity despite underlying genetic heterogeneity. These later discoveries clearly show that the genetic revolution created in the laboratory was based upon the disruption of a coadapted genetic system that was polymorphic in the ancestral population.

Although *aa* and head shape are totally different phenotypes, there are many resemblances between them. First, both have the same basic genetic architecture: a major segregating unit coupled with many epistatic modifiers. In Templeton's (67, 69) terminology, this is a type II architecture. Barton and Charlesworth claim in the accompanying article that there is ambiguity about the genetic architecture for head shape, citing Charlesworth et al (23) as their source. This paper is not a primary source and cites Templeton (62) as its source, but Templeton's paper does not support the claims made in the citation. Hence, with regard to primary sources, there is no ambiguity. It is also interesting that in both examples the major segregating unit was on the X-chromosome. Because of male hemizygosity, the founder effect is always more intense for an X-linked locus than for an autosomal one, particularly when the founding population consists of a single, multiple-inseminated female, the most likely situation for the Hawaiian *Drosophila.*

Lande (42) argues that this type of genetic architecture is not very important in natural populations because major genes have many deleterious pleiotropic effects. Hence, very strong selection on the advantageous pleiotropic effects is needed, and Lande regards such strong selective forces as rare under natural conditions. Lande views the many natural examples of this architecture—e.g. sickle-cell anemia in humans, industrial melanism, and insecticide resistance—as artificial situations induced by human activities. It is difficult to see, however, how such an argument could be used on *aa,* since the stressful environment in that case is weather fluctuations (75). Moreover, this type of genetic architecture is very common, as can be seen in the references given by Templeton (67) and in recent reviews in the plant literature (5, 36). Obviously, Lande's predictions are frequently violated.

The difficulty with Lande's arguments is his assumption that natural selective forces are weak. This assumption is widespread in evolutionary theory, but the primary impetus for it is mathematical convenience, not biological reality. For example, one of us (ART) has written many theoretical papers that assume weak selection. After the field studies with *aa* revealed that selective forces in natural populations are extremely intense, however, Templeton (72) was forced to admit that assumption of weak selection in his previous models was quite inappropriate and biologically unrealistic. Thus, natural selection can be intense, as has been documented with *aa,* and Lande's (42) model therefore clearly shows that type II genetic architectures are of major importance.

Type II architectures can also assume major importance in speciation even when intense selective forces are relatively rare. One feature of such architectures is the extensive network of pleiotropic effects associated with them. Hence, as Lande argues, a type II architecture is more likely to produce unforeseen evolutionary consequences than the usual polygenic architecture of many minor genes [a type I architecture (61)]. These unforeseen effects can then induce even more evolutionary changes in the modifier systems (64). Because it induces more extensive genetic changes, a type II genetic response under intense selection is more likely to result in speciation than a type I response under mild selection. Suppose, for example, that only 1% of all evolutionary responses to natural selection involve a type II architecture, but a type II response is 99 times more likely to result in speciation than a type I response. Then the frequency of type II architectures in speciation events will be 50%. Therefore, what appears unimportant at the intrapopulation level can be important at the speciation level.

Lande (42) has advanced one more argument against type II architectures: They are difficult to maintain as polymorphic systems. This argument leads to the second feature shared by *aa* and the sexually selected coadapted complexes: Despite evidence for strong selective forces and phenotypic uniformity, both systems are polymorphic in natural populations. Obviously, Lande's predictions do not hold up against these examples (many other examples exist as well, e.g. 56). The reason for this failure is related to the fundamental nature of the quantitative genetic models being used. Under the classical models of quantitative genetics, a trait under selection exhausts its additive genetic variance, and subsequent genetic variance is maintained by a balance between selection and mutation. Pleiotropy is either ignored entirely, assumed to operate only on traits of no selective value (i.e. indirect selection), or held to be a constant, nonevolving component of the system (which Lande assumes in 42). This last assumption is particularly unrealistic since the *major* role of the modifier system in traits like *aa* is to adjust the pleiotropic effects (71). Recently, Rose (59, 60) has been developing genetic models that actively incorporate pleiotropy into the fundamental theoretical framework. The implications of the resulting genetic models are quite different from those of the classic mutation-selection balance models favored by Lande. Rose's conclusion that pleiotropy favors the maintenance of polymorphism—which is in direct opposition to Lande (42)—is especially important. In light of the examples of polymorphic type II systems mentioned above, the Rose formulation is far more biologically realistic than the mutation-selection balance models.

Another important feature of the Rose model is that high values of additive genetic variance in fitness components can be maintained in the population at equilibrium, even though there is little or no additive variance for total fitness. The fitness components are held in equilibrium by the balance between antagonistic pleiotropic effects. If any event occurs to perturb that balance, however, a large additive variance in total fitness will be immediately realized. (*Note:* We are using *additive* as a quantitative genetic measure applicable to a reproducing population. This definition does not imply that gene action at the individual level is additive. As noted in 70, epistatic genes at the individual level can result in additive effects at the population level.) One such perturbing factor is a founder event. Thus, a founder event not only perturbs coadapted gene complexes, but it can release much additive variation, so that a strong selective response is possible. The prediction of a release of additive variation after a founder event has recently been demonstrated empirically (44).

The Origins of Coadapted Gene Complexes

In the previous section, we discussed some of the details of type II coadapted gene complexes, but we did not address the problem of their evolutionary origins. To oversimplify somewhat, there are two extreme views on the origin of such complexes: (*a*) that they are the product of a gradual (in a population-genetic sense) and sequential buildup and (*b*) that their origin is transilient. According to the sequential buildup hypothesis, a major gene is selected for strongly enough to increase in frequency—either to fixation or to a quasi equilibrium determined by the antagonistic balance of its beneficial vs deleterious pleiotropic effects (70). Since the evolutionary response at this stage occurs primarily at a single locus, the measured fitness response at the popula-tion level is necessarily an additive one. Once this gene has increased in frequency and reached a quasi equilibrium, the fitness effects of its epistatic modifiers become additive and can respond to selection. As a result of modifier evolution, the pleiotropic balances are altered and induce further evolution at the major loci. This secondary evolution at the major loci is limited by modifier evolution, however. As a result, although the system is characterized by intense fitness epistasis at the individual level, it appears to be an additive system at the population level (70). By going through one or more of these sequential buildups of modifier systems, a highly epistatic gene complex can evolve principally through additive genetic effects at the population level. Such additive evolution can be seen in classic examples of natural selection like industrial melanism and insecticide resistance (69, 70).

The second type of origin is the transilient mode, in which the major gene frequencies are altered so rapidly that the quasi equilibrium phase is never established. Many evolutionary forces could change the frequencies of the alleles at the major loci, but we will focus only upon one—founder events. We now turn our attention to the three major models of transilient origins of coadapted gene complexes induced by founder events.

THREE MODELS FOR TRANSILIENT ORIGINS VIA FOUNDING EVENTS

Genetic Revolution

Mayr (46) has proposed a theory of drastic speciation in peripheral founder populations, called genetic revolution. He has recently renamed this speciation mechanism peripatric speciation (48) to emphasize that the founders are de-rived from peripheral populations of the ancestral species. This proposal is primarily motivated by biogeographical evidence (48), but in the earlier paper (46) he borrowed heavily from the writings of Sewall Wright to outline a potential mechanism. First, Mayr takes Wright's view of epistatic gene com-plexes with their multiple adaptive peaks. The existence of multiple adaptive peaks for the same external environment implies that the genetic environment defined by the epistatic interactions between genes is of great evolutionary importance. Second, Mayr utilizes Wright's idea that genetic drift can cause a reduction in levels of genetic variability and lead to changes in a coadapted gene complex impossible in a large population. Mayr then combines the two elements of genetic environment and drift to produce a mechanism for what he terms genetic revolution. The basic idea is that the genetic drift accompanying the founding event will result in a large drop in the amount of genetic variation (Figure 3). More genetic variation will be lost following the founder event owing to continued small population size.

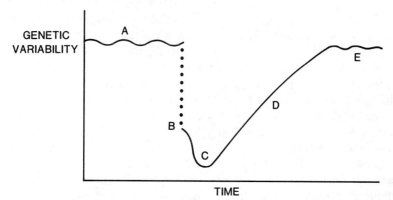

Figure 3   Loss and gradual recovery of genetic variation in a founder population according to Mayr (46). The founders (*B*) have only a fraction of the genetic variation of the parental population (*A*), and further genes are lost during the ensuing genetic revolution (*B* to *C*). Variation is gradually recovered (*D*) if the population can find a suitable niche, until a new level (*E*) is reached. (From 46.)

Mayr (46, 47) considers reduction in the level of genetic variation the most important effect of the founder event. He argues that the selective forces altered by increased homozygosity "may affect all loci at once," thereby triggering a "genetic revolution" (46, p. 170; 47, p. 533) that breaks up the old coadapted gene complexes. Mayr envisions that few populations will survive such a revolution. The ones that do, however, will be free of their previous epistatic constraints and can therefore move to a different adaptive peak as new genetic variation is accumulated and a new, coadapted gene complex is acquired (presumably through the sequential, "additive" mode discussed earlier). Speciation arises as a consequence of the acquisition of the new coadapted gene complex.

Founder–Flush

Carson (12, 14, 18) has proposed a second mechanism by which founder events can induce speciation. In this model, a founder population is established from an outcrossing, polymorphic, but coadapted, ancestral population. The genetic drift associated with the founder event begins to disrupt the ancestral coadapted gene complex. This disorganization phase continues as the founder population establishes itself in its new environment under relaxed ecological and selective conditions. Hence, the founding event is followed by a period of rapid population growth—the "flush" phase. Because of the rapid population growth immediately following the founding event, little of the ancestral genetic variation is lost (52). Moreover, the relaxed selective conditions allow recombination to occur that produces variation normally selected against. Carson argues that populations with attributes (e.g. in their genomic structure or system of mating) that promote recombination would be most sensitive to this continued disorganization during the flush phase.

As the population leaves the flush phase, it has high levels of additive genetic variation due to recombination and altered pleiotropic balances. As it saturates the environment, the selective forces reappear, perhaps even causing a population crash. The population can respond efficiently and rapidly to this selection because of the high levels of additive variation. Moreover, because of the disruption of the previous coadapted complex, the population is free to respond to this selection by moving to an alternative adaptive peak. As with the genetic revolution model, the movement to this alternative peak leads to speciation.

Genetic Transilience

The third mechanism for founder-induced speciation is genetic transilience (64). This theory requires that the ancestral population be relatively outcrossed and polymorphic for coadapted gene complexes centered around major loci. If the founder event is associated with an inbreeding effective size that is small, stochastic alterations of the frequencies of one or a few major alleles, including fixation, are probable. These chance alterations at the major loci result in a drastic fitness reweighting of the pleiotropic effects associated with the major genes. Hence, strong selective forces are created shortly after the founding event by the altered genetic environment determined by one or a few major loci. The founder population must have large amounts of genetic variation at numerous modifier loci, however, in order to respond to these altered selective forces. Thus, the same conditions for carrying-over and augmenting genetic variation in the founder population as are discussed in the founder-flush model are applicable here. In essence, these and other factors detailed (in 64) result in a large variance effective size.

When this combination of small inbreeding effective size and large variance effective size occurs, conditions are optimal for a shift to a new adaptive peak defined by the major gene systems perturbed by the founder event. Speciation is regarded as a consequence of this peak shift.

One Model or Three?

Opponents (23, 24) and some proponents (49) of the above models have tended to lump them together, thereby regarding an argument for or against any one of them as an argument for or against all of them. This approach is not justified, as we show in Table 1, which presents a summary of the salient features of the three models. One of the more obvious differences between the genetic revolution model and the other two is the role played by genetic variation. Genetic revolution requires a significant reduction in genetic variation in the founder population. In contrast, Carson's and Templeton's models require high levels of genetic variation in the founder population and predict little chance of speciation if a significant drop in levels of genetic variation occurs. Thus, founding events that could lead to genetic transilience or founder-flush specia-

Table 1   A comparison of three models of founder-induced speciation

Feature	Genetic revolution	Founder-flush	Genetic transilience
Ancestral population	Peripheral	Outcrossed and polymorphic	Outcrossed and polymorphic
Primary impact of the founding event	Great increase in level of homozygosity	Disruption of co-adapted complex through drift	Disruption of co-adapted complex through drift on major genes
Genetic events following the founding event	Continued loss of genetic variation due to small population size	Flush, recombination, and altered pleiotropic balance. Carry-over and release of genetic variation	Flush, recombination, and altered pleiotropic balance. Carry-over and release of genetic variation
Major source of selection	Genetic environment: homozygosity	External and environmental	Genetic environment: altered frequencies of major genes
Period of strongest selection	When homozygosity is maximal	After flush (relaxed during flush)	During flush, shortly after founding event
Genetic response	Most loci (later revised downward)	Polygenic, but most loci unaffected	A few major genes and their modifiers

tion would exclude genetic revolution and vice versa. Genetic revolution is not only different from the other two models, it is incompatible with them.

The founder-flush and genetic transilience models also differ significantly. The first focuses upon the ecological context of the founder population by stressing the importance of relaxed or stringent selective conditions as a function of population density. The second emphasizes the importance of the genetic environment as defined by epistatic networks and pleiotropy. (This difference is a matter of emphasis, as the authors of both models acknowledge the role of both external and genetic environments.) As a result, the flush phase is the period of most intense selection in the genetic transilience model [despite an erroneous statement to the contrary by Charlesworth & Smith (24)], whereas it is a period of relaxed selection in the founder-flush model.

Although the genetic transilience and founder-flush models are different, they are not incompatible with one another. Both require similar conditions with respect to the ancestral population, the genetic role played by the founder event, and the population dynamic and recombinational events following the founding of a population. Because of these similarities, a single founder population can selectively respond both to an altered genetic environment in the flush phase and to an altered ecological environment in the postflush phase.

Hence, although the underlying selective mechanisms are different, founder-flush and genetic transilience are compatible and can occur together; both are incompatible with genetic revolution, however.

CRITIQUE OF FOUNDER–INDUCED SPECIATION MODELS

Theoretical Critiques

GENETIC REVOLUTION   Lewontin (43) criticizes Mayr's genetic revolution theory, arguing that although a founder event can have a great effect on the frequency of a particular allele or genotype, it generally has little impact on any measure of variation (such as average heterozygosity) that weights genotypes by their frequencies. Consequently, contrary to Mayr's claims (see Figure 3), the founder event will not significantly increase the amount of homozygosity in the founder population. Therefore, the altered genetic environment needed to trigger genetic revolution will not occur. Lande (40) has strongly reiterated this criticism. Mayr (46), however, also predicted that genetic variation would be lost after the founding event owing to the continued small population size. Hence, the requisite high levels of homozygosity can be achieved if the founder population remains small for many generations after the founding event.

In his critique, Lande (40) also emphasizes that quantitative genetic variability is replenished rapidly through mutation, even in populations with variance effective sizes on the order of 100 individuals. Substantial and geologically rapid phenotypic evolution is therefore possible in these founder populations. This argument actually supports Mayr's theory, however. The principal role of genetic revolution is to break down the old coadapted gene complex, and it does not necessarily create a new one. The creation of the new adaptive complex occurs after the genetic revolution has freed the population from its previous epistatic constraints. Thus, Lande's (40) model actually shows the validity of the postrevolutionary events that Mayr outlined.

More serious criticisms of genetic revolution have been raised (69). Genetic revolution requires a significant increase in homozygosity relative to the ancestral condition. By emphasizing that the founders come primarily from peripheral demes, however, Mayr makes it more difficult to satisfy this requirement in the many species in which peripheral demes are already characterized by inbreeding and increased homozygosity. More damaging is the fact that a population's ability to respond to intense selection is directly proportional to the amount of genetic variation it has. Yet the genetic revolution model demands a rapid and effective response to selection precisely when genetic variation is at a minimum—conditions that make a rapid and effective response impossible. Thus, Mayr's genetic revolution model is based upon mutually contradictory population-genetic conditions.

We are not arguing that peripheral populations are unimportant in speciation for they are, but through mechanisms other than genetic revolution (68, 73). In cases involving founder events from highly inbred peripheral populations, the primary role of the founding event is to establish a geographically isolated population, and there is nothing about the founding event *per se* that *induces* speciation (64, 65, 67). This review is only concerned with founder-induced speciation, and genetic revolution is a most unlikely explanation for this type of speciation.

FOUNDER–FLUSH AND GENETIC TRANSILIENCE   Because these two modes of founder-induced speciation are compatible and share many of the same requirements, we discuss them together, even though their underlying genetic mechanisms are distinct.

These two models do not suffer from the major population-genetic incompatibility found in the genetic revolution model: In both models, the response to selection occurs during a phase with much genetic variability. The genetic impact of founder events in these models is also compatible with the comments made by Lewontin (43) and Lande (40). Charlesworth & Smith (24), however, have raised quantitative objections on the basis of computer simulations, even while recognizing that their "conclusions agree qualitatively" with the predictions made by Carson (14) and Templeton (68, 69). From the quantitative point of view, their simulations have two serious flaws.

First, despite their stated purpose, Charlesworth & Smith fail to simulate either Carson's founder-flush model or genetic transilience. Recall that the sequence of events in the founder-flush model is as follows: (*a*) a founder event that initiates the disorganization of ancestral coadapted gene complexes, (*b*) a flush phase of relaxed selection that accentuates the drift-initiated disorganization through recombination, and (*c*) the reimposition of stringent selective conditions after the flush phase that results in the evolution of a new coadapted gene complex. In contrast, Charlesworth & Smith simulated the following sequence of events: (*a*) a flush phase of relaxed selective conditions in the *ancestral* population for n generations, (*b*) a founder event at generation $n + 1$, and (*c*) reimposition of selection at normal intensities at generation $n+1$ and thereafter. To put it bluntly, Charlesworth & Smith put the cart (relaxed selection in the flush phase) before the horse (the founder event). They apparently confused the founder-flush model with a related, but distinct, model of Carson's—that of *in situ* reduction of population size or the flush-crash model (12, 18). In this model, relaxed ecological and selective conditions allow a population flush, but when the ecological conditions become more restrictive, a population crash follows. Carson argues that if several of these flush-crash cycles occur, speciation is likely (since Charlesworth & Smith simulated only one cycle, their simulation is biased even against this model).

This error has important quantitative implications. As is well-known, large populations have considerable genetic inertia. By having the relaxed selection occur in the large, ancestral population and by only having one flush-crash cycle, Charlesworth & Smith minimize the possibility that relaxed selection will disorganize the ancestral coadapted complex and exclude the possibility that relaxed selection will accentuate an initial disruption induced by drift. Moreover, by failing to incorporate *new* selective forces *induced* by the founder event, they fail to simulate the genetic transilience model.

The second flaw in these simulations is the type of genetic architecture used. Charlesworth & Smith only examine the case of two equally important, epistatic loci. They do argue (without any quantitative proof) that neither having several independent pairs of epistatic loci nor lowering recombination between the pair of loci to approximate a single segregating block would greatly improve the situation. Although Carson's writings do not give many genetic architectural details, he has communicated to Charlesworth that the simulated genetic architectures do not correspond to what he had in mind. Moreover, Templeton (64, 67, 69) has explicitly argued that the primary architecture is type II. This type of architecture has many unique properties, including the ability to induce rapid and drastic phenotypic transitions (28, 70, 75), the potential to have its additive variance increased by a founder event, and the capacity to induce many cascading evolutionary processes through strong pleiotropic effects (64, 72, 75). None of these properties are considered by Charlesworth & Smith. Moreover, as Templeton (64) discusses, fixation of a major segregating unit can often be crucial in triggering genetic transilience. Yet when fixation of either locus occurred in the simulations, Charlesworth & Smith terminated the run because fixation events were outside the bounds of the adaptive transition they considered. Hence, they specifically excluded the types of genetic events and architectures that should be most important quantitatively in founder-induced speciation.

A similar criticism can be leveled against the analytical treatment in Barton & Charlesworth's review. Their critique stems from quantitative comparisons based on Wright's adaptive landscape concept. They point out that this concept has many unrealistic biological assumptions, but argue that it "is still a very useful metaphor." Templeton (69, 70) has also shown how useful it is, but he strictly limits the use of this metaphor to gaining qualitative insights. There is no justification for using a "metaphor" that explicitly violates essential biological features to generate detailed quantitative comparisons.

Barton & Charlesworth's review also suffers from serious misunderstandings of the genetic transilience model. First, they say that the trigger for transilience is a deviation from Hardy-Weinberg genotype frequencies. Although factors causing such deviations influence the chances for transilience (64), the critical trigger is a reweighting of fitness components owing to drift-induced shifts in *allele* frequencies at one or more major loci that have many pleiotropic effects (64, pp. 1015–16). A second apparent misrepresentation is their claim that inbreeding effective size and variance effective size are virtually the same. This view rests upon "the cumulative effect of a whole sequence of generations with restricted numbers of individuals." Under these conditions, their claim is true, but an *essential* feature of genetic transilience is that the founder event is immediately followed by a large increase in the number of individuals. Consequently, Barton & Charlesworth's argument is only valid under conditions that violate the assumptions of the genetic transilience model.

Critiques of Experimental Examples

Several experimental systems have been presented as evidence for founder-induced speciation models. One of the first was the use of the parthenogenetic capacity present in natural populations of *Drosophila mercatorum* to simulate the most extreme founder event possible from outcrossed, diploid ancestors (63, 76). As mentioned earlier, the results of these experiments were supportive of the founder-induced speciation models and also indicated the importance of type II genetic architectures. Charlesworth et al have criticized this experimental system by stating that "the relevance of observations on totally homozygous lines to speciation in relatively outbred, sexual populations is obscure" (23, p. 483). This criticism is valid (with an exception to be mentioned shortly) but ignores the intent of these experiments. Obviously, the experimental design creates an extreme situation that is unlikely ever to be realized in natural sexual populations, as was explicitly acknowledged (63). The purpose of these experiments, however, was to accentuate the founder effects in order to make them more amenable to rigorous genetic analysis. As a result, these experiments can be (and were) used to gain insights and detailed knowledge about founder-induced speciation. As has been specifically admitted, these experiments only helped "inspire" the genetic transilience model but they "do not contribute to substantiating its validity" (64, p. 1030).

There are other experimental systems, however, that can be used to substantiate the validity of these speciation models. Powell (57) has tested the founder-flush theory using populations of *Drosophila pseudoobscura* and found that premating barriers evolved in some lines. Charlesworth et al (23) criticize this experiment because the founding flies were derived from hybrids of different stocks. This criticism would be valid if Powell had failed to institute the proper controls, but he did so and showed that the effects were attributable to the founder-flush cycle, not to either the hybrid origin of the ancestral flies or inbreeding effects (also see 58).

The criticism of hybrid origin is not applicable to Ahearn's (1) or Arita & Kaneshiro's (4) experiments that demonstrate the ease with which founder-flush cycles can induce isolating barriers in laboratory populations of Hawaiian *Drosophila*. The results of Ahearn's experiments have been attributed to inbreeding depression on the basis that "inspection of her data suggests that the males from the bottleneck stock had a lower mating success than males from the outbred stock" (23, p. 482). But this overall difference in mating success could easily have been predicted because of the mating asymmetries commonly found in Hawaiian *Drosophila* (31). If there were truly an inbreeding depression in male mating ability, the derived (i.e. "inbred") males should have difficulty in mating with derived females as well as ancestral females. Ahearn's data do not show this pattern and reveal precisely the same type of asymmetric isolation as found in comparative studies of natural populations of Hawaiian *Drosophila* (31). Consequently, not only do her data show that the effects are not due to an inbreeding depression, but Charlesworth et al's (23) entire criticism is based upon the fact that her experimental populations yield results similar to those obtained with natural populations. We regard this similarity as evidence of the validity of her experimental design, not as a weakness.

Finally, Charlesworth et al state that "strong evidence against the concept that genetic revolutions are often induced by population bottlenecks comes from the usual lack of any indication of incipient speciation in domestic and laboratory plants and animals" (23, p. 483). This criticism ignores the fact that all three models of founder-induced speciation predict that most founder events will *not* induce speciation, and the conditions under which speciation will not occur have been explicitly stated (64). In fact, Templeton (64) used the lack of speciation under certain conditions as a major inspiration and source of support for the genetic transilience model.

In particular, and as mentioned earlier, most colonizing species are not expected to display founder-induced speciation. Most domesticated species are "weedy" colonizing species. Moreover, there is a strong bias toward choosing good colonizer species as laboratory organisms because they are generally easier to maintain and because they do not change drastically when put into a laboratory environment (H. D. Stalker, personal communication). Thus, traditional laboratory species such as *Drosophila melanogaster* have a host of attributes that insure that founder-induced speciation will be unlikely (64). It is interesting to note that most of the experimental evidence for founder-induced speciation came from experiments using nonstandard *Drosophila* species (1, 4, 63).

A similar situation is now arising with respect to the criticism based upon "domestic" species. Due to extensive habitat destruction, many nondomestic species are now existing only (or nearly so) as captive populations in zoos. Such species are often subject to extreme founder events and suffer from severe inbreeding depression (77). The only (but not optimal) option to save many of these species is to alter their genetic structure deliberately, so as to eliminate the inbreeding depression. In order to maximize the population's chances of survival, this genetic alteration has to be done rapidly and effectively—the same requirements as for founder-flush and genetic transilience. The theory of genetic transilience has been explicitly applied to the case of Speke's gazelle (*Gazella spekei*), a rare African gazelle whose captive populations were founded from one male and three females (77). The breeding program was extremely successful, demonstrating that a founder population can indeed respond to intense selection rapidly and effectively shortly after the original founding event and during the flush phase (77). A follow-up analysis specifically designed to study the major factors responsible for this selective response and to separate these effects from inbreeding effects confirmed the predictions made from genetic transilience theory (78). Although small sample sizes in other captive populations prevent the type of detailed analysis possible with Speke's gazelle, a survey of many other captive populations (K. Ralls, J. Ballou & A. R. Templeton, unpublished manuscript) reveals results completely consistent with the genetic transilience hypothesis.

When the results obtained from nonstandard laboratory populations and captive zoo populations are taken into account, the failure of founder-induced speciation to occur in domestic and standard laboratory organisms actually provides strong support for the founder-flush and genetic transilience theories. The failures as well as the successes confirm the predictions of these theories, a fact not appreciated by Charlesworth et al (23).

Critiques of Natural Examples

The strongest case for founder-induced speciation in nature comes from studies on Hawaiian *Drosophila*. Charlesworth et al note that there is no evidence for a reduction in isozyme heterozygosity in these species, and hence, "little direct evidence that genetic revolutions have happened in this group" (23, p. 484). This evidence does indeed run counter to Mayr's theory of genetic revolution, but as Charlesworth et al (23) acknowledge, this pattern is expected under the founder-flush and genetic transilience models. Nevertheless, Charlesworth & Smith still used the isozyme data to question the validity of these two models by asking, "If loci affecting courtship behavior can become fixed due to genetic drift caused by founder effects, why should enzyme loci have remained apparently immune?" (24, p. 235). This question stems from a long-standing misconception about the role of genetic drift in evolution. Wright (84) stresses the interactions of genetic drift and natural selection and emphasizes the role of drift in *adaptive,* not neutral, evolution. Unfortunately, his writings have been seriously distorted, particularly in England, where the term "Sewall Wright effect" was coined to refer to the role of genetic drift in neutral evolution. As Wright has explicitly argued (86), his theories about genetic drift have very little to do with the Sewall Wright effect. Unfortunately, the role of genetic drift as a force in adaptive evolution is still not appreciated as widely as it should be (70).

The founder-flush and genetic transilience theories both emphasize the role of genetic drift in *interacting* with natural and sexual selection and not as a force yielding fixation of neutral variants. Thus, the answer to the question Charlesworth & Smith raise is straightforward: It is the distinction between selected vs neutral loci. This distinction is evident in Lande's (41) models, which show that founder events can induce strong sexual selection upon loci influencing male and female courtship behaviors and displays. Yet, there is no reason why founder events should play a similar role for loci that are unrelated to these courtship phenotypes. This distinction is also evident in Templeton's (63) experimental system in which the loci underlying the abnormal abdomen complex are subject to intense selection in establishing a new parthenogenetic strain, but the isozyme loci are apparently neutral. Charlesworth & Smith (24) fail to recognize that both the founder-flush and genetic transilience theories predict that different loci or types of genetic variation will behave differently during founder and subsequent events.

As a result of this differential behavior, Templeton (64) emphasizes that the support for founder-induced speciation from natural examples should not be based upon only one type of genetic variation but rather on the joint *pattern* defined by different types of genetic variation. For example, both genetic transilience and founder-flush theories generate the prediction that speciation will be more likely when cross-over suppressors are eliminated or fixed in the founder population. When comparing different species of Hawaiian *Drosophila,* isozyme polymorphisms should be carried over with high probability from the ancestral to the derived species, but inversion polymorphisms should not; this is precisely the observed pattern (64). Also, mitochondrial DNA is effectively inherited as a maternal haploid in *Drosophila.* Consequently, although nuclear genes are minimally affected by founder events involving a single gravid female with multiple insemination, mitochondrial genes behave as in the extreme founder event empirically modeled by the parthenogenetic *Drosophila mercatorum*—a single haploid genome. Recent studies have therefore been performed on mitochondrial DNA evolution in the *planitibia* subgroup of Hawaiian *Drosophila* (30; unpublished data), a group noted for its extreme conservatism in isozyme evolution (15, 16). As predictions based on the founder-flush and genetic transilience theories would lead us to expect, the lack of significant evolution for isozyme genes was accompanied by radical evolution for mitochondrial restriction–endonuclease sites. This pattern of variation clearly lends support to the conclusion that founder events (as supported primarily by the mitochondrial DNA data), population flushes (the isozyme data), and free recombination (the chromosomal inversion data) are all involved in the speciation of Hawaiian *Drosophila.* Although alternative hypotheses could explain one type of variation (e.g. the isozyme data could be explained by simply assuming there were no founder events at all), no alternative has yet been presented to explain the joint *pattern*.

MACROEVOLUTIONARY IMPORTANCE

One of the most hotly debated topics in current evolutionary biology is the theory of punctuated equilibrium. This theory was explicitly formulated as the paleontological implications of Mayr's genetic revolution theory (32). At the outset, we emphasize that there are many ways of achieving the phenomena explained by punctuated equilibrium (stasis interrupted by rapid evolution on a geological time scale) that do not involve founder events (70). Consequently, we will not deal with the phenomenon per se, but only with the proposed mechanism.

One inference derived from the genetic revolution theory is that evolutionary events occurring within a species are much less important in defining the overall evolutionary process than the genetic revolution itself (32, 61). Interestingly, Mayr (49) has opposed this view, and for good reasons. Recall that Mayr stresses that the primary role of genetic revolution is to free the species from its previous epistatic constraints. The evolution of a new coadapted gene complex (the event actually associated with the development of a new species) generally occurs *after* the genetic revolution, and it occurs via the normal operation of selection, mutation, drift, and so on within a single breeding population. The inference that microevolutionary processes are unimportant in speciation because of genetic revolution is totally unfounded.

In light of the difficulties mentioned above, genetic revolution is a poor foundation for a major macroevolutionary theory. This problem can be circumvented by substituting the founder-flush or genetic transilience theories for genetic revolution. Even so, however, difficulties are encountered. Remember that these theories predict that many founder events will not induce speciation. As Templeton (66) has pointed out, many of the groups whose fossil record has been interpreted as supportive of punctuated equilibrium are unlikely to satisfy the conditions required for founder-induced speciation (see also 73). Consequently, founder-induced speciation models do not provide either a general theory of macroevolution or a general interpretative framework for the fossil data.

Nevertheless, these models can be used legitimately in making macroevolutionary interpretations, but these interpretations must be made carefully, and they must be limited in scope. As Templeton (64) asserts, the conditions required for founder-induced speciation include both external factors facilitating the creation of founder events and intrinsic factors dealing with the fundamental biology and genetic organization of the group under consideration.

Because of these dependencies, founder-induced speciation will often be common in certain groups or situations. An excellent example of this is Hawaiian *Drosophila*. We do not know how often the appropriate conditions for founder-induced speciation occur among continental *Drosophila*, but the external and intrinsic conditions are optimal for founder-induced speciation in Hawaii (64). Moreover, such founder events have greatly accelerated the rate of speciation there; founder-induced speciation accounts for at least a quarter of the entire genus, even if it does not occur at all among continental *Drosophila*. In addition, such speciation is often associated with radical evolutionary transformations. As a result, the Hawaiian *Drosophila* display unique and extreme morphology, behavior, and ecology. Because of the hierarchical nature of evolutionary processes, founder-induced speciation has had a large quantitative and qualitative impact on the macroevolution of the genus *Drosophila*, even under the extreme assumption that it almost never occurs in continental situations.

Another example of the macroevolutionary amplification of this "rare" mode of speciation is provided by work on speciation in shallow-water marine organisms (37, 38, 82). The advantages of studying this type of community are the excellent fossil record and the existence of modern species, which gives us some insight into such biological features as population structure. In particular, one important contrast is between those species that have planktonic larvae (primarily nearshore species) and those that do not (primarily offshore species). Planktonic species should be characterized by extensive dispersal and gene flow, whereas nonplanktonic species have very limited dispersal capabilities and tend to show a subdivided population structure. As Templeton (64, 65, 67) states, a subdivided population structure greatly facilitates many speciation mechanisms, but it is a major barrier to founder-induced speciation, which is most probable when the ancestral population is extensively outcrossing. Moreover, the nearshore physical environment is more conducive to the formation of isolated founder populations than the more homogeneous offshore environment. Hence, the nearshore species are more prone to founder-induced speciation than the offshore species.

As a result of these contrasting population structures and physical environments, offshore assemblages should have a higher rate of speciation and extinction; the nearshore communities should be relatively resistant to speciation and extinction by virtue of their planktotrophic mode of development and the resulting population structure. When the latter do speciate, however, radical changes associated with founder-flush and/or genetic transilience are probable. This difference in the mode of speciation explains an otherwise puzzling pattern of macroevolution for these marine communities: The offshore communities display higher rates of species-level evolution, but most of the major new community types appeared first in nearshore settings and then expanded into offshore settings. Under this interpretation, the relatively rare mode of founder-induced speciation has played the major role in defining both the nearshore and offshore communities.

CONCLUSIONS

Much biogeographical, ecological, and systematic evidence implies that founder events are often a precondition for the genetic changes leading to speciation. Of three distinct mechanisms suggested to explain founder-induced speciation, genetic revolution demands contradictory population-genetic conditions and is an unlikely mode. Genetic transilience and founder-flush, although distinct, are compatible, and they are consistent with theoretical considerations and experimental and natural examples. Although the required conditions are restrictive, both modes could contribute to a single founder-induced speciation process. This relatively rare mode of speciation can have great macroevolutionary significance in certain groups of organisms or situations.

ACKNOWLEDGMENTS

The work of the authors has been supported by NSF grant DEB–7926692 to the University of Hawaii (HLC) and by NSF grant DEB–7908860 and NIH grants 1 RO1 GM2702101 and 1 RO1 AG02246 to Washington University (ART).

Literature Cited

1. Ahearn, J. N. 1980. Evolution of behavioral reproductive isolation in a laboratory stock of *Drosophila silvestris*. *Experientia* 36:63–64
2. Deleted in proof
3. Alvarez, L. W., Alvarez, W., Asaro, F., Michel, H. V. 1980. Extraterrestrial cause for the Cretaceous-Tertiary extinction. *Science* 208:1095–1108
4. Arita, L. H., Kaneshiro, K. Y. 1979. Ethological isolation between two stocks of *Drosophila adiastola* Hardy. *Proc. Hawaii Entomol. Soc.* 12:31–34
5. Bachmann, K. 1983. Evolutionary genetics and the genetic control of morphogenesis in flowering plants. *Evol. Biol.* 16:157–208
6. Baker, H. G. 1965. Characteristics and modes of origin of weeds. In *The Genetics of Colonizing Species*, ed. H. G. Baker, G. L. Stebbins, pp. 147–72. New York: Academic
7. Bush, G., Case, S. M., Wilson, A. C., Patton, J. 1977. Rapid speciation and chromosomal evolution in mammals. *Proc. Natl. Acad. Sci. USA* 74:3942–46

8. Carlquist, S. 1974. *Island Biology*. New York: Columbia Univ. Press. 660 pp.
9. Carr, G. D., Kyhos, D. W. 1981. Adaptive radiation in the Hawaiian silversword alliance (Compositae-Madiinae). I. Cytogenetics of spontaneous hybrids. *Evolution* 35:543–56
10. Carson, H. L. 1959. Genetic conditions which promote or retard the formation of species. *Cold Spring Harbor Symp. Quant. Biol.* 24:87–105
11. Carson, H. L. 1967. Selection for parthenogenesis in *Drosophila mercatorum*. *Genetics* 55:157–71
12. Carson, H. L. 1968. The population flush and its genetic consequences. In *Population Biology and Evolution*, R. C. Lewontin, pp. 123–37. New York: Syracuse Univ. Press
13. Carson, H. L. 1971. Speciation and the founder principle. *Univ. Mo. Stadler Genet. Symp.* 3:51–70
14. Carson, H. L. 1975. The genetics of speciation at the diploid level. *Am. Nat.* 109:83–92
15. Carson, H. L. 1976. Inference of the time

of origin of some *Drosophila* species. *Nature* 259:395–96

16. Carson, H. L. 1978. Speciation and sexual selection in Hawaiian *Drosophila*. In *Ecological Genetics*, ed. P. F. Brussard, pp. 93–107. New York: Springer-Verlag

17. Carson, H. L. 1982. Evolution of *Drosophila* on the newer Hawaiian volcanoes. *Heredity* 48:3–25

18. Carson, H. L. 1982. Speciation as a major reorganization of polygenic balances. In *Mechanisms of Speciation*, ed. C. Barigozzi, pp. 411–33. New York: Liss

19. Carson, H. L. 1983. Chromosomal sequences and interisland colonizations in Hawaiian *Drosophila*. *Genetics* 103: 465–82

20. Carson, H. L. 1983. Speciation and the founder effect on a new oceanic island. In *Tropical Pacific Biogeography*, ed. P. Raven, F. Radovsky, S. Sohmer. Honolulu: Bishop Mus. In press

21. Carson, H. L., Hardy, D. E., Spieth, H. T., Stone, W. S. 1970. The evolutionary biology of the Hawaiian Drosophilidae. In *Essays in Evolution and Genetics in Honor of Theodosius Dobzhansky*, ed. M. K. Hecht, W. C. Steere, pp. 437–543. New York: Appleton-Century-Crofts

22. Carson, H. L., Kaneshiro, K. Y. 1976. *Drosophila* of Hawaii: Systematics and ecological genetics. *Ann. Rev. Ecol. Syst.* 7:311–46

23. Charlesworth, B., Lande, R., Slatkin, M. 1982. A neo-Darwinian commentary on macroevolution. *Evolution* 36:474–98

24. Charlesworth, B., Smith, D. B. 1982. A computer model of speciation by founder effects. *Genet. Res. Cambridge* 39:227–36

25. DeSalle, R., Templeton, A. R. 1983. Molecular basis of the abnormal abdomen phenotype in *Drosophila mercatorum*. *Genetics* 104:s21 (Abstr.)

26. Fernald, M. L. 1925. Persistence of plants in unglaciated areas of boreal America. *Mem. Am. Acad. Arts Sci.* 15:239–342

27. Fosberg, R. 1948. Derivation of the flora of the Hawaiian Islands. In *Insects of Hawaii*, ed. E. C. Zimmerman, Introd. 1:107–19. Honolulu: Univ. Hawaii Press

28. Frankham, R., Nurthen, R. K. 1981. Foraging links between population and quantitative genetics. *Theor. Appl. Genet.* 59:252–63

29. Giddings, L. V., Carson, H. L. 1982. Behavioral phylogeny of populations of *Drosophila crucigera*. *Genetics* 100:s26 (Abstr.)

30. Giddings, L. V., DeSalle, R. 1983. Mitochondrial DNA variation in populations of *Drosophila silvestris*. *Genetics* 104:s27 (Abstr.)

31. Giddings, L. V., Templeton, A. R. 1983. Behavioral phylogenies and the direction of evolution. *Science* 220:372–78

32. Gould, S. J. 1982. Darwinism and the expansion of evolutionary theory. *Science* 216:380–87

33. Gould, S. J., Eldredge, N. 1977. Punctuated equilibria: The tempo and mode of evolution reconsidered. *Paleobiology* 6:119–30

34. Hardy, D. E. 1965. *Insects of Hawaii*, Vol. 12, *Diptera: Cyclorrhapha II, Series Schizophora, Section Acalypterae. I. Family Drosophilidae*. Honolulu: Univ. Hawaii Press. 814 pp.

35. Hedberg, O. 1969. Evolution and speciation in a tropical high mountain flora. *Biol. J. Linn. Soc.* 1:135–48

36. Hilu, K. W. 1983. The role of single-gene mutations in the evolution of flowering plants. *Evol. Biol.* 16:97–128

37. Jablonski, D., Bottjer, D. J. 1983. Soft-bottom epifaunal suspension-feeding assemblages in the late Cretaceous: Implications for the evolution of benthic paleocommunities. In *Biotic Interactions in Recent and Fossil Benthic Communities*, ed. M. J. S. Tevesz, P. L. McCall, pp. 747–812. New York: Plenum

38. Jablonski, D., Sepkoski, J. J. Jr., Bottjer, D. J., Sheehan, P. M. 1983. Onshore-offshore patterns in the evolution of Phanerozoic shelf communities. *Science* 222:1123–25

39. Johnston, R. F., Selander, R. K. 1971. Evolution in the house sparrow. II. Adaptive differentiation in North American populations. *Evolution* 25:1–28

40. Lande, R. 1980. Genetic variation and phenotypic evolution during allopatric speciation. *Am. Nat.* 116:463–79

41. Lande, R. 1981. Models of speciation by sexual selection on polygenic traits. *Proc. Natl. Acad. Sci. USA* 78:3721–25

42. Lande, R. 1983. The response to selection on major and minor mutations affecting a metrical trait. *Heredity* 50:47–65

43. Lewontin, R. C. 1965. Comment. See Ref. 6, pp. 481–84

44. Lints, F. A., Bourgeois, M. 1984. Population crash, population flush and genetic variability in cage populations of *Drosophila melanogaster*. In *Genetique, Selection, Evolution*. In press

45. Macdonald, G. A., Abbott, A. T. 1970. *Volcanoes in the Sea*. Honolulu: Univ. Hawaii Press. 441 pp.

46. Mayr, E. 1954. Change of genetic environment and evolution. In *Evolution as a Process*, ed. J. Huxley, A. C. Hardy,

130   CARSON & TEMPLETON

E. B. Ford, pp. 157–80. London: Allen & Unwin
47. Mayr, E. 1970. *Populations, Species and Evolution.* Cambridge, Mass: Belknap. 453 pp.
48. Mayr, E. 1982. Processes of speciation in animals. See Ref. 18, pp. 1–19
49. Mayr, E. 1982. *The Growth of Biological Thought: Diversity, Evolution and Inheritance.* Cambridge, Mass: Belknap. 974 pp.
50. McDougall, I., Swanson, D. A. 1972. Potassium-argon ages of lavas from the Hawi and Pololu volcanic series, Kohala Volcano, Hawaii. *Geol. Soc. Am. Bull.* 83:3731–38
51. Mueller-Dombois, D., Bridges, K. W., Carson, H. L. 1981. *Island Ecosystems.* Stroudsburg, Pa: Hutchinson Ross. 583 pp.
52. Nei, M., Maruyama, T., Chakraborty, R. 1975. The bottleneck effect and genetic variability in populations. *Evolution* 29:1–10
53. Nelson, G., Rosen, D. E. 1981. *Vicariance Biogeography.* New York: Columbia Univ. Press. 593 pp.
54. Nur, A., Ben-Avraham, Z. 1981. Lost Pacifica Continent: A mobilistic speculation. See Ref. 53, pp. 351–58
55. Paik, Y. K., Sung, K. C. 1981. Genetics of island populations of exotic *Drosophila:* Inversions in *Drosophila immigrans.* See Ref. 51, pp. 455–66
56. Parsons, P. A. 1980. Isofemale strains and evolutionary strategies in natural populations. *Evol. Biol.* 13:175–217
57. Powell, J. R. 1978. The founder-flush speciation theory: An experimental approach. *Evolution* 32:465–74
58. Powell, J. R., Morton, L. 1979. Inbreeding and mating patterns in *Drosophila pseudoobscura. Behav. Genet.* 9:425–29
59. Rose, M. R. 1982. Antagonistic pleiotropy, dominance, and genetic variation. *Heredity* 48:63–78
60. Rose, M. R. 1983. Further models of selection with antagonistic pleiotropy. In *Population Biology,* ed. H. I. Freedman, C. Strobeck, pp. 47–53. Berlin: Springer-Verlag
61. Stanley, S. M. 1979. *Macroevolution: Pattern and Process.* San Francisco: Freeman. 332 pp.
62. Templeton, A. R. 1977. Analysis of head shape differences between two interfertile species of Hawaiian *Drosophila. Evolution* 31:630–41
63. Templeton, A. R. 1979. The unit of selection in *Drosophila mercatorum.* II. Genetic revolutions and the origin of coadapted genomes in parthenogenetic strains. *Genetics* 92:1265–82

64. Templeton, A. R. 1980. The theory of speciation *via* the founder principle. *Genetics* 94:1011–38
65. Templeton, A. R. 1980. Modes of speciation and inferences based on genetic distances. *Evolution* 34:719–29
66. Templeton, A. R. 1980. Review of "Macroevolution: Pattern and Process", by S. M. Stanley. *Evolution* 34:1224–27
67. Templeton, A. R. 1981. Mechanisms of speciation—a population genetic approach. *Ann. Rev. Ecol. Syst.* 12:23–48
68. Templeton, A. R. 1982. The prophecies of parthenogenesis. In *Evolution and Genetics of Life Histories,* ed. H. Dingle, J. P. Hegmann, pp. 75–101. New York: Springer-Verlag
69. Templeton, A. R. 1982. Genetic architectures of speciation. See Ref. 18, pp. 105–21
70. Templeton, A. R. 1982. Adaptation and the integration of evolutionary forces. In *Perspectives on Evolution,* ed. R. Milkman, pp. 15–31. Sunderland, Mass: Sinauer
71. Templeton, A. R. 1983. Natural and experimental parthenogenesis. In *The Genetics and Biology of Drosophila,* ed. M. Ashburner, H. L. Carson, J. N. Thompson, 3C:343–98. New York: Academic
72. Templeton, A. R. 1983. The evolution of life histories under pleiotropic constraints and *K*-selection. See Ref. 60, pp. 64–71
73. Templeton, A. R. 1984. A population growth overview of speciation in mammals. *Acta Zool. Fenn.* In press
74. Templeton, A. R., Carson, H. L., Sing, C. F. 1976. The population genetics of parthenogenetic strains of *Drosophila mercatorum.* II. The capacity for parthenogenesis in a natural bisexual population. *Genetics* 82:527–42
75. Templeton, A. R., Johnston, J. S. 1982. Life history evolution under pleiotropy and *K*-selection in a natural population of *Drosophila mercatorum.* In *Ecological Genetics and Evolution: The Cactus-Yeast-Drosophila Model System,* ed. J. S. F. Barker, W. T. Starmer, pp. 225–39. New York: Academic
76. Templeton, A. R., Rankin, M. R. 1978. Genetic revolutions and control of insect populations. In *The Screwworm Problem,* ed. R. H. Richardson, pp. 81–111. Austin: Univ. Tex. Press
77. Templeton, A. R., Read, B. 1983. The elimination of inbreeding depression in a captive herd of Speke's gazelle. In *Genetics and Conservation: A Reference for Managing Wild Animal and Plant Populations,* ed. C. M. Schonewald-

Cox, S. M. Chambers, B. MacBryde, L. Thomas, pp. 241–61. Reading, Mass: Addison-Wesley

78. Templeton, A. R., Read, B. 1984. Factors eliminating inbreeding depression in a captive herd of Speke's gazelle (*Gazella spekei*). *Zoo Biol.* Submitted for publication

79. Throckmorton, L. H. 1966. The relationships of the endemic Hawaiian Drosophilidae. *Stud. Genet., Univ. Tex. Publ.* 6615:335–96

80. Timofeef-Ressovsky, N. 1940. Mutation and geographical variation. In *The New Systematics*, ed. J. Huxley, pp. 73–136. Oxford: Oxford Univ. Press

81. Val, F. C. 1977. Genetic analysis of the morphological differences between two interfertile species of Hawaiian *Drosophila. Evolution* 31:611–29

82. Valentine, J. W., Jablonski, D. 1983. Speciation in the shallow sea: General patterns and biogeographic controls. In *Evolution, Time and Space: The Emergence of the Biosphere*, ed. R. W. Sims, J. H. Price, P. E. S. Whalley, pp. 201–26. New York: Academic

83. Wallace, B. 1968. *Topics in Population Genetics*. New York: Norton. 481 pp.

84. Wright, S. 1932. The roles of mutation, inbreeding, crossbreeding, and selection in evolution. *Proc. 6th Int. Congr. Genet., Ithaca, NY.* 1:356–66

85. Wright, S. 1978. *Evolution and the Genetics of Populations.* Vol. 4, *Variability Within and Among Natural Populations.* Chicago: Univ. Chicago Press. 474 pp.

86. Wright, S. 1980. Genic and organismic selection. *Evolution* 34:825–43

87. Zimmerman, E. C. 1948. *Insects of Hawaii.* Vol. 1, *Introduction.* Honolulu: Univ. Hawaii Press. 206 pp.

88. Zimmerman, E. C. 1960. Possible evidence of rapid evolution in Hawaiian moths. *Evolution* 14:137–38

89. Zimmerman, E. C. 1978. *Insects of Hawaii.* Vol. 9, *Microlepidoptera.* Honolulu: Univ. Hawaii Press. 1903 pp.

90. Zimmerman, E. C. 1980. Variations on a setal theme in Hawaiian *Hyposmocoma* moths (Lep., Gelechiidae). *Entomol. Mon. Mag.* 116:125–28

Ann. Rev. Ecol. Syst. 1984. 15:133–64
Copyright © 1984 by Annual Reviews Inc. All rights reserved

GENETIC REVOLUTIONS, FOUNDER EFFECTS, AND SPECIATION

N. H. Barton

Department of Genetics and Biometry, The Galton Laboratory, University College, London NW1 2HE, England

B. Charlesworth

School of Biological Sciences, University of Sussex, Brighton BN1 9QG, England

INTRODUCTION

Are new species formed in rare catastrophes, distinct from the normal processes of phyletic evolution? Or does reproductive isolation evolve gradually, as a by-product of the divergence of gene pools? Mayr (120–124) has argued the former, holding that speciation usually results from genetic revolutions triggered by founder effects: An isolated population, small in numbers and in geographic extent, colonizes a new area. Both changes in selection pressures and genetic drift result in the rapid shift of many genes to a new, coadapted combination, which is reproductively isolated from the ancestral population. Carson (27, 29, 31) and Templeton (175–180), among others, have put forward similar models.

This cluster of theories is woven from many strands; we will try to tease these apart in order to find out precisely which processes may be involved in speciation by founder effect. By placing them in the context of other models, we will argue that, although founder effects may cause speciation under sufficiently stringent conditions, they are only one extreme of a continuous range of possibilities. Complete geographic isolation is unnecessary; absolute coadaptation between "closed" systems of alleles is unlikely; and divergence may be driven in a variety of ways, without the need for drastic external changes. Reproductive isolation is most likely to be built up gradually, in a

133

0066-4162/84/1120-0133$02.00

series of small steps. Inference from nature or from laboratory experiments is difficult, and much of the evidence that has been used to support founder effect models seems ambiguous.

SPECIES AND SPECIATION

It is important to define just what we mean by the terms *species* and *speciation*. We will keep strictly to Mayr's (119, p. 120) definition of biological species as "groups of actually or potentially interbreeding natural populations which are reproductively isolated from other such groups". In other words, a biological species is formed when a population evolves genetic differences that prevent it from exchanging genes with other populations. Although speciation is, strictly speaking, the evolution of reproductive isolation, it usually involves the splitting of the ancestral population into populations located at different equilibria under selection. When such divergent populations meet, selection will maintain their incompatible equilibria, and when adjacent demes exchange a limited proportion of migrants, they will retain their distinctive combinations of selected alleles (89). When the populations meet in a continuous habitat, a smooth, stable hybrid zone usually forms (12, 15, 16). This zone impedes gene flow at neutral loci linked to the selected genes, and some degree of reproductive isolation results (9, 13). Conversely, almost all models of reproductive isolation imply the possibility of alternative equilibria. Although we will not *define* species as populations in distinct states [or, equivalently, as "harmonious, well-integrated gene complexes" (121, p. 295)], we will nevertheless regard speciation as requiring the evolution of new, stable equilibria.

In discussing this process, we will use Wright's (193) "adaptive landscape." This is a graph of the mean fitness of a population plotted against possible states of that population—allele frequencies, means of continuous characters, and so on. When the fitness of each genotype is constant and linkage disequilibrium is negligible, populations climb up the nearest "adaptive peak." Hybrids between populations at different peaks are necessarily less fit, and so the populations are to some extent reproductively isolated. From this perspective, the problem of speciation is reduced to the problem of how a single population can split into two populations on different peaks (see Figure 1). The adaptive landscape can be used to classify different models of speciation, and it can provide quantitative estimates of rates of divergence and levels of reproductive isolation (see the section on What Drives Divergence?).

The adaptive landscape has been criticized on the grounds that it almost never describes evolution accurately (e.g. 130, 181). If there is significant linkage disequilibrium or frequency-dependent selection, mean fitness does not tend to a maximum (155). For a system involving only one variable, some gradient function can always be found (199, 201), but this usually is not possible for multidimensional systems (2): Such a system may fall into a stable limit cycle (74). Nevertheless, Wright's concept is still a very useful metaphor, and it can easily be modified to describe a general multidimensional system in which a population is attracted towards one of a number of stable states—e.g. simple equilibria, limit cycles, or even chaotic "strange attractors." Hybrids between populations in different states will not necessarily be less fit; one may not be able to give any sensible meaning to *fitness*. Gene flow between genetically differentiated populations will still be impeded, however. Arguments about the likelihood of shifts between adaptive peaks carry over to the more general case of shifts between stable states. Because of its familiarity and simplicity, we will use Wright's illustration, while bearing in mind the above limitations.

FOUNDER EFFECTS AND GENETIC REVOLUTIONS

Mayr's Theory

Mayr's theory of genetic revolutions (120–124) is based on the idea that a species in some sense constitutes a genetic unit that is hard to change radically. He argues that this is partly because gene flow helps maintain genetic uniformity in the face of spatially varying selection pressures: "The stabilizing effect of gene flow is best documented by phenotypic uniformity or at least not more than clinal variation over wide areas" (121, p. 521). More importantly perhaps, epistatic interactions in the fitness effects of different loci, together with developmental homeostasis, are held to act as strongly conservative agents that prevent radical change in large panmictic populations. "Gene flow is not nearly strong enough to make these species anywhere near panmictic. It is far more likely that all populations share a limited number of highly successful epigenetic systems and homeostatic devices which place a severe restraint on genetic and phenotypic change" (121, p. 523).

Mayr's solution to this problem is to assume that speciation is usually initiated in a completely isolated population. He argues, however, that isolation alone is not enough: If the environment of the isolate is similar to that of the parental population, homeostatic mechanisms will greatly impede genetic divergence if the population is large (121, p. 528). Nonetheless, there are many cases of an isolate's diverging strongly from its putative parental population, in the absence of obvious environmental differences (119, 121, 149, 150); for example, the New Guinea kingfisher, *Tanysiptera galatea,* is more or less uniform across all of New Guinea but has differentiated on adjacent islands. Mayr concludes that speciation must involve rapid genetic change that is independent of environmentally caused alterations in selection pressures. He assumes that a temporary bottleneck in population size, associated with the foundation of a new isolate by a small number of individuals, will perturb the population from its equilibrium under selection and lead to a transition from "one well-integrated and stable condition through a highly unstable period to another period of balanced integration" (121, p. 538). He emphasizes the loss of heterozygosity due to genetic drift during the phase of reduced population size that follows foundation of the isolate. This loss supposedly favors alleles that have a selective advantage as homozygotes. Changes in the genetic background induced by drift cause changes in the net fitness of genotypes at loci under selection (120).

Carson's Theory

Prompted by his extensive studies of the remarkably speciose Hawaiian *Drosophila*, Carson (27, 29, 31) has proposed a similar model. Carson is more explicit in his account of the genetic system needed for a genetic revolution; it consists of two classes of loci, the "open" system and the "closed" system. The former consists of variants with more or less independent effects on fitness, which can therefore "float easily in natural populations and be available to the selective process" (29, p. 85). They are responsible for "geographical and clinal genetic variability," and for the "relatively minor modifications of phenotype" caused by artificial selection (29, p. 85). The closed system consists of loci that have strong epistatic interactions with each other and hence are in strong linkage disequilibrium. Without some opposing force, such systems would rapidly move to fixation; they would not, therefore, be able to generate reproductive isolation during founder events. "A significant number, however, may be expected to achieve relational balances which confer heterozygote superiority despite their vulnerability to disintegration by crossing over" (29, p. 87). Carson does not, however, specify the nature of the genes that might be involved in the two systems, although he does suggest—with reference to the striking elaboration of courtship rituals amongst the Hawaiian *Drosophila*—that sexual selection may be important (31).

Carson proposes that drastic events are needed to disorganize the highly conservative, closed system; after this disorganization, new balances that are incompatible with their ancestors are slowly restored. He places a somewhat different emphasis than does Mayr on the various processes that operate during founder events, stressing the relaxation of selection pressures during population flushes and crashes as an agent of disorganization.

Templeton's Theory

Templeton (176, 180) has written at length on the conditions under which founder events can be expected to cause speciation; he refers to the passage of a population through an unstable intermediate state as a "transilience." Although Templeton, like Mayr and Carson, discusses the change in genetic background that results from a bottleneck (176), he also suggests that the excess of homozygotes over the Hardy-Weinberg expectation arising from extreme inbreeding will alter selection pressures enough to tip the population from one equilibrium to another (176, 179). Such effects of Hardy-Weinberg deviations on genotypes at a single locus must be distinguished from the effects of a general loss of variation and hence from homozygosity of the genetic background [which seems to be what Mayr refers to as "inbreeding" (see discussion below)].

As well as postulating slightly different mechanisms to drive divergence, Templeton envisages a genetic system different from either Mayr's or Carson's. He suggests that reproductive isolation is much more likely to evolve through changes at a few major loci, followed by coadaptations at modifier loci, rather than in a single "genetic revolution" (176, 178). Random changes in a polygenic system will tend to average out over the many genes involved, whereas "if there are a few major genes, the stochastic effects of a founder event cannot be ignored" (176, p. 1015).

Kaneshiro's Theory

Kaneshiro suggests that ethological isolation evolves in small founding populations through the loss of components of male courtship behavior because of drift (85–87). Females in the derived, isolated population will be selected to respond to courtship by such males, whereas there will be no such selection on females of the ancestral population. Ancestral females will therefore discriminate against males from the derived population, but derived females will not. Kaneshiro and Giddings & Templeton (66) claim that there is abundant evidence for such asymmetrical sexual isolation. We shall evaluate this view below.

THEORY

Reproductive isolation may arise in a wide variety of ways. Indeed, given the range of possibilities open to a population, it is hard to imagine that different populations will not eventually evolve in incompatible directions. To allow the founder effect theories to be set in this wider context, we have roughly classified speciation mechanisms (Table 1). Although this classification is inevitably somewhat arbitrary, it will help us keep in mind the features on which we base our discussion.

What Drives Divergence?

Suppose that selection can be described by an adaptive landscape with several local fitness peaks. A new peak can be reached in two ways. First, the landscape may change in such a way that the initial equilibrium either dis-

Table 1 Some speciation models classified according to the features described in the text

Features	Peripatric[a]	Founder-flush[b]	Genetic transilience[c]
Mechanism driving divergence	Background homozygosity; drift	Relaxation of selection; drift	Deviation from Hardy-Weinberg; drift
Genetic basis of isolation:			
Incompatibility per step	Substantial	Moderate	Moderate
Number of genes per step	Many	Many	A few
Type of variation	Epistatic polymorphisms	Epistatic polymorphisms	Major gene + modi‹
Geographic relations:			
During divergence	Allopatric	Allopatric	Allopatric
During spread	*	*	*
Mechanism of spread	*	*	*
Types of genetic system	*	Mating behavior	*

[a] See 120, 124.
[b] See 27, 29.
[c] See 176.

appears or becomes unstable (Figure 1a). Deterministic selection pressures will then push the population to a neighboring peak. This sequence can be triggered by a variety of changes in either the external environment or the genetic system (96). Alternatively, stochastic forces such as sampling drift can knock the population from one peak to another (71, 72, 193, 195, 196, 203, 204), as shown in Figure 1b. At the risk of oversimplification, we will call these two modes of divergence Fisherian and Wrightian respectively.

This distinction may seem identical to Templeton's division of speciation mechanisms into divergence and transilience modes (176, 177, 179, 180): "A speciation mechanism will be classified as a transilience if the isolating barriers depend on a genetic discontinuity characterized by extreme instability of the intermediate stages. . . . It is characterized by overcoming some selective barrier" (177, p. 720). We feel, however, that the term *transilience* is not clearly defined and should be avoided. First, the word suggests that the transition between equilibria is necessarily rapid: "Genetic transilience is defined as a rapid shift in a multilocus complex" (176, p. 1013). But stochastically driven changes are likely to be as slow or slower than deterministic ones. The rate of change is set by the strength of selection pressures (i.e. the steepness of the adaptive landscape) and the effective population size. To take an extreme example, the random fixation of a new neutral mutation takes $4N_e$ generations on average (N_e stands for effective population size) (93), which is a very long time in a large population. In contrast, the fixation of a strongly advantageous

Table 1 (cont.)

Classic allopatric[d]	Parapatric[e]	Stasipatric[f]	Shifting-balance[g]	Sympatric[h]
Changing selection; accumulation of different mutations	Spatial variation in selection	Drift; meiotic drive	Drift; fluctuations in the adaptive landscape	Disruptive selection
Weak to moderate	*[i]	Moderate	Weak to moderate	Strong
One or a few	One or a few	One	A few	One or a few
New mutations; polygenic	*	New mutations	Polygenic	Frequency-dependent polymorphism
Allopatric	Parapatric	Parapatric	Allopatric	Sympatric
*	Parapatric	Parapatric	Parapatric	Sympatric
*	Spread into sympatry	Moving hybrid zone	Extinction-recolonization	*
*	*	Chromosomal	*	Habitat choice

[d] See, for example, 133.
[e] See 45, 58, 64, 102.
[f] See 188, 189.
[g] See 193.
[h] See 116.
[i] An asterisk indicates that the feature is irrelevant, ambiguous, or unspecified.

mutation under selection may take only a few tens of generations. Second, it is not clear what is meant by *discontinuity*. Templeton is not referring to macromutations or to saltatory evolution generating phenotypic discontinuity between successive generations. Finally, he appears to include a series of different models under the heading of transilience, most of which are not driven by stochastic forces and do not require selective barriers to be overcome (179, 180).

Mayr, Carson, and Templeton each place a somewhat different emphasis on the various mechanisms that may be involved in founder effect speciation: loss of variability in the genetic background, relaxation of selection during population expansion, and deviations of genotypic frequencies from Hardy-Weinberg expectations, as well as the stochastic peak shifts just mentioned. These make up only a small portion of the possibilities mapped out in Table 1, and we will show that the probability that strong reproductive isolation evolves in a single founder event is low under most circumstances.

HOMOZYGOSITY OF THE GENETIC BACKGROUND   Mayr (120–124) strongly emphasizes the loss of variability at many loci during a founder event. This loss is the result of sampling drift (192), although Mayr avoids using this term and believes that "the two processes are entirely independent of each other" (124).

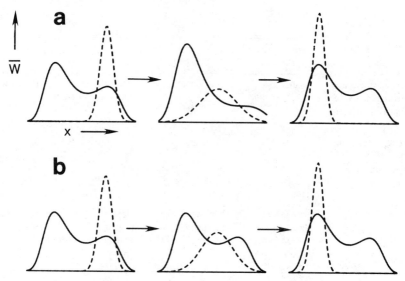

Figure 1 Disruptive selection acts on a character, x, so as to give a bimodal individual fitness of W = exp $[0.5x^2(2-x^2)-0.4x(1-x^2/3)]$. Genetic variability is based on 10 additive loci and is maintained by a mutational variance of 10^{-4} per generation; it evolves according to Felsenstein's and Kirkpatrick's model (62, 96). The solid lines show population mean fitness as a function of the character mean, making allowance for the effects of mutation and selection on the variance. The distribution of x in the population is shown by a dotted line.

In (*a*) a change in selection allows the population to evolve to the higher peak. The adaptive landscape then reverts to its original form. The reproductive isolation produced can be measured by the ratio between $\overline{W}$ at the lower peak and $\overline{W}$ of an F_1 hybrid population (87%) or $\overline{W}$ of an F_2 hybrid population (96%). The distance between the initial peak and the saddle is 2.2 standard deviations (s.d.), and the total shift is 10 s.d.

In (*b*), random drift knocks the population from one peak to the other. If N_e is reduced from a large number to 10 individuals, the expected time for a shift to occur will be very long: 303,000 generations (Equation 2).

As we show below, however, this factor is unlikely to be very important, since severe loss of variability can occur only in populations so small that they are likely to go extinct and in which genetic divergence is impeded by reductions in variability.

In his figure illustrating the effect of a founder event on genetic variability (121, Figure 17.3), Mayr assumes that the population bottleneck lasts only a short time and is followed by a rapid return to large numbers. The deleterious effects of prolonged inbreeding due to restricted population size (192, 204) and the high chance of stochastic extinction of small populations (8, 104) mean that successful founder populations will indeed usually be those that rapidly return to large numbers. The effect of a founder event on variability (as measured by the heterozygosity at neutral loci or the additive genetic variance of neutral quantitative characters) can be quite small under these circumstances (35, 100,

108, 138, 176). Consider, for example, a hermaphroditic organism with a binomial variance in offspring number. Let N_n be the number of individuals and H_n the heterozygosity at a neutral locus (i.e. the expected proportion of heterozygous individuals), n generations after the foundation of the isolate. In the absence of new mutations, we have (52)

$$H_{n+1} = \left(1 - \frac{1}{2N_0}\right)\left(1 - \frac{1}{2N_1}\right) \cdots \left(1 - \frac{1}{2N_n}\right) H_0 \qquad 1.$$

where H_0 is the heterozygosity in the ancestral population. When $N = 2$ (corresponding to foundation by a single fertilized female), the limiting values of the ratio of the final to the initial heterozygosity as $n \to \infty$ are 0.58, 0.45 and 0.06 if the rates of population increase (N_{n+1}/N_n) are 2, 1.5, and 1.1 respectively. In other words, only a very slow rate of increase after a founder event will significantly reduce the heterozygosity of the population. A similar result holds for the additive genetic variance of quantitative characters (24, 52), but here mutational variability should restore normal levels of variance after a few hundred generations (100), unless the population persists for a very long time at much reduced numbers.

It is therefore doubtful whether a founder event that succeeds in establishing a successful isolate will often alter significantly the action of selection at a locus because of the increased background homozygosity (100), particularly as loci under selection tend to resist the stochastic effects described above [as has been experimentally substantiated (159, 172)]. This resistance may be due either to direct effects of selection or to an apparent heterozygote advantage (associative overdominance) induced by stochastically caused linkage disequilibria with selected loci (143, 168, 169).

Even if genetic variability were substantially reduced, it is hard to see why this would, in itself, knock the population from one equilibrium to another. Under both two-locus (43) and multilocus models of founder effects (Figure 1; N. H. Barton and B. Charlesworth, unpublished manuscripts), reduction in variability markedly *reduces* the frequency of stochastic peak shifts. This reduction occurs because genetic change under both drift and selection requires variation. It is therefore not at all clear that increased homozygosity can increase the chance of a peak shift. We suggest that the effect, if it exists at all, is not very significant in evolution.

RELAXATION OF SELECTION   In Carson's model of founder-flush speciation, "the disorganization of the closed system of variability is thought to be accomplished through a permissive populational condition wherein natural selection is temporarily relaxed" (29, p. 88). It is assumed that epistatic selection pressures, which normally eliminate unfavorable gametes produced by recombination between loci held in linkage disequilibrium, are relaxed during phases of population growth and/or crashes, so that rare gamete types increase in frequency. When the population crashes and passes through a bottleneck of small size, stochastic transitions to new equilibria are supposedly facilitated. Charlesworth & Smith (43) have studied the simplest formalization of this model in which selection acts in such a way that two alternative, polymorphic equilibria can be simultaneously stable in a two-locus genetic system. (Models of this sort have been widely studied deterministically (22, 75, 109).) They found that several generations with relaxed selection prior to a founder event are necessary if a peak shift with a significant effect on reproductive isolation is to occur with probability greater than 10%. (Populations that went to fixation at either locus were not counted as having experienced a peak shift.) Charlesworth & Smith assumed, however, that selection continues to act during the bottleneck. Since relaxation of selection leads to a much greater rate of variational loss, as well as increasing the chance that a polymorphic system will respond to drift, the frequency of peak shifts is unlikely to be much affected by this assumption. This expectation has been confirmed in simulations (B. Charlesworth, unpublished).

It could be argued that populations that have become fixed for one or both loci have the potential to evolve onto a new peak when new mutations arise after the founder event. [This may be what Carson has in mind when he refers to the "phase of reorganization" that supposedly follows the founder event (31).] But both Carson (31, p. 424) and Templeton (176, p. 1016) think that speciation by founder events does *not* involve much fixation of polymorphic loci. Indeed, loss of heterozygosity at enzyme loci is not observed in the rapidly speciating Hawaiian *Drosophila* (50, 156). Fixation of strongly selected alleles would greatly reduce the mean fitness of the population and hence its chances of survival. Furthermore, a mutational process of this sort cannot explain the supposed laboratory analogues of founder effect speciation. Finally, the steady accumulation of new mutations in large populations forms the basis for the classic theory of allopatric speciation by gradual divergence (133) and is hard to distinguish from a process of reorganization after a founder effect.

Even if many generations of population expansion usually precede founder events, as postulated in one version of Carson's theory (29, Figure 2), it is not clear that selection will necessarily be relaxed during such a flush. It is certainly to be expected that selection pressures may be drastically different during population expansion (37, p. 51; 154, p. 311), but it is hard to see why they should be weaker. There will still be variation in fitness, even though this variation may stem more from differences in fecundity than in viability.

All that can be said with much certainty is that the adaptive landscape is likely to change during a population expansion or crash, and this may well trigger peak shifts. Many factors—such as climatic changes, new mutations, and competitive pressures—may alter selection patterns, however, and there seems little reason to attach special importance to population flushes and crashes. The same argument applies to Mayr's idea (120, 121) that changes in background variability are an important cause of a distorted adaptive landscape.

DEVIATIONS FROM HARDY–WEINBERG PROPORTIONS   Templeton has suggested that the most favorable condition for a peak shift following a founder event is when "the founder event causes a rapid accumulation of inbreeding *without* a severe reduction in genetic variability" (176, p. 1011). This condition is allegedly most likely when the inbreeding effective number (N_{ef}) of the founder population is low but the variance effective number (N_{ev}) is high. The relevance of the distinction between variance and effective numbers is unclear to us. N_{ev} for a given generation determines the sampling variance in gene or gamete frequencies in the transition to the next generation and is related to the number of individuals in the progeny generation; N_{ef} determines the relationship between inbreeding coefficients in successive generations (equivalent to $1-H$ in Equation 1) and is related to the current number of individuals (52). But, as brought out by Equation 1, the process of genetic drift during a founder event involves the cumulative effect of a sequence of generations with restricted numbers of individuals, and the ultimate reduction in average heterozygosity and variance of gene frequencies both depend in the same way on the sequence of numbers $N_0, N_1, \ldots$ Templeton's conclusion is based on consideration of only a single generation involved in the process—the initial founder generation.

Furthermore, Templeton's suggestion that increased homozygosity at a single locus is created by reduced population size and can produce a shift in the direction of selection (176, Equation 5) confuses the effect of drift, which causes an increase in the average frequency of homozygotes across a set of populations, with the deviation from Hardy-Weinberg frequencies within a single population. As is well-known, the latter effect operates in the direction of an *increased* frequency of heterozygotes and is of the order of $1/N$ for a population of size N (24, p. 220). It will therefore dissipate rapidly as population size increases and is unlikely to be of much evolutionary significance.

It is possible, of course, for a founder population to change its mating system in the direction of a higher frequency of consanguineous matings than would be expected if mating were random, with a consequent change in the effects of selection at single loci (179, p. 21). This change is not necessarily connected with any possible reduction in effective population size, however. An effect of this sort is most likely for hermaphroditic or monoecious species, where the chance of successful colonization is greatest for self-fertile individuals (5, 6). An association between geographic and ecological marginality and self-fertilization is well established in flowering plants (110). Lewis (106, 107) has documented a corresponding association with chromosomal speciation in *Clarkia*, and such a relationship should be anticipated on theoretical grounds (76, 99, 195, 196) because of the smaller effects of rearrangement heterozygosity on fertility when homozygosity is increased by inbreeding (see below). A population's ability to change its mating system *without* the operation of founder effects is illustrated by populations of *Primula vulgaris* (the primrose) in southwest England; they contain high frequencies of the self-fertile homostyle form, which has a selfing rate of greater than 90%, in contrast to the normally self-incompatible heterostyled plants found elsewhere (51; J. A. Piper, B. Charlesworth & D. Charlesworth, unpublished data). In summary, there are no grounds for expecting a strong association between change in mating system and founder events, except in self-compatible hermaphrodites.

The Basis of Reproductive Isolation

The relative importance of each mechanism of divergence will depend on the genetic and selective basis of reproductive isolation, which determines the shape of the adaptive landscape the population must traverse. Indeed, the rate of divergence may depend more on the form of the landscape than on the particular mechanism that pushes the population from state to state. Here we will show that the chances of achieving peak shifts and reproductive isolation both depend in a simple way on the depth of the saddle separating the two peaks and that this allows us to make the general argument that genetic revolutions producing strong isolation in a single founder event are most unlikely.

THE PROBABILITY OF PEAK SHIFTS   The key parameters needed to describe a given isolating barrier are the number of independent steps by which it evolved, the number of loci that changed at each step, whether the changes were between polymorphic loci or between alleles close to fixation, and the selection pressures that acted at each step. This last feature is the most important and needs some clarification. We need two measures: first, the strength of the barrier to gene flow between populations at different peaks and, second, the rate of transition between the peaks. Remarkably, if the dynamics can be described by an adaptive landscape, these two measures are both given by the ratio between the mean fitnesses of a population in the saddle between the two peaks and at one of the peaks ($\overline{W}_s/\overline{W}_p$ in Equation 2). There are of course two such ratios, corresponding to the two peaks; each ratio is related to the rate of gene flow into populations at the appropriate peak and to the chance of moving out of that peak's domain of attraction.

The flow of neutral genes between two populations depends on the balance between the transfer of genes from one background to another by recombination in hybrids and the elimination of hybrids by selection. The barrier to gene flow presented by differentiation at a single locus has been calculated for exchange between two demes (18, 166) and for a continuous population (9). When many loci have differentiated, the barrier to gene flow is much stronger (11). In general, the rate of gene flow into a population is proportional to $(\overline{W_s}/\overline{W_p})^{1/r}$, where r is the harmonic mean recombination rate between the neutral gene and the selected loci (N. H. Barton & B. O. Bengtsson, unpublished). Therefore, populations will not be reproductively isolated unless the total selection pressure is at least of the same order as the harmonic mean recombination rate. Consider, for example, an organism with a dispersal range of 1 km, and a total map length of 10 Morgans. If $(\overline{W_s}/\overline{W_p}) = 0.5$ as a result of selection on 10 loci, the barrier to gene flow is equivalent to a distance of only 25km: If selection acts at 1000 loci, the barrier is equivalent to 7700 km (13, p. 352).

The probability of a peak shift as a result of sampling drift is proportional to $(\overline{W_s}/\overline{W_p})^{2N_e}$ under rather general conditions. Wright (192, 194) showed that the probability that a population will be found at a given point on an adaptive landscape, with respect to a single locus or a set of independent loci, is proportional to $\overline{W}^{2N_e}$; Lande (98, 100) has extended this formula to continuous characters. The probability of a peak shift must have the same form as the equilibrium probability distribution, since it depends on the chance that a population will be found at or beyond the saddle. It is possible to make this argument more precise: The chance per generation of escaping from the attraction of an adaptive peak is

$$\frac{\lambda_s}{2\pi} \sqrt{\frac{|D_p|}{|D_s|}} \left(\frac{\overline{W_s}}{\overline{W_p}}\right)^{2N_e} \qquad\qquad 2.$$

where λ_s is the leading eigenvalue at the saddle (i.e. the rate at which the population moves away from the unstable equilibrium at the saddle) and D_p, D_s are the determinants of the stability matrices at the peak and saddle respectively [26, Equation (44)].

Just how likely is a drastic peak shift? The most intensively studied example is heterozygote disadvantage, which is often caused by chromosome rearrangements (19, 76, 99, 163, 186, 196). A reciprocal translocation causing 50% sterility would have a probability of fixation of only 0.00082 in a population with $N = 10$ (19). White (188, 189) has suggested that either meiotic drive or selection favoring one homozygote will increase the chance of a peak shift (76, 163). Such asymmetries inevitably weaken the level of isolation, however, since they reduce the depth of the trough separating the peaks. We are left with the conclusion that significant isolation is unlikely to evolve in a single step by this mechanism (186).

A similar conclusion was obtained from simulations of a model of two interacting loci (see above and 43). Factors that facilitate peak shifts (e.g. loose linkage between the loci, weak selection) tend to lower the degree of isolation. The same is found in studies of epistasis between loosely linked loci and of disruptive selection on continuously varying characters of the sort shown in Figure 1 (98, 100; N. H. Barton and B. Charlesworth, unpublished manuscripts). It seems unlikely that substantial reproductive isolation can evolve in a single step through genetic drift in a wide variety of genetic models.

We can take this argument further by showing that reproductive isolation is much more likely to evolve in a series of small steps than in a single genetic revolution. The number of peak shifts required to achieve a given level of isolation varies inversely with the size of each step, but the probability of each step occurring decreases exponentially; the exponential term will dominate. Walsh (186) has presented this argument for the case of heterozygote disadvantage, but it is not difficult to generalize it using Equation 2.

This argument does not indicate anything directly about the number of genes that might be involved in each step, and even a weak peak shift could involve a large number. Templeton's claim (176, 178, 180) that the genetic system most favorable to a peak shift, involves one or a few major genes interacting with a large number of modifiers. This claim is not based on any quantitative model and seems inconsistent with the argument just presented; a strongly selected locus is the one *least* likely to respond to genetic drift. Furthermore, it is hard to see how modifier loci with slight phenotypic effects can be held in strong linkage disequilibrium with the major loci by selection (59, Chaps. 6, 7), as is required for a peak shift with a significant effect on reproductive isolation.

Templeton's (177, 180) and Barton & Hewitt's (12) argument that random changes in a polygenic system tend to average out and that multilocus shifts are therefore unlikely is wrong; this can be seen by considering the effect of random drift on an additive polygenic character (100). The variance in means among isolated populations is independent of the number of loci and at most equals twice the additive genetic variance in the ancestral population (24, 198). It is unlikely, however, that the mean of an isolate will diverge from the ancestral mean by more than twice the standard deviation of the isolates. Thus, the probability that the mean of the founder population will be more than one phenotypic standard deviation from that of the original population is low, even with a heritability as high as one half. A founder event can produce a dramatic perturbation of the population mean only if the genetic variance of the ancestral population is a product of low frequency recessive genes (52, 152). As Lande (100) noted, however, such low frequency recessives are likely to be deleterious, and selection will oppose their spread. The cumulative joint effects of drift and mutation over a long period can lead to a much greater stochastic change in the population mean than a single founder event (100).

QUASI–NEUTRAL MODELS   We have shown that there is an inverse relationship between the strength of the reproductive isolation set up by a peak shift and the rate at which a population will make the shift. This proposition is apparently contradicted by models of the kind proposed by Nei (137), Bengtsson & Christiansen (20), and Nei et al (139). Mutant alleles may arise at each of two loci and have deleterious effects only when they are present at both. In the simplest case, there is only selection against double homozygotes. Since there is no great obstacle to the fixation of mutant alleles at either locus on its own, two isolated populations may fix mutant alleles at different loci. Yet when two such populations hybridize, $1/16$ of the F_2 will be double homozygotes and $9/16$ will have some mutant alleles at each locus, so that there may be a substantial initial reduction in hybrid fitness. Selection, however, will soon eliminate both mutant alleles from the hybrid population, and the level of isolation will soon fall to the feeble level expected from the shallow saddle between the two equilibria. (This raises the point that there may often be a considerable difference between the degree of isolation observed in the F_1 or F_2 between two populations and the final level of reproductive isolation established between two hybridizing populations in nature.)

Wills (191) and Nei et al (139) have proposed a formally similar model to explain ethological isolation. Males homozygous (aa) at one locus are unable to mate with females homozygous (bb) at a second locus. In a large population, both a and b will be kept rare by selection, but genetic drift can lead to the fixation of a in some isolates and b in others, thereby producing sexual isolation. No isolation between the ancestral and derived isolates is expected in either of these models, however, which has a significant impact on interpretations of the observational evidence (see the section on Ethological Isolation).

Lande (101) and Kirkpatrick (95) have analyzed models of the coevolution of a male phenotype or genotype and the level of female mating preference it engenders. Genetic variability in both traits results in a line of neutral equilibria with respect to the mean male phenotype and mean female mating preference, such that populations converge to the line but are free to drift along it without any opposition from selection. Lande points out that a small population may move so far along the line that it acquires a combination of male phenotype and female preference that effectively isolates it from the ancestral population or from other isolates. It seems unlikely, however, that a strong degree of isolation could be produced by this process in a single founder event. In Lande's model, the population evolves neutrally only along the line, so that the scope for a large founder effect is even more limited than for the neutral evolution of a quantitative character, which has been discussed above.

Finally, we stress that—in both this and the Nei/Wills model—gradual divergence between populations of moderate size can be much more effective than founder events in promoting isolation, given sufficient time, since the depletion of variability by drift can come into balance with its generation by mutation (100, 101, 139). As in the cases considered in the previous section, there is no logical reason for assigning a special evolutionary role to founder events.

GENETIC CONSTRAINTS ON EVOLUTIONARY CHANGE   Speciation is therefore more likely to involve many shallow peak shifts than a few strong ones. But Mayr's primary argument in favor of genetic revolutions is that the "cohesion of the genotype" rules out gradual change in large populations. Although there is much evidence for epistatic interactions in the fitness effects of genes segregating in natural populations (41, 53, 54, 164, 174), they need not place strong constraints on the potentiality for evolutionary change, as assumed in Mayr's concept of the "unity of the genotype" (121, Chap. 5). Epistasis (i.e. deviations from strict additivity of the effects on fitnesses of different loci) simply means that genes are selected for their ability to increase the fitness of their carriers on the current genetic background contributed by other loci (23, 132, 133, 192, 204).

Given sufficiently close linkage between loci that interact in their fitness effects, nonrandom associations of alleles at different loci (i.e. linkage disequilibria) can be maintained by selection, in opposition to recombination (23, 38, 59, 64, 77). This possibility, however, does not preclude evolutionary responses of genotypic frequencies as a result of changed selection pressures (24, 36, 115, 117).

Geographic Relations

Modes of speciation are usually classified as allopatric, parapatric, and sympatric. These terms are primarily descriptions of distribution patterns: Two forms are allopatric if they are never found together, parapatric if they meet and coexist in a region not greatly wider than their dispersal range, and sympatric if they coexist as distinct forms over a substantial area. Key (90, 91) discusses these distinctions clearly, stating that, as applied to speciation mechanisms, "these terms mean simply speciation taking place to completion under the respective geographical conditions" (91, pp. 430–32). Unfortunately, it is not simple to apply them in practice, since geographical relations usually change during evolution toward a new equilibrium (58). It is necessary to distinguish relations during the establishment of a new type from those that prevail during its spread. We will argue that the distinction between allopatry and parapatry is unclear at both stages and that speciation does not require complete isolation during a founder event.

Mayr (124) distinguishes between the "dumbbell" model of allopatric speciation (in which two large, disjunct areas diverge) and the "peripatric" model (in which a small isolate diverges). He argues that peripatric speciation predominates because widely separated populations of a continuously distributed species often diverge less than highly isolated portions. But Mayr and others accept that divergence need not occur only at the edge of the species range; it may also occur in an "internal periphery," a central region that is isolated from the surrounding population (29, 124). This possibility makes it hard to distinguish between allopatry and parapatry. Furthermore, even in a spatially homogeneous, continuous but finite population, there is always a chance that a region will not exchange migrants with its neighbors. Thus, any finite population could be divided into a mosaic of temporarily "allopatric" areas.

It is also clear that gene flow may have much less effect in reducing divergence and preventing speciation than the proponents of founder effect models suppose. Extensive theoretical studies have shown that subdivided populations can differentiate as a result of either deterministic or stochastic forces. Many studies of spatial variation in selection pressures (11, 45, 58, 64, 65, 70, 73, 82, 102, 135, 136, 160–162, 192) show that a stable cline in genotype frequencies can be maintained if selection varies in direction over a region larger than some critical scale. The scale is set by the variance in distance between the places of birth of parent and offspring (σ^2), and by the typical strength of selection (s). The critical scale will be some small multiple of the dispersal distance ($l \cong \sigma/\sqrt{s}$), and weak selection pressures can maintain substantial geographic differentiation. Consequently, weak spatial changes in the adaptive landscape can induce peak shifts and hence the evolution of reproductive isolation, even in a continuous population (64, 102).

The effects of stochastic forces such as sampling drift are more complicated, but divergence *is* possible despite gene flow. In a two-dimensional population divided into discrete demes that exchange individuals with their neighbors, the importance of random drift depends on the number of individuals exchanged per generation (61, 94, 112). In continuous, two-dimensional populations, it is controlled by the *neighborhood size*, $N = 4\pi\sigma^2 d$, proportional to the dispersal rate σ^2 and population density d (112, 197). The variance of allele frequency in a polymorphism under selection is inversely related to N and does not depend strongly on the strength of selection.

Can a new peak establish itself by drift within subdivided or continuous populations? First, consider a continuous population. If the population moves towards a new peak over too small a region ($<l$), then gene flow will swamp it. If it can establish itself over an area greater than some critical size ($\cong l$), however, it may spread outwards as a result of some intrinsic advantage or local population expansion. The chance that a new peak will arise and spread therefore depends on the chance that drift will establish it in a region sufficiently large to resist swamping. Surprisingly, the chance of divergence depends primarily on neighborhood size and only very weakly on the strength of selection (N. H. Barton, unpublished manuscript). A weakly selected shift is

more likely to occur within a given area than a strongly selected one, but it is less able to resist the swamping effects of gene flow once it is established. Models with discrete demes behave similarly: Provided that the number of individuals exchanged between demes is small, alleles subject to heterozygote disadvantage can become established and then spread by local extinction and recolonization (99, 163, 195). The general conclusion is that complete isolation is no more *necessary* for stochastic than for deterministic divergence. Provided that the neighborhood is small, peak shifts may occur without the aid of physical barriers to migration. As Wright has repeatedly emphasized in his shifting-balance model, the existence of numerous local populations within a large species range provides the opportunity for many independent stochastic trials and increases the chance that the species as a whole will eventually experience a peak shift (193, 200, 202–205).

EVIDENCE

The Genetic Structure of Populations

Carson's and Templeton's models require the existence of multilocus polymorphisms with epistatic interactions in fitness, so that strong linkage disequilibrium is maintained. Although there is abundant experimental evidence for such epistatic interactions, it is questionable whether these effects are likely to be strong enough to generate a significant degree of reproductive isolation in a single founder event. The fitness depression of the F_1 hybrid between populations located at two different equilibria under this form of selection is related to the variance in fitness generated by departures from linkage equilibrium. Thus, if recombination and epistasis are high enough to produce significant "disorganization" (31, p. 424) during a founder event, there will necessarily be an excessive variance in fitness in the original population. Consider, for example, a two-locus system in which the equilibrium frequencies of the gametes AB, Ab, aB, and ab are x_1, x_2, x_3, x_4; the corresponding marginal gametic fitnesses are W_1, W_2, W_3, W_4 (52, p. 217). In the absence of linkage disequilibrium, the marginal fitnesses are all equal. In general, the component of the total genetic variance in fitness ascribable to linkage disequilibrium is $\sigma_g^2 = 2 \Sigma x_i (W_i - \overline{W})^2$, where $\overline{W}$ is the population mean fitness ($\overline{W} = \Sigma x_i W_i$). In many two-locus models, there are two equilibria with similar gene frequencies at each locus but that differ in the direction of linkage disequilibrium. It is easily shown that in these cases the fitness of the F_1 hybrid formed by combining gametes at random from a pair of populations located at each equilibrium is reduced by no more than $\sigma_g^2/(r\overline{W}^2)$ below that of the parental populations, where r is the frequency of recombination between the two loci (B. Charlesworth, unpublished manuscript).

Mukai and his group (131) have measured $\sigma_g^2/r\overline{W}^2$ for egg-to-adult viability in

Drosophila melanogaster and arrived at a value of about 0.01 to 0.02. Similar values have been obtained for adult longevity and female fecundity (39, 153). Even if this variance were contributed by a single pair of loci, a substantial reduction in F_1 fitness would require tight linkage, which reduces the chance of a peak shift (43). If the variance is distributed over several sets of interacting loci, the chance of a peak shift with a substantial effect on reproductive isolation is severely reduced (43). Furthermore, the selection pressure for devices that restrict recombination, such as inversions, is proportional to σ_g^2 (38, 40). Therefore, strongly epistatic systems of this sort are most likely to evolve restricted recombination *before* founder effects occur and hence to have a low chance of undergoing peak shifts. Inversion polymorphisms are, of course, a prominent feature of the Hawaiian *Drosophila* fauna (28, 32, 33), from which much of the biogeographic data for founder effect speciation has been obtained.

Genetic constraints on peak shifts involving quantitative characters subject to multiple selective peaks are not so tight. All that is needed is a supply of genetic variance. There is, however, no evidence at present that such multiple peaks actually exist. Similarly, Wills's (191) and Lande's (101) ethological isolation models simply require variation in phenotypes affecting male mating success and in female preference for them. This condition seems plausible, although there is little direct evidence as far as female preference is concerned (but see 141).

Genetic Constraints

In the section on Genetic Constraints on Evolutionary Change, we showed that there are no strong theoretical reasons for expecting genetic constraints to inhibit evolutionary change. Several lines of experimental and observational evidence support this conclusion. Selection for chloramphenicol resistance in asexually reproducing populations of *Escherichia coli* (34), for proflavine resistance in bacteriophage T4 (113), and for adaptation to chemostat conditions in yeast (144) has proved effective in the face of strong epistatic interactions between the genes concerned. Mimicry in butterflies provides a good example of how adaptive evolution can proceed when a given locus's effect on fitness depends strongly on its genetic background (182, 183). In the case of Batesian mimicry, frequency-dependent epistatic selection often results in the maintenance of strong linkage disequilibrium between tightly-linked, polymorphic genes (44, 47, 48, 182), but this has not prevented the evolution of detailed mimetic resemblances. With Müllerian mimicry, selection is directional and causes allelic replacements rather than polymorphism, but considerable geographic differentiation within mimetic species can evolve by convergence to different mimicry rings (182, 183). Other examples of the evolutionary adjustment of gene effects on fitness by epistatically interacting modifiers come from observations on pesticide resistance in sheep blowflies (126) and copper tolerance in *Mimulus* (111).

Epistasis is, of course, not universally observed or expected, and many quantitatively varying characters show little evidence of strong epistasis between the genes controlling them. Numerous artificial selection experiments have shown that populations are able to respond to selection (60). Given sufficiently large population size and moderate selection pressure, so that variability is not rapidly exhausted, such responses may continue almost indefinitely as new variability is generated by mutation and exploited by selection (78, 206). Selection can be effective even on characters that are strongly developmentally canalized and normally show little variation, such as scutellar bristle number in *Drosophila* (146, 190), petal number in *Linanthus* (79), and spine number in sticklebacks (69). For these reasons, it seems unlikely that the geographic uniformity of some species, emphasized by Mayr, could be due to any inherent inability to respond to diversifying selection pressures. It probably reflects similar selection pressures across the species' range, coupled with the homogenizing effects of gene flow (42).

Genetic Differences Between Species

We have argued that speciation by the random crossing of adaptive valleys is most likely to involve several relatively small successive steps, rather than a single event. The same is probably true of Fisherian, purely adaptive, divergence (64, Chap. 2). Hence, we would normally expect the genetic basis of reproductive isolation to involve several loci or chromosome rearrangements. If, however, the Carson/Templeton mode of founder effect speciation (27, 29, 31, 176, 177, 180) were predominant, one might expect isolating mechanisms to have relatively simple genetic bases; Templeton (178, 180) has argued that founder effects are unlikely to cause the evolution of multifactorial barriers. But strong inferences about the mechanism of speciation cannot be made from the genetics of species differences. As we showed earlier, founder effect speciation *can* involve large numbers of genes, and there are examples of adaptive evolution where major genes and epistatic modifiers have been important (103, 111, 182, 183).

Surveys by Muller (133), Stebbins (167), Dobzhansky (55), and Maynard Smith (118) have shown that postzygotic isolating barriers are usually controlled by many loci. What little is known about the genetics of ethological isolation seems to support this conclusion (57, 207). Templeton (178) has argued, however, that isolation may often be based on a few major factors. While there is obviously a continuum of effects, we judge that multifactorial modes predominate. Much of the evidence comes from experimental crosses, which tend to underestimate the number of factors that contribute significantly to reproductive isolation for two reasons. First, recombination during the few generations studied in such experiments cannot break up linkages effectively (60, p. 201). For example, although the hybrid inviability of the F_1 between *Drosophila melanogaster* and *D. simulans* may depend on only 9 loci (178), it may depend on many more; Pontecorvo's (147) ingenious experiment could only measure the effects of whole chromosomes. In any case, it is hard to draw a clear line between a large number of minor genes and a few major factors with modifiers; for example, the difference in head shape between *D. heteroneura* and *D. silvestris* (173, 184) has been variously attributed to the first (42) or the second (178) category. A second problem is that minor genetic factors may contribute significantly to isolating mechanisms in nature [11; Equation (2)], since a given amount of selection forms a more effective barrier to gene flow when it is spread over many loci. The effects of minor factors can easily be obscured by the segregation of a few major factors in experimental crosses.

Fortunately, our knowledge about the genetic basis of reproductive isolation is not confined to laboratory crosses. Populations that hybridize in nature often differ extensively; for example, an average of 20% of the electrophoretic loci have diverged across a sample of 21 hybrid zones (13). Morphological, chromosomal, and behavioral differences are also common. Several lines of evidence indicate that hybrid zones often involve substantial reproductive isolation and that this is usually multifactorial. Researchers have not been able to identify any single difference that is responsible for the maintenance of any zone; in particular, meiotic nondisjunction in chromosomal heterozygotes does not usually cause a significant proportion of the isolation (10, 49, 140, 145, 157, 158). Hybrid zones are often wide relative to the dispersal range, even when hybrids are very unfit (10, 13, 158). This breadth indicates that, although total selection is strong, selection at each individual locus is weak, so that many loci must be involved. Finally, hybrid zones are often sharply stepped (7, 21, 81, 129, 170, 171), a pattern produced by linkage disequilibrium between clines at many loci (11, 13).

The evidence from both experimental crosses and hybrid zones shows that reproductive isolation usually involves many genes; this excludes Goldschmidtian macromutations as a mode of speciation (42, 118) and suggests (but does not prove) that isolation is built up in a series of small steps.

Ethological Isolation

Kaneshiro (85–87) has proposed a model of founder effects in which males of the founder population lose elements of their courtship behavior and so are discriminated against by females of the ancestral population. Males from the ancestral population are not discriminated against and may even be preferred by females from the derived population. Kaneshiro, as well as Giddings & Templeton (66), claim that asymmetrical isolation of this kind is consistent both with established phyletic relations between populations, particularly in the

Hawaiian *Drosophila*, and with laboratory experiments on founder effects (1, 3, 148).

We note first that this model is based on the assumption that deleterious genes are fixed by drift and therefore requires the maintenance of a small population for such a long time that weakly selected loci should be severely affected. There is no evidence for such effects on the enzyme loci of the Hawaiian *Drosophila* (50, 156). Second, there is a serious methodological difficulty with much of the data on which the claim of asymmetrical isolation is based. Most of the relevant experiments involve "male-choice" tests; if two strains or species, A and B, are being tested, mating chambers are set up with trios of either A males with A and B females or B males with both types of females. The numbers of matings by the two sorts of females are recorded. But, as first pointed out 30 years ago by Bateman (14) and commented on by some later authors (114), these experiments confound the relative mating propensities of A and B females with mating preferences. Differences in the frequencies of homogamic matings (A × A and B × B) between the tests with A and B males, which Kaneshiro attributes to differences in female preference, may arise if one type of female simply mates more frequently than the other.

Such tests often show a significant excess of *heterogamic* (A × B) matings in one class of trio and an excess of *homogamic* matings in the other (1, 14, 88). This result is unintelligible in terms of female preference, even with Kaneshiro's asymmetry model, and is most simply explained by a difference in mating propensity between A and B females. Most of the experiments have been done with inbred laboratory stocks, and we know that there is extensive variation in female and male mating success among inbred lines of *D. melanogaster* (185). Ahearn (1) marked one of the two types of females she was studying with paint; the pattern of asymmetry she describes is consistent with lowered mating success for the marked females.

Thus, there is no firm basis for the asymmetry model. It may, of course, seem odd that there is an apparent association between phylogeny and female mating success; the above interpretation would require the ancestral females to have lower mating success than the derived ones. But, in at least some of the cases used to support the asymmetry model, the direction of evolution has not been unequivocally established on independent grounds (88, 142). The number of Hawaiian *Drosophila* populations with a known ancestor-descendant relationship tested in these studies is too small to provide firm statistical support for an association. Furthermore, asymmetries exist in cases where there is no obvious correlation with geographical distribution (as in Dobzhansky & Streisinger's (14, 56) results with *D. prosaltans*) or when plausible alternative explanations can be proposed, such as with isolation between the races of *D. mojavensis* (187, 208) or between the Canadian populations of the stickleback *Gasterosteus aculeatus* that have black and red males (127, 128).

It could still be argued that experimental founder effects have been shown to induce ethological isolation (1, 3, 148). Only Powell's data and the unpublished data of J. Ringo, D. Wood, R. Rockwell, and M. Dowee (on *D. simulans*) seem to withstand the methodological criticism made above. But, as Charlesworth et al (42) noted, they used stocks of geographically mixed origin, so it is not clear whether their results are representative of what might happen in a natural population. Furthermore, Powell's lines that were passed through bottlenecks did not show much isolation from the ancestral stock, although they were isolated from each other to some extent. This finding is more consistent with Wills's (191) model than with founder effect models that require overcoming a selective barrier.

Although evidence for partial sexual isolation between inbred lines of *Drosophila* (92, 97) certainly exists, there are also data indicating that isolation can develop between large populations exposed to different selective regimes (92). It does not seem to us that there is good evidence that small population size plays a special role in promoting ethological isolation.

Geographical Patterns

Mayr's observation that species are often quite uniform across the central part of their ranges yet greatly divergent in peripheral isolates led him to propose his theory of founder effect speciation, and it is one of his main arguments in its favor (121, Chaps. 11, 16; 125). Perhaps the most striking example of speciation involving isolates is the massive radiation of the Hawaiian Drosophilidae, which has stimulated Carson's and Templeton's theories. Geographic patterns and examples of rapid speciation may be the best evidence we have concerning mechanisms of divergence. However, although they do suggest that the foundation of new populations is frequently accompanied by speciation, at least in the Hawaiian *Drosophila* (28, 31–33), we do not think that they support models of drastic genetic change caused by population bottlenecks.

For example, there is no direct evidence that the postulated bottlenecks have had striking genetic effects, since the proportion of polymorphic enzyme loci and their average heterozygosity seems to be just as high in Hawaiian *Drosophila* species of recent origin as in continental species (50, 156, 176). The two recent species *D. heteroneura* and *D. silvestris* are very similar with respect to both electrophoretic allele frequencies (50) and DNA sequences (80), which is inconsistent with the hypothesis of drastic bottleneck effects (35). Since the loss of heterozygosity in a small population and the transition to a new adaptive peak are both caused by the single process of random drift, population bottlenecks small enough to cause peak shifts will inevitably cause a substantial and prolonged loss of variability at neutral loci (35, 138).

Thus, although the Hawaiian Drosophilidae have speciated very rapidly and their evolution has clearly involved numerous founder events, the causal link is unclear. The fact that most species are endemic to particular islands does not imply that speciation is triggered by founder events but simply that speciation is fast relative to the rate of interisland colonization. There are several other factors that might cause rapid speciation. The flora and fauna of the Hawaiian islands are quite different from those in the habitat of continental *Drosophila*, and the flies have evolved unusual adaptations to cope with these novel selective pressures (30, 33). Their ecological requirements are often highly specific, allowing many species to live in sympatry; the importance of niche availability is shown by the fact that species colonize new islands more often than older ones, which are already fully occupied (28, 32). Even in the absence of ecological differences between habitats, the foundation of a new isolated population may allow divergence from the ancestral population through the chance accumulation of different mutations by drift or directional selection, without any drastic bottlenecks (132, 133). In the absence of isolation, such mutations would spread throughout the population, so that uniformity would result. In the case of the Hawaiian *Drosophila*, strong sexual selection promotes both morphological and behavioral divergence (151, 165). Although it is possible to include the effects of sexual selection in founder effect models (31, 66, 85–87, 101, 175), sexual selection itself could easily be the primary factor promoting divergence (64, 101, 102, 151).

In other cases of rapid divergence, there seems no reason to invoke founder effects. For example, land snails of the genus *Partula* on Moorea have evolved a complex pattern of reproductive isolation, which mostly fits a parapatric interpretation (46, 134). Some populations are isolated from the rest of the species range and diverge with respect to electrophoretic loci, as would be expected from founder effects; however, no reproductive isolation or morphological divergence is seen in these cases (84). A similar example is provided by the Venetian population of the marine snail *Littorina saxatilis*, which is divergent and genetically depauperate with respect to enzyme loci but is morphologically identical with Northern European populations (Janson, unpublished data). In one case in *Partula*, the likely mechanism of divergence can be inferred: Dextral coiling of *P. suturalis* has evolved in response to sympatry with sinistrally coiled *P. mooreana*. Since snails with opposite coiling are less likely to fertilize one another, dextrality has probably been selected in *P. suturalis* in order to reduce the frequency of matings with *P. mooreana* (46, 83).

In general, founder events should be associated with differences in ecology and selection pressures. This principle applies to "peripatric" patterns on continents as well as to islands (68, 183) and makes it difficult to distinguish mechanisms involving genetic bottlenecks from those involving responses to environmental change. Furthermore, it is often hard to be sure that present-day distribution patterns reflect the geographic relations existing at the time of divergence. For example, it is frequently argued that hybrid zones are best explained by secondary contact between populations expanding out from allopatric refugia (4, 81, 121, 158, 183). This view has been challenged by Clarke (45), Endler (58), and White (189), who have shown that clines in multiple characters can arise and persist in populations that have always been in contact. Even if the majority of present-day clines are the result of secondary contact, however, the differences did not necessarily arise first in allopatry. Provided that divergence is slow relative to population expansion and contraction, one would expect secondary contact to determine geographic patterns, even if the differences had arisen in parapatry.

The relation between population structure and speciation has also been inferred from comparisons between groups that speciate at different rates (17, 25, 105). There is a broad correlation between rates of speciation and of chromosomal evolution, which has led to the suggestion that the random fixation of chromosome rearrangements in small populations plays an important causal role in speciation (25, 105). As Lande (99), Bengtsson (17), and Charlesworth et al (42) point out, alternative interpretations can be proposed for such wide comparisons. Such problems can be avoided to some extent by comparing closely related species. Gold (67) found no evidence for a correlation in cyprinid fishes, whereas Ferris et al (63) conclude that speciation is associated with a faster rate of loss of duplicate gene expression in the tetraploid Catastomidae. The picture is therefore somewhat unclear.

CONCLUSIONS

It is very difficult to make inferences about the causes of a historical process like speciation from present-day information such as species distribution patterns (58, 189). We have therefore relied heavily on theoretical arguments in our criticisms of founder effect models, but we have also attempted to show that the available empirical evidence does not provide strong support for a major role for founder effects in speciation. Our conclusions are:

1. There are no empirical or theoretical grounds for supposing that rapid evolutionary divergence usually takes place in extremely small populations, that complete geographic isolation is necessary for speciation, or that epistatic fitness effects or developmental constraints frequently impede the effectiveness of selection in large populations.

2. Theoretical models of founder-effect speciation show that, under a wide variety of assumptions, the probability that a founder population will undergo a stochastic transition to a new selective equilibrium, causing significant reproductive isolation from its ancestral population, is low. A prolonged process of divergence at moderate population size can produce the same final degree of isolation as in a small population over a shorter period; it is, indeed, easier to

achieve a given level of isolation by crossing a series of small selective barriers than by crossing a single large one. Even with purely neutral divergence, as in the Lande/Kirkpatrick model of sexual preference, a prolonged period of genetic drift is probably necessary for strong isolation to evolve. There is, therefore, no reason to assign special significance to founder effects, even if stochastic forces are envisaged as promoting speciation.

3. The strongest biogeographic evidence for founder effect speciation comes from a general correlation between peripheral isolation and phenotypic divergence, and in particular from the Hawaiian *Drosophila,* in which colonization of new islands is almost invariably associated with speciation. The generally small chance of achieving reproductive isolation or marked phenotypic change in a single founder event means that founder effects themselves probably do not provide the explanation. It is impossible to separate the effects of isolation, environmental differences, and continuous change by genetic drift from the impact of population bottlenecks in these cases. Since all of these factors promote divergence by a variety of processes, it is not clear that the additional influence of founder events need be invoked.

ACKNOWLEDGMENTS

We thank Professor B. C. Clarke, Dr. K. Janson, Dr. R. Rockwell, Dr. M. J. van den Berg, and Dr. R. D. Ward for making their work available before publication. Dr. S. J. Arnold and Dr. L. Partridge provided us with useful references, and Dr. D. Charlesworth, Dr. J. M. Szymura, and Dr. J. S. Jones critically reviewed the manuscript.

Literature Cited

1. Ahearn, J. N. 1980. Evolution of behavioral reproductive isolation in a laboratory stock of *Drosophila silvestris.* *Experientia* 36:63–64
2. Akin, E. 1979. *The Geometry of Population Genetics, Lect. Notes Biomath.,* ed. S. Levin, No. 31. Berlin: Springer-Verlag
3. Arita, L. H., Kaneshiro, K. Y. 1979. Ethological isolation between two stocks of *Drosophila adiastola* Hardy. *Proc. Hawaii. Entomol. Soc.* 13:31–34
4. Arntzen, J. W. 1978. Some hypotheses on postglacial migration of the fire-bellied toad *Bombina bombina* (L) and the yellow-bellied toad *Bombina variegata* (L). *J. Biogeogr.* 5:339–45
5. Baker, H. G. 1955. Self-compatibility and establishment after long-distance dispersal. *Evolution* 9:347–48
6. Baker, H. G. 1967. Support for Baker's rule—as a rule. *Evolution* 21:853–56
7. Baker, R. J. 1981. Chromosome flow between chromosomally characterized taxa of the volant mammal *Uroderma bilobatum* (Chiroptera: Phyllostomatidae). *Evolution* 35:296–305
8. Bartlett, M. S. 1960. *Stochastic Population Models.* London: Methuen
9. Barton, N. H. 1979. Gene flow past a cline. *Heredity* 43:333–39
10. Barton, N. H. 1980. The fitness of hybrids between two chromosomal races of the grasshopper *Podisma pedestris.* *Heredity* 45:49–61
11. Barton, N. H. 1983. Multilocus clines. *Evolution* 37:454–71
12. Barton, N. H., Hewitt, G. M. 1980. Hybrid zones and speciation. In *Evolution and Speciation: Essays in Honour of M. J. D. White,* ed. W. R. Atchley, D. Woodruff, pp. 109–45. Cambridge: Cambridge Univ. Press
13. Barton, N. H., Hewitt, G. M. 1983. Hybrid zones as barriers to gene flow. In *Protein Polymorphism: Adaptive and Taxonomic Significance,* ed. G. S. Oxford, D. Rollinson, pp. 341–60. London: Academic

14. Bateman, A. J. 1949. Analysis of data on sexual isolation. *Evolution* 3:174–77

15. Bazykin, A. D. 1969. Hypothetical mechanism of speciation. *Evolution* 23: 683–87

16. Bazykin, A. D. 1973. Population-genetic analysis of disruptive and stabilizing selection. II. Systems of adjacent populations and populations with a continuous area. *Genetika* 9:156–66. (In Russian with English summary)

17. Bengtsson, B. O. 1980. Rates of karyotype evolution in placental mammals. *Hereditas* 92:37–47

18. Bengtsson, B. O. 1984. The flow of genes through a genetic barrier. In *Evolution—Essays in Honour of John Maynard Smith*, ed. P. J. Greenwood, M. Slatkin. Cambridge: Cambridge Univ. Press. In press

19. Bengtsson, B. O., Bodmer, W. F. 1976. On the increase of chromosomal mutations under random mating. *Theor. Popul. Biol.* 9:260–81

20. Bengtsson, B. O., Christiansen, F. B. 1983. A two-locus mutation-selection model and some of its evolutionary implications. *Theor. Popul. Biol.* 24:59–77

21. Blackwell, J. H., Bull, C. M. 1978. A narrow hybrid zone between the Western Australian frog species, *Ranidella insignifera* and *R. pseudinsignifera:* The extent of introgression. *Heredity* 40:13–25

22. Bodmer, W. F., Felsenstein, J. 1967. Linkage and selection: Theoretical analysis of the deterministic two locus random mating model. *Genetics* 57:237–65

23. Bodmer, W. F., Parsons, P. A. 1962. Linkage and recombination in evolution. *Adv. Genet.* 11:1–100

24. Bulmer, M. G. 1980. *The Mathematical Theory of Quantitative Genetics*. Oxford: Oxford Univ. Press

25. Bush, G. L., Case, S. M., Wilson, A. C., Patton, J. L. 1977. Rapid speciation and chromosomal evolution in mammals. *Proc. Natl. Acad. Sci. USA* 74:3942–46

26. Caroli, B., Caroli, C., Roulet, B., Gouryet, J. F. 1980. A WKB treatment of diffusion in a multidimensional bistable potential. *J. Stat. Phys.* 22:515–36

27. Carson, H. L. 1968. The population flush and its genetic consequences. In *Population Biology and Evolution*, ed. R. C. Lewontin, pp. 123–37. Syracuse, NY: Syracuse Univ. Press

28. Carson, H. L. 1970. Chromosome tracers of the origin of species. *Science* 168: 1414–18

29. Carson, H. L. 1975. The genetics of speciation at the diploid level. *Am. Nat.* 109:73–92

30. Carson, H. L. 1982. Evolution of Drosophila on the newer Hawaiian volcanoes. *Heredity* 48:3–25

31. Carson, H. L. 1982. Speciation as a major reorganization of polygenic balances. In *Mechanisms of Speciation*, ed. C. Barigozzi, pp. 411–33. New York: Liss

32. Carson, H. L. 1983. Chromosomal sequences and inter-island colonizations in the Hawaiian Drosophila. *Genetics* 103:465–82

33. Carson, H. L., Kaneshiro, K. Y. 1976. *Drosophila* of Hawaii: Systematics and ecological genetics. *Ann. Rev. Ecol. Syst.* 7:311–46

34. Cavalli, L., Maccacaro, G. 1952. Polygenic inheritance of drug resistance in the bacterium *E. coli*. *Heredity* 6:311–31

35. Chakraborty, R., Nei, M. 1977. Bottleneck effects on average heterozygosity and genetic distance with the stepwise mutation model. *Evolution* 31: 347–56

36. Charlesworth, B. 1976. Recombination modification in a fluctuating environment. *Genetics* 83:181–95

37. Charlesworth, B. 1980. *Evolution in Age-Structured Populations*. Cambridge: Cambridge Univ. Press

38. Charlesworth, B. 1983. Models of the evolution of some genetic systems. *Proc. R. Soc. London Ser. B* 219:265–79

39. Charlesworth, B. 1984. The evolutionary genetics of life histories. In *Evolutionary Ecology*, ed. B. Shorrocks, pp. 117–33. Oxford: Blackwell

40. Charlesworth, B., Charlesworth, D. 1973. Selection of new inversions in multi-locus genetic systems. *Genet. Res. Cambridge* 21:167–83

41. Charlesworth, B., Charlesworth, D. 1975. An experiment on recombination load in *Drosophila melanogaster*. *Genet. Res. Cambridge* 25:267–74

42. Charlesworth, B., Lande, R., Slatkin, M. 1982. A neo-Darwinian commentary on macroevolution. *Evolution* 36:474–98

43. Charlesworth, B., Smith, D. B. 1982. A computer model of founder effect speciation. *Genet. Res. Cambridge* 39:227–36

44. Charlesworth, D., Charlesworth, B. 1975. Theoretical genetics of Batesian mimicry. II. Evolution of supergenes. *J. Theor. Biol.* 55:305–24

45. Clarke, B. C. 1966. The evolution of morph-ratio clines. *Am. Nat.* 100:389–402

46. Clarke, B. C., Murray, J. 1969. Ecological genetics and speciation in land snails of the genus *Partula*. *Biol. J. Linn. Soc.* 1:31–42

47. Clarke, C. A., Sheppard, P. M. 1960. The evolution of mimicry in the butterfly *Papilio dardanus*. *Heredity* 14:163–73

48. Clarke, C. A., Sheppard, P. M. 1960. Supergenes and mimicry. *Heredity* 14: 175–85
49. Craddock, E. M. 1975. Intraspecific karyotypic differentiation in the Australian phasmatid *Didymuria violescens* (Leach). I: the chromosome races and their structural and evolutionary relationships. *Chromosoma* 53:1–24
50. Craddock, E. M., Johnson, N. E. 1979. Genetic variation in Hawaiian *Drosophila*. V. Chromosomal and allozymic diversity in *Drosophila silvestris* and its homosequential species. *Evolution* 33: 137–55
51. Crosby, J. L. 1949. Selection of an unfavourable gene-complex. *Evolution* 3: 212–30
52. Crow, J. F., Kimura, M. 1970. *An Introduction to Population Genetics Theory*. New York: Harper & Row
53. Dobzhansky, T. 1946. Genetics of natural populations. XIII. Recombination and variability in populations of *Drosophila pseudoobscura*. *Genetics* 31:269–90
54. Dobzhansky, T. 1955. A review of some fundamental concepts and problems of population genetics. *Cold Spring Harbor Symp. Quant. Biol.* 20:1–15
55. Dobzhansky, T. 1970. *The Genetics of the Evolutionary Process*. New York: Columbia Univ. Press
56. Dobzhansky, T., Streisinger, G. 1944. Experiments on sexual isolation in *Drosophila*. II. Geographic strains of *D. prosaltans*. *Proc. Natl. Acad. Sci. USA* 30:340–45
57. Ehrman, L., Parsons, P. A. 1981. *Behavior Genetics and Evolution*. New York: McGraw-Hill
58. Endler, J. A. 1977. *Geographic Variation, Speciation, and Clines*. Princeton: Princeton Univ. Press
59. Ewens, W. J. 1979. *Mathematical Population Genetics*. Berlin: Springer-Verlag
60. Falconer, D. S. 1981. *Introduction to Quantitative Genetics*. London: Longman
61. Felsenstein, J. 1976. The theoretical population genetics of variable selection and migration. *Ann. Rev. Genet.* 10:253–80
62. Felsenstein, J. 1979. Excursions along the interface between disruptive and stabilizing selection. *Genetics* 93:773–95
63. Ferris, S. D., Portnoy, S. L., Whitt, G. S. 1979. The roles of speciation and divergence time in the loss of duplicate gene expression. *Theor. Popul. Biol.* 15:114–39
64. Fisher, R. A. 1930. *The Genetical Theory of Natural Selection*. Oxford: Clarendon
65. Fisher, R. A. 1950. Gene frequencies in a cline determined by selection and diffusion. *Biometrics* 6:353–61
66. Giddings, L. V., Templeton, A. R. 1983. Behavioral phylogenies and the direction of evolution. *Science* 220:372–77
67. Gold, J. R. 1980. Chromosomal change and rectangular evolution in North American Cyprinid fishes. *Genet. Res. Cambridge* 35:157–64
68. Haffer, J. 1969. Speciation in Amazonian forest birds. *Science* 165:131–37
69. Hagen, D. W., Blouw, D. M. 1983. Heritability of dorsal spines in the fourspine stickleback *(Apeltes quadracus)*. *Heredity* 50:275–82
70. Haldane, J. B. S. 1930. A mathematical theory of natural and artificial selection. VI. Isolation. *Proc. Cambridge Philos. Soc.* 26:220–30
71. Haldane, J. B. S. 1931. A mathematical theory of natural and artificial selection. VIII. Metastable populations. *Proc. Cambridge Philos. Soc.* 27:838–44
72. Haldane, J. B. S. 1932. *The Causes of Evolution*. London: Longman
73. Haldane, J. B. S. 1948. The theory of a cline. *J. Genet.* 48:277–84
74. Hastings, A. 1981. Stable cycling in discrete time genetic models. *Proc. Natl. Acad. Sci. USA* 78:7224–25
75. Hastings, A. 1981. Simultaneous stability of $D = 0$ and $D \neq 0$ for multiplicative viabilities at two loci. *J. Theor. Biol.* 89:69–81
76. Hedrick, P. W. 1981. The establishment of chromosomal variants. *Evolution* 35: 322–32
77. Hedrick, P. W., Jain, S. K., Holden, L. R. 1977. Multilocus systems in evolution. In *Evolutionary Biology*, ed. M. K. Hecht, W. C. Steere, B. Wallace, 11: 101–82. New York: Plenum
78. Hill, W. G. 1982. Predictions of response to artificial selection from new mutations. *Genet. Res. Cambridge* 40: 255–78
79. Huether, C. A. 1968. Exposure of natural genetic variability underlying the pentamerous corolla constancy in *Linanthus androsaceus* ssp. *androsaceus*. *Genetics* 60:123–46
80. Hunt, J. A., Carson, H. L. 1983. Evolutionary relationships of four species of Hawaiian Drosophila as measured by DNA reassociation. *Genetics* 104:353–64
81. Hunt, W. G., Selander, R. K. 1973. Biochemical genetics of hybridization in

European house mice. *Heredity* 31:11–33

82. Jain, S. K., Bradshaw, A. D. 1966. Evolutionary divergence among closely adjacent plant populations. I. The evidence and its theoretical analysis. *Heredity* 21:407–41

83. Johnson, M. S. 1982. Polymorphism for direction of coil in *Partula suturalis:* Behavioural isolation and positive frequency dependent selection. *Heredity* 49:145–51

84. Johnson, M. S., Murray, J., Clarke, B. C. 1984. Genetic divergence and modes of speciation in *Partula*. *Evolution*. In press

85. Kaneshiro, K. Y. 1976. Ethological isolation and phylogeny in the *planitibia* subgroup of Hawaiian *Drosophila*. *Evolution* 30:740–45

86. Kaneshiro, K. Y. 1980. Sexual isolation, speciation and the direction of evolution. *Evolution* 34:437–44

87. Kaneshiro, K. Y. 1983. Sexual selection and direction of evolution in the biosystematics of Hawaiian Drosophilidae. *Ann. Rev. Entomol.* 28:161–78

88. Kaneshiro, K. Y., Kurihara, J. S. 1981. Sequential differentiation of sexual isolation in populations of *Drosophila silvestris*. *Pac. Sci.* 35:177–83

89. Karlin, S., McGregor, J. 1972. Application of method of small parameters to multi-niche population genetic models. *Theor. Popul. Biol.* 3:186–209

90. Key, K. H. L. 1968. The concept of stasipatric speciation. *Syst. Zool.* 17:14–22

91. Key, K. H. L. 1981. Species, parapatry, and the morabine grasshoppers. *Syst. Zool.* 30:425–58

92. Kilias, G., Alahiotis, S. N., Pelecanos, M. 1980. A multifactorial genetic investigation of speciation theory using *Drosophila melanogaster*. *Evolution* 34:730–37

93. Kimura, M. 1970. The length of time required for a selectively neutral mutant to reach fixation through random frequency drift in a finite population. *Genet. Res. Cambridge* 15:131–33

94. Kimura, M., Weiss, G. H. 1964. The stepping stone model of population structure and the decrease of genetic correlation with distance. *Genetics* 49:561–76

95. Kirkpatrick, M. 1982. Sexual selection and the evolution of female choice. *Evolution* 36:1–12

96. Kirkpatrick, M. 1982. Quantum evolution and punctuated equilibria in continuous genetic characters. *Am. Nat.* 119:833–48

97. Koref Santibanez, S., Waddington, C. H. 1958. The origin of sexual isolation between different lines within a species. *Evolution* 12:485–93

98. Lande, R. 1976. Natural selection and random genetic drift in phenotypic evolution. *Evolution* 30:314–34

99. Lande, R. 1979. Effective deme sizes during long-term evolution estimated from rates of chromosomal rearrangement. *Evolution* 33:234–51

100. Lande, R. 1980. Genetic variation and phenotypic evolution during allopatric speciation. *Am. Nat.* 116:463–79

101. Lande, R. 1981. Models of speciation by sexual selection on polygenic traits. *Proc. Natl. Acad. Sci. USA* 78:3721–25

102. Lande, R. 1982. Rapid origin of sexual isolation and character divergence in a cline. *Evolution* 36:213–23

103. Lande, R. 1983. The response to selection on major and minor mutations affecting a metrical trait. *Heredity* 50:47–66

104. Leigh, E. G. 1981. The average lifetime of a population in a varying environment. *J. Theor. Biol.* 90:213–39

105. Levin, D. A., Wilson, A. C. 1976. Rates of evolution in seed plants: Net increase in diversity of chromosome numbers and species numbers through time. *Proc. Natl. Acad. Sci. USA* 73:2080–90

106. Lewis, H. 1962. Catastrophic selection as a factor in speciation. *Evolution* 16:257–71

107. Lewis, H. 1973. The origin of diploid neospecies in *Clarkia*. *Am. Nat.* 107:161–70

108. Lewontin, R. C. 1965. Discussion of paper by Dr. Howard. In *The Genetics of Colonizing Species*, ed. H. G. Baker, G. L. Stebbins, p. 481. New York: Academic

109. Lewontin, R. C., Kojima, K. 1960. The evolutionary dynamics of complex polymorphisms. *Evolution* 14:458–72

110. Lloyd, D. G. 1980. Demographic factors and mating patterns in angiosperms. In *Demography and Evolution in Plant Populations*, ed. O. T. Solbrig, pp. 67–88. Oxford: Blackwell

111. MacNair, M. R., Christie, P. 1983. Reproductive isolation as a pleiotropic effect of copper tolerance in *Mimulus guttatus? Heredity* 50:295–302

112. Malécot, G. 1969. *The Mathematics of Heredity*. San Francisco: Freeman

113. Malmberg, R. L. 1978. The evolution of epistasis and the advantage of recombination in populations of bacteriophage T4. *Genetics* 86:607–21

114. Markow, T. A. 1981. Mating preferences are not predictive of the direction of

evolution in experimental populations of *Drosophila*. *Science* 213:1405–7

115. Mather, K. 1943. Polygenic inheritance and natural selection. *Biol. Rev.* 18:32–64

116. Maynard Smith, J. 1966. Sympatric speciation. *Am. Nat.* 100:637–650

117. Maynard Smith, J. 1980. Selection for recombination in a polygenic model. *Genet. Res. Cambridge* 35:269–77

118. Maynard Smith, J. 1983. The genetics of stasis and punctuation. *Ann. Rev. Genet.* 17:11–25

119. Mayr, E. 1942. *Systematics and the Origin of Species.* New York: Columbia Univ. Press

120. Mayr, E. 1954. Change of genetic environment and evolution. In *Evolution as a Process*, ed. J. S. Huxley, A. C. Hardy, E. B. Ford, pp. 156–80. London: Allen & Unwin

121. Mayr, E. 1963. *Animal Species and Evolution.* Cambridge, Mass: Harvard Univ. Press

122. Mayr, E. 1970. *Populations, Species and Evolution.* Cambridge, Mass: Harvard Univ. Press

123. Mayr, E. 1976. *Evolution and the Diversity of Life: Selected Essays.* Cambridge, Mass: Harvard Univ. Press

124. Mayr, E. 1982. Processes of speciation of animals. See Ref. 31, pp. 1–19

125. Mayr, E. 1982. *The Growth of Biological Thought: Diversity, Evolution and Inheritance.* Cambridge, Mass: Harvard Univ. Press

126. McKenzie, J. A., Whitten, M. J., Adena, M. A. 1982. The effect of genetic background on the fitness of diazinon resistance genotypes of the Australian sheep blowfly, *Lucilia cuprina*. *Heredity* 49:1–10

127. McPhail, J. D. 1969. Predation and the evolution of a stickleback *(Gasterosteus)*. *J. Fish. Res. Board Can.* 26:3183–3208

128. Moodie, G. E. E. 1982. Why asymmetric mating preferences may not show the direction of evolution. *Evolution* 36:1096–97

129. Moran, C., Wilkinson, P., Shaw, D. D. 1980. Allozyme variation across a narrow hybrid zone in the grasshopper *Caledia captiva*. *Heredity* 44:69–81

130. Moran, P. A. P. 1964. On the nonexistence of adaptive topographies. *Ann. Hum. Genet.* 27:383–93

131. Mukai, T., Nagano, S. 1983. The genetic structure of natural populations of *Drosophila melanogaster*. XVI. Excess additive genetic variance of viability. *Genetics* 105:115–34

132. Muller, H. J. 1939. Reversibility in evolution considered from the standpoint of genetics. *Biol. Rev.* 14:261–80

133. Muller, H. J. 1940. Bearings of the *Drosophila* work on systematics. In *The New Systematics*, ed. J. S. Huxley, pp. 185–268. London: Syst. Assoc.

134. Murray, J., Clarke, B. C. 1980. The genus *Partula* on Moorea: Speciation in progress. *Proc. R. Soc. London Ser. B* 211:83–117

135. Nagylaki, T. 1978. The geographical structure of populations. In *Studies in Mathematical Biology. II. Populations and Communities*, ed. S. Levin, pp. 588–624. Washington DC: Am. Math. Assoc.

136. Nagylaki, T. 1982. Geographical invariance in population genetics. *J. Theor. Biol.* 99:159–72

137. Nei, M. 1976. Models of speciation and genetic distance. In *Population Genetics and Ecology*, ed. S. Karlin, E. Nevo, pp. 723–65. New York: Academic

138. Nei, M., Maruyama, T., Chakraborty, R. 1975. The bottleneck effect and genetic variability in populations. *Evolution* 29:1–10

139. Nei, M., Maruyama, T., Wu, C. I. 1983. Models of evolution of reproductive isolation. *Genetics* 103:557–79

140. Nevo, E. 1982. Speciation in subterranean mammals. See Ref. 31, pp. 191–218

141. O'Donald, P., Majerus, M. E. N., Weir, J. 1982. Female mating preference is genetic. *Nature* 300:521–23

142. Ohta, A. T. 1978. Ethological isolation and phylogeny in the *grimshawi* species complex of Hawaiian *Drosophila*. *Evolution* 32:485–92

143. Ohta, T., Kimura, M. 1971. Behaviour of neutral mutants influenced by associated overdominant loci in finite populations. *Genetics* 63:229–38

144. Paquin, C. E., Adams, J. 1983. Relative fitness can decrease in evolving asexual populations of *S. cerevisiae*. *Nature* 306:368–71

145. Patton, J. L., Sherwood, S. W. 1983. Chromosome evolution and speciation in rodents. *Ann. Rev. Ecol. Syst.* 14:139–58

146. Payne, F. 1918. An experiment to test the nature of the variation on which selection acts. *Ind. Univ. Stud.* 36:1–45

147. Pontecorvo, G. 1943. Viability interactions between chromosomes of *Drosophila melanogaster* and *Drosophila simulans*. *J. Genet.* 45:51–66

148. Powell, J. R. 1978. The founder-flush speciation theory: An experimental approach. *Evolution* 32:465–74

149. Rensch, B. 1939. Typen der Artbildung. *Biol. Rev.* 14:180–222

150. Rensch, B. 1959. *Evolution Above the Species Level.* New York: Columbia Univ. Press

151. Ringo, J. M. 1977. Why 300 species of Hawaiian *Drosophila?* The sexual selection hypothesis. *Evolution* 31:694–96

152. Robertson, A. 1952. The effects of inbreeding on the variation due to recessive genes. *Genetics* 37:189–207

153. Rose, M. R., Charlesworth, B. 1981. Genetics of life history in *Drosophila melanogaster.* I. Sib analysis of adult females. *Genetics* 97:173–86

154. Roughgarden, J. 1979. *Theory of Population Genetics and Evolutionary Ecology: An Introduction.* New York: Macmillan

155. Sacks, J. M. 1967. A stable equilibrium with minimum average fitness. *Genetics* 56:705–8

156. Sene, F. M., Carson, H. L. 1977. Genetic variation in Hawaiian *Drosophila.* IV. Allozymic similarity between *D. silvestris* and *D. heteroneura* from the island of Hawaii. *Genetics* 86:187–98

157. Shaw, D. D., Coates, D. 1983. Chromosomal variation and the concept of the coadapted genome—a direct cytological approach. In *Kew Chromosome Conference,* ed. P. E. Bradham, M. D. Bennett, 2:207–16. London: Allen & Unwin

158. Shaw, D. D., Moran, C., Wilkinson, P. 1981. Chromosomal reorganisation, geographic differentiation, and the mechanism of speciation in the genus *Caledia.* In *Insect Cytogenetics,* ed. R. L. Blackman, G. M. Hewitt, M. Ashburner, pp. 171–94. Oxford: Blackwell

159. Sing, C. F., Brewer, G. J., Thirtle, B. 1973. Inherited biochemical variation in *Drosophila melanogaster:* Noise or signal? I. Single-locus analyses. *Genetics* 75:381–404

160. Slatkin, M. 1973. Gene flow and selection in a cline. *Genetics* 75:733–56

161. Slatkin, M. 1975. Gene flow and selection in a two-locus system. *Genetics* 81:209–22

162. Slatkin, M. 1978. Spatial patterns in the distribution of polygenic characters. *J. Theor. Biol.* 70:213–28

163. Slatkin, M. 1981. Fixation probabilities and fixation times in a subdivided population. *Evolution* 35:477–88

164. Spiess, E. B. 1959. Effects of recombination on viability in *Drosophila. Cold Spring Harbor Symp. Quant. Biol.* 23: 239–50

165. Spieth, H. T. 1974. Mating behaviour and evolution of the Hawaiian *Drosophila.* In *Genetic Mechanisms of Speciation in Insects,* ed. M. J. D. White. Sydney: Aust. & N. Z. Book Co.

166. Spirito, F., Rossi, C., Riggoni, M. 1982. Reduction of gene flow due to the partial sterility of heterozygotes for a chromosome mutation. I. Studies of a "neutral" gene not linked to the chromosome mutation in a two-population model. *Evolution* 37:785–97

167. Stebbins, G. L. 1950. *Variation and Evolution in Plants.* New York: Columbia Univ. Press

168. Sved, J. A. 1968. The stability of linked systems of loci with a small population size. *Genetics* 59:543–63

169. Sved, J. A. 1972. Heterosis at the level of the chromosome and at the level of the gene. *Theor. Popul. Biol.* 3:491–506

170. Szymura, J. 1976. New data on the hybrid zone between *Bombina bombina* and *B. variegata. Bull. Acad. Pol. Sci.* 24:355–63

171. Szymura, J. 1976. Hybridization between discoglossid toads *Bombina bombina* and *Bombina variegata* in southern Poland as revealed by the electrophoretic technique. *Z. Zool. Syst. Evolutionsforsch.* 14:227–36

172. Tantawy, A. O., Reeve, E. C. R. 1956. Studies in quantitative inheritance. IX. The effects of inbreeding at different rates in *Drosophila melanogaster. Z. Indukt. Abstamm. Vererbungsl.* 87:648–67

173. Templeton, A. R. 1977. Analysis of head shape differences between two interfertile species of Hawaiian *Drosophila. Evolution* 31:630–42

174. Templeton, A. R. 1979. The unit of selection in *Drosophila mercatorum.* II. Genetic revolution and the origin of coadapted genomes in parthenogenetic strains. *Genetics* 92:1256–82

175. Templeton, A. R. 1979. Once again, why 300 species of Hawaiian *Drosophila? Evolution* 33:513–17

176. Templeton, A. R. 1980. The theory of speciation via the founder principle. *Genetics* 94:1011–38

177. Templeton, A. R. 1980. Modes of speciation and inferences based on genetic distances. *Evolution* 34:719–29

178. Templeton, A. R. 1981. Mechanisms of speciation—a population genetic approach. *Ann. Rev. Ecol. Syst.* 12:23–48

179. Templeton, A. R. 1982. Adaptation and the integration of evolutionary forces. In *Perspectives on Evolution,* ed. R. Milkman, pp. 15–31. Sunderland, Mass: Sinauer

180. Templeton, A. R. 1982. Genetic architectures of speciation. See Ref. 31, pp. 105–21

181. Turner, J. R. G. 1970. Changes in mean

fitness under natural selection. In *Mathematical Topics in Population Genetics,* ed. K. Kojima, pp. 32–78. Berlin: Springer-Verlag

182. Turner, J. R. G. 1977. Butterfly mimicry: The genetical evolution of an adaptation. See Ref. 77, 10:163–206

183. Turner, J. R. G. 1981. Adaptation and evolution in *Heliconius:* A defense of neo-Darwinism. *Ann. Rev. Ecol. Syst.* 12:99–121

184. Val, R. C. 1977. Genetic analysis of the morphological differences between two interfertile species of Hawaiian *Drosophila. Evolution* 31:611–29

185. Van den Berg, M. J., Thomas, G., Hendriks, H., van Delden, W. 1984. A reexamination of the negative assortative mating phenomenon and its underlying mechanism in *Drosophila melanogaster. Behav. Genet.* 14:45–61

186. Walsh, J. B. 1982. Rate of accumulation of reproductive isolation by chromosome rearrangements. *Am. Nat.* 120:510–32

187. Wasserman, M., Koepfer, H. R. 1980. Does asymmetrical mating preference show the direction of evolution? *Evolution* 34:1116–26

188. White, M. J. D. 1968. Models of speciation. *Science* 158:1065–70

189. White, M. J. D. 1978. *Modes of Speciation.* San Francisco: Freeman

190. Whittle, J. R. S. 1969. Genetic analysis of the control of number and pattern of scutellar bristles in *Drosophila melanogaster. Genetics* 63:167–81

191. Wills, C. J. 1977. A mechanism for rapid allopatric speciation. *Am. Nat.* 111:603–5

192. Wright, S. 1931. Evolution in Mendelian populations. *Genetics* 16:97–159

193. Wright, S. 1932. The roles of mutation, inbreeding, crossbreeding and selection in evolution. *Proc. 6th Int. Congr. Genet.* 1:356–66

194. Wright, S. 1937. The distribution of gene frequencies in populations. *Proc. Natl. Acad. Sci. USA* 23:307–20

195. Wright, S. 1940. Breeding structure of populations in relation to speciation. *Am. Nat.* 74:232–48

196. Wright, S. 1941. On the probability of fixation of reciprocal translocations. *Am. Nat.* 75:513–22

197. Wright, S. 1943. Isolation by distance. *Genetics* 28:114–38

198. Wright, S. 1951. The genetical structure of populations. *Ann. Eugen.* 15:323–54

199. Wright, S. 1955. Classification of the factors of evolution. *Cold Spring Harbor Symp. Quant. Biol.* 20:16–24

200. Wright, S. 1965. Factor interaction and linkage in evolution. *Proc. R. Soc. London Ser. B* 162:80–104

201. Wright, S. 1967. "Surfaces" of selective value. *Proc. Natl. Acad. Sci. USA* 58:165–72

202. Wright, S. 1978. *Evolution and the Genetics of Populations. IV. Variability Within and Among Natural Populations.* Chicago: Univ. Chicago Press

203. Wright, S. 1978. Modes of speciation. *Paleobiology* 4:373–79

204. Wright, S. 1980. Genic and organismal selection. *Evolution* 34:825–43

205. Wright, S. 1982. Character change, speciation, and the higher taxa. *Evolution* 36:427–43

206. Yoo, B. M. 1980. Long term selection for a quantitative character in large replicate populations of *Drosophila.* I. Response to selection. *Genet. Res. Cambridge* 35:1–17

207. Zouros, E. 1981. The chromosomal basis for sexual isolation in two sibling species of *Drosophila: D. arizonensis* and *D. mojavensis. Genetics* 97:703–18

208. Zouros, E., d'Entremont, C. J. 1980. Sexual isolation among populations of *Drosophila mojavensis:* Response to pressure from a related species. *Evolution* 34:421–30

Ann. Rev. Ecol. Syst. 1984. 15:165–89

THE EVOLUTION OF EUSOCIALITY

Malte Andersson

Department of Zoology, University of Gothenburg, P.O. Box 25059,
400 31 Gothenburg, Sweden

INTRODUCTION

Eusociality is characterized by overlapping adult generations, cooperative brood care, and more or less nonreproductive workers or helpers (96, 151). Evolutionists have long recognized that the presence of sterile workers is difficult to explain by traditional individual selection (128). Darwin said of the workers in eusocial insects that they posed "one special difficulty, which at first appeared to me insuperable, and actually fatal to my whole theory . . . [because] being sterile, they cannot propagate their kind" (44, p. 236). Darwin proposed that workers evolve through selection at the level of the colony, a mechanism that is probably important in several respects (39, 109). Processes at the colony level, however, do not fully explain why individuals give up reproduction and hence the possibility of passing on their genes via offspring. Williams & Williams (149) and Hamilton (63, 65) showed that genetic relatedness between the workers and the reproductives they help raise may resolve the apparent paradox.

The most extreme cases of eusociality are species with highly specialized, permanently sterile castes of workers and soldiers. Sterile castes have evolved in the Hymenoptera (ants, bees, and wasps), Isoptera (termites), and apparently in one rodent (the naked mole-rat). Eusociality has arisen many times independently in Hymenoptera (albeit not always to the point of permanent worker sterility), and most of the discussion of the evolution of eusociality has focused on that order.

This review begins with an evaluation of the various hypotheses proposed to explain the evolution of eusociality in Hymenoptera and Isoptera. A comparative analysis of preconditions that are important for the evolution of insect

165

0066-4162/84/1120-0165$02.00

eusociality then follows. Finally, the occurrence of these preconditions is surveyed among vertebrates as a test of their generality. It turns out that the crucial preconditions have been similar in eusocial insects and eusocial-like vertebrates.

Another conclusion is that no single trait suffices to explain the evolution of eusociality, which depends on a combination of several factors. The haplodiploid sex determination of Hymenoptera has sometimes been overemphasized; hymenopterans have several important ecological preconditions in common with termites and eusocial-like vertebrates. Two conditions that have been crucial for the evolution of eusociality are: parental care, including the defense and feeding of offspring in a nest or other protected cavity; and low success of young adults or solitary pairs that attempt to reproduce.

HYMENOPTERA

Haplodiploidy and Kin Selection: The 3/4 Relatedness Hypothesis

Eusociality has arisen at least 11 different times in Hymenoptera and perhaps only once—in termites—among other insects (150–152, but see 54). Hamilton's (63) celebrated explanation is that haplodiploid sex determination in Hymenoptera makes sisters share three quarters of their genes, whereas a daughter only receives half her genome from her mother. Hymenopteran females may therefore propagate their genes better by helping to raise reproductive sisters than by raising daughters of their own. Haplodiploidy therefore should make the evolution of nonreproductive female workers particularly likely among the Hymenoptera. This and other stimulating ideas of Hamilton's started a revolution in the study of social behavior, particularly of the role of kin selection (91, 102).

Several entomologists have warned against overemphasis on the 3/4 relatedness hypothesis, and they have pointed to other factors important in the evolution of eusociality. (e.g. 4, 21, 39, 48, 54, 75, 87, 96, 99, 144, 146). Hamilton (63, 65) and Wilson (151) also noted that haplodiploidy alone cannot explain eusociality in Hymenoptera. Such reservations were often forgotten, however, and the 3/4 hypothesis came to dominate many textbook and popular accounts. For example, in his comprehensive review of social behavior in animals, Wilson stated that "the key to Hymenopteran success is haplodiploidy" (152, p. 415) and that "nothing but kin selection seems to explain the statistical dominance of eusociality by the Hymenoptera" (152, p. 418). A long list of similar evaluations of the 3/4 relatedness hypothesis by other authors could be cited.

The main empirical evidence in favor of the 3/4 hypothesis is that eusociality seems to have arisen many more times in the haplodiploid Hymenoptera than in other insects (21, 151). This evidence initially appeared impressive, but several recent findings indicate that haplodiploidy and 3/4 relatedness between sisters may have been of limited importance for the evolution of eusociality. Other factors have clearly been involved, and it seems possible that haplodiploidy has even been insignificant compared to these factors. At least five lines of evidence cast doubt on the overwhelming importance sometimes ascribed to haplodiploidy.

1. TAXONOMIC DISTRIBUTION OF HAPLODIPLOIDY   Because most eusocial species belong to Hymenoptera, there is a bias towards haplodiploidy among eusocial animals, but the association is far from universal. All forms so far examined among the one quarter of a million or so species in Hymenoptera are haplodiploid; yet only 6–8% of them are eusocial (124). Moreover, most mites and ticks, many thrips, whiteflies, scale insects, and some beetles are also haplodiploid (Table 1), as are most rotifers and certain nematodes (15, 28, 148). Yet none of these groups have eusocial species. Furthermore, eusociality has evolved in the diploid termites and certain vertebrates as well (see below).

2. MULTIPLE MATINGS   If the queen is inseminated by several males, a female on average might be more closely related to daughters of her own than to her sisters (63, 137), and the queen does mate with several males in many eusocial ants and bees (see references in 35, 110). For a given number of matings by the queen, however, the relatedness between sisters is higher in Hymenoptera than in diploid species (134). In addition, multiple matings occur mainly in highly social species with morphologically specialized, sterile workers. Multiple matings in these species may have evolved at a late stage, perhaps as an adaptation for maintaining large colonies (35), and need not necessarily be evidence against the importance of haplodiploidy in the origin of eusociality (35, 65, 110, 134).

3. MULTIPLE QUEENS   The relatedness between workers and the reproductive females they help raise is reduced even more if there are several breeding queens in the nest (e.g. 63, 137, 146, 153). Routes to sociality in which several mated females establish a nest together present problems for the 3/4 hypothesis (146, 153). If there are several breeding queens, workers on average will be less closely related to the reproductives they help raise than they would be to offspring of their own. If, on the other hand, only one foundress reproduces, the nonreproductive foundresses are related at most with $r = 3/8$ to her offspring. (If the other foundresses are not sisters of the queen, their relatedness to her offspring will of course be lower still.) So, in neither case is the relatedness as high as 3/4 between the nonreproductive females and the repro-

ductives that they raise. Yet, eusociality has apparently evolved along a multiple-foundress route in some bees (97) and wasps (e.g. 146).

Even though a nonreproductive foundress sister is related only with $r = \frac{3}{8}$ to the queen's offspring, the corresponding relatedness among diploid organisms is lower still ($r = \frac{1}{4}$). For this reason, haplodiploidy may have facilitated the evolution of eusociality (134, 144, 146)—to the extent that it has progressed along the semisocial route—with multiple foundresses that are related. Under these conditions, the facilitating effect of haplodiploidy is independent of the sex ratio (146), contrary to the case of matrifilial colonies discussed below.

4. WORKER–QUEEN CONFLICT OVER SEX RATIOS   The 3/4 relatedness between sisters may not promote eusociality unless workers bias the production of reproductive siblings towards sisters because worker relatedness to brothers is only $\frac{1}{4}$. If the queen enforces a sex ratio of 1:1, which is optimal for her, the average relatedness of a worker to her siblings is $\frac{1}{2}$, the same situation as in diploid species, and the 3/4 relatedness mechanism cannot work (31, 134).

Trivers & Hare (134; also see 32) pointed out that the optimal sex ratio among reproductives is not the same for the queen and the workers. Whether actual sex ratios in eusocial Hymenoptera are closer to 1:1 or 3:1 (as predicted with "queen control" and "worker control," respectively) is a matter of controversy; the ratio varies strongly among species (5, 134; reviewed in 153). In at least some wasps and ants, the ratio is close to 1:1, and workers help raise siblings to which they are not more closely related than they would be to offspring of their own (84a, 93, 106). If males are produced by laying workers, the optimal colony sex ratio would also approach 1:1 for workers (134), but laying workers were not observed in these species (93, 106). In certain species of ants, the sex ratio comes close to 3:1, as one would expect if workers control the sex ratio. A female-biased sex ratio is also predicted if there is inbreeding and local mate competition among males (64), which may partially explain the female-biased sex ratios in certain Hymenoptera (5, 64, reviewed in 153).

5. RESULTS FROM GENETIC MODELS   Using exact genetic models, Craig (36, 37) found that high relatedness among sisters need not favor the evolution of nonreproductive workers, even if there is a strong bias towards females in the brood. When the proportion of males declines, a brother becomes increasingly efficient at propagating the genes that he carries, since he will mate on average with more than one female (36, 70, 134). In the dynamics of gene frequency change, this development will counteract the greater relatedness among sisters. Haplodiploidy combined with female-biased sex ratios, therefore, need not make the evolution of eusociality easier than in diploid organisms (36, 37, 73).

Charnov (31), Craig (36), and Wade (136) showed that the initial invasion of a rare allele for altruism is favored by a female-biased sex ratio. But when the proportion of nonreproductive workers increases, the female bias loses its advantage and prevents further spread of the allele (37; also see 2). Craig (37) also found that if all females in the population produce the same sex ratio, eusociality does not evolve more easily in haplodiploid than in diploid species. There are cases, however, where the sex ratio differs among broods (e.g. 111, 141), which under certain circumstances can favor the evolution of eusociality (134).

Contrary to previous suggestions, Craig (38) also concluded that, from a purely genetic point of view, male workers should evolve as easily in haplodiploid as in diploid species. The fact that males usually are not workers may have little to do with the special coefficients of relatedness in the Hymenoptera. Instead, the main reason may be that only females perform parental care in this group, and males would make poor workers (e.g. 63, 65, 87). The sting in Hymenoptera is a modified ovipositor, so males possess none. They therefore lack this formidable weapon for hunting and defense and are less fit than females to serve the colony as workers (4, 144).

Craig's results seem to undermine Hamilton's (63) hypothesis as an explanation for eusociality and female workers in Hymenoptera. The conclusions appear to be sensitive to the assumptions one makes, however, and genetic models based on somewhat different assumptions support the importance of haplodiploidy. For example, laying workers, which occur in many eusocial Hymenoptera (54, 63, 65, 87, 141, 152), can alter the picture. Aoki & Moody (8), Iwasa (70), and Bartz (13) confirmed Hamilton's (63) and Trivers & Hare's (134) proposal that the evolution of eusociality is facilitated by haplodiploidy if workers add unfertilized eggs of their own, especially if workers substitute sons (with $r = \frac{1}{2}$) for brothers ($r = \frac{1}{4}$). Moreover, Bartz (13) concluded that the presence of laying workers creates degrees of relatedness that favor either female workers or male workers, but not both; the restrictive conditions of relatedness favoring male workers seem unlikely to be met in haplodiploid species.

The life-history pattern may also be important. Modeling the optimal sex ratio in bivoltine haplodiploid species that overwinter as mated adults, Seger (120) found evidence of the importance of haplodiploidy. If some males born in the spring survive to mate in the autumn as well, the optimal sex ratio for females reproducing in the summer becomes female-biased. This is because a number of males from the spring generation will still be alive. Workers born in the spring owing to the female bias in the summer generation will have high average relatedness to the reproductives they help raise. A somewhat similar sex-ratio effect can also occur if workers lay unfertilized male eggs (70). Because of the asymmetrical overlap of generations with different optimal sex ratios, these models are not subject to the problems that Craig (36, 37) and Crozier (39, 41) discussed. Seger (120) concluded that the evolution of eusociality is enhanced by haplodiploidy in species conforming to his assumptions. The occurrence of eusociality within partially bivoltine Hymenoptera provides supportive evidence: Most temperate, primitively eusocial species belong to taxa in which solitary species overwinter as mated females (21, 120). Another possible interpretation is that the factors that select for overwintering as a mated adult also select for social nesting (see 21).

To summarize, genetic models yield partly contradictory results with respect to Hamilton's (63) 3/4 relatedness hypothesis. The outcome is sensitive to several of the assumptions, the realism of which is difficult to evaluate at present; further empirical and theoretical work is needed. The evidence seems less conclusive today than it did some years ago, and genetic theory alone cannot establish whether haplodiploidy has been important for the frequent evolution of eusociality in Hymenoptera (37, 39, 41).

This tentative conclusion does not imply that indirect kin selection (i.e. the propagation of genes through relatives other than offspring; see 26) has been unimportant. On the contrary, as Sturtevant (128), Williams & Williams (149), Hamilton (63, 65), West-Eberhard (144, 146), and Wilson (152) have suggested, there is much evidence that relatedness among colony members, together with other preconditions and mechanisms, has influenced eusociality in Hymenoptera. Thus, indirect kin selection seems to have been important in the evolution of eusociality in wasps, bees, and ants, no matter what role haplodiploidy will eventually turn out to have played.

Two recent discoveries strongly support the importance of indirect kin selection for the evolution of sterile castes, although the cases do not fully fit the definition of eusociality. Nonreproductive soldiers have been found in cyclically parthenogenetic aphids, defending their siblings against predatory insects (9, 10). A form of soldiers also occurs in polyembryonic wasps (42). In these parasitic wasps, the egg does not develop into a single adult. Instead, it forms a cell mass that gives rise to numerous larvae within the host. The first few larvae, which hatch several weeks before the others, differ morphologically from their later siblings and die without pupating. Experiments indicate that these early larvae constitute a defender morph; they improve the survival of their siblings by attacking and eliminating competing parasites of other species within the same host (42). The soldiers in both these cases are probably genetically identical to the siblings they protect, so they are at the upper endpoint of the relatedness scale. Their evolution seems to represent an extreme of indirect selection, and it would be hard to explain without kinship theory.

Mutualism

The importance of mutualism in the evolution of eusociality among Hymenoptera has long been emphasized by Michener (e.g. 87, 95–98), Evans (53, 54), and West-Eberhard (144–146). They pointed out that associations of nesting females may be adaptive in reducing predation and parasitism upon the brood, a suggestion supported by numerous field studies (1, 57, 59, 72, 86, 89, 99). Such mutualistic associations may be important steps along the route to eusociality.

Mutualism alone, however, does not seem to explain the evolution of nonreproductive workers, which cannot gain any mutualistic reproductive advantage if they have no chance of laying eggs; indirect kin selection or manipulation appears necessary (39).

Manipulation

There is abundant evidence that the queen in some species manipulates her daughters or other females into foregoing reproduction (4, 99; reviewed in 18, 21, 153). In some wasps and bees, she controls workers by behavioral dominance, even eating any eggs that they may lay. Another means of manipulation is restricting the food provided to developing offspring, which limits their size as adults. A small daughter has poor reproductive capacity, and her best choice may be to help raise near relatives rather than breed herself. By suitable restriction of food rations, the queen might therefore induce some of her daughters to become workers. In some advanced eusocial bees and ants, the queen controls the workers through pheromones; if the queen dies and her chemical control ceases, workers may start reproducing.

Craig (36) and Crozier (39) pointed out that manipulation alone is not sufficient for the evolution of nonreproductive workers. In some species where several foundresses start a nest together, one of them becomes the reproductive queen after a period of conflict (e.g. 107, 143). Alexander (4) suggested that the other foundresses are her sisters, being manipulated by their mother to remain with their sister queen as workers. But the fact that subordinate foundresses do remain as workers can hardly be explained solely through manipulation by their mother, as they are no longer in contact with her (36, 39). For an unrelated female that has no chance of laying eggs in the nest now or in the future, the best option should be to leave and attempt to reproduce elsewhere. There are several reasons why she might benefit from staying. First, the subordinate foundress may be related to the queen and gain inclusive fitness from helping raise her offspring. This alternative can be better than attempting to found a new nest alone if the success of a solitary nest is low (e.g. 76, 94, 107, 127). Second, the present queen may die or lose her position, enabling a subordinate foundress to become queen (e.g. 88, 89, 127, 145). Third, in many species a subordinate female may actually lay eggs (reviewed in 21). It seems that either indirect kin selection or the possibility of future reproduction is required to explain subordinate foundress behavior (36, 41, 94, 107, 127, 145, 147).

Parental manipulation is not a mutually exclusive alternative to kin selection (92, 102). By definition, parental manipulation involves close relatives, so it is inextricably tied to kin selection. (In genetic models, parental manipulation is usually represented by an allele in the mother enforcing worker behavior in daughters). In some cases, parental manipulation may lead to evolution of nonreproductive workers where pure kin selection on the offspring would not suffice (i.e. when the allele for altruism occurs in offspring; see 30, 31, 36). But the close relatedness between the manipulated workers and the siblings that they help to raise weakens the selection on workers to escape from manipulation, so indirect kin selection is inevitably at work (also see 36).

TERMITES

Termites (Isoptera) are thought to have originated from cockroach-like ancestors and seem to have evolved eusociality about 100 million years ago; all of the 2,200 known extant species have sterile castes (151). It is usually assumed that isoptera evolved eusociality once [a conclusion that Evans (54) has questioned]. Termites are diploid and have workers of both sexes. At least three hypotheses have been offered to explain their eusociality.

The Symbiont Hypothesis

Termites depend on symbiotic intestinal protozoa for digestion of cellulose. Cleveland et al (33) suggested that this symbiotic relationship predisposes termites to social life. The protozoa are lost together with the lining of the hindgut wall at each molt, so growing termites must obtain new protozoa through transfer from other colony members. Hence, the symbiosis with protozoa necessitates social life until adulthood. The symbiotic relationship need not select for advanced sociality, however (12, 82, 126); extended parent-offspring contact alone could apparently suffice for transfer of symbionts. A similar symbiotic relationship with protozoa is maintained in cryptocercid wood roaches with a much simpler social system: extended parental care for offspring in log galleries (105, 119).

To conclude, the symbiosis with protozoa seems to select for at least simple forms of sociality that permit inoculation with symbionts after each molt, but this does not explain why termites evolved eusociality.

The Inbreeding Hypothesis

Hamilton (65, 66) and Bartz (12) suggested that inbreeding may lead to high degrees of relatedness within colonies and hence facilitate eusociality (also see 25). Lacy (82) objected, pointing out that inbreeding increases the genetic similarity to offspring as well as to other relatives, so it would not change the relative merits of raising other relatives as compared to offspring. Genetic models indicate, however, that inbreeding can facilitate the evolution of altruism, at least if the altruism is not extreme (102, 138). Moreover, termite societies often contain replacement or secondary reproductives that are recruited from the offspring of the founding couple and hence are closely related to each other (12, 132). The resulting inbreeding can make relatedness in subsequent broods higher among sibs than between parents and offspring (12, 56). Perhaps the ancestors of termites were limited to the special habitat of decaying wood and were strongly inbred, which might facilitate evolution of eusociality (12, 65, 66).

For these reasons, it would be interesting to analyze the degree of inbreeding in cryptocercid wood roaches, which are believed to resemble the ancestors of termites in morphology as well as ecology (105, 119).

The Chromosomal Linkage Hypothesis

In several termites, part of the genome is propagated as a sex-linked complex (129). Lacy (82) pointed out that this increases the relatedness among brothers and among sisters above 0.5; brother-sister relatedness is reduced below 0.5. He suggested that termite workers may bias their care toward their own sex, which might facilitate the evolution of sterile workers.

Leinaas (84) objected that sex-biased investment may not be feasible. Moreover, an allele disposing its carrier to raise reproductive siblings of its own sex may neither increase in frequency nor make the evolution of eusociality easier (84). If the colony sex ratio is 1:1 and the allele is distributed at random with respect to sex, the total investment in reproductives in the colony will also yield an expected sex ratio of 1:1. The average relatedness of a worker to its reproductive siblings then is 0.5, the same as for workers that raise reproductives without regard to their sex. Chromosomal linkage therefore may not facilitate the evolution of eusociality in termites (but see 82a). Gene frequency modeling and field tests of the presumed same-sex bias in the allotment of care by workers should clarify the issue.

PRECONDITIONS FAVORING EUSOCIALITY AMONG INSECTS

Comparative Approach

The previous review of selective mechanisms does not explain why eusociality has only evolved in Hymenoptera and Isoptera among arthropods or why eusociality has arisen independently in the former order so many times. What preconditions favored eusociality in insects? A taxonomic analysis should help provide an answer. If eusociality is consistently associated with certain traits that clearly evolved before eusociality and that rarely occur without it, then these traits may have promoted the evolution of nonreproductive workers. One indication that a trait has evolved before eusociality is if the trait also occurs in nonsocial members of the same and related taxa.

Table 1 Taxonomic distribution among arthropods of some traits in relation to eusociality

Taxonomic group	Eusociality[a]	Haplo-diploidy[b]	Offspring raised in a nest[c]	Repeated food pro-visioning[c]	Poisonous weapon	Selected references
Araneae			(Cocoon)	Some	All	29
Acarina		Many				
Orthoptera			Some	Some		142
Isoptera	All		All	All	Some	151
Dermaptera			Some	Some		83
Thysanoptera		Many				
Homoptera						
Aleurodidae		Some				
Coccidae		Many				
Coleoptera						
Carabidae			Some		Some	16, 131
Silphidae			Some	Some		104
Staphylinidae			Some			19
Passalidae			Some[d]			62
Scarabeidae			Some			61
Scolytidae		Some	Some[d]			51, 118
Micromalthidae		One[e]				117
Hymenoptera	Many	All	Many	Many	Many	97, 151

[a] Wilson (151).
[b] White (148).
[c] Eickwort (48).
[d] Feeding and reproducing in wood galleries.
[e] Genus contains one species.

But association is not enough. Eusociality is nicely correlated with the possession of 12 antennal segments in females, as most aculeate Hymenoptera belong to families with this trait (e.g. 23). However, it probably has not been crucial for the evolution of eusociality! A behavioral or other mechanism is also needed to explain why the trait promotes the evolution of non-reproductive workers. Such mechanisms may still be operating, making their presumed consequences open to testing by observation and experiment.

One way to estimate the importance of various preconditions, therefore, is to list the occurrence of eusociality and compare its taxonomic distribution with that of associated, previously evolved traits of potential importance. The proposed mechanisms through which the associated traits may facilitate the evolution of eusociality also need to be tested. Table 1 provides a summary; two excellent, recent reviews by Eickwort and Brockmann survey sociality among insects in general (48) and hymenopteran eusociality in particular (21).

Nest Building and Repeated Food Provisioning

Two traits that have apparently been crucial for the evolution of eusociality among Hymenoptera are the construction of a nest and transport of food to the offspring in it (21, 48, 52, 53, 87, 95, 96). Within Hymenoptera (Table 2), species in the suborder Symphyta (sawflies) feed on plants and build no nests. The other suborder, Apocrita, contains the divisions Parasitica and Aculeata. In aculeates, the ovipositor is modified into a sting; nest building and eusociality in Hymenoptera are restricted to certain aculeates. These traits have never been found in the Symphyta or Parasitica, although the two groups contain the majority of hymenopteran species (80).

One reason why eusociality is associated with nest building and the feeding of larvae may simply be that these traits provide opportunities for workers to help (21, 39, 65). Another likely reason is that nest building requires a great deal of work, and females can save time and energy by sharing a nest (140). Lone females in a sweat bee, *Lasioglossum zephyrum*, took almost twice as long to make their first cells as did colonies of four females; hence, colonial females may have a shorter generation time (100). Females in many species reuse nests (reviewed in 48). Reuse sometimes leads to incidental nest sharing, which may set the stage for the evolution of joint nesting if this becomes beneficial for some reason (22).

Progressive provisioning of food to each larva in the nest may also have facilitated the evolution of eusociality (54). Progressive provisioning is not a necessary precondition, however, as eusociality has evolved repeatedy among bees that "mass provision" their offspring with food before the eggs hatch (96, 97). Moreover, nest building and repeated food provisioning are not sufficient for eusociality to evolve. Both traits occur in many nonsocial insects—aculeate Hymenopterans as well as certain crickets and mole crickets, earwigs and beetles (Table 1).

Table 2 Taxonomic groups within the order Hymenoptera[a]

Suborder Division Superfamily	Examples
Symphyta	Phytophagous sawflies and horntails
Apocrita	
Parasitica	Parasitoids
Aculeata	"Stinging Hymenoptera"
Bethyloidea	Parasitoids
Sphecoidea	Digger wasps, sand wasps, mud-daubers, and bees
Vespoidea	Wasps, ants, and parasitoids

[a] Based on Brothers (23).

Since all extant termites are eusocial, the preconditions that favored sociality are even harder to determine in Isoptera than in Hymenoptera. Like eusocial Hymenoptera, termites seem to have evolved from ancestors with extended parental care in a nest or protected cavity. These ancestors probably lived under conditions in which it was difficult to found new nests because of the patchy distribution of decaying logs (e.g. 65, 66). Similar ecological preconditions therefore seem to have characterized the ancestors of both termites and eusocial Hymenoptera. In addition, the special conditions of relatedness created by inbreeding may have facilitated the evolution of eusociality in termites (12, 65, 66). Several families of beetles, however, also live in wood galleries (Table 1); passalid beetles (62) seem particularly similar in their ecology to *Cryptocercus* wood roaches. It is not clear why prototermites gave rise to eusocial lines, while none of these groups of beetles did.

Defense of Offspring

Another aspect of eusociality that Michener (95, 97, 98), Lin (86, 87), and Evans and West-Eberhard (54, 55) emphasized is the importance of communal nest defense against parasites and predators by blocking the nest entrance and biting or stinging. Lin & Michener proposed that "a major selective factor favoring the establishment of colonies . . . is almost surely parasite and predator pressure" (87, p. 143), and Evans (54) expressed similar views. Females nesting together may run a lower risk than solitary females of having their brood destroyed. In a colony, one or more females are likely to be present and deter enemies (e.g. 47, 57, 88, 89), whereas a solitary female must leave her nest unguarded while she forages. Nest guarding reduces predation and parasitism (116), and in many species, solitary nests run a higher risk of failure than communal nests (e.g. 1, 57, 59, 89, 94, 140). Moreover, the average productivity per female can be higher for multiple foundresses than for a lone female (59, 107, 127, 140, 150), and in some cases the inclusive fitness may be higher for a subordinate foundress in a communal nest than for a solitary foundress (94, 107, 127).

Mutualistic benefits may occur not only in a communal nest but also in aggregations of single-female nests (e.g. 3, 14, 103, 115, 133). An alternative hypothesis to account for clumped nest distributions is a limitation of suitable habitat. This explanation may often be correct, but in some studies groupings of nests apparently were not fully explained by habitat limitation alone (97, 101, 133). Michener (98) suggested that differences in predator pressure among habitats may partly explain the variation in the degree of sociality among certain bees (*Ceratina* and Halictini).

It may be important that probably all eusocial Hymenoptera evolved from species with a poisonous sting (Table 1). In animals with poor defense capability, clumping usually attracts predators (e.g. 43). In species with efficient weapons, however, communal defense can reduce the risk of nest predation (7). Good defense capability may help explain why eusociality has evolved so frequently among aculeate Hymenoptera: They have strong mandibles and excellent flight ability, and their main distinguishing feature is the poisonous sting. Comparable weapons are only found in certain termites (46). Thus, other nest-building insects probably have less potential for communal defense (Table 1). Certain eusocial aculeates—numerous ants and the meliponine bees—have secondarily lost the sting. Many of them have evolved other means of defense instead, which often involve biting and distributing sticky or poisonous substances (e.g. 97, 122). The reasons for such shifts in defense methods are poorly understood and merit further study.

Overlapping Adult Generations

Another precondition that should facilitate the evolution of eusociality is temporal overlap between successive adult generations, which would make it possible for daughters to help their mother raise later broods (e.g. 52, 95, 151). The importance of this precondition is not clear, however. Lin & Michener (87) and Alexander (4) emphasized that there is usually no temporal overlap between mother and daughters in solitary Hymenoptera (but see 97 and 146). This lack of overlap and other factors led Lin and Michener (87, 95) to suggest that eusociality has sometimes arisen through associations among females of the same generation (see discussion in 39).

Chromosome Numbers

Sherman (121a) and Seger (120a) independently noted that eusocial insects tend to have higher chromosome numbers than their nonsocial relatives. This observation applies to hymenopterans as well as termites. Sherman proposed the following explanation: The variance in relatedness between siblings increases as the number of chromosomes decreases (121a, 130a). Thus, a worker in a species with low chromosome numbers might be able to assist preferentially those sibs that are genetically most similar to the worker itself. Such favoritism would probably reduce the overall reproductive success of the queen. Eusociality might therefore evolve more easily in species with high chromosome numbers, where the variance in relatedness among sibs is lower and hence the reproductive interests between the workers and their mother more similar (45, 121a, 130a). Additional reasons for the correlation between eusociality and high chromosome numbers have been suggested (130a).

Seger (120a) proposed a radically different explanation. He showed that linkage disequilibrium between two loci can make selection favor (a) phenotypic responses to relative heterozygosity and (b) increased recombination between the loci involved. He also suggested that social insects' population structure is favorable to such selection. As the rate of recombination should increase with the number of chromosomes, these considerations may explain why the number seems to be higher in eusocial insects than in their solitary relatives.

The available evidence does not lend itself to a test between these two hypotheses, as Seger (120a) pointed out, because both make similar comparative predictions. For this and other reasons, it is difficult at present to judge the relative importance of high chromosome numbers as compared to other preconditions that facilitate the evolution of eusociality.

The Large Number of Species in Hymenoptera

One additional condition that may have contributed to the frequent evolution of eusociality in Hymenoptera is their sheer numbers. Brothers (23) recognized 38 families in Aculeata, containing on the order of 50,000 species (125), most of which build nests. Hence, the evolutionary "raw material" of species possessing the crucial precondition of nest building has probably been much larger in Hymenoptera than in any other group. For purely statistical reasons, one would therefore expect that ways of life conducive to eusociality have arisen more often in Hymenoptera than in other animals.

VERTEBRATES

Helpers Among Birds and Mammals

To summarize the discussion so far: It seems likely that kin selection, mutualism, and parental manipulation have all been important in the evolution of eusociality in Hymenoptera, albeit in proportions that vary from one species to another. The same processes may have been involved in Isoptera, but they are less well-known (40). Among the preconditions that have promoted the evolution of eusociality, parental care for offspring in a nest or other protected cavity seems the most important.

Do these conclusions also apply to vertebrates with helper systems? It has often been emphasized that eusociality has arisen independently at least 11 times in Hymenoptera and only once elsewhere—in termites—but Evans (54) argued that these numbers are highly uncertain. There is also increasing evidence that the Hymenoptera are not as unique as they once seemed. Eusocial-like systems with temporarily nonbreeding helpers also occur in vertebrates (123). Helpers have apparently evolved independently many times in birds (25, 49; see also Table 3) and mammalian carnivores (90; see also Table 4), and once in a rodent, the naked mole-rat (*Heterocephalus glaber*) (71). This rodent, which lives underground in burrows and feeds on bulbs and tubers, is unique among vertebrates in having castes with large and apparently lifelong differences in body size. One female and several males breed in each colony. The nonreproducing animals form two or three castes: The "non workers"

Table 3 Occurrence of helpers among birds

Variable	Altricial species	Precocial species
Proportion of bird fauna[a]	80–90%	10–20%
No. of families or subfamilies with helpers[b]	40	1

[a] Lack (81), Ar & Yom-Tov (11).
[b] Grimes (60), Rowley (114), Woolfenden (154), Zahavi (155); more recently discovered cases of helpers do not substantially alter this distribution.

are the largest mole-rats in the colony; the "frequent workers" and the "infrequent workers" forage and dig out the burrow system and usually remain considerably smaller than the reproductives and nonworkers throughout their lives. Eusociality in *Heterocephalus* shows several parallels to that in termites (71). Less advanced cooperative breeding is also found in primates (69), sciurids (68), capybaras (89a), and some fishes (130).

In birds and carnivores, helpers may breed later in life, even if many of them probably never will. This situation is similar to conditions among certain eusocial wasps (e.g. 146). In many eusocial Hymenoptera, workers lay unfertilized male eggs, and some even mate and produce offspring of both sexes (review in 21). As Vehrencamp (135) pointed out, there are thus great similarities between certain eusocial Hymenoptera and cooperatively breeding birds and carnivores. The main difference seems to be that the most advanced eusocial insects have morphologically specialized castes that are permanently incapable of reproduction. The reasons for this difference are poorly understood. Haplodiploidy and 3/4 relatedness among sisters has not necessarily

Table 4 Nonreproductive helpers in relation to parental care in carnivores[a]

	No. of species		Location where offspring are raised				
Family	Total	Studied in detail	With non-reproductive helpers	The open	Natural cavity	Nest or den prepared by another species	Nest or den prepared by parents	Parents bring prey to nest
Canidae	35	12	9		x	x	x	x
Ursidae	9	4	0	x			x	
Procyonidae	18	3	1	x	x	x		x
Mustelidae	70	10	0		x	x	x	x
Viverridae	75	2	2		x	x	x	x
Hyaenidae	4	4	1		x	x	x	x
Felidae	36	10	0	x	x	x		x

[a] Reviews in Walker et al (139) and Macdonald & Moehlman (90).

been important for the evolution of even the most advanced ants, as the most complex caste systems occur in the diploid termites (109). Differences between insects and vertebrates in ecology and life spans may be pivotal (78). For example, helpers in vertebrates often survive more than one season, but this is less likely for insect workers. In addition, nest construction takes relatively longer in many hymnenopterans than in vertebrates (78). Therefore, the chances that an insect worker will successfully found a nest are probably lower than for a vertebrate helper. The expected benefits of retaining breeding capacity, therefore, should be higher for helpers in vertebrates than in insects.

Provisioning of Offspring in a Nest or Den

The main precondition for eusociality in vertebrates, as in insects, seems to be raising the young in a nest or other protected cavity, where they are provisioned repeatedly with food (26, 135). This behavior characterizes not only the solitary relatives of eusocial Hymenoptera and termites (Table 1) but also birds (Table 3) and carnivores (Table 4). Out of about 41 families or subfamilies of birds in which helpers are known to occur (60, 114, 154, 155), all except one (Rallidae) are altricial, feeding their young in a nest. Yet precocial species, with chicks that leave the nest and gather their own food soon after hatching, make up 10–20% of most bird faunas (11, 81). Hence, helpers in birds occur disproportionately often among species that raise their young in a nest (Table 3). Woolfenden (154) remarked that cooperative breeding in birds appears to be closely linked with active nest defense.

Some of the figures for carnivores in Table 4 are rough approximations, but it seems clear that helpers occur particularly often among species that repeatedly provide prey to the offspring in a nest or den. Compare for instance bears (Ursidae), where prey provisioning is rare or absent and where helpers have not been recorded, with canids, which bring prey to the den over a long period and in which helpers often occur (90).

Constraints on Breeding

Although repeated provisioning for offspring in a nest facilitates the evolution of helpers, it is not a sufficient condition (135). Many nest-building insects, birds, and mammals with repeated food provisioning are solitary. In carnivores, helping behavior is often well developed in species feeding on large prey whose capture requires cooperation by many hunters (e.g. the lion *Panthera leo,* the hunting dog *Lycaon pictus,* the wolf *Canis lupus,* and probably also the dhole *Cuon alpinus;* reviewed in 90). Among birds, species with helpers seem to be characterized by a lack of potential mates for one of the sexes, by habitat saturation (e.g. 24, 26, 77, 121), or by low levels of food availability in stable (108) as well as fluctuating and unpredictable environments (50).

In both birds and mammals, helper systems are thus associated with difficulties for young animals in gaining breeding status or for solitary pairs in reproducing on their own. Whereas the reasons are not identical, a similar situation also applies to eusocial insects, where solitary breeders are often disfavored by the long time required for nest building and by the high risks of parasitism and predation (54, 87, 97, 126). Similar ecological conditions therefore seem to underlie the evolution of nonreproductive helpers among insects, birds, and mammals (26, 78, 135). It is hard to test, however, whether difficulties for young adults in gaining breeding status have preceded the evolution of eusociality. Such obstacles are probably partly a consequence of the eusocial system, which often makes it harder for young or solitary pairs to reproduce (85). The likelihood of successful breeding need not have been lower in species that later evolved sociality than in species that did not. The important factor seems to be that helpers must have the opportunity to increase their inclusive fitness by helping rather than by attempting the breed.

The Role of Kin Selection

What role has indirect kin selection played in the evolution of helper systems? Helpers are usually related to the reproductive adults in eusocial-like canids, and the average coefficient of relatedness has been estimated at 0.125–0.5 for groups of hunting dogs, foxes, jackals, and lions—i.e. ranging roughly from cousins to full siblings (90). In birds helpers are also often closely related to the reproductives (25), as is usually the case in eusocial Hymenoptera (40), and indirect kin selection seems to be important in many helper systems (26, 135). In a few species, helpers are often unrelated to the reproductives (87, 113); long-term mutualism may be important for the coherence of these groups (25, 27, 58, 85).

High relatedness among helpers and reproductives today need not imply that indirect kin selection was as important during the evolution of eusociality as it now seems to be. In birds and mammals that are cooperative breeders and in some of the eusocial insects, helpers are not permanently sterile but may breed later in life (reviewed in 21, 25). In these cases, the prospects of future reproduction could make it favorable for temporary nonbreeders to remain in the group. There is also selection for kin recognition and for sharing of benefits with close relatives in mutualistic groups if this raises the inclusive fitness of the participants. Close kinship between workers and reproductives could therefore be a secondary feature (107, 144).

EXPLANATIONS OF EUSOCIALITY: A REFLECTION

The main conclusion from this review is that eusociality in insects and similar social systems among vertebrates has evolved in taxa sharing constellations of behavioral and ecological preconditions that provided favorable starting points.

The most crucial precondition seems to be parental care, including feeding and defense of offspring in a nest or other protected cavity. In the processes leading to eusociality, it seems likely that mutualism, parental manipulation, and indirect kin selection have all been important.

Given the persistent evidence that Hamilton's (63) 3/4 relatedness hypothesis at most provides part of the explanation for the predominance of hymenopterans among eusocial species, why did the hypothesis have such a strong impact on the secondary literature, where it often overshadows the importance of other factors? The main reason is probably that Hamilton's novel, elegant, and simple 3/4 relatedness mechanism seemed eminently plausible, especially when combined with female-biased sex ratios (134). Not until gene frequency changes were simulated exactly was it discovered that the mechanism may be less powerful than it intuitively appears to be (36, 37).

A second reason for the popularity of this hypothesis could be the appeal of quantitative genetic arguments as compared to a morass of morphological, behavioral, and ecological preconditions, the relative importance of which are hard to quantify. (This may be a scientific version of "the street lamp temptation," which overwhelmed the couple who dropped their key one night when unlocking the door. After searching in vain in the darkness, they noticed the bright light beneath the street lamp and so searched there instead.)

Third, there has been an emphasis in sociobiology on "strong inference" (112) based on alternative rather than complementary hypotheses (e.g. 152, 153). When feasible, this approach has much to say for it, but clearcut answers in terms of a single hypothesis often are not possible in evolutionary biology. Strong inference works excellently when the hypotheses are qualitative and mutually exclusive, as for example, when a physicist asks "Do or don't W particles with such and such properties exist?" It may be enough for the physicist to record five identical specified events on her detector to proclaim the existence of W particles and hence confirm a fundamental similarity between the electromagnetic and weak forces. But the situation is very different for biologists who try to explain the evolution of a trait such as eusociality. Here, different hypotheses usually are not mutually exclusive, and the main problem is to identify and estimate the relative importance of each factor (27, 109). The time scale over which evolution occurs often precludes experimental tests, and the comparative method—with all its pitfalls—may be the only tool available (34). In addition, there can be joint causation by several necessary factors, making it impossible even in theory to label one as more important than another (e.g. 67, 74). Nevertheless, there is a great need for approaches that attempt to evaluate the relative importance of different preconditions and mechanisms for the evolution of eusociality and helper behavior (26, 135, 144). Based on the approach Hamilton pioneered (63), Brown (27) has developed a life-history formulation of inclusive fitness in helper systems with overlapping generations, which should make it easier to estimate the relative magnitude of direct and indirect fitness components. Their importance also needs to be evaluated relative to conditions that influence the scope for and consequences of aid giving. Recent work has shed light on such ecological and behavioral conditions and has widened our understanding of the evolution of eusociality and helper systems.

SUMMARY

Eusociality—with overlapping adult generations, cooperative brood care, and nonreproductive workers—has evolved often in Hymenoptera but perhaps only once among other insects (termites). Hamilton's 3/4 relatedness hypothesis suggests that the haplodiploid sex determination in Hymenoptera has been the main reason for this difference. It makes a female more closely related ($r = 3/4$) to her sisters than to daughters of her own ($r = 1/2$). A female might therefore propagate her genes better by helping raise reproductive sisters than by breeding.

Other factors have also been essential, and haplodiploidy may actually have been less important than certain other traits. Doubts about the 3/4 relatedness hypothesis are based on: (*a*) the occurrence of haplodiploidy in large groups of solitary Hymenoptera and other arthropods; (*b*) multiple matings and egg-laying by several queens in the colony, both of which reduce the relatedness among nest mates; (*c*) queen-controlled sex ratios (close to 1:1), which prevent workers from capitalizing on the high relatedness among sisters; and (*d*) evidence from gene frequency models. The conclusions derived from such models are contradictory and differ depending on the assumptions that are made; additional theoretical and empirical work seems necessary. Mutualism and manipulation have probably been involved in the evolution of eusociality, but they do not seem sufficient to explain nonreproductive workers. Relatedness among colony members also appears necessary. Hence mutualism, manipulation, and kin selection may all have contributed to the evolution of eusociality in Hymenoptera.

These factors do not explain, however, why sterile workers have evolved so often in that order. Parental care for offspring in a nest or other protected cavity has probably been crucial; it is found in more species in Hymenoptera than in any other group. Nest building and the feeding of larvae provide possibilities for workers to help. Building requires considerable work, so females can save time and energy by sharing a nest. The presence of several females enables them to improve offspring survival by sharing duty as nest guards and by attacking parasites and predators communally. Their poisonous stings can make such attacks particularly effective.

Similar preconditions characterize both termites and the eusocial-like vertebrates. Of the about 41 families of birds with helpers, all except one are altricial, feeding their young in a nest. Yet 10–20 % of bird faunas are made up of species that do not raise their young in a nest. Similarly, among carnivores nonreproductive helpers mainly occur in species that bring prey to their offspring in a nest or den.

Among insects as well as vertebrates, the evolution of helpers apparently has been promoted by conditions that make it difficult for young adults to gain breeding status or for solitary pairs to reproduce successfully on their own. Similar behavioral and ecological preconditions therefore seem to underlie the evolution of nonreproductive helpers among insects, birds, and mammals.

ACKNOWLEDGMENTS

I thank Jane Brockmann, Jerram Brown, Frank Götmark, Charles Michener, Linda Partridge, Jon Seger, and David Winkler for very useful suggestions on the manuscript; David Macdonald for comments on the vertebrate section; Robert Holt and John Wenzel for specific comments; and Brian Charlesworth, Bengt Gunnarson, Paul Harvey, John Maynard Smith, and Kjell Wallin for their discussions. Dr. Brockmann sent me her excellent review (21) at an early stage, which was very helpful. Dr. Seger kindly clarified several aspects of his and other genetic models. This work was supported by the Swedish Natural Sciences Research Council.

Literature Cited

1. Abrams, J., Eickwort, G. C. 1981. Nest switching and guarding by the communal sweat bee *Agapostemon virescens* (Hymenoptera, Halictidae). *Insectes Soc.* 28:105–16

2. Abugov, R. 1981. Non-linear benefits and the evolution of eusociality in the Hymenoptera. *J. Theor. Biol.* 88:733–42

3. Alcock, J. 1975. Social interactions in the solitary wasp *Cerceris simplex* (Hymenoptera: Sphecidae). *Behaviour* 54:142–52

4. Alexander, R. D. 1974. The evolution of social behavior. *Ann. Rev. Ecol. Syst.* 4:325–83

5. Alexander, R. D., Sherman, P. W. 1977. Local mate competition and parental investment in social insects. *Science* 196:494–500

6. Alexander, R. D., Tinkle, D. W., eds. 1981. *Natural Selection and Social Behavior: Recent Research and New Theory.* New York: Chiron. 532 pp.

7. Andersson, M., Wiklund, C. G. 1978. Clumping versus spacing out: Experiments on nest predation in fieldfares (*Turdus pilaris*). *Anim. Behav.* 26:1207–12

8. Aoki, K., Moody, M. 1981. One- and two-locus models of the origin of worker behavior in Hymenoptera. *J. Theor. Biol.* 89:449–74

9. Aoki, S. 1977. *Colophina clematis* (Homoptera, Pemphigidae), an aphid species with "soldiers". *Kontyû* 45:276–82

10. Aoki, S. 1982. Soldiers and altruistic dispersal in aphids. See Ref. 17, pp. 154–58

11. Ar, A., Yom-Tov, Y. 1978. The evolution of parental care in birds. *Evolution* 32:655–69

12. Bartz, S. H. 1979. Evolution of eusociality in termites. *Proc. Natl. Acad. Sci. USA* 76:5764–68

13. Bartz, S. H. 1982. On the evolution of male workers in the Hymenoptera. *Behav. Ecol. Sociobiol.* 11:223–28

14. Batra, S. W. T. 1978. Aggression, territoriality, mating and nest aggregation of some solitary bees (Halictidae, Megachilidae, Colletidae, Anthophoridae). *J. Kans. Entomol. Soc.* 51:547–59

15. Borgia, G. 1980. Evolution of haplodiploidy: Models for inbred and outbred systems. *Theor. Popul. Biol.* 17:103–28

16. Brandmayr, P. 1977. Ricerche etolo-

giche e morfofunzionali sulle cure parentali in Carabidi Pterostichini (Coleoptera: Carabidae, Pterostichinae). *Redia* 60:275–316

17. Breed, M. S., Michener, C. D., Evans, H. E., eds. 1982. *The Biology of Social Insects*. Boulder, Colo: Westview. 419 pp.

18. Brian, M. V. 1980. Social control over sex and caste in bees, wasps and ants. *Biol. Rev.* 55:379–415

19. Bro Larsen, E. 1952. On subsocial beetles from the saltmarsh, their care of progeny and adaptation to salt and tide. *Trans. 11th Int. Congr. Entomol., Amsterdam, 1951,* 1:502–6

20. Deleted in proof

21. Brockmann, H. J. 1984. The evolution of social behavior in insects. See Ref. 79

22. Brockmann, H. J., Dawkins, R. 1979. Joint nesting in a digger wasp as an evolutionarily stable preadaptation to social life. *Behaviour* 71:203–45

23. Brothers, D. J. 1975. Phylogeny and classification of the aculeate Hymenoptera, with special reference to Mutillidae. *Univ. Kans. Sci. Bull.* 50:483–648

24. Brown, J. L. 1969. Territorial behavior and population regulation in birds. *Wilson Bull.* 81:293–329

25. Brown, J. L. 1978. Avian communal breeding systems. *Ann. Rev. Ecol. Syst.* 9:123–55

26. Brown, J. L. 1983. Cooperation—a biologist's dilemma. *Adv. Study Behav.* 13:1–37

27. Brown, J. L. 1984. The evolution of helping behavior—an ontogenetic and comparative perspective. In *The Evolution of Adaptive Skills: Comparative and Ontogenetic Approaches,* ed. E. S. Gollin. New York: Academic. In press

28. Bull, J. J. 1981. Coevolution of haplodiploidy and sex determination in the Hymenoptera. *Evolution* 35:568–80

29. Buskirk, R. E. 1981. Sociality in the Arachnida. See Ref. 66a, pp. 281–367

30. Charlesworth, B. 1978. Some models of the evolution of altruistic behaviour between siblings. *J. Theor. Biol.* 72:297–319

31. Charnov, E. L. 1978. Evolution of eusocial behavior: Offspring choice or parental parasitism? *J. Theor. Biol.* 75:451–65

32. Charnov, E. L. 1978. Sex-ratio selection in eusocial Hymenoptera. *Am. Nat.* 112:317–26

33. Cleveland, L. R., Hall, S. R., Sanders, E. P., Collier. J. 1934. The wood-feeding roach *Cryptocercus,* its Protozoa, and the symbiosis between Protozoa and roach. *Mem. Am. Acad. Arts Sci.* 17:185–342

34. Clutton-Brock, T. H., Harvey, P. H. 1984. Comparative approaches to investigating adaptation. See Ref. 79

35. Cole, B. J. 1983. Multiple mating and the evolution of social behaviour in the Hymenoptera. *Behav. Ecol. Sociobiol.* 12:191–201

36. Craig, R. 1979. Parental manipulation, kin selection, and the evolution of altruism. *Evolution* 33:319–34

37. Craig, R. 1980. Sex ratio changes and the evolution of eusociality in the Hymenoptera: Simulation and games theory studies. *J. Theor. Biol.* 87:55–70

38. Craig, R. 1982. Evolution of male workers in Hymenoptera. *J. Theor. Biol.* 94:95–105

39. Crozier, R. H. 1979. Genetics of sociality. In *Social Insects,* ed. H. R. Hermann, 1:223–86. New York: Academic

40. Crozier, R. H. 1980. Genetical structure of social insect populations. In *Evolution of Social Behavior: Hypotheses and Empirical Tests,* ed. H. Markl, pp. 129–46. Weinheim, West Germany: Chemie

41. Crozier, R. H. 1982. On insects and insects: Twists and turns in our understanding of the evolution of eusociality. See Ref. 17, pp. 4–9

42. Cruz, Y. P. 1981. A sterile defender morph in a polyembryonic hymenopterous parasite. *Nature* 294:446–47

43. Curio, E. 1976. *The Ethology of Predation*. Berlin: Springer-Verlag. 250 pp.

44. Darwin, C. 1859. *The Origin of Species by Natural Selection*. London: Murray. 502 pp.

45. Dawkins, R. 1982. *The Extended Phenotype*. Oxford: Freeman. 307 pp.

46. Deligne, J., Quennedey, A., Blum, M. S. 1981. The enemies and defense mechanisms of termites. See Ref. 66a, pp. 1–76

47. Eberhard, W. G. 1972. Altruistic behavior in a sphecid wasp: Support for kin-selection theory. *Science* 175:1390–91

48. Eickwort, G. C. 1981. Presocial insects. See Ref. 66a, pp. 199–280

49. Emlen, S. T. 1978. Cooperative breeding. In *Behavioural Ecology: An Evolutionary Approach,* ed. J. R. Krebs, N. B. Davies, pp. 245–81. Oxford: Blackwell

50. Emlen, S. T. 1982. The evolution of helping. I. An ecological constraints model. *Am. Nat.* 119:29–39

51. Entwistle, P. F. 1964. Inbreeding and arrhenotoky in the ambrosia beetle *Xyleborus compactus* (Eichh.) (Coleoptera: Scolytidae). *Proc. R. Entomol. Soc. London Ser. A* 39:83–88

52. Evans, H. E. 1958. The evolution of so-

cial life in wasps. *Proc. 10th Int. Congr. Entomol. Montreal, 1956,* 2:449–57
53. Evans, H. E. 1973. Burrow sharing and nest transfer in the digger wasp *Philanthus gibbosus* (Fabricius). *Anim. Behav.* 21:302–8
54. Evans, H. E. 1977. Extrinsic versus intrinsic factors in the evolution of insect sociality. *BioScience* 27:613–17
55. Evans, H. E., West-Eberhard, M. J. 1970. *The Wasps.* Ann Arbor: Univ. Mich. Press. 265 pp.
56. Flesness, N. 1978. Kinship asymmetry in diploids. *Nature* 276:495–96
57. Gamboa, G. J. 1978. Intraspecific defence: Advantage of social cooperation among paper wasp foundresses. *Science* 199:1463–65
58. Gaston, A. J. 1978. The evolution of group territorial behavior and cooperative breeding. *Am. Nat.* 112:1091–1100
59. Gibo, D. L. 1978. The selective advantage of foundress associations in *Polistes fuscatus* (Hymenoptera: Vespidae): A field study of the effects of predation on productivity. *Can. Entomol.* 110:519–40
60. Grimes, L. G. 1976. Cooperative breeding in African birds. *Proc. 16th Int. Ornithol. Congr., Canberra 1974,* pp. 667–73
61. Halffter, G. 1977. Evolution of nidification in the Scarabaeinae (Coleoptera: Scarabaeidae). *Quaest. Entomol.* 13: 231–53
62. Halffter, G. 1982. Evolved relations between reproductive and subsocial behaviors in Coleoptera. See Ref. 17, pp. 164–70
63. Hamilton, W. D. 1964. The genetical evolution of social behaviour. I and II. *J. Theor. Biol.* 7:1–52
64. Hamilton, W. D. 1967. Extraordinary sex ratios. *Science* 156:477–88
65. Hamilton, W. D. 1972. Altruism and related phenomena, mainly in social insects. *Ann. Rev. Ecol. Syst.* 3:193–232
66. Hamilton, W. D. 1978. Evolution and diversity under bark. In *Diversity of Insect Faunas, Symp. R. Entomol. Soc. London,* ed. L. A. Mound, N. Waloff, 9:154–75
66a. Hermann, H. R., ed. 1981. *Social Insects,* Vol. 2. New York: Academic. 491 pp.
67. Hilborn, R., Stearns, S. C. 1982. On inference in ecology and evolutionary biology: The problem of multiple causes. *Acta Biotheor.* 31:145–64
68. Hoogland, J. L. 1983. Black-tailed prairie dog coteries are cooperatively breeding units. *Am. Nat.* 121:275–80
69. Hrdy, S. B. 1976. The care and exploitation of non-human primate infants by conspecifics other than the mother. *Adv. Study Behav.* 6:101–58
70. Iwasa, Y. 1981. Role of sex ratio in the evolution of eusociality in haplodiploid social insects. *J. Theor. Biol.* 93:125–42
71. Jarvis, J. U. M. 1981. Eusociality in a mammal: Cooperative breeding in naked mole-rat colonies. *Science* 212:571–73
72. Jeanne, R. 1979. Construction and utilization of multiple combs in *Polistes canadensis* in relation to the biology of a predacious moth. *Behav. Ecol. Sociobiol.* 4:293–310
73. Kasuya, E. 1982. Factors governing the evolution of eusociality through kin selection. *Res. Popul. Ecol.* 24:174–92
74. Kempthorn, O. 1978. Logical, epistemological and statistical aspects of nature-nurture data interpretation. *Biometrics* 34:1–23
75. Kennedy, J. S. 1966. Some outstanding questions in insect behaviour. In *Insect Behaviour, Symp. R. Entomol. Soc. London,* ed. P. T. Haskell, 3:97–112
76. Klahn, J. E. 1979. Philopatric and non-philopatric foundress association in the social wasp *Polistes fuscatus. Behav. Ecol. Sociobiol.* 5:417–24
77. Koenig, W. D., Pitelka, F. H. 1981. Ecological factors and kin selection in the evolution of cooperative breeding in birds. See Ref. 6, pp. 261–80
78. Krebs, J. R., Davies, N. B. 1981. *An Introduction to Behavioural Ecology.* Oxford: Blackwell. 292 pp.
79. Krebs, J. R., Davies, N. B., eds. 1984. *Behavioural Ecology: An Evolutionary Approach.* Oxford: Blackwell. 2nd ed. In press.
80. Krombein, K. V., Hurd, P. D., Smith, D. R., Burks, B. D. 1979. *Catalog of Hymenoptera in America North of Mexico.* Washington DC: Smithsonian Inst. 2209 pp.
81. Lack, D. 1968. *Ecological Adaptations for Breeding in Birds.* London: Methuen. 409 pp.
82. Lacy, R. C. 1980. The evolution of eusociality in termites: A haplodiploid analogy? *Am. Nat.* 116:449–51
82a. Lacy, R. C. 1984. The evolution of termite eusociality: Reply to Leinaas. *Am. Nat.* 123:876–78
83. Lamb, R. J. 1976. Parental behavior in the Dermaptera with special reference to *Forficula auricularia* (Dermaptera: Forficulidae). *Can. Entomol.* 108:609–19
84. Leinaas, H. P. 1983. A haplodiploid analogy in the evolution of termite eusociality? Reply to Lacy. *Am. Nat.* 121: 302–4
84a. Lester, L. J., Selander, R. K. 1981. Genetic relatedness and the social orga-

nization of *Polistes* colonies. *Am. Nat.* 117:147–66

85. Ligon, J. D. 1983. Cooperation and reciprocity in avian social systems. *Am. Nat.* 121:366–84

86. Lin, N. 1964. Increased parasitic pressure as a major factor in the evolution of social behavior in halictine bees. *Insectes Soc.* 11:187–92

87. Lin, N., Michener, C. D. 1972. Evolution of sociality in insects. *Q. Rev. Biol.* 47:131–59

88. Litte, M. 1977. Behavioral ecology of the social wasp, *Mischocyttarus mexicanus. Behav. Ecol. Sociobiol.* 2:229–46

89. Litte, M. 1981. Social biology of the polistine wasp *Mischocyttarus labiatus:* Survival in a Colombian rain forest. *Smithson. Contrib. Zool.* 327:1–27

89a. Macdonald, D. W. 1981. Dwindling resources and the social behaviour of Capybaras, (*Hydrochoerus hydrocaeris*) (Mammalia). *J. Zool.* 194:371–91

90. Macdonald, D. W., Moehlman, P. D. 1982. Cooperation, altruism, and restraint in the reproduction of carnivores. In *Perspectives in Ethology,* ed. P. G. Bateson, P. H. Klopfer, 5:433–67. New York: Plenum

91. Maynard Smith, J. 1964. Group selection and kin selection. *Nature* 201:1145–47

92. Maynard Smith, J. 1982. The evolution of social behaviour—a classification of models. In *Current Problems in Sociology,* ed. King's Coll. Sociobiol. Group, pp. 29–44. Cambridge: Cambridge Univ. Press

93. Metcalf, R. A. 1980. Sex ratios, parent-offspring conflict, and local competition for mates in the social wasps *Polistes metricus* and *Polistes variatus. Am. Nat.* 116:642–54

94. Metcalf, R. A., Whitt, G. S. 1977. Relative inclusive fitness in the social wasp *Polistes metricus. Behav. Ecol. Sociobiol.* 2:353–60

95. Michener, C. D. 1958. Evolution of social behavior in bees. *Proc. 10th Int. Congr. Entomol., Montreal, 1956,* 2:442–47

96. Michener, C. D. 1969. Comparative social behavior of bees. *Ann. Rev. Entomol.* 14:229–342

97. Michener, C. D. 1974. *The Social Behavior of the Bees: A Comparative Study.* Cambridge, Mass: Harvard Univ. Press. 404 pp.

98. Michener, C. D. 1984. From solitary to eusocial: Need there be a series of intervening species? *Fortschr. Zool.* In press

99. Michener, C. D., Brothers, D. J. 1974. Were workers of eusocial Hymenoptera initially altruistic or oppressed? *Proc. Natl. Acad. Sci. USA* 71:671–74

100. Michener, C. D., Brothers, D. J., Kamm, D. R. 1971. Interactions in colonies of primitively social bees: Artificial colonies of *Lasioglossum zephyrum. Proc. Natl. Acad. Sci. USA* 68:1241–45

101. Michener, C. D., Lange, R. B., Bigarella, J. J., Salamuni, R. 1958. Factors influencing the distribution of bees' nests in earth banks. *Ecology* 39:207–17

102. Michod, R. E. 1982. The theory of kin selection. *Ann. Rev. Ecol. Syst.* 13:23–55

103. Miller, R. C., Kurczewski, F. E. 1973. Intraspecific interactions in aggregations of *Lindenius* (Hymenoptera: Sphecidae, Crabroninae). *Insectes Soc.* 20:365–78

104. Milne, L. J., Milne, M. 1976. The social behavior of burrowing beetles. *Sci. Am.* 235 (2):84–89

105. Nalepa, C. A. 1982. Colony composition of the woodroach *Cryptocercus punctulatus.* See Ref. 17 p. 181

106. Noonan, K. M. 1978. Sex ratio of parental investment in colonies of the social wasp *Polistes fuscatus. Science* 199:1354–56

107. Noonan, K. M. 1981. Individual strategies of inclusive-fitness-maximizing in *Polistes fuscatus* foundresses. See Ref. 6, pp. 18–44

108. Orians, G. H., Orians, C. E., Orians, K. J. 1977. Helpers at the nest in some Argentine blackbirds. In *Evolutionary Ecology,* ed. B. Stonehouse, C. Perrins, pp. 137–51. London: Univ. Park

109. Oster, G. F., Wilson, E. O. 1978. *Caste and Ecology in the Social Insects.* Princeton: Princeton Univ. Press. 352 pp.

110. Page, R. E., Metcalf, R. A. 1982. Multiple mating, sperm utilization, and social evolution. *Am. Nat.* 119:263–81

111. Pamilo, P., Rosengren, R. 1983. Sex ratio strategies in *Formica* ants. *Oikos* 40:24–35

112. Platt, J. R. 1964. Strong inference. *Science* 146:347–53

113. Rood, J. P. 1978. Dwarf mongoose helpers at the den. *Z. Tierpsychol.* 48:277–87

114. Rowley, I. 1976. Co-operative breeding in Australian birds. *Proc. 16th Int. Ornithol. Congr., Canberra, 1974,* pp. 657–66

115. Rozen, J. G. Jr., Eickwort, K. R., Eickwort, G. C. 1978. The bionomics and immature stages of the cleptoparasitic bee genus *Protepeolus* (Anthophoridae, Nomadinae). *Am. Mus. Novit.* 2640:1–24

116. Sakagami, S. F., Maeta, Y. 1977. Some presumably presocial habits of Japanese

Ceratina bees, with notes on various social types in Hymenoptera. *Insectes Soc.* 24:319–43

117. Schedl, K. E. 1958. Breeding habits of arboricole insects in Central Africa. *Proc. 10th Int. Congr. Entomol., Montreal, 1956,* 1:183–97

118. Schmitz, R. F. 1972. Behavior of *Ips pini* during mating, oviposition, and larval development (Coleoptera: Scolytidae). *Can. Entomol.* 104:1723–28

119. Seelinger, G., Seelinger, U. 1983. On the social organisation, alarm and fighting in the primitive cockroach *Cryptocercus punctulatus* Scudder. *Z. Tierpsychol.* 61:315–33

120. Seger, J. 1983. Partial bivoltinism may cause alternating sex-ratio biases that favour eusociality. *Nature* 301:59–62

120a. Seger, J. 1983. Conditional relatedness, recombination, and the chromosome number of insects. In *Advances in Herpetology and Evolutionary Biology,* ed. A. G. J. Rhodin, K. Miyata, pp. 596–612. Cambridge, Mass: Harvard Univ. Press. 725 pp.

121. Selander, R. K. 1964. Speciation in wrens of the genus *Campylorhynchus.* *Univ. Calif. Berkeley Publ. Zool.* 74:1–305

121a. Sherman, P. W. 1979. Insect chromosome numbers and eusociality. *Am. Nat.* 113:925–35

122. Skaife, S. H., Ledger, J. 1979. *African Insect Life.* London: Country Life. 279 pp.

123. Skutch, A. F. 1961. Helpers among birds. *Condor* 63:198–226

124. Snelling, R. R. 1981. Systematics of social Hymenoptera. See Ref. 66a, pp. 369–453

125. Spradbery, J. P. 1973. *Wasps.* Seattle: Univ. Wash. Press. 408 pp.

126. Starr, C. K. 1979. Origin and evolution of insect sociality: A review of modern theory. In *Social Insects,* ed. by H. R. Hermann, 1:35–79. New York: Academic

127. Strassmann, J. E. 1981. Wasp reproduction and kin selection: Reproductive competition and dominance hierarchies among *Polistes annularis* foundresses. *Fla. Entomol.* 64:74–88

128. Sturtevant, A. H. 1938. Essays on evolution. II. On the effects of selection on social insects. *Q. Rev. Biol.* 13:74–76

129. Syren, R. M., Luykx, P. 1977. Permanent segmental interchange complex in the termite *Incisitermes schwarzi.* *Nature* 266:167–68

130. Taborsky, M., Limberger, D. 1981. Helpers in fish. *Behav. Ecol. Sociobiol.* 8:143–45

130a. Templeton, A. R. 1979. Chromosome number, quantitative genetics, and eusociality. *Am. Nat.* 113:937–41

131. Thiele, H.-U. 1977. *Carabid Beetles in Their Environments.* Berlin: Springer-Verlag. 369 pp.

132. Thorne, B. 1982. Multiple primary queens in termites: Phyletic distribution, ecological context, and a comparison to polygyny in Hymenoptera. See Ref. 17, pp. 206–11

133. Thorp, R. W. 1969. Ecology and behavior of *Anthophora edwardsii* (Hymenoptera: Anthophoridae). *Am. Midl. Nat.* 82:321–37

134. Trivers, R. L., Hare, H. 1976. Haplodiploidy and the evolution of social insects. *Science* 191:249–63

135. Vehrencamp, S. L. 1979. The roles of individual, kin and group selection in the evolution of sociality. In *Handbook of Behavioral Neurobiology.* Vol. 3, *Social Behavior and Communication,* ed. P. Marler, J. G. Vandenbergh, pp. 351–94. New York: Plenum

136. Wade, M. 1979. The evolution of social interactions by family selection. *Am. Nat.* 113:399–411

137. Wade, M. 1982. The effect of multiple insemination on the evolution of social behaviors in diploid and haplo-diploid organisms. *J. Theor. Biol.* 95:351–68

138. Wade, M. J., Breden, F. 1981. Effect of inbreeding on the evolution of altruistic behavior by kin selection. *Evolution* 35:844–58

139. Walker, E. P., Warnick, F., Hamlet, S. E., Lange, K. I., Davis, M. A., Uible, H. E., Wright, P. F. 1964. *Mammals of the World,* Vol. 2. Baltimore: John Hopkins Univ. Press. 1500 pp.

140. Waloff, N. 1957. The effect of the number of queens of the ant *Lasius flavus* (Fab.) (Hym., Formicidae) on their survival and on the rate of development of the first brood. *Insectes Soc.* 4:391–408

141. Ward, P. S. 1983. Genetic relatedness and colony organization in a species complex of ponerine ants. I. Patterns of sex ratio investment. *Behav. Ecol. Sociobiol.* 12:301–7

142. West, M. J., Alexander, R. D. 1963. Sub-social behavior in a burrowing cricket *Anurogryllus muticus* (DeGeer) Orthoptera: Gryllidae. *Ohio J. Sci.* 63:19–24

143. West-Eberhard, M. J. 1969. The social biology of polistine wasps. *Misc. Publ. Mus. Zool. Univ. Mich.* 140:1–101

144. West-Eberhard, M. J. 1975. The evolution of social behavior by kin selection. *Q. Rev. Biol.* 50:1–33

145. West-Eberhard, M. J. 1978. Temporary

queens in *Metapolybia* wasps: Nonreproductive helpers without altruism? *Science* 200:441–43

146. West-Eberhard, M. J. 1978. Polygyny and the evolution of social behavior of wasps. *J. Kans. Entomol. Soc.* 51:832–56

147. West-Eberhard, M. J. 1981. Intragroup selection and the evolution of insect societies. See Ref. 6, pp. 3–17

148. White, M. J. D. 1973. Animal Cytology and Evolution. Cambridge: Cambridge Univ. Press. 961 pp.

149. Williams, G. C., Williams, D. C. 1957. Natural selection of individually harmful social adaptations among sibs with special reference to social insects. *Evolution* 11:32–39

150. Wilson, E. O. 1966. Behaviour of social insects. See Ref. 75, pp. 81–95

151. Wilson, E. O. 1971. *The Insect Societies.* Cambridge, Mass: Harvard Univ. Press. 548 pp.

152. Wilson, E. O. 1975. *Sociobiology: The New Synthesis.* Cambridge, Mass: Harvard Univ. Press. 697 pp.

153. Wittenberger, J. F. 1981. *Animal Social Behavior.* Boston, Mass: Duxbury. 722 pp.

154. Woolfenden, G. E. 1976. Cooperative breeding in American birds. *Proc. Int. Ornithol. Congr., Canberra, 1974,* pp. 674–84

155. Zahavi, A. 1976. Cooperative breeding in Eurasian birds. *Proc. Int. Ornithol. Congr. Canberra, 1974,* pp. 685–94

Ann. Rev. Ecol. Syst. 1984. 15:191–232

LIFE HISTORY PATTERNS AND THE COMPARATIVE SOCIAL ECOLOGY OF CARNIVORES

Marc Bekoff and Thomas J. Daniels

Department of Environmental, Population, and Organismic Biology, University of Colorado, Boulder, Colorado 80309

John L. Gittleman

Department of Zoological Research, National Zoological Park, Smithsonian Institution, Washington, DC 20008

INTRODUCTION

The mammalian order Carnivora is characterized by a great range of behavioral, ecological, and morphological adaptations, as well as substantial intraspecific variability (i.e. behavioral scaling; see 324). For example, in wolves (see Table 1 for scientific names), body size ranges from 31 to 78 kg, litter size varies from 1 to 11, home-range size differs 50–100 fold, populations are found in every vegetational zone except tropical forests and arid deserts, and individuals may live alone, in pairs, or in large packs (124, 204, 332).

Despite such widespread variation, comparative analyses indicate that there also is remarkable consistency (86, 105) in the ways many diverse carnivores adapt to their habitats. Therefore, it is possible to highlight trends in the phylogeny of behavior and life history characteristics by drawing on data from numerous disciplines, including anatomy, physiology, taxonomy, behavior, and ecology (16, 54, 83, 84, 92, 93, 97, 128, 191, 196, 199, 242, 243, 313, 331).

Due to space limitations, we will primarily review field studies focusing on the variation in behavior, body size, and life histories and emphasize data collected on identified individuals that have been observed directly (sometimes supplemented by radio-tracking) over long periods of time. Such studies are

191

0066-4162/84/1120-0191$02.00

Table 1   Scientific and common names of Carnivores referred to in the text

Family Genus and species	Common name
Canidae	
Dusicyon culpaeus	South American fox (culpeo)
Dusicyon griseus	Chico gray fox
Fennecus zerda	Fennec fox
Lycaon pictus	African wild dog
Otocyon megalotis	Bat-eared fox
Alopex lagopus	Arctic fox
Canis lupus	Wolf
Canis latrans	Coyote
Canis aureus	Golden jackal
Canis mesomelas	Black-backed (silverbacked) jackal
Nyctereutes procyonoides	Raccoon dog
Vulpes vulpes	Red fox
Cuon alpinus	Dhole or red dog
Cerdocyon thous	Crab-eating fox
Speothos venaticus	Bush dog
Chrysocyon brachyurus	Maned wolf
Procyonidae	
Procyon lotor	Raccoon
Nasua narica	Coati
Ursidae	
Ursus americanus	Black bear
Thalarctos maritimus	Polar bear
Helarctos malayanus	Malayan sun bear
Ailuridae	
Ailurus fulgens	Red panda
Ailuropodidae	
Ailuropoda melanoleuca	Giant panda
Mustelidae	
Mustela nivalis	Least weasel
Mustela erminea	Stoat or ermine
Mustela vison	American mink
Mustela frenata	Long-tailed weasel
Meles meles	European badger
Lutra lutra	European otter
Enhydra lutris	Sea otter
Martes americana	American marten
Martes pennanti	Fisher
Gulo gulo	Wolverine
Mephitis mephitis	Striped skunk
Viverridae	
Helogale parvula	Dwarf mongoose
Suricata suricatta	Meerkat

Table 1   *(Continued)*

Family Genus and species	Common name
Hemigalus derbyanus	Banded palm civet
Ichneumia albicauda	White-tailed mongoose
Mungos mungo	Banded mongoose
Nandinia binotata	Palm civet
Osbornictus piscivorous	Fishing genet
Genetta genetta	Common genet
Fossa fossa	Fanaloka
Hyaenidae	
Hyaena hyaena	Striped hyena
Hyaena brunnea	Brown hyena
Crocuta crocuta	Spotted hyena
Proteles cristatus	Aardwolf
Felidae	
Herpailurus jagouaroundi	Jaguarondi
Leopardus geoffroyi	Geoffroy's cat
Puma concolor	Mountain lion or cougar
Felis margarita	Sand cat
Prionailurus bengalensis	Bengal cat
Panthera leo	African lion
Panthera tigris	Tiger
Panthera pardus	Leopard
Panthera onca	Jaguar
Acinonyx jubatus	Cheetah
Leopardus pardalis	Ocelot
Lynx canadensis	Lynx
Lynx rufus	Bobcat
Caracal caracal	Caracal

limited in number and comparative breadth, so information from shorter field studies and data on captive animals will also be used as a supplement. Comparisons of artificially and nonartificially fed groups must be treated carefully, however, since food resources have strong direct effects on social behavior (263). In addition, comparisons of exploited and (relatively) unexploited populations must take the food source into account (156). Finally, we will briefly discuss some practical aspects of data collection and analysis.

Our approach is pluralistic in two respects. First, rather than focusing on either intraspecific or interspecific variation, we emphasize the reciprocal exchange of information between both levels of analysis. Comparative studies generate hypotheses that often can only be verified in single-species studies, and the generality of functional explanations of a species' trait rests with comparative analyses among related taxa. Second, because confounding variables involving size constraints and phylogenetic effects are frequently associated with behavioral and ecological diversity (62, 64, 112, 113a), we include allometric and phylogenetic trends in our discussion of carnivore behavioral ecology.

Both descriptive studies and field experiments are still sorely needed for most carnivores. Textbooks, reviews, and popular articles frequently exaggerate our knowledge of certain phenomena (71). As Dunbar (75) stressed, long-term field studies must be performed if we are ever to learn about the evolution of the behavior and social ecology of long-lived "higher" species, in which intraspecific variation is so obvious.

GENERAL CHARACTERISTICS OF CARNIVORA

Broadly speaking, members of the order Carnivora are distinguished from other mammals by their carnassial dentition and the high proportion of vertebrates in their diets (*Carnivora* is derived from the Latin *caro: carnis* meaning "flesh" and *voro,* "to devour"). As in other mammalian orders, however, there are many interesting exceptions to these general characteristics. For example, the white-tailed mongoose and the bat-eared fox are insectivorous; the red panda and the giant panda feed primarily on bamboo; and the black bear maintains itself on a catholic herbivorous/frugivorous diet. Besides their dentition and diet, carnivores' other pronounced traits (93, 140, 242, 243, 271) include: (*a*) a jaw joint that is a transverse hinge, which facilitates biting and cutting but does not permit grinding action by the teeth; (*b*) a vertebral column that is strong and flexible and a long tail; (*c*) a brain that is relatively large, particularly in comparison to herbivores and insectivores; (*d*) anal and forehead scent glands that are well-developed and are used in marking, social recognition, and defense; (*e*) a walking gait that ranges from plantigrade to digitigrade; and (*f*) in most species, soft fur covered by longer guard hairs.

The order Carnivora is divided into two superfamilies, Canoidea and Feloidea, and seven polytypic families—Canidae, Ursidae, Procyonidae, Mustelidae, Viverridae, Hyaenidae, and Felidae. Although there is continuing controversy, growing evidence suggests that the red panda and giant panda belong in two monotypic families, Ailuridae and Ailuropodidae, respectively, rather than in the Procyonidae or Ursidae, as was previously thought (83, 251).

CANIDAE   The Canidae, with 36 species divided among 16 genera, is composed of small- to medium-sized carnivores (1–60 kg) distinguished by their cursorial mobility and strong jaws and cheek muscles (60, 226, 227, 285). Canids live in a wide variety of habitats. More behavioral and ecological information is available for Canidae than for any other carnivore family because they are typically diurnal and include an unusually large number of group-living species (4, 16, 24, 97, 105, 124, 157, 159, 191, 204, 331, 332). In many of these species, both parents provision and protect their young (158). Much of our discussion below centers on patterns in the Canidae.

PROCYONIDAE   Closely related to Canidae is Procyonidae, which includes 18 species in 6 genera. Procyonids are comparatively small carnivores (0.8–12 kg) confined to the New World, and all are semiarboreal, prefer temperate and tropical vegetational zones (144), and have a plantigrade gait. Most of them are solitary, although raccoons are frequently seen in extended family groups (102; Seidensticker, personal communication) and coatis live in female-banded groups with as many as 10 adults (143, 266, 267).

AILURIDAE AND AILUROPODIDAE   Both red pandas and giant pandas (placed in Ailuridae and Ailuropodidae, respectively) are from central Chinese provinces; red pandas also occur in Nepal, Sikkim, and northern Burma. These two species feed primarily on bamboo and are solitary except during the breeding season (155, 251).

URSIDAE   The family Ursidae (bears) consists of 7 species divided among 6 genera, which (with the exception of the Malayan sun bear) inhabit the Northern Hemisphere and northern South America. Their dentition reflects a shift away from carnivory toward herbivory: The anterior premolars are small and the carnassials are nonsectorial and have lost most of their shearing character. Ursids' walking gait is plantigrade. They are relatively solitary; adults only remain together during breeding or when food is abundant in patches. Maternal care may last for an extended length of time during harsh seasons (93, 129, 130), which, combined with small litter sizes and long interbirth intervals, results in comparatively low reproductive rates.

MUSTELIDAE   The last family included in the superfamily Canoidea is Mustelidae. It is divided into the following subfamilies: Mustelinae (weasels, polecats, fisher, martens, wolverine, tayra), Mellivorinae (badgers), Mephitinae (skunks), and Lutrinae (otters). Mustelidae consists of 67 species in all, distributed among 27 genera, and they are found everywhere except Australia, Antarctica, and Madagascar. Mustelids are small- to medium-sized carnivores (.025–45 kg) with long bodies, short legs, and usually medium to long tails. The aquatic Lutrinae and the omnivorous Mephitinae and Mellivorinae are recent radiations that diverged from the basic carnivorous characteristics of the family. Most species are solitary and live in forest habitats (89, 90, 296), although group living is found in European badgers (164, 165) and sea otters (147). Most species in this large family have only been studied in captivity.

VIVERRIDAE   The oldest lineage within the superfamily Feloidea is Viverridae. This family has retained many of the features thought to be representative of ancestral carnivores (Miacidae). The Viverridae is the largest carnivore family, is comprised of 70 species distributed among 39 genera, and is confined to the Old World tropics and subtropics. It has six subfamilies: Viverrinae (civets, genets, linsangs), Paradoxurinae (palm civets, binturong), Hemigalinae (fanaloka, banded palm civet), Galadinae (Madagascar mongooses), Herpestinae (mongooses, suricate), and Cryptoprocinae (fossa). Viverrids are small carnivores (0.45–14 kg) and generally have long tails and relatively short limbs that are semiplantigrade or digitigrade (the claws are partly retractile in digitigrade species). The family is a somewhat heterogeneous assemblage and is more ecologically diverse than any other carnivore group, ranging from solitary, forest-living species (common genet) to open savannah, group-living animals (banded and dwarf mongooses). Their dietetic affinities vary from insectivorous (banded palm civets) to frugivorous (palm civet) to piscivorous (fishing genet). Viverridae have not been studied in as much detail as other families (2, 110, 135, 152, 316).

HYAENIDAE   The family Hyaenidae contains three species of hyenas (in two genera) and the aardwolf. Hyenas are found in Africa, the Middle East, and central and south Asia; aardwolves live in southern Africa. Hyenas are fairly large animals (8–70 kg) with particularly large heads and powerful jaw and neck muscles. Their teeth are large and the carnassials are well developed. Their forequarters are heavier than their hindquarters, their legs are thick, and they have a digitigrade gait. Behaviorally and ecologically, the hyenas are diverse (212, 213, 216, 230). The striped hyena is omnivorous, foraging on fruits, insects, and small mammals (as well as scavenging); it lives in acacia savannah and open grassland habitats (163, 250). Brown hyenas are also omnivorous and live in small family groups of 4 to 14 individuals in open shrub or woodland savannah habitats (213–216, 229, 232). The spotted hyena is strictly carnivorous, with larger carnassials than the other hyenas, and it lives in "clans" of up to 80 individuals in Ngorongoro Crater or in temporary associations in the Serengeti (161, 162). The aardwolf is different from the hyenas, though it is sometimes confused with the striped hyena because of its similar appearance. It has a reduced dentition and a small head and shoulders, feeds mainly on termites, and is solitary (166).

FELIDAE   The last carnivore family, Felidae (cats, caracal, puma, ocelot, jaguarondi, lynx, leopard, jaguar, lion, cheetah), contains 37 species and 21 genera. The classification of Felidae remains somewhat uncertain, particularly with regard to the genera *Felis* and *Panthera* (227). We have adopted Ewer's taxonomic scheme (93). Felidae are found on all continents except Australia. They are small- to large-sized carnivores (1.5–300 kg), short faced, with highly sectorial carnassials and sharp retractile claws (except for cheetahs). The tongue is covered with curved, backwardly directed, horny papillae. Felids are highly carnivorous, and although they are found in every type of habitat, they most often reside in woodland and fringe-woodland terrain. All felids are solitary (157), except for lions, which live in prides of 2 to 18 adult females (usually related to one another) and 1 to 7 adult males (32, 33, 37, 50, 87, 157, 182, 273). In the central Kalahari, female pride mates are frequently unrelated; prides periodically disband and some females join "foreign" prides, especially when prey densities are low (M. J. Owens, personal communication). Male breeding coalitions appear to contain nonrelatives more frequently than was previously thought (234); males are typically unrelated to pride females. [Comparative data can be found in Eaton (77–80).]

LIFE HISTORY PATTERNS

Interspecific Comparisons

Cross-species comparisons of life history traits in Carnivora reveal extensive differences in rates and modes of reproduction. Undoubtedly, much of this variation may be accounted for by body size differences (83, 319), and such allometric effects must be incorporated into comparative studies. Even taxonomically related and similarly sized species differ markedly, however. In the Viverridae, the fanaloka and common genet both weigh an average of 1.7 kg, but their life histories are quite different. The fanaloka's gestation period is about 85 days, it usually has a litter size of one, and it weans its young about 52 days after birth (1, 2, 316), whereas the common genet's gestation period is about 72 days, its litter size is 3–4, and it weans its young around 175 days after birth (93, 253, 308, 316).

Because of the great variation within the order, carnivores are an excellent group for studying the adaptive patterns of life history traits. Furthermore, although some variables are difficult to measure (e.g. weaning age) and certain taxa are overrepresented in the available data (e.g. Canidae), the following variables have been consistently measured in both captive and natural populations: gestation length, birth weight, litter size, age at weaning, age at sexual maturity (i.e. at first reproduction), age of independence (i.e. at dispersal from natal territory or establishment of foraging independence), interbirth interval, and longevity. In the following sections, we discuss the variation in these life histories in relation to size constraints, taxonomy, food habits, and ecological characteristics. [Discussions of sexual dimorphism can be found in Ralls (245, 246), Eisenberg (83), and Gittleman (105)].

Body and Brain Size

A close relationship between adult body or brain size and life histories is widely found in eutherian mammals (83, 210, 211, 319). All of the above life history variables correlate significantly with carnivore's adult body weight and brain weight (Table 2), and a common slope can be drawn for the whole order (see 105). Many of the slopes of life history traits are similar to those found in other mammalian groups, supporting the hypothesis that the same physiological and energetic factors produce constants in scaling throughout the Mammalia class (51, 202, 319).

Because some error and uncertainty of dependence occurs among these variables, the slopes in Table 2 were calculated using major axis analysis, rather than standard regression techniques (126). The slope for gestation length on body weight for 93 carnivore genera is 0.10, which is similar to Western's (319) figures for artiodactyls (0.16), primates (0.14), and carnivores (0.12), and to Kihlström's (148) calculations for 208 species of eutherian mammals (0.17).

Birth weight also scales similarly across diverse mammalian taxa. Using regression analysis, Leitch et al (179), Leutenegger (180), and Millar (210) independently calculated exponents between 0.63 and 0.83 for a wide array of bats and other small mammals, carnivores, and primates. The exponent for 62 genera of Carnivora, the largest representative sample thus far assembled, is 0.81, similar to other exponential values.

Litter weight; ages at weaning, independence, and sexual maturity; and longevity also scale to both body and brain size (105). Although physical models cannot explain the precise functional causes of such allometric scaling (51, 203, 319), comparative evidence on carnivore life histories warrants the inclusion of size factors in analyses of both inter- and intraspecific variation among these traits.

Although life history patterns across Carnivora reveal constant scaling, differences among families are evident if one compares deviations from a line of best fit for each life-history trait plotted against either body or brain weight (14 out of 16, or 88%, are statistically significant; see 105). Particularly marked familial differences are observed for relative birth weight, litter weight, age at independence and at sexual maturity, and interbirth interval (Table 2).

Thus, in comparing carnivores with other mammalian orders (210, 319), it is essential to recognize that variation within the order is significant among families for a number of life history traits; researchers need to account for this variation before suggesting functional explanations. Therefore, in the following discussion we focus on behavioral and ecological relationships with life histories *within* families.

Table 2 Statistical relationships between 8 life history traits and adult body and brain weight across the order Carnivora (from 105)

Life history trait	Compared to body/brain weight	No. of genera	No. of families	Correlation coefficient (r)	Coefficient of determination (r^2)	Slope of major axis line across order	Heterogeneity across families F(df)
Gestation length	Body	93	7	0.85	0.73	0.10	2.8 (6, 86)[b]
	Brain	93	7	0.82	0.68	0.15	1.4 (6, 86)[e]
Birth weight	Body	62	6	0.87	0.75	0.81	5.7 (6, 54)[d]
	Brain	62	6	0.91	0.83	1.35	10.45 (6, 54)[d]
Litter weight	Body	59	6	0.84	0.71	0.86	6.1 (6, 52)[d]
	Brain	59	6	0.88	0.78	1.63	6.3 (6, 52)[d]
Weaning age	Body	62	6	0.62	0.38	0.23	2.5 (6, 55)[a]
	Brain	62	6	0.58	0.34	0.37	2.4 (6, 55)[a]
Age at independence	Body	26	4	0.84	0.71	0.41	4.6 (4, 23)[c]
	Brain	26	4	0.88	0.77	0.76	6.7 (4, 23)[d]
Age at sexual maturity	Body	54	7	0.81	0.65	0.37	5.8 (6, 47)[d]
	Brain	54	7	0.77	0.60	0.58	9.4 (6, 47)[d]
Interbirth interval	Body	54	5	0.67	0.45	0.10	4.9 (5, 49)[b]
	Brain	54	5	0.64	0.41	0.13	5.2 (5, 49)[d]
Longevity	Body	48	7	0.75	0.57	0.12	2.4 (6, 41)[a]
	Brain	48	7	0.76	0.58	0.22	2.2 (6, 41)[e]

[a] $p < 0.05$.
[b] $p < 0.025$.
[c] $p < 0.01$.
[d] $p < 0.001$.
[e] $p > 0.05$.

Life History Patterns and Feeding Ecology

Carnivores are well-known for their dietetic preferences, and thus, it is not surprising that variation in carnivore life histories is associated with food habits. In the following comparisons (summarized from 105), relationships among different species have been scaled to maternal body size (indicated by our use of the term *relatively*). We divided carnivores into two groups: (*a*) omnivores, species in which meat constitutes less than 60% of the diet (as determined from the available literature) and (*b*) carnivores, species in which meat accounts for over 60% of the diet (105).

Among Canidae, omnivorous species such as black-backed jackals, fennec foxes, and crab-eating foxes have relatively heavier offspring, longer gestation periods, and lengthier periods of dependence prior to weaning, as compared to more exclusively carnivorous species such as wolves, coyotes, African wild dogs, arctic foxes, and red foxes. There is weaker evidence indicating that omnivorous canids reach sexual maturity earlier than carnivorous canids.

The differences between omnivorous and carnivorous canids may be explained in two ways. First, omnivores can choose from a wider variety of foods, and in terms of nutritional requirements, reproduction in omnivores is probably less risky. When the food supply fluctuates, a species such as the black-backed jackal, which feeds on beetles, termites, fallen fruits, berries, snakes, and various birds and mammals (40, 259, 260, 284), can maintain itself more easily than a strict carnivore. Thus, an omnivorous canid may have the energetic capacity to endure an extended period of gestation, followed by a long lactation period for relatively larger offspring.

Carnivorous species, in contrast, tend to have shorter gestation periods and give birth to lighter young, perhaps to minimize vulnerability to prey fluctuations and, consequently, to reduce maternal energetic cost of feeding offspring. Population changes seem congruent with these comparative trends. For example, during a five-year decline in coyotes' major source of prey—the snowshoe hare, *Lepus americanus*—in Alberta, Canada, the pregnancy rate declined over 25% and the mean litter size fell from an average of 4.9 ± 1.1 to 3.6 ± 1.0 (301, 302). A similar decline in reproductive rates was observed in lynx inhabiting the same area (43).

Postweaning experiences are also markedly different in carnivores and omnivores. Carnivores progress towards the age of independence more slowly than do omnivores. Among the larger canids (wolves and coyotes) and felids (African lions, tigers, leopards, and cheetahs), which generally feed on larger-sized prey, "teaching" the young how to hunt is probably more critical in strict carnivores than in omnivores. At 12 to 15 months of age, juvenile lions, tigers, and leopards begin to make foraging excursions with their mothers, and they gradually become more successful hunters (182, 272, 273). Mothers may also bring their young maimed prey so they can learn stalking methods and killing techniques. Lionesses may even distinguish between serious hunts, with only adults participating, and training hunts, with juveniles following and watching adults in pursuit (276).

Carnivores' relatively slow progress toward independence may therefore be related to their need for more hunting experience prior to dispersal. Except for hyenas, however, this trend only seems to apply to larger Carnivora. Smaller felids, Mustelidae, and Viverridae—all strict carnivores that feed on small rodents—rely more on rushing and/or ambushing prey, rapidly clasping them with the forepaws, and delivering a swift killing bite (85, 182), rather than on the elaborate stalking procedures observed in larger species.

BODY SIZE AND DIET   Haldane (117) wrote, "The most obvious differences between different animals are differences in size, but for some reason the zoologists have paid singularly little attention to them." Although this situation has improved in recent years (51, 63, 237, 287a, 288, 322), few authors of single- or cross-species studies on carnivores have analyzed behavioral and ecological differences in relation to size. Not only is size important in all species, however, but the range of body sizes found in the Carnivora is unparalleled in any other mammalian order (105, 270) (see Figure 1a). Such broad differences may be due to ecological effects, phylogenetic history, or both. Phylogenetic effects are frequently difficult to detect, and with a less than complete fossil record, they are almost impossible to explain, at least at the family level.

Available fossil specimens suggest that diet may have influenced carnivore body size (287). Fossils indicate that the first ursids, dating back to the early Miocene, were probably small forest dwellers like their miacid ancestors (130, 168). The Ursidae may have increased in body size as they entered more open habitats and expanded the proportion of fruits and vegetation in their diets (83).

The Canidae are cursorial predators, adapted to either running down prey over considerable distances or pinning it down with large forepaws and delivering a killing bite. Selection for increased size in Canidae, as in Hyaenidae and some felids, probably occurred with the exploitation of large prey in open country habitats (83, 93, 198). In Mustelidae and Viverridae, smaller individuals were adapted for exploiting small rodent prey and invertebrates, as well as to make the predator less conspicuous in open vegetation (93). Small size in felids may also have been selected to promote arboreality (157).

Dietetic factors were probably especially important in the phylogeny of size among predatory species, where body size limits the range of available prey. Comparative data on extant species support this idea. The recent accumulation of measurements on diet and body size in carnivores permits a more comprehensive examination of this relationship (105). Prey are categorized into four size classes depending on the weight of prey that compose at least 50% of the

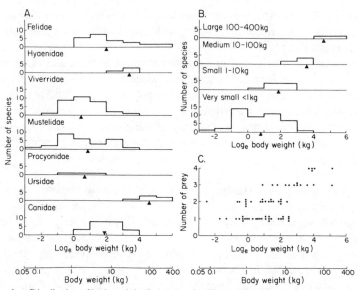

Figure 1a Distribution of body weight (kg) across families within Carnivora (from 105). Number of species shown on ordinate. Arrow indicates median value for each family.

Figure 1b Distribution of body weight (kg) across categories of prey size (from 105). Number of species shown on ordinate. Arrow indicates median value for each prey size category.

Figure 1c Number of prey (see text for definition) plotted against body weight (kg) for different carnivore species (105).

diet: (*a*) very small (less than 1 kg, e.g. the brush vole, *Lagurus curtatus*); (*b*) small (1–10 kg, e.g. the snowshoe hare, *Lepus americanus*); (*c*) medium (10–100 kg, e.g. the white-tailed deer, *Odocoileus virginianus*); (*d*) large (more than 100 kg, e.g. the wildebeest, *Connochaetes taurinus*). A direct relationship between carnivore body weight and prey size is observed across 62 species (105; Figure 1*b*). In some respects, this trend is hardly surprising. A stoat or genet would not have the strength or energetic capacity to track, pull down, and consume a large ungulate; similarly, an African lion may not move swiftly enough to follow and pounce on a vole, nor would it be energetically efficient for it to do so habitually. Yet, small mammals may be important in the diet of large carnivores—as studies of lions living in the Kalahari suggest (M. J. Owens, personal communication).

Nevertheless, the empirical association between carnivore weight and prey weight suggests that the margin of variance in the relationship is rather slight. Aside from marked differences in prey size for extremely large or small carnivores, species such as the South American fox or bobcat are restricted, *on the average*, to prey of a certain size. Nonetheless, morphological characters such as long canines and claws (93) and social hunting patterns (53, 228) may extend the limits of prey size in some cases (e.g. mountain lion, spotted hyena, wolves, African wild dog, coyotes, African lion).

Prey diversity and body size   As theoretical models would lead us to expect (278, 322), diversity of prey may also be related to differences in body size. Prey diversity can be measured in a number of ways. In carnivores, however, detailed, species-by-species descriptions of dietary contents are only available for the well-studied canids, felids, hyaenids, and mustelids, and comparative studies are therefore restricted to a crude index of diversity based on measurement of size classes. Carnivores may be classified according to the size of both "typical" (i.e. composing at least 50% of the diet) and less common prey.

By categorizing species in this fashion (Figure 1c), we find that the number of prey size classes in the diet increases with body weight ($r_{69} = 0.64$, $p < 0.01$). There are many possible causal factors involved. Larger carnivores tend to have greater home-range areas (107), and therefore the absolute chance of coming across a variety of preys is greater. Also, larger species probably have stronger forelimbs and wider jaws, enabling them to eat both small and large prey easily. Conversely, small species are morphologically limited to a smaller range of food and live in smaller home ranges.

The availability and abundance of prey also varies across different size classes. Smaller carnivores (e.g. gray fox, stoat, European wild cat) can specialize on small rodents such as *Peromyscus* or *Microtus* that reach high population densities and remain fairly common. In contrast, it would be difficult, if not impossible, for species like the African lion, mountain lion, or spotted hyena to specialize on large ungulates the size of water buffalo (*Syncerrus caffer*) or elk (*Cervus canadensis*), which have lower population densities and turnover rates. Larger carnivores must maintain flexible dietetic preferences in order to increase the absolute number of potential foods.

The advantages of increased prey diversity for large carnivores have been poorly studied. At southern latitudes in mainland Chile where the South American fox and the Chico gray fox are sympatric and differ in size, the former is larger and feeds on a greater variety of rodent species than the latter, independent of prey distribution (103). Thus, in some instances, diversity may be just as salient a factor as prey size in the selection of carnivore body size. A causal relationship has not been established, however, and diversity may be the result of increased body size, not the factor selecting for it.

Life History Patterns and Habitat

Interspecific differences in carnivore life histories are also associated with vegetation or habitat types. In Canidae, open grassland species (African wild dogs, arctic foxes) have relatively lighter birth weights than open grassland or woodland species (South American foxes, crab-eating foxes) or forest dwellers (dholes, red foxes). In Felidae, species living in sparse woodlands (mountain lions, Geoffrey's cats, leopards) have lighter total litter weights (litter size times birth weight) than do forest-living species (bengal cats, ocelots, lynx, tigers, jaguars). In explaining absolute differences in litter size between Canidae and Felidae, Kleiman & Eisenberg (157) argued that carnivores living in more forested habitats have a diversified food base, including both arboreal and terrestrial prey, that is not available in open grassland or sparse woodland terrain.

Spatial heterogeneity is also reflected in contrasting environments: Forest habitats may present more perceptual and ranging complexity than open grassland or savannah areas (83). In a study of a relatively small sample of Mustelidae, Gittleman (105) found that females of forest-living species (stoats, American minks, American martens, fishers) and aquatic forms (European otters, sea otters) reach sexual maturity later than open grassland/woodland species (long-tailed weasels, European badgers) and open grassland/forest species (wolverines, striped skunks).

Few detailed data are available on postnatal development, maternal care, natal dispersal, or establishment of territory in species of *Mustela* or *Martes* (291, 292, 296) because of their solitary existence. Behavioral observations on the aquatic European otters and sea otters, however, indicate that in addition to maternal care, both species have unusually long and attentive periods of adult care (269). Sea otter pups are usually born in the water and are supported on the mother's chest as she swims on her back (146, 147). The demands of foraging for scarce foods in oceanic waters and of locating areas protected from heavy seas may select for delayed sexual maturity to ensure familiarity and effective utilization of the environment so that young may be reared successfully (146, 277).

Life History Patterns and Social Organization

As important as ecological effects on life history patterns is the social environment in which life histories evolve (19, 20, 24, 106, 193). In many carnivores (wolves, coyotes, dwarf mongoose, lion), subadults and adults assist in provisioning, guarding, and socializing young born into the social group. Helping (alloparental) behavior or communal rearing may select for life history patterns that diverge from the traits found in species in which maternal or even biparental care is prevalent. Across Canidae—after controlling for size—birth weight, litter weight, age at which females first breed, and age at independence all increase with the degree of sociality, which is defined as the average number of individuals associating during the rearing of young (24). Moreover, across all Carnivora, species in which communal rearing is common reach sexual maturity later and have heavier litters than species in which maternal or biparental care is the modal pattern (106).

The causal mechanisms underlying these comparative trends are difficult to pinpoint. However, recent detailed field studies on the effects of helping (see below) may further our understanding of the relationship between intra- and interspecific variation in the association between life history traits and patterns of care-giving. In various canids, for example, individuals other than parents sometimes contribute directly to feeding young or parents and to guarding the den site. In black-backed and golden jackals, helpers—usually nonbreeding adults 11–20 months old—spend about 55% of their time at dens protecting pups, regularly contribute food to the young and to the lactating mother by regurgitation, and defend carcasses for utilization by the entire pack (222; for comparative data see 27, 29, 123, 195, 214, 217, 229, 231).

In some species, food provisioning does not appear to be a major part of helping (29, 171, 254–257), which primarily involves den sitting. Either food provisioning or den sitting may provide the necessary energetic requirements to select for differences in birth weight, litter size, or natal dispersal (29, 106). How helping actually affects life history patterns is unknown, however, and deserves careful attention in future research programs.

SOCIAL ECOLOGY AND SOCIAL ORGANIZATION

Methods of Study and Data Analysis: Observation and Radio–Tracking

The first step in obtaining detailed information about patterns of social organization is the compilation of ethograms, or behavioral repertoires, based on direct observation. This process involves describing basic motor patterns unambiguously, so that observers can reliably record the repeated occurrence of a particular behavior and other people can know exactly what was observed. Comparative analyses of social interaction patterns and the behavior involved rely on these basic descriptive data, especially when frequencies or rates of occurrence are used to differentiate closely related species (for data on carnivores, see 13, 15, 17, 25, 38, 156, 157).

Although many carnivores are difficult to study under field conditions, social interaction patterns and spatial relationships among individuals must be rigorously analyzed and quantitatively assessed (65, 115, 127, 178, 188, 201, 236, 304). For example, quantitative measures of association patterns (178, 188, 236), dominance relationships (14, 57), and territorial behavior (on the defense function, see 208; on the index of defendability, see 219; see also 145) have been used successfully for other animals, but they are just as applicable to analyses of carnivore social patterns.

Multivariate statistical procedures can also provide detailed information about social organization and basic behavioral patterns (14, 17, 18, 29, 65,

127, 136, 314, 315). As Hilborn & Stearns (132) pointed out, evolutionary biologists must look for multiple causes; single factor studies may delay progress when multicausal factors are actually operating—as they usually are (see 83). Statistical analyses at any level, however, cannot substitute for inadequate investigation and description of the "deep structure" of the problem at hand (109).

It is clearly important to gather observational data in any study of social organization. Most carnivores are difficult to observe or even to locate on command, however; many species are nocturnal and/or very secretive, and they move about rapidly. In these cases, observation must be supplemented by radio-tracking (for reviews of this technique, see 7, 59, 207). Radiotelemetry allows us to track individual animals over long distances and often for long periods of time. Biological attraction points (74) that influence movement patterns can also be detected. In addition, radio-tracking permits researchers to locate animals that cannot be seen and to assess spatial relationships among individuals with some degree of accuracy. As Macdonald (191) stressed, though, accurately and reliably interpreting the social significance of move-ments of radio-tagged animals (in his case, red foxes) poses many problems. In addition to not knowing what the interacting animals are doing to one another, it is difficult to assess whether their movements are independent or correlated (76).

One important methodological problem with radio-tracking free-ranging animals is that for many species, home range (territory) estimates are positively correlated with the number of radio locations gathered (for a review, see 26). Simulations relating area estimates to sample size (Figure 2) and analyses of the available literature (175) indicate that in many instances, the samples used to estimate space use have been too small. The number of locations that constitute an adequate sample may vary with a species' size, the season (e.g. presence of snow), habitat, sex, age, social status (resident, transient, or group member), reproductive condition, food resources, and the presence or absence of conspe-cifics or other species; what is sufficient for one individual may be inadequate for another. Field workers relying heavily on telemetry need to show that sample size is not a confounding variable. Radio-tracking can be a valuable tool in the study of social ecology and social organization when the effects of sample size and the accuracy with which signals can be located are taken into account.

Assessing Kin Relationships

In order to conduct detailed analyses of carnivore social organization genetic (kin) relationships (Figure 3) must also be determined whenever possible in order to assess the relative roles of kin selection (39, 118, 119, 321, 324) and

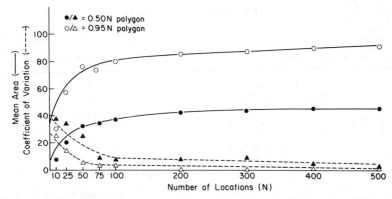

Figure 2   The relationship between estimated mean home range size (indicated by solid line) and variability (indicated by slashed line) (coefficient of variation) and number of locations. "0.50N polygon" and "0.95N polygon" refer to the convex polygons that encompass 50% and 95% of the total number of locations (*N*). The results of the present simulation indicate that, on average, field workers should obtain between 100 and 200 locations in order to estimate the home range area (*N* within which mean area and variability level off) reliably. Sampling differences for small species occupying relatively small home ranges indicate that fewer locations may be sufficient, and variability in social status, age, sex, and reproductive condition also have to be considered, as do food resources, habitat (terrain, plains, forests) and differences in sampling and analytical methods (see 26 and 30).

other possible pathways—e.g. reciprocity, mutualism, and indirect effects (323)—in the evolution of different forms of social behavior (9, 19, 20, 29, 39, 45, 46, 49, 58, 83, 96, 194, 234, 248, 257, 279, 306). For some carnivores exhibiting complex patterns of social organization, kin selection alone may be inadequate to account for observed behavioral interactions [e.g. dwarf mongooses (257), African wild dogs (248), coyotes (29), lions (234), and brown hyenas (231)]. The relative importance of different mechanisms may vary among individuals as well.

Kin relationships within social groups—as measured by the coefficient of genetic relationship, *r*—vary among species of social carnivores (*r* equals 0.125–0.5 in African wild dogs, 0.38 in red foxes, 0.5 in Arctic foxes and in jackals, 0.15–0.22 in lions, and 0.03 in spotted hyenas; data from 194). These coefficients may differ with age and sex (34, 37) as well, as a result of differences in patterns of dispersal and philopatry. In many instances, the kin relationships are not actually known, and they are inferred from social interaction patterns with varying degrees of accuracy. The importance of determining kin relationships is demonstrated by an often cited example of kin selection from an excellent field study of wild turkeys (*Meleagris gallapavo;* 312) that may be flawed because relationships among the birds were not known (11).

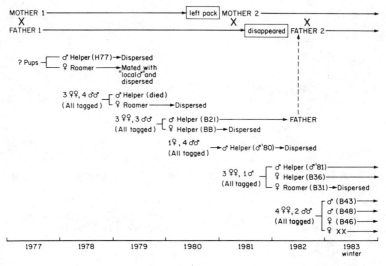

Figure 3   Social groups of carnivores often are comprised mostly of genetically related individuals (extended families). This figure presents a pedigree for a pack of coyotes observed in the Grand Teton National Park, outside of Jackson, Wyoming, from 1977 to 1983 (28, 29, 314, 315). Young of each year that are accounted for either dispersed or died before they were about 9 months of age. After the original pack mother (Mother 1) left the group in late 1980, a new and unrelated female (Mother 2) joined the pack and mated with the original pack father (Father 1) in 1981. Then, after he left the pack in spring 1981, his son, male helper B21, mated with the new female in 1982 and 1983. In 1982, the help that B21 provided to male '81 and female B36 was reciprocated when B21 and the new female's pups were born. The new pack mother was the only unrelated coyote to join the pack in 6 years (see 29 for details).

CARNIVORE SOCIAL GROUPS

In addition to the widespread but predictable differences in relative life history patterns among carnivores, there is considerable variation in social organization. Groups vary in size from lone individuals to packs of over 10 animals, age and sex ratios differ greatly, and group members may or may not be closely related. Although some species are typically considered to be "solitary" outside of the breeding season and care-giving period (e.g. felids other than lions and most mustelids and viverrids), Leyhausen (182, p. 257) has correctly stated that "the only mammal one could conceivably speak of as being socially indifferent is a dead one." Field data clearly show that members of solitary species communicate with one another using olfactory, auditory, and even visual signals, and through this exchange of information, they are able to avoid contact and maintain nonoverlapping home ranges or territories (whose boundaries are actively defended).

In some cases, intraspecific variation in social organization, due primarily to differences in food resources and habitat, is at least as pronounced as inter-specific differences [coyotes (29), wolves (124), Kalahari lions (M. J. Owens, personal communication), brown hyenas (232), golden jackals (190), striped hyenas (163, 189), red foxes (186, 187, 191, 192, 331), raccoons (102), spotted hyenas (161), and African wild dogs (100, 248)]. Although extreme intraspecific variation appears to be rare among carnivores in general, it may be misleading to characterize some species' social organization as having a modal group size (solitary or social), a typical group composition, or active defense of territorial boundaries.

Nonetheless, Altmann & Altmann (6) pointed out that animals' behavior and social organization often exhibit a range of species-specific stereotypy that is much narrower than one would expect based on experimentally demonstrated capacities for plasticity. Stressing that behavior is an important aspect of the environment, they argue that consistency in social interaction patterns among group members is made possible by the heritable component in some life history patterns. Although there is intraspecific variation in social organization among conspecifics living in different areas within a given habitat, where food resources do not drastically change, successive generations do adopt similar social organizations [coyotes (27, 29), African wild dog (100), and lion (32, 34–36, 120, 273)] with respect to most or all of the following variables: group size, group composition, patterns of emigration and space use, and mating and care-giving habits.

How Social Groups are Formed and Maintained

Understanding the many interrelated behavioral processes involved in social group formation and maintenance requires at least: (a) analyzing the develop-ment and maintenance of social bonds and relationships among offspring and between younger and older individuals (15, 20, 23, 134); (b) studying social interaction patterns between older group members and "outsiders"; (c) deter-mining movement patterns for individuals identified by sex and age; and, in the best of all possible worlds, (d) learning enough about the behavioral patterns of dispersers and nondispersers to make educated guesses about how individual movement patterns influence reproductive fitness—e.g. Do resident animals help rear offspring born into the group? Do nondispersing individuals eventual-ly inherit a portion of their natal area? Do dispersers breed at an earlier age than individuals remaining in their natal territory? Do dispersers suffer higher mortality rates than nondispersers? (see 20, 100, 120, 194, 311 for detailed discussions).

DEVELOPMENTAL ANALYSES OF SOCIAL BEHAVIOR   Comprehensive stud-ies of behavior must consider at least four major topics: evolution, adaptation, causation, and development (299, 300). While it may be beyond the ability of any one researcher to provide substantive answers to questions in all four areas, a complete analysis ultimately rests on the generation of data in all of them. Because every population is age structured (56), developmental questions are important to consider in evolutionary and ecological research in social biology (23).

Ontogenetic trajectories   A general concept applicable to both carnivores and other taxa is Wiley's (320) model of ontogenetic trajectories, in which movements of individuals through different social positions are described. This analysis can be used to trace longitudinally how individuals become incorporated into social groups (Figure 3) and to determine if variations in the natal environment affect development and later behavior (122). Wiley (320) related the evolution of different mating and social systems to ontogenetic profiles that differed in the age at sexual maturity, and he stressed that delayed reproduction is not a secondary consequence of social organization.

Wiley also pointed out that the evolution of complex social organizations is inextricably linked to the evolution of an optimal allocation of effort among reproduction, growth, and maintenance throughout an individual's lifetime. His model of ontogenetic trajectories highlights the complex trade-offs among demographic variables such as age at sexual maturity, fecundity, mortality, and developmental pathways (see also 56, 298). Clearly, life history and demographic analyses are important in studying the relationship between behavioral development and the evolution of sociality (8, 24, 94).

Development and social organization   A major difficulty in relating patterns of behavioral development to social organization is that developmental data are extremely difficult to gather under field conditions. Nonetheless, limited field data and information from captive animals concerning the ontogeny of social relationships indicate that even among closely related species (15, 17, 20, 38), interspecific differences in social organization may be associated with variations in behavioral development. Similar results have been reported for rodents (121, 325) and Bighorn sheep [*Ovis canadensis* (31)].

Apart from the problems of gathering developmental field data, intraspecific variability in social organization is highly pronounced in some taxa, making interspecific comparisons difficult. It is not known whether these intraspecific differences are reflected in variations in early social development in carnivores, as they are in Bighorn sheep (31) and lemurs [*Lemur* spp. (295)]. Field observations of coyote pups born in different habitats and into different types of social groups do not reveal any major variations in development during early life (M. Bekoff, unpublished data). But it is entirely possible that observed differences in social interaction patterns and spatial relationships among older siblings and between parents and offspring in coyote populations displaying different organizational patterns (solitary individuals, mated pairs, packs) are associated with varying developmental profiles, as perhaps are interspecific differences in social organization among some canids (20, 38). In polar bears, for example, population density influences association patterns among age and sex classes, and as a result, social organization varies (173).

To the best of our knowledge, there are no longitudinal field data for carnivores that can be used to support or refute with any degree of certainty hypotheses about the relationship between development and intraspecific variations in social organization. While there may be differences in a single species' behavioral repertoire that are dependent on habitat and parallel observed interspecific variations, this pattern apparently is not found in canids, at least. A fruitful area for further comparative study would be the analysis of differences in the frequency (or rate) of occurrence of shared behavioral patterns among carnivores (156, 157), similar to the studies done among different species of canids (see 15, 17, 25 and references therein).

Social play behavior   Variations in early social play experience may influence the development and maintenance of social bonds within a group of animals and consequently affect social interaction patterns among all group members, individual movement patterns, and the social organization in general (15, 21, 22, 88, 94, 174, 332). Detailed field data relating individual patterns of social play to later behavior are lacking and are needed to verify what appear to be robust findings based on captive animals. Information also needs to be collected on age and sex differences, variations related to social rank (15, 21, 192), and the distribution of different types of social play according to the sex, age, rank, and genetic relatedness of the participants.

SOCIAL INTERACTION PATTERNS BETWEEN OLDER GROUP MEMBERS AND OUTSIDERS   Another perspective on group formation and maintenance can be gained by studying how resident group members interact with outsiders that may be incorporated into the group. Limited comparative data indicate that some organized carnivore groups (i.e. not temporary aggregations) are relatively closed to outsiders (e.g. coyotes, jackals, and wolves); new members are rarely accepted into what are essentially extended family groups (29, 204; see also 303). In a six-year study of coyotes (29), a strange adult female was the only individual incorporated into a stable coyote pack. She was accepted after the breeding female left the group and subsequently mated with the only breeding male in the group. After he departed, she successfully bred two times with his son, who had previously been a helper (male B-21 in 29; see Figure 3). In dwarf mongooses (257) and Kalahari lions (M. J. Owens & D. D. Owens, personal communication), however, males and females commonly transfer between groups.

The permeability of relatively closed societies, as well as the openness of other types of social groups, may be influenced by age and sex differences in the degree of aggression toward outsiders or by the loss of social cohesion due to environmental stresses (M. J. Owens & D. D. Owens, personal communication). Individuals of different social rank may also vary in their attitude toward admitting new group members (48, 98). Young animals are usually more accepting of new group members than older, high-ranking individuals. The heightened level of aggression among members of the same sex also appears to influence group composition in diverse carnivores (32, 87, 100, 218, 222, 230, 240, 248, 294, 328).

SOCIAL ECOLOGY, DISPERSAL, AND PHILOPATRY   Dispersal (i.e. the movement of individuals away from their place of birth) and natal philopatry (i.e. the continued residence of animals at their birthplace past the age of independence; see 311) directly influence social organization and breeding patterns (6, 15, 20, 27, 29, 100, 104, 114, 120, 194, 222, 223a, 311). The dispersal patterns of both young and older individuals—primary and secondary emigration, respectively (100)—need to be studied further; they may vary with age, sex, and social rank. The benefits and risks associated with remaining in the natal area (possibly inheriting part or all of the natal home range or territory, providing help in rearing individuals born into the group, delaying breeding in groups where only one male and female typically breed) and dispersing (leaving a group in which breeding is unlikely or in which an individual is socially incompatible with other group members, the possibility of pairing and breeding, the possibility of increased mortality) must be considered in each specific instance.

Among wide-ranging carnivores, the difficulty of analyzing movement patterns is obvious. Radio-tracking combined with direct observation is essential, and individuals must be marked as early in life as possible. In many instances, young animals leave their natal groups and change morphologically (size, coat characteristics) between sightings. Individuals may also leave their natal area and return at a later date (29, 194), or they may leave alone and then be seen with other animals with which they may not have been previously associated. The fates of littermates that disperse together and of different-aged siblings that depart at different times also needs to be documented (100).

Many interrelated factors such as population density, food resources, and the cumulative effects of social interactions undoubtedly influence an individual's own "decisions" about whether or not to leave its natal area. These same factors probably result in involuntary dispersal by some individuals that might otherwise have remained; departure can be caused by aggression between littermates, siblings, older group members, parents and offspring, and/or adults and unrelated younger animals (15, 32, 142, 151, 154, 204, 330).

In lions, Elliott & McTaggart Cowan (87) reported a positive relationship between population density and the age at which subadults are expelled from the pride (265, 273). Bertram (32) concluded that pride size in lions is regulated by density-dependent recruitment or expulsion of subadult females. Although lion population density is influenced by food availability and social behavior, expulsion of females often occurred when food did not appear to be in short supply. He stressed the behavioral regulation of lion pride size (which may also apply to wolves), suggesting that this type of regulatory mechanism may be influenced by a pride's experience of past food shortages or its assessment of probable future shortages (see 286 for a discussion of prospective assessments of food resources). Pride size did not change in response to short-term changes in food availability. Kalahari lion pride members, however, do disband and disperse during periodic, severe droughts when large antelope prey become scarce (M. J. Owens & D. D. Owens, personal communication).

Food resources seem to have a major influence on dispersal in other carnivores (29, 88, 102, 123, 153, 185, 191, 192, 225, 232, 309, 311, 330), and differences in food availability—in quality, quantity, and distribution—are primarily responsible for intraspecific variation in social organization. When food is in short supply, dispersal is usually more pronounced. Animals may leave because they are hungry (309), or they may be driven out by more dominant individuals.

Age, social rank, and differential dispersal   Age and social rank are positively related in many carnivores. Consequently, youngsters may be unable to compete for necessary resources, including mates, and may be more disposed than older animals to leave their natal group (32, 87, 101, 185, 192, 309).

In African wild dogs living on the Serengeti Plains in northern Tanzania, individuals (usually females) commonly leave their natal pack when there is a dominant breeder of the same sex (usually a parent) present (100). Reich (247, 248) described a unique situation in which a female wild dog supplanted her mother and then mated with her father and older brother in two successive years.

In wolves, young socially subordinate individuals may get smaller rations from each kill (also observed in lions; 32), and hunger may increase subordinates' tendency to split off from their natal pack and form small hunting groups (309 and references therein). Wolves may also leave large packs, however, when there is no indication that food is in short supply. Therefore, as in lions, leopards (294), and probably most other carnivores, social and ecological factors need to be considered together (204–206, 233, 309, 330). Fritts & Mech (101) also noted that subordinate wolves tend to occupy the edge of their pack's territory or travel outside the territorial boundaries, perhaps obtaining information on the potential for colonization outside or at the edge of their natal pack's territory. Furthermore, wolves approaching breeding age often disperse early and attempt to form new social units, rather than trying to breed within their natal packs.

In addition to studying the association between social rank and age, the development of dominance relationships among littermates must be considered in a discussion of dispersal and philopatry (15, 20, 222, 330). Differential dispersal among littermates is common and may be related to relative dominance status. Because dispersal is risky (29, 32, 87, 120, 185, 192, 222, 289, 330), dominant individuals should be likely to exercise their prerogative and drive out subordinate littermates. Aggression between siblings of the same sex may be heightened further (222).

Both dominant and subordinate individuals may be predisposed to leave their natal groups because of an inability to develop strong social bonds, though for different reasons (15, 20). Based on data on captive coyotes, it has been postulated that dominant animals may disperse because other littermates avoid or constantly challenge them, whereas subordinate animals may disperse because they are scapegoats and actively avoid their siblings (15).

Golightly's (108) data on the relationship between social rank and energetics in juvenile coyotes indicate that high- and low-ranking coyotes have substantially higher daily metabolic rates than middle-ranking individuals, probably due to the regular challenges to which dominant individuals are subjected and to continued harassment of the subordinate coyote by other group members; the subordinate individual spends considerable effort avoiding interactions. Unfortunately, there are no field data on the comparative dispersal patterns of individuals with different social ranks or on rank-related energetics. But, as Golightly pointed out, the possible energetic advantages to dispersing must be considered along with other potential benefits (also see 48).

Sex differences in dispersal   Some researchers argue that among mammals, males tend to disperse more and further than females do (114), which appears to be related to a polygamous mating system (the predominant mammalian pattern). While this sex bias may hold true for some carnivores and other mammals, there are notable exceptions in at least mountain lions (280), brown hyenas (214), dwarf mongooses (257), and African wild dogs (99, 100; but see 248). In brown hyenas, both males and females disperse, apparently in order to increase their chances of breeding. Males leave their natal group and become nomadic (males living in groups were not observed to mate), while females disperse to seek a vacated territory [see Mills (214); his data were collected in the southern Kalahari]. In the central Kalahari, dispersal by female brown hyenas is less common than male emigration (231). In dwarf mongooses, males and females commonly transfer between groups (257). Female dispersal appears to predominate in African wild dogs (99, 100), but the pattern may vary from one habitat to another (248). In lions, new prides may be formed when subadult females leave their parent pride and settle in an adjacent area (120; M. J. Owens & D. D. Owens, personal communication). For most other carnivores, sufficient data are not available for any categorical generalizations to be made.

The complex relationships among behavioral, ecological, genetic, and possibly energetic factors are extremely difficult to tease apart under field conditions, and intraspecific and interspecific differences should be anticipated. Whether or not specific dispersal patterns have evolved because they favor outbreeding (i.e. reduce potential inbreeding depression) (100, 223a, 280) is not clear, because data for inbreeding depression in wild populations of carnivores are, at best, scanty and inconclusive (248, 281, 282, 329). Wolves, for example, may be highly inbred (204, 238, 329) and still not suffer from inbreeding depression. Eisenberg (82) noted that the structure of many natural populations may be a mosaic of highly inbred subpopulations.

WHY DO SOME CARNIVORES LIVE IN GROUPS?

Although animals that live in groups have attracted a disproportionate amount of attention, group living has evolved in only 10% to 15% of carnivore species (105). Comparative information on typically solitary species such as felids (37, 182, 274, 280, 293, 294) and a canid [the maned wolf (73)] has also increased our understanding of the evolution of social groups in other taxa, and it has become clear that solitary does not mean asocial (181).

Our discussion will concentrate on the possible benefits of grouping, but there may also be costs associated with living in a group. These costs include an increased probability of detection by potential competitors and predators, decreased food availability per individual, increased transmission of ectoparasites and diseases (55, 191), and increased probability of aggression (and injury) among group members. The costs and benefits of group living are not necessarily additive (200) and may differ for each individual. (General comparative reviews of the evolution of sociality and group living can be found in the following references: 3, 36, 37, 42, 48, 69, 70, 81, 83, 105, 157, 209, 264, 317, 318, 322, 326.)

A Classification of Carnivore Groups

Gittleman (105) has classified carnivore groups into four major types: (*a*) population groups—individuals sharing a common home range; (*b*) feeding groups—individuals utilizing the same food resources at a given time; (*c*) foraging groups—individuals banding together while searching for food or hunting; and (*d*) breeding groups—individuals forming a reproductive unit. Population, feeding, and foraging groups are identical for most carnivores that have been studied. In some species, population group size is related to feeding group size but not to foraging group size (wolves, spotted hyenas, brown hyenas, lions, coyotes); populations may break up into smaller hunting groups and then regroup when feeding.

OPTIMAL GROUP SIZE   Group size varies both inter- and intraspecifically, but there appears to be an optimal range within which individual group members benefit the most (37, 48, 53, 70, 91, 111, 157, 161, 169, 170, 172, 228, 252, 273), especially with respect to the exploitation of food resources and individual energy budgets. Caraco & Wolf (53) determined that lions achieve highest individual efficiency (capture success, food availability/lion, food intake/lion) in groups of two, regardless of prey size, whereas Nudds's (228) analysis showed that optimal pack size for wolves is a function of prey size.

For both lions and wolves, groups often exceed the size calculated as optimal for hunting efficiency. For wolves at least, the addition of individuals to an optimally sized pack appears to be less costly in terms of individuals' energy returns than a reduction below the optimum (228). Rodman (252) pointed out that the genetic gain through relatedness (i.e. increases in inclusive fitness) may offset the disadvantage to individual group members when surplus members are kin. Thus, despite possible costs to individual fitness, groups may evolve because of the increase in total relatedness. Competition may also promote large groups among predators that defend their kills against scavengers (37, 91, 169).

Despite possible genetic gains and the increased ability to catch prey and defend kills in larger than optimally sized groups, an upper limit on group size is imposed by: (*a*) decreased hunting success (70, 273), (*b*) reduced food intake per individual because of increased within-group competition, and (*c*) the need for small- to medium-sized predators to reduce competitive pressure by hunting and feeding inconspicuously (169), among other factors. Foraging groups that are smaller or larger than expected may form, however, when prey loss needs to be minimized (169).

Group size may also be limited by small carnivores' need to decrease the likelihood of detection by potential predators (110, 256, 257). Suffice it to say, what is optimal in one setting may not be in another, even for conspecifics. Furthermore, in determining the optimal group size for a given species, researchers must take into account all possible variables, not just food exploitation (169, 252).

INFLUENCE OF HABITAT ON CARNIVORE SOCIAL GROUPS   Comparisons of closely related species living in similar and different habitats provide additional information on the evolution of carnivore social groups (105, 110, 157, 228, 248, 256, 272–274, 280, 293). In general, species living in open grassland or in both open grassland and woodland habitats have larger population groups than forest dwellers (105), but the nature of food resources (73, 141, 204) and competition can alter group sizes on the local level.

For predators, an open environment usually increases the chance of being detected by prey, and some form of group hunting is required for successful predation, especially on prey larger than the individual hunters. For smaller carnivores, such as mongooses, the risk of predation is greater in open habitats, and larger groups afford more protection and increased vigilance (79, 256); increased detection by predators could impose an upper limit on group size in the absence of behavioral regulation. Food resources are usually readily available to the many small carnivores for which antipredatory defense is the major force behind sociality (105, 310). Most smaller carnivores displaying group defense are insectivores, and Waser's (310) model indicated that a mongoose excluding a single competitor from its foraging range would only gain about 1% in prey density. Thus, prey distribution and abundance may be a precondition for the evolution of grouping in response to the need for group vigilance (105).

A comparison of lions—the only highly social felid—with a closely related, morphologically similar solitary species such as the tiger illustrates nicely the role of habitat in the evolution of felid sociality (293). In Nepal, tigers live in dense, broken terrain, where small-sized prey are scattered and difficult to locate. Successful predation requires stalking and ambushing techniques. The necessity for crypticity along with the maintenance of an almost exclusively meat diet appears to have favored a solitary existence among tigers and other felids occupying similar habitats, such as mountain lions (280), jaguars, and leopards (274). Sociality in lions, on the other hand, seems related to increased hunting and feeding efficiency and the necessity of defending large kills in open savannah habitats (36, 53, 273). Similar selective pressures (open habitat and ungulates that were rapidly evolving into large-bodied, long-legged, fast-moving prey) also appear to have favored the evolution of sociality in wolves (157, 228).

COMPARATIVE SOCIAL ECOLOGY: THE PERVASIVE INFLUENCE OF FOOD

As we have repeatedly stressed above, the way in which food resources are exploited (located, hunted, scavenged, defended) and the nature of the available food (quality, quantity, spatial distribution) strongly influence interspecific relationships and result in marked intraspecific variations in social organization.

Interspecific Relationships

Radinsky's (242–244) studies of the evolution of skull shape in carnivores provide a paleoethological perspective on interspecific competition suggesting that the partitioning of prey resources by size—due to anatomical differences—may have been a factor in the initial radiation of carnivores. This finding is consistent with Rosenzweig's (258) suggestion that sympatric carnivores can coexist in "hunting sets" because they specialize on prey of different sizes. There are comparative data on differential prey size selection by sympatric carnivores for viverrids [white-tailed mongooses, large spotted genets, and African civets (137)], hyaenids (12), and canids and felids (37, 167, 172, 239).

Even if different species share the same prey, variations in hunting methods, prey selection, activity patterns (time-sharing), and mutual avoidance can foster coexistence (12, 37, 137, 139, 172, 224, 297). Bertram (37) has reviewed patterns of coexistence among five carnivores that inhabit the Serengeti Plains (cheetah, leopard, lion, spotted hyena, and wild dog), all of which feed on ungulates. Differences in hunting methods (stalking or pursuing), preferred prey, activity patterns, and prey selection (in terms of age/sex and health) all interact to reduce competition and promote coexistence. Social hunters (lions, wild dogs, spotted hyenas), which use various methods to pursue prey, tend to live in long-lasting social groups, whereas cheetahs and leopards, which predominantly hunt alone, live in groups containing only one adult. Cheetahs, which suffer extreme competition from spotted hyenas (224) and lions (172), actively avoid these and other competitors and eat fast at kills. They are too light and fragile to defend kills successfully, and they can be displaced from food by any animal larger than a jackal (172).

Intraspecific Variation in Social Organization

With few exceptions, intraspecific variation in carnivore social organization as a response to local food resources is the rule among species in which group living is associated with the exploitation of food resources by group members (canids: 28, 29, 41, 72, 88, 95, 100, 101, 123, 131, 160, 190, 192, 193, 204–206, 222, 248, 261, 262, 309, 330, 332; felids: 32, 35, 36, 120, 138, 183, 235, 273; brown hyenas: 215, 230, 232; striped hyenas: 163, 189; spotted hyenas: 161, 162, 184). It is also prevalent in species in which the availability of ample food permits aggregations to form, but in which exploitation of food is not typically a group affair (mustelids: 149, 164, 165; procyonids: 102; ursids: 10, 66, 67, 177, 225; brown and striped hyenas may also fit into this category, see 163, 214, 232). For example, when abundant food is clumped and economically defendable (44, 111), coyotes (29, 41, 52, and Figures 3 & 4), golden jackals (190), Kalahari desert lions (M. J. Owens & D. D. Owens, personal communication), domestic cats (138), and striped hyenas (189) show increases

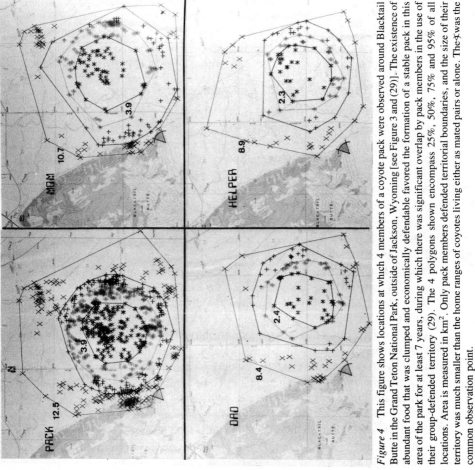

Figure 4 This figure shows locations at which 4 members of a coyote pack were observed around Blacktail Butte in the Grand Teton National Park, outside of Jackson, Wyoming [see Figure 3 and (29)]. The existence of abundant food that was clumped and economically defendable favored the formation of a stable pack in this area of the park for at least 7 years, during which there was significant overlap by pack members in the use of their group-defended territory (29). The 4 polygons shown encompass 25%, 50%, 75% and 95% of all locations. Area is measured in km². Only pack members defended territorial boundaries, and the size of their territory was much smaller than the home ranges of coyotes living either as mated pairs or alone. The X was the common observation point.

in population and feeding group size and decreases in territorial area. They may also defend food resources and territorial boundaries if intruder pressure increases.

Group and territorial sizes can also vary independently. Range size and configuration may be determined by the pattern of dispersion of food resources (and the influence of topography on the economics of defense), while group size may be determined by the richness of food patches (131). For brown hyenas living in the southern Kalahari, for example, territorial size was affected by the distribution of food, whereas group size was influenced by the quality of food in the territory (i.e. the size of available carcasses). Thus, both territorial and group size varied but were not correlated (215).

Mating Patterns, Parental Behavior, and Helping

Mating patterns, care-giving behavior, and food resources are, not surprisingly, closely related to one another (61, 153, 249, 268, 327). Here we look specifically at the relationships among resource availability, monogamy (a prolonged association and essentially exclusive mating relationship between one male and a single female; see 327), and helping behavior [alloparental behavior (324), i.e. care (food, protection) provided to young animals by individuals other than parents; (for reviews and comparative data see 29, 46, 47, 49, 123, 194, 195, 222, 229, 231, 249, 255, 257, 311)].

Although monogamy is rare in mammals [it has been documented in about 3% of species studied (153)], it appears to have evolved as the predominant mating strategy in the majority of canids for which there are sufficient data (but see 125) but not in any felids (153, 327; for a review of the available data, see 105). Harrington et al (125) stressed the difficulty of determining mating patterns under field conditions (mainly due to problems with actually observing identified individuals that are copulating), and their cautious approach deserves close attention. In many instances, researchers base their conclusions about mating patterns on circumstantial evidence—e.g. close association of a pair outside of the breeding season, absence of other adults in the pair's home range.

In general, monogamy is associated with increased male parental care (153, 158). In addition, monogamy (and possibly helping; see 105, 229, 249) has apparently evolved in environments in which food resources are energetically costly to collect and in species in which the litter size is comparatively large, as in canids (61, 153).

Kleiman (153) differentiated between two general types of monogamy— facultative and obligate. Facultative monogamy occurs when the necessary resources are thinly and patchily distributed and when only a single member of the opposite sex is available for mating. Paternal care and helping may be lacking. The only carnivore for which facultative monogamy has been described under field conditions is the maned wolf (73).

Obligate monogamy occurs when a single female cannot rear a litter without aid from conspecifics, perhaps because the carrying capacity of the habitat is insufficient to allow more than one female to breed within the same home range at a given time. [Wittenberger & Tilson (327) de-emphasize the importance of carrying capacity in the development of monogamy.] Although canids (with the possible exception of maned wolves) would be classified as obligately monogamous, the *need* for care from the father or helpers is probably not *essential* for successful rearing of young (327). The presence of helpers, and in some cases paternal care, is positively correlated with pup survival in a variety of canids, however, though the relationship is not always statistically significant (29, 194, 195, 220–222; see also 223). But, as a result of prey scarcity, pairs may produce more surviving young than do groups with one or more potential helpers; this phenomenon has been observed in wolves (123). In red foxes, if food is scarce, the presence of helpers may temporarily reduce the reproductive success of breeding individuals (305).

A detailed review of helping behavior is beyond the scope of this paper. (Recent summaries can be found in 29, 105, 106, 123, 194, 195, 222, 249, and 255). Kin selection and reciprocity, often working together, account for most instances of helping in carnivores (29, 194, 195, 222, 231, 255, 257). Among canids, helpers frequently are nonbreeding and nondispersing older siblings of the young for which they care, and an individual that previously helped his younger siblings (or half siblings) may be reciprocated in kind if he or she later breeds and if the recipients of his or her help have not dispersed (29, 231; see also Figure 3). Helpers and other nondispersing individuals may also inherit part or all of their natal area and subsequently breed there.

CONCLUSIONS

The diversity of extant carnivores makes them an exemplary group for studying the evolution of life history patterns and social behavior and ecology from a comparative perspective. Although some general explanations of carnivore social ecology appear to have broad applicability, most species have not even been considered in these syntheses. We need to know more about the *how*'s before we flood the market with grandiose theories about the *why*'s (113). Methods of collecting and analyzing data must be carefully considered in interpreting the results of a given study, and comparisons of two or more research projects should be undertaken with caution. Long-term field projects on unexploited populations are needed; researchers and funding agencies need to make the necessary commitments, despite the obvious risks (75), which are far outweighed by the potential rewards.

From a practical point of view, comparative data on social ecology may be as useful for the development of nature reserves (197) as they are in management programs. Furthermore, the study of social carnivores may shed some light on the evolution of human behavior (150, 161, 275).

ACKNOWLEDGMENTS

Marc Bekoff thanks the National Science Foundation, the National Institute of Mental Health, the Harry Frank Guggenheim Foundation, the John Simon Guggenheim Memorial Foundation, and the University of Colorado for financial support of his research, and all of the people who have participated in his research programs, especially Michael C. Wells, for their unrelenting efforts. Tom Daniels was supported by a University Fellowship from the University of Colorado (Boulder) and John Gittleman was supported by a Smithsonian Institution Post-Doctoral Fellowship during the preparation of this chapter. Joseph Ortega, Beth Bennett, Joel Berger, John Byers, Douglas Conner, Devra Kleiman, Mark Owens, Delia Owens, and John Seidensticker provided comments on an earlier draft of the manuscript. The Owenses kindly provided unpublished information from their long-term studies on lions and brown hyenas in the central Kalahari. Jan Logan drew Figures 1, 2, and 3 and Muriel Sharp and Bay Roberts helped to compile and organize the references. Mary Marcotte and Jeanie Cavanagh kindly typed numerous color-coded drafts of this paper.

Literature Cited

1. Albignac, R. 1972. The carnivores of Madagascar. In *Biogeography and Ecology of Madagascar*, ed. R. Batistina, G. Richard-Vindard, pp. 21–35. The Hague: Junk
2. Albignac, R. 1973. Mammiferes carnivores. *Faune de Madagascar*. No. 36. 206 pp.
3. Alexander, R. D. 1974. The evolution of social behavior. *Ann. Rev. Ecol. Syst.* 5:325–83
4. Allen, D. 1979. *Wolves of Minong: Their Vital Role in a Wild Community*. Boston: Houghton-Mifflin. 499 pp.
5. Altmann, J. 1974. Observational study of behavior. *Behaviour* 49:227–67
6. Altmann, S. A., Altmann, J. 1979. Demographic constraints on behavior and social organization. In *Primate Ecology and Human Origins*, ed. I. S. Bernstein, E. O. Smith, pp. 47–63. New York: Garland. 362 pp.
7. Amlaner, C. J., Macdonald, D. W., eds. 1980. *A Handbook on Biotelemetry and Radio Tracking*. Oxford: Pergamon
8. Armitage, K. B. 1981. Sociality as a life-history tactic of ground squirrels. *Oecologia* 48:36–49
9. Armitage, K. B. 1982. Social dynamics of juvenile marmots: Role of kinship

and individual variability. *Behav. Ecol. Sociobiol.* 11:33–36
10. Ballard, W. B., Miller, S. D., Spraker, T. H. 1982. Home range, daily movements and reproductive biology of brown bear in southcentral Alaska. *Can. Field. Nat.* 96:1–5
11. Balph, D. F., Innis, G. S., Balph, M. H. 1980. Kin selection in Rio Grande turkeys: A critical assessment. *Auk* 97:854–60
12. Bearder, S. K. 1975. Inter-relationship between hyaenas and their competitors in the Transvaal lowveld. *Publ. Univ. Pretoria* 97:39–49
13. Bekoff, M. 1974. Social play and play-soliciting by infant canids. *Am. Zool.* 14:323–40
14. Bekoff, M. 1977. Quantitative studies of three areas of classical ethology: Social dominance, behavioral taxonomy, and behavioral variability. See Ref. 127, pp. 1–46
15. Bekoff, M. 1977. Mammalian dispersal and the ontogeny of individual behavioral phenotypes. *Am. Nat.* 111:715–32
16. Bekoff, M., ed. 1978. *Coyotes: Biology, Behavior, and Management*. New York: Academic. 384 pp.
17. Bekoff, M. 1978. Behavioral develop-

ment in coyotes and eastern coyotes. See Ref. 16, pp. 97–126

18. Bekoff, M. 1978. A field study of the development of behavior in Adelie penguins: Univariate and numerical taxonomic approaches. In *The Development of Behavior: Comparative and Evolutionary Aspects,* ed. G. Burghardt, M. Bekoff, pp. 177–202. New York: Garland. 429 pp.

19. Bekoff, M. 1981. Vole population cycles: Kin selection or familiarity? *Oecologia* 48:131

20. Bekoff, M. 1981. Mammalian sibling interactions: Genes, facilitative environments, and the coefficient of familiarity. In *Parental Care in Mammals,* ed. D. J. Gubernick, P. H. Klopfer, pp. 307–46. New York: Plenum. 459 pp.

21. Bekoff, M. 1984. Social play behavior. *BioScience* 34:228–33

22. Bekoff, M., Byers, J. A. 1981. A critical reanalysis of the ontogeny and phylogeny of mammalian social and locomotor play: An ethological hornet's nest. In *Behavioral Development: The Bielefeld Interdisciplinary Conference,* ed. K. Immelmann, G. W. Barlow, L. Petrinovich, M. Main, pp. 296–337. New York: Cambridge Univ. Press. 754 pp.

23. Bekoff, M., Byers, J. A. 1985. The development of behavior from evolutionary and ecological perspectives. *Evol. Biol.* 19: In press

24. Bekoff, M., Diamond, J., Mitton, J. B. 1981. Life history patterns and sociality in canids: Body size, reproduction, and behavior. *Oecologia* 50:388–90

25. Bekoff, M., Hill, H. L., Mitton, J. B. 1975. Behavioral taxonomy in canids by discriminant function analysis. *Science* 190:1223–25

26. Bekoff, M., Mech, L. D. 1984. Simulation analyses of space use: Home range estimates, variability, and sample size. *Behav. Res. Methods Instrum. Comput.* 16:32–37

27. Bekoff, M., Wells, M. C. 1980. The social ecology of coyotes. *Sci. Am.* 242: 130–48

28. Bekoff, M., Wells, M. C. 1981. Behavioural budgeting by wild coyotes: The influence of food resources and social organization. *Anim. Behav.* 29:794–801

29. Bekoff, M., Wells, M. C. 1982. Behavioral ecology of coyotes: Social organization, rearing patterns, space use, and resource defense. *Z. Tierpsychol.* 60: 281–305

30. Bekoff, M., Wieland, C., Lavender, W. A. 1982. Space-Out: Graphics programs

to study and to simulate space use and movement patterns. *Behav. Res. Methods Instrum.* 14:34–36

31. Berger, J. 1979. Social ontogeny and behavioural diversity: Consequences for Bighorn sheep *Ovis canadensis* inhabiting desert and mountain environments. *J. Zool.* 118:251–66

32. Bertram, B. C. R. 1973. Lion population regulation. *East Afr. Wildl. J.* 11:215–25

33. Bertram, B. C. R. 1975. The social system of lions. *Sci. Am.* 232:54–65

34. Bertram, B. C. R. 1976. Kin selection in lions and in evolution. In *Growing Points in Ethology,* ed. P. P. G. Bateson, R. A. Hinde, pp. 281–301. New York: Cambridge Univ. Press

35. Bertram, B. C. R. 1978. *Pride of Lions.* New York: Scribner's

36. Bertram, B. C. R. 1978. Living in groups: Predators and prey. In *Behavioural Ecology: An Evolutionary Approach,* ed. J. R. Krebs, N. B. Davies, pp. 64–96. Sunderland, Mass: Sinauer. 494 pp.

37. Bertram, B. C. R. 1979. Serengeti predators and their social systems. In *Serengeti: Dynamics of an Ecosystem,* ed. A. R. E. Sinclair, M. Norton-Griffiths, pp. 221–48. Chicago: Univ. Chicago. 389 pp.

38. Biben, M. 1983. Comparative social ontogeny of social behaviour in three South American canids, the maned wolf, crab-eating fox and bush dog: Implications for sociality. *Anim. Behav.* 31:814–26

39. Boorman, S. A., Levitt, P. R. 1980. *The Genetics of Altruism.* New York: Academic. 459 pp.

40. Bothma, J. du. 1971. Food of *Canis mesomelas* in South Africa. *Ann. Transvaal Mus.* 27:15–26

41. Bowen, W. D. 1981. Variation in coyote social organization: The influence of prey size. *Can. J. Zool.* 59:639–52

42. Bradbury, J. W., Vehrencamp, S. L. 1976. Social organization and foraging in Emballonurid bats. I. Field studies; II. A model for the determination of group size. *Behav. Ecol. Sociobiol.* 1:337–81; 383–404

43. Brand, C. J., Keith, L. B. 1979. Lynx demography during a snowshoe hare decline in Alberta. *J. Wildl. Manage.* 43:27–48

44. Brown, J. L. 1964. The evolution of diversity of avian territorial systems. *Wilson Bull.* 76:160–69

45. Brown, J. L. 1974. Alternate routes to sociality in jays—with a theory for the evolution of altruism and communal breeding. *Am. Zool.* 14:63–80

46. Brown, J. L. 1978. Avian communal breeding systems. *Ann. Rev. Ecol. Syst.* 9:123–55
47. Brown, J. L. 1980. Fitness in complex avian social systems. In *Evolution of Social Behavior: Hypotheses and Empirical Tests,* ed. H. Markl, pp. 115–28. Weinheim, West Germany: Chemie. 253 pp.
48. Brown, J. L. 1982. Optimal group size in territorial animals. *J. Theor. Biol.* 95:793–810
49. Brown, J. L. 1983. Cooperation—a biologist's dilemma. *Adv. Study Behav.* 13:1–37
50. Bygott, J. D., Bertram, B. C. R., Hanby, J. P. 1979. Male lions in large coalitions gain reproductive advantage. *Nature* 26:839–41
51. Calder, W. A. III. 1983. Ecological scaling: Mammals and birds. *Ann. Rev. Ecol. Syst.* 14:213–30
52. Camenzind, F. J. 1978. Behavioral ecology of coyotes on the National Elk Refuge, Jackson, Wyoming. See Ref. 16, pp. 267–94
53. Caraco, T., Wolf, L. L. 1975. Ecological determinants of group sizes of foraging lions. *Am. Nat.* 109:343–52
54. Chapman, J. W., Feldhamer, G. A., eds. 1982. *Wild Mammals of North America: Biology, Management, and Economics.* Baltimore: Johns Hopkins Univ. Press. 1147 pp.
55. Chapman, R. C. 1978. Rabies: Decimation of a wolf pack in arctic Alaska. *Science* 201:365–67
56. Charlesworth, B. 1980. *Evolution in Age-Structured Populations.* Cambridge: Cambridge Univ. Press. 300 pp.
57. Chase, I. D. 1974. Models of hierarchy formation in animal societies. *Behav. Sci.* 19:374–82
58. Chase, I. D. 1980. Cooperative and noncooperative behavior in animals. *Am. Nat.* 115:827–57
59. Cheeseman, C. L., Mitson, R. B., eds. 1982. Telemetric studies of vertebrates. *Symp. Zool. Soc. London* 49:1–368
60. Chiarelli, A. B. 1975. The chromosomes of the Canidae. See Ref. 97, pp. 40–53
61. Clutton-Brock, T. H., Harvey, P. H. 1978. Mammals, resources and reproductive strategies. *Nature* 17:191–95
62. Clutton-Brock, T. H., Harvey, P. H. 1979. Comparison and adaptation. *Proc. R. Soc. London Ser. B* 205:547–65
63. Clutton-Brock, T. H., Harvey, P. H. 1983. The functional significance of variation in body size among mammals. See Ref. 84, pp. 632–63
64. Clutton-Brock, T. H., Harvey, P. H. 1984. Comparative approaches to investigating adaptation. In *Behavioural Ecology: An Evolutionary Approach,* ed. J. R. Krebs, N. B. Davies, pp. 7–29. Oxford: Blackwell
65. Colgan, P., Ed. 1978. *Quantitative Ethology.* New York: Wiley. 364 pp.
66. Craighead, F. C. Jr. 1979. *The Track of the Grizzly.* San Francisco: Sierra Club. 261 pp.
67. Craighead, J. J., Mitchell, J. A. 1982. Grizzly bear. See Ref. 54, pp. 515–56
68. Crook, J. H. 1964. The evolution of social organization and visual communication in the weaver birds (Ploceinae). *Behaviour Suppl.* 10:1–78
69. Crook, J. H., Ellis, J. E., Goss-Custard, J. D. 1976. Mammalian social system: Structure and function. *Anim. Behav.* 24:261–74
70. Curio, E. 1976. *The Ethology of Predation.* New York: Springer-Verlag. 250 pp.
71. Dagg, A. I. 1983. *Harems and Other Horrors: Sexual Bias in Behavioral Biology.* Waterloo, Canada: Otter. 125 pp.
72. Deutsch, L. A. 1983. An encounter between bush dog (*Speothos venaticus*) and paca (*Agouti paca*). *J. Mammal.* 64:532–33
73. Dietz, J. M. 1984. *Ecology and social organization of the maned wolf* (Chrysocyon brachyurus), *Smithson. Contrib. Zool.* No. 392
74. Don, B. A. C., Rennolls, K. 1983. A home range model incorporating biological attraction points. *J. Anim. Ecol.* 52:69–81
75. Dunbar, R. I. M. 1983. Social systems as optimal strategy sets. In *Environment and Population: Problems of Adaptation,* ed. J. B. Calhoun, pp. 4–6. New York: Praeger Sci.
76. Dunn, J. E. 1979. A complete test for dynamic territorial interaction. *Proc. 2nd Int. Conf. Wildl. Biotelem., Laramie, Wyo.,* pp. 159–69
77. Eaton, R. L. 1973. *The World's Cats: Ecology and Conservation.* Seattle: Feline Res. Group. 349 pp.
78. Eaton, R. L. 1974. *The World's Cats: Biology, Behavior and Management of Reproduction.* Seattle: Feline Res. Group. 260 pp.
79. Eaton, R. L. 1976. *The World's Cats: Contributions to Biology, Ecology, Behavior and Evolution.* Seattle: Carnivore Res. Inst. 179 pp.
80. Eaton, R. L. 1977. *The World's Cats: Contributions to Breeding Biology, Behavior and Husbandry.* Seattle: Carnivore Res. Inst. 144 pp.
81. Eisenberg, J. F. 1966. The social organization of mammals. *Handb. Zool.* VIII (10/7). Berlin: De Gruyter

82. Eisenberg, J. F. 1977. Comments. In *The Biology and Conservation of the Callitrichidae*, ed. D. G. Kleiman, pp. 220. Washington DC: Smithsonian Inst. 354 pp.

83. Eisenberg, J. F. 1981. *The Mammalian Radiations: An Analysis of Trends in Evolution, Adaptation, and Behavior*. Chicago: Univ. Chicago Press. 610 pp.

84. Eisenberg, J. F., Kleiman, D. G., eds. 1983. Advances in the study of mammalian behavior. *Am. Soc. Mammal., Spec. Publ. No. 7*. 753 pp.

85. Eisenberg, J. F., Leyhausen, P. 1972. The phylogenesis of predatory behavior in mammals. *Z. Tierpsychol.* 30:59–93

86. Eisenberg, J. F., Muckenhirn, N. A., Rudran, R. 1972. The relation between ecology and social structure in primates. *Science* 176:863–73

87. Elliott, J. P., McT. Cowan, I. 1978. Territoriality, density, and prey of the lion in Ngorongoro Crater, Tanzania. *Can. J. Zool.* 56:1726–34

88. Englund, J. 1980. Population dynamics of the red fox (*Vulpes vulpes* L., 1758) in Sweden. See Ref. 331, pp. 107–21

89. Erlinge, S. 1974. Distribution, territoriality and numbers of the weasel *Mustela nivalis* in relation to prey abundance. *Oikos* 25:308–14

90. Erlinge, S. 1977. Spacing strategy in stoat *Mustela erminea*. *Oikos* 28:32–42

91. Estes, R. D., Goddard, J. 1967. Prey selection and hunting behavior of the African wild dog. *J. Wildl. Manage.* 31:52–70

92. Ewer, R. F. 1968. *Ethology of Mammals*. New York: Plenum. 418 pp.

93. Ewer, R. F. 1973. *The Carnivores*. New York: Cornell Univ. Press. 494 pp.

94. Fagen, R. 1981. *Animal Play Behavior*. Oxford: Oxford Univ. Press. 684 pp.

95. Ferguson, J. W. H., Nel, J. A. J., de Wett, M. J. 1983. Social organization and movement patterns of black-backed jackals *Canis mesomelas* in South Africa. *J. Zool.* 199:487–502

96. Fisler, G. F. 1969. Mammalian organizational systems. *Nat. Hist. Mus. Los Angeles Cty. Sci. Bull.* 167:1–32

97. Fox, M. W., ed. 1975. *The Wild Canids*. New York: Van Nostrand-Reinhold. 508 pp.

98. Fox, M. W., Lockwood, R., Shideler, R. 1974. Introduction studies in captive wolves. *Z. Tierpsychol.* 35:38–48

99. Frame, L. H., Frame, G. W. 1976. Female African wild dogs emigrate. *Nature* 263:227–29

100. Frame, L. H., Malcolm, J. R., Frame, G. W., van Lawick, H. 1979. Social organization of African wild dogs (*Lycaon pictus*) on the Serengeti Plains, Tanzania 1967–1978. *Z. Tierpsychol.* 50:225–49

101. Fritts, S. H., Mech, L. D. 1981. Dynamics, movements, and feeding ecology of a newly protected wolf population in northwestern Minnesota. *Wildl. Monogr.* 80:1–79

102. Fritzell, E. K. 1978. Habitat use by prairie raccoons during the waterfowl breeding season. *J. Wildl. Manage.* 42:118–27

103. Fuentes, E. R., Jaksić, F. M. 1979. Latitudinal size variation of Chilean foxes: Tests of alternative hypotheses. *Ecology* 60:43–47

104. Gaines, M. S., McClenaghan, L. R. 1980. Dispersal in small mammals. *Ann. Rev. Ecol. Syst.* 11:163–96

105. Gittleman, J. L. 1984. *The behavioural ecology of carnivores*. PhD thesis. Univ. Sussex, Brighton, England

106. Gittleman, J. L. 1984. Functions of communal care in mammals. In *Evolution: Essays in Honour of John Maynard Smith*, ed. P. J. Greenwood, M. Slatkin. Cambridge: Cambridge Univ. Press. In press

107. Gittleman, J. L., Harvey, P. H. 1982. Carnivore home-range size, metabolic needs and ecology. *Behav. Ecol. Sociobiol.* 10:57–63

108. Golightly, R. 1981. *The comparative energetics of two desert canids: The coyote (Canis latrans) and the kit fox (Vulpes macrotis)*. PhD thesis. Ariz. State Univ., Tempe, Ariz.

109. Good, I. J. 1983. The philosophy of exploratory data analysis. *Philos. Sci.* 50:283–95

110. Gorman, M. L. 1979. Dispersion and foraging of the small Indian mongoose, *Herpestes auropunctatus* (Carnivora: Viverridae) relative to the evolution of social viverrids. *J. Zool.* 187:65–73

111. Gosling, L. M., Petrie, M. 1981. The economics of social organization. In *Physiological Ecology: An Evolutionary Approach*, ed. C. R. Townsend, P. Calow, pp. 315–45. Oxford: Blackwell

112. Gould, S. J. 1977. *Ontogeny and Phylogeny*. Cambridge, Mass: Harvard Univ. Press. 501 pp.

113. Gould, S. J. 1983. Genes on the brain. *NY Rev. Books* 30(11):5–10

113a. Gould, S. J., Lewontin, R. C. 1979. The spandrels of San Marco and the panglossian paradigm: A critique of the adaptationist programme. *Proc. R. Soc. London Ser. B* 205:581–98

114. Greenwood, P. J. 1980. Mating systems, philopatry and dispersal in birds and mammals. *Anim. Behav.* 28:1140–62

115. Greenwood, R. J. 1982. Nocturnal activ-

ity and foraging of prairie raccoons *(Procyon lotor)* in North Dakota. *Am. Midl. Nat.* 107:238–43

116. Guggisberg, C. A. W. 1975. *Wild Cats of the World.* New York: Taplinger

117. Haldane, J. B. S. 1956. On being the right size. In *The World of Mathematics,* ed. J. R. Newman, pp. 952–57. New York: Simon & Schuster

118. Hamilton, W. D. 1963. The evolution of altruistic behavior. *Am. Nat.* 97:354–56

119. Hamilton, W. D. 1964. The genetical evolution of social behaviour. I and II. *J. Theor. Biol.* 7:1–52

120. Hanby, J. P., Bygott, J. D. 1979. Population changes in lions and other predators. In *Serengeti: Dynamics of an Ecosystem,* ed. A. R. E. Sinclair, M. Norton-Griffiths, pp. 249–62. Chicago: Univ. Chicago Press. 389 pp.

121. Happold, M. 1976. The ontogeny of social behaviour in four conilurine rodents (Muridae) of Australia. *Z. Tierpsychol.* 40:265–78

122. Harcourt, A. H., Stewart, K. J. 1981. Gorilla male relationships: Can differences during immaturity lead to contrasting reproductive tactics in adulthood? *Anim. Behav.* 29:206–10

123. Harrington, F. H., Mech, L. D., Fritts, S. H. 1983. Pack size and wolf pup survival: Their relationship under varying ecological conditions. *Behav. Ecol. Sociobiol.* 13:19–26

124. Harrington, F. H., Pacquet, P. C., eds. 1982. *Wolves of the World: Perspectives of Behavior, Ecology, and Conservation.* Park Ridge, NJ: Noyes. 474 pp.

125. Harrington, F. H., Pacquet, P. C., Ryon, J., Fentress, J. C. 1982. Monogamy in wolves: A review of the evidence. See Ref. 124, pp. 209–22

126. Harvey, P. H., Mace, G. M. 1982. Comparisons between taxa and adaptive trends: Problems of methodology. In *Current Problems in Sociobiology,* ed. King's Coll. Sociobiol. Group, pp. 343–61. Cambridge: Cambridge Univ. Press. 394 pp.

127. Hazlett, B. A., ed. 1977. *Quantitative Methods in the Study of Animal Behavior.* New York: Academic. 222 pp.

128. Hennemann, W. W., Thompson, S. D., Konecky, M. J. 1983. Metabolism of crab-eating foxes, *Cerdocyon thous:* Ecological influences on the energetics of canids. *Physiol. Zool.* 56:319–24

129. Herrero, S. 1972. Aspects of evolution and adaptation in American black bears *(Ursus americanus* Pallus) and brown and grizzly bears *(Ursus arctos* Linn.) of North America. In *Bears—Their Biology and Management, IUCN Publ. No. 23,*

pp. 221–31. Morges, Switzerland: Int. Union Conserv. Nat. & Nat. Res. (IUCN)

130. Herrero, S. 1978. A comparison of some features of the evolution, ecology and behavior of black and grizzly/brown bears. *Carnivore* 1:7–17

131. Hersteinsson, P., Macdonald, D. W. 1982. Some comparisons between red and arctic foxes, *Vulpes vulpes* and *Alopex lagopus,* as revealed by radio tracking. *Symp. Zool. Soc. London* 49: 259–89

132. Hilborn, R., Stearns, S. C. 1982. On inference in ecology and evolutionary biology: The problem of multiple causes. *Acta Biotheor.* 31:145–64

133. Hinde, R. A. 1970. *Animal Behaviour.* New York: McGraw-Hill. 876 pp.

134. Hinde, R. A. 1974. *Biological Bases of Human Social Behaviour.* New York: McGraw-Hill. 462 pp.

135. Hinton, H., Dunn, A. 1967. *Mongooses: Their Natural History and Behavior.* London: Oliver & Boyd. 144 pp.

136. Hughes, A. L. 1983. Kin selection of complex behavioral strategies. *Am. Nat.* 122:181–90

137. Ikeda, H., Ono, Y., Baba, M., Doi, T., Iwamoto, T. 1982. Ranging and activity patterns of three nocturnal viverrids in Omo National Park, Ethopia. *Afr. J. Ecol.* 20:179–86

138. Izawa, M., Doi, T., Ono, Y. 1982. Grouping patterns of feral cats *(Felis catus)* living on a small island in Japan. *Jpn. J. Ecol. Nippon Seitai Gakkaishi* 32:373–82

139. Jaksić, F. M., Schlatter, R. P., Yanez, J. L. 1980. Feeding ecology of central Chilean foxes, *Dusicyon culpaeus* and *Dusicyon griseus. J. Mammal.* 61:254–60

140. Jerison, H. J. 1973. *Evolution of the Brain and Intelligence.* New York: Academic

141. Johnsingh, A. J. T. 1982. Reproductive and social behaviour of the dhole, *Cuon alpinus* (Canidae). *J. Zool.* 198:443–63

142. Jonkel, C. J., McTaggart Cowan, I. 1971. The black bear in the spruce-fir forest. *Wildl. Monogr.* 27:1–57

143. Kaufmann, J. H. 1962. Ecology and social behavior of the coati, *Nasua nasua,* on Barro Colorado Island, Panama. *Univ. Calif. Berkeley Publ. Zool.* 60:95–222

144. Kaufmann, J. H. 1982. Raccoon and allies. See Ref. 54, pp. 567–85

145. Kavanagh, M. 1981. Variable territory among Tantalus monkeys in Cameroon. *Folia Primatol.* 36:76–98

146. Kenyon, K. W. 1969. The sea otter in the

eastern Pacific. *North Am. Fauna* 68:1–352

147. Kenyon, K. W. 1982. Sea otter. See Ref. 54, pp. 704–10

148. Kihlström, J. E. 1972. Period of gestation and body weight in some placental mammals. *Comp. Biochem. Physiol.* 43:674–79

149. King, C. M. 1975. The home range of the weasel *(Mustela nivalis)* in an English woodland. *J. Anim. Ecol.* 44:639–68

150. King, G. E. 1980. Alternative uses of primates and carnivores in the reconstruction of early hominid behavior. *Ethology Sociobiol.* 1:99–109

151. King, J. A. 1973. The ecology of aggressive behavior. *Ann. Rev. Ecol. Syst.* 4:117–38

152. Kingdon, J. 1977. *East African Mammals.* Vol. 3A, *Carnivores.* New York: Academic

153. Kleiman, D. G. 1977. Monogamy in mammals. *Q. Rev. Biol.* 52:39–69

154. Kleiman, D. G. 1979. Parent-offspring conflict and sibling competition in a monogamous primate. *Am. Nat.* 114: 753–59

155. Kleiman, D. G. 1983. Ethology and reproduction of captive giant pandas *(Ailuropoda melanoleuca). Z. Tierpsychol.* 62:1–46

156. Kleiman, D. G., Brady, C. A. 1978. Coyote behavior in the context of recent canid research: Problems and perspectives. See Ref. 16, pp. 163–88

157. Kleiman, D. G., Eisenberg, J. F. 1973. Comparisons of canid and felid social systems from an evolutionary perspective. *Anim. Behav.* 21:637–59

158. Kleiman, D. G., Malcolm, J. 1981. The evolution of male parental investment in mammals. In *Parental Care in Mammals,* eds. D. J. Gubernick, P. H. Klopfer, pp. 347–87. New York: Plenum

159. Klinghammer, E., ed. 1979. *The Behavior and Ecology of Wolves.* New York: Garland. 588 pp.

160. Koop, K., Velimirov, B. 1982. Field observations on activity and feeding of bat-eared foxes *(Otocyon megalotis)* at Nxai Pan, Botswana. *Afr. J. Ecol.* 20:23–27

161. Kruuk, H. 1972. *The Spotted Hyena.* Chicago: Univ. Chicago Press. 335 pp.

162. Kruuk, H. 1975. Functional aspects of social hunting by carnivores. In *Function and Evolution in Behaviour: Essays in Honour of Professor Niko Tinbergen,* ed. G. Baerends, C. Beer, A. Manning, pp. 119–41. New York: Oxford Univ. Press. 393 pp.

163. Kruuk, H. 1976. Feeding and social behaviour of the striped hyaena *(Hyaena*

vulgaris Desmarest). *East Afr. Wildl. J.* 14:91–111

164. Kruuk, H. 1978. Foraging and spatial organisation of the European badger, *Meles meles* L. *Behav. Ecol. Sociobiol.* 4:75–89

165. Kruuk, H., Parish, T. 1981. Feeding specialization of the European badger *(Meles meles). J. Anim. Ecol.* 50:773–88

166. Kruuk, H., Sands, W. A. 1972. The aardwolf *(Proteles cristatus* Sparrman 1783) as a predator of termites. *East Afr. Wildl. J.* 10:211–27

167. Kruuk, H., Turner, M. 1967. Comparative notes on predation by lion, leopard, cheetah and wild dog in the Serengeti area, East Africa. *Mammalia* 31:1–27

168. Kurtén, B. 1971. *The Age of Mammals.* New York: Columbia Univ. Press

169. Lamprecht, J. 1978. The relationship between food competition and foraging group size in some larger carnivores. *Z. Tierpsychol.* 46:337–43

170. Lamprecht, J. 1978. On diet, foraging behaviour and interspecific food competition of jackals in the Serengeti National Park, East Africa. *Z. Saeugetierk.* 43:210–23

171. Lamprecht, J. 1979. Field observations on the behaviour and social system of the bat-eared fox, *Otocyon megalotis* Desmarest. *Z. Tierpsychol.* 49:260–84

172. Lamprecht, J. 1981. The function of social hunting in larger terrestrial carnivores. *Mammal Rev.* 11:169–79

173. Latour, P. B. 1981. Interactions between free-ranging, adult male polar bears *(Ursus maritimus* Phipps): A case of adult social play. *Can. J. Zool.* 59:1775–83

174. Latour, P. B. 1981. Spatial relationships and behavior of polar bears *(Ursus maritimus* Phipps) concentrated on land during the ice-free season of Hudson Bay. *Can. J. Zool.* 59:1763–74

175. Laundré, J. W., Keller, B. L. 1984. Home range of coyotes: A critical review. *J. Wildl. Manage.* 48:127–39

176. Lawton, J. 1979. A sociobiologist's new testament? *Nature* 278:193–94

177. Lecount, A. L. 1982. Characteristics of a central Arizona black bear population. *J. Wildl. Manage.* 46:861–68

178. Lehner, P. N. 1979. *Handbook of Ethological Methods.* New York: Garland. 403 pp.

179. Leitch, I., Hytten, F. E., Billewicz, W. Z. 1959. The maternal and neonatal weights of some Mammalia. *Proc. Zool. Soc. London* 13:11–28

180. Leutenegger, W. 1976. Allometry of neonatal size in eutherian mammals. *Nature* 263:229–30

181. Leyhausen, P. 1965. The communal

organization of solitary mammals. *Symp. Zool. Soc. London* 14:249–63

182. Leyhausen, P. 1979. *Cat Behavior: The Predatory and Social Behavior of Domestic and Wild Cats.* New York: Garland. 340 pp.

183. Liberg, O. 1980. Spacing patterns in a population of free roaming domestic cats. *Oikos* 35:336–49

184. Lindeque, M., Skinner, J. D. 1982. A seasonal breeding in the spotted hyena (*Crocuta crocuta* Erxleben), in Southern Africa. *Afr. J. Ecol.* 20:271–78

185. Lindström, E. 1982. *Population ecology of the red fox (Vulpes vulpes L.) in relation to food supply.* PhD thesis. Univ. Stockholm, Stockholm, Sweden

186. Lloyd, H. G. 1980. *The Red Fox.* London: Batsford. 320 pp.

187. Lloyd, H. G. 1980. Habitat requirements of the red fox. See Ref. 331, pp. 7–25

188. Lott, D. F., Minta, S. C. 1983. Random individual association and social group instability in American bison *(Bison bison).* Z. *Tierpsychol.* 61:153–72

189. Macdonald, D. W. 1978. Observations on the behaviour and ecology of the striped hyaena, *Hyaena hyaena,* in Israel. *Isr. J. Zool.* 27:189–98

190. Macdonald, D. W. 1979. The flexible social system of the golden jackal, *Canis aureus. Behav. Ecol. Sociobiol.* 5:17–38

191. Macdonald, D. W. 1980. *Rabies and Wildlife: A Biologist's Perspective.* New York: Oxford Univ. Press. 151 pp.

192. Macdonald, D. W. 1980. Social factors affecting reproduction amongst red foxes (*Vulpes vulpes* L., 1758). See Ref. 331, pp. 123–75

193. Macdonald, D. W. 1983. The ecology of carnivore social behaviour. *Nature* 301:379–84

194. Macdonald, D. W., Moehlman, P. D. 1982. Cooperation, altruism, and restraint in the reproduction of carnivores. *Perspect. Ethology* 5:433–67

195. Malcolm, J., Marten, K. 1982. Natural selection and the communal rearing of pups in African wild dogs (*Lycaon pictus*). *Behav. Ecol. Sociobiol.* 10:1–13

196. Maloiy, G. M. O., Kamau, J. M. Z., Shkolnik, A., Meir, M., Arieli, R. 1982. Thermoregulation and metabolism in a small desert carnivore: The fennec fox *(Fennecus zerda)* (Mammalia). *J. Zool.* 198:279–91

197. Margules, C., Higgs, A. J., Rafe, R. W. 1982. Modern biogeographic theory: Are there any lessons for nature reserve design? *Biol. Conserv.* 24:115–28

198. Matthews, W. D. 1930. The phylogeny of dogs. *J. Mammal.* 11:117–38

199. Maurel, D., Boissin, J. 1983. Compara-

tive mechanisms of physiological, metabolical and eco-ethological adaptation to the winter season in two wild European mammals: The European badger (*Meles meles* L.) and the red fox (*Vulpes vulpes* L.). In *Plant, Animal, and Microbial Adaptations to Terrestrial Environments,* ed. N. S. Magaris, M. Arianoutsou-Faraggitaki, R. J. Reiter, pp. 219–33. New York: Plenum

200. Maynard Smith, J. 1982. The evolution of social behaviour—a classification of models. In *Current Problems in Sociobiology,* ed. King's Coll. Sociobiol. Group, pp. 29–44. New York: Cambridge Univ. Press. 394 pp.

201. McBride, G. 1976. The study of social organizations. *Behaviour* 59:96–115

202. McNab, B. K. 1980. Food habits, energetics, and the population biology of mammals. *Am. Nat.* 116:106–24

203. McNab, B. K. 1983. Ecological and behavioral consequences of adaptation to various food resources. See Ref. 84, pp. 664–95

204. Mech, L. D. 1970. *The Wolf: The Ecology and Behavior of an Endangered Species.* New York: Doubleday. 384 pp.

205. Mech, L. D. 1977. Population trend and winter deer consumption in a Minnesota wolf pack. In *Proc. 1975 Predator Symp.,* ed. R. L. Phillips, C. Jonkel, pp. 55–83. Missoula: Univ. Mont. Press

206. Mech, L. D. 1977. Productivity, mortality, and population trends of wolves in northeastern Minnesota. *J. Mammal.* 58:559–74

207. Mech, L. D. 1983. *Handbook of Animal Radio-Tracking.* Minneapolis: Univ. Minn. Press. 107 pp.

208. Melemis, S. M., Falls, J. B. 1982. The defense function: A measure of territorial behaviour. *Can. J. Zool.* 60:495–501

209. Michener, G. R. 1983. Kin identification, matriarchies, and the evolution of sociality in ground-dwelling sciurids. See Ref. 84, pp. 528–72

210. Millar, J. S. 1977. Adaptive features of mammalian reproduction. *Evolution* 31:370–86

211. Millar, J. S. 1981. Pre-partum reproductive characteristics of eutherian mammals. *Evolution* 35:1149–63

212. Mills, M. G. L. 1978. Foraging behaviour of the brown hyaena (*Hyaena brunnea* Thinberg, 1820) in the southern Kalahari. *Z. Tierpsychol.* 48:113–41

213. Mills, M. G. L. 1978. The comparative socio-ecology of the hyaenidae. *Carnivore* 1:1–7

214. Mills, M. G. L. 1982. The mating system of the brown hyaena, *Hyaena brunnea,* in

the southern Kalahari. *Behav. Ecol. Sociobiol.* 10:131–36
215. Mills, M. G. L. 1982. Factors affecting group size and territory size of the brown hyaena, *Hyaena brunnea,* in the southern Kalahari. *J. Zool.* 198:39–51
216. Mills, M. G. L. 1983. Behavioural mechanisms and group maintenance of the brown hyaena, *Hyaena brunnea,* in the southern Kalahari. *Anim. Behav.* 31:503–10
217. Mills, M. G. L. 1983. Mating and denning behaviour of the brown hyaena *Hyaena brunnea* and comparisons with other hyaenidae. *Z. Tierpsychol.* 64:331–42
218. Mills, M. G. L., Gorman, M. L., Mills, M. E. J. 1980. The scent marking behaviour of the brown hyaena, *Hyaena brunnea. S. Afr. J. Zool.* 15:240–48
219. Mitani, J. C., Rodman, P. S. 1979. Territoriality: The relation of ranging pattern and home range size to defendability, with an analysis of territoriality among primate species. *Behav. Ecol. Sociobiol.* 5:241–51
220. Moehlman, P. D. 1979. Jackal helpers and pup survival. *Nature* 277:382–83
221. Moehlman, P. D. 1981. Reply to Montgomerie. *Nature* 289:825
222. Moehlman, P. D. 1983. Socioecology of silverbacked and golden jackals (*Canis mesomelas* and *Canis aureus.*) See Ref. 84, pp. 423–53
223. Montgomerie, R. D. 1981. Why do jackals help their parents? *Nature* 289:824–25
223a. Moore, J., Ali, R. 1984. Are dispersal and inbreeding avoidance related? *Anim. Behav.* 32:94–112
224. Myers, N. 1977. The cheetah's relationships to the spotted hyena: Some implications for a threatened species. See Ref. 205, pp. 191–200
225. Novick, H. J., Stewart, G. R. 1982. Home range and habitat preferences of black bears in the San Bernardino mountains of Southern California. *Calif. Fish Game* 68:21–35
226. Nowak, R. M. 1978. Evolution and taxonomy of coyotes and related *Canis.* See Ref. 16, pp. 3–15
227. Nowak, R. M., Paradiso, J. L. 1983. *Walker's Mammals of the World.* Baltimore: Johns Hopkins Univ. Press. 4th ed.
228. Nudds, T. D. 1978. Convergence of group size strategies by mammalian social carnivores. *Am. Nat.* 112:957–60
229. Owens, D. D., Owens, M. J. 1979. Communal denning and clan associations in brown hyenas (*Hyaena brunnea* Thunberg) of the central Kalahari Desert. *Afr. J. Ecol.* 17:35–44

230. Owens, D. D., Owens, M. J. 1979. Notes on social organization and behavior in brown hyenas (*Hyaena brunnea*). *J. Mammal.* 60:405–8
231. Owens, D. D., Owens, M. J. 1984. Helping behavior in brown hyenas. *Nature* 308:843–45
232. Owens, M. J., Owens, D. D. 1978. Feeding ecology and its influence on social organization in brown hyenas (*Hyaena brunnea*) of the central Kalahari. *East Afr. Wildl. J.* 16:113–35
233. Packard, J. M., Mech, L. D. 1980. Population regulation in wolves. In *Biosocial Mechanisms of Population Regulation,* ed. M. N. Cohen, R. S. Malpass, H. G. Klein, pp. 135–50. New Haven: Yale Univ. Press
234. Packer, C., Pusey, A. E. 1982. Cooperation and competition within coalitions of male lions: Kin selection or game theory. *Nature* 40:740–42
235. Parker, G. R., Maxwell, J. W., Morton, L. D., Smith, G. E. D. 1983. The ecology of the lynx (*Lynx canadensis*) on Cape Breton Island. *Can. J. Zool.* 61:770–86
236. Pearl, M. C., Schulman, S. R. 1983. Techniques for the analysis of social structure in animal societies. *Adv. Study Behav.* 13:107–46
237. Peters, R. H. 1983. *The Ecological Implications of Body Size.* Cambridge: Cambridge Univ. Press. 329 pp.
238. Peterson, R. O. 1977. *Wolf Ecology and Prey Relationships on Isle Royale, Natl. Park. Serv. Monogr. Ser. No. 11.* 210 pp.
239. Pienaar, U. D. V. 1969. Predator-prey relationships amongst the larger mammals of the Kruger National Park. *Koedoe* 122:108–76
240. Powell, R. A. 1979. Mustelid spacing patterns: Variations on a theme by *Mustela. Z. Tierpsychol.* 50:153–65
241. Powell, R. A. 1982. *The Fisher: Life History, Ecology, and Behavior.* Minneapolis; Univ. Minn. Press. 217 pp.
242. Radinsky, L. 1981. Evolution of skull shape in carnivores. 1. Representative modern carnivores. *Biol. J. Linn. Soc.* 15:369–88
243. Radinsky, L. 1981. Evolution of skull shape in carnivores. 2. Additional modern carnivores. *Biol. J. Linn. Soc.* 16:337–55
244. Radinsky, L. 1982. Evolution of skull shape in carnivores. 3. The origin and early radiation of the modern carnivore families. *Paleobiology* 8:177–95
245. Ralls, K. 1976. Mammals in which females are larger than males. *Q. Rev. Biol.* 51:245–76
246. Ralls, K. 1977. Sexual dimorphism in mammals: Avian models and un-

answered questions. *Am. Nat.* 111:917–38

247. Reich, A. 1978. A case of inbreeding in the African wild dog *Lycaon pictus* in the Kruger National Park. *Koedoe* 21:119–23

248. Reich, A. 1981. *The behavior and ecology of the African wild dog* (Lycaon pictus) *in the Krueger National Park.* PhD thesis. Yale Univ., New Haven, Conn.

249. Riedman, M. L. 1982. The evolution of alloparental care and adoption in mammals and birds. *Q. Rev. Biol.* 57:405–35

250. Rieger, I. 1979. A review of the biology of striped hyaenas, *Hyaena hyaena* (Linne, 1758). *Saeugetierkd. Mitt.* 2: 81–95

251. Roberts, M. S., Gittleman, J. L. 1984. *Ailurus fulgens. Mamm. Species* No. 243

252. Rodman, P. S. 1981. Inclusive fitness and group size with a reconsideration of group sizes in lions and wolves. *Am. Nat.* 118:275–83

253. Roeder, J. J. 1979. Reproduction die la Genette (*G. genetta*) en captivité. *Mammalia* 43:531–42

254. Rood, J. P. 1974. Banded mongoose males guard young. *Nature* 248:176

255. Rood, J. P. 1978. Dwarf mongoose helpers at the den. *Z. Tierpsychol.* 48:277–87

256. Rood, J. P. 1983. Banded mongoose rescues pack member from eagle. *Anim. Behav.* 31:1261–62

257. Rood, J. P. 1983. The social system of the dwarf mongoose. See Ref. 84, pp. 454–88

258. Rosenzweig, M. L. 1966. Community structure in sympatric carnivora. *J. Mammal.* 47:602–12

259. Rowe-Rowe, D. T. 1976. Food of the black-backed jackal in nature conservation and farming areas in Natal. *East Afr. Wildl. J.* 14:345–48

260. Rowe-Rowe, D. T. 1978. The small carnivores of Natal. *Lammergeyer* 25: 1–48

261. Rowe-Rowe, D. T. 1982. Home range movements of black-backed jackals in an African montane region. *S. Afr. J. Wildl. Res.* 12:79–84

262. Rowe-Rowe, D. T. 1983. Black-backed jackal diet in relation to food availability in the Natal Drakensberg. *S. Afr. J. Wildl. Res.* 13:17–23

263. Rowell, T. 1972. *Social Behaviour of Monkeys.* Baltimore: Penguin. 203 pp.

264. Rubenstein, D. I. 1978. On predation, competition, and the advantages of group living. *Perspect. Ethology* 3:205–31

265. Rudnai, J. A. 1973. *The Social Life of the Lion.* Wallingford, Pa: Washington Square. 122 pp.

266. Russell, J. K. 1983. Timing of reproduction by coatis (*Nasua narica*) in relation to fluctuations in food resources. In *Seasonal Rhythms in a Tropical Forest Ecosystem: Barro Colorado*, ed. E. G. Leigh, pp. 413–31. Washington DC: Smithsonian Inst.

267. Russell, J. K. 1983. Altruism in coati bands: Nepotism or reciprocity? In *Social Behavior of Female Vertebrates*, ed. S. K. Wasser, pp. 263–90. New York: Academic

268. Rutberg, A. T. 1983. The evolution of monogamy in primates. *J. Theor. Biol.* 104:93–112

269. Sandegren, F. E., Chu, E. W., Vandevere, J. E. 1973. Maternal behavior in the California sea otter. *J. Mammal.* 54:668–79

270. Savage, R. J. G. 1977. Evolution in carnivorous mammals. *Palaeontology* 20: 237–71

271. Scapino, R. 1981. Morphological investigation into functions of the jaw symphysis in carnivorans. *J. Morphol.* 167:339–75

272. Schaller, G. B. 1967. *The Deer and the Tiger.* Chicago: Univ. Chicago Press

273. Schaller, G. B. 1972. *The Serengeti Lion.* Chicago: Univ. Chicago Press. 480 pp.

274. Schaller, G. B., Crawshaw, P. G. 1980. Movement patterns of jaguar. *Biotropica* 12:161–68

275. Schaller, G. B., Lowther, G. R. 1969. The relevance of carnivore behavior to the study of early hominids. *Southwest. J. Anthropol.* 25:307–41

276. Schenkel, R. 1966. Play, exploration and territoriality in the wild lion. *Symp. Zool. Soc. London* 18:11–22

277. Schneider, K. B. 1978. *Sex and Age Segregation of Sea Otters, US Dep. Fish & Game Final Rep.*

278. Schoener, T. W. 1969. Models of optimal size for solitary predators. *Am. Nat.* 103:277–313

279. Schulman, S. R., Rubenstein, D. I. 1983. Kinship, need, and the distribution of altruism. *Am. Nat.* 121:776–88

280. Seidensticker, J. C., Hornocker, M. G., Wiles, W. V., Messick, J. P. 1973. Mountain lion social organization in the Idaho primitive area. *Wildl. Monogr.* 35:1–60

281. Shields, W. M. 1982. *Philopatry, Inbreeding, and the Evolution of Sex.* Albany: State Univ. NY Press. 245 pp.

282. Shields, W. M. 1983. Genetic considerations in the management of the wolf and other large vertebrates: An alternative view. In *Wolves in Canada and Alaska, Can. Wildl. Serv. Rep. Ser. No. 45*, ed.

L. N. Carbyn, pp. 90–92. Ottawa: Can. Wildl. Serv.

283. Smith, P. K. 1982. Does play matter? Functional and evolutionary aspects of animal and human play. *Behav. Brain Sci.* 5:139–84

284. Smithers, R. H. N. 1971. The mammals of Botswana. *Nat. Mus. Rhodesia No. 4*

285. Stains, H. J. 1975. Distribution and taxonomy of the Canidae. See Ref. 97, pp. 3–26

286. Stamps, J. A., Tollestrup, K. 1984. Prospective territorial defense in a territorial species. *Am. Nat.* 123:99–114

287. Stanley, S. M., van Valkenburg, B., Steneck, R. S. 1983. Coevolution and the fossil record. In *Coevolution*, ed. D. J. Futuyma, M. Slatkin, pp. 328–49. Sunderland, Mass: Sinauer

287a. Stearns, S. C. 1983. The influence of size and phylogeny on patterns of covariation among life-history traits in the mammals. *Oikos* 41:173–87

288. Stearns, S. C. 1984. The effects of size and phylogeny on patterns of covariation in the life history traits of lizards and snakes. *Am. Nat.* 123:56–72

289. Stirling, I., Latour, P. B. 1978. Comparative hunting abilities of polar bear cubs of different ages. *Can. J. Zool.* 56:1768–72

290. Storm, G. L., Andrews, R. D., Phillips, R. L., Bishop, R. A., Siniff, D. B., Tester, J. R. 1976. Morphology, reproduction, dispersal, and mortality of midwestern red fox populations. *Wildl. Monogr.* 49:1–82

291. Strickland, M. A., Douglas, C. W., Nowak, M., Hunziger, N. P. 1982. Fisher. See Ref. 54, pp. 586–98

292. Strickland, M. A., Douglas, C. W., Novak, M., Hunziger, N. P. 1982. Marten. See Ref. 54, pp. 599–612

293. Sunquist, M. E. 1981. The social organization of tigers *(Panthera tigris)* in Royal Chitawan National Park, Nepal. *Smithson. Contrib. Zool. No. 336*

294. Sunquist, M. E. 1983. Dispersal of three radiotagged leopards. *J. Mammal.* 64:337–41

295. Sussman, R. W. 1977. Socialization, social structure, and ecology of two sympatric species of *Lemur*. In *Primate Bio-Social Development*, ed. S. Chevalier-Skolnikoff, F. E. Poirier, pp. 515–28. New York: Garland

296. Svendsen, G. E. 1982. Weasels. See Ref. 54, pp. 613–28

297. Terborgh, J. 1983. *Five New World Primates: A Study in Comparative Ecology.* Princeton, NJ: Princeton Univ. Press 260 pp.

298. Thorne, B. L. 1981. Genetic consequences of variation in sib maturation schedules. *Acta Biotheor.* 30:219–27

299. Tinbergen, N. 1951. *The Study of Instinct.* New York: Oxford Univ. Press

300. Tinbergen, N. 1963. On aims and methods of ethology. *Z. Tierpsychol.* 20:410–33

301. Todd, A. W., Keith, L. B. 1983. Coyote demography during a snowshoe hare decline in Alberta. *J. Wildl. Manage.* 47:394–404

302. Todd, A. W., Keith, L. B., Fischer, C. A. 1981. Population ecology of coyotes during a fluctuation of snowshoe hares. *J. Wildl. Manage.* 45:629–40

303. van Ballenberghe, V. 1983. Extraterritorial movements and dispersal of wolves in southcentral Alaska. *J. Mammal.* 64: 168–71

304. van Schaik, C. P., van Hooff, J. A. R. A. M. 1983. On the ultimate causes of primate social systems. *Behaviour* 85:91–117

305. von Schantz, T. 1981. Evolution of group living, and the importance of food and social organization in population regulation. A study on the red fox *(Vulpes vulpes).* PhD thesis. Lund Univ., Lund, Sweden

306. Vehrencamp, S. L. 1979. The roles of individual, kin, and group selection in the evolution of sociality. In *Handbook of Behavioral Neurobiology*, ed. P. Marler, J. G. Vanderbergh, 3:351–94. New York: Plenum

307. Viljoen, S., Davis, D. H. S. 1973. Notes on stomach contents analyses of various carnivores in South Africa. *Ann. Transvaal Mus.* 28:353–63

308. Volf, J. 1965. Trente-deux jeunes de la genette. *Mammalia* 28:658–59

309. Walters, C. J., Stocker, M., Haber, G. C. 1981. Simulation and optimization models for a wolf-ungulate system. In *Dynamics of Large Mammal Populations*, ed. C. W. Fowler, T. D. Smith, pp. 317–37. New York: Wiley

310. Waser, P. M. 1981. Sociality or territorial defense? The influence of resource renewal. *Behav. Ecol. Sociobiol.* 8:231–37

311. Waser, P. M., Jones, W. T. 1983. Natal philopatry among solitary mammals. *Q. Rev. Biol.* 58:355–90

312. Watts, C. R., Stokes, A. W. 1971. The social order of turkeys. *Sci. Am.* 224:112–18

313. Webster, W. G. 1977. Territoriality and the evolution of brain asymmetry. *Ann. NY Acad. Sci.* 43:213–21

314. Wells, M. C., Bekoff, M. 1981. An observational study of scent-marking in

coyotes, *Canis latrans*. *Anim. Behav.* 29:332–50

315. Wells, M. C., Bekoff, M. 1982. Predation by wild coyotes: Behavioral and ecological analyses. *J. Mammal.* 63:118–27

316. Wemmer, C. 1977. Comparative ethology of the large-spotted genet *(Genetta tigrina)* and some related viverrids. *Smithson. Contrib. Zool.* No. 239

317. West-Eberhard, M. J. 1975. The evolution of social behavior by kin selection. *Q. Rev. Biol.* 50:1–34

318. West-Eberhard, M. J. 1979. Sexual selection, social competition, and evolution. *Proc. Am. Philos. Soc.* 123:222–34

319. Western, D. 1979. Size, life history and ecology in mammals. *Afr. J. Ecol.* 17:185–204

320. Wiley, R. H. 1981. Social structure and individual ontogenies: Problems of description, mechanism, and evolution. *Perspect. Ethology* 4:105–33

321. Williams, G. C. 1966. *Adaptation and Natural Selection*. Princeton, NJ: Princeton Univ. Press. 307 pp.

322. Wilson, D. S. 1975. The adequacy of body size as a niche difference. *Am. Nat.* 109:769–84

323. Wilson, D. S. 1980. *The Natural Selection of Populations and Communities*.

Menlo Park, Calif.: Benjamin/Cummings. 186 pp.

324. Wilson, E. O. 1975. *Sociobiology: The New Synthesis*. Cambridge, Mass: Harvard Univ. Press. 697 pp.

325. Wilson, S., Kleiman, D. G. 1974. Eliciting and soliciting play. *Am. Zool.* 14: 341–70

326. Wittenberger, J. F. 1981. *Animal Social Behavior*. Boston: Duxbury. 722 pp.

327. Wittenberger, J. F., Tilson, R. L. 1980. The evolution of monogamy: Hypotheses and evidence. *Ann. Rev. Ecol. Syst.* 11:197–232

328. Wolf, K., Schulman, S. R. 1984. Male response to "stranger" females as a function of female reproductive value among chimpanzees. *Am. Nat.* 123:163–74

329. Woolpy, J. H., Eckstrand, I. 1979. Wolf pack genetics. A computer simulation with theory. In *The Behaviour and Ecology of Wolves*, ed. E. Klinghammer, pp. 206–24. New York: Garland

330. Zimen, E. 1976. On the regulation of pack size in wolves. *Z. Tierpsychol.* 40:300–41

331. Zimen, E., ed. 1980. *The Red Fox: Symposium on Behaviour and Ecology*. The Hague: Junk. 285 pp.

332. Zimen, E. 1981. *The Wolf: A Species in Danger*. New York: Delacorte. 373 pp.

Ann. Rev. Ecol. Syst. 1984. 15:233–58

MORPHOGENETIC CONSTRAINTS ON PATTERNS OF CARBON DISTRIBUTION IN PLANTS

Maxine A. Watson

Department of Biology, Indiana University, Bloomington, Indiana 47405

Brenda B. Casper

Department of Biology, University of Pennsylvania, Philadelphia, Pennsylvania 19104

Phenotypic plasticity is an important element in the response repertoire of plants (25). It may be manifested as changes in organ (e.g. leaf) morphology as well as in patterns of biomass distribution. Morphological plasticity throughout growth is possible because plant development is modular in form (77, 184, 185). Growth results from the reiteration of basic morphological subunits produced by meristems. Because meristems may develop into either reproductive or vegetative structures, patterns of biomass distribution reflect developmental decisions (121, 172, 180).

In this essay, we review data suggesting that at least in certain cases, architectural constraints affect the range of morphological plasticity that can be expressed. Because carbon is frequently viewed as a critical currency of allocation (76, 165), we focus on how these constraints influence assimilate (i.e. carbon) production and utilization. We postulate that when these constraints are present, plants consist not only of morphological subunits (185) but, as Adams (4) first suggested, of physiological subunits as well. We call these *integrated physiological units* (IPUs) and propose that they are made up of identifiable arrays of morphological subunits that together function as relatively autonomous structures with respect to the assimilation, distribution, and utilization of carbon. If such units of physiological integration exist,

233

0066-4162/84/1120-0233$02.00

they may place additional constraints on developmental decisions that determine the fates of meristems.

Many of the patterns we describe, when considered in isolation as they are here, serve no obvious adaptive function. This may be true for at least two equally plausible reasons: (a) because examining these patterns in isolation obscures their adaptive purpose; or (b) because the patterns are architectural remnants of the phyletic heritage of the plants and constitute one set of constraints in which, as Gould & Lewontin assert, "the constraints themselves become much the most interesting aspect of evolution" (67, p. 594). Our purpose is simply to describe a number of morphological and physiological patterns that seem to constrain the range of expression of phenotypic plasticity in some plants. Attempting to explain the evolutionary origins and possible adaptive roles of such patterns is premature, although these are fundamental questions that should be explored.

While we recognize the importance of other factors such as mineral nutrients, water, and plant growth regulators in limiting and controlling patterns of biomass distribution, we confine our discussion to carbon. Specifically, we ask: (a) To what extent do morphological subunits (i.e. branches) produce the assimilate needed for their own growth and maintenance? (b) To what degree is assimilate freely moved among organs within a branch? and (c) To what extent is assimilate moved among branches?

PHOTOSYNTHESIS

Photosynthetic Capacity and Sink Strength

The strength of assimilate demand can affect the rate of photosynthesis in source leaves (64, 66, 179). While fruit development may be accompanied by an overall increase in total plant photosynthesis [e.g. in cucumber (12)], an increase in leaf photosynthesis is frequently restricted to leaves subtending or close to strong developing sinks (20, 54, 58, 59). Artificial manipulation of source-sink relationships by defoliation or sink removal can lead to changes in leaf photosynthesis. When some of the leaves are removed during active periods of pod and seed development in *Glycine* (soybean) and *Phaseolus* (bean), photosynthetic rates in the remaining leaves increase (21, 64, 168, 179). Similarly, the net photosynthetic rate of potato plants decreases following the removal of tubers (27). Photosynthetic capacity does not always decrease as sink strength decreases, however (e.g. 188).

The photosynthetic rates of leaves in tree canopies also respond to changes in sink strength. Leaves of some fruiting trees exhibit higher photosynthetic rates than those of nonfruiting trees of the same species (103). Following partial defoliation of the canopy in oak and maple (*Quercus rubra* L. and *Acer rubrum* L.), the remaining leaves exhibited so large an increase in their net photosynthesis rates that they actually overcompensated for the foliage removed (81).

Thus, leaves may photosynthesize submaximally, even in environments where resources are abundant. Increased sink demand can be met by an increase in the photosynthetic rates of source tissues. Although common, this pattern is not universal (38, 75, 142, 143). In cotton (38) and tobacco (143), for example, leaf photosynthesis does not vary with the strength of adjacent sinks. Leaves of differing final sizes and located at different node positions all have similar patterns of light-saturated photosynthesis. These patterns vary only with the absolute age of the leaf, although individual leaves achieve comparable absolute ages both in the presence and in the absence of strong developing sinks.

Other data also suggest that the availability of adequate assimilate is not necessarily limited by potential leaf photosynthetic capacity. Comparisons of nearly isogenic lines of rice that differ only in the duration of growth demonstrate that the photosynthetic production of the short-season line is adequate to fill the available sinks (i.e. grains) completely (44, 45). Similarly, soybean lines that differ in total leaf photosynthesis do not differ in seed yield, crop growth rate, harvest index, or duration of seed fill (60). These data must be viewed with caution. Gross measurements of the total leaf photosynthesis of single leaves fail to take into account differences among lines in their respiratory responses. Moreover, measurements on single leaves may be misleading, depending upon the relationship of the subject leaf to the developing sinks within the plant, a relationship we will examine below.

The Photosynthetic Capacity of Reproductive Structures

The reproductive structures themselves may contribute a portion of the carbon required for flower and fruit production. Positive net photosynthesis in plant reproductive structures is rarely observed, however, because in most cases gross photosynthesis nearly balances respiration (13, 14, 16, 36, 42, 48, 106, 113, 115, 131, 135, 139, 181). Nevertheless, fruit photosynthesis may be a significant source of assimilate. For example, in *Ambrosia trifida,* 72% of the total carbon required for maturation of male flowers and 34% of that needed for maturation of female reproductive structures is expended in respiration. Male flowers assimilate 41% and female flowers and fruits assimilate 57% of their own carbon requirements. Thus, positive net photosynthesis is observed in female reproductive structures but not in male flowers (13). Estimates of how much flowers and fruits in 15 woody species contribute to their own carbon budget range from 2.3% to 64.5%, but positive net photosynthesis occurs only in *Acer platanoides* and *Tilia platyphyllos fastigiata* (14). In many cereal crops, photosynthesis by the ear more than offsets respiration and, depending on the species and variety, may contribute up to 40% of the structural carbon in the grain (22, 54, 105, 166, 167).

The shape of the fruit or its associated structures seems to be an important determinant of its photosynthetic capacity. Leaf-shaped structures, such as the winged carpels of *Acer* and the bracts of *Tilia,* produce a greater proportion of their carbon budget than do round fruits or those with small wings (14, 16, 106, 113). Similarly, the maturing inflorescences of awned varieties of wheat produce more photosynthate than do those of awnless varieties (54). Thus, fruits with small surface to volume ratios probably are not effective as photosynthetic organs, whereas thin, flat structures such as large sepals may be. A fruit's shape and structure is, of course, the product of many selective factors, including those imposed by seed dispersal requirements. The fruit design that results may be one that precludes the fruit from functioning effectively as a photosynthetic structure (90, 91).

Some fruits internally reassimilate the CO_2 given off in respiration (42, 59, 105, 113, 131, 139). For example, some leguminous pod fruits reassimilate the respiratory CO_2 from their developing seeds (42, 59, 115, 131, 139). Thus, fruits may make a meaningful contribution to their own carbon budget, even when there is no net carbon uptake from the external atmosphere. It is also possible that chlorophyllous embryos reassimilate respiratory CO_2 (90, 91).

Few studies examine fruit photosynthesis in an ecological context; most focus on crop plants, and measurements of photosynthesis are often made under conditions that simulate full sunlight. Since flowers and fruits may be borne within the foliage and not exposed to full sunlight, photosynthesis by reproductive structures must be measured in their natural light environment before conclusions about the degree to which they contribute to their own carbon requirements can be made.

BRANCH AUTONOMY

The Modular Structure of Plants

Plants may be viewed as assemblages of metamers (70, 77, 79, 102), each consisting of a section of stem made up of the internode complex, including its leaf or leaves and their associated lateral (i.e. axillary) meristems (Prevost 1967 cited in 185; 137). Plant growth results from the reiterated production of metamers by meristems. Metamers may be assembled into nested levels of morphological organization, among them the module and branch. The module is an aggregate of metamers formed by a single apical meristem (70, 79, 137). Since a lateral meristem forms a new apex once it begins to differentiate, a simple branch resulting from the development of this meristem is also a module. Vegetative branches form from the reiteration of the basic metameric unit, while flowers and inflorescences form from the reiteration of metamers bearing modified leaves. Whether complex or simple, the aggregate of branches forms the recognizable plant body, or model unit (70, 184, 185).

Because the reiteration of metamers and modules results from the differentiation of meristems, a particular plant morphology is the product of developmental decisions about the number and types of metamers, and thus branches, produced. Is carbon availability critical to such decisions, and if so, at what level of morphological organization is its availability assessed?

Metamer and Branch Autonomy

Recent studies show that cohorts of metamers differ both in their demographic performance (e.g. fecundity) (3, 15, 33, 118) and in their physiological properties (50, 112, 126). Lovett Doust & Eaton (118) examined changes in yield in *Phaseolus vulgaris* when pods were removed as they reached maturity and found that total plant yield increased. This change was due to an increase in the number of metameric units (and hence flowers) produced, rather than to an increase in the yield components (i.e. marketable pods) made by the individual metameric units. Because the pods were removed at maturity, it is unlikely that removing them freed resources for use in subsequent metamer production. Instead, this treatment may have prevented production of the hormonal signal that halts metamer production. The yield of a particular metamer was a function of its birth date and its position within the plant, rather than the cultivation technique applied. Thus, metameric units in *Phaseolus*—consisting of a section of stem, a trifoliate leaf, and the associated, laterally borne reproductive branch—function as internally integrated and relatively autonomous physiological units. They correspond exactly to Adams's (4) "nutritional units."

Other physiological studies of legumes attest to the tight physiological linkage within metameric units (i.e. between leaves and their associated pods) (17, 23, 58, 59, 132, 169, 186). In *Pisum*, the translocation of photoassimilate between carpel and seeds is relatively independent of the rest of the plant (115). Approximately 66% of the carbon required by *Pisum* seeds ripening at the lowest reproductive node is supplied by the leaf and pod at that node (59). In *Phaseolus*, assimilate from the terminal leaf moves exclusively to subtending pods, while assimilate from leaves on branches moves only into pods within the branch (186); similarly, in soybean, exogenously supplied growth regulators are transported primarily to pods at the axil of the fed leaf and at the second axil below the fed leaf (23, 95, 163).

In cotton, the subtending leaf is the primary source of assimilate for the developing boll (8). Developing tomato fruits are almost entirely dependent on imported assimilates (175). Initially, the first truss (i.e. fruit-bearing branch) is a major sink for assimilates from all leaves, regardless of the leaf's position above or below the developing truss (98–100). As the truss matures, however, the leaves closest to it become its most important suppliers (83, 98). In

Capsicum (159) and groundnut (97), the branches themselves function semi-autonomously. Young fruits on axillary branches receive assimilate from the main axis only early in their development (159). As the branch increases in size, it becomes independent of the main stem, and maturing fruits receive assimilate only from leaves on the same branch.

The flag leaves of grasses contribute the remainder of the assimilate needed for grain development that is not supplied by the ear itself (22, 28, 54, 167). In wheat, assimilation by the flag leaf equals 110 to 120% of the final grain weight of the ear; while in barley, flag-leaf photosynthesis accounts for about half of the ear's grain weight (28, 167). Fifty percent of the carbon assimilated by the flag leaf is used in grain formation (28); ear photosynthesis provides the remaining carbon required for grain fill (28, 54, 167).

The extent to which assimilate can be transported among the tillers of clonal grasses varies. During early tiller development, there is significant transloca-tion of assimilate from the parent into daughter tillers (101, 171). In many species, daughter tillers quickly lose their assimilate dependence on the parent (cf 9, 145). Once tillers mature, what little translocation occurs often takes place from tiller to parent rather than the reverse (7, 34, 61). There are, however, several grass species in which connected tillers never cease to import assimilate from the parent (5, 144, 145). Moreover, in some grasses manipula-tion of tillers by either temporary shading or defoliation frequently leads to a renewal of assimilate movement among tillers (cf 9, 145). How long this transport will continue appears to be related, in part, to the severity of the stress that the clone experiences (133, 145). Not all grass clones, though, are capable of reinitiating translocation among tillers (cf 9).

Nongraminaceous clonal plants also vary in the extent to which they move assimilate among ramets. Diverse species such as *Ranunculus repens* (Ginzo & Lovell 1973 cited in 9), two violets (130), and the moss *Polytrichum alpinum* (Collins & Oechel 1974 cited in 9) appear to be able to move assimilate freely among ramets. Two perennial forest forms, however, differ significantly in their capacity to move assimilate (9). As in grasses, manipulation by defolia-tion or partial shading can increase this capacity—e.g. in *Maianthemum canadense* (cited in 9) and *Solidago canadensis* (80). In the latter study, translocation from parent to daughter ramet was inferred rather than measured directly. The data may as easily be interpreted as demonstrating movement of water rather than of carbon, as suggested in a related study (88).

There are limited but conflicting data regarding the extent to which photo-synthate is distributed to different parts of woody plants. In grape, the degree to which assimilate moves among branches is strongly dependent upon the growth stage (69). The pattern is similar to that observed in herbaceous species. Axillary buds are supplied by subtending leaves until they mature one or two leaves, after which the young branch exports small amounts of assimilate back to the parent stem. Once plants reach the prebloom and bloom stages, there normally is no transport among adjacent branches on a single spur (141).

At least two experiments suggest that there is widespread redistribution of assimilate in trees from year to year. In young apple trees, stored reserves in the trunk contribute to growth mainly in the early spring (74). The movement of assimilate seems to be unrestricted, since labeled carbohydrate produced in the autumn is not confined to the same vascular strands the next spring and new foliage has a uniform distribution of radioactivity (140). In a study on the effects of herbivory in *Eucalyptus,* half of each tree was sprayed weekly with insecticide for one growing season. Over the next two years, the sprayed branches grew more than the unsprayed ones on the treated trees, but both grew more than those on the controls (127). The underlying mechanism for this response is not evident, but it is possible that untreated branches on the treated trees benefited directly by importing carbon from the remainder of the tree or indirectly through increased root growth.

A different pattern was observed in a study on the effects of artificial browsing in sagebrush, *Artemisia tridentata* (39). Plants received one of two treatments over a three-year period: Either all of the current growth was removed from half of the canopy or half of the current growth was removed from every twig. In the first case, the harvested half—including the roots—died, while the intact half grew more vigorously than the untreated plants. In the latter case, the plants produced less total biomass than the controls, but only small twigs or branches died. Branches that did not produce assimilate apparently were not supported by the remainder of the plant. Furthermore, the data suggest that pathways of assimilate movement join discrete portions of the canopy with discrete portions of the root system.

Branches of trees seem to be relatively autonomous in their use of carbon within a growing season. Much of the photosynthate necessary for fruit production on a branch is provided by the branch itself (46, 57, 72, 104, 152). In apple trees, nearby leaves supply much of the carbon required for fruit development (73) and more translocation occurs from leaves on fruiting branches than from those on nonfruiting branches (72). When the fruiting branches are shaded, nearly all of the fruits on those branches abscise (152). The starch content of fruiting spurs (i.e. short shoots) in apples drops well below that of nonfruiting spurs during early fruit development (104). During fruit filling in pecans, fruiting branches use reserves built up within the branch during early summer, while nonfruiting branches continue to store carbon (57).

Additional evidence for branch autonomy in trees comes from studies designed to simulate herbivory. Totally or partially defoliating a fruiting branch negatively affects the fruit crop on that branch by increasing fruit abortion and/or decreasing the seed number per fruit or the seed weight (18, 89, 160). This pattern is evident in Kentucky coffee trees, even when less than 0.1% of the total leaf area of the tree is removed and many branches are nonfruiting (89).

Whether a meristem develops into a vegetative or a reproductive structure seems to be determined, at least in part, by conditions within the branch. Defoliating limbs of pecans decreases both the number of female flower buds initiated and the percentage of buds on those limbs that develop fruit the following year (157). Similarly, the tendency for alternate-year bearing in pistachio may be explained by local competition for carbon between inflorescence buds and developing fruits (164). Carbon limitation, however, may not be the only cause of biennial bearing. For example, in strictly biennial varieties of apple, no new floral buds are initiated on branches that bear fruit that year. At least in some of these varieties, the failure to form new flower buds is due to the presence on the branch of developing seeds rather than developing fruits. Individual spurs of a parthenocarpic (i.e. seedless) variety of apple, that normally bears annually, for example, are made to flower biennially by pollinating the flowers (30). Pollination results in seed formation. No flower buds are initiated on spurs occupied by fruits that develop from pollinated flowers. Growth regulators produced by the developing seeds, rather than the local depletion of carbon reserves by the developing fruits, may be responsible for suppressing the initiation of new floral buds (120). The ratio of male to perfect flowers on a twig of olive varies inversely with the ratio of leaves to flower buds on that twig (173). When entire leaves or portions of leaves are removed from the twig three months prior to flowering, the floral sex ratio is adjusted through abortion of the pistils in potentially perfect flowers. In some species, branches growing in the sun may produce more fruit than those in shade (18, 149); and particular branches in some species may be vegetative in some years, even though parts of the tree flower every year (89, 155). Branches of at least some tropical trees flower and fruit asynchronously (6, 43, 52).

Positional effects are also found within inflorescences or branches. The probability that a fruit will mature is often related to its location within a module or its time of pollination relative to that of nearby flowers (161, 189). The fruit or seed weight or the number of ovules maturing may also vary with flower position (29, 123, 161, 174, 189). Usually flowers located closest to the plant axis or to the leaves receive more resources (123). In *Cryptantha flava*, most flowers mature a single seed, but a few produce two. The likelihood that a flower will produce two seeds is correlated with its proximity to the main stalk of the inflorescence (29).

These data suggest that either carbon balance within certain metamers or branches is tightly regulated or that pathways of assimilate movement within and between certain morphological subunits may be restricted, or both. The degree of restriction seems to be in part a function of the growth stage. Apparently, restrictions are not caused by insufficient availability of vascular tissue per se (66, 71, 87, 136, 177; but also see 92, 111, 183).

SECTORIAL TRANSPORT

In this section, we provide further evidence that assimilate movement both within and between morphological subunits may be restricted (i.e. sectorial). Sectorial patterns contrast with those of uniform distribution in that the latter reflect a capacity to translocate assimilate with equal facility to all parts of the plant.

Changes in Patterns of Translocation Due to Age and Life Stage

Patterns of assimilate movement vary dynamically both with plant age and with life stage (65). Young leaves initially import assimilate (26, 46, 65, 103, 104, 190), but by the time most leaves reach one half to one third of their adult size, they have begun to export as well as import (69, 103, 104, 176). Once a leaf begins exporting, the destination of the assimilate varies with the position of that leaf relative to the shoot apex (23, 28, 46, 50, 69, 72, 93, 97–100, 103, 112, 141, 144, 145, 148, 154, 159, 163, 176). Early in shoot growth in *Populus*, for example, leaves tend to export to two major sinks, the roots and the shoot apex, but transport quickly becomes directional (112). Upper leaves typically send assimilate to the developing shoot apex, lower leaves export to the roots, and middle leaves export bidirectionally. Because the relative position of a leaf along the stem changes over time, the export pattern also changes. This changing pattern of assimilate flow also characterizes herbaceous shoots and branches (176).

In plants with an elongate growth axis, there is normally little transport of assimilate into mature leaves (176), while the same is not necessarily true of rosette plants—e.g. lettuce, swiss chard, radish (170). This unusual pattern ceases with the loss of the rosette habit at the onset of flowering. Another unusual pattern of assimilate movement is reported in tomato, in which upper leaves transport downward (98, 146). This pattern was initially thought to be an artifact of the short (24 hr) interval used between application of label and harvest (176) because in tomato, as in sugarcane (176) and bean (128, 176), assimilate entering the main stem from a leaf must move downward to the next node before it can move upward (Bonnemain 1965, 1966, cited in 176). More recent work on the Solanaceae (which includes tomato), however, suggests that the reported pattern accurately describes translocation in this family (146). The pattern seems to be caused by the presence of both internal and external phloem that usually transport assimilate in opposite directions (Bonnemain 1968, 1969, cited in 146). The two phloem types are developed to different degrees in leaves of different ages, making translocation pattern dependent on the developmental stage of the labeled leaf as well as on that of the plant. These findings underscore the potential importance of understanding vascular anatomy for properly measuring and interpreting patterns of assimilate movement.

Patterns of assimilate movement are also altered in the transition from vegetative to reproductive growth, when flowers and fruits become sinks that can compete effectively with the apex and roots for assimilate (12, 23, 24, 65, 66, 69, 97, 104, 134, 164, 176, 186). As shown above, most of the assimilate for flower and fruit development originates locally. Not all leaves close to a node, however, contribute equally to the development of fruit at that node (e.g. 23, 186), as we will discuss in the next section.

Evidence for Restricted Pathways of Translocation Related to Phyllotaxy

Autoradiography studies of the transport of labeled carbon within plants following photosynthetic assimilation of $^{14}CO_2$ by specific leaves indicate that restricted movement of assimilate into specific sectors of modules, branches, or ramets is common. The patterns appear to reflect the leaf phyllotaxy, that is, the geometric arrangement of leaves on the stem (129).

Studies examining the pathways of photosynthate movement from mature to young leaves in willows (84), eastern cottonwood (110), and plum and apple (11) show that at least over short distances (i.e. within branches and/or spurs), the patterns of translocation in trees are sectorial. This phenomenon is clearly seen in apple, where label applied to a single source leaf is transported into predictable and localized regions of the growing stem (Figure 1) (11). Over short intervals, the pattern of transport may be followed on a leaf by leaf basis. Leaves that lie superimposed in vertical ranks and that are located closest to the fed leaf take up label over their entire laminae. Most leaves in adjacent ranks acquire label in all but a small basal sector on one side of the midrib, while those in more distant ranks have the reverse pattern. Leaves on the opposite side of the stem from the fed leaf receive no labeled assimilate. Similar restrictions on patterns of assimilate movement are observed between leaves and the developing fruits of apple (73).

Numerous examples of sectorial transport exist for herbaceous species, among them: soybean (23, 95, 163), bean (186), tobacco (93, 154), cotton (37), tomato (146), and the domesticated sunflower (138). Additional evidence of sectorial transport in herbs comes from studies of the transmission of floral stimuli (e.g. 53, 107, 151) and hormones (94). In both soybean (23) and bean (186), sectorial patterns of assimilate movement are found between leaves along the stem and between leaves at certain nodes and developing pods at more distant nodes. When label was applied to a soybean leaf at a node midway down the stem, the greatest quantity of labeled assimilate appeared in pods at alternately numbered younger nodes; pods at intervening nodes acquired little or no label from that source leaf (23).

The pattern of assimilate movement in sunflower is especially intriguing. The cultivated sunflower is, ideally, a monopodial stem terminated by an

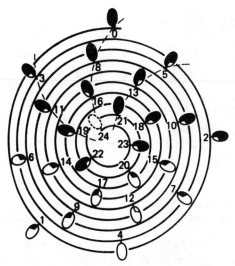

Figure 1 Leaf arrangement round the stem of apple. The source leaf is marked 0, and successive leaves upward (i.e. younger leaves) are numbered 1–24. Source leaves were fed $^{14}CO_2$ for 1–2 hours. Whole leaf autoradiography one to several days after treatment showed the distribution pattern of ^{14}C. It generally agrees with the vascular phyllotaxis of 3/8 that is characteristic of apple. The dashed lines indicate the 3 sets of leaves that should have the most direct vascular connections with the labeled leaf. Leaf 24 was not examined. (From 11)

inflorescence. It is tempting to assume that the monopodial stem acts as an integrated organ system that provides assimilate to the inflorescence. But experiments following the transport of ^{14}C-labelled acetate applied to individual leaves indicate that transport is highly sectorial, with particular leaves supplying specific files of seeds within the inflorescence (Figure 2) (138). Removing all the leaves in neighboring vertical ranks results in the asymmetric development of the compound inflorescence (138). In contrast, removing 25% of each of the plant's leaves has no significant effect on the inflorescence's ability to mature seed (150). This sectorial pattern may not be as surprising as first thought. The domestic sunflower was derived from branched ancestors relatively recently (82), and genetic control of the monopodial growth form involves comparatively few genes (85). Perhaps the pronounced sectorialization reflects an architectural constraint that is a remnant of this branched ancestry.

Sectorial translocation to reproductive structures also occurs in *Lupinus luteus*. Flowers in basal whorls are more likely to mature fruits and produce more seeds per fruit than those more distal (174). These patterns may occur in part because flowers in distal whorls have vascular tissue that is not as well developed at the time of fertilization, but many distal flowers will mature fruit if the basal whorls are removed. The resulting arrangement of mature fruits

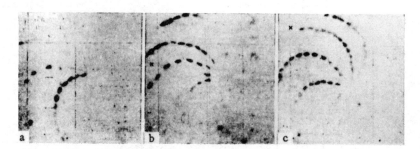

Figure 2   Autoradiographs of single files of seeds in sunflower heads (variety, Karlik Štepnoi). Heads were harvested from plants in which single leaves located at different levels in the leaf canopy had been fed ^{14}C-acetate. Label was applied for 15 consecutive days beginning 15–20 days following anthesis to (*a*) an upper leaf, (*b*) a middle leaf, or (c) a lower leaf. Crosses denote the location of the treated leaf relative to the labeled seeds in the head. (From 138)

depends, however, on the pattern of flower removal. If all of the flowers within vertical columns are removed, fewer of the remaining flowers will mature fruits than if an equal number of flowers is removed in whorls or in a spiral pattern. Thus, developing fruits seem to affect the flowers above them more than they affect those occupying the same whorl. These fruits may be competing among themselves for resources. We show below that competition may be most intense among flowers lying in common vertical columns, presumably due to constraints on assimilate movement that are imposed by vascular architecture.

Grass tillers exhibit more internal integration than the subunits of dicotyledonous trees and forbs (e.g. 122, 124, 148). As in monocotyledonous trees such as palms (190, 191), this pattern appears to be due to the large number of vascular traces and the extensive vascular anastomes at nodes, which result in the repeated reintegration of leaf traces from many different nodes of entry into the main stem (Kumazawa, 1961 cited in 51). In contrast, the reproductive ears of grasses function sectorially (41). Despite the apparent freedom of movement by assimilate within grass tillers (i.e. ramets), the degree of movement among tillers, as previously mentioned, may be related to life stage. The pattern of movement, however, appears to reflect the extent of vascular connections among daughter ramets (10, 34).

The Role of Phyllotaxy in Determining Patterns of Assimilate Movement

What underlies sectorial transport? Long-distance movement of assimilate occurs in the phloem and is predominantly oriented longitudinally. Work on soybean and tobacco indicates that there is a strict limitation on the lateral movement of assimilate away from the phloem (176). Similarly, there is little lateral movement from the vasculature of petioles, stems, and rhizomes of plants as diverse as bracken fern, peppermint, cotton, rape, rice, squash, sugarbeet, and tomato (176). Recent studies of bean (128), cotton (37, 38), and eastern cottonwood (110) provide further evidence that phyllotactic constraints on vascular architecture restrict the pathways of assimilate movement toward developing organs (129, 176).

The relationship between the vascular system of the stem and phyllotaxy is established through the leaf traces (51). Leaves are connected to the main stem by one or more leaf traces, which may vary in length. The traces of one leaf may traverse one or more nodes within the stem and interconnect it with one or several other leaves in doing so. Which leaves are directly interconnected is determined by their phyllotaxy. Phyllotaxy can be expressed as a fraction in the Fibonacci series (e.g. 1/2, 1/3, 2/5, 3/8), where the denominator is the number of leaves that develop before a direct linear overlap between two leaves occurs and the numerator is the number of turns around the stem necessary before this happens. In a plant with 3/8 phyllotaxy, for example, it takes three turns about the stem, involving eight leaves, to obtain a direct overlap between leaf n and leaf $n + 8$. Leaves that lie directly above one another form vertical ranks called orthostichies.

According to Larson (109), these divergence fractions provide the best description of the organization of the vascular strands. For a given phyllotactic ratio, the denominator of the fraction is either equal to or a multiple of the number of main vascular bundles—the sympodia—that traverse the stem. Thus, the number of sympodia bears a simple relationship to the number of orthostichies, or vertical ranks of leaves—e.g. apple has a 3/8 phyllotaxy and contains 8 sympodia per branch (Figure 1). The underlying vascular relationships are graphically illustrated for seedlings of *Populus deltoides,* which have 5 sympodia (2/5 phyllotaxy; see Figure 3) (109). *Populus* leaves are each supplied by 3 traces, with each trace arising from a different sympodium. The central leaf traces of a particular sympodium terminate in leaves belonging to a specific orthostichy. Right and left lateral traces also diverge at regular intervals.

The anatomy underlying sectorial transport is now apparent (see also 129). Leaves arrayed in a common orthostichy, or rank, are connected to each other by their central leaf traces, which occupy a common sympodium. Translocation within a sympodium is direct. Because leaves may be joined to the stem by more than one leaf trace, transport of carbon is not necessarily confined to a single orthostichy, but may follow a wedge-shaped pattern, as observed in apple (Figure 1). Leaf traces supply different areas of the leaf, and thus the sectorial pattern extends into the leaf as well. The pattern of vascular supply to branches and/or inflorescences from the main stem is similarly ordered (51). Hence, developing branches and/or inflorescences may differ in their access to assimilate from different source leaves, as occurs in soybean (23) and sunflower (138), mentioned above. Plants exhibit either right- or left-handed phyllo-

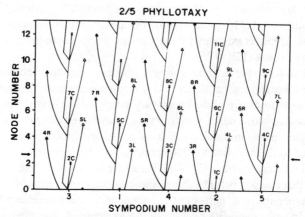

2/5 PHYLLOTAXY

Figure 3   Diagrammatic representation of the 2/5 vascular phyllotaxis of *Populus deltoides*. The vascular system has been displayed as if it were unrolled from the stem and laid flat. x, central traces; closed triangles, right traces; open triangles, left traces. Each of the three traces serving a leaf arises from a different sympodium. C-traces are the perpetuating members of the phyllotactic system; e.g., 6C is the parent for 11C and also for 8R and 9L. (From 109)

taxy. Strangely, in some species right-handed plants have a higher reproductive output (19, 47).

There is far from complete agreement on the importance of vascular architecture in determining patterns of assimilate movement (66). Furthermore, predominant pathways of translocation can be altered by stress or by changes in the relative strength of sinks (40, 66, 141). The phenomenon has been particularly well studied in soybean (17, 49, 169), where there are rapid changes in translocation patterns following severe manipulation of source-sink relationships (56). The extent of the response to manipulation may be age related, requiring the production of lateral vascular connections (e.g. rays in trees via secondary growth) before lateral transport can occur (41, 69, 104, 190, 191). Better information may be gained by examining changes in patterns of translocation that result from more realistic types of manipulations, such as changes induced by applying plant growth regulators (e.g. 153, 178). Nevertheless, sinks related directly by phyllotaxy to particular sources seem more effective at obtaining assimilate from these sources than sinks that are not so related (37, 41).

THE PLANT AS AN ASSEMBLAGE OF SEMIAUTONOMOUS PHYSIOLOGICAL UNITS

When it is possible to examine the pathway of assimilate movement among specific sources and sinks, we have shown that the pathway is frequently restricted by one or more of the following: (*a*) branches that function as autonomous or semiautonomous structures receiving assimilate only from localized regions of the main body of the plant; (*b*) a decrease in the amount of assimilate movement among various structures (e.g. branches within a model unit or ramets within a clone) with increasing maturity; and (*c*) constraints in the vascular architecture that limit patterns of translocation.

Clearly, not all plants are equally subdivided; rather, they exist somewhere on a continuum from total integration to highly localized sectorialization. The level of morphological complexity at which physiological integration occurs defines the semiautonomous, integrated physiological unit (IPU). The number of IPUs within a particular taxon depends on the vascular organization of the taxon and the individual's stage of maturity. Completely integrated plants, such as young seedlings or many monocots, may consist of a single IPU. We illustrate the dynamic character of the IPU by looking at a hypothetical plant in which the vascular architecture imposes pronounced sectorialization (Figure 4). In this figure, we indicate hypothetical pathways of assimilate movement; the structures among which assimilate is predominantly translocated are considered to belong to the same IPU. Larson (personal communication) indicates that it would be possible to reconstruct accurately such a plant from his extensive data on *Populus* (see 108, 112), but this has not been done yet.

Early in development, a young shoot consisting of only one or a few metamers should be completely integrated (Figure 4*a*). As the number of metamers increases, particular leaves and their associated meristems become linked into separate IPUs, which are defined by the vascular architecture (Figure 4*b*). The number of IPUs should correspond to the number of vascular sympodia (i.e. leaf orthostichies). Lateral meristems may differentiate into branches. The developing branch is initially parasitic on the main stem. The branch derives most of its assimilate from those leaves that lie within its orthostichy (Figure 4*c*). Once the young branch produces several leaves, it becomes photosynthetically competent. At this time, the branch may be internally integrated (i.e. it may be a single IPU), and it may or may not be integrated with the main stem. Figure 4*d* shows an internally integrated young branch that exports carbon to young leaves above its axil of origin on the main stem; the IPU consists of the branch and these young leaves. As the branch matures, it also may subdivide into a number of IPUs that reflect the vascular architecture of the branch (Figure 4*e*). In a large complex plant, the IPU may extend over large distances and may include a subset of branch leaves, a sector of the main stem (which receives assimilate from the leaves within the IPU), and a portion of the root mass (Figure 4*f*). Within this extensive IPU, assimilate is manufactured, stored, redistributed, and utilized for shoot and root growth and, in trees for example, for wood formation. All carbon for further growth or response to damage is obtained from within the IPU.

Reproductive structures may be treated similarly. Those flowers and in-

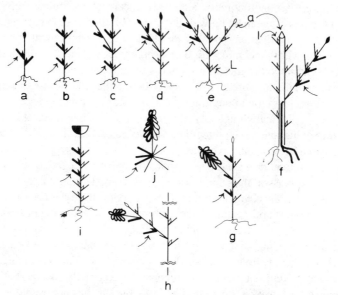

Figure 4 Integrated Physiological Units (IPUs): Hypothesized patterns of assimilate movement when the leaf indicated by the arrow is fed $^{14}CO_2$. L indicates a leaf; l, a lateral bud; and a, an apical bud. Heavy lines denote the location of label in leaves and buds; label in stems is not indicated except in *f*. Vegetative IPUs (*a–f*): (*a*) seedling module composed of two metamers; (*b*) simple module with six metamers expanded; (*c*) genet composed of two modules, one of which is a young branch; fed leaf is on main stem; (*d*) genet composed of two modules, one of which is a young branch; fed leaf is on branch; (*e*) genet composed of three modules, two of which are mature branches; label is confined to branch and is translocated only to those branch leaves lying in a common orthostichy; and (*f*) genet composed of two modules, one of which is a branch; an IPU includes the leaves lying in a common orthostichy on the branch, a sector of stem, and a portion of the root system. Reproductive IPUs (*g–j*): (*g*) genet composed of two modules, one of which is a reproductive branch or inflorescence; (*h*) spur or short shoot (i.e. vegetative module) bearing a reproductive module (i.e. an inflorescence or fruits); (*i*) genet consisting of one module terminated by a reproductive structure; and (*j*) inflorescence of a biennial rosette species.

florescences that arise directly from lateral meristems should be supported by leaves lying in a common orthostichy with their subtending leaf (Figure 4*g*). As it matures, the reproductive branch may function as a single autonomous IPU or as an assemblage of them, or it may remain partially dependent upon specific files of leaves on the main stem. Alternatively, if the reproductive branch bears enough leaves, it may be a net assimilate exporter. It can export as a single IPU or as an aggregate of IPUs in which only some components are capable of export. Fruit on a reproductive branch should be supported by leaves within the same IPU (Figure 4*h*). Terminal inflorescences should behave like lateral reproductive branches, gaining assimilate from leaves belonging to the same IPU (Figure 4*i*). Similarly, the supply of assimilate to different portions of the inflorescence of rosette species should come from leaves located in specific sectors of the rosette (Figure 4*j*).

If carbon balance is computed independently within different physiological units, then the persistence of these units—insofar as carbon availability limits their persistence—can also be determined separately. In the beginning of this paper, we equated differences in patterns of biomass allocation with the differential development of meristems. We emphasize that developmental decisions affecting the fates of meristems are made locally. For example, where alternate meristematic products differ both in life-history function and in their ability to supply their own carbon requirements, developmental programs that control their relative production may have evolved, at least in part, to yield structures that *ultimately* permit the maintenance of a positive carbon balance within the IPU. A particular developmental decision may be based on such an evolved constraint, rather than on the current availability of assimilate. Because quiescent meristems do not act as sinks, the commitment of a meristem to a particular function precedes competition by that meristem for resources. Furthermore, vegetative meristems may be equivalent to reproductive meristems in sink strength (147) and initial cost, since at least in some cases, the growth of different branch types has no effect on potential metamer production rates by the parent (1, 180). Reproductive branches may differ from vegetative branches, however, in the extent to which they contribute assimilate back to the IPU (i.e. the branch may or may not be an overproducer of assimilate). When reproductive and vegetative branch formation compete for the same meristems, the development of reproductive branches will reduce the number of vegetative branches that can form (180). Even though reproductive branches may be able to support themselves, their formation may ultimately reduce the amount of assimilate that accumulates within the IPU because successful reproductive branches are unlikely to produce excess assimilate. Thus, reproduction may influence carbon balance indirectly.

POTENTIAL APPLICATIONS TO CURRENT ECOLOGICAL AND EVOLUTIONARY QUESTIONS

It is clear that most of the literature we have reviewed deals with agricultural and horticultural species. Few data are available for undomesticated plants in native or experimental settings. Is the concept of the IPU relevant to ecological and evolutionary studies? We examine this question below.

1. *Is there a correlation between patterns of biomass allocation and habitat?* Numerous studies of life-history patterns in plants have been conducted (31, 32, 55, 187) by measuring the proportion of dry weight in reproductive and vegetative structures (78, 165). Recent reviewers, however, have found that the once commonly accepted, simple predictions of the relationship between a

plant's pattern of allocation and its environment (e.g. 63) have so many exceptions that they are of marginal utility (96, 158, 165, 187). The preponderant use of dry weight as a measure of allocation reflects an underlying assumption that assimilate is the critical currency limiting plant growth. Its adequacy as currency has been questioned recently (e.g. 2, 96, 116, 117, 119, 165, 180; J. Silvertown & D. Rabinowitz, personal communication), and there is a growing sense that its death knell has been sounded.

We believe that using fixed carbon as a currency is valid. Plants clearly require carbon for growth, and their future productive capacity reflects decisions about how to utilize assimilate in the construction of tissues differing in life-history function and photosynthetic capacity (125). Problems in its use are related to the indiscriminate application of a life-history theory (35), which maintains that sexual reproduction inhibits nonreproductive growth owing to the high costs of production and maturation of sexual structures (63, 76, 184, 187). This assumption may be valid for animals with a determinate growth form, in which all usable carbon is obtained through a single opening, the mouth, and then apportioned among competing life processes. It is far less appropriate for plants, in which the elaboration of an above-ground structure yields branches that have the potential to pay the carbon costs of their own production wholly or in part. Plants can increase the number of their carbon-acquiring structures (i.e. "mouths"); but depending on the type of structure produced and the timing of its production, the efficiency of the structure as a mouth can vary (118, 126, 180). Thus, predictions about a plant's behavior in different environments will vary according to the time period in which carbon availability acts as a developmental constraint, the degree to which translocation of assimilate is architecturally constrained, and the extent to which structures differ in their capacity to be carbon-autonomous. That is, there may be constraints on the developmental plasticity of many plants.

2. *What determines when a biennial will flower?* Flowering in biennials is frequently correlated with rosette size; the correlation may be related to the availability of stored assimilate to support maturation of the inflorescence (68, 182; but also see 117). Alternatively, the correlation may reflect the development of a sufficient number of leaves within an IPU to support the inflorescence with current assimilate.

3. *What type of branch should a plant produce?* Some argue that reproductive branches are more expensive to produce than vegetative branches (63, 76, 184, 187). Sohn & Policansky (156) assert that this is why reproductive shoots arise from larger rhizomes in mayapple, *Podophyllum peltatum*, although they provide no direct evidence that the initial cost to the parent differs from one shoot type to the other. The two shoots do, however, have distinct morphologies. The reproductive shoot has two leaves and almost twice the leaf area of the vegetative shoot, which only has one leaf (156). Shoot architecture seems to

have evolved to support the life-history functions of the two shoot types. This hypothesis is supported by the observation that new rhizomes produced by either vegetative shoots or reproductive shoots that successfully mature fruit are equivalent in size, while reproductive shoots that abort their fruit produce significantly larger new rhizomes (156, p. 1369). The old rhizomes appear to function as temporary storage organs for current assimilate (62); larger old rhizomes may be able to store more current assimilate than smaller ones. Whether the relation between old rhizome size and shoot type is due to differences in the initial costs of the two shoot types or to differences in their temporary storage requirements needs to be determined experimentally. In either case, the IPU consists of the old rhizome, the new rhizome, and the shoot.

4. *Does ovule position within a fruit affect its chance of maturing into a seed?* Not all fertilized ovules mature into seed. This variation may be due to genetic inviability or to competitive differences among genetically unique ovules. In certain plants, however, the ovule's location within a fruit affects its chance of maturing into a seed (86, 114). One hypothesis is that there is a competitive hierarchy among ovules based on their fertilization sequence (18, 86, 114). Another, possibly complementary explanation is that maturation success is influenced by vascular patterns, i.e. the presence of IPUs within the fruit; such units are reported in the flowering heads of grasses (41, 92, 183). A comparison of ovule abortion patterns among several *Cryptantha* species indicates that in perennial outcrossers the location of successful ovules within the fruit is random, while in autogamous annual species the position is always the same (29). If differential vascularization of ovules is important for their maturation success, then species either differ in their patterns of vascularization (i.e. IPUs) or vascular constraints are overridden by genetically determined differences in sink strength.

5. *Can plants compensate for herbivore foraging?* If a plant is made up of autonomous IPUs, the effect of herbivory on its reproductive success should depend strongly on the pattern of defoliation. If the damage to foliage is concentrated within an IPU (e.g. an integrated branch), fruit production on that branch may be severely affected owing to the lack of integration with other foliated branches (162). If, in contrast, the same amount of damage is spread evenly over the plant, the remaining leaves may compensate for the loss of leaf surface area by increasing photosynthesis, and there may be no discernible decrease in reproduction.

ACKNOWLEDGMENTS

Work on this essay began while we were both at the University of Utah. The ideas developed out of a thesis preliminary exam question to BC from MW. Our early thoughts benefited from conversations with J. Ehleringer, K. Harper,

I. McNulty, and D. Wiens. We are grateful for the detailed comments of D. Janzen, P. Larson, and J. White, as well as the helpful suggestions of J. Lovett Doust, C. Nelson, S. Poethig, A. Stephenson, and D. Waller on an earlier draft. While the manuscript has improved markedly from their advice, errors in fact or interpretation remain our own. The work was funded, in part, by NSF DEB 79–23435 to MW and NSF BSR 830–7888 to BC. MW also wishes to thank the Department of Biology, Indiana University, for creating the time and providing the environment in which this work could flourish.

Literature Cited

1. Abou-Haidar, S. S., Burger, D. W. 1981. Floral induction does not alter the growth parameters of *Fuchsia*. *Am. J. Bot.* 68:1278–81

2. Abrahamson, W. G., Caswell, H. 1982. On the comparative allocation of biomass, energy, and nutrients in plants. *Ecology* 63:982–91

3. Abul-Fatih, H. A., Bazzaz, F. A., Hunt, R. 1979. The biology of *Ambrosia trifida* L. III. Growth and biomass allocation. *New Phytol.* 83:829–38

4. Adams, M. W. 1967. Basis of yield component compensation in crop plants with special reference to the field bean, *Phaseolus vulgaris*. *Crop Sci.* 7:505–10

5. Allesio, M. L., Tieszen, L. L. 1975. Patterns of carbon allocation in an arctic tundra grass, *Dupontia fischeri* (Graminae), at Barrow, Alaska. *Am. J. Bot.* 62:797–807

6. Alvim, P. de T. 1964. Tree growth periodicity in tropical climates. In *The Formation of Wood in Forest Trees*, ed. M. H. Zimmerman, pp. 479–95. New York: Academic. 562 pp.

7. Anderson-Taylor, G., Marshall, C. 1983. Root-tiller interrelationships in spring barley (*Hordeum distichum* [L.] Lam.). *Ann. Bot.* 51:47–58

8. Ashley, D. A. 1972. ^{14}C-labelled photosynthate translocation and utilization in cotton plants. *Crop Sci.* 12:69–74

9. Ashmun, J. W., Thomas, R. J., Pitelka, L. F. 1982. Translocation of photoassimilates between sister ramets in two rhizomatous forest herbs. *Ann. Bot.* 49:403–15

10. Aspinall, D. 1963. The control of tillering in the barley plant. II. The control of tiller-bud growth during ear development. *Aust. J. Biol. Sci.* 16:285–304

11. Barlow, H. W. B. 1979. Sectorial patterns in leaves on fruit tree shoots produced by radioactive assimilates and solutions. *Ann. Bot.* 43:593–602

12. Barrett, J. E. III, Amling, H. J. 1978. Effects of developing fruits on production and translocation of ^{14}C-labeled assimilates in cucumber. *HortScience* 13:545–47

13. Bazzaz, F. A., Carlson, R. W. 1979. Photosynthetic contribution of flowers and seeds to reproductive effort of an annual colonizer. *New Phytol.* 82:223–32

14. Bazzaz, F. A., Carlson, R. W., Harper, J. L. 1979. Contribution to reproductive effort by photosynthesis of flowers and fruits. *Nature* 279:554–55

15. Bazzaz, F. A., Harper, J. L. 1977. Demographic analysis of the growth of *Linum usitatissimum*. *New Phytol.* 78:193–208

16. Bean, R. C., Porter, G. G., Barr, B. K. 1963. Photosynthesis and respiration in developing fruits. III. Variations in photosynthetic capacities during color change in citrus. *Plant Physiol.* 38:285–90

17. Belikov, I. F., Pirskii, L. I. 1966. Violation of the local distribution of assimilates in soybean. *Sov. Plant Physiol.* 13:361–64

18. Bertin, R. I. 1982. The ecology of sex expression in red buckeye. *Ecology* 63:445–56

19. Bible, B. B. 1976. Nonequivalence of left-handed and right-handed phyllotaxy in tomato and pepper. *HortScience* 11:601–2

20. Bidwell, R. G. S., Turner, W. B. 1966. Effect of growth regulators on CO_2 assimilation in leaves and its correlation with the bud break response in photosynthesis. *Plant Physiol.* 41:267–70

21. Binnie, R. C., Clifford, P. E. 1980. Effects of some defoliation and decapitation treatments on the productivity of french beans. *Ann. Bot.* 46:811–13

22. Biscoe, P. V., Gallagher, J. N., Littleton, E. J., Monteith, J. L., Scott, R. K. 1975. Barley and its environment. IV. Sources of assimilate for the grain. *J. Appl. Ecol.* 12:295–318

23. Blomquist, R. V., Kust, C. A. 1971.

Translocation pattern of soybeans as affected by growth substances and maturity. *Crop Sci.* 11:390–93

24. Bradbury, I. K., Hofstra, G. 1977. Assimilate distribution patterns and carbohydrate concentration changes in organs of *Solidago canadensis* during an annual developmental cycle. *Can. J. Bot.* 55:1121–27

25. Bradshaw, A. D. 1965. Evolutionary significance of phenotypic plasticity in plants. *Adv. Genet.* 13:115–55

26. Brar, G., Thies, W. 1977. Contribution of leaves, stem, siliques, and seeds to dry matter accumulation in ripening seeds of rapeseed, *Brassica napus* L. *Z. Pflanzenphysiol.* 82:1–13

27. Burt, R. L. 1964. Carbohydrate utilization as a factor in plant growth. *Aust. J. Biol. Sci.* 17:867–77

28. Carr, D. J., Wardlaw, I. F. 1965. The supply of photosynthetic assimilates to the grain from the flag leaf and ear of wheat. *Aust. J. Biol. Sci.* 18:711–19

29. Casper, B. B. 1982. *Ecological studies of ovule abortion and seed dispersal in Cryptantha (Boraginaceae).* PhD thesis. Univ. Utah, Salt Lake City. 197 pp.

30. Chan, B. G., Cain, J. C. 1967. The effect of seed formation on subsequent flowering in apple. *Proc. Am. Soc. Hortic. Sci.* 91:63–68

31. Charlesworth, B. 1980. *Evolution in Age-Structured Populations.* Cambridge: Cambridge Univ. Press. 300 pp.

32. Charnov, E. L. 1982. *The Theory of Sex Allocation.* Princeton, NJ: Princeton Univ. Press. 355 pp.

33. Clark, S. C. 1980. Reproductive and vegetative performance in two winter annual grasses, *Catapodium rigidum* (L.) C. E. Hubbard and *C. marinum* (L.) C. E. Hubbard. 2. Leaf-demography and its relationship to the production of caryopses. *New Phytol.* 84:79–93

34. Clifford, P. E., Marshall, C., Sagar, G. R. 1973. The reciprocal transfer of radiocarbon between a developing tiller and its parent shoot in vegetative plants of *Lolium multiflorum* Lam. *Ann. Bot.* 37:777–85

35. Cody, M. L. 1966. A general theory of clutch size. *Evolution* 20:174–84

36. Constable, G. A., Rawson, H. M. 1980. Photosynthesis, respiration and transpiration of cotton fruit. *Photosynthetica* 14:557–63

37. Constable, G. A., Rawson, H. M. 1980. Carbon production and utilization in cotton: Inferences from a carbon budget. *Aust. J. Plant Physiol.* 7:539–53

38. Constable, G. A., Rawson, H. M. 1980. Effect of leaf position, expansion, and

age on photosynthesis, transpiration, and water use efficiency of cotton. *Aust. J. Plant Physiol.* 7:89–100

39. Cook, C. W., Stoddart, L. A. 1960. Physiological responses of big sagebrush to different types of herbage removal. *J. Range Manage.* 13:14–16

40. Cook, M. G., Evans, L. T. 1978. Effect of relative size and distance of competing sinks on the distribution of photosynthetic assimilates in wheat. *Aust. J. Plant Physiol.* 5:495–509

41. Cook, M. G., Evans, L. T. 1983. The roles of sink size and location in the partitioning of assimilates in wheat ears. *Aust. J. Plant Physiol.* 10:313–27

42. Crookston, R. K., O'Toole, J., Ozbun, J. L. 1974. Characterization of the bean pod as a photosynthetic organ. *Crop Sci.* 14:708–12

43. Cruden, R. W., Hermann-Parker, S. M. 1977. Temporal dioecism: An alternative to dioecism? *Evolution* 31:863–66

44. Dat, T. V., Peterson, M. L. 1983. Performance of near-isogenic genotypes of rice differing in growth duration. I. Yields and yield components. *Crop Sci.* 23:239–42

45. Dat, T. V., Peterson, M. L. 1983. Performance of near-isogenic genotypes of rice differing in growth duration. II. Carbohydrate partitioning during grain filling. *Crop Sci.* 23:243–46

46. Davis, J. T., Sparks, D. 1974. Assimilation and translocation patterns of carbon-14 in the shoot of fruiting pecan trees, *Carya illinoensis* Koch. *J. Am. Soc. Hortic. Sci.* 99:468–80

47. Davis, T. A. 1963. The dependence of yield on asymmetry in coconut palms. *J. Genet.* 58:186–215

48. Dickmann, D. I., Kozlowski, T. T. 1970. Photosynthesis by rapidly expanding green strobili of *Pinus resinosa.* *Life Sci.* 9:549–52

49. Egli, D. B., Gossett, D. R., Leggett, J. E. 1976. Effect of leaf and pod removal on the distribution of ^{14}C labeled assimilate in soybeans. *Crop Sci.* 16:791–94

50. English, S. D., McWilliam, J. R., Smith, R. C. G., Davidson, J. L. 1979. Photosynthesis and partitioning of dry matter in sunflower. *Aust. J. Plant Physiol.* 6:149–64

51. Esau, K. 1965. *Vascular Differentiation in Plants.* New York: Holt, Rinehart & Winston. 160 pp.

52. Evans, G. C. 1972. *The Quantitative Analysis of Plant Growth, Studies in Ecology,* Vol. 1. Berkeley: Univ. Calif. Press.

53. Evans, L. T. 1971. Flower induction and

the florigen concept. *Ann. Rev. Plant Physiol.* 22:365–94
54. Evans, L. T., Rawson, H. M. 1970. Photosynthesis and respiration by the flag leaf and components of the ear during grain development in wheat. *Aust. J. Biol. Sci.* 23:245–54
55. Evenson, W. E. 1983. Experimental studies of reproductive energy allocation in plants. In *Handbook of Experimental Pollination Biology*, ed. C. E. Jones, R. J. Little, pp. 249–74. New York: Sci. & Acad. Ed. 558 pp.
56. Fellows, R. J., Egli, D. B., Leggett, J. E. 1979. Rapid changes in translocation patterns in soybeans following source-sink alterations. *Plant Physiol.* 64:652–55
57. Finch, A. H., Van Horn, C. W. 1936. The physiology and control of pecan nut filling and maturity. *Ariz. Agric. Exp. Stn. Tech. Bull.* 62:421–72
58. Flinn, A. M. 1974. Regulation of leaflet photosynthesis by developing fruit in the pea. *Physiol. Plant.* 31:275–78
59. Flinn, A. M., Pate, J. S. 1970. A quantitative study of carbon transfer from pod and subtending leaf to the ripening seeds of the field pea (*Pisum arvense* L.). *J. Exp. Bot.* 21:71–82
60. Ford, D. M., Shibles, R., Green, D. E. 1983. Growth and yield of soybean lines selected for divergent leaf photosynthetic ability. *Crop Sci.* 23:517–20
61. Forde, B. J. 1966. Translocation in grasses. I. Bermuda grass. *NZ J. Bot.* 4:479–95
62. Frye, D. M. 1977. *Seasonal changes in the morphology and physiology of the subterranean portions of* Podophyllum peltatum *L.* PhD thesis. Rutgers Univ., New Brunswick, NJ
63. Gadgil, M., Solbrig, O. T. 1972. The concept of *r*- and *K*-selection: Evidence from wild flowers and some theoretical considerations. *Am. Nat.* 106:14–31
64. Geiger, D. R. 1976. Effects of translocation and assimilate demand on photosynthesis. *Can. J. Bot.* 54:2337–45
65. Geiger, D. R. 1979. Control of partitioning and export of carbon in leaves of higher plants. *Bot. Gaz.* 140:241–48
66. Gifford, R. M., Evans, L. T. 1981. Photosynthesis, carbon partitioning, and yield. *Ann. Rev. Plant Physiol.* 32:485–509
67. Gould, S. J., Lewontin, R. C. 1979. The spandrels of San Marco and the Panglossian paradigm: A critique of the adaptationist programme. *Proc. R. Soc. London Ser. B* 205:581–98
68. Gross, K. L. 1981. Predictions of fate from rosette size in four "biennial" plant species: *Verbascum thapsus, Oenothera*

biennis, Daucus carota, and *Tragopogon dubius. Oecologia* 48:209–13
69. Hale, C. R., Weaver, R. J. 1962. The effect of developmental stage on direction of translocation of photosynthate in *Vitis vinifera. Hilgardia* 33:89–131
70. Hallé, F., Oldeman, R. A. A., Tomlinson, P. B. 1978. *Tropical Trees and Forests: An Architectural Analysis.* Berlin: Springer-Verlag. 441 pp.
71. Hanif, M., Langer, R. H. M. 1972. The vascular system of the spikelet in wheat (*Triticum aestivum*). *Ann. Bot.* 36:721–27
72. Hansen, P. 1967. ^{14}C-studies on apple trees. I. The effect of the fruit on the translocation and distribution of photosynthates. *Physiol. Plant.* 20:382–91
73. Hansen, P. 1969. ^{14}C-studies on apple trees. IV. Photosynthate consumption in fruits in relation to the leaf-fruit ratio and to leaf-fruit position. *Physiol. Plant.* 22:186–98
74. Hansen, P. 1971. ^{14}C-studies on apple trees. VII. The early seasonal growth in leaves, flowers and shoots as dependent upon current photosynthates and existing reserves. *Physiol. Plant.* 25:469–73
75. Hanson, W. D., West, D. R. 1982. Source-sink relationships in soybeans. I. Effects of source manipulation during vegetative growth on dry matter distribution. *Crop Sci.* 22:372–76
76. Harper, J. L. 1977. *Population Biology of Plants.* London: Academic. 892 pp.
77. Harper, J. L., Bell, A. D. 1979. The population dynamics of growth form in organisms with modular construction. In *Population Dynamics, 20th Symp. Br. Ecol. Soc.,* ed. R. M. Anderson, B. D. Turner, L. R. Taylor, pp. 29–52. Oxford: Blackwell. 434 pp.
78. Harper, J. L., Ogden, J. 1970. The reproductive strategy of higher plants. I. The concept of strategy with special reference to *Senecio vulgaris* L. *J. Ecol.* 58:681–98
79. Harper, J. L., White, J. 1974. The demography of plants. *Ann. Rev. Ecol. Syst.* 5:419–63
80. Hartnett, D. C., Bazzaz, F. A. 1983. Physiological integration among intraclonal ramets in *Solidago canadensis. Ecology* 64:779–88
81. Heichel, G. H., Turner, N. C. 1983. CO_2 assimilation of primary and regrowth foliage of red maple (*Acer rubrum* L.) and red oak (*Quercus rubra* L.). Response to defoliation. *Oecologia* 57:14–19
82. Heiser, C. B. 1976. *The Sunflower.* Norman: Univ. Okla. Press. 198 pp.
83. Ho, L. C. 1978. *Translocation of assimilates in the tomato plant, Opportunities*

for chemical plant growth regulation, *Proc. Jt. Br. Crop Prot. Counc. (BCPC) & Br. Plant Growth Regul. Group (BPGRG) Symp.*, pp. 159–66. Droitwich, England: Br. Crop Prot. Counc. 222 pp.

84. Ho, L. C., Peel, A. J. 1969. Transport of ^{14}C-labelled assimilates and ^{32}P-labelled phosphate in *Salix viminalis* in relation to phyllotaxis and leaf age. *Ann. Bot.* 33:743–51

85. Hockett, E. A., Knowles, P. F. 1970. Inheritance of branching in sunflowers, *Helianthus annuus* L. *Crop Sci.* 10:432–36

86. Horovitz, A., Meiri, L., Beiles, A. 1976. Effects of ovule positions in fabaceous flowers on seed set and outcrossing rates. *Bot. Gaz.* 137:250–54

87. Housley, T. L., Peterson, D. M. 1982. Oat stem vascular size in relation to kernel number and weight. I. Controlled environment. *Crop Sci.* 22:259–63

88. James, D. B., Hutto, J. M. 1972. Effects of tiller separation and root pruning on the growth of *Lolium perenne* L. *Ann. Bot.* 36:485–95

89. Janzen, D. H. 1976. Effect of defoliation on fruit-bearing branches of the Kentucky coffee tree, *Gymnocladus dioicus* (Leguminosae). *Am. Midl. Nat.* 95:474–78

90. Janzen, D. H. 1982. Cenizero tree (Leguminosae: *Pithecellobium saman*) delayed fruit development in Costa Rican deciduous forests. *Am. J. Bot.* 69:1269–76

91. Janzen, D. H. 1982. Ecological distribution of chlorophyllous developing embryos among perennial plants in a tropical deciduous forest. *Biotropica* 14:232–36

92. Jenner, C. F., Rathjen, A. J. 1972. Limitations to the accumulation of starch in the developing wheat grain. *Ann. Bot.* 36:743–54

93. Jones, H., Martin, R. V., Porter, H. K. 1959. Translocation of [14]carbon in tobacco following assimilation of [14]carbon dioxide by a single leaf. *Ann. Bot.* 23:493–508

94. Jones, H. G., Lamboll, D. 1980. Investigation of sectorial patterns in apple shoots using abscisic acid. *Ann. Bot.* 46:815–17

95. Karpov, E. A., Kholupenko, I. P. 1976. Comparative study of transport of radioactive isotopes of [32]phosphorus and [14]carbon from soybean leaves during the period of seed plumping. *Sov. Plant Physiol.* 23:852–58

96. Kawano, S. 1981. Trade-off relationships between some reproductive characteristics in plants with special reference to life history strategy. *Bot. Mag. Tokyo* 94:285–94

97. Khan, A. A., Akosu, F. I. 1971. The physiology of groundnut. I. An autoradiographic study of the pattern of distribution of [14]carbon products. *Physiol. Plant.* 24:471–75

98. Khan, A. A., Sagar, G. R. 1966. Distribution of ^{14}C-labelled products of photosynthesis during the commercial life of the tomato crop. *Ann. Bot.* 30:727–43

99. Khan, A. A., Sagar, G. R. 1967. Translocation in tomato: The distribution of the products of photosynthesis of the leaves of a tomato plant during the phase of fruit production. *Hortic. Res.* 7:61–69

100. Khan, A. A., Sagar, G. R. 1969. Changing patterns of distribution of the products of photosynthesis in the tomato plant with respect to time and to the age of a leaf. *Ann. Bot.* 33:763–79

101. Kirby, E. J. M., Jones, H. G. 1977. The relations between the main shoot and tillers in barley plants. *J. Agric. Sci.* 88:381–89

102. Kobayashi, S. 1975. Growth analysis of plant as an assemblage of internodal segments—A case of sunflower plants in pure stands. *Jpn. J. Ecol.* 25:61–70

103. Kozlowski, T. T., Keller, T. 1966. Food relations of woody plants. *Bot. Rev.* 32:293–382

104. Kramer, P. J., Kozlowski, T. T. 1979. *Physiology of Woody Plants*. New York: Academic. 811 pp.

105. Kriedemann, P. 1966. The photosynthetic activity of the wheat ear. *Ann. Bot.* 30:349–63

106. Kriedemann, P. E. 1968. Observations on gas exchange in the developing sultana berry. *Aust. J. Biol. Sci.* 21:907–16

107. Lang, A. 1965. Physiology of flower initation. In *Encyclopedia of Plant Physiology*, ed. W. Ruhland, 15(1):1380–1536. Berlin: Springer-Verlag

108. Larson, P. R. 1980. Interrelations between phyllotaxis, leaf development and the primary-secondary vascular transition in *Populus deltoides*. *Ann. Bot.* 46:757–69

109. Larson, P. R. 1983. Primary vascularization and the siting of primordia. In *The Growth and Functioning of Leaves*, ed. J. E. Dale, F. L. Milthorpe, pp. 25–51. Cambridge: Cambridge Univ. Press. 540 pp.

110. Larson, P. R., Dickson, R. E. 1973. Distribution of imported ^{14}C in developing leaves of eastern cottonwood according to phyllotaxy. *Planta* 111:95–112

111. Larson, P. R., Richards, J. H. 1981. Lateral branch vascularization: Its circu-

larity and its relation to anisophylly. *Can. J. Bot.* 59:2577–91

112. Larson, P. R., Isebrands, J. G., Dickson, R. E. 1980. Sink to source transition of *Populus* leaves. *Ber. Dtsch. Bot. Ges.* 93:79–90

113. Laval-Martin, D., Farineau, J., Diamond, J. 1977. Light versus dark carbon metabolism in cherry tomato fruits. I. Occurrence of photosynthesis. Study of the intermediates. *Plant Physiol.* 60: 872–76

114. Lee, T. D. 1984. Patterns of fruit maturation: A gametophytic competition hypothesis. *Am. Nat.* In press

115. Lovell, P. H., Lovell, P. J. 1970. Fixation of CO_2 and export of photosynthate by the carpel in *Pisum sativum*. *Physiol. Plant.* 23:316–22

116. Lovett Doust, J. 1980. A comparative study of life history and resource allocation in selected umbelliferae. *Biol. J. Linn. Soc.* 13:139–54

117. Lovett Doust, J. 1980. Experimental manipulation of patterns of resource allocation in the growth cycle and reproduction of *Smyrnium olusatrum* L. *Biol. J. Linn. Soc.* 13:155–66

118. Lovett Doust, J., Eaton, G. W. 1982. Demographic aspects of flower and fruit production in bean plants, *Phaseolus vulgaris* L. *Am. J. Bot.* 69:1156–64

119. Lovett Doust, J., Harper, J. L. 1980. The resource costs of gender and maternal support in an andromonoecious umbellifer, *Smyrnium olusatrum* L. *New Phytol.* 85:251–64

120. Luckwill, L. C. 1970. The control of growth and fruitfulness of apple trees. In *Physiology of Tree Crops*, ed. L. C. Luckwill, C. V. Cutting, pp. 237–54. London: Academic. 382 pp.

121. Maillette, L. 1982. Structural dynamics of silver birch. I. The fates of buds. *J. Appl. Ecol.* 19:203–18

122. Marshall, C., Sgar, G. R. 1968. The interdependence of tillers in *Lolium multiflorum* Lam—A quantitative assessment. *J. Exp. Bot.* 19:785–94

123. Maun, M. A., Cavers, P. B. 1971. Seed production and dormancy in *Rumex crispus*. II. The effects of removal of various proportions of flowers at anthesis. *Can. J. Bot.* 49:1841–48

124. Milthorpe, F. L., Moorby, J. 1969. Vascular transport and its significance in plant growth. *Ann. Rev. Plant Physiol.* 20:117–38

125. Mooney, H. A. 1972. The carbon balance of plants. *Ann. Rev. Ecol. Syst.* 3:315–46

126. Mooney, H. A., Field, C., Gulmon, S. L., Bazzaz, F. A. 1981. Photosynthetic capacity in relation to leaf position in desert versus old-field annuals. *Oecologia* 50:109–12

127. Morrow, P. A., LaMarche, V. C. Jr. 1978. Tree ring evidence for chronic insect suppression of productivity in subalpine *Eucalyptus*. *Science* 201:1244–46

128. Mullins, M. G. 1970. Transport of ^{14}C-assimilates in seedlings of *Phaseolus vulgaris* L. in relation to vascular anatomy. *Ann. Bot.* 34:889–96

129. Murray, B. J., Mauk, C., Noodén, L. D. 1982. Restricted vascular pipelines (and orthostichies) in plants. *What's New in Plant Physiol.* 13:33–36

130. Newell, S. J. 1982. Translocation of ^{14}C-photoassimilate in two stoloniferous *Viola* species. *Bull. Torrey Bot. Club* 109:306–17

131. Oliker, M., Poljakoff-Mayber, A., Mayer, A. M. 1978. Changes in weight, nitrogen accumulation, respiration and photosynthesis during growth and development of seeds and pods of *Phaseolus vulgaris*. *Am. J. Bot.* 65:366–71

132. Olufajo, O. O., Daniels, R. W., Scarisbrick, D. H. 1982. The effect of pod removal on the translocation of ^{14}C photosynthate from leaves of *Phaseolus vulgaris* L. cv. Lochness. *J. Hortic. Sci.* 57:333–38

133. Ong, C. K., Marshall, C. 1979. The growth and survival of severely-shaded tillers in *Lolium perenne* L. *Ann. Bot.* 43:147–55

134. Pate, J. S., Flinn, A. M. 1977. Fruit and seed development. In *The Physiology of the Garden Pea*, ed. J. F. Sutcliffe, J. S. Pate, pp. 431–68. London: Academic. 500 pp.

135. Pate, J. S., Sharkey, P. J., Atkins, C. A. 1977. Nutrition of a developing legume fruit: Functional economy in terms of carbon, nitrogen, water. *Plant Physiol.* 59:506–10

136. Peterson, D. M., Housley, T. L., Luk, T. M. 1982. Oat stem vascular size in relation to kernel number and weight. II. Field experiment. *Crop Sci.* 22:274–78

137. Prévost, M-F. 1978. Modular construction and its distribution in tropical woody plants. In *Tropical Trees as Living Systems*, ed. P. B. Tomlinson, M. H. Zimmermann, pp. 223–31. Cambridge: Cambridge Univ. Press. 675 pp.

138. Prokofyev, A. A., Zhdanova, L. P., Sobolev, A. M. 1957. Certain regularities in the flow of substances from leaves into reproductive organs. *Sov. Plant Physiol.* 4:402–8

139. Quebedeaux, B., Chollet, R. 1975. Growth and development of soybean

(*Glycine max* [L.] Merr.) pods: CO_2 exchange and enzyme studies. *Plant Physiol.* 55:745–48

140. Quinlan, J. D. 1969. Mobilization of ^{14}C in the spring following autumn assimilation of $^{14}CO_2$ by apple rootstock. *J. Hortic. Sci.* 44:107–10

141. Quinlan, J. D., Weaver, R. J. 1970. Modification of pattern of the photosynthate movement within and between shoots of *Vitis vinifera* L. *Plant Physiol.* 46:527–30

142. Rawson, H. M., Constable, G. A. 1980. Carbon production of sunflower cultivars in field and controlled environments. I. Photosynthesis and transpiration of leaves, stems and heads. *Aust. J. Plant Physiol.* 7:555–73

143. Rawson, H. M., Hackett, C. 1974. An exploration of the carbon economy of the tobacco plant. III. Gas exchange of leaves in relation to position on the stem, ontogeny and nitrogen content. *Aust. J. Plant Physiol.* 1:551–60

144. Rawson, H. M., Hofstra, G. 1969. Translocation and remobilization of ^{14}C assimilated at different stages by each leaf of the wheat plant. *Aust. J. Biol. Sci.* 22:321–31

145. Rogan, P. G., Smith, D. L. 1974. Patterns of translocation of ^{14}C-labelled assimilates during vegetative growth of *Agropyron repens* (L). Beauv. *Z. Pflanzenphysiol.* 73:405–14

146. Russell, C. R., Morris, D. A. 1983. Patterns of assimilate distribution and source-sink relationships in the young reproductive tomato plant (*Lycopersicon esculentum* Mill.). *Ann. Bot.* 52:357–63

147. Ryle, G. J. A. 1972. A quantitative analysis of the uptake of carbon and the supply of ^{14}C-labelled assimilates to areas of meristematic growth in *Lolium temulentum*. *Ann. Bot.* 36:497–512

148. Ryle, G. J. A., Powell, C. E. 1972. The export and distribution of ^{14}C-labelled assimilates from each leaf on the shoot of *Lolium temulentum* during reproductive and vegetative growth. *Ann. Bot.* 36: 363–75

149. Ryugo, K., Marangoni, B., Ramos, D. E. 1980. Light intensity and fruiting effects on carbohydrate contents, spur development, and return bloom of 'Hartley' walnut. *J. Am. Soc. Hortic. Sci.* 105:223–27

150. Sackston, W. E. 1959. Effects of artificial defoliation on sunflowers. *Can. J. Plant Sci.* 39:108–18

151. Sawhney, S., Sawhney, N., Nanda, K. K. 1978. Studies on the transmission of floral effects of photoperiod and gibberellin from one branch to the other in *Impatiens balsamina*. *Biol. Plant.* 20: 344–50

152. Schneider, G. W. 1977. Studies on the mechanism of fruit abscission in apple and peach. *J. Am. Soc. Hortic. Sci.* 102:179–81

153. Shindy, W., Weaver, R. J. 1967. Plant regulators alter translocation of photosynthetic products. *Nature* 214:1024–25

154. Shiroya, M., Lister, G. R., Nelson, C. D., Krotkov, G. 1961. Translocation of C^{14} in tobacco at different stages of development following assimilation of $C^{14}O_2$ by a single leaf. *Can. J. Bot.* 39:855–64

155. Singh, L. B. 1948. Studies on biennial bearing. III. Growth studies in the "on" and "off" year trees. *J. Pomol. Hortic. Sci.* 24:123–48

156. Sohn, J. J., Policansky, D. 1977. The costs of reproduction in the mayapple *Podophyllum peltatum* (Berberidaceae). *Ecology* 58:1366–74

157. Sparks, D., Brack, C. E. 1972. Return bloom and fruit set of pecan from leaf and fruit removal. *HortScience* 7:131–32

158. Stearns, S. C. 1977. The evolution of life history traits: A critique of the theory and a review of the data. *Ann. Rev. Ecol. Syst.* 8:145–71

159. Steer, B. T., Pearson, C. J. 1976. Photosynthate translocation in *Capsicum annuum*. *Planta* 128:155–62

160. Stephenson, A. G. 1980. Fruit set, herbivory, fruit reduction, and the fruiting strategy of *Catalpa speciosa* (Bignoniaceae). *Ecology* 61:57–64

161. Stephenson, A. G. 1981. Flower and fruit abortion: Proximate causes and ultimate functions. *Ann. Rev. Ecol. Syst.* 12:253–79

162. Stephenson, A. G. 1982. The role of the extrafloral nectaries of *Catalpa speciosa* in limiting herbivory and increasing fruit production. *Ecology* 63:663–69

163. Stephenson, R. A., Wilson, G. L. 1977. Patterns of assimilate distribution in soybeans at maturity. I. The influence of reproductive developmental stage and leaf position. *Aust. J. Agric. Res.* 28: 203–9

164. Takeda, F., Ryugo, K., Crane, J. C. 1980. Translocation and distribution of ^{14}C-photosynthates in bearing and nonbearing pistachio branches. *J. Am. Soc. Hortic. Sci.* 105:642–44

165. Thompson, K., Stewart, A. J. A. 1981. The measurement and meaning of reproductive effort in plants. *Am. Nat.* 117: 205–11

166. Thorne, G. N. 1963. Varietal differences in photosynthesis of ears and leaves of barley. *Ann. Bot.* 27:155–74

167. Thorne, G. N. 1965. Photosynthesis of ears and flag leaves of wheat and barley. *Ann. Bot.* 29:317–29

168. Thorne, J. H., Koller, H. R. 1974. Influence of assimilate demand on photosynthesis, diffusive resistances, translocation, and carbohydrate levels of soybean leaves. *Plant Physiol.* 54:201–7

169. Thrower, S. L. 1962. Translocation of labelled assimilates in the soybean. II. The pattern of translocation in intact and defoliated plants. *Aust. J. Biol. Sci.* 15:629–49

170. Thrower, S. L. 1977. Translocation into mature leaves—the effect of growth pattern. *New Phytol.* 78:361–64

171. Tietma, T. 1980. Ecophysiology of the sand sedge, *Carex arenaria* L. II. The distribution of ^{14}C assimilates. *Acta Bot. Neerl.* 29:165–78

172. Tuomi, J., Niemelä, P., Mannila, R. 1982. Resource allocation on dwarf shoots of birch (*Betula pendula*): Reproduction and leaf growth. *New Phytol.* 91:483–87

173. Uriu, K. 1956. Pistil abortion in the olive. *Calif. Agric.* 10:13–14

174. Van Steveninck, R. F. M. 1957. Factors affecting the abscission of reproductive organs in yellow lupins (*Lupinus luteus* L.). I. The effect of different patterns of flower removal. *J. Exp. Bot.* 8:373–81

175. Walker, A. J., Ho, L. C., Baker, D. A. 1978. Carbon translocation in the tomato: Pathways of carbon metabolism in the fruit. *Ann. Bot.* 42:901–9

176. Wardlaw, I. F. 1968. The control and pattern of movement of carbohydrates in plants. *Bot. Rev.* 34:79–105

177. Wardlaw, I. F. 1976. Assimilate partitioning: Cause and effect. In *Transport and Transfer Processes in Plants,* ed. I. F. Wardlaw, J. B. Passioura, pp. 381–91. New York: Academic. 484 pp.

178. Wareing, P. F. 1978. Growth regulators and assimilate partition. In *Plant Regulation and World Agriculture Plenum,* ed. T. K. Scott, pp. 309–17. New York: Plenum. 575 pp.

179. Wareing, P. F., Khalifa, M. M., Treharne, K. J. 1968. Rate-limiting processes in photosynthesis at saturating light intensities. *Nature* 220:453–57

180. Watson, M. A. 1984. Developmental constraints: Effect on population growth and patterns of resource allocation in a clonal plant. *Am. Nat.* 123:411–26

181. Werk, K. S., Ehleringer, J. R. 1983. Photosynthesis by flowers in *Encelia farinosa* and *Encelia californica* (Asteraceae). *Oecologia* 57:311–15

182. Werner, P. A. 1975. Predictions of fate from rosette size in teasel (*Dipsacus fullonum* L.). *Oecologia* 20:197–201

183. Whingwiri, E. E., Kuo, J., Stern, W. R. 1981. The vascular system in the rachis of a wheat ear. *Ann. Bot.* 48:189–201

184. White, J. 1979. The plant as a metapopulation. *Ann. Rev. Ecol. Syst.* 10:109–45

185. White, J. 1984. Plant metamerism. In *Perspectives in Plant Population Ecology,* ed. R. Dirzo, J. Sarukhan, pp. 15–47. Sunderland, Mass: Sinauer

186. Wien, H. C., Altschuler, S. L., Ozbun, J. L., Wallace, D. H. 1976. ^{14}C-assimilate distribution in *Phaseolus vulgaris* L. during the reproductive period. *J. Am. Soc. Hortic. Sci.* 101:510–13

187. Willson, M. F. 1983. *Plant Reproductive Ecology.* New York: Wiley. 282 pp.

188. Wittenbach, V. A. 1983. Effect of pod removal on leaf photosynthesis and soluble protein composition of field-grown soybeans. *Plant Physiol.* 73:121–24

189. Wyatt, R. 1982. Inflorescence architecture: How flower number, arrangement, and phenology affect pollination and fruit-set. *Am. J. Bot.* 69:585–94

190. Zimmermann, M. H. 1971. Transport in the phloem. In *Trees: Structure and Function,* ed. M. H. Zimmerman, C. L. Brown, pp. 221–79. New York: Springer-Verlag. 336 pp.

191. Zimmermann, M. H., McCue, K. F., Sperry, J. S. 1982. Anatomy of the palm *Rhapis excelsa.* VIII. Vessel network and vessel-length distribution in the stem. *J. Arnold Arbor. Harv. Univ.* 63:83–95

Ann. Rev. Ecol. Syst. 1984. 15:259–78

MIMICRY AND DECEPTION IN POLLINATION

A. Dafni

Institute of Evolution, Haifa University, Haifa 31999, Israel

TERMINOLOGY AND DEVELOPMENT OF IDEAS

Deception

Many flowers produce rewards, usually nectar or pollen or both. C. K. Sprengel (165), the founder of modern floral biology, was the first to report that flowers do not always reward pollinators; his observations were on the genus *Orchis*. Subsequent authors (7; 51; 66; 97; 99; 105; 186, p. 21; 195) have discussed various aspects of deceitful flowers that attract pollinators by resembling rewarding flowers or other objects.

Mimicry

There are two main types of mimicry relevant to pollination ecology: Batesian (12) and Müllerian (138). In spite of the current controversy over whether the second type represents real mimicry or just convergence (155; 200, pp. 83–85; 201), I accept Vane Wright's (187, 188) view and include it as a form of mimicry. Proctor & Yeo (152, p. 375) defined *Müllerian mimicry* as those cases in which a number of species of similar character and behavior and at comparable levels of abundance evolve a common "advertising style" to their mutual advantage. In contrast, an organism that obtains a one-sided advantage by imitating another, often more numerous organism is engaging in *Batesian mimicry*. Most floral ecologists accept these concepts, sometimes with a few reservations (23, 26, 47, 69, 78, 79, 83, 116, 118, 119, 151, 157).

Müllerian mimicry fits into the category of synergic inviting mimicry, and Batesian floral mimicry, into that of antergic inviting mimicry (187). Wiens (201) suggested the establishment of a new category called *reproductive mimicry* to describe mimetic systems that function in gamete transfer. This

259

0066–4162/84/1120–0259$02.00

class would contain several phenomena, such as sexual response (pseudocopulation), territorial defense (pseudoantagonism), brood site selection, and imitation of the pollinator's food source. Although it would be convenient to treat all of these types from a functional viewpoint, doing so would obscure the necessary distinction between deceptive and rewarding syndromes. Some phenomena are also left out of this scheme, e.g. *rendezvous pollination* (193), which involves a sexual response but no mimicry; and *mistake pollination* (5), which refers to a mimetic phenomenon with no definite model. Pasteur (145) included some floral mimicry types [Dodsonian, Pouyannian, Bakerian (details given later)] in his term *concrete homotype*. In this case, the mimic elicits the same reaction from the duped pollinator as the model does (where the model is a definite species or a cluster of similar ones).

Mistake Pollination

Mistake pollination describes visits by a pollinator to nonrewarding pistillate flowers that resemble (i.e. mimic) the rewarding, staminate ones (5). In fact, most nonrewarding pollination syndromes are based on some sort of mistake by the pollinator; this phenomenon is not peculiar to mimics of unisexual flowers. The term *mistake pollination* is appropriate from the perspective of the pollinator but with reference to the plants, no mistake is involved, of course (15).

Thus, from the plant's viewpoint, the term *chance pollination* (167), which refers to visits, usually at a low frequency, due to deceit, is more meaningful and reflects a frequency-dependent phenomenon. This term can be used for any type of deceit where instinct-based, imitated stimuli are not involved.

Parasitism

The term *parasitism* is used in anthecological contexts to denote various unrelated phenomena. The plant is regarded as a parasite (*a*) if the pollinator is not rewarded for its services (4; 26; 66, p. 58; 123, p. 70; 126; 152, p. 248; 161; 195; 196; 201), (*b*) if the flower kills the pollinator (51), (*c*) if the species with scarce nectar benefits from blooming after a species rich in nectar (82), and (*d*) if offspring laid in error on the flower have no chance to develop into adults (128, p. 104). An animal is recognized as a parasite (*a*) if it exploits floral rewards without pollinating (7a; 66, p. 58; 68; 84, p. 141; 125) or (*b*) if the pollinator or its brood destroy parts of the flower (56, p. 145; 161; 185). Not all of these phenomena, however, fit the criteria defining *parasitism* (106, p. 3; 154, p. 5; 164, p. 3). Therefore, the term should not be used in relation to pollination; it should be replaced by *unidirectional exploitation* in order to distinguish it from *mutualism*. Unidirectional exploitation covers all nonrewarding pollination syndromes as well as floral larceny.

MÜLLERIAN MIMICRY

This group includes plant species that mutually benefit by sharing the same pollinator because of the resemblance of their flowers. These species have converged to exhibit similar signals and have developed what may be regarded as a parallel to animal Müllerian mimicry (26; 78; 151; 152, p. 375). In such associations, every participant offers the pollinator at least some reward; deception is not involved at all. About 20 cases of Müllerian floral mimicry have been suggested (19, 30, 70, 95, 112, 118, 147, 157, 209). The Müllerian system results in an increase in the "effective" flower density and probably raises the chances of pollination for all partners (83; 152, p. 375; 157). The fitness of a rare species will increase if there is another species with similar flowers and if the pollinator does not distinguish between the two (21, 83, 95, 118, 137, 173, 176). A high degree of constancy by individual pollinators is unlikely because of the circumstances that promote Müllerian floral mimicry. Thus, it is more important to attract a general type of pollinator than to ensure the visits of a species-specific one. Sharing pollinators in the same ecological context may promote floral mimicry. When the combined, convergent populations command preferential pollinator activity, the parent (i.e. nonconvergent) stock of one or both may disappear (119).

NUTRITIVE DECEPTION

Batesian Mimicry

Batesian mimicry occurs when a rare species that provides no reward mimics flowers of a more abundant species that does reward (157); this phenomenon has also been called Dodsonian mimicry (145). Two criteria are crucial in this situation: low frequency of the mimic (30, 48, 60, 117, 118, 209), a well-known characteristic of animal Batesian mimics (61, 89, 145), and the supply of a compensating reward by the more common model that subsidizes the system (45, 84, 143, 195). About 30 examples (19; 30; 60; 83; 85, p. 189; 118; 136; 147; 156, pp. 109, 207; 186, p. 99; 195; 202; 208) seem to fit into this category of floral mimicry, but few have been tested experimentally (26, 48, 198). Since morphological comparisons are insufficient to demonstrate mimicry (119), many of these cases are still open to question.

Several preconditions that can promote the evolution of floral mimicry include: (a) if both species (mimic and model) have a fairly high degree of xenogamy, (b) if both species' ecological requirements are identical, (c) if there are only a few species of pollinators and short and intense flowering (relatively inconspicuous flowers are unlikely to be either models or mimic), and (d) if the mimicking species is local in comparison to the model (30). The experimental evidence (26, 48, 198) confirms most of these ideas. The question then arises, why is flower-flower mimicry so rare? It has been argued (205, 206) that it is because plants are sessile and usually clumped, so that a pollinator can simply avoid unrewarding clumps of mimics. Thus, floral mimicry should occur in species that are closely associated with their models, e.g. in hemiparasites and vines (205, 206). The available evidence (26, 45–48, 198) does not explain either the "clumpy" nature of the mimics or any close association with the model. Since pollinators are attracted to flowers by long-distance cues, the mimic and the model do not need to be close together (26). An alternative explanation for the apparent rarity of the phenomenon is simply that floral mimicry has often been overlooked by researchers.

Integrated Müllerian and Batesian Systems

In integrated systems, a minimum of three species are convergently involved, at least two of which mimic each other (Müllerian mimicry) and one or more of which mimics both the previous ones (Batesian mimicry). Several systems of this type have been reported (1, 26, 44, 133, 136, 143, 176).

Intraspecific Batesian Mimicry

IMITATION OF MALE BY FEMALE FLOWERS   In this class, nectarless female flowers attract pollinators by mimicking the nectar-producing male flowers. Their stigmatic lobes usually resemble either the petals or anthers of the male flowers (5, 7, 71, 185, 195); this phenomenon has also been termed Bakerian mimicry (145) and partial deception (7, 195). Pollinators, anticipating the reward they get from staminate flowers, visit the pistillate flowers by "mistake" (5, 71). By definition (145), the monoecious species of this class exemplify self-mimicry, while the diecious ones represent automimicry. Bakerian mimicry is quite common (5, 7, 15, 18, 38, 71, 144, 159, 195), especially in tropical Caricaceae (15). As expected with a mimetic phenomenon, female flowers occur less frequently than male flowers (14, 15, 71).

Baker (5) suggested that the lack of nectar in the fleshy female flowers of *Carica papaya* confers a selective advantage by preventing bees and birds from inflicting damage that might compromise seed production. The resources previously utilized in nectar production can also be diverted to other components of fecundity. Intersexual competition for the pollinators may lead to selective pressures towards Bakerian mimicry (15). In *Fiscus carica,* for example, there is no reward in the female syconia, and the deception imposes selective forces on the female fig. At the receptive stage, the female syconia resemble those of the caprifigs as closely as possible (179).

AUTOMIMICRY   Automimicry occurs when the model and the mimic are different individuals within the same species; it is also called Browerian mimicry (145). Nectarless individuals (or populations) may be found within nectariferous species and exploit the pollinators of the model (47). Brown & Kodric-Brown (26) describe an anectariferous population of *Lobelia cardinalis* as an example of Batesian mimicry, while Williamson & Black (206) consider it automimicry. The use of nectariferous varieties of muskmelon (*Cucumis melo*) to attract pollinators to nectarless varieties (22) that otherwise would not be pollinated is an example of automimicry.

IMITATION OF FLOWER PARTS

Pseudopollen   The term *pseudopollen* has been used in two different ways. Van der Pijl & Dodson defined it as a "pollen-like mass of cells that results from the disintegration of multicellular trichomes" (186, p. 22). In contrast, Vogel (195) treated it as "trichomes that mimic pollen and attract pollinators by deceit." It is not only a matter of definition; two different concepts are involved: The "edible" pseudopollen is regarded as a real reward (16; 54; 56; 73; 148; 186, pp. 22–23), while the "false" pseudopollen is an enticement (47; 59, p. 220; 176; 195; 196). Both have a mimetic effect in attracting pollinators.

False anthers   Vogel (196) regards the existence of two types of stamens, the "fodder" and the "attractive," as the first stage in moving from reward to deception in pollen-producing flowers. "Persisting" attractive stamens remain yellow and swollen even after anthesis and, thus, continue to attract pollinators by deceit (185; 186, p. 62; 195).

Pseudonectaries   In some blossoms, there are organs that resemble nectaries but do not produce nectar. These pseudonectaries, which are well exposed, are explained differently depending on the extent to which they attract pollinators (50; 51; 100, p. 56; 104, pp. 71–73; 152, p. 57, 116).

REPRODUCTIVE DECEPTION

Imitation of Oviposition Substrate

Many entomophilous blossoms simulate substrates for oviposition. Such breeding-site mimics include most of the carrion and dung fly flowers (sapromyophily), dung beetle flowers (coprocantharophily), and fungus gnat flowers (mycetophily) (66, pp. 103–5; 152, pp. 301–9; 177; 181; 183; 186, pp. 30, 101–2; 189–192; 194; 197; 209).

In general, myophily (attraction of Diptera) is based on sweet odors and the provision of nectar, while sapromyophily and coprocantharophily are characterized by a fetid smell and lack of nectar (186, p. 103). Most blossoms that imitate oviposition substrates have trapping devices (183, 195, 209). Some nectariferous myophilous flowers produce a carrion-like smell without having a trap (51, 75, 76). They are deceptive regarding the olfactory attraction, but they do not cause oviposition and therefore are not regarded as sapromyophilous. Sometimes it is difficult to differentiate between myophily and sapromyophily, especially in Orchidaceae. About 3000 species have traps, apparently without imitating any known oviposition substrate (186, pp. 103–22).

MEANS OF ATTRACTION   The imitation of an oviposition substrate is mainly olfactory (28, 51, 57, 98, 190, 192, 197, 209). The deceptive odor simulates volatile substances to which the insect is innately attracted and triggers the same reactions as the real material would (66, p. 103; 192). The production of imitative odoriferous compounds has been termed chemical mimicry (201). There is a large array of imitated oviposition substrates, including dung and carrion (6; 58; 59, p. 103; 94, pp. 99–100; 98; 106, p. 115; 152, p. 299; 207), fermenting sugars and fruits (52, 120), blood (98), rotten flesh or fish (41; 54; 129, p. 195; 152, p. 300), and fungi (166, 194, 197).

The inflorescences of some sapromyophilous Araceae produce several volatiles such as ammonia, skatole, indole, trimethylamine, and other amines (31, 130, 163). Indole and ammonia help stimulate oviposition by carrion flies (11, 39) and carrion beetles (62, p. 194; 75; 76). Dung beetles locate manure by its odor of ammonia, indole, and skatole (124, p. 33). These data corroborate the view (66, p. 103; 152, p. 303; 192) that sapromyophilous blossoms produce specific attractants related to pollinators' oviposition stimulants.

Although visual cues play a secondary role (51; 57; 101, p. 489; 192; 209), the striking convergence in floral appearance among certain unrelated plants indicates their selective value, probably in guiding the insect into the trap (209). Several authors discuss the various optical means that sapromyophilous flowers use (9; 66, p. 105; 152, pp. 304–9; 181; 189, pp. 50–58; 190; 192; 209). Tactile stimuli, e.g. fungus-like structures inside some traps, may also play a role in some mycetophilous flowers (197).

THE MICROENVIRONMENT INSIDE THE TRAPS   Pollinators remain imprisoned inside the traps for one to five days (29; 98; 113; 122; 127; 128, p. 193; 139; 192). Many of the trapped insects are delicate (57, 197). A favorable microenvironment within the trap is therefore extremely important, so that the insects are kept in sufficiently good condition to complete the process of pollination by visiting another trap.

Temperature   It was Lamarck in 1778 who first noticed that the inflorescence of *Arum italicum* produced heat (130). In several Araceae, the temperature can rise 10 to 22°C above the ambient (94, p. 99; 127; 129–132). At first (53; 101, p. 491), observers thought that the heat itself attracts pollinators, but Knoll (98) has shown experimentally that the heat increases the odor's rate of volatilization. Meeuse (132) confirmed this finding, showing that heat and odor production are two manifestations of the same respiratory process. In some species (102, 103), this heat enables the plant to penetrate a snow layer; it may (134) also mimic the warmth of fresh mammalian feces or of animals bodies. Flesh flies may react to temperature as an additional stimulus for oviposition (40; 62, p. 134). Blood-sucking insects pollinate *Arum conophalloids* (98), in which maximal odor production is coupled with maximal heat and carbon dioxide production (31). These conditions are strikingly similar to the oviposition stimuli of *Lucilia sericata* (a carrion fly with congeners involved in sapromyophily) (39), which needs the combination of ammonia, and carbon dioxide to lay (40). The carbon dioxide within traps probably mimics the characteristics of carrion and dung, which are inhabited by respiring decomposers (134). The narcotic effect of carbon dioxide on insects may delay the pollinator's departure. Thus, it would not be surprising if heat production has different ecological implications in different Araceous species, depending on the nature of the pollinator's oviposition stimuli.

Humidity and aeration   High humidity and adequate aeration are essential to keep the imprisoned insects alive (57; 98; 150, p. 189; 190), as well as to promote oviposition (10). Some traps have a high rate of transpiration, which produces high relative humidity (197), while others have special tissues that provide for air exchange (66, p. 172).

Illumination   Flies are positively phototropic and thus will try to escape through the trap's entrance before reaching the reproductive organs. A common adaptation to lure the insects deep into the trap is a "window-pane"—a colorless translucent area, usually at the bottom of the trap, surrounded by darker pigmentation (28; 58; 181; 186, pp. 104, 109; 190; 192; 209).

Reward and energy balance   Although the attraction is based on false promises, some sapromyophilous blossoms do offer a real reward. Daumann (51) treated these cases as partial cheaters and mentioned the conceptual problem of defining real deceitful flowers. Both *Aristolochia* spp. (28; 52; 128, p. 136; 195; 209) and *Ceropegia* spp. (190, 209), for example, have nectariferous traps. Many Araceae produce edible stigmatic liquid (32; 42; 49; 98; 101, p. 489, 120; 121; 153; 180). It is often unclear, however, whether there is any reward and, if there is, the extent to which it is consumed by pollinators. Generally speaking, rewards in traps are rare (66, p. 105; 186, p. 102). Vogel (192) noted that while nectar is common, it does not function as an attractant but maintains pollinators during their imprisonment. Similarly, Knuth (101, p. 489) argued that the stigmatic liquid in *Arum maculatum* compensates insects for their delay. Nonetheless, in some sapromyophilous traps, young from the laid eggs develop to full maturity (91, 121, 181), and the larval food does serve as a real reward (162). Such a reward system may have evolved from a deceptive syndrome to a mutualistic one (51; 66, p. 72; 201) that does not involve deception.

There is apparently some correlation between the way captive insects are handled and the offer of a reward. When the pollinator is imprisoned for a long period, it needs energy if it is to transfer pollen to the target stigma. In rewarding flowers, there generally is a surplus reward that exceeds the effort invested in extracting it, a fact that promotes repeat visits (82, 87, 96). For the plant, however, deception in pollination may produce savings in nectar production; the resultant surplus energy (23; 26; 70; 85, p. 168) may be invested in seed production (86). At least in Araceae, it is clear that heat production consumes considerable amounts of the reserves (129, 131, 132) and that the pollination mechanism cannot be regarded as an adaptation to save energy.

EVOLUTIONARY CONSIDERATIONS   Most traps are involved in the repetitive capture and release of pollinators, which have poor powers of flight and/or little ability to make precision landings (182). Some traps may have evolved to harness beetles and flies (201), since they display predictable behavior in response to special stimuli. Others are adapted to capture typical carrion or dung flies, which follow a specific odor "expecting" to find some reward.

After several unrewarded floral visits, insects will abandon the blossoms. Traps are therefore found in connection with sapromyophilous pollination (66, p. 104; 182). They force the pollinator to stay in the blossoms longer, raising the chances of a successful visit. Thus, protogyny is an essential element in the traps' efforts to use the visitor once as a carrier of foreign pollen and a second time as a messenger bearing fresh pollen from the trap plant (51; 66, p. 101; 93, pp. 67, 97; 182; 192).

Pollinators are passive, it is the plant that dictates the timing of events from capture until release. This difference apparently explains why pollinators have not evolved coevolutionarily with the trap blossoms (201); all of the adaptations have been on the part of the plant (66, p. 103; 186, p. 102; 190; 192).

Dung beetles are preadapted to this kind of sporadic and unpredictable food source. The adults, which have long life spans, can consume substitutional food and occupy a limited area for a long time. Saprophagous flies are short-lived, but adults are highly mobile and cover large areas in short periods (43, p. 474). Dung beetle pollination could be advantageous for plants that grow densely in a specific habitat and have long flowering periods.

Dung and carrion flies might also be viewed as preadapted to pollinating plants that are highly dispersed and flower for a short time. The existing data are too scanty to test the validity of this hypothesis, however. Dung beetles have a low substrate specificity (43, p. 475) in comparison to dung flies (81, pp. 59; 104). Sapromyophilous flowers produce mixtures of odors (132; 152, p. 303; 186, p. 102; 190; 209), and it is not surprising that in some instances, both beetles and flies are found in the same trap (98; 101, pp. 488–89; 127). At least in Araceae, it is hard to distinguish between fly-pollinated and beetle-pollinated species (182).

Although our present knowledge on trap blossoms is fragmentary, at least in the Araceae the existing data permit some generalizations and speculations about the possible evolution of traps. The commonly held views on evolutionary trends in Araceae are that (3, p. 213; 92, p. 629): (a) The spathe has evolved from a poorly developed or leaf-like form to an increasingly protective and colored one. (b) The flower has developed from a bisexual type with perigon to a unisexual one without it. (c) Some parts of the spadix have become barren in the advanced forms. Taking these ideas into account, along with general views on taxonomy and phylogeny in the family (65, 90, 135), we can make some generalizations: The primitive genera (e.g. *Acorus, Gymnostachys, Orontium*) have an open spathe. In more advanced genera, it is semiopen (e.g. *Zantedeschia, Monstera, Diffenbachia*); while in the most developed genera *(Arum, Arisaema, Cryptocoryne)*, the traps are highly elaborated.

It is assumed that traps have several selective advantages over exposed inflorescences including: (a) better selection of the suitable pollinators by the creation of "excluding cages," which implies less pollen loss, and (b) better protection of the ovules because of the enclosure of the reproductive organs (92, p. 629; 121; 182). Lessening the predation pressure on pollen as well as on ovules, especially by beetles in the tropics (75, 182), permits the production of fewer flowers per inflorescence. A quick glance at some representative genera reveals a growing tendency away from bare inflorescences containing numerous flowers *(Monstera, Lysichitum, Anthurium, Photos, Cyrtosperma)* toward semiclosed spathes with fewer flowers *(Alocasia, Philodendron, Amorphophallus, Caladium)* and finally toward closed traps having few, but well-protected, flowers *(Arisaema, Cryptocoryne, Arisarum, Ambrosina)*. The reduction in the number of flowers is more pronounced in females, which are better protected at the bottom of the trap. The spathe's formation of a tubular structure concentrates the means of optical attraction at the flattened part. One can recognize two main patterns. The first consists of inflorescences with open, flattened spathes that are, in general, uniformly colored without dots or patches. They are pollinated by beetles and bees in a nonsapromyophilous mode (e.g. *Anthurium, Spathiphyllum, Monstera*). A flattened and brightly colored spathe is important deep in the rain forest, the habitat of many Araceae (see 120–122 for supporting evidence). The second is made up of closed traps in which the optical guides are concentrated mainly in the flattened part of the spathe that is elaborately spotted or mottled. Pollinators include dung and carrion beetles and flies, but never bees.

Sexual Deceit

Faegri & van der Pijl (66, p. 74) regarded sexual attraction as a *primary attraction,* which contradicts their own definition of this term as an attractant that "satisfies demands like those for food, etc." (66, p. 57). In fact, except for the rare occurrence of ejaculation during pseudocopulation (34, 64), sexual attraction is based on lack of satisfaction. Thus, one has to recognize sexual attraction as a *secondary attraction.* So far, it seems to be limited to the Orchidaceae, with the possible exception of *Guiera senegalensis* of the Combretaceae (108, p. 226). About 20 plant genera are involved (54; 167–172; 186, pp. 131–40), with various types of pollinators, such as solitary bees and wasps, beetles, and flies. The attraction of male euglossine bees to orchid flowers is excluded, since this is regarded as a case of true reward (162). In other cases, sexual attraction has been observed without copulatory behavior (20, 74, 172).

PSEUDOCOPULATION   This phenomenon occurs when flowers mimic the female insect and thereby attract males. Such reproductive mimicry (201) has been termed Pouyannian (145).

The occurrence of pseudocopulation in three continents (Eurasia, South America, and Australia) and in unrelated genera implies that this syndrome had a polyphyletic origin (2; 8; 59, p. 115; 168; 170; 195). *Cryptostylis* is regarded as an ancient and well-differentiated genus (36; p. 17; 168; 169), all of whose species have the same pollinator, *Lissopimpla semipunctata* (33–37, 168, 170). *Ophrys,* in contrast, is considered a young and still evolving genus (108, p. 301; 174) with chemical and morphological variability (17, 45, 109–111) and with much hybridization (13, pp. 184–284; 108, p. 289; 140). In some respects, the genus *Chiloglottis* resembles *Ophrys.* The pollinator (*Erione* sp.) reacts only to a specific geographical race; other races have different pollinators (169), as in the case of *Ophrys fusca* (146). The complex of different biological races in *Chiloglottis* are supposedly undergoing speciation (171). Most of the pseudocopulatory flowers at least superficially (i.e. to the human eye) resemble an insect and do not produce nectar. The olfactory stimulus is the dominant one; it is the sole factor in long-range attraction (17; 36; 108, p. 292; 108–111; 199) and works on innate pollinator behavior (108, p. 291; 195; 209). The optimal cues are secondary and act primarily at close range (17; 108, pp. 276, 294), while the tactile stimulus is important only after the pollinator lands on the labellum (108, pp. 272–75).

An interesting aspect of pseudocopulation is the proterandry of the pollinator—i.e. the time lapse between the emergence of males and of the later-appearing females—in *Ophyrs* (2, 149, 195) and *Cryptostylis* 34, 64, 199). During this period, there is commonly an excess of males, so there is considerable competition for females as they emerge. Thus, the stimulus threshold is low and males attempt to copulate even with objects only remotely resembling females (108, pp. 69–71; 151, p. 239). There is evidence, however, that males visit flowers even after the females emerge (37, 63, 72, 77, 107), although flowers *are* visited more frequently at the beginning of the flowering period (146). If flowers still attract males even when females are present, any delay by the male on a flower will reduce its chances in competing for real females. In harnessing the male's sexual behavior for pollination, it is important for the plant that the insect not delay too long on any one flower so that it will be forced to visit another one. In *Ophrys* the male may leave the flower without any satisfaction, and his sexual excitement will lead him to try again (108, p. 260); but in *Cryptostylis* ejaculation was observed during the pseudocopulation (34, 36). Whether this behavior in *Cryptostylis* influences visits to additional flowers and whether it reduces the fecundity of the cheated males are topics to be explored.

An overview of the insects involved in pseudocopulation (C. O'Toole & A. Dafni, unpublished manuscript) reveals some common features. Females' nesting patterns dictate male search behavior, e.g. if there are nest aggregations, the male tends to look for females near the nesting sites. Males that have to locate a single female should have higher olfactory sensitivity than those that react to the odor of a whole colony. It is also assumed that a single "pseudo-copulatory" flower cannot compete with the olfactory attraction of a dense female colony but can compete with the odor production of a single female. In insect species that have a diffuse nesting pattern, females are likely to be found in food foraging places, e.g. the flowers amongst which the males seek to mate (see 114 for information on solitary bees). Therefore, it is not surprising that many of the species involved in pseudocopulation have diffuse nesting patterns, which give the flowers a better chance to attract males. Plants, in turn (especially *Ophrys*), generally occur in low-density populations. It is almost a necessary precondition that the insects be able to copulate from ground level up to 50 cm above it. Paulus & Gack's (146) experiments with *Ophrys* show that a change in a flower's height dramatically influences the chances of pseudocopulation. Since females of the pollinating species emerge from the soil, flowers on this level exert the greatest attraction. The authors thus concluded that there is selection for optimal plant height. The fact that bumblebees generally copulate quite high above the ground (175) could partially explain why they are not recruited for pseudocopulation. The males that are involved in pseudocopulation are not territorial and search for females in flowers. The chances of attracting a territorial insect for pollination service are considerably lower than the probability of attracting cruising males, an additional reason for excluding bumblebees.

SYNECOLOGICAL INTERFACE

Floral deception should be regarded as a recent development (51; 66, p. 104; 182; 186, p. 32), which appears mainly in highly evolved families. Deception is common, especially in Orchidaceae and Asclepiadaceae, whose flowers share common features such as large numbers of ovules and seeds, pollen packed in compact pollinia, reduction of floral parts, and highly specialized pollination. The existence of pollinia means total gambling (196) and dictates extreme precision in pollen transfer and reception; this enables the flowers to be pollinated even though they are rarely visited (85, p. 174; 183). Every mistake in pollination represents a real risk to the plants reproductive potential (183), which may be overcome, however, by highly specialized pollination syndromes together with the provision of long-lived pollen firmly attached to the pollinator (203). The large number of ovules (especially in Orchidaceae) should also be interpreted as a compensation for the low level of pollination (51). The reduction of nectar production in Orchidaceae is probably a means of preventing visits by other animals, which may rob all of the pollinia in one visit (183).

In every type of deception, adaptation occurs entirely on the part of the flower and acts on naive pollinators or on established instinctive behavior. The result is a unidirectional exploitation of the pollinators that can by no means be regarded as coevolution (59, p. 129; 186, p. 102; 198; 201). Pollinators do not appear to develop any specialization for pollination by deception (167, 199). One exception is Müllerian mimicry, in which all partners seem to benefit from the similarity between flowers that develops as a result of convergence.

Deceptive flowers impose several constraints at the community-matrix level on plants as well as pollinators. The pollination of deceitful flowers must be subsidized by rewarding species, which concurrently provide nectar or other energy sources to compensate the pollinators (7a; 45; 84; 136; 143; 172; 184; 186, p. 21; 196). The implications of mimicry for the plant populations, as well as for animal behavior, are divided in this discussion according to the two main types of deception: nutritive and reproductive.

Nutritive Deception

In nutritive-deception systems, optical cues are decisive and are based on learned signals. Thus, the efficiency of the mimicry is largely dependent on the pollinator's acquired experience. Several authors have argued that pollinators' naiveté is a precondition for the effective pollination of nonrewarding flowers (1; 26; 27; 53; 59, p. 220; 69; 70; 83; 85, p. 168; 115; 141–143; 167; 195). After a while, the pollinator recognizes the deception and learns to avoid it (115, 167). As a result, the seed set is better at the beginning of the flowering season (141) or when the pollinators have just arrived or emerged (1, 167). There is usually a low rate of visiting (70, 195), leading to a low percentage of fruit production (1, 44, 46, 48, 115, 167).

The fact that Batesian mimics occur much less frequently than their models is almost a precondition for the evolution and maintenance of mimicry in animals

(80, p. 111; 160, p. 156; 200, p. 96) and flowers (26; 66, p. 58; 118; 133; 195; 206). Having a relatively high proportion of floral mimics (as compared to models) would reduce their efficiency considerably (44, 48, 133). The pollinators' learning ability limits the population size of the nonrewarding species and, thus, its reproductive capacity. Such a negative feedback mechanism with respect to population size may have a positive survival value for plants evolving a system of deceptive pollination (85, p. 168; 167).

Deceptive flowers may reduce the pollinators' chance of learning, and thus the pollinators avoidance of them, by maintaining high color variability (1; 44; 83; 85, p. 189). Many flowers of this type have either faint fragrance (167) or none at all, which also diminishes the pollinators' learning. Heinrich (85, p. 189) suggested that lack of scent is advantageous because the pollen vectors are forced to learn to discriminate among flowers solely by color. The result is more visits to the non-rewarding flowers before learning and avoidance are achieved.

In floral mimicry, the mimic is usually constrained to flowering concurrently with its model (30, 83, 158). Another possibility is to maintain low flower availability per unit time and extend the flowering season to include intervals between or after the flowering of sympatric rewarding species (133). A third alternative is for the mimic to bloom sporadically during the model's more extended flowering period (69, 158). Gregarious flowering of a mimic species that has no specific model may enhance the optical signal that attracts naive pollinators (141).

Reproductive Deception

In syndromes that exploit reproductive stimuli, the chemical triggers act directly on the pollinators' instinct and release a pattern of innate behavior that is not susceptible to individual experience and learning (195, 209). As a result, the deceived pollinators do not learn to avoid the nonrewarding flowers, and the deceptive plants may appear in large numbers without reducing their chances of being pollinated. This pattern is paralleled by mimicry in which the deceived animal possesses an innate reaction to the model, in which case, the number of mimics is virtually unlimited (200, p. 47).

The specific volatiles are synthesized in special floral glands (osmophores). These compounds are highly efficient even in low concentrations, and they are readily perceived by insects (191, p. 752). The olfactory specificity allows a high degree of morphological variability because the selective pressures leading to uniformity—as a means for better recognition—are relaxed. When odors become the main means of attraction, they efficiently serve as isolating agents among closely related species (88, 204). Any slight change in the chemical can limit the attraction even to one species of pollinator (55, 203). When olfactory deceit occurs, one finds monophily (186, p. 31) or oligophily, but never polyphily.

In addition to specific odors, there are morphological obstacles at the entrance to oviposition traps (28, 29, 98, 190, 209) that form "enclosure cages." Thus, the limitation on the pollinators' size guarantees that only insects with the right dimensions enter the trap. Thus, there are several mechanisms that ensure a fairly high degree of pollinator specificity (143; 152, p. 303; 186, p. 102; 190; 209).

ACKNOWLEDGMENTS

I wish to express my gratitude to the following people for their encouragement, discussions, correspondence, or comments on the manuscript: P. Bernhardt, D. Cohen, S. A. Corbet, T. B. Croat, Y. Golenberg, C. C. Heyn, B. Lavie, A. D. J. Meeuse, C. D. Michener, L. A. Nilsson, C. O'Toole, P. H. Raven, W. Stoutamire, S. Vogel, and S. R. J. Woodell.

Literature Cited

1. Ackerman, J. D. 1981. Pollination biology of *Calypso bulbosa var. occidentalis* (Orchidaceae): A food-deception system. *Madroño* 28:101–10
2. Ames, O. 1937. Pollination of orchids through pseudocopulation. *Bot. Mus. Leafl. Harv. Univ.* 5:1–24
3. Arber, A. 1925. *Monocotyledons: A Morphological Study*. Cambridge: Cambridge Univ. Press
4. Baker, H. G. 1973. Evolutionary relationships between flowering plants and animals in American and African tropical forests. In *Tropical Forests in Africa and South America—A Comparative Review*, ed. B. J. Megges, E. S. Ayensu, W. D. Duckworth, pp. 145–59. Washington DC: Smithsonian Inst.
5. Baker, H. G. 1976. "Mistake pollination" as a reproductive system with special reference to the Caricaceae. In *Tropical Trees: Variation, Breeding and Conservation*, ed. J. Burley, B. T. Styles, pp. 161–69. London: Academic
6. Baker, H. G. 1977. Non-sugar constituents of nectar. *Apidologie* 8:349–56
7. Baker, H. G. 1978. Chemical aspects of the pollination of woody plants in the tropics. In *Tropical Trees as Living Systems*, ed. P. B. Tomlinson, M. Zimmerman, pp. 57–82. New York: Cambridge Univ. Press
7a. Baker, H. G., Cruden, R. W., Baker, I. 1971. Minor parasitism in pollination biology and its community function: The case of *Ceiba acuminata*. *BioScience* 21:1127–29
8. Baker, H. G., Hurd, P. D. 1968. Intrafloral ecology. *Ann. Rev. Entomol.* 13:385–414
9. Barnes, E. 1934. Some observations on the genus *Arisaema* on the Nilgiri Hills, South India. *J. Bombay Nat. Hist. Soc.* 37:630–39
10. Barton-Browne, L. B. 1964. Water regulation in insects. *Ann. Rev. Entomol.* 9:63–82
11. Barton-Browne, L. B. 1965. The analysis of the ovipositional responses of the blowfly *(Lucilia cuprina)* to ammonium carbonate and indole. *J. Insect Physiol.* 11:1131–43
12. Bates, H. W. 1862. Contribution to an insect fauna of the Amazon Valley. Lepidoptera: Heliconidae. *Trans. Linn. Soc. London (Zoology)* 23:495–566
13. Baumann, H., Künkele, S. 1982. *Die wildwachsenden Orchideen Europas*. Stuttgart: Kosmos. 432 pp.
14. Bawa, K. S. 1977. The reproductive biology of *Cupania guatemalensis* Radlk. (Sapindaceae). *Evolution* 31:52–63
15. Bawa, K. S. 1980. Mimicry of male by female flowers and intersexual competition for pollinators in *Jacaratia dolichochaula* (D. Smith) Woodson (Caricaceae). *Evolution* 34:467–74
16. Beck, G. 1914. Die Pollennachahmung in den Blüten der Orchideen—Gattung *Eria*. *Sitzungsber. Akad. Wiss. Wien: Math. Naturwiss. Kl.* 123:1033–46 (cited in Ref. 186)
17. Bergström, G. 1978. Role of volatile

chemicals in *Ophrys*-pollinator interactions. In *Biochemical Aspects of Plant and Animal Co-evolution*, ed. J. B. Harborne, pp. 207–32. London: Academic

18. Bernhardt, P., Montalvo, E. A. 1979. The pollination ecology of *Echeandia macrocarpa* (Liliaceae). *Brittonia* 31: 64–71

19. Bierzychudek, P. 1981. *Asclepias, Lantana* and *Epidendrum:* A floral mimicry complex? *Biotropica* 13:54–58 (Suppl.)

20. Bino, R. J., Dafni, A., Meeuse, A. D. J. 1982. The pollination ecology of *Orchis galilaea* (Bornm. et Schulze) Schltr. (Orchidaceae). *New Phytol.* 90:315–19

21. Bobisud, L. B., Neuhaus, R. J. 1975. Pollinator constancy and survival of rare species. *Oecologia* 21:263–72

22. Bohn, G. W., Davis, G. N. 1964. Insect pollination is necessary for the production of muskmelons (*Cucumis melo* v. *reticulatus*) *J. Apic. Res.* 3:61–63

23. Boyden, T. C. 1980. Floral mimicry by *Epidendrum ibaguense* (Orchidaceae) in Panama. *Evolution* 34:135–36

24. Brower, L. P. 1969. Ecological chemistry. *Sci. Am.* 220:22–29

25. Brower, L. P., Brower, J. V. Z., Corvino, J. M. 1967. Plant poisons in a terrestrial food chain. *Proc. Natl. Acad. Sci. USA* 57:893–98

26. Brown, J. H., Kodric-Brown, A. 1979. Convergence, competition and mimicry in a temperate community of hummingbird pollinated plants. *Ecology* 60:1022–35

27. Brown, J. H., Kodric-Brown, A. 1981. Reply to Williamson and Black's comment. *Ecology* 62:497–98

28. Cammerloher, H. 1923. Zur Biologie der Blüte von *Aristolochia grandiflora* Swartz. *Österr. Bot. Z.* 72:180–98

29. Cammerloher, H. 1933. Die Bestäubungseinrichtungen der Blüten von *Aristolochia lindneri* Berger. *Planta* 19: 351–65

30. Carlquist, S. 1979. *Stylidium* in Arnhem Land, Australia: New species, mode of speciation on the sandstone plateau and comments of floral mimicry. *Aliso* 9:411–62

31. Chen, J., Meeuse, B. J. D. 1971. Production of free indole by some arum lilies. *Acta Bot. Neerl.* 20:627–35

32. Cleghorn, M. L. 1913. Notes on the pollination of *Colocasia antiquorum*. *Proc. Asiat. Soc. Bengal* 9:313–15

33. Coleman, E. 1927. Pollination of the orchid *Cryptostylis Leptochila* F & M. *Victorian Nat.* 44:20–22

34. Coleman, E. 1928. Pollination of an Aus-

tralian orchid by the male ichneumonid *Lissopimpla semipunctata* Kirby. *Trans. R. Entomol. Soc. London* 76:533–39

35. Coleman, E. 1929. Pollination of *Cryptostylis subulata* (Labill.) Reichb. *Victorian Nat.* 46:62–66

36. Coleman, E. 1930. Pollination of some West Australian orchids. *Victorian Nat.* 46:203–6

37. Coleman, E. 1933. Pollination of *Diuris sulphurea* R. Br. *Victorian Nat.* 50:3–8

38. Cook, C. D. K. 1982. Pollination mechanisms in the Hydrocharitaceae. In *Studies on Aquatic Vascular Plants*, ed. J. J. Symones, S. S. Hooper, P. Compére, pp. 1–15. Brussels: R. Bot. Soc. Belg.

39. Cragg, J. B. 1950. The reactions of *Lucilia sericata* to various substances placed on sheep. *Parasitology* 40:179–86

40. Cragg, J. B. 1956. The olfactory behaviour of *Lucilia* species (Diptera) under natural conditions. *Ann. Appl. Biol.* 44:467–77

41. Croat, T. B. 1975. A new species of *Draconitum* (Araceae) from Panama, with notes on the sapromyophilous pollination syndrome. *Selbyana* 1:168–71

42. Croat, T. B. 1980. Flowering of the neotropical genus *Anthurium* (Araceae). *Am. J. Bot.* 67:888–904

43. Crowson, R. A. 1981. *The Biology of the Cleoptera*. London: Academic, 802 pp.

44. Dafni, A. 1983. Pollination of *Orchis caspia*—a nectarless plant species which deceives the pollinators of nectariferous species from other plant families. *J. Ecol.* 71:467–74

45. Dafni, A. 1985. On the evolution from reward to deception in *Orchis* s.l. (Orchidaceae) and its related genera. In *Orchid Biology: Reviews and Perspectives*, ed. J. Arditti, Vol. 3. Ithaca, NY: Cornell Univ. Press. In press

46. Dafni, A., Ivri, Y. 1979. Pollination ecology of and hybridization between *Orchis coriophora* L. and *Orchis collina* Sol. ex Russ. (Orchidaceae) in Israel. *New Phytol.* 83:181–86

47. Dafni, A., Ivri, Y. 1981. The flower biology of *Cephalanthera longifolia* (Orchidaceae)—pollen imitation and facultative floral mimicry. *Plant Syst. Evol.* 137:229–40

48. Dafni, A., Ivri, Y. 1981. Floral mimicry between *Orchis israelitica* Baumann and Dafni (Orchidaceae) and *Bellevalia flexuosa* Boiss. (Liliaceae). *Oecologia* 49: 229–32

49. Daumann, E. 1930. Nektarabscheidung in der Blütenregion einiger Araceen. *Planta* 12:38–48

50. Daumann, E. 1960. On the pollination ecology of *Parnassia* flowers: A new contribution to the experimental flower ecology. *Biol. Plant.* 2:113–25
51. Daumann, E. 1971. Zum problem der Täuschblumen. *Preslia* 43:304–17
52. Daumann, E. 1971. Zur Bestäubungsökologie von *Aristolochia clematitis* L. *Preslia* 43:105–11
53. Delpino, F. 1874. Ulteriori osservazioni e considerazioni sulla dicogamia nel regno vegetale. 2(IV). Delle piante ziodifile. *Ann. Soc. Ital. Sci. Nat. Mus. Civ. Stor. Nat. Milano* 16:151–349
54. Dodson, C. H. 1962. The importance of pollination in the evolution of the orchids of tropical America. *Am. Orchid Soc. Bull.* 31:525–34; 641–49; 731–35
55. Dodson, C. H., Dressler, R. L., Hills, H. G., Adams, R. M., Williams, N. H. 1969. Biologically active compounds of orchid fragrances. *Science* 169:1243–49
56. Dodson, C. H., Frymire, G. P. 1961. Natural pollination of orchids. *Ann. Mo. Bot. Gard.* 49:133–52
57. Dormer, K. J. 1960. The truth about pollination in *Arum. New Phytol.* 59:298–301
58. Drenth, E. 1972. A revision of the family Taccaceae. *Blumea* 20:365–406
59. Dressler, R. L. 1981. *The Orchids— Natural History and Classification.* Cambridge, Mass: Harvard Univ. Press, 332 pp.
60. Ducke, A. 1901. Beobachtungen über Blütenbesuch, Erscheinumgszeit etc. der bei Parávorkommenden Bienen. *Z. Syst. Hymenopt. Dipt.* 1:25–32
61. Duncan, C. J., Sheppard, P. M. 1965. Sensory discrimination and its role in the evolution of Batesian mimicry. *Behaviour* 24:269–82
62. Engelman, F. 1970. *The Physiology of Insect Reproduction.* Oxford. Pergamon
63. Erickson, R. 1951. *Orchids of the West.* Perth, Australia: Paterson, Brokensha. 109 pp.
64. Erickson, R. 1965. Insects and Australian orchids. *Aust. Orchid Rev.* 30:172–74
65. Eyde, H. R., Nicholson, D. H., Sherwin, P. 1967. A survey of floral anatomy in Araceae. *Am. J. Bot.* 54:478–97
66. Faegri, K., van der Pijl, L. 1979. *The Principles of Pollination Ecology.* Oxford: Pergamon. 244 pp. 3rd ed.
67. Ford, E. B. 1975. *Ecological Genetics.* London: Wiley. 4th ed.
68. Fritz, R. S., Morse, D. H. 1981. Nectar parasitism of *Asclepias syriaca* by ants: Effect on nectar level, pollinia insertion, pollinia removal and pod production. *Oecologia* 50:316–19

69. Gentry, A. H. 1974. Coevolutionary patterns in Central American Bignoniaceae. *Ann. Mo. Bot. Gard.* 61:728–59
70. Gentry, A. H. 1974. Flowering phenology and diversity in tropical Bignoniaceae. *Biotropica* 6:64–68
71. Gilbert, L. E. 1975. Ecological consequences of coevolved mutualism between butterflies and plants. In *Coevolution of Animals and Plants,* ed. L. E. Gilbert, P. H. Raven, pp. 211–39. Austin: Univ. Tex. Press
72. Gölz, P., Reinhard, H. R. 1977. Weitere Beobachtungen über die Bestäbuhg von *Ophrys speculum* Link. *Die Orchidee* 28:147–48
73. Goss, G. J. 1977. The reproductive biology of the epiphytic orchids of Florida. 6: *Polystachya flavescens* (Lindley) J. J. Smith. *Am. Orchid Soc. Bull.* 46:990–94
74. Goss, G. J., Adams, R. M. 1976. The reproductive biology of the epiphytic orchids of Florida, 5: Sexually selective attraction of moths to the floral fragrance of *Epidendrum anceps* Jacquin. *Am. Orchid Soc. Bull.* 46:997–1000
75. Gottsberger, G. 1970. Beiträge zur Biologie von Annonaceen-Blüten *Österr. Bot. Z.* 118:237–79
76. Gottsberger, G. 1977. Some aspects of beetle pollination in the evolution of flowering plants. *Plant Syst. Evol.* 1: 211–26 (Suppl.)
77. Graf, V. 1977. Ein neuer *Ophrys*-Bastard aus Korfu: *O. bomblyliflora* Linx × *O. attica* (Boiss. et Orph.) Soó. *Die Orchidee* 28:149
78. Grant, K. A. 1966. A hypothesis concerning the red coloration in Californian hummingbird flowers. *Am. Nat.* 100:85–98
79. Grant, K. A., Grant, V. 1968. *Hummingbirds and Their Flowers.* New York: Columbia Univ. Press
80. Grant, V. 1963. *The Origin of Adaptations.* New York: Columbia Univ. Press. 606 pp.
81. Hammer, O. 1941. *Biological and Ecological Investigations on Flies Associated with Pasturing Cattle and Their Excrement.* Copenhagen: Bianco Lunos Bogtrykkeri A/S. 257 pp.
82. Heinrich, B. 1975. The role of energetics in bumblebee-flower interrelationships. See Ref. 71, pp. 141–55
83. Heinrich, B. 1975. Bee flowers: A hypothesis on flower variety and blooming time. *Evolution* 29:325–34
84. Heinrich, B. 1977. Pollination energetics: An ecosystem approach In *The Role of Arthropods in Forest Ecosystems,* ed. W. J. Matton pp. 41–46. New York: Springer-Verlag

85. Heinrich, B. 1979. *Bumblebee Economics.* Cambridge, Mass: Harvard Univ. Press. 245 pp.
86. Heinrich, B. 1981. The energetics of pollination. *Ann. Mo. Bot. Gard.* 68: 370–78
87. Heinrich, B., Raven, P. H. 1972. Energetics and pollination ecology. *Science* 176:597–602
88. Hills, G. H., Williams, N. H., Dodson, C. H. 1972. Floral fragrances and isolating mechanisms in the genus *Catasetum* (Orchidaceae). *Biotropica* 4:61–76
89. Holling, C. S. 1965. The functional response of predators to prey density and its role in mimicry and population regulation. *Mem. Entomol. Soc. Can.* 45:1–60
90. Hotta, M. 1971. Study on the family Araceae: General remarks. *Jpn. J. Bot.* 20:269–310
91. Hubbard, H. G. 1895. Insect fertilization of an aroid plant. *Insect Life* 7:340–45
92. Hutchinson, J. 1959. *The Families of Flowering Plants.* Vols. 1, 2. Oxford: Clarendon. 2nd ed.
93. Jaeger, P. 1961. *The Wonderful Life of Flowers.* London: Harrap
94. James, W. O., Clapham, A. R. 1935. *The Biology of Flowers.* Oxford: Clarendon
95. Kallunki, J. A. 1981. Reproductive biology of mixed-species populations of *Goodyera* (Orchidaceae) in northern Michigan. *Brittonia* 33:137–55
96. Kevan, P. G., Baker, H. G. 1983. Insects as flower visitors and pollinators. *Ann. Rev. Entomol.* 28:407–53
97. Kirchner, O. 1911. *Blumen und Insekten, ihre Anpassungen aneinander und ihre gegenseitige Abhängigkeit.* Leipzig/Berlin (cited in Ref. 51)
98. Knoll, F. 1926. Insekten und Blumen. Experimentelle Arbeiten zur Vertiefung unserer Kennthisse Über die Wechselbeziehungen zwischen Pflanzen und Tieren. IV. Die *Arum*-Blütenstände und ihre Besucher. *Abh. Zool. Bot. Ges. Wien* 12:379–482
99. Knoll, F. 1931. Bestäubung. In *Handwörterbuch der Naturwissenschaften* 1:870–908. Jena. 2nd ed. (cited in Ref. 51)
100. Knoll, F. 1956. *Die Biologie der Blüte.* Berlin: Springer-Verlag. 164 pp.
101. Knuth, P. 1909. *Handbook of Flower Pollination,* Vol. 3. Transl. J. R. Ainsworth Davis. Oxford: Oxford Univ. Press. 644 pp. (From German)
102. Knutson, R. M. 1972. Temperature measurements of the spadix of *Symplocarpus foetidus* (L.) Nutt. *Am. Midl. Nat.* 88:251–54
103. Knutson, R. M. 1974. Heat production and temperature regulation in eastern skunk cabbage. *Science* 186:746–47
104. Kugler, H. 1955. *Einführung in die Blütenökologie.* Stuttgart: Fischer. 278 pp.
105. Kugler, H. 1970. *Blütenökologie.* Stuttgart: Fischer
106. Kuijt, J. 1969. *The Biology of Parasitic Flowering Plants.* Berkeley: Univ. Calif. Press
107. Kullenberg, B. 1950. Investigations on the pollination of *Ophrys* species. *Oikos* 2:1–19 (publ. 1952)
108. Kullenberg, B. 1961. Studies in *Ophrys* pollination. *Zool. Bidr. Uppsala* 34:1–340
109. Kullenberg, B., Bergström, G. 1973. The pollination of *Ophrys* orchids. In *Chemistry in Botanical Research, Proc. 25th Nobel Symp.,* ed. G. Bendz, J. Santesson, pp. 253–58. Stockholm: Nobel Found.
110. Kullenberg, B., Bergström, G. 1975. Chemical communication between living organisms. *Endeavour* 34:59–66
111. Kullenberg, B., Bergström, G. 1976. The pollination of *Ophrys* orchids. *Bot. Not.* 129:11–19
112. Lack, A. 1976. Competition for pollinators and evolution in *Centaurea. New Phytol.* 77:787–92
113. Lindner, E. 1928. *Aristolochia lindneri* Berger und ihre Bestaübung durch Fliegen. *Biol. Zentralbl.* 48:93–101
114. Linsley, G. E. 1958. The ecology of solitary bees. *Hilgardia* 27:543–99
115. Lock, J. M., Profita, J. C. 1975. Pollination of *Eulophia cristata* (Sw.) Steud. (Orchidaceae) in southern Ghana. *Acta Bot. Neerl.* 24:135–38
116. Macior, L. W. 1968. Pollination adaptation in *Pedicularis groenlandica. Am. J. Bot.* 55:927–33
117. Macior, L. W. 1970. Pollination ecology of *Dodecatheon amethystinum* (Primulaceae). *Bull. Torrey Bot. Club* 97:150–53
118. Macior, L. W. 1971. Co-evolution of plants and animals—systematic insights from plant-insect interaction. *Taxon* 20: 17–28
119. Macior, L. W. 1974. Behavioral aspects of coadaptations between flowers and insect pollinators. *Ann. Mo. Bot. Gard.* 61:760–69
120. Madison, M. 1977. A revision of *Monstera* (Araceae). *Contrib. Gray Herb. Harv. Univ.* 207:3–100
121. Madison, M. 1979. Protection of developing seeds in neotropical Araceae. *Aroideana* 2:52–61
122. Madison, M. 1981. Notes on *Caladium*

(Araceae) and its allies. *Selbyana* 5:342–77

123. Manning, A. 1979. *An Introduction to Animal Behaviour*. London: Arnold. 329 pp. 3rd ed.

124. Markl, H. 1974. Insect behavior: Functions and mechanisms. In *The Physiology of Insects*, ed. M. Rockstein, 3:3–110. New York: Academic. 2nd ed.

125. McDade, L. A., Kinsman, S. 1980. The impact of floral parasitism in two neotropical hummingbird-pollinated plant species. *Evolution* 34:944–58

126. Meeuse, A. D. J. 1973. Co-evolution of plant hosts and their parasites as a taxonomic tool. In *Taxonomy and Ecology*, ed. V. H. Heywood, pp. 289–316. London: Academic

127. Meeuse, B. J. D. 1959. Beetles as pollinators. *Biologist* 42:22–32

128. Meeuse, B. J. D. 1961. *The Story of Pollination*. New York: Ronald. 243 pp.

129. Meeuse, B. J. D. 1966. The voodoo lily. *Sci. Am.* 215:80–88

130. Meeuse, B. J. D. 1973. Films of liquid crystals as an aid in pollination studies. In *Pollination and Dispersal*, ed. N. B. M. Brantjes, H. F. Linskens, pp. 19–20. Nijmegen. The Netherlands: Dep. Botany, Univ. Nijmegen

131. Meeuse, B. J. D. 1975. Thermogenic respiration in Aroids. *Ann. Rev. Plant Physiol.* 26:117–26

132. Meeuse, B. J. D. 1978. The physiology of some sapromyophilous flowers. See Ref. 154, pp. 97–104

133. Melampy, M. N., Hayworth, A. H. 1980. Seed production and pollen vectors in several nectarless plants. *Evolution* 34:1144–54

134. Moodie, G. E. E. 1976. Heat production and pollination in Araceae. *Can. J. Bot.* 54:545–46

135. Mookerjea, A. 1955. Cytology of different species of aroids with a view to tracing the basis of their evolution. *Caryologia* 7:221–91

136. Mosquin, T. 1970. The reproductive biology of *Calypso bulbosa* (Orchidaceae). *Can. Field Nat.* 84:291–96

137. Mosquin, T. 1971. Competition for pollinators as a stimulus for the evolution of flowering time. *Oikos* 22:338–402

138. Müller, F. 1878. Über die Vortheile der Mimikry bei Schmetterlingen. *Zool. Anz.* 1:54–55

139. Müller, L. 1926. Zur biologischen Anatomie der Blüte von *Ceropegia woodii* Schlechter. *Biol. Gen.* 2:799–814

140. Nelson, E. 1962. Gestaltwandel und Artbildung erörtert am Beispiel der Orchidaceen Europas und der Mittelmeerländer insbesondere der Gattung *Ophrys*. Chernex-Montreux, Switzerland: Nelson

141. Nilsson, L. A. 1980. The pollination ecology of *Dactylorhiza sambucina* (Orchidaceae). *Bot. Not.* 133:367–85

142. Nilsson, L. A. 1981. Pollination ecology and evolutionary processes in six species of orchids. *Abstr. Uppsala Diss. Sci.* 593

143. Nilsson, L. A. 1983. Anthecology of *Orchis mascula* (Orchidaceae). *Nord. J. Bot.* 3:157–79

144. Pannell, C. M. 1980. *Taxonomic and ecological studies in Aglaia (Meliaceae)*. PhD thesis. Oxford Univ., Oxford

145. Pasteur, G. 1982. A classificatory review of mimicry systems. *Ann. Rev. Ecol. Syst.* 13:169–99

146. Paulus, H. F., Gack, C. 1980. Beobachtungen und Untersuchungen zur Bestäubungsbiologie südspanischer *Ophrys* Arten. In *Probleme der Evolution bei europäische und mediterranen orchideen*, ed. K. Senghas, H. Sundermann. *Jahresber. Naturwiss. Ver. Wuppertal* 33:55–68. (Suppl.)

147. Pennel, F. 1948. Taxonomic significance of an understanding of floral evolution. *Brittonia* 6:301–8

148. Porsch, O. 1909. Neuere Untersuchungen über Insektenanlockungsmittel der Orchideenblüte. *Mitt. Naturwiss. Ver. Steiermark* 45:346–70

149. Pouyanne, A. 1917. La fécondation des *Ophrys* par les insectes. *Bull. Soc. Hist. Nat. Afr. Nord* 8:6–7

150. Prime, C. T. 1960. *Lords and Ladies*. London: Collins 241 pp.

151. Proctor, M. C. F. 1978. Insect pollination syndromes in an evolutionary and ecosystem context. See Ref. 155, pp. 105–16

152. Proctor, M. C. F., Yeo, P. 1973. *The Pollination of Flowers*. London: Collins. 418 pp.

153. Ramirez, B. W., Gómez, P. L. D. 1978. Production of nectar and gums by flowers of *Monstera deliciosa* (Araceae) and of some species of *Clusia* (Guttiferae) collected by New World *Trigona* bees. *Brenesia* 14/15:407–12

154. Read, C. P. 1970. *Parasitism and Symbiology*. New York: Ronald

155. Richards, A. J., ed. 1978. *The Pollination of Flowers by Insects, Symp. Linn. Soc. London No. 6*. London: Academic

156. Schelpe, E. A. C. L. E. 1966. *An Introduction to the Southern Orchids*. London: Macdonald

157. Schemske, D. W. 1980. Floral convergence and pollinator sharing in two bee-pollinated tropical herbs. *Ecology* 62:946–54

158. Schemske, D. W., Willson, M. F., Melampy, M. N., Mitker, L. J., Verner, L. et al. 1978. Flowering ecology of some spring woodland herbs. *Ecology* 59:351–66
159. Schmid, R. 1978. Reproductive anatomy of *Actinidia chinensis* (Actinidiaceae). *Bot. Jahrb. Syst.* 100:149–95
160. Sheppard, P. M. 1958. *Natural Selection and Heredity.* London: Hutchinson Univ. Libr.
161. Simpson, B. B., Neff, J. L. 1977. *Prosopis* as a niche component of plant and animal. In *Mesquite: Its Biology in Two Desert Ecosystems*, ed. B. B. Simpson, pp. 124–32. Stroudsburg, Pa: Dowden, Hutchinson & Ross
162. Simpson, B. B., Neff, J. L. 1981. Floral rewards: Alternatives to pollen and nectar. *Ann. Mo. Bot. Gard.* 68:301–22
163. Smith, B. N., Meeuse, B. J. D. 1966. Production of volatile amines and skatole at anthesis in some arum lily species. *Plant Physiol.* 41:343–47
164. Smyth, J. D. 1976. *Introduction to Animal Parasitology.* London: Hodder & Stoughton
165. Sprengel, C. K. 1793. *Das Entdeckte Geheimnis der Natur im Bau und in der Befruchtung der Blumen.* Berlin: Viewveg
166. Stoutamire, W. P. 1967. Flower biology of lady's slippers (Orchidaceae: *Cypripedium*). *Mich. Bot.* 6:159–75
167. Stoutamire, W. P. 1971. Pollination in temperate American orchids. In *Proc. 6th World Orchid Conf., Sydney, Australia, 1969*, ed. M. J. G. Corrigan, pp. 233–43. Sydney: Halstead
168. Stoutamire, W. P. 1974. Australian terrestrial orchids, thynnid wasps and pseudocopulation. *Am. Orchid Soc. Bull.* 43:13–18
169. Stoutamire, W. P. 1975. Pseudocopulation in Australian terrestrial orchids. *Am. Orchid Soc. Bull.* 44:226–33
170. Stoutamire, W. P. 1976. Pollination strategies in orchids of southern Australia. In *First Symp. Sci. Aspects Orchids, 1974*, ed. H. H. Szmant, J. Wemple, pp. 27–33. Southfield, Mich: Dep. Chem., Univ. Detroit
171. Stoutamire, W. P. 1981. Pollination studies in Australian terrestrial orchids. *Nat. Geogr. Soc. Res. Rep.* 13:591–98
172. Stoutamire, W. P. 1983. Wasp pollinated species of *Caladenia* (Orchidaceae) in south-western Australia. *Aust. J. Bot.* 31:383–94
173. Straw, R. M. 1972. A Markov model for pollinator constancy and competition. *Am. Nat.* 106:597–620
174. Sundermann, H. 1977. The genus *Ophrys*—an example of the importance of isolation for speciation. *Am. Orchid Soc. Bull* 46:825–30
175. Svensson, B. G. 1979. Patrolling behaviour of bumble bee males (Hymenoptera, Apidae) in a subalpine/alpine area, Swedish Lapland. *Zoon* 7:67–94
176. Thien, L. B., Marcks, B. G. 1972. The floral biology of *Arethusa bulbosa, Calopogon tuberosus*, and *Pogonia ophioglossoides* (Orchidaceae). *Can. J. Bot.* 23:19–25
177. Troll, W. 1928. *Organisation und Gestalt im Bereich der Blüte.* Berlin: Springer-Verlag
178. Deleted in proof
179. Valdeyron, L. G., Lloyd, D. G. 1979. Sex differences and flowering phenology in the common fig, *Fiscus carica* L. *Evolution* 33:673–85
180. Van der Pijl, L. 1937. Biological and physiological observations on the inflorescence of *Amorphophallus. Recl. Trav. Bot. Neerl.* 34:157–67
181. Van der Pijl, L. 1953. On the flower biology of some plants from Java. *Ann. Bogor.* 1:77–99
182. Van der Pijl, L. 1960. Ecological aspects of flower evolution I. *Evolution* 14:403–16
183. Van der Pijl, L. 1966. Pollination mechanisms in orchids. In *Reproductive Biology and Taxonomy of Vascular Plants*, ed. J. C. Hawkes, pp. 61–75 London: Pergamon
184. Van der Pijl, L. 1969. Evolutionary action of tropical animals on the reproduction of plants. *Biol. J. Linn. Soc.* 1:85–92
185. Van der Pijl, L. 1978. Reproductive integration and sexual disharmony in floral functions. See Ref. 155, pp. 79–88
186. Van der Pijl, L., Dodson, C. H. 1966. *Orchid Flowers: Their Pollination and Evolution.* Coral Gables, Fla: Univ. Miami Press. 214 pp.
187. Vane-Wright, R. I. 1976. A unified classification of mimetic resemblances. *Biol. J. Linn. Soc.* 8:25–56
188. Vane-Wright, R. I. 1980. On the definition of mimicry. *Biol. J. Linn. Soc.* 13:1–6
189. Vogel, S. 1954. *Blütenbiologische Typen als Elemente der Sippengliederung.* Jena: Fischer
190. Vogel, S. 1961. Die Bestäubung der Kesselfallen-Blüten von *Ceropegia. Beitr. Biol. Pflanz.* 36:159–237
191. Vogel, S. 1963. Duftdrüsen im Dienste der Bestäubung: über Bau und Funktion der Osmophoren. *Abh. Math.-Naturwiss. Kl. Aka. Mainz* 1962(10):599–763

192. Vogel, S. 1965. Kesselfallen-Blumen. *Umschau* 65:12–17
193. Vogel, S. 1972. Pollination von *Orchis papilionaceae* L. in den Schwarmbahnen von *Eucera tuberculata* F. In *Probleme der Orchideengattung Orchis,* ed. K. Senghas, H. Sundermann. *Jahresber. Naturwiss. Ver. Wuppertal* 25:67–74 (Suppl.)
194. Vogel, S. 1973. Fungus gnat flowers and fungus mimesis. See Ref. 30, pp. 13–18
195. Vogel, S. 1975. Mutualismus und Parasitismus in der Nutzung von Pollenträgern. *Ver. Dtsch. Zool. Ges.* 1975: 102–10
196. Vogel, S. 1978. Evolutionary shifts from reward to deception in pollen flowers. See Ref. 154, pp. 89–96
197. Vogel, S. 1978. Pilzmückenblumen als Pilzmimeten. I. and II. *Flora (Jena)* 167:329–66; 367–98
198. Vöth, W. 1982. Die "ausgeborten" Bestäuber von *Orchis pallens* L. *Die Orchidee* 33:196–203
199. Wallace, B. J. 1978. On *Cryptostylis* pollination and pseudocopulation. *Orchadian* 5:168–69
200. Wickler, W. 1968. *Mimicry in Plants and Animals.* Transl. R. D. Martin. London: Weidenfeld & Nicholson. 255 pp. (From German)

201. Wiens, D. 1978. Mimicry in plants. *Evol. Biol.* 11:365–403
202. Williams, N. H. 1972. A reconsideration of *Ada* and the glumaceous *Brassias* (Orchidaceae). *Brittonia* 24:93–110
203. Williams, N. H., Dodson, C. H. 1972. Selective attraction of male euglossine bees to orchid fragrances and its importance in long-distance pollen flow. *Evolution* 26:84–95
204. Williams, N. H., Dressler, R. L. 1976. Euglossinae pollination of *Spathiphyllum* (Araceae). *Selbyana* 1:350–56
205. Williamson, G. B. 1982. Plant mimicry: Evolutionary constraints. *Biol. J. Linn. Soc.* 18:49–58
206. Williamson, G. B., Black, E. M. 1981. Mimicry in hummingbird-pollinated plants? *Ecology* 62:494–96
207. Winkler, H. 1927. Über eine *Rafflesia* aus Zentralborneo. *Planta* 4:1–97
208. Yeo, P. F. 1968. The evolutionary significance of the speciation of *Euphrasia* in Europe. *Evolution* 22:736–46
209. Yeo, P. F. 1972. Floral allurements for pollinating insects. In *Insect-Plant Relationships, 6th Symp. R. Entomol. Soc. London,* ed. H. F. Van Emden, pp. 51–57. Oxford: Blackwell

Ann. Rev. Ecol. Syst. 1984. 15:279–301

MIGRATION AND GENETIC POPULATION STRUCTURE WITH SPECIAL REFERENCE TO HUMANS

E. M. Wijsman and L. L. Cavalli-Sforza

Department of Genetics, Stanford University, Stanford, California 94305

INTRODUCTION

The importance of migration as a force influencing genetic structure has long been recognized. There are a number of ways in which migration can occur, however, each with its characteristic properties, effects on genetic structure, and problems of estimation. In the following pages, we will discuss these processes and their effects on the genetic structure and makeup of populations. In our analysis of the quantitative data on this phenomenon, we will look at human populations because of the paucity of data on migration in other organisms.

For purposes of discussion, it is convenient to consider three aspects of migration: (*a*) individual migration (natal dispersal plus breeding dispersal), (*b*) exchange among genetically differentiated populations, and (*c*) demic diffusion (the cumulative effects of migration with population expansion, where the expansion accompanies the migration). These three aspects are not mutually exclusive, and they interact with other variables (selection, geographic distance, drift, etc.) to influence genetic differentiation among populations. We will consider some of these interactions in this review; we will not consider the interaction between migration and population structure [for a review of this, see Karlin (64)].

Individual migration and demic diffusion are discussed in the first and last major parts of the review, respectively. Exchange among populations is discussed in three sections in which other variables are also considered. In the

0066-4162/84/1120-0279$02.00

section on isolation by distance, we consider the effect of geographic distance on genetic differentiation, and in the discussion on gene flow, we consider the problem of estimating the amount of genetic exchange that has occurred among populations. In the section on clines, we consider the interaction of genetic exchange with selection, drift, and other factors and apply some of these results to data on migration in human populations.

INDIVIDUAL MIGRATION

Movement for the purpose of finding mates or resources causes a certain amount of gene migration. Although parent-offspring migration (or birth site to breeding site dispersal) is the parameter of primary genetic interest (123), few data on such instances exist. More commonly studied are the movements of individuals during a particular life stage or the distances between mates' birthplaces or residences.

Theory

Models of migration can be divided into three general classes, distinguished according to their assumptions about the underlying processes. The first class is based on random walks in which migrants have no memory of previous visits. The normal distribution in either one or two dimensions (depending on the distribution of the population in space) is one example of a function that describes this process for constant migration (26, 76). A second diffusion model introduced by Malecot (79)—the K-distribution—assumes that mating distance is a function of time, which has a gamma distribution.

Although plants may in fact migrate in a random fashion, few animals do. Therefore, several knowledge-based models have been introduced. The most explicit of these (23) assumes that migration (and mating) frequency at a particular distance is proportional to the number of contacts made at that distance. This number, in turn, is a function of population density and the number of visits at that distance. This model is closely related (76) to the gravitational model (27).

A third class of models which has been used to describe migration may best be called empirical because the models do not contain parameters that have a direct interpretation. These models generally follow some sort of negative exponential distribution (17, 23, 26, 32, 48, 114, 115).

Migration matrices provide an alternative approach to the study of distributions of distances (22). Such matrices show the numbers (or proportions) of individuals migrating from places i to j for $1 < i, j < k, k > 1$, where k is the number of places. For genetic purposes, the parent-offspring migration matrix is optimal. In general, the migration matrix contains more information than the distribution of distances, and it can be used to answer a variety of questions. These include the direction as well as amount of migration, the final population sizes of the various communities, and the genetic variation ultimately expected among the communities. If the communities are stationary in space, the migration matrix can also be used to indicate the equilibrium distribution of genetic distances.

Estimation[1]

In order to study individual migration, it is necessary to assume some distribution of the population in space. It is often convenient to assume a discontinuous population structure, but a second possibility is a continuous distribution of individuals in space. In fact, most populations should fall somewhere between these two extremes, although more plants may fall into a continuous, and more social animals into a discontinuous distribution.

One convenient source of data on human populations is written records. These usually provide information on birthplaces or residences of mates from which marital distances are obtainable. Parent-offspring distances are rarely directly estimatable and must be calculated by, for example, linking the birth records of parents and offspring. Records usually consist of locations identified by administrative units rather than addresses, so it is difficult to estimate short distance movements. Written records also reveal historical changes in migration rates and distances, often over several centuries (24, 39, 58, 73). Direct questioning (69, 72) and direct observation of mother-offspring distances in nonhuman populations are alternative sources of data.

A second source of data on migration is surnames. Since they are generally transmitted patrilineally, surnames usually provide information only on male migration. The number of surnames in a given sample size fits the Karlin-McGregor (66) and logarithmic distributions (46); thus, it is possible to estimate the immigration rate into an area simply by knowing the number of surnames and individuals in the area (129). The available comparisons between actual migration rates and estimates based on this procedure indicate that this method is satisfactory (129), but more comparisons are needed.

When surname data are available over two different time periods for several neighboring regions that share the same language (75), the whole migration matrix can be estimated (120). This was calculated for surnames in Sardinia; gene-frequency data could be used in a similar manner to estimate parent-offspring migration (120). An important advantage of the matrix is that it contains information on the direction as well as the distance of migration, but

---

[1]Almost all quantitative data on migration comes from human populations, so some of the following estimation techniques will be applicable only to the study of migration in humans.

estimating a migration matrix for even a moderate number of regions with these techniques requires large sample sizes.

Data

The quantitative aspects of short-distance migration have been investigated in a large number of human populations plus a few other populations. Table 1 presents the results of some of these studies in humans. Although most of the data are marital distances, a regression of mean parent-offspring distance on marital distance where both kinds of data exist indicates that parent-offspring distances equal roughly 79% of the marital distances. Except for the oceanic populations and some modern Japanese urban populations, the root mean square migration distances in a wide variety of populations are all within an order of magnitude. The two hunter-gatherer populations (Pygmies and !Kung bushmen) and the Indian populations are somewhat more mobile than others; the European populations are somewhat less so.

In many cases, there are differences in the amount of migration across economic or social classes (1–3, 19, 37, 39, 54, 61, 74, 104, 113). Marital movement among classes is often large (72, 74), however, obliterating any heterogeneity and encouraging the use of the population average migration as an estimator. Differences in migration rates as a function of social dominance have also been observed in nonhuman species (36); such complications may be widespread and may explain why it is hard to obtain good fits to some of the models presented earlier.

Problems

To understand the genetic variation of populations, it is necessary to estimate the relevant parameters of individual migration. Most studies of human populations have been concerned with marital distances; it would be more useful to collect data on parent-offspring distances. Also, the data need to be reported so that the root mean square distance is available; the migration matrix may be a suitable format, since it also contains information on the direction of migration. The study of migration in nonhuman populations may also benefit from the migration matrix approach.

ISOLATION BY DISTANCE

The term *isolation by distance* originated with Wright (124, 125), and the model currently in use was first suggested by Malecot (78). Kimura & Weiss independently proposed a similar model (68, 119) for a discontinuous population distribution under the name of the *stepping stone model*. The aim was to establish a relationship between the genetic similarity of two populations and the geographic distance separating them. The exact relationship is complicated;

Table 1 Migration distances (km) for various human populations

Population	Data[a]	Mean	l[b]	References
Pygmies (modern)	mt	53.0	68.2	32
	po	18.5	40.8	c
!Kung (modern)	po	51.2	66.2	53
Pyrenees				
(1850–1910)	mt	10.4	24.8	1
	fo	6.9	21.3	1
	mo	4.4	18.0	1
(1850–1915)	mt	5.7	16.0	2
	fo	2.5	10.9	2
	mo	1.9	8.8	2
Sweden (1800–1959)	po	2.0	14.5	7
Sweden (1900–1944)	mt	26.0	32.4	48
North				
(1861–1875)	mt	6.1	9.1	18
(1876–1890)	mt	6.1	9.2	18
(1891–1905)	mt	6.0	9.7	18
(1906–1920)	mt	6.6	11.0	18
(1900)	mt	18.8	—	19
(1930)	mt	20.3	—	19
(1960)	mt	46.6	—	19
Aland Islands				
(1800–1850)	mt	1.2	4.8	43
(1878–modern)	po	5.4	15.0	62
Britain (1920–1960)	mt	11.0	42.3	37
England (1861)	mt	4.2	4.7	23
Oxfordshire				
(1850–modern)	mt	0.3	0.8	58
(before 1850)	mt	10.8	—	73
(after 1850)	mt	39.6	—	73
Oxford				
(<1900)	mt	24.9	—	74
(>1900)	mt	31.7	—	74
Otmoor Parishes				
(1650–modern)	mt	31.0	74.4	24
Belgium				
(1935)	mt	12.1	26.3	40
(1960)	mt	16.5	32.1	40
Italy				
Parma (1958)	mt	6.2	12.2	27, 33
	fo	2.7	7.2	27, 33
	mo	3.6	8.1	27, 33
Sardinia (1870–1970)	fo	31.9	53.4	120
Malay (modern)	mt	9.3	9.9	47
Japan (modern)				
Gyoda	mt	53.1	159.1	59
	fo	39.0	155.7	59
	mo	35.7	121.6	59

Table 1 (Continued)

Population	Data[a]	Mean	l[b]	References
Hasuda	mt	107.1	300.4	59
	fo	90.5	265.8	59
	mo	60.6	198.0	59
Mine	mt	18.0	24.7	127
Ohdate	mt	10.8	16.2	127
Misima	mt	2.2	2.6	127
	mt	3.1	3.9	128
	fo	1.1	1.6	128
	mo	2.0	2.4	128
India (modern)				
Santals	mt	12.1	19.1	16
Maratha	mt	36.8	55.6	76
Chamar	mt	38.6	56.1	76
Dhangar castes	mt	26.4	49.8	77
US–New Mexico				
Hispanic (modern)	mt	33.7	54.3	38
Massachusetts (1810–1819)	mt	12.5	14.0	115
Pingelap (modern)	po	13.7	119.3	89
Mokil (modern)	po	44.9	221.1	89
New Guinea (modern)				
Kiunga speakers	mt	1.3	4.7	105
Gainj, Kalam	po	1.3	3.9	d
	mo	4.4	7.1	d
Madang	mt	22.6	103.4	105
Pidgin	mt	50.5	132.0	105
English speakers	mt	171.1	294.1	105

[a] mt = marital, po = parent-offspring, fo = father-offspring, mo = mother-offspring.
[b] l = root mean square distance.
[c] G. Zei & E. Zanardi, submitted for publication.
[d] J. W. Wood, P. E. Smouse, J. C. Long, submitted for publication.

in the case of a unidimensional population distribution, the relationship is approximately exponential:

$$\phi\ (d)\ =\ ac^{-bd}, \tag{1.}$$

where $\phi\ (d)$ is the "kinship" between two populations and d is distance. Kinship and related quantities have been estimated with a variety of formulas [for a discussion of the difficulties encountered in estimating kinship, see Jacquard (60)]. For the one-dimensional case, the two constants a and b are approximately equal to (78):

$$a\ =\ 1/[1+4N\ \sqrt{c(c+2m)}];\ b\ =\ \sqrt{2c/m}, \tag{2. & 3.}$$

where N is the effective population size; c the coefficient of recall, which is the sum of the effects of mutation, stabilizing selection (i.e. heterozygous advantage near equilibrium gene frequencies), and long-range migration (to be defined more accurately later); and m is the proportion of individuals migrating to (and from) the nearest neighbor colonies in the discrete model of isolation by distance. In the case of a continuous geographic distribution, m can be replaced by I^2, the second moment of the distance between parent and offspring. In the bidimensional case, the theory gives a substantially different prediction for a, and the exponential relationship (Equation 1) is valid only under more restricted conditions, e.g. long distances and a certain range of b values (80).

The analysis of the dependence of correlation (or alternatively, covariance or kinship) between two populations and their geographic distance can be looked at as a form of analysis of spatial autocorrelation (112). Another method is to plot the square of the difference between the gene frequencies of the two populations against their geographic distances. Near the origin, this quantity is expected to vary linearly with distance. In some mathematical (82) and geological uses, the quantity "half the squared difference" receives the name *variogram*. In genetic applications (97), dividing the square of the difference by the sum of the difference of gene frequencies provides a standardization similar to that obtained by the "Wahlund" variance of gene frequencies (30).

Applications

Equation 1 seems remarkably efficient, since it has been applied to a great variety of data (e.g. phenotypes, metric and migration data, surnames; see Morton (88), who pioneered this work).

For two reasons, however, the robustness of Equation 1 may not be as high as it may superficially seem to be. The first is that the equation estimated in practice (88) often is not Equation 1 but the three-parameter equation:

$$\phi\ (d) = (1-L)\ ac^{-bd} + L, \qquad\qquad 4.$$

where the third parameter, L, is an undefined quantity that is not justified by theory and which can absorb some of the deviations from Equation 1. Among the sources of deviation is the possible existence of selective clines, which are totally unaccounted for theoretically and which should be eliminated before model fitting. There is no satisfactory procedure for identifying and correcting for such clines. The second reason is that the goodness of fit is never actually tested; it is, in fact, difficult to test because the regression of log kinship is calculated on all possible pairs of populations; hence, the observations are not independent, which violates a basic requirement for the validity of the goodness of fit test. As a rule, only the estimates of a and b—and not the observed points—are published, so that not even a visual test of deviation from linearity is possible.

The data reviewed by Morton (88) came from local populations, usually relatively isolated ones. Piazza & Menozzi (97) performed a reasonably independent test of the relationship between genetic similarities and distances on general data using the variogram approach (see also 28). An approximately linear increase with distance was observed for most genes up to some 1000 km or more. Interestingly, the slope of the variogram averaged over all genes and continents is similar to the average b value that Morton observed in local populations. Different continents have gross differences in average slopes, and differences of some importance also exist for different genes.

Interpretation of Constants

The constants a and b of the linear relationship between log kinship and geographic distance have a theoretical interpretation, indicated by Equations 2 and 3, valid for the one-dimensional case. It would be interesting to test the validity of these theoretical formulas in cases in which the basic quantities that define them can be estimated independently.

If valid, a and b might provide relatively simple ways of estimating quantities such as c, m, and N in Equations 2 and 3. Morton (87, 88) has attempted to estimate the necessary quantities from situations in which the values of a and b were known and some demographic information was available. The theory is probably inadequate, however, when immigration is not from nearest neighbors but from more distant populations (i.e. the migration distributions have long tails, as is often the case when means of transportation that allow long distances to be covered more easily are common). Morton (87) tried to solve this problem by separating the migration distribution into two parts: One segment lay above 4 standard deviations (4σ) and the other below. The choice of 4σ was arbitrary. Migrants from distances above 4σ were counted in the "long-range" migration and those from below 4σ in the "short-range." Long-range migration was then used as an estimate of c. Even if this attempt at correcting for the long tails were satisfactory, the migrational component of c is not the only one that should be considered. Mutation is negligible, but stabilizing selection for some genes may be an interesting and not insignificant component. A more important contributor to c is migration *unrelated to distance*. In many situations, different ethnic groups or social classes or castes often coexist in the same village or city. Their gene frequencies may differ and they may exchange gametes only rarely; but even the rare exchanges will be important because of the relatively large differences in gene frequencies.

The second source of error is that N is not usually a well-defined quantity and can vary with the size and nature of the group under consideration. The N in the present case should be of a *deme*, for which there exists, in human populations, no unique definition. According to Wright (126), it should be calculated on the basis of population density δ, as $N = 4\pi l^2 \delta$. Morton arbitrarily defines N as the average area corresponding to a radius equal to σ' (σ' = short-range migration) *divided* by 10 and multiplied by δ. The resulting value is clearly a much smaller number than Wright's.

A third difficulty is that the formula for the *a* constant is from one-dimensional theory, while the data used, and hence the population densities, are two-dimensional. Moreover, three quantities *(N,m,c)* have to be calculated from two *(a,b)*.

From the regressions of the calculated *a* and *b* values on the short- and long-range (called *m* in 88) migration and *N*, Morton concluded (88): (*a*) that the values of *b* depend only on the short-range migration σ' (which roughly corresponds to our *l*); (*b*) that the *a* values (using 2-dimensional population densities) can be used to perform the magic feat of calculating *both N* and *c*. Naturally, the estimated values of *N* and *c* have a correlation of 1.

Conclusion (*a*) seems to rest on firmer ground than (*b*), although there are inherent difficulties due to the lack of direct evaluations of all of the nonnegligible components of *c*. It is difficult, however, to give credence to estimates of *N* and *c* that depend on too many unwarranted assumptions.

An additional problem is that Morton does not recognize the relativity of the concept of *N*, which is used for the somewhat abstract and elusive unit *deme*. There are far fewer problems when *N* is related to a better defined entity, like a village, an administrative unit, an ethnic group, or even a species. It is clear that in general, *N* will vary from one entity to another. Instead, Morton criticizes other authors' (25, 118) estimates of *N*, which were calculated for whole ethnic groups, and hence presumably for a sum of demes, because they conflicted with his own estimates derived for demes on the basis of a totally different definition and for entirely distinct purposes.

Problems

The shape of the relationship between kinship and distance needs to be investigated more thoroughly. The following cross validation method for testing deviation from linearity (L. Moses, personal communication) has been suggested: Take out one colony (e.g. a village or island); fit a parabola to the residual data; repeat this process after taking out all other colonies one by one from the global data; estimate, using the usual jackknifing rules (85), the quadratic coefficient and its standard error; and test the null hypothesis that the average quadratic coefficient is significantly different from zero. It is also important to estimate *a* and *b* independently for different genes when more than one gene has been tested in the same population, as any differences among estimates of *a* and *b* may be of considerable interest. The jackknife method can also provide estimates of the standard errors of *a* and *b*. Those obtained by the usual methods cannot be trusted because of the nonindependence of the pairs of villages.

Clearly, interpretations of the constants (especially *a*) are less obvious. The ethnic heterogeneity of the surrounding populations, examined gene by gene, may be an important factor. A better definition and examination of *demes*, of migration distributions, of the relations among *a*, *b*, and F_{ST} values (the standardized variance of gene frequencies among populations; see 30) may be rewarding. Analyses of gene frequencies should be carried out on a gene by gene basis.

It should be acknowledged at this stage, however, that it is clear that there is a relation between genetic difference and geographic distance. It increases monotonically (for most genes and places) up to an average over a number of genes of roughly 1000 miles, placing geographic distance among the important evolutionary factors.

GENE FLOW

Large-scale migrations of populations sometimes occur (see 6 for a review on humans). If the migrants settle in an inhabited area, intermating can produce a hybrid population. Although this type of migration may be relatively rare, it can have a marked effect on genetic structure. Gene flow can also occur in small increments over many generations when two populations live relatively close together but exchange only a limited number of migrants. Thus, we can use gene frequencies to try to estimate the amount of migration that occurred in cases where we have a specific hypothesis about the genetic origin of a population. In doing so, it may be possible to identify genes that may have been subjected to selective change (121), since their frequencies would deviate markedly from the linear combinations of gene frequencies expected on the basis of the other loci (57, 99, 100, 103, 122).

Theory

In our discussion of the theory of gene flow, we will only treat the problem of estimating the parental contributions to a dihybrid population, since no additional insight is obtained from looking at trihybrid populations. The subscripts 1 and 2 refer to the two parental populations, and *H*, to the hybrid population.

In an ideal situation with no drift, mutation, selection, or unaccounted-for-migration, the frequency of an arbitrary allele in the hybrid is: $p_H = mp_1 + (1-m)p_2$, where *m* is the fraction of genes in the hybrid derived from parental population 1. Any or all of the above assumptions may be false, however. Mutation is unlikely to provide much of a complication; errors introduced by violations of the other assumptions can be loosely divided into errors in the parental gene frequencies, which are generally assumed to be absent, or in the hybrid gene frequencies.

Estimation

Estimates of m for individual loci may vary widely (5, 20, 29, 49, 51, 98–101). An important factor in this variability is the difference in gene frequencies between the parental populations: Estimates of m are less likely to vary substantially when the gene frequencies in the parental populations differ substantially. A weighted average of the individual estimates obtained by assuming no error in the parental gene frequencies can provide an overall estimate of admixture, where the weights are inversely proportional to the variances of the individual estimates (30); such weights may include variance due to drift, sampling error, or both.

There are a number of techniques that use all loci simultaneously to estimate m. Most of them ignore all sources of error in the parental populations, an assumption that is approximately correct if sampling error and drift are of a lower order of magnitude in both parental populations than in the hybrid. These techniques include maximum likelihood estimation (29, 41, 72), weighted or unweighted least squares (41, 101, 102), and gene identity (34) which have varying abilities to take into account different amounts of gene frequency error in the hybrid due to sampling and drift.

The simplest application of (unweighted) least squares and maximum likelihood assumes that the error at all loci in the hybrid is of constant magnitude. This assumption is appropriate only under two conditions: first, if sample sizes at all loci are the same and second, if the gene frequencies are transformed using the angular transformation (31) in order to stabilize the variances. This transformation, however, often introduces a bias into the estimate of m.

Weighted least squares or maximum likelihood (41) techniques permit solutions of m that weight loci inversely proportional to the variance at each locus. This variance can be due to sampling error and/or to drift. The gene identity method (34) provides a solution incorporating the first sort of variance but not the second.

The error in the gene frequencies of parental populations is not always negligible. One method that takes this source of error into consideration calculates m by standard least squares (13) from modified Mahalanobis distances (13, 14). This method weights the distances between pairs of populations by the inverse of the pooled variances of all three populations. The variances include both sampling error and drift.

Korey (70) has compared the above methods' ability to estimate m. By conducting a limited simulation of drift in the hybrid population, he obtained the mean and variance of m for several methods. Gene identity was by far the least reliable method, probably in part because it behaves poorly for gene frequencies that are similar in the parental populations (71). Least squares and maximum likelihood were slightly superior to the remaining methods. Since the simulation did not allow for error in the parental gene frequencies, however, it is not clear that these methods would continue to perform the best under more realistic conditions.

Another approach to the problem of estimating m when error in parental gene frequencies is considered is to find the line that provides the "best" fit for the three sets of gene frequencies, since the null hypothesis is that there is a linear relationship among the gene frequencies. The estimate of m is then obtained from the relative distances between the projections of the data onto this line. A maximum likelihood criterion of a best line (116) easily allows the inclusion of drift in any of the populations, but it only readily allows for sampling error when the sample sizes are constant at all loci in a particular population. Also, with this method, bias is introduced through the angular transformation. Such a bias can be eliminated by performing the analysis in Euclidean space with a minimum variance criterion for a best line (120a). This technique readily encompasses the variance introduced by unequal sample sizes at different loci, as well as drift, but it is computationally cumbersome.

The procedures outlined above for finding admixture all require a choice of two parental populations and gene frequency estimates in all populations. When drift is specifically included in the weighting scheme, estimates of the effective population sizes are also necessary. Most methods do not take into account sampling error in the parental gene frequencies. Bias in the choice of parental populations may also be important, but it is more difficult to correct for it. Misidentification of parental populations is likely to cause deviations from the expected values for a large number of gene frequencies in the hybrid, obscuring the effects of selection. When the parental populations are not identified correctly, the use of "ranked estimates" when multiple hybrid populations exist (21) is likely to give spurious indications of selection (81).

Measuring the effective population size may be difficult or impossible when hybridization occurred many generations previously. If historical information is available, the overall effective size may be taken to be the harmonic mean of the effective sizes over time (126); the effective size at any point in time is roughly 1/3 of the census size in human populations (30).

Problems

Although studies of gene flow may be one of the best ways of investigating the selection of certain genes in humans (e.g. Hb^S, ABO, etc.), multiple studies of the origins of certain ethnic groups (20, 25, 29, 65, 87, 88, 90, 91, 96, 116, 120a) have often indicated varying contributions from presumed parental populations, which demand explanation. In many cases, the available data seem to be inadequate, especially in one or more of the parental populations; appropriate samples need to be obtained in order to resolve some of these problems.

The number of competing techniques for estimating admixture indicates the need for criteria for choosing the technique most suitable for a particular application. Korey's simulation (70) was a start, but it did not allow for errors in the parental gene frequencies, selection, or any other complicating factors. It would be useful to do extensive simulations covering the range of possible sources of error and to examine both the means and variances for the competing methods.

CLINES

A gene frequency cline consists of a gradual change in gene frequencies over geographic distance. The effects of migration on such clines can be considered from two points of view: How does migration affect the existence and stability of clines? And what impact does it have on the dynamics or persistence of clines. Even though the existence of clines has been demonstrated through ample observational data (42, 67), almost all of the work on the interaction of migration and clines is theoretical. Therefore, the following discussion will be primarily concerned with theoretical results. The related topic of selection-migration structures and the conditions for a protected polymorphism has been reviewed by Karlin (64).

Theory

In addition to migration, the parameters that affect the stability of clines include selection, genetic drift, mutation, and linkage. The simplest models of migration balanced by selection make the following assumptions: (*a*) migration is isotropic (no directional bias occurs), (*b*) no mutation or drift occurs, (*c*) only one locus is under selection, (*d*) the habitat is one-dimensional and infinite, and (*e*) migration rates are uniform over the habitat and over different genotypes. These models then describe the effects of migration vs selection on clines by considering a number of attributes of clines—their shape, position, and width. The shape of a cline is determined by the maximum slope in gene frequencies; the position is the geographic location of the maximum slope; and the width is the inverse of the maximum slope, or in measureable terms, roughly the distance over which 60% of the maximum gene-frequency change occurs (83).

The major parameter necessary to quantify migration is the square root of the second moment of the migration distances about the origin, l. The distribution is unimportant in discussions of the stability of clines and their characteristics (108). The selection function can describe a gradual or a discrete change in selective coefficients over space; the results for one-locus, two-allele models with dominance are qualitatively similar to those results obtained with models assuming only additive effects (83, 92). The selective coefficients change over distance Δ. The characteristic distance, l_c, is a parameter that combines selection with l; $l_c = l/\sqrt{s}$, where s is the selection against one homozygote in a two-allele, additive model (83, 108).

If $l_c > \Delta$, the width of a stable cline with one change of sign in the selection function is l_c (83). Without overdominance, the gene frequencies at the ends of the cline should be 1 or 0. If $\Delta > l_c$, the width is $(l_c^2 \Delta)^{1/3}$ (83). Obviously, decreasing selection widens the cline, but not in a linear fashion. A decrease from $s = 0.1$ to $s = 0.01$ produces about a three-fold increase in the cline's width (52).

The variable l_c also describes the distance over which adaptation to local conditions can occur: A pocket of selective differences whose length is less than l_c cannot produce a cline (108). Similarly, if the habitat is bounded at one end, instead of being infinite, and if selection on the infinite side is of opposite sign and stronger than that on the bounded side, the stability of the cline is only ensured if the boundary is at least $\pi l_c/(2\sqrt{2})$ from the origin, or a little more than the characteristic length (92).

Genetic drift has little qualitative effect on clines. It widens the cline, but it does not strongly affect the covariance of gene frequencies among populations (44, 110). Genotype-dependent variability in migration rates also has few qualitative effects (94).

Anisotropy in migration rates can affect both the position and the width of clines. For $l_c > \Delta$ and small anisotropy $(2x_0/(l\sqrt{s}) < 1$ where x_0 is the average migration distance from the birthplace), the displacement of the position of the cline is $d = x_0/s$ (83). For larger anisotropy, displacement is more severe until sufficient anisotropy destroys the cline (93). In either case, the width of the cline increases (83). For $\Delta > l_c$, anisotropy has little effect on the width and causes a smaller displacement of $d = (x_0/s)(\Delta/l_c)^{2/3}$.

Linkage of genes where the recombination rate is at most of the same order of magnitude as the selective coefficient both can decrease the characteristic distance by $l'_c = l_c/\sqrt{k}$ for k linked loci under parallel selection and can steepen clines (109). If selection occurs in different directions for the loci, the cline shifts for both loci if selection is comparable for both; if selection on one locus is considerably stronger, only the cline of the weaker locus will shift (109). This shift also affects neutral genes linked to selected genes (15).

Little work has been done on the dynamics of clines. Simulation indicates that the time to reach equilibrium is not strongly dependent on the initial gene frequencies in a one-locus system (12). Linkage disequilibrium resulting from nonparallel selection lasts for many generations, however, and it is dependent on initial frequencies (117).

The effect of mutation on divergence of populations (without selection) is to increase the interpopulation variance of additive quantitative characters linearly over time (35). The rate of approach to the equilibrium divergence (m/s) measured with the negative log of inter- vs intraspecific kinship is

$(8N\mu + 1)/(4N)$ for populations of effective size N, where μ is the mutation rate (95).

Estimation and Data

The theory of clines clearly establishes which parameters are necessary for investigating the interaction between the stability of clines and migration. One important parameter is l, as estimated from parent-offspring migration, but for many of the populations studied in Table 1, mating distance is the only parameter estimated. For the populations in Table 1 for which both measurements exist, mating distance tends to be larger than parent-offspring distances by a factor of 1.26. Estimating the effective population size, N, and the bias in the direction of migration may also be useful. (See the section on demic diffusion for a discussion of estimating N.)

To estimate selection from an observed cline, however, it is not sufficient to know l. According to the existing theory, linkage and multiple directions of selection can displace and steepen clines or even produce clines where none should exist. To conclude that a cline is maintained by selection requires estimates of both the direction and the strength of selection. At best, we can hope to put an upper bound on the extent or distribution of selection. Drawing on the data in Table 1, for example, for a selection coefficient of $s = 0.1$, a cline of minimum width 150–200 km might be expected for the relatively mobile hunter-gatherer populations, the pygmies and !Kung. For more sessile agricultural populations, this width might drop to 20–60 km (e.g. in Parma, Sweden, the Pyrenees) or may be as high as 120–175 km (for some of the Indian populations). The range in cline widths may vary substantially among different subpopulations within a country: The Dhangar caste-group (77) in the Indian populations ranges from 45 to 351 km for $s = 0.1$; the New Guinea populations range 12–740 km. When selection is weaker, the clines are wider; for example, for $s = 0.01$, the minimum cline widths are approximately three times as wide as when $s = 0.1$. If the selection gradients are greater than the characteristic lengths, as may often be the case, the clines also will be wider than suggested above.

Problems

There are a number of open questions about the effects of migration on clines. Most important among these is the issue of the dynamics of clines. How long can ephemeral clines last under various conditions? Is the distribution of migration distances important? What predictions can be made for ephemeral clines determined by the interaction between migration and drift? Other interesting topics include two-dimensional clines and their interactions with other clines; the effects of linkage on such clines; and the nature of circular clines that arise because of population expansion, which will be discussed in the next section.

DEMIC DIFFUSION AND POPULATION EXPANSION

The expression *demic diffusion* indicates the gene flow accompanying population expansion. It was first used to compare the expansion of a technology that is spread by users who multiply and migrate with cultural diffusion (9). Demic diffusion is likely to take place when technological changes have a positive effect on the asymptotic population density (i.e. the "carrying capacity"). The models discussed here were developed specifically to study the spread of early farming techniques, which probably caused the greatest increase in carrying capacity in human development. Population densities reached under hunting-gathering or fishing economies were usually quite low. With the beginning of farming in the early neolithic, they increased at least tenfold (10, 55).

There are two major examples of demic diffusion linked with the spread of early farming: one from the Middle East to Europe 9000–5000 years ago (8, 10) and the other from Cameroon/Nigeria to Central and South Africa (the Bantu expansion), which took place over the last 3000 years (56). Areas of agricultural development in China and Southeast Asia and in Central and South America were probably at the center of similar expansions. In Europe, the expansion can be mapped fairly accurately using radiocarbon dating of the first arrivals of neolithic farmers at the major archeological areas of Europe (8, 9). The rate of advance of this technology averaged 1 km per year, with some local variations. The conclusion that this advance was due to demic diffusion was reached: (*a*) by noting that this rate of spread is compatible with ordinary demographic performances of the human species; (*b*) by evaluating the genetic expectations of a mixed demic/cultural spread with various assumptions about the degree of relative importance of the two; and (*c*) by analyzing the geographic gene frequency distribution in Europe.

The simplest model that can predict the spread of a growing population is Fisher's "wave of advance" model (45). Although designed to show the advance of an advantageous gene, it can also predict the progress of a population multiplying logistically and migrating at a constant rate (107). According to Fisher's formula,

$$\rho = 2 \sqrt{m\alpha}, \qquad\qquad 5.$$

where ρ is the rate of advance, m is the migration rate (as in a stepping-stone model; see also below), and α is the initial growth rate in a logistic model. Archeological information with respect to α is extremely limited and may even be contradictory (11); there is none on m. Values of m, observed in comparable ethnic situations, are consistent with the observed value of ρ of 1 km/yr.

Fisher's Equation 5 was offered as a solution to the differential equation:

$$dp/dt = md^2p/dx^2 + \alpha p (1-p), \qquad\qquad 6.$$

where p is the gene frequency in the original application of an advantageous gene and the population density is expressed as a fraction of the value at saturation in our application. x is space, t is time, and m is a measure of migration equivalent to a diffusion coefficient (the square of the standard deviation of Brownian movement in continuous space). With a discrete distribution, m can be estimated, as before, by the proportion of parents born ouside the area of birth of a child. Fisher's equation was recently solved for the one-dimensional case (4). The validity of Equation 6 in the two-dimensional case has been tested by simulation (30, 106; S. Rendine, A. Piazza, and L. Cavalli-Sforza, in preparation); approximate validity was found for a substantial range of migration values.

To model the spread of farming to an area previously occupied by hunter-gatherers (106), two stepping-stone models were set up, one for hunter-gatherers (with low population numbers) and a parallel one for farmers. In the simulation, a community of farmers occupies the same area as a hunting tribe but has a carrying capacity, M_f, of one to two orders of magnitude greater than M_h, the carrying capacity of the hunter-gatherers. Farmers and hunter-gatherers reproduce and exchange members with neighboring groups independently. Exchange within the same area, however, is limited and occurs only in the direction hunter-gatherers to farmers, that is, a small fraction of the hunter-gatherers becomes acculturated and takes up agriculture. This fraction is proportional to the number of farmers in the area, the number of hunters in the area, and an acculturation constant γ. To complete the model, estimates of the initial growth rates of farmers and hunter-gatherers (α_f and α_h), and the migration rates of hunter-gatherers to neighboring hunter-gatherer groups (m_h) and of farmers to neighboring farmer groups (m_f) must be supplied.

The first stage covers the events *within* each specific area, including growth and acculturation for every generation. After one generation, N_f and N_h change to N_f' and N_h' as follows:

$$N_h' = N_h + \alpha_h (1 - N_h/M_h) - \gamma N_h N_f$$

$$N_f' = N_f + \alpha_f (1 - N_f/M_f) + \gamma N_h N_f.$$

7.

For the stage involving migration between areas, we add i and j as suffixes to indicate the location of each area. As in the stepping-stone model:

$$N_{h,ij}' = (1-m)N_{h,ij} + .25m_h (N_{h,i-1,j} + N_{h,i+1,j} + N_{h,i,j-1} + N_{h,i,j+1})$$

$$N_{f,ij}' = (1-m)N_{f,ij} + .25m_f(N_{f,i-1,j} + N_{f,i+1,j} + N_{f,i,j-1} + N_{f,i,j+1}).$$

8.

At the beginning of the simulation, the hunter-gatherers had reached saturation everywhere, while farmers had only started as an "upgraded" hunter-gatherer group in some portion of the area. They will therefore spread to the whole area at rate of $\sqrt{2m_f}\ \alpha_f$. With finite numbers in each group, all hunter-gatherers will eventually convert to farming if M_f/M_h is sufficiently large.

Gene frequencies were included in the model so that we could predict the genetic composition of the population during and after the transition to agriculture. Differentiation among the original hunter-gatherer tribes was generated by random genetic drift. One allele was used as a "population tracer," i.e. it was given a frequency of 100% in the hunter-gatherer tribe originally converted to farming and 0% in all other tribes, and it was used as a marker of the proportion of "farmers' genes" among populations at the end of the transition. Additional time was allowed to elapse to test the flattening of gradients produced after the transition was accomplished.

In the first simulation (106), a relatively small matrix of populations was used (11 × 11), and attention was focused on the gradient of farmer genes being formed. The parameter of greatest importance was clearly γ, expressing the probability of acculturation, i.e. of the transition of hunter-gatherers to farming and their assimilation into the farmers' culture (or, as it might be also called, of the gene flow of hunter-gatherers into farmers). For low values of γ, farmers essentially replace hunter-gatherers over the whole territory; with high values, farmer genes remain confined to a small area around their original location. With intermediate values of γ, continuous gene frequency gradients are established.

When real genes are examined, there is not as great an initial difference on average between hunter-gatherer and farmer gene frequencies as with the population tracer; moreover, every hunter-gatherer tribe has its own gene frequencies. One can still hope to detect a latent pattern corresponding to the proportion of farmer genes by using a multivariate technique of analysis.

Patterns of potential significance can be discerned by employing principal components analysis on real data. In a recent and fairly realistic simulation (S. Rendine, A. Piazza, and L. Cavalli-Sforza, in preparation), the actual geography of Europe was imitated, although with modest resolution. There is an obvious need to keep the number of tribes and genes reasonably low to save computer time. Twenty biallelic genes were simulated. At the same time, three demic diffusions were started from different locations and at different times, roughly corresponding to those postulated from the analysis of actual data, with their effects superimposed on one other. The geographic maps of the first three principal components separated two of the three diffusions. The second principal component did not correspond to any demic diffusion but was clearly influenced by the geographic separation between North Africa and Europe.

The analysis of real data (84) showed that the first principal component corresponds to an elliptic cline centered in the Middle East that has a geographic distribution similar to the spread of farming. The second principal component showed an East-West gradient and was interpreted as a consequence of the many migrations from Central Asia that took place in the last 3000 years. The third principal component (an elliptic cline centered between

Poland and the Ukraine) was interpreted as an indication of either the places of origin of the barbarian invasions at the end of the Roman Empire or the cultures that some archeologists and linguists (50) place at the origin of the Indo-European invasions 5000 years ago (or both). An analysis limited to the HLA data employed by Menozzi et al (84) was carried out (111) by a totally independent method—using spatial autocorrelation methods—and it confirmed the original results. Maps of principal components, however, can only provide information on the places of origin of putative population expansions, but not on timing, which can only come from historical and archeological information.

Population expansions must have been frequent events in recent human evolution. The Caucasian expansion in the last six centuries is an example recorded in recent history. The geographic expansion of populations is usually the result of increased carrying capacity (it depends more on K than on r, to use standard ecological terminology and symbols), and both r and K depend not only on technology but also on social customs. While technology can change and diffuse quickly, social customs may be conserved more carefully and will not evolve so fast. Also, the acculturation constant may be lower for the transition from hunting to farming than for most other cases of technological advance, therefore favoring demic more than cultural diffusion. The genetic clines formed after these transitions are unstable, of course, and will eventually be destroyed by regular neighbor-to-neighbor (i.e. short-range) migration. But when the clines extend over long distances and originate from fairly large initial differences, they will be remarkably stable over time, as preliminary simulations have shown.

Literature Cited

1. Abelson, A. 1978. Population structure in the Western Pyrenees: Social class, migration and the frequency of consanguineous marriage, 1850 to 1910. *Ann. Hum. Biol.* 5:165–78
2. Abelson, A. 1979. Population structure in the western Pyrenees. I. Population density, social class composition, and migration, 1850–1915. *J. Biosocial Sci.* 11:353–62
3. Abelson, A. 1980. Population structure in the western Pyrenees. II. Migration, the frequency of consanguineous marriage and inbreeding, 1877 to 1915. *J. Biosocial Sci.* 12:93–101
4. Ablowitz, M., Zeppetella, A. 1979. Explicit solutions of Fisher's equation for a special wave speed. *Bull. Math. Biol.* 41:835–40
5. Adams, J., Ward, R. H. 1973. Admixture studies and the detection of selection. *Science* 180:1137–43
6. Adams, W. Y., Van Gerven, D. P.,

Levy, R. S. 1978. The retreat from migrationism. *Ann. Rev. Anthropol.* 7: 483–532
7. Alstrom, C. H., Lindelius, R. 1966. A study of the population movement in nine Swedish subpopulations in 1800–1849 from the genetic-statistical viewpoint. *Acta Genet. Stat. Med.* 16:1–44 (Suppl.)
8. Ammerman, A. J., Cavalli-Sforza, L. L. 1971. Measuring the rate of spread of early farming in Europe. *Man* 6:674–88
9. Ammerman, A. J., Cavalli-Sforza, L. L. 1973. A population model for the diffusion of early farming in Europe. In *The Explanation of Culture Change*, ed. C. Renfrew, pp. 343–57. London: Duckworth
10. Ammerman, A. J., Cavalli-Sforza, L. L. 1985. *The Neolithic Transition and the Genetics of Populations in Europe.* Princeton, NJ: Princeton Univ. Press. 169 pp. In press
11. Ammerman, A. J., Cavalli-Sforza, L.

L., Wagener, D. K. 1976. Towards the estimation of population growth in Old World pre-history. In *Demographic Anthropology*, ed. E. Zubrow, pp. 27–61. Santa Fe: Univ. N. Mex. Press

12. Antonovics, J. 1968. Evolution in closely adjacent plant populations. VI. Manifold effects of gene flow. *Heredity* 23:507–24

13. Balakrishnan, V. 1973. Use of distance in hybrid analysis. See Ref. 86, pp. 268–73

14. Balakrishnan, V., Sanghvi, L. D. 1968. Distance between populations on the basis of attribute data. *Biometrics* 24:859–65

15. Barton, N. H. 1979. Gene flow past a cline. *Heredity* 43:333–39

16. Basu, A. 1973. A note on the distribution of marriage distance among the Santals in the neighbourhood of Giridih, Bihar. *J. Biosocial Sci.* 5:367–76

17. Bateman, A. J. 1950. Is gene dispersion normal? *Heredity* 4:353–63

18. Beckman, L. 1961. Breeding patterns of a north Swedish parish. *Hereditas* 47:72–80

19. Beckman, L., Cedergren, B. 1971. Population studies in northern Sweden. I. Variations of matrimonial migration distances in time and space. *Hereditas* 68:137–42

20. Bjarnason, O., Bjarnason, V., Edwards, J. H., Fridriksson, S., Magnusson, M. et al. 1973. The blood groups of Icelanders. *Ann. Hum. Genet.* 36:425–58

21. Blumberg, B. S., Hesser, J. E. 1971. Loci differentially affected by selection in two American black populations. *Proc. Natl. Acad. Sci. USA* 68:2554–58

22. Bodmer, W. F., Cavalli-Sforza, L. L. 1975. The analysis of genetic variation using matrices. In *Genetic Distance*, ed. J. F. Crow, C. Denniston, pp. 45–61. New York: Plenum

23. Boyce, A. J., Kuchemann, C. F., Harrison, G. A. 1967. Neighborhood knowledge and the distribution of marriage distances. *Ann. Hum. Genet.* 30:335–38

24. Boyce, A. J., Kuchemann, C. F., Harrison, G. A. 1971. Population structure and movement patterns. In *Biological Aspects of Demography*, ed. W. Brass, pp. 1–9. New York: Barnes & Noble

25. Carmelli, D., Cavalli-Sforza, L. L. 1979. The genetic origin of the Jews: A multivariate approach. *Hum. Biol.* 51:41–61

26. Cavalli-Sforza, L. L. 1958. Some data on the genetic structure of human populations. *Proc. 10th Int. Congr. Genet.* 1:389–407

27. Cavalli-Sforza, L. L. 1962. The distribution of migration distances: Models, and applications to genetics. In *Les Deplacements Humains, Proc. 1st Entret Monaco Sci. Hum.* ed. J. Sutter, pp. 139–58

28. Cavalli-Sforza, L. L. 1983. Isolation by distance. In *Human Population Genetics: The Pittsburgh Symposium*, ed. A. Chakravarti. Stroudsberg, Pa: Dowden, Hutchinson & Ross. In press

29. Cavalli-Sforza, L. L., Carmelli, D. 1979. The Ashkenazi gene pool: Interpretations. In *Genetic Diseases Among Ashkenazi Jews*, ed. R. M. Goodman, A. G. Motulsky, pp. 93–102. New York: Raven

30. Cavalli-Sforza, L. L., Bodmer, W. 1971. *The Genetics of Human Populations*. San Francisco: Freeman

31. Cavalli-Sforza, L. L., Edwards, A. W. F. 1967. Phylogenetic analysis: Models and estimation procedures. *Am. J. Hum. Genet.* 19:233–57

32. Cavalli-Sforza, L. L., Hewlett, B. 1982. Exploration and mating range in African Pygmies. *Ann. Hum. Genet.* 46:257–70

33. Cavalli-Sforza, L. L., Kimura, M., Barrai, I. 1966. The probability of consanguineous marriages. *Genetics* 54:37–60

34. Chakraborty, R. 1975. Estimation of race admixture—a new method. *Am. J. Phys. Anthropol.* 42:507–12

35. Chakraborty, R., Nei, M. 1982. Genetic differentiation of quantitative characters between populations or species. I. Mutation and random genetic drift. *Genet. Res.* 39:303–14

36. Cheney, D. L., Seyfarth, R. M. 1983. Nonrandom dispersal in free-ranging vervet monkeys: Social and genetic consequences. *Am. Nat.* 122:392–412

37. Coleman, D. A. 1977. The geography of marriage in Britain, 1920–1960. *Ann. Hum. Biol.* 4:101–32

38. Devor, E. J. 1980. Marital structure and genetic isolation in a rural Hispanic population in northern New Mexico. *Am. J. Phys. Anthropol.* 53:257–65

39. Dobson, T., Roberts, D. F. 1971. Historical population movement and gene flow in Northumberland parishes. *J. Biosocial Sci.* 3:193–208

40. Dodinval, P. A. 1973. Distribution of matrimonial migrations in Belgium. *Hum. Hered.* 23:59–68

41. Elston, R. C. 1971. The estimation of admixture in racial hybrids. *Ann. Hum. Genet.* 35:9–17

42. Endler, J. A. 1973. Gene flow and population differentiation. *Science* 179:243–50

43. Eriksson, A. W., Fellman, J. O., Workman, P. L., Lalouel, J. M. 1973. Population studies on the Åland Islands. I.

Prediction of kinship from migration and isolation by distance. *Hum. Hered.* 23:422–33

44. Felsenstein, J. 1975. Genetic drift in clines which are maintained by migration and natural selection. *Genetics* 81:191–207

45. Fisher, R. A. 1937. The wave of advance of advantageous genes. *Ann. Eugen.* 7:355–69

46. Fisher, R. A. 1943. The relation between the number of species and the number of individuals in a random sample of an animal population. Part 3. *J. Anim. Ecol.* 12:42–58

47. Fix, A. G. 1974. Neighbourhood knowledge and marriage distance: The Semai case. *Ann. Hum. Genet.* 37:327–32

48. Francarro, M. 1959. Breeding structure of human populations. An analysis of distance. *Eugen. Q.* 6:32–35

49. Franco, M. H. L. P., Weimer, T. A., Salzano, F. M. 1982. Blood polymorphisms and racial admixture in two Brazilian populations. *Am. J. Phys. Anthropol.* 58:127–32

50. Gimbutas, M. 1973. The beginning of the bronze age in Europe and the Indo-Europeans: 3500–2500 B.C. *J. Indo-Eur. Stud.* 3:163–214

51. Glass, B., Li, C. C. 1953. The dynamics of racial intermixture—an analysis based on the American Negro. *Am. J. Hum. Genet.* 5:1–19

52. Hanson, W. D. 1966. Effects of partial isolation (distance), migration, and different fitness requirements among environmental pockets upon steady state gene frequencies. *Biometrics* 22:453–68

53. Harpending, H. 1976. Regional variation in !Kung populations. In *Kalahari Hunter-Gatherers*, ed. R. B. Lee, I. DeVore, pp. 152–65. Cambridge, Mass: Harvard Univ. Press

54. Harrison, G. A., Hiorns, R. W., Kuchemann, C. F. 1971. Social class and marriage patterns in some Oxfordshire populations. *J. Biosocial Sci.* 3:1–12

55. Hassan, F. A. 1981. *Demographic Archaeology*, pp. 8:39–50. New York: Academic

56. Hiernaux, J. 1975. *The People of Africa*, pp. 175–89. New York: Scribner's

57. Hiorns, R. W. 1979. The consequences of intermixture in human populations. *Adv. Appl. Probab.* 11:5–7

58. Hiorns, R. W., Harrison, G. A., Boyce, A. J., Kuchemann, C. F. 1969. A mathematical analysis of the effects of movement on the relatedness between populations. *Ann. Hum. Genet.* 32:237–50

59. Imaizumi, Y. 1977. A demographic approach to population structure in Gyoda and Hasuda, Japan. *Hum. Hered.* 27:314–27

60. Jacquard, A. 1974. *The Genetic Structure of Populations.* New York: Springer-Verlag

61. Jeffries, D. J., Harrison, G. A., Hiorns, R. W., Gibson, J. B. 1976. A note on marital distances and movement, and age at marriage, in a group of Oxfordshire villages. *J. Biosocial Sci.* 8:155–60

62. Jorde, L. B., Workman, P. L., Eriksson, A. W. 1982. Genetic microevolution in the Aland Islands, Finland. In *Current Developments in Anthropological Genetics*, ed. M. H. Crawford, J. H. Mielke, pp. 333–66. New York: Plenum

63. Deleted in proof

64. Karlin, S. 1982. Classifications of selection-migration structures and conditions for a protected polymorphism. *Evol. Biol.* 14:61–204

65. Karlin, S., Kenett, R., Bonne-Tamir, B. 1979. Analysis of biochemical genetic data on Jewish populations. II. Results and interpretations of heterogeneity indices and distance measures with respect to standards. *Am. J. Hum. Genet.* 31:341–65

66. Karlin, S., McGregor, J. 1972. Addendum to a paper of W. Ewens. *Theor. Popul. Biol.* 3:113–17

67. Kettlewell, H. B. D., Berry, R. J. 1969. Gene flow in a cline. *Heredity* 24:1–14

68. Kimura, M., Weiss, G. H. 1964. The stepping stone model of population structure and the decrease of genetic correlation with distance. *Genetics* 49:561–76

69. Kirkland, J. R., Jantz, R. L. 1977. Inbreeding, marital movement, and genetic isolation of a rural Appalachian population. *Ann. Hum. Biol.* 4:211–18

70. Korey, K. A. 1978. A critical appraisal of methods for measuring admixture. *Hum. Biol.* 50:343–60

71. Korey, K. A. 1979. Cherokee admixture and its estimation by the gene identity method: A critique. *Am. J. Phys. Anthropol.* 50:51–56

72. Krieger, H., Morton, N. E., Mi, M. P., Azevedo, E., Freire-Maia, A., Yasuda, N. 1965. Racial admixture in northeastern Brazil. *Ann. Hum. Genet.* 29:113–25

73. Kuchemann, C. F., Boyce, A. J., Harrison, G. A. 1967. A demographic and genetic study of a group of Oxfordshire villages. *Hum. Biol.* 39:251–76

74. Kuchemann, C. F., Harrison, G. A., Hiorns, R. W., Carrivick, P. J. 1974. Social class and marital distance in Oxford City. *Ann. Hum. Biol.* 1:13–27

75. Lasker, G. W. 1978. Relationships among the Otmoor villages and surround-

ing communities as inferred from surnames contained in the current register of electors. *Ann. Hum. Biol.* 5:105–11

76. Majumder, P. P. 1977. Matrimonial migration: A review, with special reference to India. *J. Biosocial Sci.* 9:381–401

77. Majumder, P. P., Malhotra, K. C. 1979. Matrimonial distance, inbreeding coefficient and population size: Dhangar data. *Ann. Hum. Biol.* 6:17–27

78. Malécot, G. 1950. Quelques schémas probabilistes sur la variabilité des populations naturelles. *Ann. Univ. Lyon Sci. Sect. A* 13:37–60

79. Malécot, G. 1967. Identical loci and relationship. In *Proc. 5th Berkeley Symp. Math. Stat. & Probab.* Vol. 4, *Biology and Problems of Health*, ed. L. Lecam, J. Neyman, pp. 317–32. Berkeley: Univ. Calif. Press

80. Malécot, G. 1973. Génétique des populations diploïdes naturelles dans le cas d'un seul locus. *Ann. Genet. Sel. Anim.* 5: 333–61

81. Mandarino, L., Cadien, J. D. 1974. Use of ranked migration estimates for detecting natural selection. *Am. J. Hum. Genet.* 26:108–10

82. Matheron, G. 1973. The intrinsic random functions and their application. *Adv. Appl. Probabl.* 5:439–68

83. May, R. M., Endler, J. A., McMurtrie, R. E. 1975. Gene frequency clines in the presence of selection opposed by gene flow. *Am. Nat.* 109:659–76

84. Menozzi, P., Piazza, A., Cavalli-Sforza, L. L. 1978. Synthetic maps of human gene frequencies in Europeans. *Science* 201:786–92

85. Miller, R. G. 1974. The jackknife—a review. *Biometrika* 61:1–15

86. Morton, N. E., ed. 1973. *Genetic Structure of Populations.* Honolulu: Univ. Hawaii Press

87. Morton, N. E. 1977. Isolation by distance in human populations. *Ann. Hum. Genet.* 40:361–65

88. Morton, N. E. 1982. Estimation of demographic parameters from isolation by distance. *Hum. Hered.* 32:37–41

89. Morton, N. E., Harris, D. E., Yee, S., Lew, R. 1971. Pingelap and Mokil Atolls: Migration. *Am. J. Hum. Genet.* 23:339–49

90. Morton, N. E., Kenett, R., Yee, S., Lew, R. 1982. Bioassay of kinship in populations of Middle Eastern origin and controls. *Curr. Anthropol.* 23:157–67

91. Morton, N. E., Dick, H. M., Allan, N. C., Izatt, M. M., Hill, R., Yee, S. 1977. Bioassay of kinship in northwestern Europe. *Ann. Hum. Genet.* 41:249–55

92. Nagylaki, T. 1975. Conditions for the existence of clines. *Genetics* 80:595–615

93. Nagylaki, T. 1978. Clines with asymmetric migration. *Genetics* 88:813–27

94. Nagylaki, T., Moody, M. 1980. Diffusion model for genotype-dependent migration. *Proc. Natl. Acad. Sci. USA* 77:4842–46

95. Nei, M., Feldman, M. 1972. Identity of genes by descent within and between populations under mutation and migration pressures. *Theor. Popul. Biol.* 3:460–65

96. Patai, R., Patai-Wing, J. 1975. *The Myth of the Jewish Race.* New York: Scribner's

97. Piazza, A., Menozzi, P. 1984. Geographic variation in human gene frequencies. *Syst. Zool.* In press

98. Pollitzer, W. S. 1958. The Negroes of Charleston (S.C.): a study of hemoglobin types, serology and morphology. *Am. J. Phys. Anthropol.* 16:241–63

99. Reed, T. E. 1969. Critical tests of hypotheses for race mixture using Gm data on American Caucasians and Negros. *Am. J. Hum. Genet.* 21:71–83

100. Reed, T. E. 1969. Caucasian genes in American Negroes. *Science* 165:762–68

101. Roberts, D. F., Hiorns, R. W. 1962. The dynamics of racial intermixture. *Am. J. Hum. Genet.* 14:261–77

102. Roberts, D. F., Hiorns, R. W. 1965. Methods of analysis of the genetic composition of a hybrid population. *Hum. Biol.* 37:38–43

103. Saldanha, P. H. 1957. Gene flow from white into Negro populations in Brazil. *Am. J. Hum. Genet.* 9:299–309

104. Saugstad, L. F. 1977. The relationship between inbreeding, migration and population density in Norway. *Ann. Hum. Genet.* 40:331–41

105. Serjeantson, S. 1975. Marriage patterns and fertility in three Papua New Guinean populations. *Hum. Biol.* 47:399–413

106. Sgaramella-Zonta, L., Cavalli-Sforza, L. L. 1973. A method for the detection of a demic cline. See Ref. 86, pp. 128–35

107. Skellam, J. 1973. The formulation and interpretation of mathematical models of diffusionary processes in population biology. In *The Mathematical Theory and Dynamics of Biological Populations*, ed. M. Bartlett, R. Hiorns, pp. 63–86. London: Academic

108. Slatkin, M. 1973. Gene flow and selection in a cline. *Genetics* 75:733–56

109. Slatkin, M. 1975. Gene flow and selection in a two-locus system. *Genetics* 81:787–802

110. Slatkin, M., Maruyama, T. 1975. Genetic drift in a cline. *Genetics* 81:209–22

111. Sokal, R. R., Menozzi, P. 1982. Spatial autocorrelations of HLA frequencies in Europe support demic diffusion of early farmers. *Am. Nat.* 119:1–17
112. Sokal, R. R., Wartenberg, D. E. 1983. A test of spatial autocorrelation analysis using an isolation-by-distance model. *Genetics* 105:219–37
113. Spuhler, J. N., Clark, P. J. 1961. Migration into the human breeding population of Ann Arbor, Michigan, 1900–1950. *Hum. Biol.* 33:223–36
114. Sutter, J., Tran-Ngoc-Toan. 1957. The problem of the structure of isolates and of their evolution among human populations. *Cold Spring Harbor Symp. Quant. Biol.* 22:379–83
115. Swedlund, A. C. 1972. Observations on the concept of neighbourhood knowledge and the distribution of marriage distances. *Ann. Hum. Genet.* 35:327–30
116. Thompson, E. A. 1973. The Icelandic admixture problem. *Ann. Hum. Genet.* 37:69–80
117. Thomson, G., Bodmer, W. F., Bodmer, J. 1976. The HL-A system as a model for studying the interaction between selection, migration, and linkage. In *Population Genetics and Ecology*, ed. S. Karlin, E. Nevo, pp. 465–98. New York: Academic
118. Wagener, D., Barakat, R., Cavalli-Sforza, L. L. 1978. Ethnic variation of genetic disease: Role of drift for recessive lethal genes. *Am. J. Hum. Genet.* 30:262–70
119. Weiss, G. H., Kimura, M. 1965. A mathematical analysis of the stepping stone model of genetic correlation. *J. Appl. Probab.* 2:129–49
120. Wijsman, E. M., Zei, G., Moroni, A., Cavalli-Sforza, L. L. 1984. Surnames in Sardinia. II. Computation of migration matrices from surname distributions in different periods. *Ann. Hum. Genet.* 48:65–78
120a. Wijsman, E. M. 1984. Techniques for estimating genetic admixture and applications to the problem of the origin of the Icelanders and the Ashkenazi Jews. *Hum. Genet.* In press
121. Workman, P. L. 1968. Gene flow and the search for natural selection in man. *Hum. Biol.* 40:260–79
122. Workman, P. L., Blumberg, B. S., Cooper, A. J. 1963. Selection, gene migration and polymorphic stability in a U.S. white and Negro population. *Am. J. Hum. Genet.* 15:429–37
123. Wright, S. 1943. Isolation by distance. *Genetics* 28:114–38
124. Wright, S. 1946. Isolation by distance under diverse systems of mating. *Genetics* 28:114–38
125. Wright, S. 1951. The genetical structure of populations. *Ann. Eugen.* 15:322–54
126. Wright, S. 1969. *Evolution and the Genetics of Populations.* Vol. 2, *The Theory of Gene Frequencies*, pp. 210:302–3. Chicago: Univ. Chicago Press
127. Yasuda, N. 1975. The distribution of distance between birthplaces of mates. *Hum. Biol.* 47:81–100
128. Yasuda, N., Kimura, M. 1973. A study of human migration in the Mishima district. *Ann. Hum. Genet.* 36:313–22
129. Zei, G., Matessi, R. G., Siri, E., Moroni, A., Cavalli-Sforza, L. L. 1983. Surnames in Sardinia. I. Fit of frequency distributions for neutral alleles and genetic population structure. *Ann. Hum. Genet.* 47:329–52

Ann. Rev. Ecol. Syst. 1984. 15:303–28

FLOW ENVIRONMENTS OF AQUATIC BENTHOS

A. R. M. Nowell and P. A. Jumars

School of Oceanography, WB–10, University of Washington, Seattle, Washington 98195

INTRODUCTION

Appreciation of the importance of fluid and particle motions to aquatic organisms is a precept of limnology and biological oceanography. The incentives to understand the details of these motions near sea, river, and lake beds, as well as around particles settling through the water column, are growing steadily. The recruitment of microbes and invertebrate larvae to surfaces of all kinds is controlled or strongly modulated by boundary-layer flow dynamics (18, 21, 30). The fluxes of dissolved nutrients and wastes to and from attached algae depend on the details of the flow regime (55, 134). The survival and growth of suspension feeders hinge upon the near-bottom—horizontal as well as vertical—flux of particles; less obviously, the same is true of many kinds of deposit feeders (80). The ability to track scents (60, 140) or detect vibrations (53, 91) also depends upon the details of near-bed fluid motions. Furthermore, any organism protruding above the bed must withstand forces whose magnitudes and directions depend on the details of the local flow (131, 132).

The scales of these ecological phenomena range from that occupied by a single individual to that inhabited by a small population. On the smaller end of the scales we will treat, viscous flows predominate, and progress is being made, for example, in studying the important details of the flow around feeding appendages (2). On the larger end, there is a voluminous literature on geophysical flows (98). It is especially difficult, however, to find literature relevant to the intermediate scales and to extract those parts with strong ecological implications. The need to do so is critical for both designing and interpreting manipulative laboratory and field experiments—including those with microbes—which are largely conducted on an intermediate scale.

303

0066-4162/84/1120-0303$02.00

Our aim is threefold. First, we hope to provide access to the ecologically relevant fluid dynamic literature concerning the above range of scales. Second—and most importantly—we wish to emphasize the parameters that need to be measured or controlled in order to describe or simulate a natural flow regime adequately. Few ecologists would attempt to characterize a normal distribution without specifying both a mean and a variance, yet many aquatic scientists try to characterize complex flows using single velocity measurements or, at best, single Reynolds numbers. Before beginning an ecologically oriented, fluid dynamic study, it is essential to determine the number and nature of the critical flow parameters (Figure 1). Otherwise, reproducibility (i.e. dynamical similarity; see 67) is difficult to achieve. Third, we will review the literature establishing the ecological significance of benthic flow environments and microenvironments. Due to space limitations, our intent is to provide representative and selective, rather than exhaustive, coverage and to highlight fertile areas for future research.

We will begin by placing benthic microenvironments in context by summarizing what is known in general about bottom-boundary-layer structure, including flow around settling particles as a special case. We then treat the major categories of flow microenvironments within bottom-boundary layers: flow around obstacles (e.g. stones or animal tubes), flow around and through meshes (e.g. grass beds and cages), and flow over and in depressions (e.g. stingray feeding pits or feeding depressions produced by deposit feeders). For each category and for the macroenvironments, we characterize the flow, establish its ecological relevance (e.g. to processes of recruitment or feeding), and discuss both its laboratory simulation and its bearing upon past and future field manipulations. The characterizations are not fully derived herein; we refer the interested reader to the cited literature. We also caution that we follow the oceanographic rather than the engineering convention for symbols (see Appendix 1) and assume a familiarity with simple Reynolds numbers and the terms in them (130).

BOTTOM–BOUNDARY–LAYER STRUCTURE

Fully Developed, Uniform, Steady Flow

Analytic solutions of the (Navier-Stokes) equations that describe the flow near a solid boundary exist only for the simplest situation of steady (no temporal variation in the outer flow forcing), uniform (no spatial variation in the properties of the boundary), entirely laminar flow (107). Virtually all boundary-layer flows of aquatic interest are turbulent, at least over most of their regions. Even approximate solutions of the equations for turbulent cases require strict assumptions. The simplest case is based on an assumption of strictly one-dimensional flow with no temporal variation (i.e. the mean of turbulent velocity fluctuations is zero), where gradients in velocity exist only in the z direction. One length and one velocity scale are needed to describe the flow (136). Both for consistency and for ease of conversion into a measure of the shear stress acting on the bed (Figure 1), the friction velocity, u_*, is typically used as the velocity parameter. For a flat bed, the grain diameter, D, is the obvious choice for a length scale, since it is the grain roughness that extracts momentum from the fluid. These parameters combine naturally into a roughness Reynolds number, Re_*, that is fully adequate to characterize such simple flows near the bed. That flow velocity at one height alone is insufficient is illustrated by the matching velocities at 10 cm above the two beds in Figure 1, Case 1.

An alternate means of characterizing these flows comes from what is usually the most practical means of estimating u_*, namely regression. Mean velocity, measured at a series of distances from the bed, is plotted against the logarithm of z (e.g. Figure 5 in 89); the z intercept is labelled z_0 and is called the hydraulic roughness, boundary roughness, or simply roughness length. The slope of this plot can be used to estimate u_* (89). Intuitively, z_0 is the height at which the mean velocity would drop to zero if the log-linear relation between z and U held so close to the bed. Only at high Re_*'s is there a simple linear relationship between D and z_0 (Figure 1).

A sublayer next to the bed and present only for small Re_*'s is characterized by long periods of quiescent, viscous flow interrupted occasionally by injections and eruptions of turbulent fluid as the large, energetic eddies in the outer part of the turbulent boundary layer penetrate all the way to the bed (13). In the viscous sublayer, the velocity is proportional to the distance from the bed, and the stress is constant. In the lowermost portion of the viscous sublayer, there is a diffusive sublayer, and the only vertical motion in it occurs via molecular diffusion. Above the viscous sublayer or throughout boundary layers with high Re_*'s, flow is dominated by turbulent eddies, with high exchange rates of momentum and consequently high rates of mixing and diffusion. This near-bed, turbulent layer—the logarithmic region (or inertial layer; cf 119)—has a level of turbulent shear stress that decreases linearly with distance from the bed (Figure 1).

When the roughness of the boundary increases (z_0 gets larger) or the overlying flow velocity rises, the region of viscous influence shrinks and eventually disappears altogether (when $Re_* > 100$, i.e. when there is fully rough, turbulent flow). High rates of momentum and contaminant exchange predominate, and the stresses in the fluid are entirely turbulent, in contrast to the viscous stresses that dominate the momentum flux in a viscous sublayer. This hydraulically rough flow is common in fluvial boundary layers and occurs most frequently in the ocean during periods of active sediment transport. Close to the

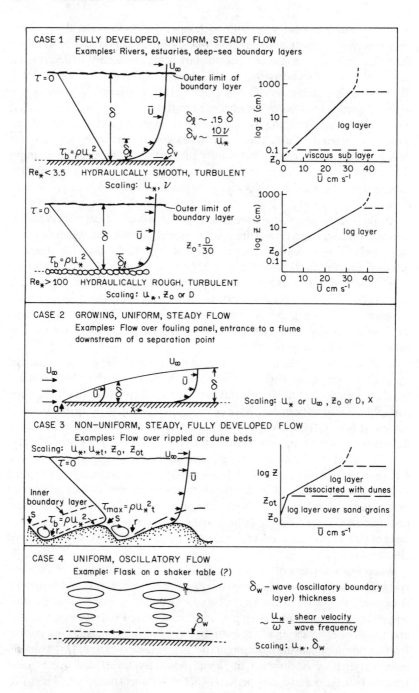

CASE 1   FULLY DEVELOPED, UNIFORM, STEADY FLOW
Examples: Rivers, estuaries, deep-sea boundary layers

$\tau = 0$

U_∞

Outer limit of boundary layer

δ

$\bar{U}$

$\delta_\ell \sim .15\,\delta$

$\delta_v \sim \dfrac{10\,\nu}{u_*}$

$\tau_b = \rho u_*^2$

δ_ℓ δ_v

$\mathrm{Re}_* < 3.5$ HYDRAULICALLY SMOOTH, TURBULENT
Scaling: $u_*,\ \nu$

1000

log $\bar{z}$ (cm)

10

0.1
z_0

log layer

viscous sub layer

0   10   20   30   40
$\bar{U}$ cm s^{-1}

$\tau = 0$

Outer limit of boundary layer

δ

$\bar{U}$

$z_0 = \dfrac{D}{30}$

$\tau_b = \rho u_*^2$

δ_ℓ

$\mathrm{Re}_* > 100$ HYDRAULICALLY ROUGH, TURBULENT
Scaling: $u_*,\ z_0$ or D

1000

log $\bar{z}$ (cm)

10

z_0
0.1

log layer

0   10   20   30   40
$\bar{U}$ cm s^{-1}

CASE 2   GROWING, UNIFORM, STEADY FLOW
Examples: Flow over fouling panel, entrance to a flume
downstream of a separation point

U_∞

U_∞

$\bar{U}$ δ

$\bar{U}$ δ

a
x

Scaling: u_* or U_∞, z_0 or D, X

CASE 3   NON-UNIFORM, STEADY, FULLY DEVELOPED FLOW
Examples: Flow over rippled or dune beds
Scaling: $u_*,\ u_{*t},\ z_0,\ z_{0t}$

$\tau = 0$ U_∞

$\bar{U}$

Inner
boundary layer $\tau_{max} = \rho u_{*t}^2$

$\tau_b = \rho u_*^2$ s

s   r

log $\bar{z}$

$\bar{U}$

log layer
associated with dunes

z_{0t}
z_0

log layer over sand grains

$\bar{U}$ cm s^{-1}

CASE 4   UNIFORM, OSCILLATORY FLOW
Example: Flask on a shaker table (?)

δ_w

δ_w – wave (oscillatory boundary
layer) thickness

$\sim \dfrac{u_*}{\omega} = \dfrac{\text{shear velocity}}{\text{wave frequency}}$

Scaling: $u_*,\ \delta_w$

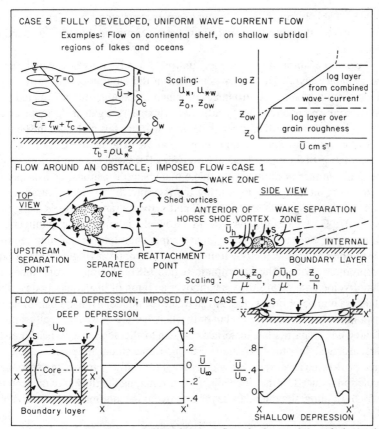

Figure 1   Examples of the five types of boundary-layer flows for the two classes of microenvironments observed in marine and freshwater environments. Note that not even in Case 1 will measurements of velocity at a single height adequately characterize the flow, especially the shear exerted on the boundary. Flow separation and reattachment (127) are observed in most aquatic boundary layers and have dramatic effects on fluxes near the bed.

bed, there are strong velocity gradients in all directions as the flow accelerates and decelerates over and around obstacles. Thus, within approximately three particle diameters ($3D$) of the boundary, the fluid is no longer characterized by a one-dimensional model, and the simple logarithmic profile cannot be expected to hold (87). To measure the boundary friction velocity under these circumstances, one must resort to velocity profiles made within the logarithmic region, using enough measurement points to achieve respectable confidence limits on the regression parameters of interest, namely u_* and z_0 (48). In gravel streams, the boulders often extend all the way through the flow depth, or the flow in riffles is less than three boulder scales thick. While there are methods for estimating u_* under such circumstances, average boundary skin friction would probably be much less relevant to any organism on a boulder than would the complex, three-dimensional flow pattern induced by the surrounding topography.

Flows of the simplest type, namely uniform, steady, fully developed, turbulent boundary layers, are easily generated in the laboratory and provide the simplest fluid dynamic environment in which to examine the postulated relations among momentum flux, particulate flux, and organism growth or behavior. Because we are dealing with the flow within a few meters of the bed, the inherent length scales are small enough to allow us to ignore Coriolis effects due to rotation of the earth. Hence, flow over a flat surface is similar to flow in a pipe. This similarity allows us to make many simplifications, as it is easy to determine u_* in a pipe by measuring the pressure drop (67). Nikuradse (86) developed the basic data set relating D and z_0 over a range of Re_*'s using pipe flow (107; C. E. Long & A. R. M. Nowell, submitted for publication). Estimating u_* over a flat surface in either the laboratory or the field requires measurements of the velocity within the viscous sublayer (Figure 1) or several measurements within the logarithmic region (48). Obviously, one measurement of velocity at an arbitrary height over a region of inadequately described roughness will be of little value. An alternate technique, using a flush-mounted sensor in the bed, works well—but only when the flow is hydraulically smooth (6) and when no particulate material is present to alter the sensor's heat-transfer characteristics. In order to simulate a one-dimensional boundary layer, the flume or flow channel must be wide enough ($> 7 \delta$) to prevent the secondary circulation induced by the sidewalls from dominating the flow and long enough ($> 50 \delta$) to allow the boundary layer to grow to its equilibrium thickness (see below).

Many empirical studies indicate that there is a correlation between Re_* and benthic community structure. Perhaps the most frequent generalization about both marine and freshwater benthos is that community structure varies with sediment type (58, 73, 84, 95). The level (and temporal pattern) of u_* determines, to a substantial extent, grain size and sorting (137). While the implicit chain of cause and effect is generally from flow regime through sediment type to community structure, there is no reason to believe that this sequence is the only one possible. The flow regime (Re_*) can affect the community structure much more directly, even in the absence of sediment transport, by altering the fluxes of dissolved materials to organisms on and in the bed (141). Organisms can, in fact, significantly change both z_0 and the response of sediments to an imposed shear stress (66), making assumptions regarding cause and effect even more risky.

Studies that succeed in establishing a mechanism for flow effects on benthos are far rarer than correlative approaches. Most notably, nutrient supply and uptake rate in attached algae have been shown to be strongly dependent (135) upon near-bed turbulent mixing rates (Re_*). Pipe flow has been used to show that there is an optimal u_* for the initial substratum-exploration behavior of barnacle cyprids and that this optimum differs among species (21). Pipe flow has also been used to simulate flow and its interactions with fouling bacteria in a heat exchanger (18). Results from pipe flow should be generalizable to one-dimensional boundary layers in the field. For bare substrata in an initially hydraulically smooth flow ($Re_* < 3.5$), bacterial population growth (i.e. immigration plus birth) is a function of u_* (18, 97). In a predator-free system, an equilibrium microbial film thickness is established (with birth and immigration balancing erosive emigration), which decreases with increasing u_* (18). It would be extremely interesting to follow up these experiments under a range of boundary roughnesses and to use techniques that resolve microbial immigration from local colony growth (9, 69).

Field manipulation of boundary-layer structure is in a primitive state. Grain diameter, D, is the easiest parameter to vary, altering z_0 and Re_* in a predictable fashion. One must alter a sufficiently large area and work far enough from the edge of the region to ensure that the flow is fully developed and uniform. We are unaware of any experiments based on deliberate alteration of the soft-sediment grain size aimed at monitoring changes in the boundary-layer structure and the consequent, flow-mediated effects on biota. Dredge-spoil disposal frequently produces such modifications, but it has many other confounding effects. Using substrata with varying surface roughnesses has been a relatively frequent practice in successional studies of aquatic fouling communities (36, 100, 115). Consequent modifications of fluid boundary-layer structure have not been monitored, and procedures have often confounded changes in chemical composition with changes in hydraulic roughness. Surface chemistry plays a strong role in the initial phases of adhesion to substrata (4). Controlled experiments that begin with a constant surface chemical composition and then vary surface roughness in order to monitor the effects on boundary-layer structure and ensuing colonization would be especially informative. Blinn et al's (7) finding that enhanced early colonization is correlated with greater surface roughness (z_0), and hence turbulent exchange, among three natural (but chemically disparate substrata) is particularly tantalizing.

Devices that modify near-bed flow velocity have also been used. The prototypes have been relatively short, open-ended, and primitively instrumented with respect to flow measurement. Nonetheless, they have provided valuable data on sediment entrainment (138) and have led to more sophisticated devices—in terms of both flow manipulation and measurement of the resulting flow structure—for in situ use (88). All of these devices use energy provided by pumps. Where current directions are reasonably predictable, it is also possible to emplace simple weirs or channels that alter the local flow's geometry and intensity. J. E. Eckman (personal communication) used such a device to increase u_* locally and produce ripple migration in a small section of an intertidal sand flat. No data have been published yet on the ecological consequences of such in situ modifications, but the potential is great.

Growing, Uniform, Steady Flow

Boundary layers require finite distances to grow to equilibrium thicknesses (107). Fouling plates with an edge facing upstream; many sections of topographically rough, natural bottoms; and short ($<$ 50 δ) flumes or flow tanks thus have boundary-layer characteristics that vary with downstream distance as well as with distance from the boundary. Hence Re_* and an additional length scale, namely, the distance (x) downstream from the point at which boundary-layer growth begins, are needed to characterize the flow. Growing, turbulent boundary layers exhibit a growth rate of $(x/\delta)^{0.8}$ (99, 124). At any fixed downstream distance within the boundary layer, the flow is characterized by the same parameters as in the one-dimensional case. The boundary shear stress decreases rapidly, however, from a maximum at the leading edge (or at the point where boundary-layer growth begins) to the equilibrium value that develops approximately 50 (equilibrium) boundary-layer thicknesses downstream.

Sudden changes in boundary roughness, from smooth to rough or vice versa, are common in marine and fluvial environments. The momentum field takes a certain distance to relax to the new boundary conditions, and the transition from smooth to rough occurs more rapidly than the converse (3, 124). In the field, a sufficient upstream distance of uniform roughness must exist before the flow can be characterized by only u_* and z_0, and this distance increases the further one is from the bed. Cummins & Lauff's (23) data require reinterpretation in light of this fact. They found that some species of stream invertebrates' apparent preference for settlement in certain particle sizes depends on the upstream-downstream order of the trays of graded sediments in their laboratory flow tank. They also ran some experiments with varying sediment structures in the cross-stream direction (silt on one side of the tray but not on the other); this pattern of varying roughness can cause subtle but important changes in near-bed fluid circulation (78).

The growing boundary layer presents problems in the laboratory. In order to measure the drag on an obstacle fixed on the bed—the drag equals the integral of the momentum extracted by the obstacle—the ratio of the obstacle height to the boundary-layer thickness must be similar to that found in the field. In the architectural aerodynamics literature, examples abound where the drag on simulated high-rise buildings has been miscalculated because an adequate ratio of boundary layer thickness to obstacle height was not maintained (cf comments by Gartshore (40) on Lee & Soliman (72)). When a flume less than 50δ

long is used (1, 26), the results of a one-dimensional characterization are not reliable quantitatively.

Growing boundary layers have the most immediate ecological applications to the design and interpretation of fouling experiments. The effect of area on species diversity is an important case in point. It often has been assumed that the only difference among fouling plates of different dimensions is the area available for colonization (93). Small fouling plates, however, may never develop a boundary layer of equilibrium thickness, and boundary-layer edge effects will cover a proportionately large area (61). One ecologically important consequence is that microbes or suspension feeders near the leading edge of a plate will not experience diffusional limitations of dissolved or particulate nutrients, while organisms hidden within a fully developed diffusive sublayer may (51).

A fouling plate dipped down in the flow (i.e. whose downstream edge is higher than its upstream edge) at angles less than 10° off the horizontal will produce a negative pressure gradient and accelerate the flow within the boundary layer over the plate. The boundary layer will thin, and the shear stress on the upper surface of the plate will be greater than that on a plate parallel to the flow. If the plate tips up at shallow angles, there will be a deceleration, but as this angle gets larger the flow will tend to separate; no simple boundary layer model can be used to describe such a momentum regime (17). Such arbitrarily complex and ill-characterized flow regimes and the sedimentation patterns they produce are likely to be more important (77, 114) than the substratum angle per se (52) in determining directions and rates of colonization. Orientation effects (i.e. of bare substrata relative to the flow) are well established in the periphyton literature (82, 125), but the degree to which streamwise variations in boundary-layer thicknesses are involved has yet to be determined.

A less obvious but perhaps ecologically more important implication of growing boundary layers is related to the cost-benefit trade-offs of microbes' attachment to particles or surfaces. Without *extreme* measures (e.g. water jacketing and the exclusion of any open surfaces that might lead to evaporation), convective circulations are set up in most containers of fluids. A boundary layer grows from the point at which a circulation cell meets the wall of the container to the point at which it detaches, but due to the small size of most containers and of the even smaller convection cells, this boundary layer is generally thin. Thus, microbes attached to the walls enjoy enhanced rates of nutrient supply and waste diffusion. We suspect that this essentially passive means of continuing gain explains the major portion of the so-called solid surface effect in aquatic microbiology (62). Similarly, microbes attached to particles can experience relatively large fluxes of nutrients and wastes (96). Flow past a small, isolated particle (e.g. one that is settling or that projects into fully rough, turbulent flow) can generate only a thin boundary layer, but particles of the size and specific gravity of bacteria have virtually no gravitational or inertial motion with respect to the surrounding fluid, and their own swimming velocities usually are not high enough to yield substantial nutrient gains over those provided by molecular diffusion (103).

A related problem plagues the interpretation of field results from benthic flux chambers or bell jars. Such chambers are generally equipped with pumps or stirring motors to prevent stagnation (139). The motors again set up circulation cells that are small compared with the distance required for growth of a boundary layer of equilibrium thickness over the bottom, biasing the measured fluxes upward. The relative insensitivity of measured fluxes to stirring speed (105, Figure 7) is thus a weak argument for accuracy. Designs for flux chambers that produce more natural boundary-layer structure are now available (88), but they have yet to be applied to geochemical flux measurements.

Nonuniform, Steady, Fully Developed Flow

As the flow field becomes more complicated over a more complex boundary, the number of independent scaling parameters rises. Hence, the number of measurements required to characterize such a complex environment adequately also increases. In recent years, considerable theoretical work has been devoted to describing nonuniform flow fields, which are common in nature. The flow over sand waves is a well-studied case of nonuniform turbulent flow, and in many fluvial and marine environments, the additional conditions of steady and fully developed flow are adequately met (111). The extra scaling parameters come from increased resistance by the large obstacles to flow. Flow over dunes separates from the bottom. The lee region has a recirculating flow (Figure 1), and this separation increases the drag on the flow. At the same time, the individual grains of the bed still extract momentum, and thus two roughness lengths and two shear velocities can be identified. Far away from the bed, the resistance to flow is produced primarily by the large-scale separation flow over the dunes, and hence we find large u_{*t} values. Within the 1000 grain diameters of the bed the velocity profile will yield a shear velocity, u_*, that is determined, as in simpler boundary layers, by skin friction on the sediment grains. By using spatial averages over the dominant topographic scale (say the wavelength of a dune), the flow field can be simplified (111).

Flow over dune fields oriented normal to the flow exhibits predictable velocities and momenta. The boundary shear stress profiles over dunes have been investigated in detail, though the structure of the diffusive sublayer has not been. The abundances of megafauna (123), macrofauna (28) and meiofauna (54, 106) vary spatially with a periodicity comparable to that of the local bedforms, which implies that they have some ecological link with bedform (42) or flow dynamics, but cause and effect have not been clearly identified and verified.

Ambühl (1) used flow visualization techniques that resolve the u and w components of velocity over objects and series of objects mounted on the bottom of a flume. Qualitatively, his results resemble those for flows over dunes. His quantitative results are difficult to interpret or generalize, however, because the flume was not long enough to allow equilibrium boundary-layer development.

Fully Developed, Uniform, Oscillatory Flow

The simplest case of this class of flow is a wave boundary layer developed over a flat bottom under a train of monochromatic waves (15, 43). The extra length scale that turns up is the thickness of the oscillatory boundary layer. For typical high frequency waves, say 5-sec waves and commonly observed shear velocities of 1 cm sec^{-1}, the total oscillatory boundary layer is only 2 mm thick. Thus, there will be strong velocity gradients, high shear stresses, and high rates of sediment resuspension. Purely oscillatory boundary layers rarely exist in the field, however. If one thinks of the tidal boundary layer as an oscillatory flow with a frequency of 2×10^{-4} sec^{-1}, the boundary layer will be of order 20 m thick, and hence, the near-bed flow can be treated as quasi-steady (48).

One well-quantified study of the effects of oscillatory flow (19) examined the effects of wave surge on algal spore adhesion. A more common laboratory application of oscillatory flow is the so-called shaker table. The container's size and shape may keep the boundary layer from full development, but a maximum boundary-layer thickness can be calculated for a given oscillation frequency. If we assume that the motion is purely oscillatory at a frequency of 75 min^{-1} (11) and that u_* equals 0.1 cm sec^{-1}, the *entire* boundary layer over the bottom of the flask can not be any more than 400 μm thick. Attached microbes will thus experience much greater nutrient flux than freely floating ones in the same kind of container, providing a facile fluid dynamic explanation of Bright & Fletcher's (11) results. It is certainly an oversimplification to regard flow in a flask on a shaker table as purely oscillatory, but this technique is used so frequently that it would be worth the effort to characterize such flows more accurately. Quantification of oscillatory flows may be easier, however, in simple apparatuses of alternative design (118).

Fully Developed, Uniform, Wave-Current Flow

In the presence of a steady mean flow and an oscillatory component due to either surface or internal waves, the momentum field exhibits a complex form. Within the current boundary layer, there is a thinner boundary layer created by the oscillatory flow. This wave boundary layer increases the roughness (i.e. by extracting momentum from the flow) felt by the outer part of the flow, and thus the velocity profiles are segmented, with differing slopes showing the differing momentum flux regions (Figure 1). Obviously, if there are too few measurement heights to detect the complex shape of the profile or if an inappropriate time scale is used for averaging, the actual boundary friction velocity (u_*) cannot be evaluated. The modeling techniques and measurement difficulties in dealing with such flows have been reviewed by Grant & Madsen (43, 44).

This advance in modeling wave-current boundary layers has occurred too recently to have been adopted in ecological applications, but some historical information suggests that such applications will be profitable. Shelford (110, p. 27), for example, notes that faunas of shallow lake bottoms exposed to strong, wave-generated oscillatory flows (plus unspecified but probably slow mean flows) closely resemble those of swiftly flowing streams. It would be interesting to compare the net u_* values in these kinds of environments to see whether they are roughly equal.

Particle Behavior in Bottom Boundary Layers

Proper analysis and dynamic scaling of two-phase flows—made up of fluid plus particles, where the particles may be resuspended sediments, food for suspension feeders, bacteria, or larvae—is inherently more difficult and involves more parameters than examination of fluid flows alone (112). A thorough treatment of the problem would fill a textbook (41, 79, 137), but much ecologically useful information can be gained from examining one simple nondimensional scaling, i.e. the still-water particle settling velocity, w_s, divided by u_* (65).

Suspended load and particle behavior in the lowermost 5% of the bottom boundary layer has not been studied extensively. While theories of suspended load transport have been verified (137), most require measurement of particle concentration at some reference height, usually set arbitrarily at 5% of the boundary layer thickness above the bottom. These theories do not predict particle behavior below the reference height. Suspended load in the region occupied by suspension feeders thus has not been well characterized in either the theoretical or practical literature on sediment transport; the only exception is the elegant pair of papers by Sumer and his coworkers (116, 117). Another recent, ecologically important contribution is by Nielsen (85), who showed that particles can be trapped readily and hence concentrated in vortical circulation patterns, such as those existing in steady flow over dunes (Figure 1). The degree of concentration depends on the angular rotation rate of the vortex relative to w_s. Local concentrations of particles produced by this mechanism may well be important for both passive and active suspension feeders.

Despite the paucity of information on near-bed particle behavior, Grant & Madsen (44) have recently made a major contribution to ecologically relevant sediment transport theory. They focused on unsteady, oscillatory flow that was produced near the bottom by gravity waves passing overhead. They showed that w_s/u_* is still an excellent index of particle behavior under waves or combined currents and waves. Specifically, they calculated the response times of sediment particles to wave-produced unsteadiness and found them, for all practical purposes, to be instantaneous. Thus, while waves add a great deal to the magnitude of u_*, particles resuspend no differently than one would expect in a steady flow of sufficient velocity to produce an equivalent u_*. Net horizontal flux is determined by net horizontal velocity, coupled with the concentration gradient; but the flux of import to a passive, benthic suspension feeder will lie somewhere between this net horizontal flux and the gross flux calculated by taking the scalar sum of steady and oscillatory components.

In summary, although much can be said about patterns over the entire boundary layer, there is a major gap in knowledge of suspended sediment transport in a region of primary ecological concern, i.e. the near bottom layer controlling nutrition of suspension and deposit feeders and the settlement of larvae and bacteria. This gap is not crucial to field scientists interested in sediment transport itself, who can now predict net rates of sediment transport without these details (because the bulk of net transport occurs well above the bottom in major transport events). Rapid progress in closing this gap will require the attention of ecologists concerned with the near-bottom region. Workers interested in sediment transport *can* be relied upon to identify the minimal suite of particle and flow characteristics that need to be measured in the field and controlled in the dynamic scaling of laboratory analog systems. In the simplest abiotic case, this set includes δ, u_*, ρ, μ D, ρ_s, and g (112). In designing measurements, it must also be borne in mind that the strong shear within the bottom boundary layer can both significantly affect the particle aggregation of cohesive materials (74) and bias sediment-trap catches (49).

MICROENVIRONMENTS WITHIN THE BOTTOM BOUNDARY LAYER

Flow and Sedimentation Around Objects on the Bottom

It might at first seem logical to take the vast literature on flow around spheres and cylinders (17) and use it directly in predicting flow patterns around objects on the bottom. The problem is that this literature is largely based upon flow around objects far from any boundary. Objects within a boundary layer experience a gradient of incident velocities, producing much more complex flow patterns than one might at first suppose. This complication, for example, invalidates the otherwise elegant approach by O'Neill (92), who used airfoil theory to infer flow patterns around sand dollars. Only two simple shapes of objects attached to a bed have been investigated in any quantitative detail, namely erect cylinders (29, 31) and hemispheres (10, 94). The ecological equivalents of these shapes are, respectively, animal tubes or macrophyte stalks and sediment mounds or protruding stones.

Qualitatively, simple dye studies or other visualization techniques will reveal distinct regions in the flow about an arbitrarily shaped object attached to the bottom (Figure 1). Upstream-downstream asymmetry grows in concert with imposed Re_*. Qualitative patterns of bottom shear stress about such objects are similar for cylinders and hemispheres and accord with one's intuition from observing flow and scour about a piling. Quantitatively, however, pilings provide poor analogs of shorter objects attached to the bottom. Pilings generally extend through and beyond the entire boundary layer, while animal tubes and mounds usually reach only the lowest part of the logarithmic layer. The flow velocity gradient is steepest close to the boundary, making small changes in the heights of objects exceedingly important in that region. For the same reason, the structure of the boundary layer incident on the object must be well specified for any quantitative study.

The full range of parameters for cylinders has not been investigated. Eckman & Nowell (31) explored only hydraulically smooth flow incident on a cylinder, selecting this case because a cylinder is likely to have the strongest effect when it is the only object protruding through the diffusive sublayer. Likewise, only a narrow range of parameters has been investigated for hemispheres (94). Flows about objects of more complex geometries, about flexible objects (e.g. seagrasses), and about any class of objects in an unsteady bottom boundary-layer flow setting remain essentially uncharacterized. One generalization emerging from the measurements to date is that an isolated, rigid object that protrudes through the viscous sublayer increases the spatially integrated shear stress on the immediately surrounding bed. In terms of the effects on bed structure, one can expect scour and deposition to reflect these patterns of bed shear stress (31).

When objects are sparsely distributed on the bottom, their effects on the flow will remain essentially isolated, and group effects can be treated as the sum of individual flow disturbances. For cylinders that project well into the boundary layer, the minimal spacing to avoid significant flow interactions is approximately 2 cylinder diameters upstream, 5 cross-stream, and 20 downstream. When arbitrarily shaped objects are more closely spaced, flow within and around the cluster of objects cannot be characterized without detailed measurements (10). Flow interactions among frames or plates plague interpretations in many fouling studies (128). If the rigid objects in a group attached to the bed are similar in height, are randomly or evenly dispersed over the bottom, and the group is sufficiently expansive to allow the boundary layer to equilibrate with it (i.e. to develop fully), then some bulk properties of the flow, if not the local details, can be specified. In this situation, turbulence levels in the fluid among the objects are higher than those prevailing in the absence of the structure, up to a density of objects that equals approximately one twelfth of the cover of the bed's plan area (87). Natural densities of tube builders usually fall below this threshold, ruling out hydrodynamic stabilization (32) as a cause of the frequently observed correlation (34) between the presence of the tube builders and sediment stability. At areal densities above the threshold, a skimming flow (sensu 81) is observed; the fluid "finds it easier" to flow over rather than through the field of structures, and both flow velocities and turbulence levels among the objects decrease.

Colonization rates and species compositions differ among the various portions of the flow structure produced by objects attached to the bed (37, 110), but the mechanisms accounting for these differences have rarely been identified and substantiated experimentally. In one simple but elegant experiment on caddis-fly larvae, Edington (33) showed (by moving the obstacle to flow) that the larvae behaviorally select—on a species-specific basis—particular flow environments about rocks. Periphyton colonization experiments employing bare substrata generally show enhanced initial colonization of the upper leading and trailing edges of objects attached to the bed (70, 82). The roles of direct interception (104), variation in boundary-layer thickness, vortex concentration (85) of propagules, and wake recirculation remain to be unraveled experimentally.

There are other incentives for understanding the details of flow around and among objects protruding from the bed. Bacterial swimming speeds typically fall near 50 μm sec^{-1} (101), while those of invertebrate larvae generally fall below 10 mm sec^{-1} (Hsueh-tze Lee, personal communication). Their likelihoods of reaching particular locations on the bottom and of being able to maintain their positions once having arrived must therefore be influenced strongly by the details of flow about natural biogenous or abiotic topography. J. E. Eckman (personal communication) has recently demonstrated that bacterial population growth on initially sterile substrata is enhanced in the horseshoe vortex region surrounding isolated tubes, supporting his (32) hypothesis that tubes may enhance sediment stability by stimulating microbial growth and its contribution to adhesion between sedimentary grains. The importance of this process in the field and to the next trophic level has also been established (D. Thistle, unpublished manuscript). Some preliminary data suggest that the flow pattern in the lee of a facultatively suspension-feeding polychaete enhances the flux of suspended particles to its feeding tentacles, which are situated near the top of its nearly cylindrical tube (14). The way that capitellid polychaete larvae are trapped in vortices and caught in sediment traps indicates that they apparently behave as passively settling particles (50). A similar vortex concentration mechanism may be used by passive suspension feeders (M. Patterson & K. L. Sebens, personal communication). Experiments designed to dissect the contributions of passive and active particle behavior in settlement as well as in feeding would be amply repaid. We do not know at present, for example, whether larvae tend to "sediment" or stay preferentially in any of the various subregions of flow about an obstruction (Figure 1).

Laboratory simulation of flow about objects on the bed must be undertaken carefully. First, the boundary layer incident upon the organism or structure must be tailored to the situation of interest, i.e. a sufficient upstream expanse ($> 50\,\delta$) of the proper Re_* is necessary to ensure that an equilibrium boundary-layer thickness has evolved. Next, the flow must be deep enough; if it is less than three times the object's height, its field will be strongly affected by induced deflections of the free surface. Perhaps the most insidious problem in laboratory modeling of flow about an object or group of objects is flow blockage. If the object or group of objects occupies more than 25% of the channel width, then the fluid will have a greater tendency to go over (rather than around) the object(s) than in a comparable field situation without sidewalls, making interpretation of results difficult, if not impossible (35). Similarly, if objects are brought within about 5 diameters of a sidewall or if interacting clusters of objects are brought within about 3 cluster diameters of the sidewall, the circulation patterns induced by the object(s) and the sidewall are likely to interact. These considerations place serious constraints on the sizes of the objects or arrays that can be investigated quantitatively in existing flow facilities (30).

The clearest implications of fluid dynamics in community ecology have come from intentional reproduction in the field of flow regimes that have been characterized in the laboratory. At present, such manipulations are limited to individual cylinders and hemispheres or to expansive arrays of objects. Eckman (28) clearly demonstrated that isolated cylinders less than a millimeter in diameter and protruding less than a millimeter into the water column of an intertidal sand flat significantly enhance the recruitment of some species to the cylinder's immediate vicinity (28). By producing large arrays of artificial tubes spanning the threshold for skimming flow, he (30) showed that a wide variety of recruits increase over time in a pattern that is consistent with their passive sedimentation (109) into the tube arrays. Equally importantly, he identified a few species that do not fit this pattern.

The inclusion of proper fluid dynamic controls in manipulative experiments also reveals the effects of flow on community structures. Rigorous attention to such controls allows one to evaluate the potential importance of fluid and sediment dynamic effects before deciding whether it is possible or necessary to characterize them. Eckman's (28) results led Gallagher et al (38) to avoid any unnaturally protruding structures in their manipulations. Fluid dynamic controls were still necessary, however, in order to determine whether the effects that implanted tube builders have on succession are due to the effects of their tubes on flow and sedimentation or (at least in part) to first-order biological interactions. Artificial tubes were implanted in the same spatial configurations and areal densities as the manipulated tube builders. While the fluid and sediment dynamic effects of such arbitrary arrays are difficult to characterize, these simple controls did demonstrate that much of the tube builders' facilitative effect on succession in this sand flat community was due to the physical effects of the tubes.

Dean & Hurd (25) similarly produced inanimate mimics of early colonists on hard substrata and found that they have strong effects on subsequent successional events. We suggest that these effects, which Dean (24) labels simply structural, are predominantly mediated by induced changes in fluid and particle fluxes to and from the settling plate, much as J. E. Eckman (personal communication) demonstrated for bacterial recruitment at the bases of simulated tubes in soft substrata. It is difficult, in fact, to conceive of a mechanism whereby a new recruit could perceive or respond to this "structure" except via modification of flow patterns and shear stresses on the plate. This idea could be tested by attaching isolated objects whose fluid dynamic effects have been characterized (i.e. erect cylinders or hemispheres) to large (> 50 δ) settling plates oriented parallel to the flow.

Thus, extreme caution should be used in interpreting data from litter bags (5), settling chambers (27), and sediment boxes or baskets (45, 68, 133), or from areas near other solid obstructions (16, 76) set on the bottom. These structures exert arbitrary, and to date uncharacterized, effects on near-bed flow. Current findings (28, 38) suggest that the flow artifacts can still be large if *any* artificial structure remains exposed. The techniques of implanting abiotic cores flush with the sediment surface (38) or of producing an area devoid of macrofauna via asphyxiation (J. F. Grassle, personal communication) are much more acceptable from a fluid dynamics viewpoint.

Flow Through Meshes

Meshes' dominant effects on flow are easily visualized: They do not allow flow structures (e.g. vortices) larger than the mesh size to pass through unaffected. Conversely, they produce flow structures of scales comparable to the mesh size and to the "wire" diameter of the mesh material. The geometry and spacing of natural "meshes" vary considerably, as the simple contrast between gorgonians (47) and grass beds (12) demonstrates.

Meshes are used for a wide variety of purposes in laboratory flows. At the upstream end of flow tanks or flumes, they are utilized to break up undesired cross-stream or vertical circulation in order to produce well-characterized, steady flows. The most effective meshes here are bundles of tubes or "hexcell" material aligned with the stream (108). Conversely, meshes are often used to generate fully three-dimensional turbulence by oscillating a grid or screen in a tank (121). This technique produces well-characterized, turbulent flow fields that have dominant temporal and spatial scales (71). We know of only one study that quantifies the effects of flow about and through meshes attached to the bed (i.e. a snow fence; see 63), though there are qualitative appraisals in aquatic environments (22, 129). In the sparsest of meshes, individual vertical elements should act much like isolated tubes, while the effects of denser meshes will approach the flow blockage and flow separation effects induced by similarly shaped solid bodies (Figure 1).

By far the most frequent use of meshes in ecological field manipulation is in predator inclusion and exclusion experiments. To institute a control including fluid dynamic effects, the standard procedure is to use topless or sideless cages (20). A long history of attempts to devise traps for bedload transport estimation (56) indicates that these controls are woefully inadequate. Friction with any object, even one with an open weave, slows the flow through it and accelerates the flow around it. Upstream-facing (i.e. front-less) "cages" produce underestimates of the gravel transport rate of as much as 50% (8); gravel hops over and around them.

Cages, in turn, have been used most frequently on hard substrata at shallow depths. Here, due to surface waves, near-bed flows usually have a strong oscillatory component. Hence cages provide a crude analog of the laboratory turbulence grid. If the sides are removed in a control, the orientation of the remaining meshes to the oscillatory flow will be of paramount importance. Fortunately, due to the rough texture of many hard substrata and the rapid accelerations produced by oscillatory flow, the fluid surrounding the cages also tends to be turbulent. In such settings, the flow artifact of caging should be relatively small. On hard substrata where or when flow is steady and weak, cages may, however, act primarily to impede prevailing currents, thus increasing the residence times of water parcels in regions under cages. Thus, cages on hard substrata do produce artifacts (74).

On soft substrata, the artifacts should vary widely in direction and magnitude. Where oscillatory flow components are strong (e.g. in shallow water on exposed coasts), turbulence generation may be the prevailing effect, producing scour in and around the cage. Where currents are steady and weak, there should be enhanced sedimentation (perhaps including larvae) under a cage. If oscillatory and steady flows are very weak, however, as may occur in regions of lakes and ponds, cages may impede passive recruitment (122). Unless the cage protrudes above the level of the grass or has an extremely close weave, relatively small caging artifacts would be expected in grass beds where flow already encounters "meshes." The greatest problems should occur in areas with otherwise low Re_*'s, i.e. in flat and featureless bottoms in low-energy environments. Here, warnings about caging artifacts must be taken seriously (28, 38, 57).

Exclusion and inclusion experiments are so useful that it would be foolish to recommend their discontinuation. There is an abundant and obvious need, however, to carry out laboratory modeling and field measurements with proper dynamic scaling for the purposes of (a) characterizing flow and sediment transport within and about cages, (*b*) designing cages with minimal flow perturbations, and (most importantly) (*c*) designing controls that closely simulate the flow modifications produced by cages or at least err in the conservative direction of accentuating the flow artifacts.

Flow and Sedimentation in a Depression

Qualitatively, a depression has the opposite effect from an object protruding above the bottom. Since there is more cross-sectional area for it to fill, the flow must slow down as it flows over and into a depression. Consequently, a depression is an area of reduced shear stress and enhanced deposition or residence time of suspended material. Once again, there are comparatively few quantitative studies (64, 113) and even fewer generalizations.

Pits are common in nature. They result primarily from the foraging activities of animals feeding either on sediments or on animals or plants buried within them. Subsurface deposit feeding in a localized area often leads to slumping. In some species, this slumping leads to characteristically shaped depressions and the functional group label of "funnel feeders" (83). Surface-deposit feeding species that place fecal material outside their feeding areas must also produce pits (90). A diverse group of more mobile animals feeds on buried plant (39) or animal (46) material. Its members make pits ranging in spatial scales from the beak widths of shorebirds to the jaw dimensions of bottom-feeding whales. The ecological importance of pits produced in deposit feeding is clear: Sedimentation, especially of flocculent and presumably nutritious aggregates, is enhanced because of reduced shear, increasing both the quantity and quality of food that is available to the deposit feeder without moving. Removal of plants or animals from the deposit and the associated production of a pit alter the path of succession by making both space and food resources available (120).

Pits have received little ecological attention in the laboratory. Nowell et al (90), working with an ampharetid polychaete, and D. M. Dauer (personal communication), working with a spionid polychaete, found that these animals have a greater propensity to move to a new feeding area if they are placed in an artificial laboratory situation excluding sediment transport. When sediment transport is reproduced in a laboratory flume, each species remains sessile, making use of material sedimenting into its feeding pit. Both of these studies are anecdotal, partly because the flumes presently available to benthic ecologists and oceanographers cannot accurately stimulate a steady supply rate of natural flocs. Material is generally forced through a pump or other region of unnaturally high shear in existing recirculation systems. Thus, the relationships among pit morphology, flow patterns, and the sedimentation rates of particulate food or of new recruits still need to be quantified. One of a number of interesting questions that remain to be answered concerns the degree to which the sedimentation process into pits is selective for particles of differing characteristics (e.g. varying w_s).

Field manipulation of depressions, on the other hand, has primarily been limited to creating analogs of those produced by fish feeding and has been aimed at successional questions. The importance of pits in affecting directions and rates of succession varies from one environment to the next (102, 126). VanBlaricom (126) emphasized active selection of pits by the recruits of pit opportunists as an important component of the successional pattern observed. A viable alternative, however, is that recruits have transport characteristics (e.g. w_s; see 49, 50) similar to those of the food resources they utilize. In VanBlaricom's (126) study area, flow was strongly oscillatory due to swell, while at Reidenauer & Thistle's (102) site, it was not. How fluid flow and particulate (including larval) transport differ in these two flow regimes and pit morphologies is not known, again highlighting the need for parameterization and characterization of fluid and particle trajectories in and around pits.

CONCLUSIONS AND PROSPECTUS

The importance of flow to the ecology of aquatic benthos is unquestionable. Fluid dynamic parameters are important in microbial and metazoan recruitment and in the supply of particulate food to both deposit and suspension feeders. At a minimum, fluid dynamics must be considered in the design of experiments and controls on benthic community structure and processes, even when biological interactions are the prime targets of the experiments. Despite the weight of the evidence, mechanisms responsible for ecologically important flow effects remain poorly identified, poorly parameterized, and largely unquantified. Given this state of affairs, experiments with well-characterized boundary layers (one of the categories noted above) and geometrically simple flows seem much more promising than deployments of experimental devices of arbitrary design in poorly characterized flows. The sequential elucidation of flow effects in well-controlled laboratory models, coupled with subsequent manipulation and evaluation of their ecological significance in the field, will continue to be profitable.

Quantitatively, our perusal of the literature suggests that boundary-layer flows in rivers, streams, and oceans have been relatively well parameterized and characterized by the simple formulations we have summarized. Some of these formulations (e.g. those for wave-current interactions) have been derived comparatively recently and have opened the way for a great deal of new, ecologically relevant, and exciting work. Far less is known about bottom-boundary-layer structure in lakes, and the existing data suggest that fluid dynamics are often dominated by episodic events (59) and therefore may not be susceptible to the simple (i.e. steady or periodic) modeling and measuring approaches we have outlined.

Flow microenvironments within boundary layers are inherently difficult to characterize quantitatively because they result from perturbations of already strong gradients in velocity. It is therefore imperative to determine the number and nature of measurements required to quantify adequately the flow in the microenvironment of ecological concern. Given adequate parameterization, there is an impressive list of questions in physiological ecology, population dynamics, and community dynamics that await refined measurements of benthic flow microenvironments.

ACKNOWLEDGMENTS

This review (contribution no. 1388 from the School of Oceanography, University of Washington) was supported by NSF grant OCE–8117397. We thank D. C. Miller, D. Penry, R. F. L. Self, and R. Strathmann for their contributions and revisions.

APPENDIX 1

a	Attachment point (start of boundary layer)
D	Grain diameter, cm
g	Gravitational acceleration, cm sec^{-2}
r	Reattachment point
Re_*	Roughness Reynolds number, u_*D/ν
s	Separation point
u_*	Shear or friction velocity, cm sec^{-1}
u_{*t}	Shear velocity associated with topography, cm sec^{-1}
$\overline{U}$	Mean velocity at one elevation, cm sec^{-1}
U_h	Mean velocity at height of obstacle, cm sec^{-1}
U_∞	Free-stream velocity, or velocity at water surface, cm sec^{-1}
w_s	Settling velocity of a particle, cm sec^{-1}
x,y,z	Coordinate system with z vertical, x along stream, and y across stream
z_0	Roughness length (due to sediment grains or surface texture), cm
z_{0t}	Roughness length due to bedforms, cm
z_{0w}	Roughness length due to oscillatory current, cm
δ	Boundary-layer thickness, distance at which $\overline{U} = .99\ U_\infty$, cm
δ_c	Boundary-layer thickness generated by mean current, cm
δ_l	Logarithmic-layer thickness, cm
δ_ν	Viscous-sublayer thickness, cm
δ_w	Wave boundary-layer thickness, cm
μ	Dynamic viscosity, g cm sec^{-1}
ν	Kinematic viscosity, μ/ρ, cm^2 sec^{-1}
ρ	Fluid density, g cm^{-3}
ρ_s	Particle density, g cm^{-3}
τ_b	Shear stress on the bed, g cm^{-1} sec^{-2}
τ_c	Shear stress generated by mean current, g cm^{-1} sec^{-2}
τ_w	Shear stress generated by oscillatory current, g cm^{-1} sec^{-2}

Literature Cited

1. Ambühl, H. 1959. Die Bedeutung der Strömung als ökologischer Factor. *Schweiz. Z. Hydrol.* 20:133–264

2. Andrews, J. C. 1983. Deformation of the active space in the low Reynolds number feeding current of calanoid copepods. *Can. J. Fish. Aquat. Sci.* 40:1293–1302

3. Antonia, R. A., Luxton, R. E. 1972. The response of a turbulent boundary layer to a step change in surface roughness. Part 2. Rough to smooth. *J. Fluid Mech.* 53: 737–57

4. Baier, R. E. 1980. Substrate influences on adhesion of microorganisms and their resultant new surface properties. In *Adsorption of Microorganisms to Surfaces*, ed. G. Bitton, K. C. Marshall, pp. 59–104. New York: Wiley. 439 pp.

5. Bärlocher, F., Schweizer, M. 1983. Effects of leaf size and decay rate on colonization by aquatic hyphomycetes. *Oikos* 41:205–10

6. Bellhouse, B. J., Schultz, D. L. 1966. Determination of mean and dynamic skin friction, separation and transition in low-speed flow with a thin-film heated element. *J. Fluid Mech.* 24:379–400

7. Blinn, D. W., Fredericksen, A., Korte, V. 1980. Colonization rates and community structure of diatoms on three different rock substrata in a lotic system. *Br. Phycol. J.* 15:303–10

8. Bogardi, J. 1978. *Sediment Transport in Alluvial Streams.* Budapest: Akademiai Kiado. 826 pp.

9. Bott, T. L., Brock, T. D. 1970. Growth and metabolism of periphytic bacteria: Methodology. *Limnol. Oceanogr.* 15: 333–42

10. Brayshaw, A. C., Frostick, L. E., Reid, I. 1983. The hydrodynamics of particle clusters and sediment entrainment in coarse alluvial channels. *Sedimentology* 30:137–43

11. Bright, J. J., Fletcher, M. 1983. Amino acid assimilation and respiration by attached and free-living populations of a marine *Pseudomonas* sp. *Microb. Ecol.* 9:215–26

12. Burke, R. W. 1982. *Free surface flow through salt marsh grass.* PhD thesis. Woods Hole Oceanogr. Inst./MIT, Woods Hole, Mass. 252 pp.

13. Cantwell, B. J. 1981. Organized motion in turbulent flow. *Ann. Rev. Fluid Mech.* 13:457–515

14. Carey, D. A. 1983. Particle resuspension in the benthic boundary layer induced by flow around polychaete tubes. *Can. J. Fish. Aquat. Sci.* 40:301–8 (Suppl. 1)

15. Carstens, T. 1968. Wave forces on boundaries and submerged bodies. *Sarsia* 34:37–60

16. Chandler, G. T., Fleeger, J. W. 1983. Meiofaunal colonization of azoic estuarine sediment in Louisiana: Mechanisms of dispersal. *J. Exp. Mar. Biol. Ecol.* 69:175–88

17. Chang, P. K. 1970. *Separation of Flow.* London: Pergamon. 777 pp.

18. Characklis, W. G. 1981. Microbial fouling: A process analysis. In *Fouling of Heat Transfer Equipment*, ed. E. F. C. Somerscales, J. G. Knudsen, pp. 251–91. Washington DC: Hemisphere. 743 pp.

19. Charters, A. C., Neushul, M., Coon, D. 1973. The effect of water motion on algal spore adhesion. *Limnol. Oceanogr.* 18: 884–96

20. Connell, J. H. 1961. Effects of competition, predation by *Thais lapillus,* and other factors on natural populations of the barnacle *Balanus balanoides*. *Ecol. Monogr.* 31:61–104

21. Crisp, D. J. 1955. The behaviour of barnacle cyprids in relation to water movement over a surface. *J. Exp. Biol.* 32:569–90

22. Crowley, P. H., Pierce, C. L., Johnson, D. M., Bohanan, R. E. 1983. An enclosure for experimental manipulation of lentic littoral and benthic communities. *J. Freshwater Ecol.* 2:59–66

23. Cummins, K. W., Lauff, G. H. 1969. The influence of substrate particle size on the microdistribution of stream macrobenthos. *Hydrobiologia* 34:145–81

24. Dean, T. A. 1981. Structural aspects of sessile invertebrates as organizing forces in an estuarine fouling community *J. Exp. Mar. Biol. Ecol.* 53:163–80

25. Dean, T. A., Hurd, L. E. 1980. Development of an estuarine fouling community: The influence of early colonists on later arrivals. *Oecologia* 46:295–301

26. Décamps, H., Capblanq, J., Hirigoyen, J. P. 1972. Étude des conditions d'écoulement près du substrat en canal expérimental. *Verh. Int. Ver. Limnol.* 18: 718–25

27. Desbruyères, D., Bervas, J. Y., Khripounoff, A. 1980. Un cas de colonisation rapide d'un sédiment profond. *Oceanol. Acta* 3:285–91

28. Eckman, J. E. 1979. Small-scale patterns and processes in a soft-substratum, intertidal community. *J. Mar. Res.* 37:437–57

29. Eckman, J. E. 1982. *Hydrodynamic effects exerted by animal tubes and marsh grasses and their importance to the ecol-*

ogy of soft-substratum marine benthos. PhD thesis. Univ. Wash., Seattle. 275 pp.

30. Eckman, J. E. 1983. Hydrodynamic processes affecting benthic recruitment. Limnol. Oceanogr. 28:241–57

31. Eckman, J. E., Nowell, A. R. M. 1984. Boundary skin friction and sediment transport about an animal-tube mimic. Sedimentology. In press

32. Eckman, J. E., Nowell, A. R. M., Jumars, P. A. 1981. Sediment destabilization by animal tubes. J. Mar. Res. 39:361–74

33. Edington, J. M. 1968. Habitat preferences in net-spinning caddis larvae with special reference to the influence of water velocity. J. Anim. Ecol. 37:675–92

34. Fager, E. W. 1964. Marine sediments: Effects of a tube-building polychaete. Science 143:356–59

35. Fonseca, M. S., Fisher, J. S., Zieman, J. C., Thayer, G. W. 1982. Influence of the seagrass, Zostera marina L., on current flow. Estuarine Coastal Shelf Sci. 15:351–62

36. Foster, M. S. 1975. Regulation of algal community development in a Macrocystis pyrifera forest. Mar. Biol. 32:331–42

37. Fritsch, F. E. 1929. The encrusting algal communities of certain fast flowing streams. New Phytol. 28:165–96

38. Gallagher, E. D., Jumars, P. A., Trueblood, D. D. 1983. Facilitation of soft-bottom benthic succession by tube builders. Ecology 64:1200–16

39. Garbisch, E. W. Jr., Waller, P. B., McCallum, R. J. 1975. Salt Marsh Establishment and Development, US Army Corps of Eng. Tech. Memo. 52. 110 pp.

40. Gartshore, I. S. 1977. Discussion of B. E. Lee and B. F. Soliman. J. Fluids Eng. 99:510

41. Graf, W. H. 1971. Hydraulics of Sediment Transport. New York: McGraw-Hill. 513 pp.

42. Grant, J. 1983. The relative magnitude of biological and physical sediment reworking in an intertidal community. J. Mar. Res. 41:673–89

43. Grant, W. D., Madsen, O. S. 1979. Combined wave and current interaction with a rough bottom. J. Geophys. Res. 84:1797–1808

44. Grant, W. D., Madsen, O. S. 1982. Movable bed roughness in unsteady oscillatory flow. J. Geophys. Res. 87: 469–81

45. Grassle, J. F. 1977. Slow recolonization of deep-sea sediment. Nature 265:618–19

46. Gregory, M. R., Ballance, P. F., Gibson, G. W., Ayling, A. M. 1979. On how some rays (Elasmobranchia) excavate feeding depressions by jetting water. J. Sediment. Petrol. 49:1125–30

47. Grigg, R. W. 1972. Orientation and growth form of sea fans. Limnol. Oceanogr. 17:185–92

48. Gross, T. F., Nowell, A. R. M. 1983. Mean flow and turbulence in a tidal boundary layer. Cont. Shelf Res. 2:109–26

49. Hannan, C. A. 1984. Initial settlement of marine invertebrate larvae: The role of passive sinking in a near-bottom turbulent flow environment. PhD thesis. Woods Hole Oceanogr. Inst./MIT, Woods Hole, Mass. 517 pp.

50. Hannan, C. A. 1984. Planktonic larvae act like passive particles in turbulent near-bottom flows. Limnol. Oceanogr. In press

51. Hargrave, B. T., Phillips, G. A. 1977. Oxygen uptake of microbial communities on solid surfaces. In Aquatic Microbial Communities, ed. J. Cairns Jr., pp. 445–587. New York: Garland. 624 pp.

52. Harris, L. G., Irons, K. P. 1982. Substrate angle and predation as determinants in fouling community succession. In Artificial Substrates, ed. J. Cairns Jr., pp. 131–74. Ann Arbor, Mich: Ann Arbor Sci. 279 pp.

53. Hawkins, A. D., Rasmussen, K. J. 1978. The calls of gadoid fish. J. Mar. Biol. Assoc. UK 58:891–911

54. Hogue, E. W., Miller, C. B. 1981. Effects of sediment microtopography on small-scale spatial distributions of meiobenthic nematodes. J. Exp. Mar. Biol. Ecol. 53:181–91

55. Horner, R. R., Welch, E. B. 1981. Stream periphyton development in relation to current velocity and nutrients. Can. J. Fish. Aquat. Sci. 38:449–57

56. Hubbell, D. W. 1964. Apparatus and Techniques for Measuring Bedload. US Geol. Surv. Water Supply Pap. 1748. 74 pp.

57. Hulberg, L. W., Oliver, J. S. 1980. Caging manipulations in marine soft-bottom communities: Importance of animal interactions or sedimentary habitat modifications. Can. J. Fish. Aquat. Sci. 37: 1130–39

58. Hynes, H. B. N. 1970. The Ecology of Running Waters. Liverpool: Liverpool Univ. Press. 555 pp.

59. Imboden, D. M., Lenmin, U., Joller, T., Shurter, M. 1983. Mixing processes in lakes: Mechanisms and ecological relevance. Schweiz. Z. Hydrol. 45:11–44

60. Ingram, C. L., Hessler, R. R. 1983. Distributional and behavioral patterns of

scavenging amphipods from the central North Pacific. *Deep-Sea Res.* 30:683–706

61. Jackson, J. B. C. 1977. Habitat area, colonization, and development of epibenthic community structure. In *Biology of Benthic Organisms*, ed. B. F. Keegan, P. O. Ceidigh, P. J. S. Boaden, pp. 349–58. New York: Pergamon. 630 pp.

62. Jannasch, H. W., Pritchard, P. H. 1972. The role of inert particulate material in the activity of aquatic microorganisms. *Mem. Ist. Ital. Idrobiol. Dott Marco de Marchi Pallanza Italy* 29:289–308 (Suppl.)

63. Jensen, M. 1954. *Shelter Effect.* Copenhagen: Danish Tech. 74 pp.

64. Johnston, J. P. 1960. On the three-dimensional turbulent boundary layer generated by secondary flow. *ASME J. Basic Eng.* 82:233–48

65. Jumars, P. A., Nowell, A. R. M. 1984. Fluid and sediment dynamic effects on marine benthic community structure. *Am. Zool.* 24:45–55

66. Jumars, P. A., Nowell, A. R. M. 1984. Effects of benthos on sediment transport: Problems with functional grouping. *Cont. Shelf Res.* 3:115–30

67. Kay, J. M. 1963. *An Introduction to Fluid Mechanics and Heat Transfer,* Cambridge: Cambridge Univ. Press. 327 pp. 2nd ed.

68. Khalaf, G., Tachet, H. 1980. Colonization of artificial substrata by macroinvertebrates in a stream and variations according to stone size. *Freshwater Biol.* 10:475–82

69. Kieft, T. L., Caldwell, D. E. 1983. A computer simulation of surface microcolony formation during microbial colonization. *Microb. Ecol.* 9:7–13

70. Korte, V. L., Blinn, D. W. 1983. Diatom colonization on artificial substrata in pool and riffle zones studied by light and scanning electron microscopy. *J. Phycol.* 19:332–41

71. Laws, E. M., Livesy, J. L. 1978. Flow through screens. *Ann. Rev. Fluid Mech.* 10:247–66

72. Lee, B. E., Soliman, B. F. 1977. An investigation of the forces on three dimensional bluff bodies in rough wall turbulent boundary layers. *J. Fluids Eng.* 99:503–9

73. Macan, T. T. 1974. *Freshwater Ecology,* New York: Wiley. 343 pp. 2nd ed.

74. Marshall, J. J., Rowe, F. W. E., Fisher, R. P., Smith, D. F. 1980. Alterations to the relative species-abundance of ascidians and barnacles in a fouling community due to screens. *Aust. J. Mar. Freshwater Res.* 31:147–53

75. McCave, I. N. 1984. Size spectra and aggregation of suspended particles in the deep ocean. *Deep-Sea Res.* 31: In press

76. McClatchie, S., Juniper, S. K., Knox, G. A. 1982. Structure of a mudflat diatom community in the Avon-Heathcote Estuary, New Zealand. *NZ J. Mar. Freshwater Res.* 16:299–309

77. McDougall, K. D. 1943. Sessile marine invertebrates of Beaufort, North Carolina. *Ecol. Monogr.* 13:321–74

78. McLean, S. R. 1981. The role of non-uniform roughness in the formation of sand ribbons. *Mar. Geol.* 42:49–74

79. Middleton, G. V., Southard, J. B. 1978. *Mechanics of Sediment Movement.* Lect. Notes for Short Course No. 3, Sponsored by East. Sect. Soc. Econ. Paleontol. & Mineral, Binghamton, NY, March 29–30, 1977. 231 pp.

80. Miller, D. C., Jumars, P. A., Nowell, A. R. M. 1984. Effects of sediment transport on deposit feeding: Scaling arguments. *Limnol. Oceanogr.* In press

81. Morris, H. M. 1955. A new concept of flow in rough conduits. *Trans. Am. Soc. Civ. Eng.* 120:373–98

82. Munteanu, N., Maly, E. J. 1981. The effect of current on the distribution of diatoms settling on submerged glass slides. *Hydrobiologia* 78:273–82

83. Myers, A. C. 1977. Sediment processing in a marine subtidal bottom community. *J. Mar. Res.* 35:609–47

84. Newell, R. C. 1970. *Biology of Intertidal Animals.* London: Elek. 555 pp.

85. Nielsen, P. 1984. On the motion of suspended sand particles. *J. Geophys. Res.* 89:616–26

86. Nikuradse, J. 1933. Laws of flow in rough pipes. *Nat. Advis. Comm. Aeronaut. Tech. Memo. 1292.* 62 pp. (Transl. from German, 1950)

87. Nowell, A. R. M., Church, M. A. 1979. Turbulent flow in a depth-limited boundary layer. *J. Geophys. Res.* 84:4816–24

88. Nowell, A. R. M., Hess, F. R., Zaneveld, J. R. V., Bartz, R., McCave, I. N., Winget, C. L. 1984. Instrumentation in HEBBLE. *Mar. Geol.* In press

89. Nowell, A. R. M., Jumars, P. A., Eckman, J. E. 1981. Effects of biological activity on the entrainment of marine sediments. *Mar. Geol.* 42:133–53

90. Nowell, A. R. M., Jumars, P. A., Fauchald, K. 1984. Foraging strategy of a subtidal and deep-sea deposit feeder. *Limnol. Oceanogr.* 29:645–49

91. Ockelmann, K. W., Vahl, O. 1970. On the biology of the polychaete *Glycera alba,* especially its burrowing and feeding. *Ophelia* 8:275–94

92. O'Neill, P. L. 1978. Hydrodynamic

analysis of feeding in sand dollars. *Oecologia* 34:157–74

93. Osman, R. W. 1977. The establishment and development of a marine epifaunal community. *Ecol. Monogr.* 47:37–63

94. Paola, C. 1983. *Flow and skin friction over artificial rough beds.* PhD thesis. Woods Hole Oceanogr. Inst./MIT, Woods Hole, Mass. 347 pp.

95. Parsons, T. R., Takahashi, M., Hargrave, B. 1977. *Biological Oceanographic Processes.* Oxford: Pergamon. 332 pp.

96. Pasciak, W. J., Gavis, J. 1975. Transport limited nutrient uptake rates in *Ditylum brightwellii*. *Limnol. Oceanogr.* 20:604–17

97. Pedersen, K. 1982. Factors regulating microbial biofilm development in a system with slowly flowing seawater. *Appl. Environ. Microbiol.* 44:1196–1204

98. Pedlosky, J. 1979. *Geophysical Fluid Dynamics.* New York: Springer-Verlag. 624 pp.

99. Peterson, E. W. 1969. Modification of mean flow and turbulent energy by a change in surface roughness under conditions of neutral stability. *Q. J. R. Meteorol. Soc.* 95:561–75

100. Pomerat, C. M., Weiss, C. M. 1946. The influence of texture and composition of surface on the attachment of sedentary marine organisms. *Biol. Bull.* 91:57–65

101. Purcell, E. M. 1977. Life at low Reynolds number. *Am. J. Phys.* 45:3–11

102. Reidenauer, J. A., Thistle, D. 1981. Response of a soft-bottom harpacticoid community to stingray (*Dasyatis sabina*) disturbance. *Mar. Biol.* 65:261–67

103. Roberts, A. M. 1981. Hydrodynamics of protozoan swimming. In *Biochemistry and Physiology of Protozoa*, ed. M. Levandowsky, S. Hutner. 4:5–66. New York: Academic. 213 pp.

104. Rubenstein, D. I., Koehl, M. A. 1977. The mechanisms of filter feeding: Some theoretical considerations. *Am. Nat.* 111:981–94

105. Santschi, P. H., Bower, P., Nyffeler, U. P., Azevedo, A., Broecker, W. S. 1983. Estimates of the resistance to chemical transport posed by the deep-sea boundary layer. *Limnol. Oceanogr.* 28:899–912

106. Schafer, C. T. 1971. Sampling and spatial distribution of benthic Foraminifera. *Limnol. Oceanogr.* 16:944–51

107. Schlichting, H. 1968. *Boundary-Layer Theory.* New York: McGraw-Hill. 658 pp. 6th ed.

108. Schubauer, G. B., Spangenberg, W. G., Klebanoff, P. 1950. Aerodynamics of damping screens. *NACA Tech. Note 2001.* 16 pp.

109. Scoffin, T. P. 1970. The trapping and binding of subtidal carbonate sediments by marine vegetation in Bimini Lagoonn, Bahamas. *J. Sediment. Petrol.* 40:249–73

110. Shelford, V. E. 1918. Conditions of existence. In *Fresh-Water Biology*, ed. H. B. Ward, G. C. Whipple, 2:21–60. New York: Wiley. 1111 pp.

111. Smith, J. D., McLean, S. R. 1977. Spatially averaged flow over a wavy boundary. *J. Geophys. Res.* 82:1735–46

112. Southard, J. B., Boguchwal, L. A., Romea, R. D. 1980. Test of scale modeling of sediment transport. *Earth Surf. Process.* 5:17–23

113. Squire, H. B. 1956. Note on the motion inside a region of recirculation (cavity flow). *J. R. Aeronaut. Soc.* 60:203–5

114. Stevenson, R. J. 1983. Effects of current and conditions stimulating autogenically changing microhabitats on benthic diatom immigration. *Ecology* 64:1514–24

115. Straughan, D. 1972. Ecological studies of *Mercierella enigmata* Fauvel (Annelida: Polychaeta) in the Brisbane River. *J. Anim. Ecol.* 41:93–136

116. Sumer, B. M., Deigaard, R. 1981. Particle motions near the bottom in turbulent flow in an open channel. Part 2. *J. Fluid Mech.* 109:311–37

117. Sumer, B. M., Oğuz, B. 1978. Particle motions near the bottom in an open channel. *J. Fluid Mech.* 86:109–27

118. Svoboda, A. 1970. Simulation of oscillating water movement in the laboratory for cultivation of shallow water sedentary organisms. *Helgol. Wiss. Meeresunters.* 20:676–84

119. Tennekes, H., Lumley, J. L. 1972. *A First Course in Turbulence.* Cambridge, Mass: MIT Press. 300 pp.

120. Thistle, D. T. 1981. Natural physical disturbances and communities of marine soft bottoms. *Mar. Ecol. Prog. Ser.* 6:223–28

121. Thompson, S. M., Turner, J. S. 1975. Mixing across an interface due to turbulence generated by an oscillating grid. *J. Fluid Mech.* 67:349–68

122. Thorp, J. H., Bergey, E. A. 1981. Field experiments on responses of a freshwater, benthic macroinvertebrate community to vertebrate predators. *Ecology* 62:365–75

123. Thum, A. B., Allen, J. C. 1975. Distribution and abundance of the lamp urchin *Echinolampas crassa* (Bell) 1880 in False Bay, Cape. *Trans. R. Soc. S. Afr.* 41:359–73

124. Townsend, A. A. 1966. The flow in a turbulent boundary layer after a change in surface roughness. *J. Fluid Mech.* 26:255–66

125. Tuchman, M. L., Stevenson, R. J. 1980. Comparison of clay tile, sterilized rock and natural substrate diatom communities in a small stream in southeastern Michigan, U.S.A. *Hydrobiologia* 75:73–79

126. VanBlaricom, G. R. 1982. Experimental analyses of structural regulation in a marine sand community exposed to oceanic swell. *Ecol. Monogr.* 52:283–305

127. Van Dyke, M. 1982. *An Album of Fluid Motion*. Stanford, Calif: Parabolic. 176 pp.

128. Vandermeulen, H. DeWreede, R. E. 1982. The influence of orientation of an artificial substrate (transite) on settlement of marine organisms. *Ophelia* 21:41–48

129. Virnstein, R. W. 1977. The importance of predation by crabs and fishes on benthic infauna in Chesapeake Bay. *Ecology* 58:1199–1217

130. Vogel, S. 1981. *Life in Moving Fluids*. Boston: Grant. 352 pp.

131. Wainwright, S. A., Koehl, M. A. R. 1976. The nature of flow and the reaction of benthic Cnidaria to it. In *Coelenterate Ecology and Behavior*, ed. G. O. Mackie. pp. 5–21. New York: Plenum. 579 pp.

132. Welch, P. S. 1935. *Limnology*. New York: McGraw-Hill. 471 pp.

133. Welch, P. S. 1948. *Limnological Methods*. New York: McGraw-Hill. 381 pp.

134. Whitford, L. A. 1960. The current effect and growth of freshwater algae. *Trans. Am. Microsc. Soc.* 79:302–9

135. Whitford, L. A., Shumacher, G. J. 1961. Effect of current on mineral uptake and respiration by a freshwater alga. *Limnol. Oceanogr.* 6:423–25

136. Yaglom, A. M. 1979. Similarity laws for constant-pressure and pressure-gradient turbulent wall flows. *Ann. Rev. Fluid Mech.* 11:505–40

137. Yalin, M. S. 1977. *Mechanics of Sediment Transport*. New York: Pergamon, 307 pp. 2nd ed.

138. Young, R. A., Southard, J. B. 1978. Erosion of fine-grained sediments: Seafloor and laboratory experiments. *Geol. Soc. Am. Bull.* 89:663–72

139. Zeitzschel, B. 1980. Sediment-water interactions in nutrient dynamics. In *Marine Benthic Dynamics*, ed. K. R. Tenore, B. C. Coull, pp. 195–218. Columbia: Univ. SC Press. 451 pp.

140. Zimmer-Faust, R. K., Case, J. F. 1983. A proposed dual role of odor in foraging by the California spiny lobster. *Panulirus interruptus* (Randall). *Biol. Bull.* 164:341–53

141. Zimmerman, P. 1961. Experimentelle Untersuchungen über die ökologische Wirkung der Stromgeschwindigkeit auf die Lebensgemeinschaften des fliessenden Wassers. *Schweiz, Z. Hydrol.* 23:1–81

Ann. Rev. Ecol. Syst. 1984. 15:329–51

THE EVOLUTION OF FOOD CACHING BY BIRDS AND MAMMALS

C. C. Smith and O. J. Reichman

Division of Biology, Kansas State University, Manhattan, Kansas 66506

INTRODUCTION

We limit our discussion of food storage, or caching, to the movement of potential food items from one location to another for eating at some later time. This activity occurs exclusively in those animals that bring food to their offspring, and not in those that bring their offspring to food (i.e. that lay their eggs near food in a favorable microhabitat). Provisioning offspring is limited taxonomically to mammals, most birds, and some Hymenoptera. Moving food to a favorable microhabitat was apparently the first transitional step in the evolution of more complex systems of provisioning offspring in hymenopterans (72). Although all species that cache food provision their offspring, the converse is not true. Relatively few of the animals that provision their young with food also cache food, and caching food has no obvious connection to the glandular secretion of milk, which probably initiated offspring provisioning in the evolutionary history of mammals. Repeated traveling from a foraging area to dependent young seems to precondition animals for caching food. One of the goals of our review will be to determine what other conditions among species of birds and mammals and their food favor the evolution of caching. We limit our discussion to birds and nonhuman mammals because of our own backgrounds and because experimental studies have recently been done on these vertebrates (4, 14, 20, 48a, 62, 101a, 103–105, 116, 117, 128), although earlier investigations started with wasps (121).

A corollary of provisioning offspring with food is a stable, relatively high metabolic rate (even in the Hymenoptera), which ensures that parents can maintain a steady flow of food to the relatively few, "expensive" offspring

329

0066-4162/84/1120-0329$02.00

Table 1 Information on taxa that are known to cache in the wild

Taxonomic group	Food stored	Cache type[a]	Period[b]	References
Birds				
Goshawk	Small vertebrates	S	D	97
Sparrow hawk	Small vertebrates	S	D	6, 19
Snowy owl	Lemmings	L	D	89
Some small owls	Small vertebrates	S?[c]	D	88
Some woodpeckers	Large seeds	L, S	M	10, 54, 56, 70, 80
Some crows	Vertebrate flesh, seeds, marine invertebrates	S	D	37, 48a
Some jays	Large seeds	S	M	14, 20a, 37, 62, 126, 130
Gray jays	Vertebrate flesh, seeds	S	M	96
Nutcrackers	Large seeds	S	M	119, 123–126, 128–130
Some tits	Seeds, insects	S	D	20, 35, 39, 103
Nuthatches	Seeds, insects	S	D	57, 115
Shrikes	Small vertebrates, large insects	S	D	17
Mammals				
Shrews	Mice, large invertebrates, seeds	L	D	48, 74, 90, 114, 122
Moles	Earthworms	L	D	33
Pikas	Dried vegetation	L	M	29, 32, 107, 108
Tree squirrels	Large seeds	S	M	30, 36, 116, 117
Southern flying squirrels	Large seeds	S	M	81
Pine squirrels	Conifer cones	L	M	109, 110
Chipmunks	Seeds	L + S	M	23, 58, 100, 120, 133
Some ground squirrels	Seeds, fungi	L	?	25, 63
Beavers	Bark, twigs	L	M	2, 52, 84, 106
Large kangaroo rats	Small seeds	L	M	76, 77, 131
Small Heteromyidae	Small seeds	L?	M?	9, 28, 75, 98, 127
Pocket gophers	Tubers, vegetation	L	?	40
Mole rats	Tubers, bulbs, stems	L	?	51
Some Hystricomorphs	Large seeds	S	M	78, 113
Some Cricetidae	Seeds, vegetation	L	M?	1, 27, 29, 34, 38, 46, 64, 82, 87
Some Muridae	Seeds	L	?	7, 26, 31, 91
Canidae	Vertebrates	S	D	43, 45, 68, 69, 71
Brown bears	Large vertebrates	L	D	22, 68
Some Mustelidae	Small vertebrates	S	D	42, 68
Leopards	Large vertebrates	L	D	59, 68
Some Hyaenidae	Large vertebrates	L	D	59, 68

[a] L = larder, S = scatterhoard.
[b] D = days, M = months.
[c] A question mark indicates uncertainty about that aspect of the taxon.

during their helpless period of growth. Ensuring a steady supply of energy in a variable environment is one of the main benefits of storing food, although this can also be accomplished by storing energy as fat. Furthermore, when not raising young, energy can be saved by escaping the high costs of a variable environment through migration or the use of torpor (93). A second goal of our review will be to contrast the relative advantages of responding to food shortage by storing food, storing fat, migrating, or reducing metabolic demands through torpor.

Caching food presumably preserves it for later use by the individuals that store it; defense of food territories serves the same function without expending the time or effort needed to move food. This contrast highlights a third goal of the review—to analyze the efficiency of caching behavior as a means of defending food.

At first glance, food caching seems contrary to the tenets of natural selection, i.e. to the process of maximizing the contribution to future generations by rapidly converting environmental resources into offspring. Why delay the use of resources by storing them? If there is an advantage to storing energy, why not store it as fat reserves in the body? It is not surprising that the complex behavior of food storage coevolved along with provisioning offspring with food, which seems to preadapt animals behaviorally and metabolically for caching.

THE TYPES OF FOOD STORED

All major categories of terrestrial food are stored by some birds and mammals (Table 1), with the exception of nectar and pollen, which are stored by some hymenopterans. No marine birds or mammals are known to cache food. The absense of food storage in marine habitats is probably related to their low environmental variability and to the advantages of storing fat for thermal insulation in the aquatic medium, which also provides buoyancy.

In terrestrial environments, food storage is more common at higher latitudes, where the environment is more variable, than at lower latitudes. A season of freezing temperatures can help preserve stored food, as well as make storage useful. Nonetheless, some tropical rodents do cache seasonally available seeds over a period of months (36, 78, 113), and some carnivores (59, 68, 71) at low latitudes cache excess food from kills for short periods of time. Male MacGregor's bower birds in New Guinea make short-term fruit caches near their bower to reduce foraging time away from the bower during peak periods of display activity (M. A. Pruett-Jones, S. G. Pruett-Jones, submitted).

Seeds are the food cached most commonly for the longest times, presumably because their adaptations to dormancy make them ideal for surviving storage and their concentrated energy reserves make them efficient to carry. Subterranean tubers may have somewhat the same properties, but they are readily available only to fossorial (40, 51) or semifossorial (34) animals. Vegetation, cambium, fruit, fungi, and animal flesh must be dried or frozen to be stored for more than a few days; thus animals develop behaviors that are specific to the food items and the microhabitats where they are stored.

CACHING BEHAVIOR

Species that cache extensively as adults appear to develop the appropriate behavior early in their development. In hamsters, the initial food carrying, pouch filling, and burying can occur as early as 2–4 weeks of age (60). During the first week they leave the nest, pine squirrels cache fungi, and they steadily improve in their efficiency of storage over a period of more than 6 weeks (112). Fox pups begin obvious but unpracticed caching attempts early in their lives (68). In several rodent species that use cached food as part of parental care, adult females almost always cache more frequently and greater amounts than males (24, 28, 47).

The season for caching coincides with the period when excess supplies are available and is just prior to when they are required. At higher latitudes, caching animals take advantage of summer and fall productivity to lay in their caches for the winter (2, 10, 14, 20a, 23, 46, 62, 64, 74, 81, 95, 107, 109, 123, 129, 131). In tropical forests that are seasonally dry, concentrated nut fall in the dry season provides caching material for use at the end of the wet season when fruit fall is at its lowest (36, 113). In contrast, northwestern crows, which forage in the intertidal, cache on land during low tide to have a food supply during high tide (202), and MacGregor's bower birds (see above) also tend to use caches in a diurnal cycle.

A number of species have evolved structures that help them to carry large quantities of food items to storage locations. Epithelial-lined pockets in the lateral wall of the buccal cavity have evolved independently in at least 4 major taxa [3 genera of murid rodents, 4 genera of cricetid hamsters, 2 genera of sciurid chipmunks, and some ground squirrels (132)]. External, fur-lined, cheek pouches are found in all heteromyid and geomyid rodents, which probably have a common phyletic origin, and they are not found in any other species (66). External cheek pouches may be advantageous for carrying food excavated from the soil by keeping grit out of the mouth where it could wear down the cheek teeth or by preventing potentially dangerous mycotoxins in food from entering the buccal cavity (92, 94a). Animals with internal cheek pouches may gather food for storage from vegetation, which would be less gritty than food gathered from the soil. During the evolution of internal cheek pouches, even slight expansions of the cheeks would be advantageous; the intermediate stages of external cheek-pouch formation confer no obvious advantage (66). The apparent difficulty in evolving external cheek pouches

(66) may explain why several genera of gerbils and several families of fossorial rodents (83) lack the cheek pouches their ecological equivalents among heteromyids and geomyids possess. Canids' ability to regurgitate food may be functionally analogous to cheek pouches, but it is most frequently used to feed offspring, rather than to carry food back to a cache (68).

Among birds, nutcrackers have a sublingual pouch (13) that can hold up to 90 large pine seeds (130). Piñon jays, Steller's jays (130), European jays (14) and blue jays (20a) have expandable esophagi that allow them to carry up to 56 large pine seeds, 18 large pine seeds, 6 acorns, and 5 acorns, respectively. Crows *(Corvus)* also have throat pouches that increase the load they can carry to a storage site (37). As with mammalian internal cheek pouches, these adaptations for transporting food could easily evolve in gradual stages with only minor anatomical modifications (13, 130).

THE DEFENSE OF STORED FOOD FROM MICROBES

Most of the important details concerning caching involve evolutionary pressures brought to bear on caching individuals by the need to prevent theft of their stored food. One group of competitors caching animals cannot avoid are microbes, which Janzen (49) suggests have evolved spoilage strategies as a competitive mechanism against vertebrates. Because microorganisms are ubiquitous and occur in most cache locations, food destined for caches may have to be "treated" (e.g. dried), and caches may have to be placed in locations where the potential impact of decomposers is minimized (e.g. dry, cold locations). Pine squirrels carry fresh fungi up into trees to dry them in the sun (109) before moving some of them to the nest cavity (16). Pikas dry vegetation ("hay") on tops of rocks before storing it under the rocks (107). Canids bury meat underground, which may limit the activity of aerobic microbes as well as protect it from other large carnivores and flies (68). Foxes seek sunny sites for caching fresh kills, presumably to inhibit microbial action (68). Half the caches made by bears (22) were covered with *Sphagnum*, a plant with antiseptic properties. Some species circumvent the problems of spoilage by caching live material. Moles are known to bite earthworms, leaving them alive but unable to crawl from the storage locations (33). Shrews sometimes store live snails in underground corrals before consuming them (48), and the poisonous bite of short-tailed shrews is thought to leave their prey immobilized (74, 122) in a manner analogous to parasitic wasps' sting. Shrews cache seeds before insects and both before mice, apparently in order of resistance to spoilage (74). The large seed caches of desert heteromyids (77, 101) are maintained in areas with 100% relative humidity (53) and apparently will rapidly succumb to molds and bacteria if they are not actively managed by the rodents (92, 94a). Although we do not know the details of the management procedures, heteromyids may take advantage of the beneficial aspects of the mold/seed/caching relationship by actually preferring slightly moldy seeds to unmoldy and very moldy seeds (92, 94a).

THE DEFENSE OF STORED FOOD FROM INTERSPECIFIC COMPETITORS

Competition from other species of animals influences both the microhabitat in which food items are cached and the spacing of the caches within the habitat. Morris (78) coined the term *scatter hoarding* to describe the formation of many caches, each with one or a few food items, scattered through much of the caching individual's home range. *Larder hoarding* involves storing most of the food in one concentration in a small part of the home range. Larder hoards appear to be used when an animal can defend its caches from interspecific competitors, while scatter hoards are more appropriate when an animal cannot. By collecting food from a concentrated site (e.g. nuts on a tree) and scattering its caches, an owner can increase the dispersion of a resource occurring in a rich patch to the point where it is less economical for a competitor to steal from a cache than to forage for food.

The importance of the placement and spacing of caches in protecting them from competitors is illustrated by the guilds of birds and mammals that feed on beechnuts and acorns in the eastern deciduous forests of North America. Squirrels of the genus *Sciurus* scatter hoard nuts primarily on the forest floor (18, 116, 117). Chipmunks form a larder hoard in their nesting burrow (23) where they sleep and/or hibernate (133), and they also form scatter hoards on the forest floor (23, 100). Southern flying squirrels scatter hoard nuts in leaf litter, tree branches, and bark (81). White-footed and deer mice usually eat acorns as they find them (118), but they have been reported to make scattered caches when given large numbers of pine seeds from a feeder (1) or to make larder hoards of acorns in nest boxes with small openings (46).

Among birds, red-headed woodpeckers larder hoard pieces of acorn or the whole nut in the bark and cavities of one or a few large trees, which they defend aggressively from all seed eaters (54, 55, 80). Red-bellied woodpeckers form scatter hoards in the bark and crevices of many large trees in winter territories (56). Blue jays scatter hoard nuts just under the surface of the ground on forest edges and in brushy areas (20a) in a manner similar to European jays (14). Black-capped chickadees (39, 101a) and whitebreasted nuthatches (57) scatter hoard nuts in the bark and branch forks in their home ranges, but only for periods of a few hours to a few days, rather than the weeks or months characteristic of the other six species discussed. White-tailed deer, raccoons, turkeys, passenger pigeons, and Carolina parakeets (73) all fed on large concentrations of acorns, beechnuts, and chestnuts in presettlement deciduous forests, and they must also be considered selective forces in the evolutionary history of caching behaviors in these guilds. Each larder hoarder has a specific microhabitat in which it can defend its caches against all competing species. In some cases, the defense involves placing the caches where competitors cannot reach them. For example, chipmunks can enter burrows too narrow for tree squirrels. Furthermore, because some chipmunks do not enter torpor (133) and those that do periodically rouse, it is dangerous for mice to attempt theft.

The interactions among the competing species—and their effect on caching strategies—are complex and multidimensional. For example, red-headed woodpeckers are able to drive other acorn feeding birds away from their cache trees (54, 55, 80), and the woodpeckers' loud calls may drive squirrel thieves away by attracting potential predators' attention. Flying squirrels are the one competing species woodpeckers cannot control because of the squirrels' nocturnal habits. Although red-headed woodpeckers do co-occur with flying squirrels in these forests, they are found mainly on prairie edges (11) where they are beyond the range of flying squirrels. In this habitat, they often cache acorns in exposed, dead trees distant from the source of acorns, which might attract competitors to their caches. Lewis's and acorn woodpeckers, which larder hoard, also live outside the range of flying squirrels. Red-bellied woodpeckers, a scatter-hoarding, melanerpine species, overlap extensively with the range and habitat of flying squirrels. These woodpeckers may protect their scatter hoards to some degree by placing them in deep crevices in trees, from which their long tongues allow them to remove the acorns where other competitors might fail (56). Chickadees and nuthatches, which cache food for short intervals, seem to rely on dispersed spacing to protect their caches (102). *Sciurus* species and flying squirrels probably cannot defend larder hoards on the ground from deer, raccoons, and turkeys. Squirrels' burial of single nuts reduces the foraging efficiency of larger mammals and eliminates competition from birds, which lack a good sense of smell. Blue jays (20a), which would have problems with larder hoards similar to squirrels' (14, 116), place their caches outside mature forest, thus reducing the chances that squirrels will find them.

In other habitats, cachers encounter similar problems. Pine squirrels can larder hoard conifer seeds in their cones because woody tissue in the cones makes them indigestible if eaten whole by larger animals and the cones are difficult to open for smaller mammals such as mice. When pine squirrels dry fungi in trees for preservation (109), however, they scatter hoard them, probably because flying squirrels could easily steal the fungi at night if they were larder hoarded. Once dried, fungi are sometimes larder hoarded in the squirrel's nest cavity where the fungi can be defended at night (16).

Pikas dry vegetation and larder hoard it in piles (107). In North American species, the vegetation is usually dried on talus slopes and later cached under large boulders where large herbivores cannot steal it. On the Eurasian steppes, an area apparently devoid of potential thieves such as large herbivores (29; but see 108), voles and pikas dry vegetation in unprotected hay piles.

In North American desert communities of heteromyid rodents, it is the largest species that larder hoard large quantities of seeds in their elaborate burrows (44, 76, 77, 131). Some intermediate-sized heteromyids make hundreds of small surface caches near their burrow entrances. Shaw (101) suggests that the rodents were "curing" the seeds before taking them underground for caching. Smaller, sympatric species of heteromyids may store several days worth of food; perhaps they are limited by an inability to defend a larder hoard, although their small burrows might preclude large thieves from entering and stealing seeds (38).

In western North American forests, nutcrackers scatter hoard seeds within the territory of a mated pair or in communal cache areas used by several pairs (123, 129). Piñon jays cache in a communal area used by all of the members of a socially cohesive flock that defends a large territory (5, 62). The communal caches cover a small part of the caching birds' range, although they do not constitute a larder hoard. Seeds are cached in the soil in small groups, presumably to prevent them from being attractive for exploitation by sympatric rodents. Communal cache areas are formed on wind-swept, south-facing slopes where the snow melts first and where plants, their seeds, and seed predators are likely to be rare.

Carnivores may form scatter or larder hoards. When carnivores capture a large prey item that they cannot consume immediately, the size of the prey itself serves as a larder hoard. Leopards move prey into trees, reducing competition from other large carnivores and at least some scavengers (59). Brown bears rake litter over carcasses until they are ready to feed again (22). Although the food is not moved to a new cache site in accordance with our definition of caching, covering the food does, in effect, change its position and preclude its use by some competitors. Canids, with their ability to regurgitate food, may carry food from a kill and bury it in several small caches (43, 69), thus depending on spacing rather than a safe location to protect their food.

Nesting northern shrikes (17) scatter hoard caches of small mammals, birds, and large insects, but they do not cache within 50 m of their nest, apparently to keep from attracting thieves to the nest area, where they might also prey on nestlings. Small raptors scatter hoard their prey around their hunting areas for brief periods (6, 19, 88, 97), but snowy owls, which can defend their caches from all but the largest mammals, accumulate large piles of lemmings around their nests (89).

Beavers' placement of caches is related less to defense than to access during the winter. At the northern extreme of their range, beavers live below ice for up to eight months of the year and therefore place their caches of sticks in the pond bottom just outside the entrance to the lodge (2, 84). Beavers at northern latitudes also build "rafts" under which they place some of their winter caches. The rafts become frozen into the ice in the winter, making the material in them unavailable for consumption during that period. The top of the rafts consist of items the beavers cannot or do not like to eat (106).

SOCIAL ORGANIZATION AND FOOD CACHING

Andersson & Krebs (3) argue that selection should always favor behavior in individuals that reduces the chances that conspecifics will pilfer their caches. Communal caching always invites cheaters who use caches without contributing to them, although this may be less of a problem if the cheaters are close relatives. Nel (82) equated caching behaviors with a solitary existence when comparing the hoarding behaviors of nine species of Kalahari Desert rodents.

Larder hoards can easily be defended from cheaters by using a system of individual territories, as has evolved in pine squirrels (109), pikas (107), large kangaroo rats (99, 131), and wood rats (27, 64). In addition to insuring the cacher sole use of the cache, individual territories require 0.71 times as much time and energy for carrying food to the cache as do territories exploited by a pair (109). In contrast, beavers have family territories, which allow males to help feed the female during lactation and provision the young during the time between weaning and independence (84). The joint engineering projects shared by a pair of beavers and the male's provision of some parental care presumably compensate for the inefficiency of forming a joint cache.

Chipmunks (23) and red-headed woodpeckers (55, 80) defend their larder hoards, but they do not defend the areas where the items in their caches are gathered, apparently because so many larger birds and mammals use the same areas to obtain seeds that defense would be impossible. Some mated pairs of Lewis's woodpeckers produce winter caches in their breeding territories (10). Such caching may have been a step in the evolution of the permanent group territories noted in acorn woodpeckers in which an extended family group defends an area containing the group's larder hoard and most of the trees from which it is gathered. As in beavers, the advantages of cheating on other members of the social group are offset by the benefits of assisting close genetic relatives (3, 70).

Tree squirrels scatter small caches in order to minimize interspecific theft, thus also precluding an effective intraspecific territorial defense of the caches (116). Although tree squirrels have broadly overlapping home ranges, each squirrel tends to concentrate caches around its nest tree, thereby minimizing cache overlap between neighbors. This strategy maintains cache densities below a level at which theft is economical (116). Nuthatches and tits tend to store the same food when several individuals of each taxon forage together in mixed-feeding flocks during the winter. Their caches, however, are scattered in different areas by different individuals (20, 102, 103). Individual or mated pairs of red-bellied woodpeckers scatter hoard throughout defended winter territories of two or more acres (56). The winter social organization of most jays is made up of pairs (14, 37, 41) or family units (37, 96) caching in mutually exclusive areas that may (37) or may not (14, 37) be actively defended. A mated pair of nutcrackers may defend a territory containing scattered caches while also contributing to communal caches (123, 129). In other regions, all caching may be limited to individual territories (119). Canid family groups scent mark and defend mutually exclusive territories within which each group scatters its caches (68, 69). Even for those species that do scatter hoard within their own territories, it is doubtful if the function of the territory is to defend the cached food from conspecifics. These species all maintain territories for breeding, and the territorial spacing makes central-place foraging the most efficient type during the progressive provisioning of young.

MEMORY AND CACHE LOCATION

A critical consideration in the evolution of caching is for the caching individual to have a better chance of relocating the cache than other individuals (4, 14, 102, 104, 105, 128). This condition will obviously be met for a single, defended larder hoard, but remembering the specific location of scattered individual caches would be a great advantage for a scatter-hoarding individual. The ability of almost all birds and mammals to return from foraging to provision their offspring indicates that they have significant spatial memory. Experiments have demonstrated the existence of a memory for specific cache sites over a period of 24 h in black-capped chickadees (101a), 24 h in marsh tits (104), 72 days in European jays (14), 24 days in nutcrackers (128), and several days in foxes (68). These studies did not provide any information about the decay of memory over time or the period over which effective retrieval could occur. The importance of memory in finding the exact location of a cache was demonstrated in an experiment where equivalent supplemental caches were placed within 3 m of caches made by foxes (68); the animals often ignored these while rapidly locating their own caches. Experiments with marsh tits produced a similar response, even when the experimental caches were only 1 m from the naturally cached items (20). D. B. McQuade, E. H. Williams & H. E. Eichenbaum (in preparation) found that gray squirrels in a controlled environment use spatial and specific visual cues, but not olfaction, to learn food locations. Gray (61) and fox squirrels (18) unfamiliar with the location of buried nuts can find them by olfaction, but only from close range. Regardless of an animal's olfactory acumen, visual cues are probably more easily located from a distance than olfactory ones.

European jays (14) also use their visual orientation toward specific landmarks near a buried cache in order to proceed directly to it. For an animal to return quickly to a cache, it would have to know its own position relative to the cache site. In other words, it would have to be able to recognize landmarks and know how far they are from the caches, both of which European jays (14) and nutcrackers (128) can do. Edges between different types of substrate (4, 14, 128) frequently serve as landmarks, and we know that European jays orient more strongly to sticks in vertical than in horizontal positions (14). Both jays (14) and nutcrackers (4, 128) have been found to cache preferentially near landmarks, but there was no significant relationship between nutcrackers' success in retrieving their caches and the caches' distance from landmarks (4). Experiments with marsh tits show that both site preference and memory are involved in retrieving cached food (105). The use of site preference in nutcrackers means that conspecific thieves will search preferentially in the same type of site the cacher used; thus, thieves have a better chance of finding a cache than if they searched all sites with equal probability (128).

All published experiments designed to test spatial memory in birds and mammals have yielded positive results. Nonetheless, spatial memory need not be a prerequisite for the evolution of scatter hoarding. If cache sites are protected from individuals of competing species (e.g. mammals burying food where they can smell it but birds cannot see it), selection could still favor caching behaviors. The ability to use spatial memory would only enhance an already effective behavior.

OPTIMAL FORAGING TO FILL CACHES

Gathering food for storing presents different problems depending on whether it is gathered for a larder hoard or a scatter hoard. For larder hoards, there is an important difference between food gathered within a defended territory or items collected outside it. Animals collecting food for larder hoards within defended territories have the greatest control over their resources. If territorial defense is completely effective, the maximum amount of food can be stored by gathering items near the larder first, with successively more distant resources being harvested subsequently. Even though the territory holder may have exclusive control over the food resources in its territory, it may still have to form a larder hoard so it can have access to the food during the winter (2, 52, 84). If the territory is just large enough to encompass the necessary food reserve and there is a threat of theft from neighboring territory holders, peripheral food resources should be harvested first to prevent competing neighbors from securing the resource. The caching individual should then harvest progressively toward the larder. Larder hoarding in pine squirrels fits the latter situation (109, 110). Pine squirrels can only use conifer seeds effectively when they remain in the cone.

Different species of cones open in a predictable sequence, with larger species tending to open first. Starting with the largest and earliest-opening cones, squirrels trim hundreds of cones from trees, come to the ground, and carry them one at a time to a central mixed-species larder. The squirrels harvest all of the largest species first, working in from the periphery toward the cache and burying them in moist ground to prevent the cones from opening and losing their seeds. The pattern is subsequently repeated with those cone species opening next and containing the next largest amount of energy. Theoretically, there should be a trade-off between the energy content of the food items (the benefit) and the distance they are carried to a larder hoard (the cost) such that the net energy gain from small items near the cache is approximately equivalent to that from large items far from the caches. This relationship suggests that there will be greater variation in quality (e.g. size) in the items taken near the cache than in those far from the cache, as was found for beavers (52).

If only the larder hoard—and not the territory containing the food source—can be defended, then its placement becomes critical to the efficiency of acquiring food for storage. Defensibility of the larder site should be the larder owner's first consideration. Given a variety of safe sites for larder hoards, the best position is the one that minimizes the amount of time required to secure food for the cache. This position will depend on the natural distribution of food and on the abundance and distribution of conspecifics' hoards. Placing a larder hoard in an area of undefended food resources becomes a problem of minimizing the travel distances by taking the nearest food first or using an optimum itinerary (85). As in uncontested territories, there should be trade-offs between travel distance and the amount of food carried each trip. When undefended food is cached, there is the additional cost of repeated sampling of the food distribution to estimate the effect of resource depression by competitors.

Optimal foraging to fill scatter hoards is dependent on the density and spacing of cache sites. Stapanian & Smith (116) argue that there is an optimum density of caches at which the difference between the cost of making the cache and the benefit of being able to retrieve it is maximized. This cache density will be optimal, in a two-dimensional system, if the distance food is carried from a central concentration for caching is proportional to the square root of the area of burial (i.e. if the area is proportional to the square of a linear dimension) and if the probability of survival of the caches is proportional to the area of burial. The optimal cache density is the density below which cache thieves cannot afford to consider the caches as a food source when they forage. Sherry et al (103) cached seeds at five experimental densities in sites similar to those used by marsh tits. The survival of seeds in caches increased as the cache density decreased down to the median-density sites, but it did not change appreciably at lower densities. Marsh tits cached seeds at the median density, as Stapanian and Smith had predicted they would (109). Fox squirrels' consumption of walnuts was also directly proportional to the density of the nut burial down to the density at which squirrels bury nuts, but experiments were not performed at densities lower than the apparent optimum (116).

The quality of the food items also influences optimum scatter-hoard density, with higher value items requiring lower cache densities to be protected from potential cache thieves (116, 117). Fox squirrels, for example, cached nuts with a lower energy content at a higher density (117). Experimental grids of the lower energy nuts had higher survival rates when placed at the same density as the higher energy species, and grids with the two species alternating positions exhibited intermediate survival rates (117). In addition, the optimum density of caching of the lower quality nut species was higher than for the higher quality nuts, as predicted.

If the probability of being detected by thieves is different for scatter hoards within different microhabitats, caching individuals should benefit from being able to respond appropriately to variations in the theft rate among microhabitats. Cowie et al (20) found that individual tits had different preferences for caching in five microhabitats and that an individual's preferences change over time. Using experimentally placed caches (103), researchers found that of five microhabitats used for caching by marsh tits, the ones closest to the ground—where rodents forage—had the highest cache loss rate. Sherry (102) quotes Stevens as finding that tits store the fewest seeds in microhabitats where they had lost the most seeds to thieves in the past. Perhaps as thieves learn the microhabitats tits use for caching, the tits change their microhabitat preferences in order to keep ahead of the thieves (103).

Based on cost-benefit models of scatter hoarding food, Stapanian & Smith (116) predict a uniform density of caches in all directions around a central source, with the earliest caches being made close to the source to minimize the cost of carrying food. Their observations of squirrels caching walnuts revealed no pattern in the distances successive caches were carried; each squirrel formed most of its caches in a small arc of a circle around the central source of nuts. Marsh tits, however, tended to carry items for early caches further than items for later caches (20), contrary to Stapanian & Smith's predictions (116). As with the squirrels, each marsh tit used only a small arc in the circle around the food source for its caches. The natural distribution of food encountered by individual animals is likely to be a product of the interaction among food from several sources, each representing a different nut tree, for example. Thus, the problem an individual faces may not be spreading caches around a single source but filling an individual cache area from several concentrated resource locations. Minimizing carrying distance may be less critical than maintaining the optimum density of caches. For marsh tits, successive caches are significantly closer to each other than would be expected from random placement, suggesting a systematic filling of the available caching space (20).

One problem with maintaining an optimum density of scattered caches is that given certain resource and competitor densities, very few caches will be feasible in individually exclusive cache areas. Some populations of *Sciurus* live at such densities that only a fraction of their winter diet can be cached at the optimum density (116). Establishing buried cache densities greater than the optimum would protect more seeds from birds but also increases the possibility of exploitation by pilfering squirrels. Therefore, the scattered caches probably serve more as an emergency food reserve than as a complete and reliable winter food supply. Scattered caches in tits, nuthatches, crows, and canids may serve the same general function, but because they are generally retrieved within a few days, storing at higher than optimum densities is unlikely to occur.

European jays, piñon jays, nutcrackers, and perhaps some other jays are the only species that scatter hoard enough food to support themselves through the winter, and only the first three have enough left in the spring to feed nestlings and/or fledglings (14, 20a, 62, 119, 130). The extensive food stores are possible because of the large territories pairs of European jays (14) and nutcrackers (119) maintain and the large, exclusive areas used by communal flocks of piñon jays (5, 62). The dry, exposed, relatively barren communal cache areas used by piñon jays and nutcrackers may allow greater optimum densities of caches than other parts of the piñon jay's and the nutcrackers' territories because these areas support lower densities of potential rodent thieves. Within these areas, birds do not cache right next to patches of living vegetation, perhaps because such sites are more susceptible to thievery (123). Nutcrackers may have smaller territories on islands because they can place their caches closer together where there are no rodents (119).

Many factors associated with the large number of caches these communal cachers require affect the foraging and anatomy of these birds. For example, the isolated communal cache areas may be up to 10 km and 22 km from the harvest sites for piñon jays and nutcrackers, respectively. Such distances seemingly select for the birds' ability to fly faster and carry more seeds per trip than other North American jays (130). These communally caching corvids can also distinguish viable seeds from aborted seeds when foraging, and they may bury two or more seeds per cache (130). Piñon jay reproduction is finely tuned to the supply of cached food; gonad development is initiated by the sight of pine seeds, rather than by photoperiod (62). European jays increase their foraging efficiency by choosing long narrow acorns to hold in their esophagi and large spherical ones to carry in their bills, allowing them to transport more acorns (14).

In addition to scattering caches at some optimum density, individual cachers usually conceal their caches under the surface of the soil, in bark, or under lichens or pieces of wood (1, 6, 18, 39, 56, 57, 68, 81, 113, 123, 124). Even red-headed woodpeckers, which defend their larder hoards, minimize the chances of theft by breaking up acorns and dispersing and covering the pieces

(54, 80). European jays prefer to cache acorns in rough ground, perhaps because it is easier to insert the acorn into the soil (14) or because disturbances in rough ground are less detectable by cache thieves. Caged nutcrackers use such disturbances as a clue for finding each other's caches (128). Gray jays can cache food in a great variety of locations because their modified salivary glands (12) allow them to make boluses of food that can be stuck onto vertical surfaces (21).

THE SEQUENCE OF CACHE USE

Having larder hoards should allow the owner to maintain a balanced diet while using up its cached food. Scatter hoarders, however, should eat their most valuable food first, assuming that all of their caches have an equal probability of being stolen. Even scatter hoarders should maintain a more balanced diet than animals that do not cache, at least to the extent that scatter hoarders have more control of their cached food supply than animals that do not cache. In one experiment, both deer mice and pocket mice were given a 10-day supply of seeds. The pocket mice, which are more likely to cache food in the lab than the deer mice, ate a more balanced diet of five species of seeds of differing quality (94). Scatter-hoarding black-capped chickadees allowed to store two species of seeds in a controlled environment revisited and spent more time at caches of the preferred seed species than at those of the less preferred seed species, as would be expected of a scatter hoarder (101a).

Factors other than the ability to defend caches may influence the order of their use. If seeds cached in humid burrows tend to germinate or decompose, then the first cached seeds should be consumed first. Large kangaroo rat species make extensive underground caches in humid burrows, and the seeds may become moldy. The rodents may manage their caches to take advantage of the beneficial by-products produced by the molds and avoid the liabilities. Thus, laboratory experiments show that kangaroo rats prefer slightly moldy seeds over both very moldy (i.e. toxic) seeds and unmoldy, moist seeds (92, 94a). Chipmunks bite off the tip of the embryo in germinating beech seeds in their larder hoards and thus maintain their food value (23).

Gray and fox squirrels gain energy faster from acorns than from walnuts, but they must consume almost twice the amount of acorns to get the same energy as the more lipid-rich, highly digestible walnut kernels yield (112). This difference makes acorns a preferred food in the winter when rapid feeding allows squirrels to spend more time in the nest and walnuts a preferred food in the spring and fall when a light gut content may benefit active squirrels involved in mating and caching, respectively (112). The seasonal variation helps explain why both types of nuts are cached and why the different species are used during different seasons (117). Other squirrels that cache two or more species of food should also feed from their caches in an appropriate sequence, perhaps by eating the species that provides the most efficient food intake during the worst weather. Black-capped chickadees, which in a laboratory setting store food of two different qualities, remember the location of the food and its quality and search more often in the sites where they stored the higher quality food (101a).

To retrieve caches efficiently, animals must remember which caches have been emptied and which have not. Controlled laboratory experiments demonstrate that black-capped chickadees have this ability, whether the cache was emptied by the bird or randomly by the investigator (101a). Red foxes (45) and wolves and coyotes (43) keep an inventory of which caches they deplete by urinating on the cache hole immediately after retrieving the food. These canids search in areas for pilfered caches that they had not been able to scent mark much longer than in areas where they had retrieved the cached items and marked the location (43, 45). Individual flying squirrels (81) and chipmunks (133) can distinguish between nuts they have cached and those cached by other individuals, apparently on the basis of individually specific secretions from sebaceous glands on the lips and cheeks. They prefer to eat nuts from caches made by other individuals (81, 133), thus preserving their own caches.

EVOLUTIONARY FEEDBACK FROM STORED FOOD

For the most part, cached items are prey that have lost in an ecological predator-prey interaction. The evolutionary effect on cached items, if any, should not be different from any other predator-prey relationship: The prey that are easiest to catch leave fewer genes to the next generation, making that generation more difficult to catch. The large species of seeds that are scatter hoarded ("planted") in the ground by jays, nutcrackers, squirrels, and hystricomorph rodents, however, are probably represented more heavily in future generations than other seeds of the same species (14, 18, 20a, 36, 62, 78, 113, 116, 123, 125, 126, 129).

The traits of individual trees or seeds that promote their use by soil scatter hoarders will appear more frequently because dispersal away from the parent plant is usually beneficial. In piñon pines, color differences between full and aborted seeds allow piñon jays and nutcrackers to identify appropriate seeds to cache quickly, and the vertical orientation of the pine cone with seeds in pockets in the scales gives birds rapid access to the seeds before many drop to the ground (130). Piñon pines at lower elevational extremes of all pines and white-barked and limber pines at the high elevational extremes are adapted to dry soils by their low bushy shape and large seeds, both of which also inhibit effective wind dispersal (125, 129). They were probably preadapted to dispersal by scatter hoarding animals. The short, bushy growth form of the piñon tree displays the entire cone crop for easy assessment by flying birds (130). In addition, communal cachers, which choose dry, exposed soils to avoid cache loss to rodents (123, 125), "plant" the seeds in an ideal germination location.

Limber and white-barked pines growing in dense stands lose most of their cones to tree squirrels (8, 123), creating further selection for adaptation to barren slopes where fewer seed predators reside.

The long, narrow shape of acorns in the subgenus of the white oaks *(Lepidobalanus)* allows jays to carry more of them, which in turn promotes long distance flights (14). It is the acorns of this subgenus that germinate in the fall, using up an appreciable portion of the cotyledonary energy reserve by the following spring when the birds return to the caches (30). European jays check developing oak seedlings to find the remains of buried acorns and eat any remaining tissue (14). The petioles of the cotyledons have apparently been selected to minimize losses to the jay by having natural abscission layers that break when the acorn is pulled, leaving the seedling intact (30). Gray squirrels bite off the tip of the embryo before caching many white oak acorns, thus killing the embryo and preserving the full food value of their cache (30); and adult squirrels kill a larger proportion of the seeds in their caches than naive juvenile squirrels. The large annual variation in oak mast crops may be advantageous for trees by generating a greater proportion of naive juveniles in the consumer populations. The large annual variation in piñon pine seed crops means that nutcrackers (129) and piñon jays (62) store food far in excess of their needs during mast years and leave many unretrieved caches to germinate. Temperate species of nuts in the Fagaceae and Juglandaceae scatter hoarded by squirrels provide no food other than the embryo and endosperm. In contrast, in the New World tropics the nuts from palms and legumes that are scatter hoarded ("planted") by rodents have a pulpy layer around the nut that is rich in lipids or carbohydrates and is used by rodents for food. Researchers disagree as to whether selection for the pulpy layer came from rodents that could store more nuts if they got a quick meal (111) or from an extinct megafauna that ate the entire fruit and defecated the seeds before rodents extracted them from dung piles for their scatter hoards (50).

The optimal spacing of scatter-hoarded nuts is determined by their value to cache thieves (see above). The value of the nut is determined by the rate at which it provides food to the thief, which is influenced by the size of the kernel and the thickness of the shell. Selection acting on the thickness of black walnut shells may result in the optimum cache density of one nut per 100 m^2 detected in field studies (116), which coincides with the spacing used for maximum production in commercial orchards.

DISCUSSION

Several circumstances seem to preadapt animals to caching. All birds and mammals known to cache also progressively provision their young, as do the advanced Hymenoptera, where caching was initiated as provisioning for offspring. Bringing food back to a safe place to eat ("security eating"; 25, 26) or to an "anvil" for processing (95) may also have preadapted animals to caching. All of these activities would be easier for animals with a good spatial memory, which is extremely advantageous for using caches as well. A variable food supply is also associated with the evolution of caching, providing a benefit during periods of food shortage in return for the cost of caching when food is plentiful.

The phylogenetic distribution of caching species seems to demonstrate the opportunistic nature of the evolutionary process. Many factors can interact to promote the evolution of caching, including a competitive environment, the type of food available and eaten, resource spacing, climate, and body size. For example, the right climate, seasonally scarce food, and microhabitats in which caches can be kept from interspecific competitors provide the opportunity for larder hoarding to evolve. If the food cannot be defended against interspecific competitors, scatter hoarding will still decrease the probability that a competitor will get it and increase the chance that the cacher will retrieve it, especially if memory is involved.

Because one or more of the factors mentioned above can be critical in determining the evolution of caching, overall patterns among taxa that cache are not obvious or abundant. One that does emerge is that although almost every type of food is cached by birds and mammals, seeds are probably the most commonly stored food items, and seed eaters, the most common cachers. The concentrated energy reserves of seeds coupled with their adaptations to dormancy make them ideal for collection, transport, and storage.

The type of caching and body size are also related. For example, granivores that store food for weeks or months range in size from about 100 g to 3 kg. The size of animals using seeds as a sole source of stored energy while staying active over a period of months is probably limited by seeds' relatively small size. The largest desert rodents that larder hoard small seeds (76, 131) weigh about 150 g while in temperate and tropical forests squirrels (1 kg) and agoutis (3 kg), which store much larger seeds (113, 116), lie at the upper extreme. Corvids and woodpeckers that depend on stored seeds for several months also lie near the upper size limit for seed-eating birds. Smaller birds and mammals that store seeds may not be able to defend a larder large enough to support them for several months. In addition, because of their higher metabolic rates, they need to store more food per gram of body weight (R. P. Balda, personal communication). For tits and nuthatches, these factors may prevent them from making caches large enough to last the winter. Because cheek pouch volume is proportional to body weight in small heteromyid rodents (79), smaller rodents may have to spend more time foraging than a larger species in order to amass a large enough cache.

Geographically, it appears that caching is more common in temperate regions than in tropical areas. This pattern may be related to the greater predictability of food supplies in tropical areas, which reduces the selection pressures favoring its evolution. Tropical habitats may also promote spoilage because of their relatively high temperatures and humidity, reducing the efficacy of caching as a means of surviving through periods of poor production.

Caching is only one way of avoiding food shortages during periods of unreliable or low productivity; migration, torpor, and fat storage are alternative strategies. Migration is inappropriate, however, for animals with low vagility, such as most mammals. Torpor is usually effective for small homeotherms, which do not have large body masses that must be cooled and rewarmed as they enter and exit torpor. Many small mammals that do use winter torpor also cache food to ingest when they periodically rouse to excrete wastes, often relying on the small gauge of their burrows to protect their small larders from thieves. The alternative to storing energy as food is storing it as body fat. This strategy is unsuitable for animals that must remain active throughout the period of cache use, as it can significantly reduce mobility. In addition, the physiological conversion of energy to fat and back to usable energy is wasteful, since a considerable percentage is "lost" in the conversion process.

Analyses of the relative advantages of different responses to seasonal food shortages are best performed on organisms that express a wide range of behavioral responses to variation in food supply—e.g. hamsters (67) and chipmunks (23, 100). Eastern chipmunks make both larder and scatter hoards (23, 100), apparently choosing the type that optimizes foraging under each set of conditions (58). In controlled laboratory conditions, individuals vary their levels of fat deposition (15, 86, 100), the amount of stored food (15, 65), and their storing method (in larder or scatter hoards; 65) in response to environmental variables. The length of torpor varies among individuals (15, 86, 133). It is longer in females than in males (133) on the average, although members of both sexes may opt for no torpor at all. Southern (120) and northern (23) populations may differ in all of these characteristics, offering the prospect that both genetic and facultative differences are salient.

Literature Cited

1. Abbott, H. G., Quink, T. F. 1970. Ecology of eastern white pine seed caches made by small forest mammals. *Ecology* 51:271–78

2. Aleksiuk, M. 1970. The seasonal food regime of Arctic beavers. *Ecology* 51:264–70

3. Andersson, M., Krebs, J. 1978. On the evolution of hoarding behaviour. *Anim. Behav.* 26:707–11

4. Balda, R. P. 1980. Recovery of cached seeds by a captive *Nucifraga caryocatactes*. *Z. Tierpsychol.* 52:331–46

5. Balda, R. P., Bateman, G. C. 1971. Flocking and annual cycle of the pinon jay, *Gymnorhinus cyanocephalus*. *Condor* 73:287–302

6. Balgooyen, T. G. 1976. Behavior and ecology of the American kestrel (*Falco sparverius* L.) in the Sierra Nevada of California. *Univ. Calif. Berkeley Publ. Zool.* 103:1–83

7. Begg, R. J., Dunlop, C. R. 1980. Security eating, and diet in the large rock-rat, *Zygomys woodwardi* (Rodentia:Muridae). *Aust. Wildl. Res.* 7:63–70

8. Benkman, C. W., Balda, R. P., Smith, C. C. 1984. Adaptation for seed dispersal and the compromises due to seed predation in limber pine. *Ecology* 45:632–42
9. Blair, W. F. 1937. The burrows and food of the prairie pocket mouse. *J. Mammal.* 18:188–91
10. Bock, C. E. 1970. The ecology and behavior of the Lewis woodpecker *(Asyndesmus lewis)*. *Univ. Calif. Berkeley Publ. Zool.* 92:1–100
11. Bock, C. E., Lepthien, L. W. 1975. A Christmas count analysis of woodpecker abundance in the United States. *Wilson Bull.* 87:355–66
12. Bock, W. J. 1961. Salivary glands in the gray jays *(Perisoreus)*. *Auk* 78:355–65
13. Bock, W. J., Balda, R. P., Vander Wall, S. B. 1973. Morphology of the sublingual pouch and tongue musculature in Clark's nutcracker. *Auk* 90:491–519
14. Bossema, I. 1979. Jays and oaks: An eco-ethological study of symbiosis. *Behaviour* 70:1–117
15. Brenner, F. J., Lyle, P. D. 1975. Effects of previous photoperiodic conditions and visual stimulation on food storage and hibernation in the eastern chipmunk *(Tamias striatus)*. *Am. Midl. Nat.* 93:227–34
16. Buller, A. H. R. 1920. The red squirrel of North America as a mycophagist. *Trans. Br. Mycol. Soc.* 6:355–62
17. Cade, T. J. 1967. Ecological and behavioral aspects of predation by the northern shrike. *Living Bird* 6:43–86
18. Cahalane, V. H. 1942. Caching and the recovery of food by the western fox squirrel. *J. Wildl. Manage.* 6:338–52
19. Collopy, M. W. 1977. Food caching by female American kestrels in winter. *Condor* 79:63–68
20. Cowie, R. J., Krebs, J. R., Sherry, D. F. 1981. Food storing in marsh tits. *Anim. Behav.* 29:1252–59
20a. Darley-Hill, S., Johnson, W. C. 1981. Acorn dispersal by the Blue Jay *(Cyanocitta cristata)*. *Oecologia* 50:231–32
21. Dow, D. D. 1965. The role of saliva in food storage by the gray jay. *Auk* 82:139–54
22. Elgmork, K. 1982. Caching behavior of brown bears *(Ursus arctos)*. *J. Mammal.* 63:607–12
23. Elliott, L. 1978. Social behavior and foraging ecology of the eastern chipmunk *(Tamias striatus)* in the Adirondack Mountains. *Smithson. Contrib. Zool.* 265:1–107
24. Etienne, A. S., Emmanelli, E., Zinder, M. 1982. Ontogeny of hoarding in the golden hamster: The development of motor patterns and their sequential coordination. *Dev. Psychobiol.* 15:33–45
25. Ewer, R. F. 1965. Food burying in the African ground squirrel, *Xerus erythropus* (E. Geoff.). *Z. Tierpsychol.* 22:321–27
26. Ewer, R. F. 1967. The behaviour of the African giant rat *(Cricetomys gambianus* Waterhouse). *Z. Tierpsychol.* 24:6–79
27. Finley, R. B. 1958. The wood rats of Colorado: Distribution and ecology. *Univ. Kansas Publs., Mus. Nat. Hist.* 10:213–552
28. Fleming, T. H., Brown, G. J. 1975. An experimental analysis of seed hoarding and burrowing behavior in two species of Costa Rican heteromyid rodents. *J. Mammal.* 56:301–15
29. Formozov, A. N. 1966. Adaptive modifications of behavior in mammals of the Eurasian steppes. *J. Mammal.* 47:208–23
30. Fox, J. F. 1982. Adaptation of gray squirrel behavior to autumn germination by white oak acorns. *Evolution* 36:800–9
31. Fulk, G. W. 1977. Food hoarding of *Bandicota bengalensis* in a rice field. *Mammalia* 41:539–41
32. Fulk, G. W., Khokhar, D. R. 1980. Observations on the natural history of a pika *(Ochotona refescens)* from Pakistan. *Mammalia* 44:51–58
33. Funmilayo, O. 1979. Food consumption, preferences and storage in the mole. *Acta Theriol.* 24:379–89
34. Gates, J. E., Gates, D. M. 1980. A winter food cache of *Microtus pennsylvanicus*. *Am. Midl. Nat.* 103:407–8
35. Gibb, J. A. 1960. Populations of tits and goldcrests and their food supply in pine plantations. *Ibis* 102:163–208
36. Glanz, W. E., Thorington, R. W. Jr., Giacalone-Madden, J., Heaney, L. R. 1982. Seasonal food use and demographic trends in *Sciurus granatensis*. In *The Ecology of a Tropical Forest: Seasonal Rhythms and Long-term Changes*, ed. E. G. Leigh Jr., A. S. Rand, D. M. Windsor, pp. 239–52. Washington DC: Smithsonian Inst.
37. Goodwin, D. 1976. *Crows of the World.* Ithaca, NY: Cornell Univ. Press. 354 pp.
38. Grinnell, J., Orr, R. T. 1934. Systematic review of the *californicus* group of the rodent genus *Peromyscus*. *J. Mammal.* 15:210–20
39. Haftorn, S. 1974. Storage of surplus food by the boreal chickadee *Parus hudsonicus* in Alaska, with some records on the mountain chickadee *Parus gambeli* in Colorado. *Ornis Scand.* 5:145–61
40. Hansen, R. M. 1960. Pocket gophers in Colorado. *Colo. State Univ. Exp. Stn. Bull. 508-S.* 26 pp.
41. Deleted in proof

42. Harper, R. J., Jenkins, D. 1982. Food caching in European otters *(Lutra lutra). J. Zool.* 197:297–98
43. Harrington, F. H. 1981. Urine-marking and caching behavior in the wolf. *Behaviour* 76:280–88
44. Hawbrecker, A. C. 1940. The burrowing and feeding habits of *Dipodomys venustus. J. Mammal.* 21:388–96
45. Henry, J. D. 1977. The use of urine marking in the scavenging behavior of the red fox *(Vulpes vulpes). Behaviour* 61:82–106
46. Howard, W. E., Evans, F. C. 1961. Seeds stored by prairie deer mice. *J. Mammal.* 42:260–63
47. Imaizumi, Y. 1979. Seed storing behavior of *Apodemus speciosus* and *Apodemus argenteus. Zool. Mag. Zool. Soc. Jpn.* 88:43–49
48. Ingram, W. M. 1942. Snail associates of *Blarina brevicauda talpoides* (Say). *J. Mammal.* 23:255–58
48a. James, P. C., Verbeek, N. A. M. 1983. The food storage behaviour of the northwestern crow. *Behaviour* 85:276–91
49. Janzen, D. H. 1977. Why fruits rot, seeds mold, and meat spoils. *Am. Nat.* 111: 691–713
50. Janzen, D. H., Martin, P. S. 1982. Neotropical anachronisms: The fruits the gomphotheres ate. *Science* 215:19–27
51. Jarvis, J. U. M., Sale, J. B. 1971. Burrowing and burrow patterns of East African mole-rats *Tachyoryctes, Heliophobius,* and *Heterocephalus. J. Zool.* 163: 451–79
52. Jenkins, S. H. 1980. A size-distance relation in food selection by beavers. *Ecology* 61:740–46
53. Kay, F. R., Whitford, W. G. 1978. The burrow environment of the banner-tailed kangaroo rat, *Dipodomys spectabilis,* in south-central New Mexico. *Am. Midl. Nat.* 99:270–79
54. Kilham, L. 1958. Sealed-in winter stores of red-headed woodpeckers. *Wilson Bull.* 70:107–13
55. Kilham, L. 1958. Territorial behavior of wintering red-headed woodpeckers. *Wilson Bull.* 70:347–58
56. Kilham, L. 1963. Food storing in red-bellied woodpeckers. *Wilson Bull.* 75: 227–34
57. Kilham, L. 1974. Covering of stores by white-breasted and red-breasted nuthatches. *Condor* 76:108–9
58. Kramer, D. L., Nowell, W. 1980. Central place foraging in the eastern chipmunk, *Tamias striatus. Anim. Behav.* 28:772–78
59. Kruuk, H. 1972. Surplus killing by carnivores. *J. Zool.* 166:233–44
60. Launay, M. 1982. Ontogénèse du comportement d'amassement chez le hamster. *Biol. Behav.* 7:1–15
61. Lewis, A. R. 1980. Patch use by gray squirrels and optimal foraging. *Ecology* 61:1371–79
62. Ligon, J. D. 1978. Reproductive interdependence of piñon jays and piñon pines. *Ecol. Monogr.* 48:111–26
63. Linsdale, J. M. 1946. *The California Ground Squirrel.* Berkeley: Univ. Calif. Press. 475 pp.
64. Linsdale, J. M., Tevis, L. P. Jr. 1951. *The Dusky-Footed Wood Rat.* Berkeley: Univ. Calif. Press. 664 pp.
65. Lockner, F. R. 1972. Experimental study of food hoarding in the red-tailed chipmunk, *Eutamias ruficaudus. Z. Tierpsychol.* 31:410–18
66. Long, C. A. 1976. Evolution of mammalian cheek pouches and a possibly discontinuous origin of a higher taxon (Geomyoidea). *Am. Nat.* 110:1093–97
67. Lyman, C. P. 1954. Activity, food consumption and hoarding in hibernators. *J. Mammal.* 35:545–52
68. MacDonald, D. W. 1976. Food caching by red foxes and some other carnivores. *Z. Tierpsychol.* 42:170–85
69. MacDonald, D. W. 1979. Some observations and field experiments on the urine marking behaviour of the red fox, *Vulpes vulpes* L. *Z. Tierpsychol.* 51:1–22
70. MacRoberts, M. H., MacRoberts, B. R. 1976. Social organization and behavior of the acorn woodpecker in central coastal California. *Ornithol. Monogr.* 21:1–115
71. Malcolm, J. R. 1980. Food caching by African wild dogs *(Lycaon pictus). J. Mammal.* 61:743–44
72. Malyshev, S. I. 1968. *Genesis of the Hymenoptera and the Phases of Their Evolution.* London: Methuen. 319 pp. (From Russian)
73. Martin, A. C., Zim, H. S., Nelson, A. L. 1951. *American Wildlife and Plants, A Guide to Wildlife Food Habits.* New York: McGraw-Hill. 500 pp.
74. Martin, I. G. 1984. Factors affecting food hoarding in the short-tailed shrew *Blarina brevicauda. Mammalia* 48:65–71
75. Matson, J. O., Christian, D. P. 1977. A laboratory study of seed caching in two species of *Liomys* (Heteromyidae). *J. Mammal.* 58:670–71
76. Monson, G. 1943. Food habits of the banner-tailed kangaroo rat in Arizona. *J. Wildl. Manage.* 7:98–102
77. Monson, G., Kessler, W. 1940. Life history notes on the banner-tailed kangaroo rat, Merriam's kangaroo rat, and the white-throated wood rat in Arizona and New Mexico. *J. Wildl. Manage.* 4:37–43

78. Morris, D. 1962. The behaviour of the green acouchi *(Myoprocta pratti)* with special reference to scatter hoarding. *J. Zool.* 139:701–32

79. Morton, S. R., Hinds, D. S., MacMillen, R. E. 1980. Cheek pouch capacity in heteromyid rodents. *Oecologia* 46:143–46

80. Moskovits, D. 1978. Winter territorial and foraging behavior of red-headed woodpeckers in Florida. *Wilson Bull.* 90:521–35

81. Muul, I. 1968. Behavioral and physiological influences on the distribution of the flying squirrel, *Glaucomys volans. Misc. Publ. Mus. Zool. Univ. Mich.* 134:1–66

82. Nel, J. A. J. 1975. Aspects of the social ethology of some Kalahari rodents. *Z. Tierpsychol.* 37:322–31

83. Nevo, E. 1979. Adaptive convergence and divergence of subterranean mammals. *Ann. Rev. Ecol. Syst.* 10:269–308

84. Novakowski, N. S. 1967. The winter bioenergetics of a beaver population in northern latitudes. *Can. J. Zool.* 45:1107–18

85. Orians, G. H., Pearson, N. E. 1979. On the theory of central place foraging. In *Analysis of Ecological Systems,* ed. D. J. Horn, G. R. Stairs, R. D. Mitchell, pp. 155–77. Columbus: Ohio State Univ. Press

86. Panuska, J. A. 1959. Weight patterns and hibernation in *Tamias striatus. J. Mammal.* 40:554–66

87. Pettifer, H. L., Nel, J. A. J. 1977. Hoarding in four Southern African rodent species. *Zool. Afr.* 12:409–18

88. Phelan, F. J. S. 1977. Food caching in the screech owl. *Condor* 79:127

89. Pitelka, F. A., Tomich, P. Q., Treichel, G. W. 1955. Ecological relations of jaegers and owls as lemming predators near Barrow, Alaska. *Ecol. Monogr.* 25:85–117

90. Platt, W. J. 1976. The social organization and territoriality of short-tailed shrew *(Blarina brevicauda)* populations in old-field habitats. *Anim. Behav.* 24:305–18

91. Rao, A. M. K. M. 1980. Demography and hoarding among the lesser bandicoot rat *Bandicota bengalensis* in rice fields. *Saeugetierkd. Mitt.* 28:312–14

92. Rebar, C., Reichman, O. J. 1983. Ingestion of moldy seeds by heteromyid rodents. *J. Mammal.* 64:713–15

93. Reichman, O. J., Brown, J. H. 1979. The use of torpor by *Perognathus amplus* in relation to resource distribution. *J. Mammal.* 60:550–55

94. Reichman, O. J., Fay, P. 1983. Comparison of the diets of a caching and a noncaching rodent. *Am. Nat.* 122:576–81

94a. Reichman, O. J., Rebar, C. 1985. Seed preferences by desert rodents based on levels of mouldiness. *Anim. Behav.* In press

95. Roberts, R. C. 1979. The evolution of avian food-storing behavior. *Am. Nat.* 114:418–38

96. Rutter, R. J. 1969. A contribution to the biology of the gray jay *(Perisoreus canadensis). Can. Field Nat.* 83:300–16

97. Schnell, J. H. 1958. Nesting behavior and food habits of goshawks in the Sierra Nevada of California. *Condor* 60:377–403

98. Schreiber, R. K. 1978. Bioenergetics of the Great Basin pocket mouse, *Perognathus parvus. Acta Theriol.* 23:469–87

99. Schroder, G. D. 1979. Foraging behavior and home range utilization of the bannertail kangaroo rat *(Dipodomys spectabilis). Ecology* 60:657–65

100. Shaffer, L. 1980. Use of scatterhoards by eastern chipmunks to replace stolen food. *J. Mammal.* 61:733–34

101. Shaw, W. T. 1934. The ability of the giant kangaroo rat as a harvester and storer of seeds. *J. Mammal.* 15:275–86

101a. Sherry, D. 1984. Food storage by black-capped chickadees; memory for the location and contents of caches. *Anim Behav.* 32:451–64

102. Sherry, D. F. 1985. Foraging for stored food. In *Foraging: Sixth Harvard Symposium on Quantitative Analysis of Behavior.* In press

103. Sherry, D. F., Avery, M., Stevens, A. 1982. The spacing of stored food by marsh tits. *Z. Tierpsychol.* 58:153–62

104. Sherry, D. F., Krebs, J. R., Cowie, R. J. 1981. Memory for the location of stored food in marsh tits. *Anim. Behav.* 29:1260–66

105. Shettleworth, S. J., Krebs, J. R. 1982. How marsh tits find their hoards: The roles of site preference and spatial memory. *J. Exp. Psychol. Anim. Behav. Process.* 8:354–75

106. Slough, B. G. 1978. Beaver food cache structure and utilization. *J. Wildl. Manage.* 42:644–46

107. Smith, A. T. 1981. Territoriality and social behaviour of *Ochotona princeps.* In *Proc. World Lagomorph Conf.,* ed. K. Myers, C. D. MacInnes, pp. 310–23. Guelph, Canada: Guelph Univ. Press

108. Smith, A. T. 1981. Population dynamics of pikas. See Ref. 107, pp. 572–86

109. Smith, C. C. 1968. The adaptive nature of social organization in the genus of tree squirrels *Tamiasciurus. Ecol. Monogr.* 38:31–63

110. Smith, C. C. 1970. The coevolution of pine squirrels *(Tamiasciurus)* and conifers. *Ecol. Monogr.* 40:349–71

111. Smith, C. C. 1975. The coevolution of seeds and seed predators. In *Coevolution of Animals and Plants,* ed. L. E. Gilbert, P. H. Raven, pp. 53–77. Austin: Univ. Tex. Press

112. Smith, C. C., Follmer, D. 1972. Food preferences of squirrels. *Ecology* 53:82–91

113. Smythe, N., Glanz, W. E., Leigh, E. G. Jr. 1982. Population regulation in some terrestrial frugivores. See Ref. 36, pp. 227–38

114. Sorenson, M. W. 1962. Some aspects of water shrew behavior. *Am. Midl. Nat.* 68:445–62

115. Stallcup, P. L. 1968. Spatio-temporal relationships of nuthatches and woodpeckers in ponderosa pine forests of Colorado. *Ecology* 49:831–43

116. Stapanian, M. A., Smith, C. C. 1978. A model for seed scatterhoarding: Coevolution of fox squirrels and black walnuts. *Ecology* 59:884–96

117. Stapanian, M. A., Smith, C. C. 1984. Density-dependent survival of scatterhoarded nuts: An experimental approach. *Ecology.* In press

118. Sullivan, T. P. 1978. Lack of caching of direct-seeded Douglas fir seeds by deer mice. *Can. J. Zool.* 56:1214–16

119. Swanberg, P. O. 1956. Territory in the thick-billed nutcracker *Nucifraga caryocatactes. Ibis* 98:412–19

120. Thomas, K. R. 1974. Burrow systems of the eastern chipmunk *(Tamias striatus pipilans* Lowery) in Louisiana. *J. Mammal.* 55:454–59

121. Tinbergen, N., Kruyt, W. 1938. Über die Orientierung des Bienenwolfes *(Philanthus triangulum* Fabr.) III. Die Bevorzugung bestimmter Wegmarken. *Z. Vgl. Physiol.* 25:292–334

122. Tomasi, T. E. 1978. Function of venom in the short-tailed shrew, *Blarina brevicauda. J. Mammal.* 59:852–54

123. Tomback, D. F. 1977. Foraging strategies of Clark's nutcrackers. *Living Bird* 16:123–61

124. Tomback, D. F. 1980. How nutcrackers find their seed stores. *Condor* 82:10–19

125. Tomback, D. F. 1983. Nutcrackers and pines: Coevolution or coadaptation? In *Coevolution,* ed. M. H. Nitecki, pp. 179–223. Chicago: Univ. Chicago Press

126. Turcek, F. J., Kelso, L. 1968. Ecological aspects of food transportation and storage in the Corvidae. *Commun. Behav. Biol.* 1:277–97

127. Vandermeer, J. H. 1979. Hoarding behavior of captive *Heteromys desmarestianus* (Rodentia) on fruits of *Welfia georgii,* a rainforest dominant palm in Costa Rica. *Brenesia* 16:107–16

128. Vander Wall, S. B. 1982. An experimental analysis of cache recovery by Clark's nutcracker. *Anim. Behav.* 30:84–94

129. Vander Wall, S. B., Balda, R. P. 1977. Coadaptations of the Clark's nutcracker and piñon pine for efficient seed harvest and dispersal. *Ecol. Monogr.* 47:89–111

130. Vander Wall, S. B., Balda, R. P. 1981. Ecology and evolution of food-storage behavior in conifer-seed-caching corvids. *Z. Tierpsychol.* 56:217–42

131. Vorhies, C. T., Taylor, W. P. 1922. Life history of the kangaroo rat, *Dipodomys spectabilis spectabilis* Merriam. *US Dep. Agric. Bull. 1091,* pp. 1–40

132. Walker, E. P., Paradiso, J. L. *Mammals of the World.* Baltimore: Johns Hopkins Univ. Press. 1500 pp. 3rd ed.

133. Wrazen, J. A., Wrazen, L. A. 1982. Hoarding, body mass dynamics, and torpor as components of survival strategy of the eastern chipmunk. *J. Mammal.* 63:63–72

Ann. Rev. Ecol. Syst. 1984. 15:353–91

THE ROLE OF DISTURBANCE
IN NATURAL COMMUNITIES

Wayne P. Sousa

Department of Zoology, University of California, Berkeley, California 94720

INTRODUCTION

Two features characterize all natural communities. First, they are dynamic systems. The densities and age-structures of populations change with time, as do the relative abundances of species; local extinctions are commonplace (37). For many communities, a self-reproducing climax state may only exist as an average condition on a relatively large spatial scale, and even that has yet to be rigorously demonstrated (36). The idea that equilibrium is rarely achieved on the local scale was expressed decades ago by a number of forest ecologists (e.g. 101, 168). One might even argue that continued application of the concept of climax to natural systems is simply an exercise in metaphysics (41). While this view may seem extreme, major climatic shifts often recur at time intervals shorter than that required for a community to reach competitive equilibrium or alter the geographical distributions of species (6, 21, 43, 76, 92). Climatic variation of this kind influences ecological patterns over large areas, sometimes encompassing entire continents. Other agents of temporal change in natural communities operate over a wide range of smaller spatial scales (47, 242).

Second, natural communities are spatially heterogeneous. This statement is true at any scale of resolution (242), but it is especially apparent on what is commonly referred to as the regional scale. (By *region* I mean an area that potentially encompasses more than one colonizable patch.) Across any land or seascape, one observes a mosaic of patches identified by spatial discontinuities in the distributions of populations (153, 159, 161, 231, 239, 240). Closer examination often reveals a smaller-scale patchwork of same-aged individuals (e.g. 85–87, 101, 146, 199, 204, 217–220, 235, 246).

Discrete patch boundaries sometimes reflect species-specific responses to

353

0066-4162/84/1120-0353$02.00

steep gradients in the physical environment. Such responses account for only a small part of the spatial heterogeneity found in natural communities, however. Even where background physical conditions are relatively uniform across a site, opportunities for recruitment, growth, reproduction, and survival vary spatially, reflecting variation in the intensity of biological interactions, resource availability, and microclimatological conditions. By itself, this spatial and temporal variation in the density of "safe sites" for establishment (sensu 83) may only partially explain local differences in the demography of populations. The availability of propagules sometimes limits rates of establishment (84, 94, 215). Since environmental characteristics and population parameters change with time, the mosaic patterns are themselves dynamic.

To interpret and predict the patterns observed in nature accurately, our methods of study must embrace temporal and spatial variability as essential features of population and community dynamics. There is now abundant evidence that in the absence of such variability many species would cease to exist. Inherent in this view is the recognition that traces of history are etched in the structures of many, if not most, natural communities. Often, present-day patterns can only be interpreted if the organisms themselves yield clues (e.g. fire scars) as to the identity and timing of historical events or if the assemblage has been monitored continuously for a long time.

Disturbance is both a major source of temporal and spatial heterogeneity in the structure and dynamics of natural communities and an agent of natural selection in the evolution of life histories. These roles are clearly interdependent. The differential expression of life history attributes under different regimes of disturbance produces much of the spatial and temporal heterogeneity one observes in natural assemblages. On the other hand, the heterogeneity in environmental conditions (both biological and physical) induced by disturbance is probably a key part of the "habitat templet" (194) that selects among life history variants.

This review emphasizes the impact of disturbance on the numerical abundance of populations and on the relative abundance of species in guilds and communities. Disturbance also has an important influence, however, on ecosystem-level processes such as primary and secondary production, biomass accumulation, energetics, and nutrient cycling (e.g. 17, 20, 136, 197, 222, 233). Indeed, too often studies focused at the population or community level overlook potentially significant effects of ecosystem-level processes on population dynamics. In forests, for instance, disturbance sometimes causes a net increase in the amount of soil nitrogen available to early colonizers. It is probably not coincidental that the seeds of a number of pioneer plant species germinate in response to high levels of soil nitrates (9, 157).

Assemblages of both sessile and mobile organisms are subject to disturbance. The effects of disturbance have been much more thoroughly studied in the former, however, simply because sessile organisms are easier to observe and quantify. Because of this disparity in our understanding of the role of disturbance in the two sorts of assemblages and because of apparent fundamental differences in their responses to disturbance, I have treated mobile organisms in a separate section at the end.

WHAT IS A DISTURBANCE?

Traditionally, disturbances have been viewed as uncommon, irregular events that cause abrupt structural changes in natural communities and move them away from static, near equilibrium conditions (104, 235). This definition has little utility in light of the following observations:

1. Evidence from long-term censuses suggests that few natural populations or communities persist at or near an equilibrium condition on a local scale (37). There is no clear demarcation between assemblages in an equilibrium state and those that are not.

2. The change caused by any force can vary from negligible to extreme, depending on the intensity of the force and the vulnerability of the target organisms. How does one objectively decide what degree of change along this continuum constitutes a disturbance? The response of perennial species to regular seasonal change in the physical environment is a case in point. When temperatures or rainfall oscillate close to their long-term seasonal averages, organisms respond physiologically and/or behaviorally to ameliorate possible negative effects of the change. With more extreme seasonal fluctuations in the physical environment, the limits of effective physiological or behavioral response are exceeded. At first, this may cause reductions in growth and reproduction, but if the stress becomes severe enough, organisms will die. The number killed can vary from one or just a few individuals to entire populations. Thus, the same basic phenomenon can elicit responses ranging from physiological acclimatization to population extinction, depending on the magnitude of the variation. Since lethal and sublethal responses to such changes (e.g. seasonal migrations of birds; see 103) can markedly alter the community structure, the "objective definition of a threshold at which a periodicity becomes a disturbance [is] difficult at best" (104).

Moreover, the levels of environmental fluctuation to which a present-day species responds with effective homeostatic mechanisms probably represented a far greater hazard early in the species' evolutionary history. Differential mortality and/or reproductive success in the past among individuals differing in genotype probably contributed to the evolution of the homeostatic mechanisms. Therefore, an environmental fluctuation that once caused disturbance does so no longer. This dynamic evolutionary relationship between an organism and the environmental stresses it encounters is ongoing and subject to numerous constraints; no organism can perfectly track fluctuations in its environment. The evolutionary moderation of stress is exhibited in its most extreme form by species that depend on environmental disruption for the completion of their life cycles and the persistence of their populations (227). Allen & Starr (5) argue that at this point the disruptive event ceases to be a disturbance at all.

Given the complexities discussed above, it seems wisest to adopt the view (104) that disturbance lies near one extreme of the continuum of natural perturbations that affect organisms. In the context of this review, a *disturbance* is a discrete, punctuated killing, displacement, or damaging of one or more individuals (or colonies) that directly or indirectly creates an opportunity for new individuals (or colonies) to become established.

AGENTS OF DISTURBANCE

Both physical and biological processes act as agents of disturbance. The former are the kind most often associated with the term disturbance, and their role in natural communities is the primary focus of this review. Examples include fires, ice storms, floods, drought, high winds, landslides, large waves, and desiccation stress. Agents of biological disturbance (45, 235) encompass everything from predation or grazing to nonpredatory behaviors that inadvertently kill or displace other organisms [e.g. digging by mammals and ants in grasslands (114, 163, 179) or by elasmobranchs in marine soft sediments (207, 216)].

The impact of biological agents of disturbance seems generally similar to that of physical agents. Organisms are killed, thereby creating opportunities for recruitment. The timing of biological disturbance, however, is probably subject to a somewhat more complex set of controls. Rates of predation depend on the functional, numerical, and developmental responses of the predator (139). These responses, in turn, are influenced by the physical environment, habitat complexity, presence of alternate prey, availability of prey refuges, and the impact of higher level predators and parasites. For example, the timing of insect outbreaks in forests may be simultaneously influenced by parasitic infection (8), spatial heterogeneity in forest structure (138), and the recent history of physical disturbance (181, 205, 235).

DISTURBANCE IN ASSEMBLAGES OF SESSILE ORGANISMS

A full understanding of the dynamics of populations within habitats subject to disturbance requires knowledge of the regime of disturbance and of the subsequent patterns of recolonization and succession in the disturbed patches. These patterns are a product of certain characteristics of the original disturbance and the life histories of the species available to reoccupy the disturbed site.

The Regime of Disturbance

How an investigator characterizes a regime of disturbance depends on the particular disruptive force and responses being studied. The most commonly used descriptors (e.g. 35, 62, 86, 192, 223) are listed below:

1. *Areal extent*—the size of the disturbed area
2. *Magnitude*—consists of the following two components:
    a.   *Intensity*—a measure of the strength of the disturbing force (e.g. fire temperature, wind speed, wave velocity)
    b.   *Severity*—a measure of the damage caused by the disturbing force [Both of these terms have often been used interchangeably (e.g. 33, 192). Severity seems to denote better the amount of damage caused by a disturbance.]
3. *Frequency*—the number of disturbances per unit time. Separate terms are used for the average frequency of disturbance at the local and the regional spatial scales:
    a.   *Random point frequency*—the mean number of disturbances per unit time at a random point within a region; this is often expressed as the *recurrence* or *return interval* (i.e. the average time between disturbances)
    b.   *Regional frequency*—the total number of disturbances that occur in a geographical area per unit time
4. *Predictability*—measured by the variance in the mean time between disturbances
5. *Turnover rate* or *rotation period*—the mean time required to disturb the entire area in question

Regimes of disturbance vary considerably along a number of spatial and temporal scales. For example, the well-studied forests of North America exhibit a wide range of variation in both present-day and presettlement disturbance regimes (19). Near one extreme is the mixed mesophytic cove forest of the southern Appalachian mountains (176). Scattered deaths of single trees or at most a few neighboring trees by windthrow, lightning, or glaze storms form the predominant pattern of disturbance. Fire is uncommon. According to one study, treefall gaps range up to 1490 m^2, but the average gap is only 31 m^2. For a number of sites, the average percentage of the forest canopy converted to gaps per year ranges from 0.5% to 2.0% (grand average = 1.2%), so that the rotation interval for the canopy layer is 50–200 years. The annual rate at which new gap area is generated by treefall did not vary markedly over time.

The presettlement regimes of disturbance in the conifer and hardwood forests of the Boundary Waters Canoe Area (BWCA) of northern Minnesota lie near the opposite extreme on the geographical scale of variation. Fire is the primary source of disturbance in all forest types there (85–87). The presettlement fire regime of "near boreal" conifer forests consisted of crown fires and/or severe surface fires with a return interval of 50–100 years. A typical fire in this forest type usually burned a large area, probably 400–4000 ha. Pine forests experienced a regime of moderate surface fires with a return interval of approximately 36 years, as well as an overlying regime of severe surface or crown fires that occurred about once every 180 years. The latter fires burned 40–400 ha of old growth stands. Adjacent enclaves of mixed hardwood forest were burned in 400–4000 ha patches by severe surface and crown fires (where conifers were prevalent) about every 80 years. High intensity surface and crown fires kill most of the trees in a burned stand, and regeneration initiates from dispersed or stored seeds. Extensive fire scar analyses indicate that all present-day stands in BWCA originated after fire. Smaller-scale disturbances such as windthrow were insignificant by comparison. Before the adoption of effective fire suppression procedures, an average of 0.8% of the entire BWCA study area (405,000 ha) burned per year, and the average fire rotation period was about 100 years long (85, 87).

Surprisingly, turnover rates estimated for forests in these two areas are quite comparable, despite the striking dissimilarity in the predominant agent of disturbance. The critical difference is that the Minnesota forests experience infrequent large-scale disturbances whereas the disturbances in the Appalachian cove forests are much more regular in time and smaller in area. In addition, the severity of disturbance caused by intense fires is much greater than that caused by windthrow.

I intentionally chose two rather extreme temperate forest examples to demonstrate the wide range of disturbance regimes that occur within this habitat in different geographical locations in North America. Forests in other regions of the continent—for example, the Harvard Forest Tract in New Hampshire (90)—experience a more balanced mix of disturbance by fire and windthrow.

A knowledge of long-term average climatological conditions is useful in explaining coarse differences in the geographic patterns of disturbance. For example, the average annual pattern of precipitation determines the relative importance of fire versus treefall as agents of disturbance in forest communities (29, 87, 235). Similarly, differences in winter temperatures explain why floating ice commonly scours the rocky intertidal shores of New England and eastern Canada but not those along the Pacific coast of the United States (192, 234).

Local patterns of disturbance cannot be predicted from a knowledge of large-scale climatic variation, however. The regime of disturbance at any one site depends on a multitude of local physical and biological factors. In the following sections, I briefly describe some of the more common, and better-studied, local physical agents in natural communities.

WIND   Wind is an important agent of disturbance in many temperate and tropical forests, where it creates gaps of various sizes in the forest canopy by blowing down large branches or trees (19, 22, 84, 176, 237, 238). Rates of treefall vary substantially over time and across a landscape. In most areas, treefall is seasonal. It occurs most frequently during seasons with strong gusting winds and high rainfall (e.g. 22, 232).

Spatial variation in treefall rates is attributable to a number of factors including differences in topography and soil type. Treefall is more likely to occur where there are high prevailing winds and at sites that lie in the path of hurricanes (22, 232, 236, 238). In some forests, surviving trees at the edges of existing gaps may be more likely to blow down in subsequent storms (22, 84, 237, 238); but in other forests, they are not (176). The risk of windthrow is often greater for trees that grow on steep slopes or in soils where a stable root hold cannot be established—for example, in wet, sandy, and some very fertile soils (22, 84, 176).

As trees grow older and taller, they are more likely to be blown down (22, 176). In part, this is because of the increasing forces a stationary, and relatively inflexible, object like a tree experiences as it grows larger in a moving fluid environment. This is particularly true if the flow sometimes accelerates, as when the wind gusts (224: 94). Other factors also contribute to the increasing risk of blowdown as a tree ages, including the weakening effects of insect attack, lightning strikes, disease, and physiological stress. A heavy load of epiphytes, which often develops on older rain forest trees, may contribute further to the chance that a tree will be blown down in storms (201).

Gap size is related to the manner in which a tree falls (22; G. B. Williamson, unpublished manuscript). As one would expect, gap area is positively correlated with tree size in the most common case where the tree falls laterally owing to wind stress. In contrast, some species of trees die in a standing position and gradually collapse downward. This phenomenon produces smaller gaps, and their size is usually uncorrelated with the size of the tree at the time of death. Limb fall usually creates even smaller gaps.

The very largest gaps are generated by multiple tree falls. Such gaps are created when several trees fall synchronously in "domino" fashion or when a new gap is contiguous or overlaps with an older one. In tropical rain forest, extensive liana connections may increase the rate of multiple synchronous treefall (F. E. Putz, cited in 22).

FIRE   The local intensity, frequency, and areal extent of fire in terrestrial plant communities is controlled by complex interrelations among the following six factors (29, 30, 72, 87, 227): (a) frequency and seasonality of ignition sources, (b) moisture content of the fuel (i.e. the potentially flammable living and dead plant material), (c) the rate of fuel accumulation, (d) structural and chemical characteristics of the fuel, (e) mosaic nature of the landscape, and (f) local weather conditions at the time of the fire. Since all of these factors vary over space and time, there is considerable heterogeneity in local fire regimes and consequently in the effects of fire on vegetation.

Lightning is the most significant natural source of fire (205, 226, 227). Far less common natural sources include spontaneous combustion (221), sparks from falling rocks (89, 135), and volcanic eruptions (219, 220, 227). The frequency of lightning strikes alone, however, rarely explains local patterns of fire occurrence, even where there is little anthropogenic influence on the fire regime. Only about 0.03% of lightning discharges that strike vegetated areas of the world result in wildfires (205). Thus, while it is true that fires must have a source of ignition and that seasonal patterns of thunderstorm activity may influence their timing (87), the other five factors listed above determine, in a proximate sense, the fire regime in a particular area. They determine whether or not ignition will occur, how intense the fire will be, and what area it will cover.

The moisture content of the fuel determines the likelihood of ignition and the ability of the fuel to carry a fire. The amount of moisture is influenced, in turn, by a number of local factors including the aspect of the site, the water retention properties of the underlying soil, and wind conditions (226). In certain geographical regions, there are only brief periods during a "normal" year when fuels are dry enough to ignite and carry a fire. Consequently, the most significant fires (i.e. the most intense, severe, and extensive) burn during periodic droughts that recur at intervals of 10–20 years or longer (66, 85–87, 204, 249). Substantial amounts of fuel accumulate between droughts, so that when fires occur they are exceedingly intense. Similarly, fire only invades semipermanent wetlands such as shrub bogs (29) during droughts. Fires occur much more frequently and regularly in areas that have predictable annual dry seasons as well as occasional droughts [e.g. grasslands and savanna (29, 93, 120, 144, 226), chaparral and forests of the Sierra Nevada Mountains in California (109, 111), and the dry sclerophyll forest of Australia (31)].

The rate of fuel accumulation can influence the frequency and intensity of fires because it determines how much fuel will be available for burning at any given time. Assuming that the likelihood of fire is positively related to the standing mass of fuels, fires should occur most often at sites where fuels accumulate the most quickly, all else being equal. (This assumption is critically evaluated below.) Also, the more fuel that has accumulated since the last burn, the more intense a fire will be.

The rate at which fuels accumulate equals the difference between their rates of production and decomposition (29, 30). Site conditions or characteristics of the vegetation itself that enhance the production of fuels and/or slow their decomposition result in higher rates of fuel accumulation. Insect and parasite attacks, disease, competition, windthrow, lightning strikes, and senescence convert living vegetation into dead fuels. Since dead plant material is often drier and therefore more flammable than living plant tissue, the likelihood of intense fires is greater in older stands. There are exceptions to this pattern when past fires have killed the vegetation but not consumed it completely. In this case, dead "carry over" fuels may be abundant in the early stages of stand regeneration, creating a high risk of fire even in a young stand (e.g. 66, 86). The rate at which dead fuels decompose is controlled by climatological conditions (i.e. temperature and humidity) and by the characteristics of the vegetation. Sclerophyllous foliage, for example, is decay resistant and decomposes relatively slowly.

Structural characteristics of the fuel influence fire intensity (29, 174). When fuels are distributed in widely separated strata—e.g. in savannas or frequently burned pine woods—relatively low–intensity surface or ground fires are the rule. In shrublands and long unburned coniferous forest, the vertical distribution of fuels is more continuous, and the surface to volume ratio of the vegetation is large. As a consequence, high intensity crown fires are common in these vegetation types.

Fuel flammability is also a function of the chemical composition of the foliage (29, 174, 175). Fuels rich in secondary organic compounds are ignited more easily and burn more intensely. In some plant species, the concentrations of extractable organics vary seasonally and increase with age [e.g. the chaparral shrub *Adenostoma fasciculatum* (158, 175)].

The vegetational mosaic and local topographic features strongly influence the point fire frequency and the areal extent of fires (29, 86, 227). Only a fraction of the fires that burn through the vegetation at a particular point start within the stand that includes that point. Therefore, the point fire frequency depends to a large degree on the rate at which fire encroaches from surrounding areas. This rate is influenced by the ability of neighboring phases of the vegetational mosaic to carry a fire and by the extent and orientation of natural and artificial firebreaks.

Finally, weather conditions at the time of a fire affect its intensity and size (87). Such conditions include the level of precipitation, air temperature and humidity, wind speed, and wind direction, especially with respect to the spatial distribution of fuels and the position and orientation of fire barriers. A fire's impact is also influenced by its rate of spread (30, 93), which depends on both weather conditions and the winds, convection currents, etc., that the fire itself produces. In general, rapidly moving fires consume less fuel and burn at a lower temperature.

In summary, a reciprocal relationship appears to exist between vegetation and fire. The state of the vegetation affects the fire; and the interfire interval regulates the composition, structure, and quantity of living vegetation and dead fuels. All else being equal, lightning fires are more likely to start in older stands of vegetation and to be more intense than fires that start in younger stands. In many instances, however, this description of fire behavior may be overly simplistic (86). Once a fire starts, it can spread into stands of many different ages. "All else" is rarely equal; the intensity and areal extent of any particular fire may be influenced to a considerable degree by factors other than the characteristics of the accumulated fuel.

WATER MOTION   In marine and freshwater environments, moving water exerts forces on sessile organisms, just as wind does in terrestrial environments. In addition, suspended particulate matter (e.g. sediment) or larger objects transported by the moving water (e.g. logs or cobbles) may strike and abrade the substratum over which the water flows. Aquatic organisms may also be killed either by burial under sediments that have been displaced and redeposited by moving water (e.g. 126, 192) or by exposure to air when water motion changes drainage regimes (e.g. coral reef crests; J. H. Connell, personal communication).

Detailed study of the natural regimes of disturbance caused by water motion in freshwater habitats—in streams in particular—is in its fledgling stages (e.g. 88, 113, 131, 131a); little is known about the frequency, areal extent, and intensity of such disturbances. The disruptive influence of moving water has been much more thoroughly studied in marine habitats, particularly along temperate, rocky intertidal shores (e.g. 45, 153, 189, 192) and on tropical coral reefs (e.g. 33–35, 54, 155, 247). Wave action is a major agent of disturbance in these habitats. As with wind and fire, the regime of wave-induced disturbance varies in space and time and is influenced by physical and biological components of the environment. Wave energy is maximal during seasons of high storm activity. Therefore, the disturbance it causes is highly seasonal (35, 45, 54, 153, 189, 192). The frequency of disturbance is also strongly influenced by the physiological and morphological characteristics of the organisms in question and the properties of the substratum to which the organisms are attached.

As a sessile marine organism grows larger in an environment subject to periodically accelerating water motion, i.e. in wave-swept habitats, its risk of being detached or broken by wave stress often increases (50, 117, 192). Older individuals are more likely to suffer injury or death from wave forces if weakening wounds caused by boring organisms, predators, and grazers accumulate with age (35, 192). Epiphytic or epizoic overgrowth can also increase the chance that an organism will be dislodged by wave action (192). The risk of damage or death by a given wave force varies among species that differ in shape, flexibility, or internal structure, among other factors. For example, tree or bush-like corals (e.g. *Acropora* spp.) are more susceptible to wave damage than mound or sheet-like corals (33–35). Similarly, some aggregations of sessile organisms become more vulnerable to disruption as their individual members increase in size and number. Dense, multilayered beds of the mussel *Mytilus californianus* are less stable and more likely to be torn from the rock surface by wave forces than less dense, single-layered beds (82, 151, 153).

In some aquatic habitats, the stability of the substratum directly determines the rates of disturbance. In streams and on the seashore, strong water motion overturns loose rocks, damaging the attached organisms. The frequency of this kind of disturbance declines with increasing rock size (125, 131, 131a, 150, 189).

The sizes of the areas disturbed by wave action vary considerably (35, 192). Patches cleared in mussel beds during stormy winter months and at more exposed sites are much larger on average than those created during calm summer months or at protected sites (153). Similarly, larger boulders are overturned more often during winter storms and at sites exposed to heavy wave action than at other seasons or sites (189). In both systems, the rate of disturbance differs from one year to the next, reflecting annual variation in the intensity and frequency of storms. Similarly, there was substantial spatial and temporal heterogeneity in the regime of disturbance caused by four hurricanes that struck the reef at Heron Island, Australia (35; J. H. Connell, personal communication). Though such storms are seasonal, their effects are unpredictable from year to year. The underlying causes of the variation in this case are not as clear as in the mussel bed and boulder examples.

PATTERNS OF DISTURBANCE IN ASSEMBLAGES OF SESSILE ORGANISMS
Within any particular habitat, a variety of agents operate independently or in concert to generate the overall regime of physical disturbance to which organisms respond. The heterogeneity of natural disturbance regimes is due, in part, to local variation in the intensity, timing, and spatial distribution of potentially disturbing forces. Often, however, heterogeneity in local patterns of disturbance is better explained by temporal and spatial variation in the "intrinsic vulnerability" (192) of the organisms affected by these forces. Conspecific individuals differing in age and size usually differ in their vulnerability to a particular force. Increasing size and/or age decreases the risk of damage or death from some forces (e.g. of woody vegetation from ground fires; intertidal organisms from desiccation stress; coral colonies from sediment burial) but increases vulnerability to others (e.g. of large, old trees to windthrow; older stands of trees and bushes to crown fires; larger intertidal organisms and branched corals to wave forces).

Vulnerability to a given force is also species- and assemblage-specific. Not uncommonly, the vulnerability of a community to disturbance changes over the course of succession (e.g. 191), owing to differences in the physiology, morphology, or growth habit of the species characteristic of each seral stage. The risk of disturbance by fire, for example, often changes with the succession-al age of the vegetation because seral stages differ in fuel characteristics. Such differences can influence the temporal pattern of disturbance. Some terrestrial plant communities appear to exhibit cycles of inflammability that are controlled by the rate of accumulation of combustible plant material (95, 134, 235). There may be similar cycles of disturbance on mussel-dominated shores of the Pacific Northwest coast of the United States. The rate of succession to a community dominated by mussels appears to set a lower limit of 7–8 years on the interval between successive major disturbances of assemblages occupying a particular area of substratum (153).

In many forests, gaps in the canopy are generated primarily by windthrow of isolated individual trees or small groups of neighboring trees. In these cases, gap regeneration is a small-scale, spatially and temporally asynchronous pro-cess. There are remarkable, though relatively rare, exceptions to this pattern where large tracts of forest are leveled by wind (e.g. 19, 24a, 56, 168, 170, 232, 237). To my knowledge, however, a large-scale, cyclic pattern of wind-induced tree mortality has only been rigorously demonstrated in some high-altitude balsam fir forests (170, 196, 198).

Such cyclic patterns of disturbance are certainly intriguing and have partially inspired the hypothesis that species have evolved physiological mechanisms and morphologies that promote disturbance and/or determine its characteristics (see below). The phenomenon of regular disturbance controlled by the biotic component of the environment is by no means universal, however. A signifi-cant amount of the variability in the impact of a disturbing force may be unrelated to biotic properties. In fact, it has been argued that patterns of local fires in some presettlement forests were largely random with respect to vegeta-tion type and best explained by patterns of lightning ignition and the vagaries of the winds and weather during a fire (86, p. 399). Large-scale disturbances such as mass movements caused by landslides (60, 170, 199, 217–220), earthquakes (68, 219), and volcanic eruptions (219, 220) usually occur at random with respect to the age, successional state, or species composition of the populations and communities they disrupt.

This caution aside, the regime of disturbance in many communities reflects the interplay between the properties of organisms and the characteristics of the physical forces that cause disturbance. Thus, the categories of *endogenous* and *exogenous disturbance* (sensu 19, 20) are difficult, if not impossible to apply (176, 235). It is therefore hard to unambiguously classify a particular succes-sional sequence as *autogenic* or *allogenic* (145).

Present-day disturbance regimes may be very different from the regimes characteristic of communities unaffected by humans. Fire regimes in vegeta-tion surrounding inhabited areas were surely altered when indigenous peoples began to use fire thousands of years ago as a tool for managing vegetational cover and game animal populations (29, 31, 112, 166, 226). Modern man's influence has been much more pervasive.

While the activities of indigenous peoples and early settlers may have initially increased the frequency of fire, the adoption of effective fire suppression techniques during the last century has sharply reduced its frequency in many vegetation types (e.g. 62, 112, 119, 134, 137, 173, 204, 206). In North America only the unexploited boreal forests of northern Canada and Alaska (87) and forests with extremely long natural fire cycles (66, 172) have remained relatively unaffected by this policy of fire suppression. Where fire cycles have been unnaturally lengthened, large quantities of fuel accumulate and infrequent, unusually intense fires are the result (53, 134). In other habitats, such as swamps where draining has lowered the water table (29), fires are much more frequent now than in presettlement times. These changes in the fire regime have had a marked effect on vegetation patterns (see above citations).

The effects exerted by humans are not limited to disturbance regimes caused by fire. For example, forest logging practices produce much larger clearings than those usually created by windthrow, and the temporal and spatial patterns of harvesting are often quite distinct from natural treefall. Flood control procedures combined with human modification and degradation of watersheds can lead to extreme flooding and sedimentation of abnormally long duration in coastal wetlands. This altered regime of disturbance has much more severe effects on the biota of the floodplains and estuaries than the normal, lower volume discharge of freshwater runoff during the rainy season (147, 250). Human-caused deterioration of riparian watersheds can have similar negative effects on stream communities (214).

The ubiquity of human-caused alterations in natural disturbance regimes significantly complicates evolutionary interpretation of present-day patterns of morphology, physiology, and life history in relation to physical disturbance (e.g. 109). Only in some communities, such as forests, is accurate reconstruction of presettlement regimes of disturbance possible. The same concern applies to human alterations of biotic components of the environment (47, D. Lindberg, J. Estes, K. Warheit, in preparation). Cautious consideration of such effects should precede speculation about the evolutionary mechanisms underlying present-day patterns.

The Repopulation of Disturbed Sites by Sessile Organisms

Propagules of sessile organisms, be they sexually or asexually produced (sensu 143), are rarely able to invade and become established in areas densely occupied by other organisms (22, 35, 36, 46, 74, 77–80, 83, 176, 192, 231, 252). Resident organisms inhibit recruitment from propagules by a variety of mechanisms. Occupants may consume the dispersed propagules, as in some assemblages of sessile, suspension-feeding invertebrates (35, 190, 192, 246), or the residents may simply have preempted the available space. In other cases, residents modify site conditions in ways that inhibit the germination or metamorphosis of those propagules that do reach the substratum. They may release toxic chemicals (allelopathy) or reduce the supply of essential resources such as food, light, nutrients, and water.

Physical disturbance is one of the major mechanisms that break this inhibition and generate conditions favorable for recruitment, growth, and reproduction. Disturbances not only reduce or eliminate the cover of resident organisms, thereby lessening competition for resources that are present on the site, but in some cases they indirectly replenish some of the depleted nutrients [e.g. nutrient-rich ash produced by fire (17), the accumulation of detritus in pits dug by foraging rays (216), and nutrients leached from rotting treefall debris in forest light gaps (e.g. 22, 222)]. In addition, the disturbance may eliminate toxic chemicals that have accumulated in or on the substratum [e.g. volatilization of allelopathic chemicals by fire (70); see 30 for a recent discussion of the controversial role of allelopathy, and its interaction with fire, in shrublands]. The disturbance may also temporarily reduce the density of predators or parasites of the propagules (70).

The rate and pattern of reestablishment following a disturbance depend on the following elements:

1. The morphological and reproductive traits of species that are present on the site when the disturbance occurs. Such traits determine, in part, the likelihood that these species will survive the event and rapidly reoccupy the site.
2. The reproductive biology of species that were not present on the site when it is disturbed but have occupied it previously or live within dispersal distance of it.
3. Characteristics of the disturbed patch including:
    a.   the intensity and severity of the disturbance that created it,
    b.   its size and shape,
    c.   its location and degree of isolation from sources of colonists,
    d.   the heterogeneity of its internal environment, and
    e.   the time it was created.

Below, under subheadings 3a-e, I briefly discuss the influence of the characteristics of these elements on the recolonization of a disturbed patch. This information is certainly not sufficient to predict the abundance and demography of a species within a particular habitat or even the likelihood that the species will persist there. Such properties of a population depend on the complex interaction of an organism's complete life cycle (including events that occur during disturbance-free periods) with the overall regime of disturbance in the habitat.

INTENSITY AND SEVERITY OF DISTURBANCE   The influence of the intensity of disturbance on recolonization has been most thoroughly studied in communities exposed to fire (e.g. 2, 70, 71, 108). Trees and shrubs whose aboveground buds are covered by a thick layer of heat-resistant bark are more likely to survive fires of moderate intensity, even if there is some scorching and defoliation. Hotter fires may kill all aboveground tissues, and only those species that are able to resprout from underground buds associated with rhizomes, roots, root crowns, or lignotubers will survive and regrow vegetatively.

The probability that a plant will survive a fire (even with some bud protection) may decline with increasing age and number of exposures to fire (70, 71). The degree of vegetative regeneration depends on the season in which burning occurs if such regrowth draws on a seasonally fluctuating pool of stored carbohydrates (e.g. 175). There is also the interesting case (100) in which mutualistic ant associates mediate the impact of fire on resprouting swollen-thorn acacia plants. The obligate acacia-ants clear foreign vegetation from around the base and branches of the acacia, thus reducing the chance that the plant will be killed by a fire. If the fire is hot enough to kill aboveground shoots but not the ants, the colony will move into the regenerating sucker and defend it against encroaching vegetation and herbivorous insects.

The mortality of even those species that possess subterranean buds increases with fire intensity (71). This relationship holds especially for underground fires in layers of peat or humus that directly damage tissues below the soil surface and often kill all of the vegetation on a site. Thus, the probability that a site will be repopulated by the vegetative regeneration of surviving residents decreases with the intensity of the disturbance. This same pattern seems to apply to many species of intertidal algae (191, 192) and colonial marine invertebrates (35). On coral reefs, for example, fragments of colonies that are broken loose in storms can establish themselves in clearings and fill the space by asexual reproduction (91). The survival of these asexual propagules is probably inversely related to storm intensity, since heavy wave action breaks fragments into smaller pieces and scours off much of the living tissue; this greatly reduces the number that can successfully colonize disturbed sites (116).

Repopulation of disturbed sites by other kinds of propagules, including those produced by sexual reproduction, can also be influenced by disturbance intensity. Once again, the best examples come from plant communities exposed to fire (e.g. 2, 30, 70, 71, 107–109, 252). The flowering of some plant species characteristic of fire-prone habitats is stimulated by low intensity surface fires. Other species produce seed that is stored in fruits on the plant until fire of a particular intensity triggers their dehiscence [e.g. serotinous pine cones (225–227, 251)]. Still other species release hard-coated seeds that remain dormant on or in the soil until fire, in concert with other environmental factors, stimulates germination. The cues that stimulate germination and the mechanisms involved vary among species (30). In northern conifer forests, the intensity of the fire is particularly important to patterns of seedling establishment. Hot fires consume the organic layer, including its seed bank, and expose mineral soil that is favorable to the establishment of conifers. Less intense fires leave the organic layer and its seed bank intact; thus, species whose seeds have accumulated in this layer will dominate the regenerating vegetation, and conifer seedlings will be scarce (86).

Plants vary substantially, both within and among species, in the degree to which they respond to fire with vegetative resprouting versus germination from seed (70, 108). Similarly, patterns of serotiny in pines differ markedly among populations and species (132, 156).

A buried seed strategy (20, p. 108) is also employed by some plant species of mesic or wet terrestrial environments in response to disturbances other than fire (11, 13, 22, 25, 39, 40, 78, 79, 83, 84, 130). Large quantities of dormant seed may be stored in the soil. Some of it will have been produced by resident plants—usually early successional species—mature individuals of which no longer occupy the site. The remainder will have been dispersed to the site, in many cases transported by animals (e.g. 20, 25, 40, 84, 130, 211). The creation of a large gap in the vegetation, e.g. by treefall, alters the soil environment and/or the light regime in such a way that seed germination is stimulated (9, 11, 20, 130, 157). Cues for germination vary among species, as does the viability and length of dormancy of their buried seeds. Though an equivalent recolonization mechanism may exist in marine and freshwater communities, its existence has yet to be rigorously demonstrated.

The seeds of other mesic or wet environment plant species do not accumulate in a long-lived seed pool. Instead, they germinate immediately upon dispersal or soon thereafter; germination may be briefly delayed until favorable seasonal climatic conditions develop (25, 67). This pattern is exhibited by some shade-intolerant, early successional species whose light, often wind-dispersed seeds colonize transient or frequently disturbed habitats (e.g. floodplains and river banks). It is particularly characteristic, however, of middle and late succession-al tree species in temperate and tropical closed-canopy forests (22, 25, 67, 84, 176). Such species produce relatively large seeds that often are dispersed close to the parent tree. The large seed reserve permits germination and establish-ment in the low light conditions prevalent under a closed canopy and may facilitate root penetration of dense litter on the forest floor. The seedlings are shade tolerant and persist in the understory for varying lengths of time, depending on the species. They grow slowly, if at all, until a gap opens in the canopy over them, permitting a pulse of growth. A suppressed individual rarely can recruit to the canopy layer after a single disturbance. Before it can do so, either the hole in the canopy is closed by lateral expansion of the crowns of surviving trees on the gap border or the gap becomes so choked with regenerating vegetation that the resources needed to support further growth are used up. Thus, several episodes of low intensity canopy disturbance in the immediate vicinity of the suppressed individual are usually required before it grows into the canopy (176). The disturbance must be severe enough to open a gap, but not so intense that it kills the understory plants. Suppressed individuals are often damaged by falling trees and other debris, but most survive and are able to resprout.

Following an intense disturbance—such as that produced by a landslide (60, 68, 170, 199, 217–220), a volcanic mudflow (49), a long-overturned boulder on the seashore (191, 192), erosion and redeposition of alluvial sediments on floodplains (63, 141, 200), or a receding glacier (42)—neither resident organisms nor stored dormant propagules survive. All recolonization must come from either propagules dispersed into the open patch from surrounding areas or vegetative ingrowth of neighboring individuals or clones.

PATCH SIZE AND SHAPE   The size and shape of a patch indirectly influence its repopulation in several ways. The internal physical and biological environments may vary with patch size. For example, there is an increase in light intensity and its daily duration, in mean soil and air temperature (and their ranges), and in subsurface soil moisture as the size of forest treefall gaps increases (assuming a constant severity of disturbance). The levels of these microclimatological variables are substantially higher in gaps than under a forest canopy (20, 22, 25, 51, 176, 237, 238). Conversely, humidity decreases with gap size; the initial competition for nutrients with surviving trees at the edge of the gap may also. [However, there have been no detailed studies of nutrient dynamics within a single treefall gap (222)]. A number of these patterns have also been observed in disturbed sites in shrublands (30). These size-related differences in the characteristics of the physical environment can influence the germination of seeds, resprouting from buds, and the subsequent survival and growth of seedlings and saplings within a gap.

In aquatic environments, organisms surrounding a clearing can influence the patterns of water flow in and around the open patch and thereby affect the availability of food and the density of settling propagules. These organisms may also consume incoming propagules or otherwise directly interfere with their settlement or recruitment. In intertidal habitats, however, the presence of organisms on the edge of a clearing may moderate the internal microclimate of the patch at low tide when it is exposed to the air. These influences will be more strongly felt in small clearings than in large ones (35, 192).

Small patches have a greater perimeter-to-area ratio than large patches, which has a number of consequences for patterns and rates of recolonization (133). The vegetative ingrowth of clonal organisms or the lateral encroachment of attached but semimobile solitary organisms will make a proportionately greater contribution than dispersed propagules to the recolonization of small as compared to large clearings if these organisms occur along patch edges. This phenomenon has been demonstrated in beds of intertidal mussels (153, 193, 202) and in subtidal assemblages of colonial invertebrates (35, 106, 154). Numerous factors influence the rate at which a patch of a given size is filled by vegetative ingrowth, including the morphology and rate of expansion of the invading colony, clone, or shifting assemblage of individuals.

Similarly, in some instances the rate of colonization by dispersed propagules varies with patch size. The number of nearby adults per unit patch area is usually greater for small patches than for large ones. This differential may result in a greater density of recruited propagules and therefore more rapid recolonization of small patches. This pattern will only develop when propagules are not dispersed far from their parents and when there is no overriding negative influence on recruitment by organisms that surround the patch. When the size of a clearing is large relative to the dispersal ability of potential colonists, invasion from the edge of the patch may be slow; its closure by recruitment from dispersed propagules may take several generations (94, 195, 230).

In some systems, the abundance of mobile consumer species within disturbed patches and their influence on recolonization varies predictably with the size of the patch. In intertidal mussel beds, for example, small clearings contain higher densities of grazers, particularly limpets, than do large patches (153, 193, 202). This relationship is analogous to the association between rates of vegetative ingrowth and patch size. The surrounding bed of mussels appears to serve as a refuge for small grazers from wave shock, desiccation stress, and possibly predation. These grazers migrate into the open patch from the edge to feed on colonizing algae. Since the length of edge is proportionately greater in small patches, a higher density of grazers accumulates in small clearings than in large ones. As a consequence of this variation in grazing pressure, different assemblages of macroalgae develop in patches of different size (193, 202).

This pattern is likely to occur in any system where natural enemies prefer—or are forced—to live largely within the phase(s) of the community mosaic that surrounds a patch, while their prey occupy the interior of the patch. Bartholomew's (10) demonstration that vertebrates concentrate their grazing and seed predation within a narrow zone along the edge of chaparral stands suggests that their influence on the revegetation of clearings may vary with clearing size in the same way.

Given that the physical and biological environments of patches often vary with patch size, it is not surprising that species are differentially distributed among clearings of various sizes in a number of communities. These include intertidal assemblages (192, 193), subtidal epifaunal assemblages (35, 97,

110), tropical forests (22, 51, 84, 237), temperate forests (20, 57, 176, 243), and old fields (12, 44, 74, 78).

Patch shape also influences the degree to which surrounding organisms affect within-patch dynamics. The more irregularly shaped a clearing of a given area is, the higher the ratio of perimeter length to area and the greater the influence (both positive and negative) organisms in the immediate neighborhood of the patch will have on its recolonization (192). For example, long, narrow gaps in a forest canopy allow less light to penetrate than more circular gaps (22; 149; J. Tomanek, 1960, cited in 176).

THE LOCATION OF THE PATCH AND ITS DEGREE OF ISOLATION FROM SOURCES OF PROPAGULES   The proximity of a patch to sources of colonists can greatly influence the mode and rate of colonization (84, 164, 192, 193). For example, colonization by expanding clones is only possible if they occur on or near the borders of the patch. Discrete patches of substratum, such as cobbles on the seashore, are physically isolated from other substrata. When they are denuded by a disturbance, invasion by neighboring clones is impossible, and dispersed propagules must initiate recolonization (35, 192). Patch location is particularly important for species that do not disperse their propagules very far, e.g. many terrestrial and some marine plants (20, 46, 84, 94, 152, 164, 192, 193). In contrast, the larvae of many, but not all (e.g. 69), marine invertebrates spend days, weeks, or months in the plankton before they are competent to settle; consequently they can be dispersed long distances from the parent organism. Recruitment of such species to disturbed patches rarely correlates with the abundance of propagule-releasing adults in the immediate vicinity of the clearing (35, 192, 215).

WITHIN–PATCH ENVIRONMENTAL HETEROGENEITY   The environment within a patch cleared by disturbance is seldom, if ever, homogeneous (22, 84, 176). In many cases, this internal heterogeneity influences the process of recolonization. In forest lightgaps, soil conditions and light intensity vary with position in the gap. Fallen dead plant material may significantly alter the local microclimate and distribution of soil nutrients. Conditions along the bole of a fallen tree may be very different from those under and around its crown, and both may differ from conditions near the upturned root mass where mineral soil may be exposed (see references in 176). The association of certain species with particular areas of the forest floor adjacent to a fallen tree suggests a degree of specialization to these microenvironments (149, 176, 210). The decomposing log itself may be a critical germination site for some species (e.g. 66, 218, 220). This local heterogeneity in the conditions for germination and growth can enhance the diversity of vegetation within the gap (e.g. 149, 208).

In marine communities, within-patch heterogeneity in the characteristics of the substratum may also affect patterns and rates of patch colonization (35, 192). Settlement and/or survival of propagules on hard substrata is sometimes influenced by relatively small-scale differences in surface texture or composition. Larger-scale differences in the rock surface such as cracks and crevices can provide refuges from predators and grazers.

Within-patch heterogeneity in the form of a spatial refuge afforded by a particular substratum can also be important in terrestrial communities. The perennial herb, *Lomatium farinosum*, suffers far lower mortality caused by small mammals in shallow rocky soils than in deeper, less rocky soils (209) because mammals are better able to create runways, dig burrows, and feed on plant roots in deeper soils.

TIME OF PATCH CREATION   When an open patch is created by disturbance will indirectly affect colonization if, as is commonly the case, the availability of propagules varies over time. Production of propagules is seasonal for at least some species in most habitats [e.g. tropical rain forest (63, 67, 84, 237), temperate forest (25), temperate intertidal zone (192), temperate subtidal epifauna (150, 203), coral reefs (35), and chaparral (107)]. Yearly variation in the production of propagules is often great (see also 80). Since large patches often remain open to colonization longer than small ones, temporal variation in the availability of propagules may have less of an impact on the long-term development of their assemblages (35). In some systems, the largest patches of open space are only generated in certain seasons, whereas small patches are produced year-round (35, 192). In this case, there should be less variation in the composition of colonists among large patches than among small ones (189), assuming that yearly variation in propagule availability is smaller than seasonal variation.

Temporal variation in the production of propagules may also be less important where propagules are stored for long periods in the soil or on the parent organism. Such storage effectively dampens variation in the number of propagules that are immediately available to recolonize a disturbed site. If the interval between successive disturbances is short, however, there may not be sufficient time to replenish the pool of stored propagules, particularly if this depends on the maturation of individuals that became established since the last disturbance. Under these conditions, few if any stored propagules will be available to repopulate the site, and local extinction may result (70, 251, 253). Therefore, in fire-prone habitats the dynamics of a population that relies on stored propagules for regeneration depends on the interrelation between fire frequency and fire intensity, the rate of propagule production as a function of plant age/size, and the viability of stored propagules. The relationship between inputs to and losses from a pool of dormant propagules is equally important in closed-canopy mesic forests where windthrow is the predominant form of disturbance (25).

The timing of disturbance can have a significant effect on patterns of recolonization even when reproduction is stimulated by the disturbance event itself. For example, patterns of fire-induced flowering and seed production can vary with the time of year the plant is exposed to fire (e.g. 2, 70).

CORRELATIONS AMONG THE CHARACTERISTICS OF A DISTURBANCE RE-GIME AND COMMUNITY RESPONSE   The characteristics of natural disturbances are often correlated. Disturbances that affect large areas are generally the least frequent and the most severe (i.e. leave the fewest survivors). The above correlation has been observed on coral reefs (35), rocky seashores (192), and in some terrestrial plant communities (235). The simplest explanation for this pattern is that the events that trigger massive earth movements, severe droughts, and large storms occur less frequently than those that produce smaller-scale, less intense disruptive forces. In other cases, this correlation is due to the interacting influences of the biotic and abiotic components of the system on patterns of disturbance. This sort of interaction best describes correlations among characteristics of a fire regime. If ignition is restricted to infrequent periods of severe drought, considerable fuel will have accumulated, and the resulting fire will be intense, killing much of the vegetation over a large area. Alternatively, if dry periods favorable to ignition occur annually, fires will be frequent but usually of low intensity and small areal extent because the short interval between burns will not allow much fuel to accumulate.

One general consequence of this correlation among disturbance characteristics is that there is a fairly continuous relationship between, the predominant mode by which disturbed sites are recolonized and patch size. At one extreme are large clearings with no survivors and no dormant propagules. These areas will initially be recolonized by species producing widely dispersed propagules at the time the patch is opened. Germination or metamorphosis of the propagules soon after dispersal and rapid early growth are advantageous for successful establishment (73). If they are not redisturbed subsequently, such patches often undergo a long period of succession as other species slowly invade and gradually replace earlier colonists. At the other extreme, the smallest disturbances are filled almost exclusively, and relatively quickly, by the vegetative growth of survivors living either within the clearing or on its edge. Little or no successional replacement takes place in this case.

Repopulation of patches whose size and severity of destruction falls between these two extremes occurs by one or more of the following mechanisms: (*a*) germination or metamorphosis from recently dispersed or stored propagules, (*b*) growth from a suppressed juvenile state, or (*c*) regrowth from damaged tissues. The first mechanism predominates in larger patches where conditions are more open and where the disturbance has been severe enough to kill most adults and juveniles. Quite often, mass seeding of such patches in forests produces extensive even-aged stands (85–87, 101, 141, 146, 199, 204, 217–220).

The latter two mechanisms are more important in smaller patches where some survivors remain and where competition for resources such as light and nutrients is likely to be intense. Here, larger individual size at the outset should be advantageous for procuring the limited resources available. In small light-gaps, for example, suppressed seedlings or saplings usually outcompete individuals that germinate from seed after the gap has opened (25). Similarly, in coral reef communities, attached fragments of corals probably have an advantage over metamorphosing planula larvae in small clearings on coral reefs (35). Because fragments are larger at initial colonization, they can grow to a large size more quickly. At this larger size, they are better able to hold off colonies invading from the edge of the clearing.

It is often impossible to predict a priori the relative contribution of each of these mechanisms to the process of recolonization in any particular clearing. This will depend on the environmental setting and the life histories of the particular species involved. Chance events also play a significant role in many cases.

Comments on Evolutionary Responses to Disturbance

Disturbances are clearly an important cause of local heterogeneity in the environmental conditions relevant to the recruitment, growth, survival, and reproduction of organisms. Circumstantial evidence strongly suggests that under the constraints of adaptive compromise, the differential reproductive success of individuals has resulted in specialization to particular phases of the environmental mosaic generated by disturbance (e.g. 11–13, 25, 35, 51, 52, 77–80, 164, 171).

Disturbance causes environmental heterogeneity along two dimensions—the temporal and the spatial. The temporal component derives from asynchrony in patch creation, causing the assemblages of neighboring areas to differ in age. Species may partition this temporal component of heterogeneity by colonizing and/or growing to maturity only during a particular stage of succession (159) or only in patches created at a particular time of year.

The spatial component results from variation in patch characteristics other than age, e.g. size, microclimate, and location. Species may differentially exploit patches that differ in one or more of these characteristics. Levin (124) refers to these temporal and spatial components of disturbance-induced heterogeneity as *phase difference* and *local uniqueness*, respectively. Undoubtedly, most species exploit a combination of these two forms of environmental variation. Despite the specialization described above, many species overlap considerably in the environmental conditions suitable for completion of their life cycles.

Several authors have suggested that organisms are incapable of evolving adaptive responses to disturbances that occur at long intervals relative to their generation time (e.g. 15, 83, 152, 207). As Harper states:

> I distinguish (as ends of a continuum) disasters and catastrophes. A disaster recurs frequently enough for there to be reasonable expectation of occurrence within the life cycles of successive generations . . . the selective consequence may be expected to leave relevant genetic and evolutionary memories in succeeding generations. A "catastrophe" occurs sufficiently rarely that few of its selective consequences are relevant to the fitness of succeeding generations. The selective consequence of disasters is therefore likely to be to increase short-term fitness and the consequence of catastrophes is to decrease it (83, p. 627).

I see at least two problems with this idea. The first is largely a matter of semantics. It is unfortunate that Harper chose to assign the specific terms *disaster* and *catastrophe* to opposite ends of the continuum of disturbances. In principle, it seems a poor practice to designate specific categories without providing objective criteria for distinguishing among them. How much "relevant genetic memory" must be passed to future generations for an event to be classified as a disaster and not a catastrophe? While I doubt that Harper intended these terms to be adopted in empirical research, several workers have recently applied these terms to real situations (e.g. 152, 207). Such usage is best discouraged both for the reason just discussed and because these words are so commonly, but ambiguously, used in everyday speech.

The second problem with this view is the tacit assumption that the observed "fit" of an organism to its environment is attributable largely to the optimizing process of natural selection (75). It overlooks the real and lasting influence that a "catastrophe" may exert on the genotypes and phenotypes of succeeding generations, even if its immediate effects are density independent and largely random with respect to genotype. Until we know much more about the relative influences of different evolutionary processes, it is difficult to say whether natural selection by frequent disturbance has a greater evolutionary impact than the population bottlenecks, local extinctions, and founder effects associated with rare catastrophes (228).

A final caveat seems warranted. Hypotheses concerning the adaptive nature of observed patterns of life history, morphology, and behavior are easy to propose. It has often been suggested, or even taken as fact, that certain characteristics have evolved by natural selection in response to recurring physical disturbance. Furthermore, some species that depend on particular kinds of disturbance for completion of their life cycles or whose fitness is otherwise enhanced by such disturbances exhibit morphological or physiological features that seem to increase the likelihood that these disturbances will occur. This correlation has elicited the hypothesis that the latter features, sometimes referred to as "disturbance facilitating," have evolved by natural selection (e.g. 140, 244).

While there is little question that physical disturbance can act as a potent agent of natural selection and may play a part in other evolutionary processes as well one must be careful not to invoke natural selection as an explanation without considering alternatives. For example, many features of plants classically considered to be "fire-adapted," including flammability and the ability to resprout after fire, may have evolved in response to alternative selective forces such as herbivory or drought (29, 30, 70, 71) or by a different evolutionary mechanism altogether (75). "The immediate utility of an organic structure often says nothing at all about the reason for its being" (75:593). Whether "disturbance-dependent" organisms have evolved specific mechanisms that increase the probability of particular kinds and/or patterns of disturbance favorable to their persistence is an interesting, though probably an unanswerable, question.

The Role of Disturbance in Population and Community Dynamics

WITHIN–PATCH DYNAMICS

Demography   I have mentioned a variety of factors that influence the reestablishment of populations within a disturbed patch of habitat. Subpopulations inhabiting different regions of a patch may differ in age structure, genetic composition, and life history characteristics (210). The selective effects of intra- and interspecific interaction may vary over small distances within a single patch—e.g. at its edge versus at its center. The evolutionary consequences of such variation have yet to be fully explored.

Interspecific interactions and the patterns of subsequent disturbance strongly influence the absolute and relative abundances of species within a patch, including the probability of continued local persistence.

Species diversity   The resources made available by a disturbance are soon exploited by colonists and regenerating survivors, and a successional sequence of species replacements usually ensues (36, 143, 190). In the course of most successions, one or a few competitively dominant and/or long-lived species come to monopolize the resources of the disturbed patch (e.g. 36, 45, 113, 127, 129, 151, 153, 160, 190). In such cases, the time to local extinction of early successional species depends on the characteristics of the patch—for example, its size and degree of isolation (35, 192)—and of the species participating in the succession. When late successional species are able to invade the open patch, dominance will be quickly attained. So, small patches surrounded by adults of these species should be dominated more quickly than larger and/or more isolated patches.

In situations where hierarchical competitive interactions or differential longevities [see the inhibition model of succession in (36)] will probably lead to the monopolization of patch resources, disturbance can maintain within-patch diversity by one of two mechanisms—compensatory mortality or intermediate disturbance (33, 34). *Compensatory mortality* refers to the situation in which the potential late successional dominant suffers a disproportionately high rate of disturbance-related mortality as compared to other species that it might otherwise exclude from the patch. Selective predation by the starfish *Pisaster* on the competitively dominant mussel *Mytilus californianus* (151) is a classical example of compensatory biological disturbance. Likewise, selective herbivory can maintain high local diversity in plant communities (128). Physical disturbance associated with heavy wave action can act in a similar manner on mussel-dominated intertidal shores (153) and coral reefs (33, 34); spates can have the same influence in stream communities (113).

Physical disturbance that does not cause compensatory mortality may nonetheless maintain within-patch diversity. To do so, a physical disturbance must renew resources, such as space, at a rate sufficient to allow continued recruitment and persistence of species that would otherwise be driven locally extinct. It must not occur so often or with such intensity, however, that many species are eliminated. Therefore, the disturbance must occur with some intermediate frequency and intensity/severity (33, 34), hence the term *intermediate disturbance*. These intermediate scales of disturbance allow species to accumulate within the patch but prevent it from becoming dominated by one or a few of them. The assemblage is maintained in a nonequilibrium state, and assuming that the system is open [i.e. dispersal can freely occur among patches (26)], local coexistence of species is ensured (33, 34, 88, 160, 189). The scales of disturbance that will maintain the highest within-patch diversity in any particular assemblage depend on factors such as the rate of competitive exclusion (96) and the relative rates of recruitment.

The hypotheses of disturbance-mediated coexistence "assume a transitive hierarchical ranking of competitive abilities among the species, with competitive outcomes being consistent and asymmetrical, i.e. one of a pair of competitors always winning over the other" (35). For some assemblages of sessile organisms, this assumption does not seem to hold, however. In some cases, the competitive abilities of species are about equal and the outcome of their interaction is largely stochastic (e.g. 1, 35, 73, 98, 177). The winner in any particular interaction is determined by its order of invasion, its relative size, and the angle of encounter with the competitor, among other factors. In other cases, species are arranged in a relatively intransitive competitive network defined by "the occurrence of a loop in an otherwise hierarchical sequence of interference competitive abilities" (23:223–24). The degree of intransitivity in such a network can vary, depending on the competitive symmetry of its members (35).

Even if there is an overall hierarchy, a high degree of symmetry in competitive interactions among high-ranking species or a competitive network involving such species should reduce the likelihood that within-patch diversity will decline in the later stages of succession (167). Reciprocal replacement of species, as has been documented in some forest stands (e.g. 248), might also maintain high diversity in the absence of large-scale disturbance. Where such relationships among competitors exist, disturbance might seem unnecessary to maintain within-patch diversity. In the absence of disturbance, however, within-patch diversity in these systems may eventually decline owing to differences in the growth rates and competitive abilities of the species (23) or simply because the populations of some species fluctuate to low densities and go extinct by chance. Even rare disturbances can then be crucial in the long-term maintenance of local species richness (35). If nothing else, such nonhierarchical relationships may slow the rate of competitive exclusion (23, 102, 167) and thus increase the opportunity for mechanisms such as physical disturbance to maintain local diversity.

THE REGIONAL DYNAMICS OF POPULATIONS AND COMMUNITIES

The persistence of populations   Most sessile organisms are, to varying degrees, "fugitive species." On a small enough spatial scale, probably no population persists indefinitely (37). The replacement of adults requires dispersal of propagules to sites favorable for recruitment. A population will persist in a given area only if that area provides a sufficient number of safe sites per unit time to guarantee successful recruitment during the adults' lifetimes. The minimum area needed for persistence will vary among species. A species whose offspring can survive and grow under environmental conditions similar to those experienced by the adults usually does not disperse its propagules far, and it potentially can persist in a relatively small area. In contrast, if the offspring require conditions significantly different from those found in the area occupied by the adults, propagules are often dispersed a long distance, and persistence will only occur on a relatively large spatial scale.

As noted earlier, many species depend on disturbance to create conditions favorable for the recruitment, growth, and reproduction of their offspring. The regional dynamics and abundance of such a species will reflect the interplay of its life history and the regime of disturbance. For such a species to persist, disturbances must generate colonizable space within the dispersal range of extant populations and within the period of time it takes for these populations to go extinct (e.g. 152).

Species diversity   The species diversity of a region is a function of both the number and relative abundances of species that persist there. The regional abundance of a species can depend on the rate of disturbance (i.e. the area or number of patches cleared per unit time) and its predictability (as just discussed). In addition, regional population dynamics can be influenced by the size distribution of open patches and how synchronously they are produced.

Miller (133) assumes, and Abugov (3) demonstrates theoretically (as have others), that regional diversity peaks at intermediate rates of disturbance. At high mean disturbance rates, early successional species dominate most patches, while at low mean disturbance rates, late successional species do. These studies showed, however, that the distribution of patch sizes and the degree to which disturbances were phased (i.e. how synchronously they occurred) could strongly influence the level of diversity maintained by any given level of disturbance.

Assuming that small patches are usually invaded and dominated more quickly than large ones, Miller (133) concludes that large patches should favor colonizing species (i.e. early successional species) while small patches should favor competitive species (i.e. late successional species). Given these assumptions, he shows that at a low overall disturbance rate, regional diversity will be higher if the mean size of cleared patches is large, thus favoring the continued persistence of colonizing species. When the rate of disturbance is high, regional diversity will be greater when the areas disturbed are small, since this will ensure the continued persistence of competitive species. Alternative ways in which patch size may influence regional dynamics have also been discussed (35, 192).

Abugov (3:289) has examined how the phasing of disturbance might affect regional diversity. He describes *phasing* as follows:

> . . . the disturbance of a patch is unphased if its probability of being cleared during each time interval is independent of whether any other patches are being cleared during the same time interval. Conversely, other patches may be disturbed in phase. Each time a disturbance clears one of these patches it clears them all.

In the simplest case, as phasing increases, so does the proportion of colonizing species in the assemblage. Thus, an increase in the phasing of disturbance will decrease diversity when colonizing species predominate but will increase diversity when competitively dominant species occupy most of the space. More complicated patterns are also possible if one allows for changes in the relative competitive and migratory abilities of the colonizing species. To my knowledge, these theoretical predictions have yet to be tested empirically.

Patch dynamics and landscape pattern   Recent research on the role of disturbance in natural communities has increasingly focused on patterns at the regional scale, where the diversifying influence of disturbance is easy to observe. Patchy and locally asynchronous disturbance transforms the land or seascape into a continuously changing mosaic of patches of different sizes and ages (153, 159, 172, 173, 189, 239). The spatial orientation of each of the phases of this mosaic may strongly influence regional dynamics, particularly if the dispersal range of propagules is limited (159, 192). The dynamics of this patchwork of successional stages is beginning to receive theoretical attention (e.g. 153, 184, 185).

Do within-patch, nonequilibrium conditions average to an equilibrium pattern when on considers the mean dynamics of an area containing many such patches? Bormann & Likens (19, 20) refer to this large-scale equilibrium as a "shifting-mosaic steady-state." Many workers would answer the above question affirmatively (e.g. 19, 20, 196, 198, 249); but Romme (172, 173) found strong cyclic, rather than steady-state, dynamics in the vegetation of a 73 km^2 watershed in Yellowstone National Park. Recent simulation studies (185) indicate that the likelihood of a large-scale steady state is a function of the total landscape area and the size of the individual disturbances. The larger the area affected by a single disturbance, the more extensive the landscape must be to average out its effects. Landscapes that are small in absolute area (including those that have been fragmented by human activity) or that experience typical disturbances covering many thousands of hectares (e.g. 66) are unlikely to be in equilibrium.

DISTURBANCE AND MOBILE ANIMALS

With some notable exceptions, ecologists have largely overlooked the significant role of physical disturbance in the biology of mobile animals. Appreciation of the direct and indirect influences of disturbance on mobile animals has been slow to develop for at least two reasons. First, the direct effects of disturbance on mobile organisms are not as easy to observe and measure as those on sessile organisms. Second, during the last two decades many of the most influential investigators of the ecology of mobile animals—vertebrates in particular—have been proponents of a competition-based equilibrium theory of community organization (104, 241). Once equilibrium is assumed, there is little cause to examine the influence of supposedly rare and inconsequential disturbances.

In recent years, this view has become quite controversial (178). Physical disturbance can be a major cause of local disequilibrium when it kills appreciable numbers of animals, as demonstrated by the following incomplete list of examples: cold temperatures—butterflies (24, 58, 182), coral reef fish (18), stream fish (104, 214); oxygen depletion—lake fishes (213); drought—pond fish (121), stream fish (122), salamanders (99); flooding—desert fish (32); hurricanes—coral reef fish (123); fire—vertebrates (14, 28, 31, 169), soil and litter invertebrates (4, 31, 169); miscellaneous agents—intertidal invertebrates (192).

Severe climatic conditions can indirectly exert a strong negative impact on populations of mobile organisms by eliminating or producing shortages in vital resources. For example, prolonged drought stresses or kills the food plants of butterflies and causes large declines in, or even extinctions of, local populations (59, 182). A similar risk of local extinction is faced by any parasite whose host is subject to physical disturbance (165).

Even when a climatically induced shortage in resources does not cause much mortality, organisms that remain in the area may experience increased intra- and interspecific competition. Highly mobile species may move in search of more benign sites where resources are more plentiful. Increased competition and emigration during harsh climatic periods can reduce local population densities. These effects are well documented in recent long-term experimental and observational studies of lizard (55, 186) and bird (187) assemblages.

As was true for sessile organisms, mobile animals vary considerably in their vulnerability to disturbing forces and in their ability to avoid them. Animal responses to fire are a case in point (4, 14, 31, 118, 169). Highly mobile species can escape harm from all but the most quickly spreading fires by moving just ahead of the advancing flames or by seeking refuge out of the fire's path in patches of unburnt vegetation. If fleeing animals are overtaken, they can survive (as long as the fire is not too intense) by taking refuge in burrows or under rocks. Consequently, low to medium intensity fires often have relatively little direct impact on vertebrate populations. The main effect of such fires is to alter temporarily the spatial distribution and density of these populations. The burned site is frequently reoccupied soon after the fire has passed, though population densities may not return to prefire levels until the appropriate vegetation has regenerated on the site. In contrast, relatively immobile soil and litter invertebrates have no means of escaping an advancing fire; even a low intensity burn can kill large numbers of these organisms. The rate at which the burned area is recolonized by such organisms depends on the species' dispersal abilities and the proximity of source populations in surrounding areas of unburnt vegetation.

Physical disturbances also cause short- and long-term changes in the habitat that can have major indirect effects on populations of mobile species. Some of these changes can be detrimental to mobile animal populations, as the follow- ing examples demonstrate. As a result of unusually heavy rains, large quanti- ties of sediment were deposited in a southern California coastal lagoon. These deposits reduced the low tide volume of the lagoon and eliminated much of the previously extensive cover of eel grass. Probably for lack of sufficient habitat—as opposed to direct mortality caused by the sedimentation—the den- sity of water column fishes declined significantly (148). Similar declines in fish populations have been observed on storm-damaged coral reefs (105, 229). Here too, the reduction in fish numbers was due more to storm-induced alterations in the habitat than to any direct mortality caused by storm con- ditions.

Disturbances can also have indirect positive effects on mobile animal populations. Those that cause some immediate mortality may in the long run produce a net increase in population size and vitality. Fires, for example, may kill some birds and mammals, but they also reduce the populations of parasites that afflict some species (14).

Disturbances from such agents as fire, windthrow, and water motion indirectly affect populations of mobile animals by influencing the composition and structure of sessile assemblages. Sessile organisms provide cover and/or food, thus constituting a major component of the habitat of mobile animals. As this paper has emphasized, physical disturbance transforms sessile assemblages into mosaics of different seral stages. Sessile organisms comprising each of these stages differ in many characteristics critical to the welfare of mobile animals, including structural complexity, species composition, microclimate, and the quantity and quality of the food and shelter available. Although mobile animal species vary in their habitat requirements, many prefer the productive conditions associated with areas undergoing regeneration from recent disturbance. The quantity and palatability of plant foods are often higher in earlier stages of succession (27, 65, 81, 173). Some species of birds forage preferentially for fruits or insects in forest light gaps (180, 211).

Other species exploit the tissues of dead or dying plants as a resource and preferentially colonize recently disturbed sites where this material is abundant. For example, bark beetles and other boring insects are strongly attracted to injured or dead trees (181).

The compositions of insect, bird, and mammal assemblages change with the successional stage of the vegetation (e.g. 7, 16, 28, 61, 64, 115, 142, 173, 181, 183, 188, 206, 245). The population density and species richness of mobile animals often decline in later successional stages, paralleling similar trends in the vegetation. Some species, however, become more abundant as time elapses since the last disturbance. Such patterns reflect species-specific responses to changes in vegetation structure and composition, food availability, and interspecific interactions that accompany successional shifts in the plant community. Similar changes in mobile animal assemblages occur over the course of succession in sessile communities on rocky seashores (e.g. 48). On the landscape scale, mosaic patterns in sessile assemblages generated by disturbance are likely to have a profound effect on the regional population dynamics of mobile animals (65, 81, 173, 242).

CONCLUSIONS

I have attempted to summarize some of the key themes that run through the vast and rapidly expanding literature on disturbance in natural communities. These include: (*a*) the factors that determine natural regimes of disturbance, (*b*) organismal-level responses to disturbance and their evolution, and (*c*) the influence of disturbance on population and community structure and dynamics at both the local and regional scales.

Although all natural communities probably experience disturbance at some spatial and temporal scale, historically its role in community dynamics has been largely overlooked, except by some temperate-forest ecologists. There are many reasons for this neglect, but one of the prime causes may be that major disturbances often recur at intervals longer than the duration of an average research project or even than the lifespan of the investigator (146). Thus, the effects of disturbance cannot always be directly observed, which may lead one to conclude that disturbance is unimportant. Even a very long recurrence interval does not necessarily indicate that the impact of disturbance on the community is inconsequential, however. When the affected organisms are long-lived, the "compositional effects of disturbance can persist for centuries or even millennia" (66:216)).

There is a growing realization that disturbance may play as great a role in community dynamics as do biological interactions such as competition and predation, which have received far more empirical and theoretical attention from ecologists. The interplay between disturbance and these biological processes seems to account for a major portion of the organization and spatial patterning of natural communities.

ACKNOWLEDGMENTS

I thank J. Connell, P. Frank, A. Meyer, B. Mitchell, B. Okamura, G. Roderick, and S. Swarbrick for their very helpful comments on an earlier draft of the manuscript. I also thank those authors who provided prepublication drafts of chapters from a forthcoming book on disturbance edited by S. T. A. Pickett and P. S. White (162).

Literature Cited

1. Aarssen, L. W. 1983. Ecological combining ability and competitive combining ability in plants: Toward a general evolutionary theory of coexistence in systems of competition. *Am. Nat.* 122:707–31
2. Abrahamson, W. G. 1984. Species responses to fire on the Florida Lake Wales Ridge. *Am. J. Bot.* 71:35–43
3. Abugov, R. 1982. Species diversity and phasing of disturbance. *Ecology* 63:289–93
4. Ahlgren, I. F. 1974. The effect of fire on soil organisms. See Ref. 119, pp. 47–72
5. Allen, T. F. H., Starr, T. B. 1982. *Hierarchy.* Chicago: Univ. Chicago Press. 310 pp.
6. Amundson, D. C., Wright, H. E. Jr. 1979. Forest changes in Minnesota at the end of the Pleistocene. *Ecol. Monogr.* 49:1–16
7. Andersen, D. C., MacMahon, J. A., Wolfe, M. L. 1980. Herbivorous mammals along a montane sere: Community structure and energetics. *J. Mammal.* 61:500–19

8. Anderson, R. M., May, R. M. 1980. Infectious diseases and population cycles of forest insects. *Science* 210:658–61
9. Auchmoody, L. R. 1979. Nitrogen fertilization stimulates germination of dormant pin cherry seeds. *Can. J. For. Res.* 9:514–16
10. Bartholomew, B. 1970. Bare zone between California shrub and grassland communities: The role of animals. *Science* 170:1210–12
11. Bazzaz, F. A. 1979. The physiological ecology of plant succession. *Ann. Rev. Ecol. Syst.* 10:351–71
12. Bazzaz, F. A. 1983. Characteristics of populations in relation to disturbance in natural and man-modified ecosystems. See Ref. 136, pp. 259–75
13. Bazzaz, F. A., Pickett, S. T. A. 1980. Physiological ecology of tropical succession: A comparative review. *Ann. Rev. Ecol. Syst.* 11:287–310
14. Bendell, J. F. 1974. Effects of fire on birds and mammals. See Ref. 119, pp. 73–138

15. Bleakney, J. S. 1972. Ecological implications of annual variations in tidal extremes. *Ecology* 53:933–38

16. Bock, C. E., Bock, J. H. 1978. Response of birds, small mammals, and vegetation to burning sacaton grasslands in southeastern Arizona. *J. Range Manage.* 31:296–300

17. Boerner, R. E. J. 1982. Fire and nutrient cycling in temperate ecosystems. *BioScience* 32:187–92

18. Bohnsack, J. A. 1983. Resiliency of reef fish communities in the Florida Keys following a January 1977 hypothermal fish kill. *Environ. Biol. Fish.* 9:41–53

19. Bormann, F. H., Likens, G. E. 1979. Catastrophic disturbance and the steady state in northern hardwood forests. *Am. Sci.* 67:660–69

20. Bormann, F. H., Likens, G. E. 1979. *Pattern and Process in a Forested Ecosystem.* New York: Springer-Verlag. 253 pp.

21. Botkin, D. B., Sobel, M. J. 1975. Stability in time-varying ecosystems. *Am. Nat.* 109:625–46

22. Brokaw, N. V. L. 1984. Treefalls, regrowth, and community structure in tropical forests. See Ref. 162

23. Buss, L. W., Jackson, J. B. C. 1979. Competitive networks: Non-transitive competitive relationships in cryptic coral reef environments. *Am. Nat.* 113:223–34

24. Calvert, W. H., Zuckhowski, W., Brower, L. P. 1983. The effect of rain, snow and freezing temperatures on overwintering monarch butterflies in Mexico. *Biotropica* 15:42–47

24a. Canham, C. D., Loucks, O. L. 1984. Catastrophic windthrow in the presettlement forests of Wisconsin. *Ecology* 65:803–9

25. Canham, C. D., Marks, P. L. 1984. The response of woody plants to disturbance: Patterns of establishment and growth. See Ref. 162

26. Caswell, H. 1978. Predator-mediated coexistence: A non-equilibrium model. *Am. Nat.* 112:127–54

27. Cates, R. G., Orians, G. H. 1975. Successional status and the palatability of plants to generalized herbivores. *Ecology* 56:410–18

28. Catling, P. C., Newsome, A. E. 1981. Responses of the Australian vertebrate fauna to fire: An evolutionary approach. See Ref. 72, pp. 273–310

29. Christensen, N. L. 1981. Fire regimes in southeastern ecosystems. See Ref. 137, pp. 112–36

30. Christensen, N. L. 1984. Shrubland fire regimes and their evolutionary consequences. See Ref. 162

31. Christensen, P., Recher, H., Hoare, J. 1981. Responses of open forests (dry sclerophyll forests) to fire regimes. See Ref. 72, pp. 367–93

32. Collins, J. P., Young, C., Howell, J., Minckley, W. L. 1981. Impact of flooding in a Sonoran desert stream, including elimination of an endangered fish population (*Poeciliopsis o. occidentalis,* Poeciliidae). *Southwest. Nat.* 26:415–23

33. Connell, J. H. 1978. Diversity in tropical rain forests and coral reefs. *Science* 199:1302–10

34. Connell, J. H. 1979. Tropical rain forests and coral reefs as open non-equilibrium systems. In *Population Dynamics,* ed. R. M. Anderson, B. D. Turner, L. R. Taylor, pp. 141–63. Oxford: Blackwell. 434 pp.

35. Connell, J. H., Keough, M. J. 1984. Disturbance and patch dynamics of subtidal marine animals on hard substrates. See Ref. 162

36. Connell, J. H., Slatyer, R. O. 1977. Mechanisms of succession in natural communities and their role in community stability and organization. *Am. Nat.* 111:1119–44

37. Connell, J. H., Sousa, W. P. 1983. On the evidence needed to judge ecological stability or persistence. *Am. Nat.* 121:789–824

38. Conrad, C. E., Oechel, W. C., eds. 1982. *Dynamics and Management of Mediterranean-Type Ecosystems, US Dep. Agric. For. Ser. Gen. Tech. Rep. PSW–58.* Berkeley, CA: USDA. 637 pp.

39. Cook, R. E. 1980. The biology of seeds in the soil. In *Demography and Evolution in Plant Populations,* ed. O. T. Solbrig, pp. 107–29. Oxford: Blackwell. 222 pp.

40. Cook, R. E., Lyons, E. E. 1983. The biology of *Viola fimbriatula* in a natural disturbance. *Ecology* 64:654–60

41. Cottam, G. 1981. Patterns of succession in different forest ecosystems. See Ref. 233, pp. 178–84

42. Crocker, R. L., Major, J. 1955. Soil development in relation to vegetation and surface age at Glacier Bay, Alaska. *J. Ecol.* 43:427–48

43. Davis, M. B. 1981. Quaternary history and the stability of forest communities. See Ref. 233, pp. 132–53

44. Davis, R. M., Cantlon, J. E. 1969. Effect of size area open to colonization on species composition in early old-field succession. *Bull. Torrey Bot. Club* 96:660–73

45. Dayton, P. K. 1971. Competition, disturbance and community organization: The provision and subsequent utilization

of space in a rocky intertidal community. *Ecol. Monogr.* 41:351–89

46. Dayton, P. K. 1973. Dispersion, dispersal, and persistence of the annual intertidal alga, *Postelsia palmaeformis* Ruprecht. *Ecology* 54:433–38

47. Dayton, P. K., Tegner, M. J. 1984. The importance of scale in community ecology: A kelp forest example with terrestrial analogs. In *A New Ecology: Novel Approaches to Interactive Systems*, ed. P. W. Price, C. N. Slobodchikoff, W. S. Gaud, pp. 457–81. New York: Wiley. 515 pp.

48. Dean, R. L. 1983. *The influence of marine algal succession on the invertebrate community*. PhD thesis. Univ. Calif., Santa Barbara. 222 pp.

49. del Moral, R. 1983. Initial recovery of subalpine vegetation on Mount St. Helens, Washington. *Am. Midl. Nat.* 109:72–80

50. Denny, M. W., Daniel, T. L., Koehl, M. A. R. 1984. Mechanical limits to size in wave-swept organisms. *Ecol. Monogr.* In press

51. Denslow, J. S. 1980. Gap partitioning among tropical rainforest trees. *Biotropica* 12:47–55 (Suppl.)

52. Denslow, J. S. 1984. Disturbance-mediated coexistence of species. See Ref. 162

53. Dodge, M. 1972. Forest fuel accumulation—a growing problem. *Science* 177:139–42

54. Dollar, S. J. 1982. Wave stress and coral community structure in Hawaii. *Coral Reefs* 1:71–81

55. Dunham, A. E. 1980. An experimental study of interspecific competition between the iguanid lizards *Sceloporus merriami* and *Urosaurus ornatus*. *Ecol. Monogr.* 50:309–30

56. Dunn, C. P., Guntenspergen, G. R., Dorney, J. R. 1983. Catastrophic wind disturbance in an old-growth hemlock-hardwood forest, Wisconsin. *Can. J. Bot.* 61:211–17

57. Ehrenfield, J. G. 1980. Understory response to canopy gaps of varying size in mature oak forest. *Bull. Torrey Bot. Club* 107:29–41

58. Ehrlich, P. R., Breedlove, D. E., Brussard, P. F., Sharp, M. A. 1972. Weather and the "regulation" of subalpine populations. *Ecology* 53:243–47

59. Ehrlich, P. R., Murphy, D. D., Singer, M. C., Sherwood, C. B., White, R. R., Brown, I. L. 1980. Extinction, reduction, stability and increase: The responses of checkerspot butterfly (*Euphydryas*) populations to the California drought. *Oecologia* 46:101–5

60. Flaccus, E. 1959. Revegetation of landslides in the White Mountains of New Hampshire. *Ecology* 40:692–703

61. Force, D. C. 1981. Postfire insect succession in southern California chaparral. *Am. Nat.* 117:575–82

62. Forman, R. T. T., Boerner, R. E. 1981. Fire frequency in the pine barrens of New Jersey. *Bull. Torrey Bot. Club* 108:34–50

63. Foster, R. B. 1980. Heterogeneity and disturbance in tropical vegetation. In *Conservation Biology*, ed. M. E. Soule, B. A. Wilcox, pp. 75–92. Sunderland, Mass: Sinauer. 395 pp.

64. Fox, B. J. 1982. Fire and mammalian secondary succession in an Australian coastal heath. *Ecology* 63:1332–41

65. Fox, J. F. 1978. Forest fires and the snowshoe hare—Canada lynx cycle. *Oecologia* 31:349–74

66. Franklin, J. F., Hemstrom, M. A. 1981. Aspects of succession in the coniferous forests of the Pacific Northwest. See Ref. 233, pp. 212–29

67. Garwood, N. C. 1983. Seed germination in a seasonal tropical forest in Panama: A community study. *Ecol. Monogr.* 53:159–81

68. Garwood, N. C., Janos, D. P., Brokaw, N. 1979. Earthquake-caused landslides: A major disturbance to tropical forests. *Science* 205:997–99

69. Gerrodette, T. 1981. Dispersal of the solitary coral *Balanophyllia elegans* by demersal planular larvae. *Ecology* 62:611–19

70. Gill, A. M. 1981. Adaptive responses of Australian vascular plant species to fires. See Ref. 72, pp. 243–71

71. Gill, A. M. 1981. Fire adaptive traits of vascular plants. See Ref. 137, pp. 208–30

72. Gill, A. M., Groves, R. H., Noble, I. R., eds. 1981. *Fire and the Australian Biota*. Canberra: Aust. Acad. Sci. 582 pp.

73. Goldberg, D. E., Werner, P. A. 1983. Equivalence of competitors in plant communities: A null hypothesis and a field experimental approach. *Am. J. Bot.* 70:1098–1104

74. Goldberg, D. E., Werner, P. A. 1983. The effects of size of opening in vegetation and litter cover on seedling establishment of goldenrods (*Solidago* spp.). *Oecologia* 60:149–55

75. Gould, S. J., Lewontin, R. C. 1979. The spandrels of San Marco and the Panglossian paradigm: A critique of the adaptationist programme. *Proc. R. Soc. London Ser. B* 205:581–98

76. Green, D. G. 1982. Fire and stability in the postglacial forests of southwest Nova Scotia. *J. Biogeogr.* 9:29–40

77. Grime, J. P. 1979. *Plant Strategies and Vegetation Processes.* Chichester, England: Wiley. 222 pp.
78. Gross, K. L. 1980. Colonization by *Verbascum thapsus* (Mullein) of an old-field in Michigan: Experiments on the effects of vegetation. *J. Ecol.* 68:919–27
79. Gross, K. L., Werner, P. A. 1982. Colonizing abilities of "biennial" plant species in relation to ground cover: Implications for the distributions in a successional sere. *Ecology* 63:921–31
80. Grubb, P. J. 1977. The maintenance of species richness in plant communities: The importance of regeneration niche. *Biol. Rev.* 52:107–45
81. Hansson, L. 1979. On the importance of landscape heterogeneity in northern regions for the breeding population densities of homeotherms: A general hypothesis. *Oikos* 33:182–89
82. Harger, J. R. E., Landenberger, D. E. 1971. The effect of storms as a density dependent mortality factor on populations of sea mussels. *Veliger* 14:195–201
83. Harper, J. L. 1977. *Population Biology of Plants.* London: Academic. 892 pp.
84. Hartshorn, G. S. 1978. Tree falls and tropical forest dynamics. See Ref. 212, pp. 617–38
85. Heinselman, M. L. 1973. Fire in the virgin forest of the Boundary Waters Canoe Area, Minnesota. *Quat. Res. (NY)* 3: 329–82
86. Heinselman, M. L. 1981. Fire and succession in the conifer forests of northern North America. See Ref. 233, pp. 374–405
87. Heinselman, M. L. 1981. Fire intensity and frequency as factors in the distribution and structure of northern ecosystems. See Ref. 137, pp. 7–57
88. Hemphill, N., Cooper, S. D. 1983. The effect of physical disturbance on the relative abundances of two filter-feeding insects in a small stream. *Oecologa* 58: 378–82
89. Henniker-Gotley, G. R. 1936. A forest fire caused by falling stones. *Indian For.* 62:422–23
90. Henry, J. D., Swan, J. M. A. 1974. Reconstructing forest history from live and dead plant material—an approach to the study of forest succession in southwest New Hampshire. *Ecology* 55:772–83
91. Highsmith, R. C. 1982. Reproduction by fragmentation in corals. *Mar. Ecol. Progress Ser.* 7:207–26
92. Holbrook, S. J. 1977. Roden faunal turnover and prehistoric community stability in northwestern New Mexico. *Am. Nat.* 111:1195–1208
93. Hopkins, B. 1965. Observations on savanna burning in the Olokemeji Forest Reserve, Nigeria. *J. Appl. Ecol.* 2:367–81
94. Horn, H. S. 1981. Some causes of variety in patterns of secondary succession. See Ref. 233, pp. 24–35
95. Horn, H. S. 1981. Succession. In *Theoretical Ecology,* ed. R. M. May, pp. 253–71. Sunderland, Mass: Sinauer. 489 pp. 2nd ed.
96. Huston, M. 1979. A general hypothesis of species diversity. *Am. Nat.* 113:81–101
97. Jackson, J. B. C. 1977. Habitat area, colonization, and development of epibenthic community structure. In *Biology of Benthic Organisms,* ed. B. F. Keegan, P. O. Ceidigh, P. J. S. Boaden, pp. 349–58. London: Pergamon. 630 pp.
98. Jackson, J. B. C. 1979. Overgrowth competition between encrusting cheilostome ectoprocts in a Jamaican cryptic reef environment. *J. Anim. Ecol.* 48: 805–23
99. Jaeger, R. G. 1980. Density-dependent and density-independent causes of extinction of a salamander population. *Evolution* 34:617–21
100. Janzen, D. H. 1967. Fire, vegetation structure, and the ant × acacia interaction in Central America. *Ecology* 48:26–35
101. Jones, E. W. 1945. The structure and reproduction of the virgin forest of the north temperate zone. *New Phytol.* 44:130–48
102. Karlson, R. H., Jackson, J. B. C. 1981. Competitive networks and community structure: A simulation study. *Ecology* 62:670–78
103. Karr, J. R., Freemark, K. E. 1983. Habitat selection and environmental gradients: Dynamics in the stable tropics. *Ecology* 64:1481–94
104. Karr, J. R., Freemark, K. E. 1984. Disturbance, perturbation, and vertebrates: An integrative perspective. See Ref. 162
105. Kaufman, L. S. 1983. Effects of Hurricane Allen on reef fish assemblages near Discovery Bay, Jamaica. *Coral Reefs* 2:43–47
106. Kay, A. M., Keough, M. J. 1981. Occupation of patches in the epifaunal communities on pier pilings and the bivalve *Pinna bicolor* at Edithburgh, South Australia. *Oecologia* 48:123–30
107. Keeley, J. E. 1977. Seed production, seed populations in soil, and seedling production after fire for two congeneric pairs of sprouting and nonsprouting chaparral shrubs. *Ecology* 58:820–29
108. Keeley, J. E. 1981. Reproductive cycles and fire regimes. See Ref. 137, pp. 231–77

109. Keeley, J. E., Zedler, P. H. 1978. Reproduction of chaparral shrubs after fire: A comparison of sprouting and seeding strategies. *Am. Midl. Nat.* 99:142–61

110. Keough, M. J. 1984. Effects of patch size on the abundance of sessile marine invertebrates. *Ecology* 65:423–37

111. Kilgore, B. M. 1981. Fire in ecosystem distribution and structure: Western forests and scrublands. See Ref. 137, pp. 58–89

112. Kilgore, B. M., Taylor, D. 1979. Fire history of sequoia-mixed conifer forest. *Ecology* 60:129–42

113. Kimmerer, R. W., Allen, T. F. H. 1982. The role of disturbance in the pattern of a riparian bryophyte community. *Am. Midl. Nat.* 107:370–83

114. King, T. J. 1977. The plant ecology of anthills in calcareous grasslands. I. Patterns of species in relation to anthills in southern England. *J. Ecol.* 65:235–56

115. Kirkland, G. L. 1977. Responses of small mammals to the clearcutting of northern Appalachian forests. *J. Mammal.* 58:600–9

116. Knowlton, N., Lang, J. C., Rooney, M. C., Clifford, P. 1981. Evidence for delayed mortality in hurricane-damaged Jamaican staghorn corals. *Nature* 294: 251–52

117. Koehl, M. A. R. 1982. The interaction of moving water and sessile organisms. *Sci. Am.* 247:124–34

118. Komarek, E. V. 1969. Fire and animal behavior. *Proc. Tall Timbers Fire Ecol. Conf.* 9:161–207

119. Kozlowski, T. T., Ahlgren, C. E., eds. 1974. *Fire and Ecosystems.* New York: Academic. 542 pp.

120. Kucera, C. L. 1981. Grasslands and fire. See Ref. 137, pp. 90–111

121. Kushlan, J. A. 1974. Effects of a natural fish kill on the water quality, plankton, and fish production of a pond in the Big Cypress Swamp, Florida. *Trans. Am. Fish. Soc.* 103:235–43

122. Larimore, R. W., Childers, W. F., Heckrotte, C. 1959. Destruction and reestablishment of stream fish and invertebrates affected by drought. *Trans. Am. Fish. Soc.* 88:261–85

123. Lassig, B. R. 1983. The effects of a cyclonic storm on coral reef fish assemblages. *Environ. Biol. Fish.* 9:55–63

124. Levin, S. A. 1976. Population dynamic models in heterogeneous environments. *Ann. Rev. Ecol. Syst.* 7:287–310

125. Lieberman, M., John, D. M., Lieberman, D. 1979. Ecology of subtidal algae on seasonally devastated cobble substrates off Ghana. *Ecology* 60:1151–61

126. Littler, M. M., Martz, D. R., Littler, D.
S. 1983. Effects of recurrent sand deposition on rocky intertidal organisms: Importance of substrate heterogeneity in a fluctuating environment. *Mar. Ecol. Progress Ser.* 11:129–39

127. Loucks, O. L. 1970. Evolution of diversity, efficiency, and community stability. *Am. Zool.* 10:17–25

128. Lubchenco, J. 1978. Plant species diversity in a marine intertidal community: Importance of herbivore food preference and algal competitive abilities. *Am. Nat.* 112:23–39

129. Lubchenco, J., Menge, B. A. 1978. Community development and persistence in a low rocky intertidal zone. *Ecol. Monogr.* 48:67–94

130. Marks, P. L. 1974. The role of pin cherry (*Prunus pensylvanica* L.) in the maintenance of stability in northern hardwood ecosystems. *Ecol. Monogr.* 44:73–88

131. McAuliffe, J. R. 1983. Competition, colonization patterns, and disturbance in stream benthic communities. In *Stream Ecology,* ed. J. R. Barnes, G. W. Minshall, pp. 137–56. New York: Plenum. 399 pp.

131a. McAuliffe, J. R. 1984. Competition for space, disturbance, and the structure of a benthic stream community. *Ecology* 65: 894–908

132. McMaster, G. S., Zedler, P. H. 1981. Delayed seed dispersal in *Pinus torreyana* (Torrey Pine). *Oecologia* 51:62–66

133. Miller, T. E. 1982. Community diversity and interactions between the size and frequency of disturbance. *Am. Nat.* 120: 533–36

134. Minnich, R. A. 1983. Fire mosaics in southern California and northern Baja California. *Science* 219:1287–94

135. Moir, E. McA. 1923. Natural causes of forest fires. *Emp. For. J.* 2:17–18

136. Mooney, H. A., Godron, M., eds. 1983. *Disturbance and Ecosystems.* Berlin: Springer-Verlag. 292 pp.

137. Mooney, H. A., Bonnicksen, T. M., Christensen, N. L., Lotan, J. E., Reiners, W. A., eds. 1981. *Fire Regimes and Ecosystem Properties, US Dep. Agric. For. Ser. Gen. Tech. Rep. WO–26.* Washington, DC: US For. Ser. 594 pp.

138. Morris, R. F. 1963. The dynamics of epidemic spruce budworm populations. *Mem. Entomol. Soc. Can.* 31:1–332

139. Murdoch, W. W., Oaten, A. 1975. Predation and population stability. *Adv. Ecol. Res.* 9:2–131

140. Mutch, R. W. 1970. Wildland fires and ecosystems—a hypothesis. *Ecology* 51: 1046–51

141. Nanson, G. C., Beach, H. F. 1977.

Forest succession and sedimentation on a meandering-river floodplain, northeast British Columbia, Canada. *J. Biogeogr.* 4:229–51

142. Newsome, A. E., McIlroy, J. C., Catling, P. C. 1975. The effects of extensive wildfire on populations of twenty ground vertebrates in southeast Australia. *Proc. Ecol. Soc. Aust.* 9:107–23

143. Noble, I. R., Slatyer, R. O. 1980. The use of vital attributes to predict successional changes in plant communities subject to recurrent disturbances. *Vegetatio* 43:5–21

144. Norton-Griffiths, M. 1979. The influence of grazing, browsing, and fire on the vegetation dynamics of the Serengeti. In *Serengeti*, ed. A. R. E. Sinclair, M. Norton-Griffiths, pp. 310–52. Chicago: Univ. Chicago Press. 389 pp.

145. Odum, E. P. 1971. *Fundamentals of Ecology.* Philadelphia: Saunders. 574 pp. 3rd ed.

146. Oliver, C. D. 1981. Forest development in North America following major disturbances. *For. Ecol. Manage.* 3:153–68

147. Onuf, C. P. 1984. *The Ecology of Mugu Lagoon: An Estuarine Profile.* Washington DC: US Fish & Wild. Serv. Biol. Serv. Program. In press

148. Onuf, C. P., Quammen, M. L. 1983. Fishes in a California coastal lagoon: Effects of major storms on distribution and abundance. *Mar. Ecol. Progress Ser.* 12:1–14

149. Orians, G. H. 1982. The influence of tree falls in tropical forests on tree species richness. *Trop. Ecol.* 23:255–79

150. Osman, R. W. 1977. Establishment and development of a marine epifaunal community. *Ecol. Monogr.* 47:37–64

151. Paine, R. T. 1974. Intertidal community structure: Experimental studies on the relationship between a dominant competitor and its principal predator. *Oecologia* 15:93–120

152. Paine, R. T. 1979. Disaster, catastrophe and local persistence of the sea palm *Postelsia palmaeformis.* *Science* 205:685–87

153. Paine, R. T., Levin, S. A. 1981. Intertidal landscapes: Disturbance and the dynamics of pattern. *Ecol. Monogr.* 51:145–78

154. Palumbi, S. R., Jackson, J. B. C. 1982. Ecology of cryptic coral reef communities. II. Recovery from small disturbance events by encrusting bryozoa: The influence of "host" species and lesion size. *J. Exp. Mar. Biol. Ecol.* 64:103–15

155. Pearson, R. G. 1981. Recovery and recolonization of coral reefs. *Mar. Ecol. Progress Ser.* 4:105–22

156. Perry, D. A., Lotan, J. E. 1979. A model of fire selection for serotiny in lodgepole pine. *Evolution* 33:958–68

157. Peterson, D. L., Bazzaz, F. A. 1978. Life cycle characteristics of *Aster pilosus* in early successional habitats. *Ecology* 59:1005–13

158. Philpot, C. W. 1969. *Seasonal Changes in Heat Content and Ether Extractives Content of Chamise. US Dept. Agric. For. Ser. Res. Pap. INT–61.* Ogden, Utah: Intermount. For. & Range Exp. Stn. 10 pp.

159. Pickett, S. T. A. 1976. Succession: An evolutionary interpretation. *Am. Nat.* 110:107–19

160. Pickett, S. T. A. 1980. Non-equilibrium coexistence of plants. *Bull. Torrey Bot. Club* 107:238–48

161. Pickett, S. T. A., Thompson, J. N. 1978. Patch dynamics and the design of nature reserves. *Biol. Conserv.* 13:27–37

162. Pickett, S. T. A., White, P. S., eds. 1984. *Natural Disturbance: The Patch Dynamics Perspective.* New York: Academic. In press

163. Platt, W. J. 1975. The colonization and formation of equilibrium plant species associations on badger disturbances in a tall-grass prairie. *Ecol. Monogr.* 45:285–305

164. Platt, W. J., Weis, I. M. 1977. Resource partitioning and competition within a guild of fugitive prairie plants. *Am. Nat.* 111:479–513

165. Price, P. W. 1980. *Evolutionary Biology of Parasites.* Princeton, NJ: Princeton Univ. Press. 237 pp.

166. Pyne, S. J. 1982. *A Cultural History of Wildland and Rural Fire.* Princeton, NJ: Princeton Univ. Press. 656 pp.

167. Quinn, J. F. 1982. Competitive hierarchies in marine benthic communities. *Oecologia* 54:129–35

168. Raup, H. M. 1957. Vegetational adjustment to the instability of the site. In *6th Proc. Tech. Meet. Int. Union Conserv. Nature Nat. Resour., Edinburgh, 1956,* pp. 36–48. London: Soc. Promo. Nat. Resour.

169. Recher, H. F., Christensen, P. E. 1981. Fire and the evolution of the Australian biota. In *Ecological Biogeography of Australia,* ed. A. Keast, pp. 135–62. The Hague: Junk. 805 pp.

170. Reiners, W. A., Lang, G. E. 1979. Vegetational patterns and processes in the balsam fir zone, White Mountains, New Hampshire. *Ecology* 60:403–17

171. Ricklefs, R. E. 1977. Environmental heterogeneity and plant species diversity: A hypothesis. *Am. Nat.* 111:376–81

172. Romme, W. H. 1982. Fire and landscape

diversity in subalpine forests of Yellowstone National Park. *Ecol. Monogr.* 52: 199–221

173. Romme, W. H., Knight, D. H. 1982. Landscape diversity: The concept applied to Yellowstone Park. *BioScience* 32: 664–70

174. Rundel, P. W. 1981. Structural and chemical components of flammability. See Ref. 137, pp. 183–207

175. Rundel, P. W. 1982. Successional dynamics of chamise chaparral: The interface of basic research and management. See Ref. 38, pp. 86–90

176. Runkle, J. R. 1984. Disturbance regimes in temperate forests. See Ref. 162

177. Russ, G. R. 1982. Overgrowth in a marine epifaunal community: Competitive hierarchies and competitive networks. *Oecologia* 53:12–19

178. Salt, G. W., ed. 1983. *A Round Table on Research in Ecology and Evolutionary Biology. Am. Nat.* 122:593–705

179. Schaal, B. A., Leverich, W. J. 1982. Survivorship patterns in an annual plant community. *Oecologia* 54:149–51

180. Schemske, D. W., Brokaw, N. 1981. Treefalls and the distribution of understory birds in a tropical forest. *Ecology* 62:938–45

181. Schowalter, T. D. 1984. Adaptations of insects to disturbance. See Ref. 162

182. Shapiro, A. M. 1979. The phenology of *Pieris napi microstriate* (Lepidoptera: Pieridae) during and after the 1975–77 California drought, and its evolutionary significance. *Psyche* 86:1–10

183. Shugart, H. H. Jr., James, D. 1973. Ecological succession of breeding bird populations in northwestern Arkansas. *Auk* 90:62–77

184. Shugart, H. H. Jr., Seagle, S. W. 1984. Modeling forest landscapes and the role of disturbance in ecosystems and communities. See Ref. 162

185. Shugart, H. H. Jr., West, D. C. 1981. Long-term dynamics of forest ecosystems. *Am. Sci.* 69:647–52

186. Smith, D. C. 1981. Competitive interactions of the striped plateau lizard *(Sceloporus virgatus)* and the tree lizard *(Urosaurus ornatus). Ecology* 62: 679–87

187. Smith, K. G. 1982. Drought-induced changes in avian community structure along a montane sere. *Ecology* 63:952–61

188. Smith, K. G., MacMahon, J. A. 1981. Bird communities along a montane sere: Community structure and energetics. *Auk* 98:8–28

189. Sousa, W. P. 1979. Disturbance in marine intertidal boulder fields: The nonequilibrium maintenance of species diversity. *Ecology* 60:1225–39

190. Sousa, W. P. 1979. Experimental investigations of disturbance and ecological succession in a rocky intertidal algal community. *Ecol. Monogr.* 49:227–54

191. Sousa, W. P. 1980. The responses of a community to disturbance: The importance of successional age and species' life histories. *Oecologia* 45:72–81

192. Sousa, W. P. 1984. Disturbance and patch dynamics on rocky intertidal shores. See Ref. 162

193. Sousa, W. P. 1984. Intertidal mosaics: Patch size, propagule availability, and spatially variable patterns of succession. *Ecology.* In press

194. Southwood, T. R. E. 1977. Habitat, the templet for ecological strategies? *J. Anim. Ecol.* 46:337–65

195. Spring, P. E., Brewer, M. L., Brown, J. R., Fanning, M. E. 1974. Population ecology of loblolly pine *Pinus taeda* in an old field community. *Oikos* 25:1–6

196. Sprugel, D. G. 1976. Dynamic structure of wave-regenerated *Abies balsamea* forests in the north-eastern United States. *J. Ecol.* 64:889–911

197. Sprugel, D. G. 1984. Natural disturbance and ecosystem energetics. See Ref. 162

198. Sprugel, D. G., Bormann, F. H. 1981. Natural disturbance and the steady state in high-altitude balsam fir forests. *Science* 211:390–93

199. Stewart, G. H., Veblen, T. T. 1982. Regeneration patterns in southern rata *(Metrosideros umbellata)*—kamahi *(Weinmannia racemosa)* forest in central Westland, New Zealand. *NZ J. Bot.* 20:55–72

200. Stone, E. C., Vasey, R. B. 1968. Preservation of coast redwood on alluvial flats. *Science* 159:157–61

201. Strong, D. R. 1977. Epiphyte loads, treefalls, and perennial forest disruption: A mechanism for maintaining higher tree species richness in the tropics without animals. *J. Biogeogr.* 4:215–18

202. Suchanek, T. H. 1979. *The* Mytilus californianus *community: Studies on the composition, structure, organization, and dynamics of a mussel bed.* PhD thesis. Univ. Wash., Seattle. 286 pp.

203. Sutherland, J. P., Karlson, R. H. 1977. Development and stability of the fouling community at Beaufort, North Carolina. *Ecol. Monogr.* 47:425–46

204. Tande, G. F. 1979. Fire history and vegetation pattern of coniferous forests in Jasper National Park, Alberta. *Can. J. Bot.* 57:1912–31

205. Taylor, A. R. 1974. Ecological aspects of lightning in forests. *Proc. Tall Timbers Fire Ecol. Conf.* 13:455–82

206. Taylor, D. L. 1973. Some ecological implications of forest fire control in Yellowstone National Park. *Ecology* 54:1394–96

207. Thistle, D. 1981. Natural physical disturbances and communities of marine soft bottoms. *Mar. Ecol. Progress Ser.* 6:223–28

208. Thompson, J. M. 1980. Treefalls and colonization patterns in temperate forest herbs. *Am. Midl. Nat.* 104:176–84

209. Thompson, J. M. 1983. Partitioning of variance in demography: Within-patch differences in herbivory, survival, and flowering of *Lomatium farinosum* (Umbelliferae). *Oikos* 40:315–17

210. Thompson, J. M. 1984. Within-patch dynamics of life histories, populations, and interactions: Selection over time in small spaces. See Ref. 162

211. Thompson, J. M., Willson, M. F. 1978. Disturbance and the dispersal of fleshy fruits. *Science* 200:1161–63

212. Tomlinson, P. B., Zimmerman, M. H., eds. 1978. *Tropical Trees as Living Systems*. Cambridge: Cambridge Univ. Press. 675 pp.

213. Tonn, W., Magnuson, J. J. 1982. Patterns in the species composition and richness of fish assemblages in northern Wisconsin lakes. *Ecology* 63:1149–66

214. Toth, L. A., Dudley, D. R., Karr, J. R., Gorman, O. T. 1982. Natural and man-induced variability in a silverjaw minnow *(Ericymba buccata)* population. *Am. Midl. Nat.* 107:284–93

215. Underwood, A. J., Denley, E. J. 1984. Paradigms, explanations and generalizations in models for the structure of intertidal communities on rocky shores. In *Ecological Communities: Conceptual Issues and the Evidence*, ed. D. R. Strong Jr., D. Simberloff, L. G. Abele, A. B. Thistle. 1984. Princeton, NJ: Princeton Univ. Press. 613 pp.

216. VanBlaricom, G. R. 1982. Experimental analyses of structural regulation in a marine sand community exposed to oceanic swell. *Ecol. Monogr.* 52:283–305

217. Veblen, T. T. 1979. Structure and dynamics of *Nothofagus* forests near timberline in south-central Chile. *Ecology* 60:937–45

218. Veblen, T. T. 1984. Stand dynamics in Chilean *Nothofagus* forests. See Ref. 162

219. Veblen, T. T., Ashton, D. H. 1978. Catastrophic influences on the vegetation of the Valdivian Andes, Chile. *Vegetatio* 36:149–67

220. Veblen, T. T., Donoso Z., C., Schlegel, F. M., Escobar R., B. 1981. Forest dynamics in south-central Chile. *J. Biogeogr.* 8:211–47

221. Viosca, P. Jr. 1931. Spontaneous combustion on marshes of southern Louisiana. *Ecology* 12:439–42

222. Vitousek, P. M. 1984. Community turnover and ecosystem nutrient dynamics. See Ref. 162

223. Vitousek, P. M., White, P. S. 1981. Process studies in succession. See Ref. 233, pp. 267–76

224. Vogel, S. 1981. *Life in Moving Fluids*. Boston: Grant. 352 pp.

225. Vogl, R. J. 1973. Ecology of knobcone pine in Santa Ana Mountains, California. *Ecol. Monogr.* 43:125–43

226. Vogl, R. J. 1974. Effects of fire on grasslands. See Ref. 119, pp. 139–94

227. Vogl, R. J. 1977. Fire: A destructive menace or a natural process. In *Recovery and Restoration of Damaged Ecosystems*, ed. J. Cairns Jr., K. L. Dickson, E. E. Herricks, pp. 261–89. Charlottesville: Univ. Press Va. 531 pp.

228. Vrijenhoek, R. C. 1984. Animal population genetics and disturbance: The effects of local extinctions and recolonizations on heterozygosity and fitness. See Ref. 162

229. Walsh, W. J. 1983. Stability of a coral reef fish community following a catastrophic storm. *Coral Reefs* 2:49–63

230. Watt, A. S. 1925. On the ecology of British beechwoods with special reference to their regeneration. Part II, Sect. II & III. The development and structure of beech communities on the Sussex Downs. *J. Ecol.* 13:27–73

231. Watt, A. S. 1947. Pattern and process in the plant community. *J. Ecol.* 35:1–22

232. Webb, L. J. 1958. Cyclones as an ecological factor in tropical lowland rain forest, north Queensland. *Aust. J. Bot.* 6:220–28

233. West, D. C., Shugart, H. H. Jr., Botkin, B. D., eds. 1981. *Forest Succession.* New York: Springer-Verlag. 517 pp.

234. Wethey, D. S. 1979. *Demographic variation in intertidal barnacles.* PhD thesis. Univ. Mich., Ann Arbor. 260 pp.

235. White, P. S. 1979. Pattern, process, and natural disturbance in vegetation. *Bot. Rev.* 45:229–99

236. Whitmore, T. C. 1974. *Change with Time and the Role of Cyclones in Tropical Rain Forest on Kolombangara, Solomon Islands, Commonw. For. Inst. Pap. 46, Univ. Oxford.* 78 pp.

237. Whitmore, T. C. 1975. *Tropical Rain Forests of the Far East.* Oxford: Clarendon. 282 pp.

238. Whitmore, T. C. 1978. Gaps in the forest canopy. See Ref. 212, pp. 639–55
239. Whittaker, R. H., Levin, S. A. 1977. The role of mosaic phenomena in natural communities. *Theor. Popul. Biol.* 12: 117–39
240. Wiens, J. A. 1976. Population responses to patchy environments. *Ann. Rev. Ecol. Syst.* 7:81–120
241. Wiens, J. A. 1977. On competition and variable environments. *Am. Sci.* 65:590–97
242. Wiens, J. A. 1984. Vertebrate responses to environmental patchiness in arid and semi-arid ecosystems. See Ref. 162
243. Williamson, G. B. 1975. Pattern and seral composition in an old-growth beech-maple forest. *Ecology* 56:727–31
244. Williamson, G. B., Black, E. M. 1981. High temperature of forest fires under pines as a selective advantage over oaks. *Nature*. 293:643–44
245. Wirtz, W. O. II. 1982. Postfire community structure of birds and rodents in southern California chaparral. See Ref. 38, pp. 241–46
246. Woodin, S. A. 1976. Adult-larval interactions in dense infaunal assemblages: Patterns of abundance. *J. Mar. Res.* 34: 25–41

247. Woodley, J. D., Chornesky, E. A., Clifford, P. A., Jackson, J. B. C., Kaufman, L. S. et al. 1981. Hurricane Allen's impact on Jamaican coral reefs. *Science* 214:749–55
248. Woods, K. D., Whittaker, R. H. 1981. Canopy-understory interaction and the internal dynamics of mature hardwood and hemlock-hardwood forests. See Ref. 233, pp. 305–23
249. Zackrisson, O. 1977. Influence of forest fires on the north Swedish boreal forest. *Oikos* 29:22–32
250. Zedler, J. B. 1983. Freshwater impacts in normally hypersaline marshes. *Estuaries* 6:346–55
251. Zedler, P. H. 1981. Vegetation change in chaparral and desert communities in San Diego County, California. See Ref. 233, pp. 406–30
252. Zedler, P. H. 1982. Plant demography and chaparral management in southern California. See Ref. 38, pp. 123–27
253. Zedler, P. H., Gautier, C. R., McMaster, G. S. 1983. Vegetation change in response to extreme events: The effect of a short interval between fires in California chaparral and coastal scrub. *Ecology* 64:809–18

Ann. Rev. Ecol. Syst. 1984. 15:393–425

THE ONTOGENETIC NICHE AND SPECIES INTERACTIONS IN SIZE–STRUCTURED POPULATIONS

Earl E. Werner

Kellogg Biological Station and Department of Zoology, Michigan State University, Hickory Corners, Michigan 49060

James F. Gilliam

Department of Biological Sciences, State University of New York at Albany, Albany, New York 12222

INTRODUCTION

Body size is manifestly one of the most important attributes of an organism from an ecological and evolutionary point of view. Size has a predominant influence on an animal's energetic requirements, its potential for resource exploitation, and its susceptibility to natural enemies. A large literature now exists on how physiological, life history, and population parameters scale with body dimensions (24, 131).

The ecological literature on species interactions and the structure of animal communities also stresses the importance of body size. Differences in body size are a major means by which species avoid direct overlap in resource use (153), and size-selective predation can be a primary organizing force in some communities (20, 70). Size thus imposes important constraints on the manner in which an organism interacts with its environment and influences the strength, type, and symmetry of interactions with other species (152, 207).

Paradoxically, ecologists have virtually ignored the implications of these observations for interactions among species that exhibit size-distributed populations. For instance, it has been often suggested that competing species

393

0066-4162/84/1120-0393$02.00

using the same habitats must differ in size by a factor of 2 in weight in order for them to coexist (18, 103, 153). In many taxa, however, the body weight of individuals *within* species commonly spans 1 order of magnitude and sometimes even 4 or more orders of magnitude (e.g. fish, reptiles). Thus, the body dimensions experienced ontogenetically often transcend those limits purported to isolate strongly competing species (see also 48, 85, 104, 133).

Given that resource utilization abilities and predation risk are generally related to body size, many species will undergo extensive ontogenetic shifts in food or habitat use. Such shifts create a complex fabric of ecological interactions in natural communities. Individuals face different competitors and/or predators as they grow, and in many cases even the sign of the interaction between species changes with size (see below). The size- or stage-specific nature of these interactions is critically important in shaping species life histories, the dynamics of species interactions, and the structure of the communities in which they are imbedded.

In this review, we first document the widespread existence of ontogenetic shifts in diet and habitat and then explore the consequences of such shifts for species interactions and community structure. The majority of our examples are from the lower vertebrates and invertebrates in freshwater communities, in part because this reflects our experience and in part because the phenomena are well documented in aquatic communities. Such stage-specific interactions should be abundantly represented in other taxa, however, since most exhibit strongly size-distributed populations (e.g. most invertebrate phyla, fish, amphibians, reptiles). Birds and mammals may be exceptions, but even in these taxa some have precocial young or adults and juveniles that differ in resource use (183). We have not attempted to review such interactions in plants or other sessile organisms, though they are clearly important in these groups.

In the second half of the paper, we offer a conceptual framework for predicting ontogenetic shifts and suggest some preliminary approaches for exploring their ecological and evolutionary consequences. We indicate how such life histories may be incorporated into a population dynamics framework and review the relevant findings from studies of structured population models. These are only tentative suggestions about how we might begin to develop a theory of species interactions in size-structured populations. We hope that this review will serve to highlight some important questions and stimulate broader treatments of the problem.

BODY SIZE AND THE ONTOGENETIC NICHE

For the purposes of this paper *ontogenetic niche* refers to the patterns in an organism's resource use that develop as it increases in size from birth or hatching to its maximum. We examine changes in food or habitat use almost exclusively because these are the niche dimensions that typically have been quantified with respect to body size. Many other important factors also scale with body size—e.g. predation risk and susceptibility to physical factors, which either influence fitness directly or affect patterns of resource use. We will consider some of these factors later.

There is a gradient in the form that ontogenetic changes in resource use can take. In some species, e.g. certain phytophagous insects and filter feeding organisms, few changes occur with growth. In other species food or habitat use changes continuously as individuals grow, but there is little variance in niche breadth. Thus, smaller and larger size-classes are virtually isolated, and intraspecific density dependence is manifest only over some range of sizes (e.g. 133). Alternatively, the range of prey types and the mean prey size may increase with body size, but the smallest prey sizes change little with predator size (207). Therefore, the niche of all smaller size-classes is included in that of all larger classes. Finally, ontogenetic changes in resource use can be relatively discrete. Spectacular examples of the latter are the habitat shifts occurring at metamorphosis in amphibians and holometabolous insects (see 201 for a recent review). Obviously, at the various life history stages, these organisms have markedly different competitors and predators, and intraspecific interactions occur only within particular stages. A cohort of such organisms can be viewed as navigating a landscape of ecological niches, and one of the challenging problems is to develop theory that will help us predict the course a species will take and the ecological consequences.

Discrete shifts are not limited to species that metamorphose, however, but are also approximated in many groups where morphology simply changes allometrically with growth. Among fish, for instance, ontogenetic changes in resource use are nearly universal. Size-specific shifts in food types have been documented in a great variety of species (e.g. 31, 34, 45, 67, 68, 85, 86, 92, 98, 106–108, 130, 141, 143, 156, 168, 175, 195), and these shifts are often associated with or caused by shifts in habitat. Commonly piscivorous fish undergo 3 to 4 rather abrupt shifts, e.g. the largemouth bass (*Micropterus salmoides*) switches from feeding on zooplankton to littoral invertebrates and then to fish as it grows (62). While growing only 120 mm in length, the pinfish (*Lagodon rhomboides*) progresses from carnivore to herbivore in five well-ordered stages (167, 168). These ontogenetic shifts are often correlated with discrete growth periods in the life history, in some cases attributed to the increases in food particle size (92, 106–108).

Discrete shifts are common in reptiles as well. In watersnakes of the genus *Nerodia*, the diets of a number of species exhibit changes with size that are associated with concomitant shifts in habitat use (118), e.g. *N. erythrogaster* switches from fish to frogs when it reaches about 50 cm in length. Turtles often switch from a largely carnivorous diet when small to a herbivorous diet when larger. In *Pseudemys scripta,* animal foods may comprise 80% of the diet for small individuals but drop abruptly at 60 mm plastron length to less than 10% of the diet (32). Across five different lizard families, species weighing less than 50–100 g were nearly all carnivorous; species larger than 300 g were nearly all herbivorous. In the larger herbivorous species, however, juveniles were carnivorous and became herbivorous at about 100 g (135; see also 8). In anoline lizards, ontogenetic habitat shifts have been well documented (154).

Such ontogenetic resource shifts are also common among invertebrates, including leeches (39, 209), phantom midges (50), damselflies (83), seastars (174), and spiders (181). Predatory zooplankters often switch from herbivory to carnivory as they grow (121). Stoneflies exhibit a similar pattern (58). Extensive changes in both habitat and prey type occur with increases in size in intertidal gastropods (P. G. Fairweather, in preparation). In other words, the majority of the world's species are characterized by size-structured populations; many of these species will exhibit the sort of ontogenetic changes in diet and habitat illustrated above.

THE ONTOGENETIC NICHE AND SPECIES INTERACTIONS

Ontogenetic resource shifts can vastly complicate species interactions and have important consequences for community dynamics. For example, consider the common case of a smaller species interacting with a larger one. Often the critical feature of this interaction is not how adults of these species interact but how the larger species is able to recruit through juvenile stages that are identical to the range of sizes (and therefore size-scaled niche properties) expressed in the smaller species (53, 55, 85, 119). It is remarkable how little attention ecologists have paid to this very simple observation when considering species coexistence.

A classic example is found in the history of the size-efficiency hypothesis, the major organizing framework for much zooplankton ecology over the last two decades (70). Brooks & Dodson (20) offered this hypothesis to explain the observed inverse relation between the abundance of small- and large-bodied herbivorous zooplankton in freshwater lakes. Simply stated, the hypothesis proposed that large species are more effective competitors because they can filter particulate matter as fine as that smaller species consume, but they can also utilize larger particles. Furthermore, larger species are more efficient, since the filtering rate increases at roughly the 2nd–3rd power of length whereas metabolic expenditures increased at the 1st–2nd power of length. They argued that size-selective predation counters this advantage (20, 70; see 47 for alternative hypotheses).

For nearly a decade ecologists using this idea to examine species' coexistence largely overlooked the fact that larger species have to recruit through smaller size-classes. Thus, during these early stages, they would be subject to the same disadvantages as the smaller species (but see 119). Indeed, other broad theoretical treatments of competition and body size also ignored the problem, even though these models were ostensibly patterned after species with strongly size-distributed populations (60, 152, 207).

Empirical studies have clearly documented the existence of a recruitment bottleneck. In carefully controlled microcosms, Neill (119) demonstrated that a small zooplankter, *Ceriodaphnia,* could monopolize the 3–6 μm range of food sizes and thereby affect juveniles of much larger species, such as *Pseudosida* and *Simocephalus,* and actually drive *Daphnia magna* extinct. Adult *Daphnia* introduced into these systems, in contrast, survived and produced abundant young, but not a single one of these offspring survived more than four days in the presence of normal *Ceriodaphnia* populations. When *Ceriodaphnia* was reduced by fish predation, juvenile survivorship of all species increased, but competitive interactions among adults also increased. *Ceriodaphnia* has a similar effect on the larger *Daphnia pulex,* as Lynch (101) found in experiments in small natural ponds. Lynch's study also revealed the importance of resource structure. Competitive effects changed seasonally, and the results suggested that strong indirect effects mediated through the resource community were responsible.

Within species, positive correlations between food size and body size have now been documented for a large number of taxa, e.g. fish (68, 85, 108, 123, 134, 143, 161), postmetamorphic frogs (30, 99), lizards (150, 151, 154, 155, 160), salamanders (55), seastars (128), copepods (105, 205), waterbugs (110), leeches (39, 209), crabs (76), snails (P. G. Fairweather, in preparation; 51–53), damselflies (110), ant lions (206), and a sea anemone (158), among others. These data alone would suggest that larger species often negotiate juvenile periods when individuals must obtain resources that are similar to those consumed by smaller species.

The juvenile bottleneck problem can be exacerbated if there are trade-offs among features that adapt species to alternate ontogenetic niches. These trade-offs are particularly likely to exist in species that undergo discrete shifts without metamorphosis. For instance, Werner (194) quantified trade-offs in the body plan of fish by comparing "piscivorous" and "invertebrate feeding" morphologies under laboratory conditions where both forms fed on fish and zooplankton; significant trade-offs in feeding efficiency were evident. Individuals of the piscivorous species, however, must pass through a zooplankton feeding stage when they are small, yet they are burdened with the piscivorous morphology, which hinders their ability to compete for zooplankton. In fact, during their early life history stages many fish feed on zooplankton; species that are specialized planktivores (i.e. highly efficient planktivores as adults) can have dramatic competitive effects on species that are only planktivores during their early stages, in some cases leading to competitive exclusion (34, 49, 171).

Potential juvenile bottlenecks are not limited to species that differ in adult size. Many species use similar resources when small but diverge in niche when larger. In anuran communities, small ponds attract the breeding adults of many species, and tadpoles are less segregated than the terrestrial adults (201). The diets of four species of coexisting water snakes are very similar when small but diverge as the snakes get larger (118). Similarly, in freshwater fish the young of many species are confined to macrovegetation or similar cover owing to predation risk, but they can move into more open habitats and diverge in niche as they grow larger (113, 115, 196).

Among carnivorous species, another complication may be added. Members of a larger species may initially compete with a smaller species that they will eventually prey on (though this is not necessarily restricted to larger and smaller species). Such interactions appear widespread among fish, amphibians, reptiles, crustacea, and insects. If the ontogenetic shifts in the larger species are relatively discrete, then the sign of the interaction for that species will change at a given size (see below). If resource use changes continuously with size and the niche of the smaller species is included in the larger (207), a larger species may simultaneously compete with and prey on a smaller species. Certain larval salamanders may interact in this fashion (116, 166).

A classical example of a mixed competition/predation interaction in fish is found in Paul Lake, British Columbia. This lake was fishless until it was stocked with rainbow trout (*Salmo gairdneri*) in 1909. Between 1937 and 1949 the annual catch from the lake stabilized (34). All size-classes of trout fed on a mixture of plankton, benthos, and terrestrial insects (34, 91). Just before 1950, a smaller species, the redside shiner (*Richardsonius balteatus*), entered the lake from upstream in the drainage basin. The shiner also fed on plankton and benthos and overlapped in the littoral zone with young trout. Following the shiner invasion, large trout (with fork lengths greater than 20 cm) fed on shiners and dropped plankton from the diet.

As the shiner population increased during the following 10 years, the growth of large trout increased in response to the availability of shiners as food. In contrast, the growth rate and amount of food in the stomachs of smaller trout declined markedly, and it took the trout a full year longer to reach a given size. Thus, the two life stages were affected in opposite ways by the shiner invasion. Since larger fish exhibit greater fecundity (7), this should compensate to some extent for the competitive effects experienced early in the life history. Though only catch data are available, it appears that the overall effect of the shiner invasion was to reduce the trout population in the lake.

The introduction of threadfin shad (*Dorosoma petenense*) as a forage fish to improve largemouth bass growth in California provides a parallel example. As expected, growth of bass in the piscivorous size-classes increased after shad introduction. The growth rate during early stages was often just as markedly reduced, however, presumably owing to competition with the shad for zooplankton. This competition often reduced survivorship and produced smaller year-classes of the bass (49). Largemouth bass and bluegill (*Lepomis macrochirus*) (two dominant fish species in many Eastern North American lakes) also interact in this fashion (62). Competition experiments demonstrate that the bluegill can have a considerable depressive effect on the growth and survivorship of young bass, but the bluegill forms an important component of the larger bass's diet (J. F. Gilliam, unpublished).

Analogous interactions are documented in the freshwater zooplankton. The copepod, *Cyclops bicuspidatus thomasi*, is herbivorous in the naupliar stages but becomes predatory at larger sizes. When algal production is low, the predatory stages of *Cyclops* have little effect on herbivore community composition; when it is high, however, the reduced competition with herbivores in the naupliar stage permits a greater numerical response, and the carnivorous stages then markedly affect the relative abundance of the herbivorous zooplankton in the community (121). Similar interactions may occur in communities containing *Mysis relicta*, which is herbivorous over the first 6–9 months of life and then becomes an important plankton predator.

Maly (105) provides an insightful analysis of a mixed competition-predation interaction between two calanoid copepods, *Diaptomus shoshone* and *D. coloradensis*. In high altitude ponds in Colorado *D. shoshone* reaches roughly twice the size of *D. coloradensis* and has one generation a year, whereas *D. coloradensis* may have two. As *D. shoshone* grows, it eats progressively larger food sizes and can eventually consume *D. coloradensis*. With data on food habits and laboratory studies delimiting the sizes of *D. shoshone* that can feed on *D. coloradensis* Maly constructed a graph with the sizes of the two species on the axes and delimited those regions where competition or predation between the species would be severe (i.e. if too similar in size, they would feed on the same foods; if too dissimilar, predation could occur). The first generations of *D. coloradensis* and *D. shoshone* were timed so that their relative sizes always fell outside the ranges of severe competition or predation (see also 204). Analyses of this sort on other species where the option for temporal adjustments of discrete generations is open would be extremely useful. Though detailed studies are not available, such interactions also appear to be important among nymphal odonates in small lakes and ponds (11, 35, 94).

Complex, mixed interactions mediated by ontogenetic size changes are also common in amphibian communities. Morin (116) has experimentally dissected the interaction between the newt, *Notophthalmus viridescens*, and the salamander, *Ambystoma tigrinum*, in temporary ponds. The newts are voracious predators on the eggs and developing embryos of the salamander and at high densities can exclude the salamanders from the ponds. Furthermore, the diet of the larval and adult newts overlaps with the salamander's, and competition with the newts can retard growth in the latter. The salamander, however, eventually grows larger than the newt and then becomes a significant predator on the newt larvae. Larval *A. tigrinum* can also act as either a competitor or a predator of three other species of larval *Ambystoma,* depending on the availability of alternative food resources (larval anurans) (199). Finally, it has been shown that two *Ambystoma* species may interact both as competitors and as predator and prey (166).

Alterations in such stage-specific interactions can dramatically affect community dynamics. A nice example is the experimental work on the phantom midge *Chaoborus* (121). Early instars of *Chaoborus* feed on rotifers, while later instars feed on larger herbivorous plankton. Many studies have shown that the larger herbivorous zooplankton competitively suppress rotifers (69, 120). At low algal production levels, populations of rotifers are small, and *Chaoborus* suffers high mortality in instars I and II and consequently has little effect on the composition of the zooplankton community. At high algal production levels, rotifers are released from competition, and the survivorship of *Chaoborus* through instar IV increased by two orders of magnitude. Under these conditions, the larger instars of *Chaoborus* have a pronounced impact on the structure of the zooplankton community, virtually reducing it to one species. Thus, competitive interactions among the prey species maintained a juvenile bottleneck for *Chaoborus,* which in turn limited the effect of the larger instars on this community. Although in this case the bottleneck was ameliorated by an external perturbation (nutrient addition), subtle changes in the balance of interactions structured in this way could move a system between alternate configurations.

Other investigators (46, 61) have suggested that in certain circumstances, the actual presence of *Chaoborus* is a result of indirect effects mediated through resources. In alpine ponds, the salamander *A. tigrinum* preys selectively on the larger zooplankton, allowing smaller species to increase and thereby support large *Chaoborus* populations. Dodson (46) has termed this a complementary niche. If *Ambystoma* is removed, larger plankton dominate, and *Chaoborus* is not able to invade because the smaller instars starve (61).

The preceding examples illustrate the importance of size-specific interactions for community dynamics and their complexity. Two basic types of interactions were illustrated: the juvenile bottleneck problem and the mixed competition/predation interaction. These are relevant for a wide range of taxa, but a plethora of other types of size-specific interactions can obviously be envisioned. Among sessile organisms, size-dependent competitive abilities can lead to intransitivity and frequency dependence in species interactions (23). Changes in such stage-specific interactions can set in motion a chain of events that radically alters community structure; there is also a rich potential for indirect effects. If a series of such communities were examined, the patterns in community structure might well appear random unless the underlying mechanisms responsible for community dynamics were appreciated.

A CONCEPTUAL APPROACH

The previous sections suggest that complex stage-specific interactions are widespread and important. It would be useful to formulate approaches that would enable us to predict the occurrence and nature of such interactions and to explore their consequences. A diversity of empirical and theoretical approaches will undoubtedly be required to make significant progress on this front. To date, there is little theory that even begins to examine the richness of these interactions. Our purpose in this section is to suggest some approaches that may be useful. The scope of these suggestions is rather modest; the development of a recipe for dealing with the complexities of species interactions in size-structured populations is clearly still in an embryonic stage.

A central feature of an approach to size-class interactions would be theoretical and/or empirical relationships predicting how resource-use capabilities and vital rates scale ontogenetically with body size. In many of the taxa that undergo significant ontogenetic shifts, mortality and fecundity rates are size specific. Moreover, growth rates in most invertebrates and lower vertebrates are also size specific, age effects are often minimal (4, 19, 59, 129, 139), and most species exhibit either indeterminate growth or considerable flexibility in growth rates and ultimate size. Transplant experiments with fish (e.g. 2, 22, 108) and intertidal invertebrates (112, 128, 159) illustrate that individuals quickly adopt the size-specific growth rates and ultimate size characteristic of the habitats to which they have been transferred, even though these individuals may have been "stunted" for long periods of time in a former habitat. Thus, growth is a sensitive index of available resources and probably is the parameter through which intra- and interspecific competition is expressed. For the above reasons, models of these organisms' population dynamics are often more appropriately indexed by size than by age.

In size-structured populations, individual growth rates are as important as survivorship and fecundity rates in determining population trajectories (184). Traditional age-structured demography includes only two vital rates: age-specific mortality and fecundity, from which the familiar l_x and m_x schedules of life tables are derived. This age-structured demography does not explicitly include individual growth rates, although the mortality and birth rates of an x-year-old may in fact depend upon body size and hence individual growth rates at all previous ages. A size-indexed demography, however, includes three vital rates: size-specific mortality and fecundity rates, plus individual growth rates. The explicit inclusion of individual growth rates is a natural mathematical consequence of indexing by size rather than by age. If we can determine how these vital rates scale with body size, these relationships can be used as basic submodels for the parameters of a size-based demography. A size-based demography, in turn, would permit us to explore the population-dynamic consequences of various types of interactions.

Size-specific individual growth rates may also provide the foundation for predicting discrete ontogenetic resource shifts. Because a higher growth rate reduces the time it takes to reach reproductive size, often increases survivorship (see below), and/or increases size (and therefore generally fecundity) at first reproduction, it has often been argued that there is strong selection for maximizing growth rates. Thus, it seems reasonable to hypothesize that all else being equal, niche shifts occur so as to maximize absolute growth rates or surplus energy at each size. Growth rate curves can then be plotted as a function of body size for each available habitat, or for food type if this is appropriate, and we predict that niche shifts will occur at the intersections of these curves (Figure 1). A habitat switch at size $\hat{s}$ maximizes the growth rate and minimizes the time it takes to reach each size (including reproductive size). If mortality rates are identical in each habitat and a decreasing function of size, a switch at $\hat{s}$ also maximizes the survivorship to each age. This is not necessarily true if the mortality rate increases with size over some range, but survivorship to each *size* is maximized regardless of the relationship between the mortality rate and size as shown later. We will consider the effects of habitat-specific mortality rates on predictions of ontogenetic shifts in a later section.

The Relationship Between Growth and Size

The difference between foraging gains and costs as a function of body size—or the surplus energy curve—is a quantitative measure of the scope for growth (192) or of the energy available for investment in reproduction or allied fitness-related activities at each size. In species where the range of foods or habitats changes continuously with size, quantification of the surplus energy

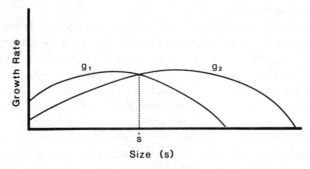

Figure 1   Absolute growth rates *(g)* for a hypothetical species in two habitats as a function of body size *(s)*. ŝ denotes the size at which a habitat switch maximizes the growth rate.

curve directly determines size-specific performance and thus, in part, the nature of intra- and interspecific size-class interactions. For discrete shifts, these curves must be determined for each habitat type in order to predict when shifts should occur as well.

The manner in which the rate of energy metabolism (E_m), or the cost, scales with body size has been studied in some detail (149). Metabolism scales by the allometric equation:

$$E_m = kM^\delta, \qquad\qquad 1.$$

where M is mass, δ is a constant that averages 0.75 across species but intraspecifically may range from 0.5 to 1.0 (79), and k is a species-specific constant that depends on factors such as temperature and activity.

How energy gain curves scale with body size is less obvious, however. It is unlikely that a general relationship will be discovered here, given the diversity of feeding modes in animals and the subtle ways in which behavior and morphology affect foraging relationships. Nonetheless, within certain taxa energy gain has been found to scale by the allometric equation as well, though with quite variable exponents (e.g. see 70, 159, 190, 207), thus giving a surplus energy (E_s) curve of

$$E_s = hM^\gamma - kM^\delta, \qquad\qquad 2.$$

where h and γ are fitted constants that will be habitat- and species-specific. Size-specific growth rates can be obtained from the surplus energy curve using conversion efficiencies. If energy intake increases at a lesser power of weight than does cost, the difference between the two (i.e. the scope for growth) rises to some maximum and then decreases. Intake and cost curves then take the form in Figure 2 and the difference between the two gives a surplus energy or growth curve as in Figure 1 (see 84 and 33 for examples of such growth curves in fish and frogs). The maximum possible size (where gains and costs balance), M_{max}, equals $(k/h)^{(1/\gamma - \delta)}$ (159, 191); and the optimal body size, M_{opt}, if defined as the size providing maximum surplus energy, equals $(\delta k/\gamma h)^{(1/\gamma - \delta)}$ (159). This second value will not be equivalent to the optimal body size based on various efficiency criteria such as energy gained/energy spent, a point that has led to confusion in the literature (188).

Equation 2 has been useful in the context of filter feeders (70, 159); and the relation between body size, collecting abilities, and metabolism has been measured most often in filter feeding organisms [e.g. zooplankton (70), mollusks (66, 111, 122, 173, 182), anemones (158, 159)]. The exponent of the intake rate γ ranged below that of metabolic rates in studies of anemones (158) and some mollusks (182), but the reverse was true in studies of other mollusks (111, 122, 137, 67) and in the zooplankton (70), conferring a greater feeding advantage on larger body sizes. In many cases, species-specific attributes of

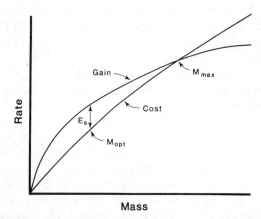

Figure 2   Schematical metabolic cost and foraging gain curves as a function of body mass. M_{max} is the maximum size attainable on the given resource(s), E_s is the surplus energy available for growth and reproduction, and M_{opt} is the size at which surplus energy is maximized.

course, will contravene these general body-size relations (57), but they are useful guidelines to initiate analyses.

Gain curves have been quantified in particulate feeding organisms less often. Ware (190) constructed a gain curve based on a Holling (77) Type II functional response in order to analyze optimal swimming speeds in fish. One promising framework for predicting energy gain—and the range of foods consumed by different sized animals—is optimal foraging theory. In our work with several centrarchid sunfishes, we quantified prey encounter rates and handling and searching costs as functions of predator size, prey size, and prey density in three different types of habitats (62, 113, 193, 194). Given simply the size-frequency distribution of prey from a specific habitat, the theory can be used to predict the optimal diet and the energy gain curves as a function of fish size. The estimates of energy gain from different habitats take the form depicted in Figure 1 (114), and they have been used to predict habitat shifts as resource levels change in different habitats (113, 197, 198). A number of studies examining instar-specific interactions of insect predators have quantified some fraction of the relations necessary to implement the above or similar approaches (72, 110, 172). Belovsky (10) extended this approach to terrestrial herbivores by using linear programming techniques. Schoener (152) attempted to predict the optimal body size for solitary predators that feed by different modes ("sit and wait" and "searching") using such a framework. He constructed a simulation model from general relations between body size and foraging costs and expenditures and developed some general notions of scale effects in ecological interactions.

If the cost and gain curves as functions of body size can be estimated by one of these methods, we can predict size-specific growth rates for different foods or habitats. For species whose resource use changes continuously with size, knowing these relationships enables us to quantify relative performance and exploitative competitive relations among size-classes intra- or interspecifically. Density dependence can also be incorporated into the analysis by using these parameters. If the ontogenetic shifts are discrete, we will be able to predict the size at which a shift should occur (Figure 1).

This size, however, can also be influenced by size-specific mortality rates in different habitats. Furthermore, whether changes in ontogenetic niches are continuous or discrete, size-specific mortality rates are crucial for considerations of both optimal body size or fitness in different environments and population dynamics.

The Relationship Among Mortality, Size, and Growth

Mortality rates may be positively or negatively correlated with size or the direction of correlation may reverse as size increases. Particularly vulnerable periods, e.g. during metamorphosis in amphibians (5), can also complicate these trends. Certainly in the zooplankton vulnerability can increase with size owing to positive size-selection by vertebrate predators (20, 70). This vulnerability, then, is one important factor that can counter the posited filtering rate advantage of larger sized zooplankters, and body size may reflect a balancing of these factors (1, 100). In most taxa, however, the mortality rates of different sized individuals are likely to be inversely related to size (189). For example, in fish it is not uncommon for 99% of a cohort's mortality to occur during the first 100 days of life (36, 54, 95) and in amphibians the early larval stages are subject to extremely high mortality rates (25, 201).

It is often argued that predators are the actual agents of most of this early life history mortality, although there is rarely unequivocal evidence to support this thesis. Vulnerability to given predators generally decreases with size (89), and the total number of potential predators is usually greater for smaller organisms. Thus, the overall risk from predators usually falls precipitously as an individual increases in size. In *Rana* tadpoles, for instance, Calef (25) found two phases in the survivorship curve; over the first 3–4 weeks mortality was extreme, but then the survivorship curve flattened out. The break in the curve occurred when tadpoles became too large for salamanders to eat them (see also 178, 179). Such size refuges from predators are also common in intertidal organisms (40, 128) and fish (82, 196), and an individual will usually reach a series of such refuges as it grows. Small size also renders individuals more susceptible to starvation and physicochemical factors, which in turn increases vulnerability to predation.

Based on the above observations, it has been argued that for many taxa, there is a selective premium for fast growth early in the life history. An increased growth rate reduces the time spent in the smaller, more vulnerable size-classes and thereby minimizes the overall risk of mortality. Any factor that reduces the growth rate (e.g. competition, an inferior genotype, temperature) protracts the time spent in more vulnerable stages where the mortality factors are concentrated, which may explain why the incidence of poor year-classes of Baltic salmon and other fish appears to be related to cooler temperatures (G. Svardson, personal communication). Growth and mortality rates are therefore intricately related so that factors affecting growth may indirectly regulate recruitment and population size.

Gilliam (62, 196) has derived a simple expression for survivorship between two sizes based on size-specific growth and mortality rates that allows us to illustrate the relation between growth and mortality quantitatively. If a change in the environment uniformly multiplies the growth rate by a factor c ($c < 1$ for a growth reduction) over some size interval, then the new survivorship (L') between the sizes of interest can be expressed as a power of the original survivorship (L, i.e. survivorship through the size interval at the original growth rate):

$$L' = L^{(1/c)}. \qquad\qquad 3.$$

For example, if the growth rate is multiplied by ½, the animal will spend twice as long in the size range, and the new survivorship is L^2. Thus, a given reduction in growth can have an enormous effect on survivorship if the latter is already low. A change in the size-specific mortality rate can be examined simultaneously. If the size-specific daily mortality rate is multiplied by a factor k, the new survivorship through the size interval is given by

$$L' = L^{(k/c)}. \qquad\qquad 4.$$

These ideas on the interaction of size-specific mortality and growth rates in determining survivorship have a tradition in the fish population-dynamics literature that goes back at least to the 1940s (12, 37, 140). Density-dependent growth has often been postulated as a primary factor regulating mortality rates in populations of fish (123, 162, 187, 189, 195, 196), amphibians (44, 177–179, 200, 201, 203), and intertidal organisms (157). Shepherd & Cushing (162) argued that marine fish populations' ability to withstand high exploitation rates on adults (often 2–5 times natural mortality) is due to the fact that this exploitation reduces the densities of young spawned; thus, these fish grow quickly through the vulnerable size-classes and have lower mortality rates than at higher densities and lower growth rates. They constructed a model of density-dependent growth in fish, holding the instantaneous mortality rate constant; therefore, density-dependent variation in overall mortality could only occur through effects on the growth rate, or the time it takes to reach a critical size. This relation has the basic form of a Beverton & Holt (12) stock-recruitment curve and clearly shows how the effects of predation and food availability (growth) interact to produce strong density-dependent relationships.

Such models illustrate the critical relation between growth and mortality when mortality rates are strongly size-specific. In many taxa with size-structured populations, these ideas may provide a reasonable first approximation of how growth, size, and mortality interrelate to affect survivorship and ultimately population dynamics.

Predicting Ontogenetic Shifts Given a Growth Rate/Predation Risk Trade–Off

Earlier we argued that the maximization of surplus energy can be a useful predictor of ontogenetic habitat shifts if habitats do not differ in mortality rates. It is unlikely, however, that size-specific mortality rates will be identical in two habitats if these habitats differ in the predators present or in their physical structure. Even within a habitat, exploitation of different resource types (e.g. sedentary vs mobile prey) may require different foraging tactics and hence different mortality risks. If mortality rates differ among habitats, how does this alter our predictions of the switching size that maximizes fitness? For example, if the mortality rates for all sizes are greater in the second habitat of Figure 1 than in the first, would fitness be increased by delaying the shift?

Gilliam (62 and manuscript submitted for publication) has approached this problem for the case of a population in a constant environment. Since growth and mortality rates are expressed in different units, they must be integrated into the common currency of fitness. Thus, habitat shifts were determined that maximized the net reproductive rate, R_0, in equilibrial populations (since a phenotype can invade only if its R_0 exceeds 1 when the resident phenotype(s) has a net reproductive rate equal to 1) or r in the Euler-Lotka equation in changing populations.

The mortality rate (μ) at age x was made a function of both size (s) and growth (g), i.e. $\mu[s(x),g(x)]$; the mortality rate is a function of the growth rate, since the latter is determined by the habitat the animal chooses at age x. If the choice is between only two habitats and we assume that the animal spends all of its time in one of the habitats, then at each size there are only two choices of $g(x)$, i.e. g_1 and g_2 as in Figure 1. More generally, $\mu[g(x),s(x)]$ can be a continuous function of both size and growth rate, and $g(x)$ can range between 0 and the maximal rate that could be achieved by ignoring mortality risk; this upper bound will be a function of body size at age x. The birth rate was made a function of body size, since fecundity increases with size in many animals (reviewed in 13, 131)—e.g. fish (7), zooplankton (65, 102), amphibians (146, 147), lizards (186) and many invertebrates (165). Reproduction was assumed to commence upon reaching a critical size and to occur continuously; available (although sparse) evidence suggests that the onset of reproduction is often better predicted by size than age (3, 132, 203).

The problem of determining optimal growth rates (through ontogenetic choice of habitats) was stated as follows: Find the series of growth rates that maximizes

$$R_0 = \int_0^\infty l(x)b(x) \, dx, \qquad\qquad 5.$$

where $l(x) = \exp[-D(x)]$,
         $b(x) = b[s(x)]$,

$$\frac{dD(x)}{dx} = \mu[s(x), g(x)],$$

$$\frac{ds(x)}{dx} = g(x),$$

$$x = \text{age}, \ D(0) = 0, \text{ and } s(0) = s_0$$

In other words, an animal is born at size s_0. Its survivorship to age x, $l(x)$, is equal to $\exp[-D(x)]$, where $D(x)$ is the integral of the mortality rates to that age, and exp denotes exponentiation of the base of the natural logarithm. The birth rate at age x, b(x), is a function of the size at x. The mortality rate, which is the rate at which $D(x)$ "accumulates," is a function of size and choice of growth rate at age x. The rate of change of the body size is the growth rate.

This set of equations was used to find the optimal growth rate (and hence mortality rate) at each age (or size), employing optimal control theory. It was determined (62; J. F. Gilliam, submitted for publication) that a necessary condition for R_0 to be maximized is that the following partial differential equation be satisfied at each age (and size):

$$\frac{\partial \mu}{\partial g} = \frac{\mu - b/v}{g}, \qquad\qquad 6.$$

where v equals $v(x)$, the reproductive value at age x, which is itself determined by the path of $g(x)$.

This equation can be greatly simplified for the case of juveniles, an important special case because many of the resource shifts described in this paper occur at prereproductive sizes. In that case, $\partial\mu/\partial g = \mu/g$, and this can be restated simply as: To maximize fitness, minimize the ratio, μ/g, at each size. This result allows a return to the graphical approach, and Figure 3 shows how the predicted optimal switching size is changed from that depicted in Figure 1. Multiple habitats are easily incorporated, and cases can be constructed in which the optimal solution involves a switch from habitat 1 to habitat 2 and then back to habitat 1, in qualitative agreement with the shifts by some planktivorous fish

(E. E. Werner, D. J. Hall, unpublished). Intuitively, the minimization of μ/g at each size minimizes the mortality cost for each increment in size and hence maximizes the probability of reaching each size (but not age), including reproductive size.

If the fitness criterion is to maximize r in the Euler-Lotka equation, the rule becomes "minimize the ratio, $(\mu + r - b/v)/g$," where r is the rate of increase afforded by the optimal series of decisions, or simply "minimize $(\mu + r)/g$" for juveniles.

This model may be viewed in an ecological or evolutionary sense. Ontogenetic shifts in a number of organisms appear to be genetically fixed with respect to occurrence and timing; those in many others are facultative. For instance, metamorphosis in many insects may be programmed to occur when they reach a particular size or at a specific season. In other species, the decision to shift may be based facultatively on conditions in the present habitat, e.g. size at metamorphosis in amphibians can be determined to a large extent by conditions in the larval habitat (201). Species that are able to sample alternative foods or habitats base their decision to switch on their assessment of these alternatives. A large literature on foraging now documents such abilities in many organisms (see 88 and the review in this volume by G. H. Pyke). Bluegill sunfish in small Michigan lakes, for instance, switch from littoral vegetation to the open water to feed when they reach a critical size (E. E. Werner, D. J. Hall, unpublished; 115). Experimental evidence suggests that this is a facultative switch and that it depends on the relative foraging rates and predation risks in these two habitats (196, 198).

The results of the above model would change, of course, if the vital rates are not time invariant (e.g. if there is seasonal reproduction or a seasonally dependent mortality rate). For example, if a summer growth season is followed

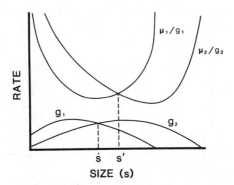

Figure 3   Growth rates *(g)* and the ratio of mortality (μ) to growth rates for two habitats as a function of body size (s). $\hat{s}$ is the size at which a habitat shift maximizes growth rate, s' denotes the optimal size to shift when predation risk is incorporated and is greater in the second habitat.

by a winter in which growth ceases and mortality rates are a declining function of size, fitness would be higher for an animal that chooses a higher growth rate (and hence higher mortality rates) in the summer than the above time-invariant solutions specify, since it would be a larger size during the winter. In amphibians, Wilbur (201, 202) has stressed the interplay of growth and mortality rate in determining ontogenetic shifts (metamorphosis) from aquatic to terrestrial habitats, but he also notes the influence of the seasonal drying of ponds.

The above analysis is also based on the assumption of a given size-specific birth rate. Of course, the birth rate can be subject to behavioral and evolutionary control, but the above analysis applies to any time-invariant birth rate function, including the optimal function. In general, the problem of ontogenetic niche shifts involving trade-offs among current growth, mortality, and birth rates is closely related to the problem of optimal life histories. Sophisticated analyses of the latter can incorporate parameters that vary over time as well as allowing vital rates to depend on both age and size or other properties (148), but such analyses have yielded few simple rules that are amenable to experimental tests. At present, the most productive path may be to view the "minimize μ/g" rule as a starting point for juveniles in time-invariant environments in which vital rates are purely size dependent and assess the impact of alternate conditions by characterizing the effects as departures from this rule.

DEMOGRAPHY AND POPULATION DYNAMICS OF SIZE-STRUCTURED POPULATIONS

Given a basis for predicting diet and habitat shifts and mechanistic bases for growth, mortality, and birth rates, we possess the essential tools to begin to model the dynamics of a population and of interacting populations. A simple case is depicted in Figure 4, where the introduction of a competitor in the first habitat utilized by a resident species lowers the resident's growth rate in that

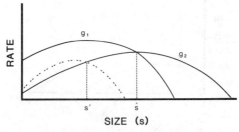

Figure 4 Growth rates *(g)* in two habitats as a function of body size *(s)*. The dashed line represents the lowered growth rate due to introduction of a competitor into habitat 1. $\hat{s}$ and s' are the optimal switching sizes for the two conditions.

habitat. This reduction, in turn, decreases the size interval over which the resident should stay in the first habitat, with the switch occurring at s' rather than $\hat{s}$. (For simplicity, mortality rates are assumed to be equal in the two habitats, and the other growth rates are held constant.) If the growth rate is sufficiently depressed, the first habitat could be dropped completely from the resident's ontogeny—either immediately, if the shifts are facultative, or in evolutionary time, if not.

The above graphical representation can be translated into equations that describe the resultant demography and dynamics of such a size-structured population, although subsequent analysis of the equations can be difficult. While population models without any size or age structure deal with the change in total numbers, i.e. dn/dt, the basic continuous time model of size-structured populations deals with changes in a size distribution. Let $n(s,t)\cdot ds$ be the density of animals between the sizes s and $s + ds$ at time t. Then the size distribution of the population, $n(s,t)$, changes according to the following partial-differential equation, usually called the balance or continuity equation (124, 125, 145):

$$\frac{\partial n(s,t)}{\partial t} = - \frac{\partial g(s,t) \cdot n(s,t)}{\partial s} - \mu(s,t) \qquad 7.$$

In other words, the rate of change of $n(s,t)$ is determined by both the net rate at which the animals grow through size s and the rate of loss (i.e. mortality) at s. The production of newborn is described by a boundary condition:

$$n(s_0,t) = \int_{s_0}^{\infty} n(s,t) \cdot b(s,t) \, ds, \qquad 8.$$

where s_0 is the size at birth. The substitution of appropriate submodels for growth, mortality, and birth rates, and possible coupling to other populations would complete the specification of the dynamics. These equations can be extended to include age or other properties in addition to size (124, 126, 127, 144, 163, 170), but doing so can introduce substantial technical difficulties. Nearly all authors have treated only age or only size, or used discrete-time matrix methods with arbitrary stage classifications (see below).

For the special case of time-invariant, density-independent vital rates, the above equations yield the demography developed by Van Sickle (184; see also 41–43, 87, 185). In this case, the time spent growing from some size s_1 to a size s_2, $\tau(s_1,s_2)$, and the survivorship through that size interval, $l(s_1,s_2)$, are given by

$$\tau(s_1, s_2) = \int_{s_1}^{s_2} \frac{1}{g(s)} \, ds; \; l(s_1,s_2) = \exp\left[- \int_{s_1}^{s_2} \frac{\mu(s)}{g(s)} \, ds\right]. \qquad 9.$$

The rate of increase, r, of a population with a stable size-distribution is defined implicitly by:

$$1 = \int_{s_0}^{\infty} \exp[-r\tau(s_0,s)] \, [l(s_0,s) \, b(s)/g(s)]ds. \qquad 10.$$

The net reproductive rate, R_0, equals the right-hand side of this equation when $r = 0$. The stable size distribution does not necessarily decline with size; "stacking" or "piling up" occurs at size s if $dg(s)/ds + \mu(s) + r < 0$.

The most conspicuous difference between this size-indexed demography and the age-indexed demography is the explicit inclusion of the individual growth rate $g(s)$. This addition is a natural mathematical consequence of the fact that the rate of change of age, dx/dt, is equal to one and need not appear in the age-indexed demography, but the rate of change of size, ds/dt, is $g(s)$ and cannot be deleted. The familiar Euler-Lotka equation ($1 = \int e^{-rx}l_x m_x dx$) is obtained from the above equation if x is substituted for s and $g(s)$ is set equal to one.

There is also a literature on discrete-time models of density-independent populations in which the organisms are classified by "stages." The stages can be defined by size, age, and/or other properties, and a projection matrix or similar mathematics describes the transitions among the stages (26, 27, 63, 64, 71, 78, 93, 96). As in the Leslie matrix, which is a special case of the above models that is age-structured, the population's finite rate of increase is given by the dominant eigenvalue of the projection matrix, and the stable stage distribution and reproductive values are given by the corresponding right and left eigenvalues, respectively.

There has been much less work on density-dependent models of size-structured populations (14–16, 56, 117, 125, 144), and we know of no published mathematical work on two interacting, explicitly size-structured populations. The work on single-species, density-dependent models demonstrates a phenomenon mentioned above: Strongly density-dependent recruitment rates to adult size can result from density-dependent growth rates, even if size-specific mortality and birth rates are completely density-independent.

Botsford's (14, 15) work constitutes the most complete treatment of a size-structured population model. He sought to explain some exploited fish and crab populations' failure to recover following cessation of exploitation. In the model, individual growth rates are density-dependent, and adults cannibalize juveniles. The model predicts two alternative stable states: (a) the population reaches a stable state at high densities and low individual growth rates and (b) the population stays at low densities with high individual growth rates. Botsford (16) suggested that the inclusion of individual growth rates can reveal optimal harvesting policies that differ substantially from those derived from less complete models. Frogner's (56) model treated an insect population in which maturation occurs after eating a given amount of food and the feeding rate is density-dependent; this situation is analogous to maturation at a critical size with a density-dependent growth rate. He investigated numerically the effect of an increasing daily mortality rate on the larvae. When the average adult density was plotted against juveniles' daily mortality rate, the resulting function first increased, then declined, increased again, and finally declined to zero. In the context of the ontogenetic niche model developed in this review, this means that an increase in the density of a predator on the small sizes of a species can either increase or decrease the average density of larger sizes, and the effect need not be monotonic.

In future theoretical work on size-structured population dynamics, ecologists should pay careful attention to the pattern of resource change with size. It was noted earlier that intraspecific resource use patterns may vary on a continuum bounded by the lack of resource partitioning among size-classes at one end and the discrete ontogenetic diet/habitat shifts where there is no direct interaction among different size-classes at the other end. These patterns of resource use will have different consequences for intraspecific population dynamics. In the first case (little partitioning among size-classes), a crucial question is whether the allometry of the intake and metabolic rates is such that the growth rate of small individuals will become negative because of resource reduction by larger individuals. If so, the population may show pulsed recruitment, with recruitment only occurring when resources are extremely abundant because of the death of the larger individuals or environmental perturbations. On the other hand, juveniles may be able to tolerate lower resource levels than larger size-classes. If so, the continued recruitment and growth of juveniles would result in a "stunted" population consisting of many small individuals and in which larger individuals would have negative growth rates (114). Finally, in the case of discrete diet/habitat shifts, small and large individuals may not directly interact at all, or they may alternate between simultaneous habitat use and complete segregation as resources change.

There is a much larger literature on density-dependent age-structured models (see 124; also 29, 97, 180). There has also been some work on two competing age-structured species (73, 136, 176) and somewhat more research on age-structured predator-prey interactions (9, 38, 74, 75, 109, 127, 208). Some of this work has also examined multiple stable equilibria and complex relations between juvenile and adult abundances (73, 180). Much of this literature describes age structure as "destabilizing" because its introduction into models increases the dimensionality of the model and introduces delays (i.e. time lags). In some cases, however, age structure can "buffer" against perturbations [see Ricker's (138) early numerical work and Botsford & Wickham's (15, 17) analytical confirmation (see also 97)]. Recent work is also forcing theoreticians to reassess the commonly held generalization that delays are destabilizing;

age-structured predator-prey models can exhibit multiple switches between local stability and instability as a delay is extended (74). The implications of this cannot be safely generalized to size-structured populations, but it should be noted that one effect of a reduced individual growth rate is to increase the time it takes to reach the next size.

In general, the results of age-structured models can be extended to size-structured populations only cautiously, if at all. In a population at equilibrium or with fixed vital rates, deterministic age-structured and size-structured demographies are interconvertible, since there is a one-to-one correspondence between age and size. If we are interested in the dynamics of populations, however, and if vital rates depend fundamentally on body size with density-dependence acting in part through the depression of individual growth rates, then purely age-structured models are inappropriate. The mortality and fecundity rates of an i-year-old simply cannot be written only as a function of age and the animal's present biotic and abiotic environment; current survival rates are functions of the animal's growth history and hence all past states in its ontogeny. Furthermore, when ontogenetic niche shifts are size based, past growth history can even affect the sign of the interaction between two age-classes: Whether an i-year-old preys upon, competes with, or has no interaction with a j-year-old of the same or another species can depend upon the animals' growth histories. As noted above, the natures of the delays also differ: The time between ages i and j is fixed, but the time between sizes i and j depends on growth rate (125). The mathematical implications of this property are not yet known. Thus, although age-structured models are often easier to analyze mathematically (because $d(\text{age})/dt = 1$ but $ds/dt = g$), the explicit inclusion of individual growth rates will be essential for many populations.

TWO EVOLUTIONARY PROBLEMS

Morphological Responses to Species Introductions

Our predictions of ontogenetic niche shifts are based on habitat- and size-specific growth and mortality curves. The curves themselves are subject to evolution, however, through changes in morphology and other phenotypic properties, and in many cases, selective pressures for appropriate morphologies at each size will be conflicting. For example, we earlier presented the problem of a fish that is piscivorous when sufficiently large but that must pass through an invertebrate-feeding stage. Since morphological traits that facilitate piscivory differ from those that facilitate other types of feeding, the evolution of the fish's morphology is presumably shaped by a balancing of selection at the various stages. How then will this balance be affected if a competitor is introduced that reduces the fish's growth rate during the invertebrate-feeding stage?

One can set this problem in an optimal control framework similar to the model of ontogenetic shifts if the constraints on the feasible morphological configurations and their ramifications can be specified. (Other mathematical frameworks also exist that describe the way in which conflicting ontogenetic attributes might be resolved; see 6, 28, 90, 142). To illustrate the problem, consider a simplified case where we assume a morphological trade-off between two juvenile stages and hold morphology at larger sizes constant. As before, the optimization criterion is to minimize the area under the μ/g curves; this maximizes the survivorship to size at first reproduction (that size, of course, is also subject to natural selection). Curves *1, 2,* and *3* of Figure 5 represent the values of μ/g in three habitats before the introduction of the competitor. The animal is assumed to switch habitats at their intersection, and for convenience it is assumed that the morphology the curves reflect is the morphology that minimizes the area under the curves, relative to other morphologies.

Let the introduction of a competitor in the first habitat be represented by replacing curve *1* with curve *A.* Curve *A* is shifted upward from curve *1* to reflect the competitor's depression of the resident's growth rate in the first habitat. For convenience, we leave curve *2* unaffected. How will the evolution of the resident's morphology respond to this change? Consider a small variation in morphology that raises the resident's growth rate in the first habitat but lowers it in the second habitat. This variant is represented by curves *A'* and *2'.* The variant would be favored by the present criterion of fitness, since the area under curves *A'–2'–3* is slightly smaller than the area under the curves *A–2–3.* If the introduction of the competitor had resulted in curve *B* rather than curve *A,* however, the variant would not have been favored, since the area under the curves *B'–2'–3* exceeds that under *B–2–3.* If the introduction of the competitor had resulted in curve *C,* a perfectly facultative switcher would immediately drop the first habitat from its ontogeny; the variant would not use the first

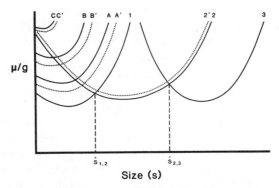

Figure 5   Ratios of mortality (μ) to growth *(g)* rates for three different habitats as a function of body size *(s).* Curves *A, B* and *C* represent the effects of introduced competitors that reduce growth rates in habitat 1. See text for details.

habitat either and the evolution of the animal's morphology would proceed independently of the first habitat.

It is notable that the response to a small reduction (A) is to become a better forager in the first habitat, but the response to a larger reduction *(B,C)* is to become a less efficient forager in that habitat. Thus, this preliminary analysis suggests that gradual coevolution with a competitor may produce character convergence (at least initially), but the response to an abrupt invasion may be character divergence.

Complex Life Cycles

Many major taxa—e.g. holometabolous insects and amphibians—have evolved complex life cycles with larvae and adults having distinct ecological life styles separated by an intervening metamorphosis. A number of authors have speculated on the evolution and maintenance of such life cycles, but there is little apparent consensus (reviewed in 201). Though metamorphosis renders such life cycles spectacular in appearance, we do not think they differ in any fundamental way from the ontogenetic niche shifts abundantly documented in other groups, and therefore they should be accessible by the same analyses.

Istock (80) was the first to offer a formal theory exploring complex life cycles, and his model remains the only systematic attempt to develop such a theory. He concluded that rates of evolution in both stages must be equal, and both environments must be saturated; or the two-stage life cycle will be unstable. This conclusion was based on his inability to construct feedback mechanisms when one phase enters a habitat promoting prolonged, density-independent, directional selection. Under these conditions, a reduction or loss of the alternative bottleneck stage is favored. Several authors (21, 164, 169) have disputed these conclusions.

Complex life cycles, however, have been stable over long periods of evolutionary time in groups such as insects and amphibians (81). Clearly the time (and therefore sizes) apportioned to different ontogenetic niches during the life cycle should be related to the distribution of selective forces operating in the various habitats or stages (80). The advantage of a habitat switch can be due to factors such as size-specific differences in growth or mortality rates in two habitats, or a size-related growth rate/predation risk trade-off. The balance in the duration of life history stages is determined by maximizing r in our model (in Istock's model, optimally balancing regulation in the two stages can sometimes lead to a reduction in r; see 164); when the components of fitness in each habitat are evaluated this way, the forces maintaining the two-stage life cycle become apparent. Furthermore, it is easy to predict the conditions under which one or the other stage of a complex life cycle will be eliminated by natural selection. In reference to Figure 3, if μ_2 increased or g_2 decreased (or vice versa in habitat 1) such that the μ_2/g_2 curve came to lie above the μ_1/g_1

curve, then selection would eliminate the second stage of the life cycle and vice versa for the first stage.

The fact that an enormous variety of complex life cycles are maintained in organisms that do not metamorphose suggests that factors like the above are important and widespread. In these cases, there would be little resistance to regression to a single-stage life history if it were advantageous. The growth rate/predation risk trade-off, of course, is only one of a large number of possible trade-offs or factors that may maintain complex life cycles. Attempts to construct a theory of ontogenetic niche shifts seems to offer considerable promise for gaining insight into the origin and maintenance of complex life cycles.

CONCLUSIONS

We have provided examples of changes in niche that are correlated with individual growth for a variety of taxa. As a fairly large body of empirical evidence indicates, such ontogenetic shifts can create a rich nexus of stage-specific interactions that have profound effects on the dynamics of interacting species and the resulting community structure. Traditional approaches to community ecology are founded on models lacking population structure. Consequently, the importance and richness of such interactions have largely been ignored. For a great many communities, we must begin to deal directly—both experimentally and theoretically—with this complexity.

It seems that a theory (or theories) of the ontogenetic niche will be a necessary feature of many approaches to the above problems. The allometries of factors such as resource use, net return rates, and predator vulnerability with body size are fundamental parameters for predicting the size-specific asymmetries in species interactions and/or ontogenetic niche shifts. Some starts have been made along these lines in studies of continuous resource change with body size (20, 152, 207). All of these studies, however, explored the implications of such allometries for interspecific interactions among "adults." More attention needs to be focused on the consequences of these allometries for changes in resource use and intraspecific competitive ability as an individual grows and the implications of this for interspecific interactions. For discrete ontogenetic shifts, our approach predicts resource switches based on two important size-specific components of fitness—the growth rate and predation risk. It is hoped that such theories will enable us to predict the range of sizes or the period in the life history over which a specific interaction will occur. Theories of the ontogenetic niche are also closely related to the critical problems of life history evolution; in many ways, they are the same problem.

Theoretical work on size-structured population dynamics is in a formative stage, and there is almost no work on interactions among such populations.

Studies of single species models, however, indicate that novel behaviors can be expected. This conclusion is reinforced by studies of age-structured models, though extrapolations of these results to size-structured populations are difficult. The empirical studies of species interactions clearly indicate that at least several broad classes of interactions—e.g. the juvenile bottleneck problem and mixed competition/predation interactions—would repay theoretical investigation. The analytical problems presented here are formidable, but there may be systematic ways to simplify these models (125). Empirical work to dissect these complex interactions and identify classes of interactions is critical in order to give direction to the theory amidst what can appear to be a daunting celebration of biological uniqueness.

ACKNOWLEDGMENTS

We are indebted to Donald Hall, Craig Osenberg, Gary Mittelbach, Mathew Leibold, Carol Folt, Scott Gleeson, Timothy Ehlinger, and Ronald Pulliam for comments on earlier versions of the manuscript. Laura Riley provided her customary expert assistance. This work was supported by NSF Grant BSR–8119258. Ralph Slatyer and Barry Osmond of the Research School of Biological Sciences, Australian National University, provided support and facilities for EEW during preparation of this paper. Contribution No. 535 of the Kellogg Biological Station.

Literature Cited

1. Allan, J. D. 1974. Balancing predation and competition in cladocerans. *Ecology* 55:622–29
2. Alm, G. 1946. Reasons for the occurrence of stunted fish populations with special regard to the perch. *Inst. Freshwater Res. Drottningholm Rep.* 25:1–146
3. Alm, G. 1959. Relation between maturity, size and age in fishes. *Inst. Freshwater Res. Drottningholm Rep.* 40:5–145
4. Andrews, R. M. 1982. Patterns of growth in reptiles. In *Biology of the Reptilia,* ed. C. Gans, H. Tough, 13:273–319. New York: Academic
5. Arnold, S. J., Wassersug, R. J. 1978. Differential predation on metamorphic anurans by garter snakes (Thamnophis): Social behavior as a possible defense. *Ecology* 59:1014–22
6. Atchley, W. R. 1984. Ontogeny, timing of development, and genetic variance-covariance structure. *Am. Nat.* 123:519–40
7. Bagenal, T. B. 1978. Aspects of fish fecundity. In *Ecology of Freshwater Fish Production,* ed. S. D. Gerking, pp. 75–101. London: Blackwell
8. Ballinger, R., Newlin, M., Newlin, S. 1977. Age-specific shift in the diet of the crevice spring lizard, *Sceloporus poinsetti* in southwestern New Mexico. *Am. Midl. Nat.* 97:482–84
9. Beddington, J. R., Free, C. A. 1976. Age structure effects in predator-prey interactions. *Theor. Popul. Biol.* 9:15–24
10. Belovsky, G. E. 1978. Diet optimization in a generalist herbivore: The moose. *Theor. Popul. Biol.* 14:105–34
11. Benke, A. C. 1978. Interactions among coexisting predators—a field experiment with dragonfly larvae. *J. Anim. Ecol.* 47:335–50
12. Beverton, R. J. H., Holt, S. J. 1957. On the dynamics of exploited fish populations. *Fish. Invest. Ministry Agric. Fish. Food (GB) Ser. II, No. 19.* 533 pp.
13. Blueweiss, L., Fox, H., Kudzma, V., Nakashima, D., Peters, R., Sams, S. 1978. Relationships between body size and some life history parameters. *Oecologia* 37:257–72
14. Botsford, L. W. 1981. The effects of increased individual growth rates on depressed population size. *Am. Nat.* 117:38–63

15. Botsford, L. W. 1981. More realistic fishery models: Cycles collapse and optimal policy. *Lect. Notes Biomath.* 40:6–20

16. Botsford, L. W. 1981. Optimal fishery policy for size-specific, density-dependent population models. *J. Math. Biol.* 12:265–93

17. Botsford, L. W., Wickham, D. E. 1978. Behavior of age-specific, density-dependent models and the northern California Dungeness crab *(Cancer magister)* fishery. *J. Fish. Res. Board Can.* 35:833–43

18. Bowers, M. A., Brown, J. H. 1982. Body size and coexistence in desert rodents: Chance or community structure? *Ecology* 63:391–400

19. Brett, J. R. 1979. Environmental factors and growth. In *Fish Physiology*, ed. W. S. Hoar, D. J. Randall, J. R. Brett, 8:599–675. New York: Academic

20. Brooks, J. L., Dodson, S. I. 1965. Predation, body size, and composition of plankton. *Science* 150:28–35

21. Bryant, E. H. 1969. A system favoring the evolution of holometabolous development. *Ann. Entomol. Soc. Am.* 62: 1087–91

22. Burnet, A. M. R. 1970. Seasonal growth of brown trout in two New Zealand streams. *NZ J. Mar. Freshwater Res.* 4:55–62

23. Buss, L. W. 1980. Competitive intransitivity and size-frequency distributions of interacting populations. *Proc. Natl. Acad. Sci. USA* 77:5355–59

24. Calder, W. A. III. 1983. Ecological scaling: Mammals and birds. *Ann. Rev. Ecol. Syst.* 14:213–30

25. Calef, G. W. 1973. Natural mortality of tadpoles in a population of *Rana aurora.* *Ecology* 54:741–58

26. Caswell, H. 1982. Optimal life histories and the maximization of reproductive value: A general theorem for complex life cycles. *Ecology* 63:1218–22

27. Caswell, H. 1982. Stable population structure and reproductive value for populations with complex life cycles. *Ecology* 63:1223–31

28. Caswell, H. 1983. Phenotypic plasticity in life-history traits: Demographic effects and evolutionary consequences. *Am. Zool.* 23:35–46

29. Charlesworth, B. 1980. *Evolution in Age-Structured Populations.* Cambridge: Cambridge Univ. Press

30. Christian, K. A. 1982. Changes in the food niche during postmetamorphic ontogeny of the frog *Pseudacris triseriata. Copeia* 1982:73–80

31. Clady, M. D. 1974. Food habits of yellow perch, smallmouth bass and large-

mouth bass in two unproductive lakes in northern Michigan. *Am. Midl. Nat.* 91:453–59

32. Clark, D. B., Gibbons, J. W. 1969. Dietary shift in the turtle *Pseudemys scripta* (Schoepff) from youth to maturity. *Copeia* 1969:704–6

33. Clarke, R. D. 1974. Postmetamorphic growth rates in a natural population of Fowler's toad, *Bufo woodhousei fowleri.* *Can. J. Zool.* 52:1489–98

34. Crossman, E. J., Larkin, P. A. 1959. Yearling liberations and change of food as effecting rainbow trout yield in Paul Lake, British Columbia. *Trans. Am. Fish. Soc.* 88:36–44

35. Crowley, P. H., Johnson, D. M. 1982. Habitat and seasonality as niche axes in an odonate community. *Ecology* 63: 1064–77

36. Cushing, D. H. 1974. The possible density-dependence of larval mortality and adult mortality in fishes. In *The Early Life History of Fish*, ed. J. H. S. Blaxter, pp. 103–11. New York: Springer-Verlag

37. Cushing, D. H., Harris, J. G. K. 1973. Stock and recruitment and the problem of density dependence. *Rapp. P. V. Reun. Cons. Int. Explor. Mer.* 164:142–55

38. Cushing, J. M., Saleem, M. 1982. A predator-prey model with age structure. *J. Math. Biol.* 14:231–50

39. Davies, R. W., Wrona, F. J., Linton, L., Wilkialis, J. 1981. Inter- and intraspecific analyses of the food niches of two sympatric species of Erpobdellidae (Hirudinoidea) in Alberta, Canada. *Oikos* 37:105–11

40. Dayton, P. K. 1971. Competition, disturbance and community organization: The provision and subsequent utilization of space in a rocky intertidal community. *Ecol. Monogr.* 41:351–89

41. DeAngelis, D. L., Coutant, C. C. 1982. Genesis of bimodal size distributions in species cohorts. *Trans. Am. Fish. Soc.* 111:384–88

42. DeAngelis, D. L., Hackney, P. A., Webb, J. C. 1980. A partial differential equation model of changing sizes and numbers in a cohort of juvenile fish. *Environ. Biol. Fish.* 5:261–66

43. DeAngelis, D. L., Mattice, J. S. 1979. Implications of a partial-differential-equation cohort model. *Math. Biosci.* 47:271–85

44. DeBenedictis, P. A. 1974. Interspecific competition between tadpoles of *Rana pipiens* and *Rana sylvatica*: An experimental field study. *Ecol. Monogr.* 44:129–51

45. Deelder, C. L. 1953. A contribution to the knowledge of the stunted growth of

perch (*Perca fluviatilis* L.) in Holland. *Hydrobiologia* 3:357–78

46. Dodson, S. I. 1970. Complementary feeding niches sustained by size-selective predation. *Limnol. Oceanogr.* 15:131–37

47. Dodson, S. I. 1974. Zooplankton competition and predation: An experimental test of the size-efficiency hypothesis. *Ecology* 55:605–13

48. Enders, F. 1976. Size, food finding, and Dyar's constant. *Environ. Entomol.* 1:1–10

49. Fast, A. W., Bottroff, L. H., Miller, R. L. 1982. Largemouth bass, *Micropterus salmoides*, and bluegill, *Lepomis macrochirus*, growth rates associated with artificial destratification and threadfin shad, *Dorosoma petenense*, introductions at El Capitan Reservoir, California. *Calif. Fish Game* 67:4–20

50. Fedorenko, A. 1975. Instar and species-specific diets in two species of *Chaoborus*. *Limnol. Oceanogr.* 20:238–49

51. Fenchel, T. 1975. Characteristic displacement and coexistence in mud snails (Hydrobiidae). *Oecologia* 20:19–32

52. Fenchel, T. 1975. Factors determining the distribution patterns of mud snails (Hydrobiidae). *Oecologia* 20:1–17

53. Fenchel, T., Kofoed, L. H. 1976. Evidence for exploitative interspecific competition in mud snails (Hydrobiidae). *Oikos* 27:367–76

54. Forney, J. L. 1976. Year-class formation in the walleye (*Stizostedion vitreum vitreum*) population of Oneida Lake, New York, 1966–73. *J. Fish. Res. Board Can.* 33:783–92

55. Fraser, D. F. 1976. Coexistence of salamanders in the genus *Plethodon*: A variation of the Santa Rosalia theme. *Ecology* 57:238–51

56. Frogner, K. J. 1980. Variable developmental period: Intraspecific competition models with conditional age-specific maturity and mortality schedules. *Ecology* 61:1099–1106

57. Frost, B. W. 1980. The inadequacy of body size as an indicator of niches in the zooplankton. In *Evolution and Ecology of Zooplankton Communities*, ed. W. C. Kerfoot, pp. 742–53. Hanover, N.H.: Univ. Press New Engl.

58. Fuller, R. L., Stewart, K. W. 1977. The food habits of stoneflies (Plecoptera) in the upper Gunnison River, Colorado. *Environ. Entomol.* 6:293–302

59. Gerking, S. D., Raush, R. R. 1979. Relative importance of size and chronological age in the life programme of fishes. *Arch. Hydrobiol. Beih. Ergebn. Limnol.* 13:181–94

60. Gerritsen, J. 1984. Size frequency recon-

sidered: A general foraging model for free-swimming aquatic animals. *Am. Nat.* 123:450–67

61. Giguere, L. 1979. An experimental test of Dodson's hypothesis that *Ambystoma* (a salamander) and *Chaoborus* (a phantom midge) have complementary feeding niches. *Can. J. Zool.* 57:1091–97

62. Gilliam, J. F. 1982. *Habitat use and competitive bottlenecks in size-structured fish populations*. PhD thesis. Michigan State University, East Lansing, Michigan. 107 pp.

63. Glasser, J. 1983. A model of the growth of populations composed of individuals whose probabilities of growth, reproduction and death are size-specific. *J. Plankton Res.* 5:305–10

64. Goodman, L. A. 1969. The analysis of population growth when the birth and death rates depend upon several factors. *Biometrics* 25:659–81

65. Green, J. 1956. Growth, size and reproduction in *Daphnia* (Crustacea: Cladocera). *Proc. Zool. Soc. London* 126:173–204

66. Griffiths, C. L., King, J. A. 1979. Some relationships between size, food availability, and energy balance in the ribbed mussel *Aulacomya ater*. *Mar. Biol.* 51:141–49

67. Griffiths, D. 1975. Prey availability and the food of predators. *Ecology* 56:1209–14

68. Grossman, G. D. 1980. Ecological aspects of ontogenetic shifts in prey size utilization in the bay goby (Pisces:Gobiidae). *Oecologia* 47:233–38

69. Hall, D. J., Cooper, W. E., Werner, E. E. 1970. An experimental approach to the production dynamics and structure of freshwater animal communities. *Limnol. Oceanogr.* 15:839–928

70. Hall, D. J., Threlkeld, S. T., Burns, C. W., Crowley, P. H. 1976. The size-efficiency hypothesis and the size structure of zooplankton communities. *Ann. Rev. Ecol. Syst.* 7:177–208

71. Hartshorn, G. S. 1975. A matrix model of tree population dynamics. In *Tropical Ecological Systems*, ed. F. B. Golley, E. Medina, pp. 41–51. New York: Springer-Verlag

72. Hassell, M. P. 1978. *The Dynamics of Arthropod Predator-Prey Systems*. Princeton: Princeton Univ. Press. 237 pp.

73. Hassell, M. P., Comins, H. N. 1976. Discrete time models for two-species competition. *Theor. Popul. Biol.* 9:202–21

74. Hastings, A. 1983. Age-dependent predation is not a simple process. I. Con-

tinuous models. *Theor. Popul. Biol.* 23: 347–62

75. Heller, R. 1978. Two predator-prey difference equations considering delayed population growth and starvation. *J. Theor. Biol.* 70:401–13

76. Hines, A. H. 1982. Coexistence in a kelp forest: Size, population dynamics, and resource partitioning in a guild of spider crabs (Brachyura, Majidae). *Ecol. Monogr.* 52:179–98

77. Holling, C. S. 1965. The functional response of predators to prey density and its role in mimicry and population regulation. *Mem. Entomol. Soc. Can. No. 45.* 60 pp.

78. Hubbell, S., Werner, P. 1979. On measuring the intrinsic rate of increase of populations with heterogeneous life histories. *Am. Nat.* 113:277–93

79. Hughes, G. M. 1977. Dimensions and the respiration of lower vertebrates. In *Scale Effects in Animal Locomotion,* ed. T. J. Pedley, pp. 57–81. New York: Academic

80. Istock, C. A. 1967. The evolution of complex life cycle phenomena: An ecological perspective. *Evolution* 21:592–605

81. Istock, C. A. 1984. Boundaries to life history variation and evolution. In *A New Ecology: Novel Approaches to Interactive Systems,* ed. P. W. Price, C. N. Slobodchikoff, W. S. Gaud, pp. 143–68. New York: Wiley

82. Jackson, P. B. N. 1961. The impact of predation especially by the tiger fish (*Hydrocynus vittatus* Cast) on African freshwater fishes. *Proc. Zool. Soc. London* 136:603–22

83. Johannsson, O. E. 1978. Co-existence of larval Zygoptera (Odonata) common to the Norfolk Broads (U.K.). *Oecologia* 32:303–21

84. Jones, R., Johnston, C. 1977. Growth, reproduction and mortality in gadoid fish species. In *Fisheries Mathematics,* ed. J. H. Steele, pp. 37–62. New York: Academic

85. Keast, A. 1977. Mechanisms expanding niche width and minimizing intraspecific competition in two centrarchid fishes. In *Evolutionary Biology,* ed. M. K. Hecht, W. C. Steere, B. Wallace, 10:333–95. New York: Plenum

86. Keast, A. 1978. Trophic and spatial interrelationships in the fish species of an Ontario temperate lake. *Environ. Biol. Fish.* 3:7–31

87. Kirkpatrick, M. 1984. Indeterminate growth, and demographic theory based on size, not age. *Ecology.* In press

88. Krebs, J. R., Stephens, D. W., Sutherland, W. J. 1983. Perspectives in optimal foraging. In *Perspectives in Ornithology,* ed. A. H. Brush, G. A. Clark, Jr., pp. 165–221. Cambridge: Cambridge Univ. Press

89. Kusano, T. 1981. Growth and survival rate of the larvae of *Hynobius nebulosus tokyoensis* Tago (Amphibia, Hynobiidae). *Res. Popul. Ecol.* 23:360–78

90. Lande, R. 1982. A quantitative genetic theory of life history evolution. *Ecology* 63:607–15

91. Larkin, P. A., Smith, S. B. 1954. Some effects of introduction of the redside shiner on the Kamloops trout in Paul Lake, British Columbia. *Trans. Am. Fish. Soc.* 83:161–75

92. Larkin, P. A., Terpenning, J. G., Parker, R. R. 1957. Size as a determinant of growth rate in rainbow trout *Salmo gairdneri. Trans. Am. Fish. Soc.* 86:84–96

93. Law, R. 1983. A model for the dynamics of a plant population containing individuals classified by age and size. *Ecology* 64:224–30

94. Lawton, J. H. 1971. Maximum and actual field feeding-rates in larvae of the damselfly *Pyrrhosoma nymphula* (Sulzer) (Odonata: Zygoptera). *Freshwater Biol.* 1:99–111

95. LeCren, E. D. 1962. The efficiency of reproduction and recruitment in freshwater fish. In *The Exploitation of Natural Animal Populations,* ed. E. D. LeCren, M. W. Holdgate, pp. 283–96. Oxford: Blackwell

96. Lefkovitch, L. P. 1965. The study of population growth in organisms grouped by stages. *Biometrics* 21:1–18

97. Levin, S. A. 1981. Age-structure and stability in multiple-age spawning populations. *Lect. Notes Biomath.* 40:21–45

98. Lister, D. B., Genoe, H. S. 1970. Stream habitat utilization by cohabiting underyearlings of chinook (*Oncorhynchus tshawytscha*) and coho (*O. kisutch*) salmon in the Big Qualicum River, British Columbia. *J. Fish. Res. Board Can.* 27:1215–24

99. Loman, J. 1979. Food, feeding rates and prey-size selection in juvenile and adult frogs, *Rana arvalis* Nilss. and *R. temporaria* L. *Ekol. Pol.* 27:581–601

100. Lynch, M. 1977. Fitness and optimal body size in zooplankton populations. *Ecology* 58:763–74

101. Lynch, M. 1978. Complex interactions between natural coexploiters—*Daphnia* and *Ceriodaphnia. Ecology* 59:552–64

102. Lynch, M. 1980. The evolution of cladoceran life histories. *Q. Rev. Biol.* 55: 23–42

103. MacArthur, R. H. 1972. *Geographical*

Ecology: Patterns in the Distribution of Species. New York: Harper & Row. 269 pp.

104. Maiorana, V. 1978. An explanation of ecological and developmental constants. *Nature* 273:375–77

105. Maly, E. J. 1976. Resource overlap between co-occurring copepods: Effects of predation and environmental fluctuation. *Can. J. Zool.* 54:933–40

106. Martin, N. V. 1952. A study of the lake trout, *Salvelinus namaycush,* in two Algonquin Park, Ontario, lakes. *Trans. Am. Fish. Soc.* 81:111–37

107. Martin, N. V. 1966. The significance of food habits in the biology, exploitation, and management of Algonquin Park, Ontario, lake trout. *Trans. Am. Fish. Soc.* 95:415–22

108. Martin, N. V. 1970. Long-term effects of diet on the biology of the lake trout and the fishery in Lake Opeongo, Ontario, *J. Fish. Res. Board Can.* 27:125–46

109. Maynard-Smith, J., Slatkin, M. 1973. The stability of predator-prey systems. *Ecology* 54:384–91

110. McArdle, B. H., Lawton, J. H. 1979. Effects of prey-size and predator-instar on the predation of *Daphnia* by *Notonecta. Ecol. Entomol.* 4:267–75

111. McLusky, D. S. 1973. The effect of temperature on the oxygen consumption and filtration rates of *Chlamys (Aequipecten) opercularis* (L.) (Bivalvia). *Ophelia* 10:141–54

112. Menge, B. A. 1972. Competition for food between two intertidal starfish species and its effect on body size and feeding. *Ecology* 53:635–44

113. Mittelbach, G. G. 1981. Foraging efficiency and body size: A study of optimal diet and habitat use by bluegills. *Ecology* 62:1370–86

114. Mittelbach, G. G. 1983. Optimal foraging and growth in bluegills. *Oecologia* 59:157–62

115. Mittelbach, G. G. 1984. Predation and resource partitioning in two sunfishes (Centrarchidae). *Ecology* 65:499–513

116. Morin, P. J. 1983. Competitive and predatory interactions in natural and experimental populations of *Notophthalmus viridescens dorsalis* and *Ambystoma tigrinum. Copeia* 1983:628–39

117. Murphy, L. F. 1983. A nonlinear growth mechanism in size structured population dynamics. *J. Theor. Biol.* 104:493–506

118. Mushinsky, H. R., Hebrard, J. J., Vodopich, D. S. 1982. Ontogeny of water snake foraging ecology. *Ecology* 63:1624–29

119. Neill, W. E. 1975. Experimental studies of microcrustacean competition, community composition and efficiency of resource utilization. *Ecology* 56:809–26

120. Neill, W. E. 1984. Regulation of rotifer densities by crustacean zooplankton in an oligotrophic montane lake in British Columbia. *Oecologia* 61:175–81

121. Neill, W. E., Peacock, A. 1980. Breaking the bottleneck: Interactions of invertebrate predators and nutrients in oligotrophic lakes. See Ref. 57, pp. 715–24

122. Newell, R. C., Johnson, L. G., Kofoed, L. H. 1977. Adjustment of the components of energy balance in response to temperature change in *Ostrea edulis. Oecologia* 30:97–110

123. Nielsen, L. A. 1980. Effect of walleye *(Stizostedion vitreum vitreum)* predation on juvenile mortality and recruitment of yellow perch *(Perca flavescens)* in Oneida Lake, New York. *Can. J. Fish. Aquat. Sci.* 37:11–19

124. Nisbet, R. M., Gurney, W. S. C. 1982. *Modeling Population Fluctuations.* New York: Wiley. 379 pp.

125. Nisbet, R. M., Gurney, W. S. C. 1983. The systematic formulation of population models for insects with dynamically varying instar duration. *Theor. Popul. Biol.* 23:114–35

126. Oster, G. 1976. Internal variables in population dynamics. *Lect. Math. Life Sci.* 8:37–68

127. Oster, G., Takahashi, Y. 1974. Models of age-specific interactions in a periodic environment. *Ecol. Monogr.* 44:483–501

128. Paine, R. T. 1976. Size-limited predation: An observational and experimental approach with the *Mytilus-Pisaster* interaction. *Ecology* 57:858–73

129. Parker, R. P., Larkin, P. A. 1959. A concept of growth in fishes. *J. Fish. Res. Board Can.* 16:721–45

130. Persson, L. 1983. Food consumption and the significance of detritus and algae to intraspecific competition in roach *Rutilus rutilus* in a shallow eutrophic lake. *Oikos* 41:118–25

131. Peters, R. H. 1983. *The Ecological Implications of Body Size.* Cambridge: Cambridge Univ. Press. 329 pp.

132. Policansky, D. 1983. Size, age and demography of metamorphosis and sexual maturation in fishes. *Am. Zool.* 23:57–64

133. Polis, G. A. 1984. Age structure component of niche width and intraspecific resource partitioning: Can age groups function as ecological species? *Am. Nat.* 123:541–64

134. Popova, O. A., Sytina, L. A. 1977. Food and feeding relations of Eurasian perch *(Perca fluviatilis)* and pikeperch *(Stizostedion lucioperca)* in various waters

of the USSR. *J. Fish. Res. Board Can.*
34:1559–70
135. Pough, F. H. 1973. Lizard energetics and diet. *Ecology* 54:837–44
136. Pruss, J. 1981. Equilibrium solutions of age-specific population dynamics of several species. *J. Math. Biol.* 11:65–84
137. Randlóv, A., Riisgard, H. U. 1979. Efficiency of particle retention and filtration rate in four species of ascidians. *Mar. Ecol. Program Ser.* 1:55–59
138. Ricker, W. E. 1954. Stock and recruitment. *J. Fish. Res. Board Can.* 14:669–81
139. Ricker, W. E. 1979. Growth rates and models. See Ref. 19, pp. 677–743
140. Ricker, W. E., Foerster, R. E. 1948. Computation of fish production: A symposium on fish populations. *Bull. Bingham Oceanogr. Collect.* 11, pp. 173–211
141. Robertson, A. I. 1980. The structure and organization of an eelgrass fish fauna. *Oecologia* 47:76–82
142. Rose, M. 1983. Theories of life-history evolution. *Am. Zool.* 23:15–24
143. Ross, S. T. 1978. Trophic ontogeny of the leopard searobin, *Prionotus scitulus* (Pisces:Triglidae). *Fish. Bull.* 76:225–34
144. Rotenberg, M. 1977. Mathematical description of holometabolous life cycles. *J. Theor. Biol.* 64:333–53
145. Roughgarden, J. 1979. *Theory of Population Genetics and Evolutionary Ecology: An Introduction.* New York: Macmillan. 634 pp.
146. Salthe, S. N. 1969. Reproductive modes and the number and sizes of ova in the urodeles. *Am. Midl. Nat.* 81:467–90
147. Salthe, S. N., Duellman, W. E. 1973. Quantitative constraints associated with reproductive mode in anurans. In *Evolutionary Biology of the Anurans*, ed. J. L. Vial, pp. 229–49. Columbia: Univ. Mo. Press
148. Schaffer, W. M. 1983. The application of optimal control theory to the general life history problem. *Am. Nat.* 121:418–31
149. Schmidt-Nielsen, K. 1977. Problems of scaling: Locomotion and physiological correlates. See Ref. 79, pp. 1–21
150. Schoener, T. W. 1967. The ecological significance of sexual dimorphism in size in the lizard *Anolis conspersus. Science* 155:474–77
151. Schoener, T. W. 1968. The *Anolis* lizards of Bimini: Resource partitioning in a complex fauna. *Ecology* 49:704–26
152. Schoener, T. W. 1969. Models of optimal size for solitary predators. *Am. Nat.* 103:277–313
153. Schoener, T. W. 1974. Resource parti-

tioning in ecological communities. *Science* 185:27–39
154. Schoener, T. W. 1977. Competition and the niche. In *Biology of the Reptilia,* ed. C. Gans, D. W. Tinkle, pp. 35–136. New York: Academic
155. Schoener, T. W., Gorman, G. C. 1968. Some niche differences in three Lesser Antillean lizards of the genus *Anolis. Ecology* 49:819–30
156. Scott, W. B., Crossman, E. J. 1973. *Freshwater Fishes of Canada.* Ottawa: Fish. Res. Board Can. 966 pp.
157. Sebens, K. P. 1979. The energetics of asexual reproduction and colony formation in benthic marine invertebrates. *Am. Zool.* 19:683–97
158. Sebens, K. P. 1981. The allometry of feeding, energetics, and body size in three sea anemone species. *Biol. Bull.* 161:152–71
159. Sebens, K. P. 1982. The limits to indeterminate growth: An optimal size model applied to passive suspension feeders. *Ecology* 63:209–22
160. Sexton, O. J., Bauman, J., Ortleb, E. 1972. Seasonal food habits of *Anolis limifrons. Ecology* 53:182–86
161. Shelbourne, J. E. 1962. A predator-prey size relationship for plaice larvae feeding on *Oikopleura. J. Mar. Biol. Assoc. UK* 42:243–52
162. Shepherd, J. G., Cushing, D. H. 1980. A mechanism for density-dependent survival of larval fish as the basis of a stock-recruitment relationship. *J. Cons. Cons. Int. Explor. Mer* 39:160–67
163. Sinko, J. W., Streifer, W. 1967. A new model for age-size structure of a population. *Ecology* 48:910–18
164. Slade, N. A., Wassersug, R. J. 1975. On the evolution of complex life cycles. *Evolution* 29:568–71
165. Spight, T. M., Emlen, J. 1976. Clutch sizes of two marine snails with a changing food supply. *Ecology* 57:1162–78
166. Stenhouse, S. L., Hairston, N. G., Cobey, A. E. 1983. Predation and competition in *Ambystoma* larvae: Field and laboratory experiments. *J. Herpetol.* 17: 210–20
167. Stoner, A. W. 1980. Feeding ecology of *Lagodon rhomboides* (Pisces: Sparidae): Variation and functional responses. *Fish. Bull.* 78:337–52
168. Stoner, A. W., Livingston, R. J. 1984. Ontogenetic patterns in diet and feeding morphology in sympatric sparid fishes from seagrass meadows. *Copeia* 1984:174–87
169. Strathmann, R. 1974. The spread of sibling larvae of sedentary marine invertebrates. *Am. Nat.* 108:29–44

170. Streifer, W. 1974. Realistic models in population ecology. In *Advances in Ecological Research,* ed. A. MacFadyen, 8:199–266. New York: Academic

171. Svardson, G. 1976. Interspecific population dominance in fish communities of Scandinavian lakes. *Inst. Freshwater Res. Drottningholm Rep.* 55:144–71

172. Thompson, D. J. 1975. Towards a predator-prey model incorporating age-structure: The effects of predator and prey size on the predation of *Daphnia magna* by *Ischnura elegans. J. Anim. Ecol.* 44:907–16

173. Thompson, R. J., Bayne, B. L. 1972. Active metabolism associated with feeding in the mussel *Mytilus edulis* L. *J. Exp. Mar. Biol. Ecol.* 9:111–24

174. Town, J. C. 1981. Prey characteristics and dietary composition in intertidal *Astrostole scabra* (Echinodermata: Asteroidea). *NZ J. Mar. Freshwater Res.* 15:69–80

175. Trautman, M. B. 1981. *The Fishes of Ohio.* Columbus: Ohio State Univ. Press. 782 pp.

176. Travis, C. C., Post, W. M., DeAngelis, D. L., Perkowski, J. 1980. Analysis of compensatory Leslie matrix models for competing species. *Theor. Popul. Biol.* 18:16–30

177. Travis, J. 1980. Phenotypic variation and the outcome of interspecific competition in hylid tadpoles. *Evolution* 34:40–50

178. Travis, J. 1983. Variation in development patterns of larval anurans in temporary ponds. I. Persistent variation within a *Hyla gratiosa* population. *Evolution* 37:496–512

179. Travis, J. 1983. Variation in growth and survival of *Hyla gratiosa* larvae in experimental enclosures. *Copeia* 1983: 232–37

180. Tschumy, W. O. 1982. Competition between juveniles and adults in age-structured populations. *Theor. Popul. Biol.* 21:255–68

181. Turner, M. 1979. Diet and feeding phenology of the green lynx spider, *Peucetia viridans* (Aranae: Oxyopidae). *J. Arachnol.* 7:149–54

182. Vahl, O. 1973. Pumping and oxygen consumption rates of *Mytilus edulis* L. of different sizes. *Ophelia* 12:45–52

183. Van Horne, B. 1982. Niches of adult and juvenile deer mice *(Peromyscus maniculatus)* in seral stages of coniferous forest. *Ecology* 63:992–1003

184. Van Sickle, J. 1977. Analysis of a distributed-parameter population model based on physiological age. *J. Theor. Biol.* 64:571–86

185. Van Straalen, N. M. 1982. Demographic analysis of arthropod populations using a continuous stage-variable. *J. Anim. Ecol.* 51:769–83

186. Vitt, L. J., Congdon, J. D. 1978. Body shape, reproductive effort and relative clutch mass in lizards: Resolution of a paradox. *Am. Nat.* 112:595–608

187. Walters, C. J., Hilborn, R., Peterman, R. M., Staley, M. J. 1978. Model for examining early ocean limitation of Pacific salmon production. *J. Fish. Res. Board Can.* 35:1303–15

188. Ware, D. M. 1975. Growth, metabolism, and optimal swimming speed of a pelagic fish. *J. Fish. Res. Board Can.* 32:33–41

189. Ware, D. M. 1975. Relation between egg size, growth, and natural mortality of larval fish. *J. Fish. Res. Board Can.* 32: 2503–12

190. Ware, D. M. 1978. Bioenergetics of pelagic fish: Theoretical change in swimming speed and ration with body size. *J. Fish. Res. Board Can.* 35:220–28

191. Ware, D. M. 1980. Bioenergetics of stock and recruitment. *Can. J. Fish. Aquat. Sci.* 37:1012–24

192. Warren, C. E., Davis, G. E. 1967. Laboratory studies on the feeding, bioenergetics, and growth of fish. In *The Biological Basis of Freshwater Fish Production,* ed. S. D. Gerking, pp. 175–214. London: Blackwell

193. Werner, E. E. 1974. The fish size, prey size, handling time relation in several sunfishes and some implications. *J. Fish. Res. Board Can.* 31:1531–36

194. Werner, E. E. 1977. Species packing and niche complementarity in three sunfishes. *Am. Nat.* 111:553–78

195. Werner, E. E. 1979. Niche partitioning by food size in fish communities. In *Predator-Prey Systems in Fisheries Management,* ed. R. H. Stroud, H. Clepper, pp. 311–22. Washington DC: Sport Fish. Inst.

196. Werner, E. E., Gilliam, J. F., Hall, D. J., Mittelbach, G. G. 1983. An experimental test of the effects of predation risk on habitat use in fish. *Ecology* 64:1540–48

197. Werner, E. E., Mittelbach, G. G. 1981. Optimal foraging: Field tests of diet choice and habitat switching. *Am. Zool.* 21:813–29

198. Werner, E. E., Mittelbach, G. G., Hall, D. J., Gilliam, J. F. 1983. Experimental tests of optimal habitat use in fish: The role of relative habitat profitability. *Ecology* 64:1549–55

199. Wilbur, H. M. 1972. Competition, predation, and the structure of the *Ambystoma-Rana sylvatica* community. *Ecology* 53:3–21

200. Wilbur, H. M. 1977. Density-dependent aspects of growth and metamorphosis in *Bufo americanus*. *Ecology* 58:196–200
201. Wilbur, H. M. 1980. Complex life cycles. *Ann. Rev. Ecol. Syst.* 11:67–93
202. Wilbur, H. M. 1984. Complex life cycles and community organization in amphibians. See Ref. 81, pp. 195–224
203. Wilbur, H. M., Collins, J. P. 1973. Ecological aspects of amphibian metamorphosis. *Science* 182:1305–14
204. Williams, E. H. 1980. Disjunct distributions of two aquatic predators. *Limnol. Oceanogr.* 25:999–1006
205. Wilson, D. S. 1973. Food size selection among copepods. *Ecology* 54:907–14

206. Wilson, D. S. 1974. Prey capture and competition in the ant lion. *Biotropica* 6:187–93
207. Wilson, D. S. 1975. The adequacy of body size as a niche difference. *Am. Nat.* 109:769–84
208. Wollkind, D., Hastings, A., Logan, J. 1982. Age structure in predator-prey systems. II. Functional response and stability and the paradox of enrichment. *Theor. Popul. Biol.* 21:57–68
209. Wrona, F. J., Davies, R. W., Linton, L., Wilkialis, J. 1981. Competition and coexistence between *Glossiphonia complanata* and *Helobdella stagnalis* (Glossiphoniidae:Hirudinoidea). *Oecologia* 48:133–37

Ann. Rev. Ecol. Syst. 1984. 15:427–47

RESTITUTION OF *r*- AND *K*-SELECTION AS A MODEL OF DENSITY-DEPENDENT NATURAL SELECTION

Mark S. Boyce

Department of Zoology and Physiology, University of Wyoming, Laramie, Wyoming 82071

INTRODUCTION

Dawkins (36, p. 293) observes that "ecologists enjoy a curious love/hate relationship with the r/K concept, often pretending to disapprove of it while finding it indispensable." Others have suggested that the model of *r*- and *K*-selection is inadequate and outmoded and does not further our understanding of life history phenomena (105, 120, 123). These views unfortunately result from frequent misuse and overgeneralization of the model. I contend that *r*- and *K*-selection may have important ramifications for our understanding of life history evolution, but for the model to be useful, it must be interpreted strictly as it was originally formulated: as a model of density-dependent natural selection.

To minimize confusion, I begin this review with an outline of *r*- and *K*-selection explicitly as a model of density-dependent natural selection. I then attempt to place this model within a historical context with other life history theory. As with recent criticisms of competition theory (109), I find that *r*- and *K*-selection suffers from a lack of true tests of its hypotheses. Although it may play a significant role in the evolution of life histories, there are few empirical studies that can identify density dependence that is independent of other potential selective forces.

427

0066-4162/84/1120-0427$02.00

THE MODEL

In the simplest case, consider a population consisting of 2 genotypes, $i = 1,2$, reproducing clonally according to logistic dynamics. The Malthusian parameter of fitness, m_i, for each genotype is a decreasing linear function of the total population size, N, where each genotype has different values of r_i, the potential rate of increase when $N = 0$, and of K_i, the carrying capacity or population size when $dN_i/dt = 0$. Thus, $m_i = dN_i/N_i dt = r_i - (r_i/K_i)N$. When N is small, the genotype with the highest r_i possesses superior Malthusian fitness, irrespective of K_i. Conversely, individuals of the genotype with highest K_i have highest fitness at large N, irrespective of r_i. [Graphical illustration of this model is presented in MacArthur's *Geographical Ecology* (73, Figure 8–15).]

Such a continuous-time model is inadequate for monitoring diploid populations because changes in allelic frequency cannot reflect corresponding changes in genotypic frequency (68). This problem is avoided in a discrete-time version of the logistic model, commonly written as a function of absolute fitness, W_{ij} (given $W_{ij} > 0$) for the ith and jth alleles segregating at a single locus (68, 80, 101): $W_{ij} = 1 + r_{ij} - (r_{ij}/K_{ij})N$. For small to moderate values of r_{ij} (41), this version has properties similar to the original differential equation model, although it potentially allows for a heterozygote advantage at any density and consequent maintenance of polymorphism (68, 99, 101).

A frequently voiced interpretation of this model is that natural selection operates on K at high densities (29, 72). This extension of the model is trivial because although natural selection still maximizes the Malthusian parameter of fitness, genotypes with maximum K possess superior fitness at large N.

Considerable confusion exists because r is frequently used to denote dN/Ndt in exponential growth models. In the logistic model, however, r is not equal to dN/Ndt except at the imaginary point where $N = 0$. Logistic r is a constant that quantifies the rate at which a population will grow in a particular environment given the absence of *any* density-dependent effects. Kozlowski (63) misunderstood this and suggested that r is a density-dependent variable. In their appendix demonstrating that maximizing K is equivalent to maximizing realized r (which equals dN/Ndt), Sibly & Calow's (107) argument is similar to the preceding paragraph, but they are not clear about the distinction between realized r and logistic r.

Logistic r is also distinct from the abstraction called r_{max}. r_{max} is the potential growth rate in a totally unlimited environment, that is, the maximum growth rate attainable under ideal conditions—e.g. in the laboratory. Logistic r, on the other hand is the growth rate that may be achieved *in a particular environment* if there are no density effects (contra 42). Evolutionarily, this value may be maximized by either increasing the birth rate or decreasing the death rate, depending upon the particular demographic environment (105). Therefore, a high r does not necessarily mean a high birth rate, and it certainly cannot be equated with the birth rate (as in 89).

The structure of the model does not imply that K should necessarily decrease as r increases or visa versa. Mutants with higher r and K will clearly have superior fitness at all densities. It still holds that at high density forms with higher K should be favored irrespective of r, and similarly, at low N natural selection favors individuals with high r irrespective of K. Thus, it does not seem at all contradictory that Luckinbill (69) should find that genotypes of *Escherichia coli* with high r and K did best and that there was no necessary trade-off between r and K. Still such a trade-off may occur for some organisms. Various characters correlated with fitness may maintain additive genetic variance through the evolution of antagonistic pleiotropism (40). Consequently, within a population we may see that characters correlated with high K will be inversely correlated genetically with characters associated with high r. Pleiotropic genes for a positive association between r and K would become genetically fixed, and the only variability that would remain would be that for antagonistic pleiotropy.

Heckel & Roughgarden (58) presented a stochastic r- and K-model where selection for increased K leads to a reduction in r. They argued that genotypes with high r-values would realize greater variance in fitness; therefore, long-term geometric mean fitness would select for reduced r. Turelli & Petry (130) argued that this result was not robust, that it depended critically on the specific manner in which stochasticity was incorporated, and that it could also be modified for nonlinear generalizations of the logistic (e.g. 49).

Robustness of the r and K Model

The logistic model is obviously too simplistic to be realistic, and r- and K-selection is consequently dismissed by some authors (78). Specific criticisms of logistic constraints are not justified because the r- and K-model is readily written in generalized forms (29, 100)—the logistic version just happens to be the simplest case. Except for the case of frequency-dependent selection, the maximization of population size under K-selection and of potential r at low N appears to be a robust prediction.

Although the logistic model is only valid if mortality and fertility rates are independent of age (29), Charlesworth (28, 29) claimed that r- and K-selection is generally valid for both discrete and overlapping generations, even when density-dependence occurs age specifically. Dingle (37) and Blau (12) argued, in contrast, that r- and K-selection is not always valid because natural populations often violate the implicit assumption of a stable age-distribution. Templeton (126) pointed out that fluctuating populations may not maximize $m = dN/Ndt$ because growth in populations that do not have a stable age-distribution is determined by all of the eigenvalues of a population projection matrix.

Nevertheless, long-term mean population growth is approximated reasonably well by mean fertility and survivorship parameters in stochastic projection matrices (34; but see 14, 129). Therefore, deviations from a stable age-distribution do not invalidate long-term growth rates as one measure of fitness (129).

Relaxing the linearity assumption does not influence the basic structure of the model (72; see also 15, 29, 45, 90, 100); indeed, selection may operate directly on the concavity of $m(N)$ (termed θ-selection) (43, 49). For example, in populations maintained at intermediate N (i.e. $< K$), genotypes able to sustain high rates of increase—in spite of N—may possess a selective advantage irrespective of r_{ij} or K_{ij}, potentially leading to a more concave $m(N)$ function. Indeed, $m(N)$ is concave for many populations of vertebrates (43).

Stochastic extensions of the deterministic model of r- and K-selection complicate the mathematics and convert outcomes to probabilistic functions, but in the case of a linear $m(N)$, they do not drastically alter the structure of the model or its interpretation (29, 58). For non-linear $m(N)$'s, however, the stochastic model's results can be influenced by the degree of concavity of $m(N)$ for each genotype (15, 130).

Complex life cycles complicate the theory because although selection for increased K may occur in one environment or stage of the life cycle, totally different selective pressures may dominate another (133). Still, r- and K-selection is valid within the relevant life cycle stages. Charlesworth (29) noted that K-selection in age-structured populations tends to maximize the number of individuals in a critical age group. But of course the critical age group or life cycle stage may vary considerably among taxa. In blowfly *(Lucilia cuprina)* experiments, the critical stage under K-selection was female fecundity (84), whereas in *Drosophila birchi* K-selection enhanced adult survivorship (6; F. Ayala, personal communication).

The r- and K-model suffers most seriously with the incorporation of frequency-dependent natural selection (68); indeed, mean fitness may decrease under these conditions (29, 68, 82, 86). This complication is perhaps best illustrated by the evolution of competitive ability (termed α-selection; 24, 47, 48)—either interspecific or intergenotypic. Here, a superior competitor may possess a selective advantage in a genetically variable population, even at the cost of lower K. Although K-selection is not the only evolutionary response possible at high density, there is considerable empirical evidence that K-selection occurs (6, 29, 62, 69, 70, 80, 84). But by definition, K-selection only occurs when selection at high density results in an increase in equilibrium population size. Thus, K-selection is not necessarily synonymous with selection for competitive ability (contra 87), and selection at high densities need not result in K-selection (47, 51).

Single-species population models with continuous, density-dependent functions (e.g. the logistic and age-structured or nonlinear extensions) may be most applicable to taxa with a constant rate of resource renewal (27). Many species, however, interact with other species that constitute their resources. The result is an interactive set of equations with much more complex dynamics than the logistic ones—e.g. predator-prey models and plant-herbivore models (27). These interactive models reveal frequency-dependent influences on evolution, and consequently, density-dependent natural selection may not always maximize population size (1).

Density-dependence is only one of a multitude of factors that can shape the evolution of life histories (8). Furthermore, density-independent selection can occur at any density (93). For example demographic selection can alter life history parameters irrespective of density. Thus clearly *r-* and *K*-selection is limited in scope. Yet, some authors have attributed virtually all life history variation to this simple model (118, 119, 123). To place these works in perspective, I review the historical context of *r-* and *K*-selection.

THE HISTORY OF *r–K* THEORY

The basic rudiments of *r-* and *K*-selection appear as early as MacArthur's 1958 *Ecology* paper (71), where he attempted to understand variation in clutch size among 5 species of parulid warblers. He noted that "in a population which has reached an equilibrium size, abundance is independent of birth and death rates. For species in equilibrium, then, a study of birth and death rates is not necessary to understand the control of the equilibrium abundance" (71, p. 610). In the same context, MacArthur quotes Darwin (35): "A large number of eggs is of some importance to those species which depend upon a fluctuating amount of food, for it allows them rapidly to increase in numbers" (71, p. 610). This paper was followed by MacArthur's genetic formalization of how natural selection maximizes *K* (72), which was heralded as one of the early efforts to integrate population genetics and ecology.

The first explicit reference to *r-* and *K*-selection appears to have been by MacArthur's student, M. Cody (33), who credits the idea to MacArthur. Cody (33) used *r* and *K* in his attempt to explain patterns of geographic variation in avian clutch size. He argued that in seasonal environments, density-independent mortality lowers the mean population size, thus reducing competition among survivors. Therefore, birds in seasonal environments have more resources to allocate to reproductive functions, including clutch size. In contrast, in less seasonal environments, populations are nearer to carrying capacity, and more resources are consequently allocated to competitive activities. Although characterizing environmental seasonality as simply imposing density-independent mortality is a gross oversimplification (15), Cody's loose interpretation of *r-* and *K*-selection may have some merit here (95). When resources are abundant, individuals with large clutch sizes may leave more offspring. But in resource-limited populations, birds with smaller clutches may have lower energetic demands—and thus a higher K. Slagsvold (111) challenged the latitudinal pattern of avian density dependence, however, so references to r- and K-selection may be invalid. Ashmole's (5, 15, 96) hypothesis, which explicitly defines the role of environmental seasonality in clutch size evolution, is greatly preferable.

MacArthur & Wilson (74, p. 149ff) described r- and K-selection as a model of density-dependent natural selection. They noted that in an initially colonized population, r-selection would predominate for a time, but ultimately the population would come under K-selection. Under r-selection, they argued, evolution promotes productivity, high rates of resource exploitation, and high reproductive output. Under K-selection, "genotypes which can at least replace themselves with a small family at the lowest food level will win, the food density being lowered so that large families cannot be fed. Evolution here favors efficiency of conversion of food into offspring" (74, p. 149). Nothing in this discussion is inconsistent with a strict interpretation of r and K as a model of density-dependent natural selection.

MacArthur & Wilson went on to suggest, however, that in seasonal environments with highly density-independent mortality, we might expect r-selection for large productivity. Unfortunately, this argument is a very loose interpretation of the consequences of seasonality (15) and implies that density-independent mortality falls evenly on all age classes. This implication certainly may not be true, and it has led to much confusion about the consequences of density-independent vs density-dependent sources of mortality. Also, many authors appear to have equated r-selection with selection in fluctuating environments, overextending the original model (87, 98). Additionally, MacArthur & Wilson were confused about the mechanism for K-selection, suggesting that it may require group selection. Roughgarden (99) clearly showed that group selection was not necessary, a point MacArthur (73) later recognized.

Pianka's (89) short note published in 1970 had a long lasting impact on the interpretation of r–K theory. Pianka implicitly assumed that increased population density will result in decreased juvenile survivorship, and consequently the optimal reproductive effort should decrease with increasing population size (60a, 90, 106). Based upon this overgeneralization, he generated a table that lists what he claims are life history correlates of r- and K-selection, e.g. r-selected taxa mature early, exhibit high levels of reproductive effort, and do not have parental care; K-selected forms should somehow have the opposite traits, such as delayed breeding, low levels of reproductive effort, and parental care. No such conclusions can be drawn from density-dependent natural selection, and it is not true that r- and K-selection collapses into demographic theory, even if one accepts Pianka's assumption of age specificity [contra Horn (60a)].

Since the appearance of Pianka's table, r-selection has been equated with high reproductive potential and K-selection with low reproductive potential

(87, 119). Why?—because *r*-selection should clearly select for high reproductive potential; and *K*-selection, for the forms with highest fitness in a competitive environment. It is easy to envisage a "good competitor" as a large, aggressive individual with a large territory (despite the consequences for *K*). Critical evaluations of the concept or explicit definitions of the model were quickly ignored. It was convenient to have a label for species with high vs low *r*; indeed, there are many useful distinctions among taxa with variable potential rates of increase (122, 123). The origin of the concepts no longer mattered; "*r*-strategists" and "*K*-strategists" became jargon that everyone could understand: small species that could reproduce quickly vs large species that could not.

Pianka's table was reproduced in many introductory ecology texts (39, 64, 91, 94); misuse of the concept became rampant. People working on life history evolution quickly recognized that Pianka's interpretation was not a general one (104, 134). MacArthur (73) apparently recognized the problem and presented an explicit model of *r*- and *K*-selection in his last work, *Geographical Ecology*. At the same time, population geneticists began to examine density-dependent natural selection (3, 28, 31, 99), but these authors were usually cautious about extrapolating their results to life histories. Life history theory became a demographic theory with a focus on schedules of reproductive effort and life table prediction (105), perhaps even avoiding density-dependence and resource-based models.

Southwood (116) developed a very different interpretation of *r*- and *K*-selection. According to it, the permanence of habitats or "habitat durational stability" can clearly influence maximum longevity and the potential for density dependence. Obviously different life histories are associated with temporary vs permanent habitats, but the association with *r*- and *K*-selection is not always a clear one. As Parry (87) argued, to imply exact synonymy between habitat durational stability and *r*- and *K*-selection can only obscure the actual selective forces that operate in different habitats. Similar difficulties arise with attempts to collapse a wide variety of stress phenomena into a single factor (52, 76a), as in Grime's (54, 55) triangular classification of life histories.

Many of the authors purporting to provide evidence in support of *r*- and *K*-selection interpreted the model very loosely or as Pianka (89) proposed in his table. For example, Solbrig's (46, 115) study of dandelions *(Taraxacum officianale)* is frequently championed as a rigorous test of *r*- and *K*-selection (101, 106). Solbrig compared the reproductive effort, size, and survivorship of three populations of dandelions, presumably of different genotypes. The so-called *r*-selected population was frequently trampled and mowed and thus thought to be experiencing high, density-independent mortality. The population claimed to be *K*-selected was in a less disturbed site with high *interspecific* competition. But the population density on the trampled and mowed site was higher than on the undisturbed site! The higher reproductive effort on the trampled and mowed site may have been a consequence of selective trampling and mowing, and the large biomass of individuals in the low density population may have been a product of interspecific competition and the resulting α-selection. Consequently, Solbrig's dandelion study is not clearly a test of r- and K-selection.

The most recent phase in studies of r and K has been prompted by Stearns (118, 119), who proposed to test life history theory by comparing Pianka's predictions of r- and K-selection and those of bet-hedging and demographic theory with results from empirical studies. Several similar "tests" of r and K have subsequently appeared (7a, 38, 125). Since Pianka (89) made rather stringent assumptions and overgeneralized, r and K does not usually fare well under such "tests." In this case, r and K is rejected in favor of demographic theory. Thus, misinterpretations of theory can lead to spurious hypotheses that have little or no bearing on the model. Since density is only one of many environmental forces shaping life histories, we cannot expect life history patterns to reflect the consequences of r- and K-selection, except in restricted cases. The only appropriate solution is to study the original model of r- and K-selection as formulated in MacArthur's (72) original paper and to consider carefully the possible life history consequences of natural selection as it is affected by density dependence.

THE IMPLICATIONS FOR LIFE HISTORIES

Demographic approaches to life history evolution attempt to predict the optimal schedule of reproductive effort, E_x, that maximizes the average m in a particular demographic environment (26, 29, 79, 104). There is no reason that such a theory should not be valid at any density; similarly, any form of density-independent selection can occur irrespective of density (93).

As Schaffer (105) emphasizes, neither r nor K can be clearly expressed as explicit life table parameters (105). Stearns (119) stated that K is not a simple population parameter but a composite of a population, its resources, and their interaction. This statement is not necessarily true, since the density-dependent mechanism determining K need not be resource based, but his point is valid. K is not linked in any predictable way either to the life table or to the schedule of reproductive effort.

Which life history traits determine variation in K? Our ability to answer this question is obscured by the fact that a variety of density-dependent mechanisms may be responsible for setting K. Even in generalized models with age structure included (29), whether density-dependence occurs via competition, predation, or dispersal is not specified. I will proceed by reviewing the possible life history consequences of each density-dependent mechanism separately.

Density–Dependent Resource Availability

We know that the relationship between dN/Ndt and resource availability is generally concave, often following Michaelis-Menton, Monod (127) or von Bertalanffy functions (15). In the last instance, we see $dN/Ndt = r^* \{1 - \exp[-Z(X-D)]\}$, where r^* is the potential rate of increase given unlimited resource availability, X; D is the resource demand ($D=X$ when $dN/Ndt = 0$); and Z is the initial slope of the function $dN/Ndt(X)$. We may easily envisage an analog to the model of r- and K-selection (illustrated in 15, Figure 1). Here natural selection favors genotypes with high r^* when resources are abundant, but genotypes with low resource demands, D, have a selective advantage when resources are scarce (see 15 for a density-dependent extension of this model). MacArthur (73) explicitly interpreted r- and K-selection with a focus on density-dependent resource availability (see his Figures 8–15) and with good justification. The ecological constraints of resource availability on reproduction and growth have been extensively documented for many groups of organisms, including plants (136, 137), mammals (102), and birds (66).

Critical to understanding r- and K-selection under density-dependent resource availability is a knowledge of the correlations between life history and K or D. As Roughgarden (99a) noted, K-selection occurs when a genotype is able to continue reproduction and sustain a positive growth rate when all other genotypes have reached carrying capacity. MacArthur (73) pointed out that one mechanism resulting in higher K is increased efficiency of resource conversion. Calder (18) showed that the efficiency of resource conversion is size independent, at least among birds and mammals, but a variety of physiological adaptations can improve this efficiency. MacArthur (73), Smith (112), and Odum & Pinkerton (85) have suggested that high reproductive output is inconsistent with high efficiency, and conversely, individuals with low reproductive output should have higher efficiency and higher K. I am unaware of the existence of adequate biological evidence to support this hypothesis. The issue is complicated by the fact that the energy conversion efficiencies involved in the production of gametes are often higher than for the production of somatic tissues—but they differ among taxa (22).

In resource-limited populations, carrying capacity is higher for genotypes with lower total-resource budgets. Therefore, K-selection should favor individuals with lower total resource or energy demands (51, 72). Conversely, when resources are abundant, increased resource acquisition can be used selfishly to enhance both fecundity and survivorship via larger size and/or increased reproductive output (19, 21). Thus, total resource budgets should increase under r-selection and decrease under K-selection.

Abrams (1) came to a similar conclusion noting that less foraging effort and thus less total resource acquisition results in a larger equilibrium population size. He assumed that an increase in foraging effort and the concomitant increase in the total resource budget is funneled to reproductive effort, but this assumption is unnecessary. Increasing foraging effort may result in greater somatic allocation with the same consequences, i.e. a larger total resource budget and a smaller equilibrium population size.

Smaller body size is an example of a life history response that may occur as a consequence of K-selection (contra 89). Merritt & Merritt (77) argued that the lower food requirements of small individuals of lower mass enhance the probability of survival for voles in restricted subnivean environments. Larger individuals appear to be less able to tolerate the long periods of acute food shortage associated with high density; Clutton-Brock & Harvey reviewed the evidence for mammals (32). Lower mass and the concomitant lower food requirements increase the probability that individuals can meet their basic metabolic requirements during food shortages and still acquire energy to allocate to reproduction. Conversely, the allocation of resources to large body size or tap roots can *enhance* survival, particularly in seasonal environments (15). Thus, when resources are abundant r-selection may favor increased growth rates and larger size. Size may also be correlated with age-specific fecundity in some taxa (118), but not all. When size is positively correlated with age at first reproduction, large size may still offer a selective advantage through enhanced survivorship (15) if adequate resources are available.

Although not a life history character, reduced metabolic rate is a physiological response that may be expected to occur as a consequence of K-selection. Paraphrasing Bennett (10), "the low metabolic rate of ectothermic vertebrates has obvious advantages in permitting survival on very little food and conversion of a large fraction of ingested food into biomass." The consequence of variation in metabolic rates is greater than was previously thought. Nagy (81) noted that if food supply limits density, lizards could maintain a population 50 times as dense as a population of mammals of the same size and diet! I am unaware of direct evidence suggesting that density-dependent natural selection has resulted in reduced metabolic rates, although the indirect evidence from lizards is tantalizing (30, 81).

DEMOGRAPHIC RESPONSE   Michod (79) evaluated the consequences of variation in resource levels for optimal reproductive effort as Charlesworth (29) did for age-structured density-dependence. Basically, increased resources may result in either an increase or a decrease in optimal reproductive effort, depending upon the relative "rewards" to increased fecundity or increased survivorship. If increased reproductive allocation does not increase fitness substantially, increased somatic allocation may be able to enhance fitness more through increased survivorship. In populations where survival is hardly affected by somatic allocation, fitness may be enhanced more by increasing reproductive effort.

Such a demographic approach has proven to be useful for interpretating many instances of phenotypic variation in life histories. For example, decreasing food supplies stimulate a greater reproductive effort in triclad flatworms because young individuals may have a better chance than adults of surviving starvation conditions (23). Decreased food results in decreased reproductive effort in *Tribolium*, in contrast, because adult mortality is affected less than juvenile mortality by food shortages (78, 121). Among fish, reduced food availability typically leads to delayed breeding at a smaller size, and the delay in breeding decreases with increasing adult mortality (121). High densities in red deer *(Cervus elaphus)* populations (2), however, result in delayed breeding at a larger body size, probably because of the seasonal constraint on breeding.

The basic premise of much life history theory is that increased reproductive effort at a particular age results in decreased survivorship and residual reproductive value (20, 105) known as the "cost of reproduction". But along a resource or density gradient, this hypothesis need not be true. Soule & Werner (115a) proposed that along an environmental gradient, there may be an optimum at which reproductive effort is maximal. Deviations from this optimum may increase the relative maintenance costs and reduce the resources available for reproduction, thus decreasing the reproductive effort. Independent of demography, allometric constraints may also shape resource allocation patterns, e.g. more massive supportive structures may be necessary to ensure reproduction and survival at large sizes (75).

Calow (20) observed that if resources were adequate to ensure all maintenance requirements, increased reproductive allocation need not entail survival costs. Because such patterns might be viewed as contradictions of demographic theory that assumes reproductive effort is inversely correlated with residual reproductive value (104, 107), Calow proposed to redefine reproductive effort as the extent to which nutrients are allocated to reproduction when they are also required "to support other aspects of metabolism" (20, p. 36). Although Calow's approach helps us understand some apparent exceptions to demographic life history theory, reproductive effort becomes even more elusive and difficult to quantify. As a consequence, it is harder to use the theory to understand the life history parameters we observe in nature. As discussed in the previous section, resources can influence the evolution of life history patterns irrespective of trade-offs between reproductive effort and residual reproductive value. Therefore, I do not find Calow's redefinition of reproductive effort necessary; but furthermore, such redefinition implies excessive focus on the cost of reproduction for determining life history optima (9a).

SEX AND GENETIC VARIABILITY   The most direct evidence for *K*-selection would appear to come from *Drosophila* spp., where *K* becomes progressively larger over many generations in laboratory populations raised at high densities, and the rate of evolution of K is highest in populations with high genetic variability (6, 17, 62). These observations, however, have been shown to be a consequence of both frequency-dependent and density-dependent natural selection (128). When resources are limiting, competition is greatest among individuals of the same genotype, whereas the fitness of other genotypes exploiting the environment differently is inversely correlated with frequency (7, 128). As in models of competitive ability (24, 47), genotypic carrying capacities may actually decrease, even if population size increases. Interestingly, positive frequency-dependent selection occurs at low densities, and negative frequency-dependent selection only occurs at high densities (114, 128).

Nevertheless, this process may still occur under K-selection when genes control genetic variability and thereby the collective niche breadth of the offspring. Genes that confer high rates of recombination or other mechanisms resulting in high genetic variability among offspring will be favored under K-selection. Note that precisely this rationale has been used in the "Tangled Bank" hypothesis to explain the evolution of sex (9). Even though asexual forms are able to achieve higher genotypic rates of increase in unlimiting environments (i.e. r-selection), in populations at K sexual forms can continue to increase because their offspring exploit the environment in a variety of ways, and therefore they possess higher K.

Williams (134a) proposed that the ratio of sexual to vegetative reproductive effort should increase with density. However, Schaffer & Gadgil (106) have proposed exactly the opposite—i.e. that vegetative reproduction should become more important at high densities [contrary to Pitelka et al's reading (92) of 106]. Empirical observations are inconsistent as well (92) and are probably confounded by varying opportunities for dispersal, density-dependent pollinator abundance, and density-dependent seed predation. The same problems complicate investigations of the optimal balance between seed size and number as a function of density (106, 113). More theoretical and empirical work is clearly needed on these problems (136, 137).

Density–Dependent Predation

Predation can be either density dependent or density independent. Although predation can clearly play a major role in shaping the evolution of life histories, in most cases it is not clear why it should make any difference whether predation is density-dependent or density-independent. Therefore, I will not attempt to review the possible consequences of predation for life histories.

One situation does seem particularly relevant, however. It has been shown that weasels *(Mustela nivalis)* develop a search image on great tit *(Parus major)* nests and thus impose density-dependent mortality (65). The probability of predation on great tit nests is a function of brood size because parent birds have difficulty keeping large broods fed, and therefore large broods make more noise and are more likely to be found by weasels (88). In other words, the predation rate is positively correlated with clutch size, resulting in selection against large clutch size.

As noted above, predator-prey (or plant-herbivore) interactions may be inadequately modeled by the logistic equation. If dynamics can be character-ized by an interactive system such as Caughley & Lawton's (27) plant-herbivore system, a decreased foraging rate increases the equilibrium density for the herbivore (i.e. the predator). Holding all else constant, an increase in the potential growth rate of the predator will ultimately result in a larger equilib-rium population size, as it would for the logistic if the rate of density-dependence (which equals $-r/K$) were maintained constant. Nevertheless, frequency-dependent selection will sometimes (but not always) confound pred-ator-prey interactions (1, 68). Predator-prey models can become quite com-plex, and the mechanisms of density-dependent selection may vary according to the parameters that are incorporated into the system. Regardless, it should be quite apparent that the life history responses to density-dependent predation may be quite different from those imposed by density-dependent resource availability (134).

Density–Dependent Migration

The population growth rate may vary in a density-dependent fashion owing to differential rates of dispersal or immigration. The genotypic growth rate is maximized under r-selection by low dispersal rates. As population density increases, individuals presumably maximize their own fitness by dispersing rather than facing low fitness opportunities in a crowded environment, particu-larly if crowding is localized (44, 53, 61). Hamilton & May (56) argued that even in stable habitats, parents should ensure some dispersal among their offspring, even when dispersal results in considerable mortality, because of competition with relatives or other conspecifics. The dispersal of prereproduc-tives can also reduce the disadvantages of inbreeding depression, yet among animals, parents can maintain the advantage of familiarity with the area which they acquired through learning (53).

But how does K-selection operate under density-dependent dispersal, and what implications does this have for life histories? The fact that philopatry may be favored under r-selection does not mean that it will not be favored under K-selection. As Greenwood & Harvey (53) pointed out, it is not clear which life history characteristics correlate with dispersal patterns or density-dependent dispersal.

The spatial and temporal patterns of resources do offer some interesting correlates with dispersal and life history. For example, Andersson (3a) showed that autocorrelated periodic environments can favor nomadism and large clutch size. Similarly, Southwood (116) noted that ephemeral habitats offer a selective advantage for high reproductive effort and dispersal ability, since opportunities for future survival may be low. But neither of these examples have any necessary relation with r- and K-selection.

Similarly, dispersal ability or propensity need not be tied to r- and K-selection (37, 108), contrary to the loose interpretations by MacArthur & Wilson (74). Dingle (37) presented data on migratory forms within species, noting that r is actually higher among low migratory populations. He argued that this is a consequence of deteriorating habitats that stimulate dispersal—where selection favors movement even if it results in delayed reproduction—and reduced fecundity. Likewise, density-intolerant rodents possess lower reproductive rates, have larger body sizes and are highly aggressive (4, 13). These dispersal genotypes certainly are not a consequence of K-selection since they can not survive at high densities—rather, these forms may be selected for dispersal ability.

Although colonizing ability is another issue often correlated with r and K-selection, there seems no justification for doing so. Simberloff (108) showed that good island colonists may be either adept at reaching the islands and/or able to persist once colonized. MacArthur & Wilson (74, Chap. 4) suggested that colonists should be at the maximal reproductive ages, postulating that the probability of founding new populations would be highest at this point. Using branching process theory, Williamson & Charlesworth (135), however, found no particular relation between the age-specific reproductive value and the probability of survival of a population founded by individuals of a particular age.

More theoretical and empirical work is clearly needed in order to ascertain whether density-dependent migration or dispersal has any consistent consequences for life history evolution. Yet, as argued above in the case of density-dependent predation, the life history consequences of migration or dispersal are probably much different from those of density-dependent resource availability.

FLUCTUATING ENVIRONMENTS

Most organisms live in fluctuating environments, and the degree and effect of these environmental fluctuations can have important consequences for life histories (15, 25, 57, 103, 106). Contrary to frequent interpretation, the original model of r- and K-selection is a deterministic one providing no direct insight into the consequences of environmental variability. Yet, if population size is not perturbed, r-selection will be a fleeting experience with populations typically undergoing K-selection. The most common view is that fluctuating environments impose density-independent mortality, which reduces population size (33, 74). One may incorporate a density-independent mortality effect into the r- and K-model, yielding $W_{ij} = 1 + r_{ij} - (r_{ij}/K_{ij})N - y$, where y is the density-independent mortality rate, independent of genotype. As y increases, equilibrium N^* decreases. With this model, we have the frequently cited situation where high, density-independent mortality results in selection for genotypes with high r; whereas low, density-independent mortality means that populations will be near K, where selection usually results in fixation of the mutants possessing high K.

The nature of density-independent mortality is critical. MacArthur & Wilson (74) and Pianka (89, 90) implicitly assumed that y is a pure rarefaction term that imposes no selection itself, i.e. y must be age-independent, resource-independent, and independent of any life history characters. This assumption is usually unrealistic. For example, Sinclair (110) showed that drought related density-independent mortality among African buffalo *(Syncerus cafer)* is much higher among juveniles than adults. Therefore, the greater the intensity of density-independent mortality, the greater the selection for the low, age-specific reproductive effort predicted by demographic approaches (e.g. 107).

Similarly, density-independent mortality can directly select for particular life history traits. For example, high overwinter mortality may fall more heavily on small individuals than on larger ones because large individuals are able to survive longer food shortage periods (15). Likewise, density-independent mortality imposed by wave action on intertidal molluscs may select for thick-shelled individuals that are able to withstand the heavy force of waves (8, 131, 132).

In reality, the effects of environmental fluctuations are much more complex than simply imposing density-independent mortality. One may envisage seasonally fluctuating carrying capacity or resource availability (15, 16), fluctuating age-specific mortality rates (29, 57, 103), or spatial patchiness where different patches vary temporally in resource abundance (3a). The precise nature of environmental variability is critical to understanding its potential impact on life history evolution; to collapse all environmental variability into one model of r- and K-selection is naive.

Caswell (25) recently proposed that the proportion of time that a population spends increasing or decreasing will have important effects on life history. Populations that are usually increasing should experience selection for high reproductive effort, early age at first breeding, semelparity, fast development, short lifespans, and little investment in individual offspring; in decreasing populations, in contrast, a premium is placed upon long lifespans, slow development, delayed breeding, iteroparity, and high investments in individual offspring. These predictions depend entirely upon explicit assumptions about the demographic environment during population growth and decline. Caswell presumed that the change in population size during population increase equals that of population decrease, irrespective of the time spent decreasing or increasing. Therefore, the magnitude of selection is potentially equal during population increase or decrease. As with density-independent mortality, the precise selective forces operating during population growth or decline determine the life history consequences of fluctuating environments, and these are potentially independent of the proportion of time a population is increasing or decreasing. For example, "catastrophic mortality" may occur in an age-specific fashion that could reverse Caswell's predictions for populations spending a majority of time increasing.

The phenotypic plasticity of life history traits in fluctuating environments may be considerable (83). Phenotypic responses to environmental change are typically expected to occur in the same direction as that favored by natural selection (16, 50, 59). But such an approach should be viewed cautiously because phenotypic responses to environmental influences can mask genetic variation (11), and virtually nothing is known about the genetic control of phenotypic plasticity.

CONCLUSIONS

To be useful, r- and K-selection must be interpreted consistently (87). Although ecological theory does not subscribe to a law of priority like that prevalent in taxonomy, historical precedence commands that r- and K-selection be viewed strictly as a model of density-dependent natural selection (72).

Generalizations of the original model of r- and K-selection to allow for discrete time intervals and to incorporate age-structures, stochasticity, and nonlinearity show it to be reasonably robust (29, 100). The model is based on the assumption, however, that natural selection maximizes mean fitness, so extensions to frequency dependence may invalidate it (68). At high densities, K-selection can only increase K, but the evolution of competitive ability via frequency-dependent natural selection (α-selection) can actually reduce genotypic K (24, 47, 76). Therefore, evolution of competitive ability is not synonymous with K-selection.

Density-dependence is only one of many environmental factors that may shape the evolution of life histories. This evolution can occur with decidedly different consequences via density-dependent competition, predation, or migration. Even so, Prout (93) demonstrated that density-independent selection can occur at any density. Additionally, frequency-dependent natural selection may also be a function of density (128). Therefore, it may be difficult to demonstrate that any particular example of life history evolution is the consequence of density-dependent natural selection, especially in field studies.

This statement does not imply that density-dependent natural selection is not an important force in the evolution of life histories. Indeed, in resource limited populations K-selection increases the efficiency of resource utilization or favors a decrease in total resource use per individual. It may be associated with decreased reproductive output or decreased somatic allocation, e.g. smaller body size. In contrast, r-selection results in the use of as many resources as can possibly enhance fitness—through increased reproductive output to enhance fecundity and/or increased growth and somatic allocation for improved survivorship (21). Additionally, when linked with frequency-dependence, *r*- and *K*-selection offers a relevant framework for the fixation of genes that promote genetic variability or niche breadth, including sexual reproduction via Ghiselin's "Tangled Bank" hypothesis (9).

Much confusion exists over the role of environmental variability in the interpretation of *r*- and *K*-selection. Although density-independent mortality can reduce *N* and thus promote *r*-selection, it may also impose selective mortality, which may shape the evolution of life histories. The precise nature of environmental variability can be extremely important in shaping life history evolution (15, 57, 103, 130), but this effect may be totally independent of *r*- and *K*-selection.

Law (67) criticized Charlesworth (29) for developing theory on *K*-selection, claiming that there is little evidence that *K*-selection in natural populations is significant. I insist that the existing tests of the theory have not been adequate to justify such a criticism, although granted that the scope of the model is narrow, i.e. restricted to density effects. Schaffer (105) implied that life history evolution entails the prediction of life tables and age-specific schedules of reproductive effort. But such an approach obscures life history parameters that may be correlated with an organism's total resource budget, e.g. reproductive output and the size of the organism (60). Life history theory will remain rudimentary until predictions can encompass variation in the original life history characters that stimulated the development of the theory initially, including clutch size, litter size, total reproductive output, body size, and plant biomass.

ACKNOWLEDGMENTS

I commenced this work while I was a NATO postdoctoral fellow in the Department of Zoology at Oxford. I thank Sir Richard Southwood, Michael Bulmer, Jonathan Roughgarden, Pat Werner, Francisco Ayala, Peter Calow, Richard Sibly, Mark Wetton, Ingrid Jensen, Bruce Don, Tom LaPoint, and Simon Levin for valuable discussions and/or reviews of the manuscript.

Literature Cited

1. Abrams, P. 1983. Life-history strategies of optimal foragers. *Theor. Popul. Biol.* 24:22–38
2. Albon, S. D., Mitchell, B., Staines, B. W. 1983. Fertility and body weight in female red deer: A density-dependent relationship. *J. Anim. Ecol.* 52:969–80
3. Anderson, W. W. 1971. Genetic equilibrium and population growth under density-dependent selection. *Am. Nat.* 105: 489–98
3a. Andersson, M. 1980. Nomadism and

site tenacity as alternative reproductive tactics in birds. *J. Anim. Ecol.* 49:175–84
4. Archer, J. 1970. Effects of population density on behavior in rodents. In *Social Behavior in Birds and Mammals*, ed. J. H. Crook, pp. 169–210. New York: Academic
5. Ashmole, N. P. 1963. The regulation of numbers of tropical oceanic birds. *Ibis* 103b:458–73
6. Ayala, F. J. 1968. Genotype, environ-

ment, and population numbers. *Science* 162:1453–59

7. Ayala, F. J. 1972. Competition between species. *Am. Sci.* 60:348–57

7a. Barclay, H. J., Gregory, P. T. 1981. An experimental test of models predicting life-history characteristics. *Am. Nat.* 117:944–61

8. Begon, M., Mortimer, M. 1981. *Population Ecology.* Sunderland, Mass: Sinauer. 200 pp.

9. Bell, G. 1982. *The Masterpiece of Nature.* Berkeley: Univ. Calif. Press. 635 pp.

9a. Bell, G. 1984. Measuring the cost of reproduction. I. The correlation structure of the life table of a plankton rotifer. *Evolution* 38:300–13

10. Bennett, A. F. 1983. Ecological consequences of activity metabolism. In *Lizard Ecology,* ed. R. B. Huey, E. R. Pianka, T. W. Schoener, pp. 11–23. Cambridge, Mass: Harvard Univ. Press. 501 pp.

11. Berven, K. A. 1982. The genetic basis of altitudinal variation in the wood frog *Rana sylvatica.* I. An experimental analysis of life history traits. *Evolution* 36:962–83

12. Blau, W. S. 1981. Latitudinal variation in life histories of insects in disturbed habitats. In *Insect Life History Patterns,* ed. R. F. Denno, H. Dingle, pp. 75–95. New York: Springer-Verlag. 225 pp.

13. Boonstra, R., Krebs, C. J. 1979. Viability of large- and small-sized adults in fluctuating vole populations. *Ecology* 60:567–73

14. Boyce, M. S. 1977. Population growth with stochastic fluctuations in the life table. *Theor. Popul. Biol.* 12:366–73

15. Boyce, M. S. 1979. Seasonality and patterns of natural selection for life histories. *Am. Nat.* 114:569–83

16. Boyce, M. S., Daley, D. J. 1980. Population tracking of fluctuating environments and natural selection for tracking ability. *Am. Nat.* 115:480–91

17. Buzzati-Traverso, A. A. 1955. Evolutionary changes in components of fitness and other polygenic traits in *Drosophila melanogaster* populations. *Heredity* 9:153–86

18. Calder, W. A. III. 1983. Ecological scaling: Mammals and birds. *Ann. Rev. Ecol. Syst.* 14:213–30

19. Calow, P. 1978. *Life Cycles.* New York: Wiley. 164 pp.

20. Calow, P. 1979. The cost of reproduction—a physiological approach. *Biol. Rev.* 54:23–40

21. Calow, P. 1982. Homeostasis and fitness. *Am. Nat.* 120:416–19

22. Calow, P. 1983. Energetics of reproduction and its evolutionary implications. *Biol. J. Linn. Soc.* 20:153–65

23. Calow, P., Woollhead, A. S. 1977. The relationship between ration, reproductive effort and age-specific mortality in the evolution of life history strategies—some observations on freshwater triclads. *J. Anim. Ecol.* 46:765–81

24. Case, T. J., Gilpin, M. E. 1974. Interference competition and niche theory. *Proc. Natl. Acad. Sci. USA* 71:3073–77

25. Caswell, H. 1982. Life history theory and the equilibrium status of populations. *Am. Nat.* 120:317–39

26. Caswell, H. 1982. Optimal life histories and the age-specific costs of reproduction. *J. Theor. Biol.* 98:519–29

27. Caughley, G., Lawton, J. H. 1981. Plant-herbivore systems. In *Theoretical Ecology,* ed. R. M. May, pp. 132–66. Oxford: Blackwell. 489 pp.

28. Charlesworth, B. 1971. Selection in density-regulated populations. *Ecology* 52:469–74

29. Charlesworth, B. 1980. *Evolution in Age-Structured Populations.* Cambridge: Cambridge Univ. Press. 300 pp.

30. Christian, K., Tracy, C. R., Porter, W. P. 1983. Seasonal shifts in body temperature and use of microhabitats by Galapagos land iguanas *(Conolophus pallidus).* *Ecology* 64:463–68

31. Clarke, B. 1972. Density-dependent selection. *Am. Nat.* 106:1–13

32. Clutton-Brock, T. H., Harvey, P. H. 1983. The functional significance of variation in body size among mammals. In *Advances in the Study of Mammalian Behavior, Am. Soc. Mammal., Spec. Publ. No. 7,* ed. J. F. Eisenberg, D. G. Kleiman, pp. 632–63. Shippensburg, PA: Am. Soc. Mammal. 753 pp.

33. Cody, M. L. 1966. A general theory of clutch size. *Evolution* 20:174–84

34. Cohen, J. E. 1977. Ergodicity of age structure in populations with Markovian vital rates. II. General states. *Adv. Appl. Probab.* 9:18–37

35. Darwin, C. R. 1859. *On the Origin of Species by Means of Natural Selection or the Preservation of Favoured Races in the Struggle for Life.* London: Murray. 502 pp.

36. Dawkins, R. 1982. *The Extended Phenotype.* Oxford: Freeman. 307 pp.

37. Dingle, H. 1981. Geographic variation and behavioral flexibility in milkweed bug life histories. See Ref. 12, pp. 57–73

38. Dunham, A. E. 1982. Demographic and life-history variation among populations of the iguanid lizard *Urosourus ornatus:* Implications for the study of life-history

phenomena in lizards. *Herpetologica* 38:208–21

39. Emmel, T. C. 1976. *Population Biology*. New York: Harper & Row. 371 pp.
40. Falconer, D. S. 1981. *Introduction to Quantitative Genetics*. Harlow, England: Longman. 340 pp. 2nd ed.
41. Felsenstein, J. 1979. *r*- and *K*-selection in a completely chaotic population model. *Am. Nat.* 113:499–510
42. Fenchel, T. 1974. Intrinsic rate of natural increase: The relationship with body size. *Oecologia* 14:317–26
43. Fowler, C. W. 1981. Density dependence as related to life history strategy. *Ecology* 62:602–10
44. Gadgil, M. 1971. Dispersal: Population consequences and evolution. *Ecology* 52:253–61
45. Gadgil, M., Bossert, W. 1970. Life historical consequences of natural selection. *Am. Nat.* 104:1–24
46. Gadgil, M., Solbrig, O. T. 1972. The concept of *r*- and *K*-selection: Evidence from wild flowers and some theoretical considerations. *Am. Nat.* 106:14–31
47. Gill, D. E. 1974. Intrinsic rate of increase, saturation density, and competitive ability. II. The evolution of competitive ability. *Am. Nat.* 108:103–16
48. Gilpin, M. E. 1974. Intraspecific competition between *Drosophila* larvae in serial transfer systems. *Ecology* 55:1154–59
49. Gilpin, M. E., Case, T. J., Ayala, F. J. 1976. θ-selection. *Math. Biosci.* 32:131–39
50. Goodman, D. 1979. Regulating reproductive effort in a changing environment. *Am. Nat.* 113:735–48
51. Green, R. F. 1980. A note on *K*-selection. *Am. Nat.* 116:291–96
52. Greenslade, P. J. M. 1983. Adversity selection and the habitat templet. *Am. Nat.* 122:352–65
53. Greenwood, P. J., Harvey, P. H. 1982. The natal and breeding dispersal of birds. *Ann. Rev. Ecol. Syst.* 12:1–21
54. Grime, J. P. 1977. Evidence for the existence of three primary strategies in plants and its relevance to ecological and evolutionary theory. *Am. Nat.* 111:1169–94
55. Grime, J. P. 1979. *Plant Strategies and Vegetation Processes*. Chichester, England: Wiley. 222 pp.
56. Hamilton, W. D., May, R. M. 1977. Dispersal in stable habitats. *Nature* 269:578–81
57. Hastings, A., Caswell, H. 1979. Role of environmental variability in the evolution of life history strategies. *Proc. Natl. Acad. Sci. USA* 76:4700–3
58. Heckel, D. G., Roughgarden, J. 1980. A

species near its equilibrium size in a fluctuating environment can evolve a lower intrinsic rate of increase. *Proc. Natl. Acad. Sci. USA* 77:7497–500
59. Hickman, J. C. 1975. Environmental unpredictability and plastic energy allocation strategies in the annual *Polygonum cascadense* (Polygonaceae). *J. Ecol.* 63:689–701
60. Hirshfield, M. F., Tinkle, D. W. 1975. Natural selection and the evolution of reproductive effort. *Proc. Natl. Acad. Sci. USA* 72:2227–31
60a. Horn, H. 1978. Optimal tactics of reproduction and life-history. In *Behavioural Ecology*, ed. J. R. Krebs, N. B. Davies, pp. 411–29. Sunderland, Mass: Sinauer. 494 pp.
61. Howe, H. F., Smallwood, J. 1982. Ecology of seed dispersal. *Ann. Rev. Ecol. Syst.* 13:201–28
62. Hutchinson, G. E. 1978. *An Introduction to Population Ecology*. New Haven, Conn: Yale Univ. Press. 260 pp.
63. Kozlowski, J. 1980. Density dependence, the logistic equation, and *r*- and *K*-selection: A critique and an alternative approach. *Evol. Theory* 5(2):89–102
64. Krebs, C. J. 1972. *Ecology*. New York: Harper & Row. 694 pp.
65. Krebs, J. R. 1970. Territory and breeding density in the great tit, *Parus major* L. *Ecology* 52:2–22
66. Lack, D. 1968. *Ecological Adaptations for Breeding in Birds*. London: Chapman & Hall. 409 pp.
67. Law, R. 1982. Book review: Evolution in age-structured populations. *Biol. J. Linn. Soc.* 18:65–66
68. Levin, S. A. 1978. On the evolution of ecological parameters. In *Ecological Genetics: The Interface*, ed. P. R. Brussard, pp. 3–26. New York: Springer-Verlag. 247 pp.
69. Luckinbill, L. S. 1978. *r*- and *K*-selection in experimental populations of *Escherichia coli*. *Science* 202:1201–3
70. Luckinbill, L. S. 1979. Selection and the *r/K* continuum in experimental populations of protozoa. *Am. Nat.* 113:427–37
71. MacArthur, R. H. 1958. Population ecology of some warblers of northeastern coniferous forests. *Ecology* 39:599–619
72. MacArthur, R. H. 1962. Some generalized theorems of natural selection. *Proc. Natl. Acad. Sci. USA* 48:1893–97
73. MacArthur, R. H. 1972. *Geographical Ecology*. New York: Harper & Row. 269 pp.
74. MacArthur, R. H., Wilson, E. O. 1967. *The Theory of Island Biogeography*. Princeton, NJ: Princeton Univ. Press. 203 pp.

75. McMahon, T. 1973. Size and shape in biology. *Science* 179:1201–4
76. Matessi, C., Jayakar, S. D. 1976. Models of density-frequency dependent selection for the exploitation of resources. I: Intraspecific competition. In *Population Genetics and Ecology,* ed. S. Karlin, E. Nevo, pp. 707–21. New York: Academic. 832 pp.
76a. Menges, E. S., Waller, D. M. 1983. Plant strategies in relation to elevation and light in floodplain herbs. *Am. Nat.* 122:454–73
77. Merritt, J. F., Merritt, J. M. 1978. Population ecology and energy relationships of *Clethrionomys gapperi* in a Colorado subalpine forest. *J. Mammal.* 59:576–98
78. Mertz, D. B. 1975. Senescent decline in flour beetles selected for early adult fitness. *Physiol. Zool.* 48:1–23
79. Michod, R. E. 1979. Evolution of life histories in response to age-specific mortality factors. *Am. Nat.* 113:531–50
80. Mueller, L. D., Ayala, F. J. 1981. Trade-off between *r*-selection and *K*-selection in *Drosophila* populations. *Proc. Natl. Acad. Sci. USA* 78:1303–5
81. Nagy, K. A. 1983. Ecological energetics. See Ref. 10, pp. 24–54
82. Nagylaki, T. 1979. The dynamics of density- and frequency-dependent selection. *Proc. Natl. Acad. Sci. USA* 76:438–41
83. Nichols, J. D., Conley, W., Batt, B., Tipton, A. R. 1976. Temporally dynamic reproductive strategies and the concept of *r*- and *K*-selection. *Am. Nat.* 110:995–1005
84. Nicholson, A. J. 1960. The role of population dynamics in natural selection. In *Evolution After Darwin,* ed. S. Tax, 1:477–520. Chicago: Univ. Chicago Press
85. Odum, H. T., Pinkerton, R. C. 1955. Time's speed regulator. *Am. Sci.* 43(2):331–43
86. Paquin, C. E., Adams, J. 1983. Relative fitness can decrease in evolving asexual populations of *S. cerevisiae. Nature* 306:368–71
87. Parry, G. D. 1981. The meanings of *r*- and *K*-selection. *Oecologia* 48:260–64
88. Perrins, C. M. 1977. The role of predation in the evolution of clutch size. In *Evolutionary Ecology,* ed. B. Stonehouse, C. M. Perrins, pp. 181–91. Baltimore: Univ. Park. 310 pp.
89. Pianka, E. R. 1970. On *r*- and *K*-selection. *Am. Nat.* 104:592–97
90. Pianka, E. R. 1972. *r*- and *K*-selection or *b* and *d* selection? *Am. Nat.* 106:581–88
91. Pianka, E. R. 1983. *Evolutionary Ecolo-*

gy. New York: Harper & Row. 416 pp. 3rd ed.
92. Pitelka, L. F., Stanton, D. S., Peckenham, M. O. 1980. Effects of light and density on resource allocation in a forest herb, *Aster acuminatus* (Compositae). *Am. J. Bot.* 67:942–48
93. Prout, T. 1980. Some relationships between density-independent selection and density-dependent population growth. In *Evolutionary Biology,* ed. M. K. Hecht, W. C. Steere, B. Wallace, 13:1–68. New York: Plenum. 301 pp.
94. Remmert, H. 1980. *Ecology.* New York/Berlin: Springer-Verlag. 289 pp. 2nd ed.
95. Ricklefs, R. E. 1977. On the evolution of reproductive strategies in birds: Reproductive effort. *Am. Nat.* 111:453–78
96. Ricklefs, R. E. 1980. Geographical variation in clutch size among passerine birds: Ashmole's hypothesis. *Auk* 97:38–49
97. Deleted in proof
98. Ricklefs, R. E. 1983. Comparative avian demography. In *Current Ornithology,* ed. R. F. Johnston, 1:1–32. New York: Plenum. 425 pp.
99. Roughgarden, J. 1971. Density-dependent natural selection. *Ecology* 52:453–68
99a. Roughgarden, J. 1977. Basic ideas in ecology. *Science* 196:51 (Book rev.)
100. Roughgarden, J. 1977. Coevolution in ecological systems: Results from "loop analysis" for purely density-dependent coevolution. In *Measuring Selection in Natural Populations,* ed. F. B. Christiansen, T. M. Fenchel, pp. 499–517. New York: Springer-Verlag. 564 pp.
101. Roughgarden, J. 1979. *Theory of Population Genetics and Evolutionary Ecology: An Introduction.* New York: Macmillan. 634 pp.
102. Sadleir, R. M. F. S. 1969. *The Ecology of Reproduction in Wild and Domestic Mammals.* London: Methuen. 321 pp.
103. Schaffer, W. M. 1974. Optimal reproductive effort in fluctuating environments. *Am. Nat.* 108:783–90
104. Schaffer, W. M. 1974. Selection for optimal life histories: The effects of age structure. *Ecology* 55:291–303
105. Schaffer, W. M. 1979. The theory of life-history evolution and its application to Atlantic salmon. *Symp. Zool. Soc. London* 44:307–26
106. Schaffer, W. M., Gadgil, M. D. 1975. Selection for optimal life histories in plants. In *Ecology and Evolution of Communities,* ed. M. L. Cody, J. M. Diamond, pp. 142–57. Cambridge, Mass: Belknap, 543 pp.

107. Sibly, R., Calow, P. 1983. An integrated approach to life-cycle evolution using selective landscapes. *J. Theor. Biol.* 102:527–47

108. Simberloff, D. 1981. What makes a good island colonist? See Ref. 12, pp. 195–205

109. Simberloff, D. 1982. The status of competition theory in ecology. *Ann. Zool. Fenn.* 19:241–53

110. Sinclair, A. R. E. 1977. *The African Buffalo.* Chicago: Chicago Univ. Press. 355 pp.

111. Slagsvold, T. 1981. Clutch size and population stability in birds: A test of hypotheses. *Oecologia* 49:213–17

112. Smith, C. C. 1976. When and how much to reproduce: The trade-off between power and efficiency. *Am. Zool.* 16:763–74

113. Smith, C. C., Fretwell, S. D. 1974. The optimal balance between size and number of offspring. *Am. Nat.* 108:499–506

114. Smouse, P. E. 1976. The implications of density-dependent population growth for frequency and density-dependent selection. *Am. Nat.* 110:849–60

115. Solbrig, O. T. 1971. The population biology of dandelions. *Am. Sci.* 59:686–94

115a. Soule, J. D., Werner, P. A. 1981. Patterns of resource allocation in plants with special reference to *Potentilla recta* L. *Bull. Torrey Bot. Club* 108:311–19

116. Southwood, T. R. E. 1977. Habitat, the templet for ecological strategies? *J. Anim. Ecol.* 46:337–65

117. Deleted in proof

118. Stearns, S. C. 1976. Life-history tactics: A review of the ideas. *Q. Rev. Biol.* 51:3–47

119. Stearns, S. C. 1977. The evolution of life history traits. *Ann. Rev. Ecol. Syst.* 8:145–72

120. Stearns, S. C. 1983. A natural experiment in life-history evolution: Field data on the introduction of mosquito fish *(Gambusia affinis)* to Hawaii. *Evolution* 37:601–17

121. Stearns, S. C. 1983. The evolution of life-history traits in mosquito fish since their introduction to Hawaii in 1905: Rates of evolution, heritabilities, and developmental plasticity. *Am. Zool.* 23:65–75

122. Stearns, S. C. 1983. The influence of size and phylogeny on patterns of covariation among life-history traits in the mammals. *Oikos* 41:173–87

123. Stearns, S. C. 1984. The effects of size and phylogeny on patterns of covariation in the life history traits of lizards and snakes. *Am. Nat.* 123:56–72

124. Deleted in proof

125. Taylor, C. E., Condra, C. 1980. *r*- and *K*-selection in *Drosophila pseudoobscura. Evolution* 34:1183–93

126. Templeton, A. R. 1980. The evolution of life histories under pleitropic constraints and *r*-selection. *Theor. Popul. Biol.* 18:279–89

127. Tilman, D. 1982. *Resource Competition and Community Structure.* Princeton, NJ: Princeton Univ. Press. 296 pp.

128. Tosic, M., Ayala, F. J. 1981. Density- and frequency-dependent selection at the *Mdh-2* locus in *Drosophila pseudoobscura. Genetics* 97:679–702

129. Tuljapurkar, S. D., Orzack, S. H. 1980. Population dynamics in variable environments. I. Long-run growth rates and extinction. *Theor. Popul. Biol.* 18:314–42

130. Turelli, M., Petry, D. 1980. Density- and frequency–dependent selection in a random environment: An evolutionary process that can maintain stable population dynamics. *Proc. Natl. Acad. Sci. USA* 77:7501–5

131. Vermeij, G. J. 1972. Intraspecific shore-level size gradients in intertidal molluscs. *Ecology* 53:693–700

132. Vermeij, G. J. 1973. Morphological patterns in high intertidal gastropods: Adaptive strategies and their limitations. *Mar. Biol.* 20:319–46

133. Wilbur, H. M. 1980. Complex life cycles. *Ann. Rev. Ecol. Syst.* 11:67–93

134. Wilbur, H. M., Tinkle, D. W., Collins, J. P. 1974. Environmental certainty, trophic level, and resource availability in life history evolution. *Am. Nat.* 108:805–17

134a. Williams, G. C. 1975. *Sex and Evolution.* Princeton, NJ: Princeton Univ. Press. 251 pp.

135. Williamson, J. A., Charlesworth, B. 1976. The effect of age of founder on the probability of survival of a colony. *J. Theor. Biol.* 56:175–90

136. Willson, M. F. 1983. *Plant Reproductive Ecology.* New York: Wiley. 282 pp.

137. Willson, M. F., Burley, N. 1983. *Mate Choice in Plants.* Princeton, NJ: Princeton Univ. Press. 251 pp.

Ann. Rev. Ecol. Syst. 1984. 15:449–78

TRADITIONAL AGRICULTURE IN AMERICA

R. S. Loomis

Department of Agronomy and Range Science, University of California, Davis, California 95616

> In agriculture nature too labours along with man.
>
> Adam Smith, 1776

INTRODUCTION

Agriculture presents a fertile field for the analysis of technological change for a number of reasons, including our strong concern about the future adequacy of food supplies. In our highly urbanized world, human beings are no less dependent on the land than they were in the past, but we now have only tenuous connections to it; it is natural to wonder whether things are being done properly out there. At present, as in the past, the merits of old and new directions in agriculture are actively discussed, and words such as *traditional, appropriate, alternative,* and *sustainable* are used to characterize views that contrast with contemporary (i.e. modern or conventional) agriculture.

Few of these terms have precise definitions. Schultz provides something of a foundation when he writes: "Farming based wholly upon the kinds of factors of production that have been used by farmers for generations can be called traditional agriculture" (83, p. 3). He had in mind the differences between a modern agriculture with inputs and management based on scientific knowledge, and primitive, labor-oriented systems without scientific knowledge. Since many "modern" practices have survived for hundreds of generations, a better definition would be based on measures of efficiency or intensity. Some of the terms represent beliefs in the need for less demanding life styles while recognizing that science is necessary in some degree. Others represent political or religious ideologies with elements of antiscience, including the proposition

449

0066-4162/84/1120-0449$02.00

that Adam Smith's *nature* extends beyond the materials and processes discernible by science to include cosmic and spiritual levels and the transmutation of elements (9).

The term *organic farming* embraces most nonconventional approaches, as well as those traditional in America before 1940. Current definitions of organic farming are generally based on the exclusion of manufactured chemicals; mined but still "natural" phosphate and potassium ores are permitted, but the use of natural nitrogen gas from the atmosphere, reduced with natural gas, is excluded (68). Legal definitions of organic farming may also allow the application of manures generated from chemically grown feeds or the use of lands that have been chemically treated in previous years (California State Law). All farming is organic, however, and since 18% of our cropland is normally planted to leguminous forages, the basic nutrient supply of conventional agriculture is also organic. In this review, the development of the "new farming systems" of Europe (i.e. legume rotations and the use of manure) is traced, presented as a concept of traditional agriculture, and compared with conventional agriculture.

In America, discussions about traditional agriculture suggest two basic issues: whether the present stewardship of the land and of our food sources are adequate for both now and the future; and whether our cultural root, the agrarian society, is healthy. This analysis deals mainly with the first question, considering its historical roots and exploring several topics of ecological interest.

WHAT IS TRADITIONAL IN AGRICULTURE?

Our concepts of traditional American agriculture rest in part on images of pioneers fulfilled, of peaceful fields in small farms, and of sage grandparents comfortable in their ability to produce from the land and to survive the vagaries of weather and markets. We are particularly impressed by our notions of the people, their knowledge, their competence in a range of skills about which we know little, and the seeming strength of their personal philosophies. Our images of present-day farmers, with their Steiger tractors, their knowledge of futures markets and discount rates, and their use of chemicals and supermarkets, appear in stark contrast to Grant Wood's Iowa landscapes. Coupled with this is the ever changing productivity of farm labor and the resultant changes in the scale of "family farming."

We need to separate the changes in farmers' life-styles from alterations in farming methods. The rapid extension of communications has integrated farm families into the mainstream of society. Radio, the telephone, and now television, automobiles and all-weather roads, unified school districts, and rural electrification are among the recent agents of these changes. Even as late as the

1930s, many Iowa farmers still rendered their own lard, made their own soap and bread, used outdoor privies, and made do with kerosene lamps and shallow, hand-pump wells. Those inconveniences are gone; farm life is no longer a backwoods proposition.

Farming methods have also changed. One might assume that our farming methods evolved mainly from practices in England, the origin of most early settlers. In the 1600s when America was settled, English farms were still largely manors operated by peasants or hired labor in "open-field" (unfenced) rotations with fallow. Wheat and barley were the principal grains, and the unimproved strains of animals were herded. Few of the early colonists arrived in America with even that kind of farming experience, however. With the necessity of subsistence, they adopted the native American maize as their grain, gradually welding it with European livestock, wheat, and rice as well as native cotton and tobacco into farming systems suited to the Atlantic coastal regions (18). Cotton and tobacco from the South were the major commercial crops (for sale or trade) and came to be grown in some cases on large plantations with slave labor. In both North and South, these woodland farmers depended heavily on forests for fuel, fencing, and shelter. The traditional English accommodation to nutrient depletion—an unvegetated fallow—was either unnecessary on the new lands or was replaced by shifting cultivation (a vegetated fallow), a practice that continued into the present century (95). Farming practices in the late 1700s were still primitive; Smith (88), for example, was appalled by the little attention to husbandry and land.

During the 17th and 18th centuries, a new farming system emerged in Europe (87). The population was decimated by plague in the 1500s, and much of the arable was abandoned. In the Low Countries, these idle lands were grazed, providing more manure and thus larger yields on the remaining arable. The development of root crops (fodder beet and turnip) for winter fodder was the key to maintaining larger animal herds. At the same time, rotation with forage legumes—a Roman technique—was rediscovered. Between 1650 and 1850, as British agriculture broke away from manorial systems, the new yeoman farmers gradually adopted closed-field systems based on legume rotations and mixed farming with livestock. Rapid improvements in sheep, swine, and cattle breeds occurred at the same time. These organic farming systems in Britain and the Low Countries produced striking increases in yield, and with continued improvements, they served as the basis of traditional farming in the developed nations until the middle of the twentieth century.

In such farming systems, the value of land lies not so much in its receipt of sunlight or water, which seldom limit production, but in the rate at which it gathers the scarce resource, nitrogen. Various types of fallow (bare or vegetated) provide nonuniform treatment of the landscape over time, so that the scarce resource accumulates. Soil resources may also be concentrated through nonuniform treatment in space—nutrients can be gathered from uncropped portions of the landscape and transferred to the arable as decayed residues ("turf manuring" with peat and muck) or as manures from grazing animals. Nonuniformity in space represents a form of "refuging" (38).

The new farming systems of Britain were not adopted extensively in the United States until the present century. Instead, farming simply spread to fertile new lands. After Independence, settlement of the woodland systems of the Ohio basin, the mid-South, and Oregon came rapidly; but the tall-grass prairie regions of Iowa, Illinois, and Minnesota, and the short-grass systems beyond, presented a number of difficulties in addition to the lack of wood. Chief among these was the difficulty of turning the grasslands with wooden plows. Lane developed the first narrow-profile steel plow in the 1820s, and Deere's and Oliver's innovations and designs followed quickly (18, 77). The introduction of barbed wire in 1874, which allowed fuller utilization of the abundant forage and grain potential of the plains without herding, was equally revolutionary.

Transport, a principal limitation on the expansion of farming, was solved by steam engines, which were used in river boats and on the railroads. Such engines also were the basis of the new "American system" for industrial mass production. Industrial skills led to another American genesis—the rapid development of farm machinery, which eased the chronic shortage of agricultural labor. Equally important, it greatly reduced the weather-related risks of farming while increasing yields by improving the timeliness and the quality of farm operations (106). Despite the importance of the tractor in this century, 1840 to 1860 was the great period of agricultural mechanization in America (77). Reapers, threshers, and stationary engines, which stand out for their revolutionary effects on the harvest of small grains, were accompanied by hundreds of new designs related to all aspects of farming.

The fertile soils of the American heartland, coupled with mechanized farming, have had far-reaching consequences. The large amount and low cost of grain allowed that area to begin to dominate world trade in those commodities at the end of the Civil War, and farming in many other regions became noncompetitive. Large areas of New England, for example, returned to forest. By 1900, England had turned toward grassland agriculture as the acreage given to wheat production by labor-intensive methods was cut in half and the rural population was sharply reduced (47). As the Atlantic coastal farmers moved west, they were joined by European immigrants with more experience in intensive agriculture (82). The new immigrants quickly adopted the corn-livestock system and local methods, including mechanization. They also brought an infusion of technology from Europe, including new genetic strains of crops. As the organic nitrogen reserves of the prairie soils declined, Dutch and German immigrants were among the first to turn to legume rotations.

The genetic introductions were particularly important. The early settlers were carried to the Missouri River by maize culture; but in Kansas and

Nebraska rainfall proved too sparse for maize, and wheat was adopted as the grain. These were traditional soft red spring wheats well suited to stone milling, but they proved highly risky in dryland agriculture. Drought risk could be reduced with winter wheat, but the available cultivars lacked sufficient winter hardiness. With periodic drought and severe winters, the early years of wheat farming in Kansas were characterized by cycles of expansion and bankruptcy (74). Russian-German immigrants who came to Kansas from the Crimea in the 1870s were more experienced in dryland agriculture. They based their fallow farming systems on hardy Crimean winter wheats, which persist today in the cultivar 'Turkey' and its descendants. These hard red wheats defied stone grinding, however, and were at first heavily discounted on the market. Steel roller mills eventually brought them to dominance as premiere breadstuffs and stabilized farming in the Great Plains.

Change has continued rapidly in the present century. The centers of cotton production have moved west to semiarid regions where yields and quality are superior to those in the old Cotton Belt. As a result, large portions of the Southeast have turned to grassland agriculture and forestry, and our black population has diffused into the North and West. Soybean, added in the 1930s as an oilcrop with useful oilcake by-products, now occupies nearly 30×10^6 ha in the Mississippi basin. Change is also evident in the tractor, in the expanding use of fertilizer, and in the improved efficiency of animals. Since the 1930s, the tractor has freed some 40×10^6 ha of land (nearly one fourth of US arable land) from the production of feed for horses and mules. Such innovations mark the obvious advances. Not so obvious are the advances that have come because farmers can now read and write and have a better understanding of farming. As much as one third of the difference between developed and developing countries in agricultural productivity can be attributed to education (81a).

Given this background to agriculture in North America, we can ask what features are really "traditional." Several characteristics emerge as long-embedded traits: the private ownership of land mostly in family units; labor shortages (and mechanization); the dominance of wheat as a food grain and maize as a feed grain for cattle and swine; commercial farming; and continuing technological change. In addition, American agriculture has developed with extensive rather than intensive management—large areas are farmed moderately without the extra inputs and effort needed to approach potential yields. The tradition of change is particularly important since it is the basis of our present adequacy of food and the means by which farmers remain competitive in a modern society.

DIVERSITY AND STABILITY

Genetic diversity in farming must be defined at several levels. At the field level, the degree of heterozygosity within cultivars, cultivar mixtures, and various forms of mixed cropping are important. Most agricultural cultivars, except for clonally propagated materials and closely bred hybrids, carry considerable genetic diversity, while land races (farmer-selected and maintained) and various types of synthetics can be quite diverse (84). Natural selection of such populations over generations can favor local adaptation, but it may also favor competitive ability (i.e. tall, rank vegetative growth) and thus work against desirable agronomic traits such as a large "harvest index" (25). In cereals, for example, the grain yields of earlier land races seldom exceeded 0.30–0.35 of total aboveground annual production. By careful breeding and selection, that fraction has now been brought to 0.45 or even 0.50 at about the same net annual production (3, 24, 80). Depending upon the mating system and selection pressure, such genetic shifts may occur rapidly or slowly (1, 25, 40, 90). In vulnerable crops, inspection and roguing of seed fields, frequent returns to foundation seed stocks, and seed certification now serve to minimize genetic drift.

Within-cultivar diversity and cultivar mixtures are sometimes viewed as improving stability by buffering environmental influences and reducing the impact of disease (32, 50, 60). In agriculture, relative variability in production (e.g. the coefficient of variation) over a number of years, rather than persistence (which is mainly a human decision), is the principal measure of stability. Yield variation is surprisingly small; coefficients of variation over a decade range from 5% to 10% in humid regions and with irrigation, and from 20% to 30% with dryland methods in semiarid regions. The evidence on the relationships between genetic diversity and variability is fragmentary and mixed. Depending upon the system and the causes of variation (e.g. disease, weather), mixtures may be marginally better or worse than pure lines. Much greater stability is more simply achieved by other genetic means—with winter rather than spring wheat in the Kansas example, with a safer cultivar differing in maturity date or having proper disease resistance, or with a less risky species. Quite often, the genetic materials are chosen so as to exploit the causes of variation. Plants with indeterminate reproductive habits represent an adaptation to environments with uncertain growth duration (or to gardens) (51). High stability (low year-to-year variation) could be achieved with a short-season, determinate plant, but this would sacrifice the opportunities for much greater yields in favorable years. Finally, management tactics with limited rainfall, for example—such as fallow, irrigation, modified land forms, and control of density (all involving nonuniformities in time or space)—have far more favorable and dramatic effects on yield and stability than does the genetic structure of the stand (57). Variation in yield does exist, but it is accommodated through storage, variations in use (e.g. increased feeding of animals), and year-to-year variations in the area and intensity of cropping.

Special problems in maintaining stability may arise with the introduction of new cultivars. The main limitation on the distribution of annual species is only that they must mature within the available growing season; as a consequence, new cultivars sometimes are adopted quickly over large geographic ranges. Such materials usually represent a sharp genetic advance with greater agronomic fitness in nearly all environments due to such simple and desirable attributes as an improved harvest index or increased efficiency in the use of water and nitrogen. In stability analyses (28, 31), there is a greater relative change in yield per unit change in environment—i.e. the new strains are less stable than earlier cultivars, but the yield each year, the mean yield over a number of years, and the yield per unit labor and capital expended all increase.

Much has been made of the "vulnerability" of farming to the general employment of superior cultivars (65). The epidemic of Southern corn leaf blight that reduced maize production by 10–15% in 1970 was due to widespread use in hybrid seed production of T-cytoplasmic male sterility, which carried susceptibility to *Helminthosporium maydis*. Although quickly corrected, that failure pointed to a danger inherent in the wide adoption of a single genetic trait. Such events have been rare, however, in part because one of the main traits contributing to wide adoption is superior disease resistance. Indeed, maize hybrids paid their dues hundreds of times through years of strong resistance to smut, stalk rots, and other diseases that plagued earlier land races. In addition, the basic concepts of new cultivars are diffused quickly into many cultivars. We now, for example, have hundreds of stiff-strawed wheats. Many are based on the Japanese 'Norin 10' stem dwarfism (79), which has shown no linkage to disease susceptibility, while others have different bases. Those wheats are sometimes referred to as "Mexican" wheats but are better described as "Indian," "Australian," or "Palouse," since they are products of largely local breeding programs. The assertion by critics (e.g. 91) that such high-yielding varieties (HYV) of wheat and rice "require" fertilizers for performance, and thus force social dislocations, is without foundation in fact. Since less photosynthate (and nitrogen) is allocated to stem tissue, more of the limiting resources are available for grain formation. HYVs can be said to have greater "nitrogen use efficiency" and typically are superior in yield at any nitrogen level.

A greater concern relating to improved genotypes is that they displace traditional cultivars and land races, and that without deliberate conservation, portions of the germplasm base may be lost. In the United States, we have developed cataloged collections with many thousands of entries for crops such as wheat (40,000). Collections for other crops, such as sugarbeet, are small, and with the exception of those for potato and tomato, the germplasm pools for horticultural species are mostly small. These living collections are costly to maintain, and the necessary periodic increases are subject to genetic drift. The alternatives include bulking into large populations and procedures for long-term storage. More serious is the lack of sufficient funding for plant breeders to work closely with the collections in order to identify major gene sources.

Disease epidemics can be countered in closely bred crops through further breeding or a return to older cultivars. Species substitution is also possible (although difficult with a dominant crop such as maize). A genetic shift toward resistance can also be expected (81), but this is generally not effective or rapid. When host and pathogen have multiple or alternative alleles representing resistance and pathogenicity, resistance or immunity to some of the pathotypes but not to others is termed vertical resistance (VR); general resistance to all pathotypes is called horizontal resistance (HR) (101). These may also be viewed as major gene and quantitative resistances, respectively. While HR may be reasonably durable, it is seldom as complete as VR. There are many situations where VR has proved exceedingly durable, and backcrossing methods can be employed to concentrate resistance genes at the necessary alleles into standard cultivars. Where resistance to only one or few pathotypes is possible, boom-and-bust cycles in production may occur as other pathotypes are selected and become dominant. These situations have been countered by sequential replacement of cultivars with appropriately chosen VR. Disease monitoring and careful management of the farmers' choices of cultivars are required.

An alternative based in epidemiological theory and employed by some plant breeders is to create multilines (i.e. mixtures of closely related VR and HR types) in an effort to slow the rate of secondary infections within fields (32, 50). The theory—that the disease may occur each year but at tolerable levels—has yet to be proven. Frey's multilines of oat (32), for example, have not been challenged by a major rust epidemic, perhaps because of the extensive eradication of the alternate host, barberry, and because cultivars of increasing resistance have been employed in source regions of the disease (creating a regional mosaic effect). Regional mosaics with "beet-free" periods in California to control beet yellows virus (26) have been highly effective.

Mosaic patterns also occur with species mixtures (within a community) or rotations of species in time and space (among communities). Species mixtures are widely used in forage crops and pastures in American agriculture. They serve as a low-cost means for accommodating habitat heterogeneity, but the main purposes are to improve the evenness of grazing (with cool- and warm-season species), to provide a nutritional balance for ruminants, and to introduce a low-cost source of nitrogen. Mixed cropping in rows and other patterns, as is done with corn and beans in the tropics, is not used in North America, despite the fact that squash-corn-bean complexes were widely employed by native Americans. The main reason, other than poor or reduced yields, is that the production costs are much greater.

A common assumption of alternative agriculturalists, derived in part from Gaussian exclusion notions (35) and arguments that "diversity lends stability" (36), has been that "overyielding" and improved stability should result from mixed cropping. Overyielding of species mixtures, as compared to the best pure stand, is rare, and yields are usually intermediate to those of the pure components (94). Overyielding requires resource differentiation in time (season extension) or, where soil resources are strongly limiting, in space (i.e. the species must differ in their ability to acquire soil resources). Space differentiation may occur in primitive systems strongly limited by nutrients so that interference for aerial resources (CO_2, light) is small because of the sparse canopy. Since higher plants all require the same basic resources, the possibilities of such differentiations are much smaller than for animals—especially in the short growing seasons of the temperate zone where the possibility of differentiation in time through differing phenology disappears.

An agriculturalist would view complementation based on differing root behavior as evidence of defective phenotypes and would look for a cultivar or species capable of exploiting the full profile. When soil resources are not limiting, a spatial basis for complementation no longer exists, regardless of rooting habit or capacity for nitrogen fixation. Under those conditions, the possibilities for differential use of aerial resources are slim to nonexistent. Monocultures are easily brought to complete cover (and 95% or more light interception) with leaf area indexes of 3–4 where variation in canopy architecture has little influence on the production rate (27). Thus, mixtures cannot exceed the gross photosynthesis of the most capable monoculture; in addition, differences in stature quickly lead to problems. Legumes, for example, disappear from forage stands because the more vigorous grasses block off the light. In an imaginary situation where maize is cultured with radish, interference would be strongly imbalanced, radish would lose, and the system would behave as a monoculture of maize. Even with similar stature, small grain mixtures are notoriously unstable in persistence over time (40, 84). We see here a corollary of the nonuniformity principle. When soil resources are nonlimiting, atmospheric factors become limiting. Other than changing the shape of the land to influence solar heating of the soil, these resources cannot be concentrated by nonuniform treatment in time or space—and uniform treatment gives the best result. Furthermore, if we consider the multiplicative increases in the difficulties of management with mixed crops, the much greater costs of harvest, the loss of rotation benefits, and the significantly higher cost of improvements through research, one cannot be optimistic about future possibilities for mixed cropping in North America.

Rotation of monocultures is a traditional and powerful tool for controlling pests and weeds, replenishing nitrogen (via legumes), and distributing labor requirements over time. It was not until the vast organic nitrogen reserves of the prairies (up to 10,000 kg N ha^{-1} in the surface 30 cm) were depleted to new equilibria in this century (46, 92) that American farmers adopted legume rotations. Between 1900 and 1940, set rotations such as maize-oat-clover and maize-wheat-oat-clover gradually appeared in the Midwest. (That was one of the few periods in American history that seems to fit Schultz's definition of traditional agriculture with little change: Maize yields remained static near 1800 kg ha^{-1}, and the United States became a net importer of agricultural produce.) Neither maize nor wheat is especially plagued by soil-borne diseases, so the principal advantages of these rotations lay with nitrogen supply, weed control, improved supplies of fiber for ruminants, and labor distribution.

With the advent of tractors, herbicides, and fertilizers, emphasis on traditional set rotations declined. Clover and oat are now less common (tractors don't eat oats), and mixed farming with cattle has declined in a shift toward commercial farming with maize and soybean. Labor saving is still a factor, since soybean is planted after maize and harvested before; but the rotations are now more ad hoc, depending upon the weather and the futures market. Soybean plays the role of an emergency crop when maize acreage is reduced by wet weather.

Surveys of world food supplies remind us that the great majority of human food is derived from a small number of species—about 66% from cereals and over 80% from just 11 species (58). Some people view this as a hazardous situation with possibilities for the kind of collapse that occurred during the Irish potato famine. Throughout its history, however, agriculture has been characterized by low species diversity. Cereals are emphasized because they can be stored and transported easily and because they have been remarkably stable over thousands of years. That maize, wheat, rice, sweet potato, potato, and sugarcane now appear together as carbohydrate sources in the Southeast United States represents a large increase in diversity for the region. The diversity of fruits and vegetables available to the nation is also much greater. While these seem a large portion of the diet and are important nutritionally, they are not our main sources of energy and protein, and only small amounts of land are involved. Less than 3% of US cropland is given to fruits and vegetables, with 1.1×10^6 ha in California (0.8% of US cropland) providing about 40% of the supply (59); the UN Food and Agriculture Organization's (FAO) statistics show that 2% of the world's cropland is devoted to fruits and vegetables (58).

Although similar to that of Europe, America's on-farm diversity of basic crops is clearly less than that in India and some other places. Three factors seem to account for this: Our temperate growing season restricts the possible number of crops; long-distance transport favors regional specialization; and grower "efficiency" is favored by low diversity. In 1870, apple trees were found in great abundance in nearly all of New York State (66). In recent times, this expanse has been reduced to two regional areas of concentration in the Hudson Valley and along the shore of Lake Ontario. Similar concentrations of orchardists are found in the Shenandoah Valley, Michigan, Ontario, and the Northwest. With cheaper transport and a market economy, apples are only grown now in regions where production and quality are high and only with growers highly skilled in apple production. In the early stages of our agricultural development, there was greater emphasis on subsistence; fruits and vegetables came mainly from local, in-season sources. Refrigerated transport (the "reefer" railcar appeared in 1870) has allowed remote regions to dominate the produce market, altering von Thunen's view (86) that perishables could only be produced close to markets. Greater on-farm diversity lowers grower efficiency because it requires increased investment in diverse types of machinery, greater knowledge of production technology and markets, increased management skills, and greater risk or smaller return. The last point may seem surprising until one considers that having chosen the crop with the best combination of lowest average risk and highest average return, the next choice cannot be as good. High risk can be tolerated only with high return.

The high diversity in Indian farming (a dozen or more crops per farm in some regions) reflects the long growing season, a great concern for labor costs, the closeness of markets (because of densely distributed populations and poor connections to distant centers), and a strong emphasis on subsistence farming. Market and home gardening become mixed with basic grain production. With little machinery, family labor may suffice for small acreages of many crops, but the peak labor requirements of larger acreages would require either cash expenditures for contract labor or larger families. What we see in America is little diversity per farm (unless the household garden is included) but great diversity for the nation.

PLANT NUTRITION

Agriculture is an extractive activity. The biomass removed as economic yield contains soil nutrients at levels characteristic of the types of plant tissues involved. Since the possible range of nutrient contents in a particular material is small, there is an almost stoichiometric relationship between yield and nutrient extraction. The yellow dent maize grain of the Corn Belt is typically 9% crude protein (1.5% N), 0.4% K, and 0.35% P on a dry basis (64). The current annual production of dry grain in Illinois is 28×10^6 t from 4.4×10^6 ha, removing annually an estimated 4.2×10^5 t N, 1.1×10^5 t K, and 1.0×10^5 t P. The mineral drains are small, yet they are more than continuing soil formation (weathering) and natural inputs (dust, precipitation) can replace. The average removal of nitrogen is 96 kg ha^{-1} yr^{-1} placing a heavy burden on recycling and external inputs for maintenance of an equilibrium in soil nitrogen.

Examination of such yields provides the clearest definition we have for the "intensity" of farming. Following FAO practice (17), the 6400 kg ha^{-1} mean yield of maize in Illinois is the "actual" production from the system and reflects the average levels of farmer skill and technology used for those climates and soils. Yields obtained by the best farmers, in the local "5-Acre Yield Tests," and by experiment stations range from 10,000 to 13,000 kg ha^{-1}; they provide an estimate of FAO's "attainable yield," reflecting a combination of the best skills, the best available technology, and to a lesser extent the best lands. Record yields and theoretical predictions provide the standard of "potential yield" (ca. 16,000 kg ha^{-1} for Illinois) for comparison. That potential crop would carry 240 kg N in the grain, while perhaps 100 kg N (equivalent to 20 t of fresh ruminant manure) would recycle to the soil in residues.

Clearly, Illinois farmers are not engaged in truly intensive production. Their practices are more intensive than those in Kenya, where 800 kg ha^{-1} is common with primitive methods, and in Mexico, where 1400 kg ha^{-1} is the national mean. Nutrient stress is the principal determinant of those differences. Natural inputs by free-living fixation and atmospheric transfer are small (5–15 kg N ha^{-1}yr^{-1}; see 56, 92), so grain yields near 1000 kg ha^{-1} (15 kg N) are near the maximum that can be achieved for continuous culture on depleted soils without external inputs (22, 33). [The seed increase rates are also low, so the seed itself provides a significant portion of the nitrogen supply (87).]

The equilibrium yields of primitive systems can be improved in several ways. We have seen that grazing animals can be used to refuge nutrients from surrounding areas. Large losses of nitrogen from manure by leaching and gasification (ca. 50%; see 5), coupled with export in animal products, creates the need for large ratios of refuging lands and animals to grain land. A quantum jump in production can also be achieved by rotating grain lands with leguminous forages. Yields can be pushed even higher by adding manure derived from forage. Alternative farms in Europe average about 3 ha in pasture and legume forages for each ha of crop (9). Such ratios strongly limit the amounts of grain that can be produced in a given region by organic methods. The effects of legumes and manures are seen most clearly in long-term rotation studies such as those at Rothamstad, England (130 yr, wheat; see 34), at the Morrow plots in Urbana, Illinois (100 yr, maize; see 105), and in recent work in New York (6). Selected data from the Morrow experiment are given in Table 1. Additions of lime (pH control) and phosphorus are generally necessary for adequate growth of the legume. Productivity of such organic systems is also limited by the facultative nature of legume symbionts. As the background nitrogen in the soil increases, fixation is repressed (2, 30), so that the legumes grow on mineral nitrogen rather than by fixation. As a result, one cannot achieve a sufficient nitrogen level to exploit the full photosynthetic potential of the grain subsequently planted on the same land. In the corn-soybean rotation used in the Midwest, the legume is actually a scavenger of residual nitrogen from the maize. The net removal of nitrogen from Illinois soils by soybean is the same as for maize (48).

A good alfalfa crop in a long-season environment (e.g. California) may sustain a fixation rate of 500 kg N ha^{-1}yr^{-1}, since that amount can be removed

TABLE 1   Maize yields, 1955–1975, in the Morrow plots, Urbana, Illinois (105)[a]

Soil treatment	Continuous maize (kg/ha)	Maize-oat-clover (kg/ha)
None	2600	4800
MLP	5600	8600
MLP–LNPK	7800	8400

[a] The experiments were begun in 1876; manure equal to fed produce, lime, and phosphorus (MLP) were added beginning in 1904; some of the MLP plots were changed to lime and NPK fertilizer (LNPK) in 1955. Soil nitrogen continued to decline during this period under continuous maize with no soil treatment (92). Attainable yield for this location is near 13,000 kg per ha (93).

annually in repeated forage harvests. The nitrogen remaining in the legume roots and stubble, however, is less than 200 kg N ha^{-1} at the end of legume culture (P. Meyerhoff, personal communication). Seasonal removals and residuals are much lower with shorter seasons (6, 43, 53). The nitrogen supply from legumes can be increased by incorporating shoot material into the soil as "green manure." That means that the entire cropping season must be given to nitrogen production, except in mild climates where *Vicia* spp. (vetch, faba bean) can be grown during the winter period without conflicting too much with economic crops. Vetch may yield 45 kg N ha^{-1} in California rotations with rice (107), but the current cost for seed alone exceeds the value of the nitrogen obtained. Vetch plus natural inputs support yields near 3000 kg of paddy rice per ha (worth $400), while the California state average using about 170 kg fertilizer N per ha (worth $85) is now near 7000 kg ha^{-1} (worth $930; J. Hill, personal communication).

The response ratio in the rice example to the last 100 kg N, 32 kg of grain kg^{-1} N (or about 10 cal combustible product per cal fossil energy expended), is not unusual for grain crops (6, 93). The use of external resources to alleviate nutrient stress converts the land from a place for gathering nitrogen to one for gathering solar energy and carbon. Stated another way, fertilizer substitutes for land. In the rice example, over 10 ha are replaced per ton N as compared to vetch. Heady (42) cites examples for developing countries ranging from 11 ha of wheat per ton N in Argentina to 53 ha of rice per ton in Peru. A given market requirement (number of people × annual consumption) can now be met through intensive farming of small hectarages. The conceptual basis of carrying capacity has changed. Through intensive management of their small land base (nearly 7 people per ha of agricultural land; 1800 mi^{-2}), the Dutch are able to produce caloric and protein foods in excess of their annual consumption (C. T. de Wit, personal communication). Despite their apparent disadvantages in agriculture, they are plagued with surpluses.

At the same ratio of 7 people ha^{-1}, the United States could support about 1.3 × 10^9 people or nearly 6 times our present population. Buringh & van Heemst (12) have analyzed that question more carefully through climate and soil

assessments for each continent. Using a conservative model of grain production, their group arrives at an estimate of maximum world production of 50 ×
10^9 t grain equivalents per year (vs 1.7 at present) by applying present technology to the maximum possible amount of land (25% of total land area compared to
the 11% now farmed); 7 × 10^9 t would be produced in North and Central
America on 620 × 10^6 ha (compared to 240 × 10^6 ha at present). The "standard
human nutrition unit" of 2 × 10^6 kcal yr^{-1} per capita is equivalent to 0.5 t grain
(1.37 kg grain, 5480 kcal, and about 140 g of protein per day) and includes
adequate allowances for seed supplies, storage losses, and alternative foods. At
0.5 t yr^{-1} per capita, 50 × 10^9 ÷ 0.5 = 100 × 10^9 people is an estimate for the
maximum carrying capacity of the earth with conventional agriculture supplied
with adequate energy, labor, and capital. Buringh & van Heemst (12) are more
cautious, however. They reduce potential grain production by half (to less than
the present average in Illinois) as the amount attainable by average farmers with
present crops and allow 0.34 of the cultivated land to be used for nongrain crops
(as at present). With those and other adjustments, they arrive at 13 × 10^9 t (i.e.
26 × 10^9 people) as a limit. Restricted to present lands (1.4 × 10^9 ha), those
methods would support about 10 × 10^9 people.

I have emphasized the restraints on production due to nitrogen. At present,
phosphorus and potassium are replaced by mining geologic sources. Supplies
of potassium from the high grade ores and recyclable sources (oceans) are
essentially infinite. There is more concern about phosphorus. Sources of high
grade ores are limited to the United States and Africa, low grade sources are
more plentiful, and recyclable sources (sewage, oceans) will require large
amounts of energy to process.

American farmers and policymakers have yet to appreciate these concepts
fully. They are surprised to see France and, on occasion, India export wheat
and to learn that Taiwan and Japan have very large reserves of rice. The
developing countries of the world have large resources of infertile soils whose
response ratios to added nutrients are much larger than remain to our farmers.
Given a conducive social organization, they could be highly competitive in
acquiring the energy resources to satisfy their own food needs and to enter
world grain markets.

ENERGY RELATIONS

Pimentel et al (71) made the first careful (and widely influential) study of the
energy costs of farm production. They expressed the result as an energy ratio
(cal in combustible product out to cal fossil fuel in). For maize production in
Iowa during the transition from traditional agriculture (1950–1970), the ratio
remained relatively constant at about 3, while the yield doubled. The fossil
energy input in 1970 was 7.2 × 10^6 kcal ha^{-1}. A new look at Iowa today would

undoubtedly reveal a larger ratio due to the greater use of fertilizer, diesel tractors, and minimum tillage. Pimentel et al (70a) consider some of these factors in calculating a ratio of 4.5 for 1980. A legacy of such studies is the feeling that the ratio should be >1.0 (as it is for the forestry and fossil fuel sectors of the economy), and this view has become a keystone among advocates of alternative agricultures.

With the aid of better data and methods, a number of points in the original Pimentel study can now be criticized. Doering (23), for example, calculated that machinery manufacture and repair costs equal only 10–15% of fuel costs, not 52%, because he limited machinery to that actually used on an average farm and credited the scrap value. Doering's ratio for a case study of maize production in Indiana was 6.3. Pimentel et al (71) concluded (their Figure 3) that the maximum energy ratio would be obtained at less than optimal nitrogen levels. That result was derived from a poor-fitting quadratic yield function, while a sharp-breaking plateau fits the original data (63) better. In line with other results (59), both yield and efficiency reach their maxima at the same input of nitrogen when the latter function is used.

A more serious problem common to both the Pimentel and Doering studies is that only the energy cost of actual labor time is counted. The entire cost of labor would include resting time, management time, replacement, retirement, and spouse support (in line with feed budgets for herds). Such accounting dramatically lowers the energy ratios for labor-intensive systems (49, 75). For subsistence agriculture, placing humans in the system actually brings the ratio to 0, since there is no output unless human capital increases. The American farm also uses external inputs for family living. In 1970, energy use per US farm household for heat, electricity, and transport was a modest 47×10^6 kcal yr^{-1}; yet added to the approximate 12×10^6 kcal used for management, it equalled nearly 43% of all energy consumed per farm (29). The inclusion of household energy has a particularly devastating effect on energy ratios for small farms. Labor and management, it seems, are costly in terms of fossil energy. Given our standard of living, fossil energy may be conserved by replacing labor with machines. De Wit & van Heemst (22) calculated a marginal substitution ratio of 30×10^6 kcal fossil energy per man-year for Dutch wheat production.

The embodiment of such indirect energy costs is greatly facilitated by input-output (I-O) accounting in each supporting sector of the economy. Using the Illinois I-O model for the American economy (45), Avlani & Chancellor (4) developed an elegant energy analysis for wheat production in California. The statewide energy ratio for production is 2.0; processed to flour and prepared as toast, the ratio drops to 0.6. Using a similar approach, Costanza (19) evaluated the entire US agricultural sector, including the embodied direct and indirect energy costs of labor and government services in all contributing sectors. Remarkably, he was able to include 87 sectors of the American economy, including agriculture, in the same linear regression of value output vs direct plus indirect energy input (his Figure 4C). The remaining five sectors (including petroleum production and coal mining), which represent net energy sources to society, fell as outliers. Agriculture thus behaves as a highly integrated part of the larger economy and not as an energy system.

A practical way of gaining some perspective on these energy budgets is to compare the numbers with everyday experiences. In fuel and electricity, the average home refrigerator uses 1×10^6 kcal yr^{-1}; my university uses 19×10^6 kcal yr^{-1} per student (or 340×10^6 per professor), one passenger hour in a DC-9 uses 0.3×10^6 kcal (KLM data), and 10,000 mi yr^{-1} in a private car at 25 mi gal^{-1} uses 15×10^6 kcal. It also helps to remember that we do not construct energy budgets for the publishing industry, although books can be burned, or for steel filing cabinets, which cannot. The surprising thing about energy analyses of agriculture is not how much energy is used in production but how little. In the US less than 3% of current energy use goes to production, including that for export; for the world, the proportion is about 5%. If all recoverable fossil fuels were devoted to food production, we could sustain the present system for many thousands of years.

It is difficult to assess how agriculture would respond to sharply curtailed supplies of external energy. A complete breakdown in energy resources would dictate a return to the "new farming systems," which use animal power. Applied to all possible lands (18% of the world's land area), farms could support about 5.3×10^9 people at 3300 original kcal day^{-1}; restricted to present lands, the labor-oriented methods would feed only 3.3×10^9 people on a minimum diet, 1.3×10^9 less than our present population (12). The United States, with a large land base, would not suffer a population loss. In a less technical analysis, Olson et al (68) reached a similar conclusion. Their approach is faulty however; their grain yields are estimated after adding a genetic advance to those obtained in 1944–1946 when no nitrogen fertilizer was used and only a small proportion of the land was in legumes. By that means they invent imaginary annual sources of nitrogen equal to ca. 15 kg ha^{-1} of maize and 10 kg ha^{-1} of wheat. That mistake is common to even the best analyses of genetic advance (14).

Having only intermediate levels of external energy would lead to a number of adjustments. Food processing, transport, and distribution may be more vulnerable to change than production since they use more energy, but that is offset somewhat by the greater efficiency of central cooking compared to home preparation. Other than to support labor, the largest amounts of the energy in production are devoted to fertilizers and fuels. During the energy shortages of the 1970s, American farmers increased their use of fertilizer (thus improving the energy ratio relative to the fixed costs of crop establishment and harvest), increased the proportion of diesel power used, and made a dramatic shift from moldboard plowing to "conservation tillage" with chisel plows. Surprisingly, there was little change in the hectarage of legume forages, even though the amount of land in production increased significantly. Several factors contributed to those trends: Natural gas prices were regulated so the cost of fertilizer remained low; rapid genetic advances in ruminants led to a continuing decline in the amount of forage needed, so hay prices remained low; and export markets, and thus prices for grains, were strong. Under different circumstances, an energy shortage should favor an increase in traditional rotations with legume forages. It is important to recognize, however, that the energy savings with legumes are not large. The energy used in crop establishment and forage harvests for alfalfa amounts to nearly 3000 kcal fossil fuel per kg N in the forage (calculated from 15). After losses in manure recycling, the cost would rise to perhaps 10,000 kcal kg^{-1}, compared to 14,000 kcal kg^{-1} in fertilizer.

Energy will have to be extremely scarce before there would be significant demechanization of US agriculture. Buringh & van Heemst (12) estimate that without machines or fertilizer, 50% of the world's population would have to be engaged in production. The large labor costs are accounted for in van Heemst et al's study (102; see Tables). Given that 100 hr t^{-1} is the time required for hand harvest and threshing of small grains (vs 80 hr with animal power and less than 1 hr by combine), it becomes clear why primitive systems depend heavily on child labor and why most Amish allow themselves a stationary engine and take pride in the modernity of their machines (67). In addition to the labor problem, perhaps a quarter of the land would be needed for feedstuffs for horses and mules. Land requirements would be smaller with oxen, but the labor requirement would be greater because of their slower pace.

The implications for child labor have not been addressed clearly by sociologists. They have recognized the economic advantages of having large farm families and the unusually high proportion of such labor in agriculture (70% of documented child labor in the United States in 1930, see 85); they often describe this as child work and as a healthy benefit of rural life. The large differential in fertility between urban and farm families is well known [0.85 more children per ever-married farm woman in 1960 (41) and 0.59 in 1981 (100)], but the effect of urban/farm differences is confounded with differences in education levels, alternative pursuits, income levels, and class. Farm fertility has declined dramatically during the past century in a pattern that would correlate closely with increases in mechanization and farm labor productivity. Some evidence for the existence of such a causal relationship is provided by the continued high fertility of agrarian societies in developing countries with little mechanization. This suggests that mechanization may be necessary for a demographic transition to a low birthrate.

Using a generous definition of *farm* (any rural place that sold something), Banks & Beale (7) found that the American farm population fluctuated between 30 × 10^6 and 32 × 10^6 persons during the traditional period of 1920–1941. Since then, the population has declined more or less steadily to 9.7 × 10^6 in

1970 (7) and 6.9 × 10^6 in 1980 (99). The net decline from 1920 to 1970 was 22.3 × 10^6; but during the same period, the excess of births over deaths in that population was 16.3 × 10^6, so the total movement from farms was 38.6 × 10^6 (7). In the stable period between 1920 and 1941 our farms yielded an annual crop of surplus people equal to 1.6% of their population. It seems a reasonable hypothesis that a population surplus is an intrinsic feature of labor-intensive family farming systems. My own experiences in Iowa conform with a study of rural Michigan (78): Rural youth grew up with the expectation of migration and looked forward to it in a positive way. Iowa, with its early emphasis on quality education, has populated urban centers from Schenectady to Los Angeles. In developing countries, the annual surplus of rural youth is poorly skilled and educated, and their exodus seems inexorably entwined with poverty and social decay. It was this situation that Schultz (83, p. 4) addressed with the remark that "a country dependent upon traditional agriculture is inevitably poor."

The net decline in farm population of 22 × 10^6 people from 1941 to 1970, although magnified by the generous definition of farms, can be attributed to changes in the structure of farming, including mechanization and fossil energy uses. The bulk of the migration occurred in the South (13.3 × 10^6) (7) and included 5 × 10^6 black people (99). That period was marked by a sharp decline (-3.4 × 10^6 during 1945–1978; see 98) in semisubsistence small farms with <20 ha of harvested crops. Mechanization of cotton harvests, which allowed higher quality Western areas to expand production, may be cited as a contributing factor, but rapid increases in the use of synthetic fibers and devastating soil erosion probably had a greater impact. Even more important were the higher wages and better quality of life in the urban centers (7); blacks in particular were released from rural poverty by more equal opportunities in education and employment in the cities. These statistics highlight a point of sharp contention between sociologists, who would enumerate most rural inhabitants as farm population regardless of their production (10), and economists, who tend to restrict their view to those who produce (96). By economists' standards, the declines in the numbers of "working" farms and farmers are much smaller.

While it is common to hear mechanization discussed in terms of "labor displacement," for the United States it is more accurate to say that machines replaced the labor that moved to the urban sector for higher wages. Even today there is a large earnings differential in favor of urban people. There simply is little chance in our society to achieve a living wage through hand harvest of maize or hand milking while others work in automated warehouses and at computers. Mechanization is a principal reason returns to rural labor have remained reasonably competitive (18, 61). As Poleman (73, p. 518) notes in relation to labor-intensive agriculture, "the noble peasant is a rich man's delusion." We can expect machines to become more efficient, to be used more efficiently (69, 97), and to become "smarter" (16). We can also expect them to continue to be a basic element of agricultural technology.

Approximately one third of the decline in farm population since 1941 is attributable to changes in the livestock industry. Automation of the poultry industry accounted for a portion of this share, but the main causes were improvements in the genetics and nutrition of dairy cattle. Annual milk production per cow has increased nearly three-fold since 1960, and the increase per unit feed is nearly as great. Pimentel et al (72) examined the constraint that limited energy would place on livestock production in the United States. Considering only one option—a switch to maximum grain export and "grass-only" ruminant production—they predicted a sharp drop in animal products, particularly of poultry and swine. They omitted grass and legume forages from the feedstuffs, and it is not clear how they thought to overwinter ruminants without hay. Hay is now produced on 57×10^6 ha with an annual yield of about 110×10^6 t dry matter and 15×10^6 t protein. Their assumption that grain would be exported rather than fed to cattle is unrealistic. If we are to export, it makes more sense to export animal products and keep the labor, capital, and manure at home. Ward (104) presented a series of options (for moderate constraints on energy) that seem to provide better clues about the response of the livestock industry.

The Pimentel et al analysis addresses the perceived "sin" of grain feeding; but to do that properly, variations in ruminant digestion should be considered. Ruminants have excellent abilities as grazing animals and for converting cellulosic roughages to human foods, but digestion efficiency and the end products vary considerably with diet (103). Grass-fed dairy cattle tend to produce high-fat milk; but with supplemental grain, the milk yield increases and fat content declines (62). Conversely, grass-fed cattle have low levels of body fat. Animal products now supply 33% of our dietary energy (72). Protein is a poor source of energy, however, so a corollary to the concept of "empty calories" is "empty protein" [see Speth & Spelman (89) for an interesting perspective on the value of animal fats in human diets].

Ruminant conversion of plant protein to meat protein is poor (ca. 5%), but conversion to milk protein is efficient (ca. 29%) (13). The recovery of plant protein depends upon the proportions of grains and roughages (20). Taking feed grains as "human edible" and roughage as "nonedible," Bywater & Baldwin (13) found that the use of human-edible grains in standard rations gave an increased return of as much as 1.8 times that amount in milk protein with dairy cattle and 1.1 times as much in meat protein with beef cattle (Table 2). With grain as an energy source, the rumen bacteria recover more of the protein from roughages, so grain feeding of cattle amplifies the human food supply.

Confinement feeding provides other advantages as well: The energy (feed) cost of foraging and disease recycling are eliminated; "least-cost" rationing (20) allows optimum use of other feedstuffs (by-products and wastes); and forage and grain crops can be grown under optimum management without trampling, so that yields and quality will be higher than in pastures. These

TABLE 2 Example calculations of inputs and relative returns for various animal production systems based on practices common in California[a]

| System | Inputs per slaughter animal | | | | Output of human edible relative to inputs of: | | | |
	E_T (10^b kcal)	E_{HE} (10^b kcal)	P_T (kg)	P_{HE} (kg)	E_T (%)	E_{HE} (%)	P_T (%)	P_{HE} (%)
Dairy	20.0	4.6	700	112	23	101	29	181
Beef								
A	21.6	1.5	750	31	4.9	73	5.8	139
B	18.7	1.4	650	31	4.3	60	5.9	126
C	20.6	1.9	820	40	5.2	57	5.3	108
D	21.1	2.9	730	64	5.1	36	6.0	68
Swine	1.5	0.6	66	29	23	29	38	86
Poultry[b]	2.3	1.1	120	48	15	31	30	75

[a] The costs of the entire life cycle including the maintenance of the herd or flock are included. The total inputs of digestible energy (E_T) and protein (P_T) represent all the materials consumed; the "human-edible" inputs (E_{HE} and P_{HE}) are the feed grains. The four beef strategies involve different ways of finishing grass-reared, yearling animals. Dairy calculation is annual budget per milk cow. Adapted from (13).
[b] Inputs per 100 birds; output as meat.

advantages have been apparent throughout our agricultural history. Societies with large population densities and/or a small proportion of grazing lands emphasize confinement feeding of swine and poultry. These animals require a much greater proportion of grain in their diets than ruminants and thus are more competitive with humans, but they have reasonable feeding efficiencies (over 30% of plant protein is converted to animal protein with 80% of the human-edible returned; see Table 2). With high reproductive rates, they effectively buffer the human food supply by consuming surplus and waste grain. At present, with low cost grain, the United States emphasizes swine and poultry as well as ruminants. Given our abundant grazing lands suitable for ruminant reproduction, a scarcity of energy might shift the balance toward ruminants with an important role for confinement feeding. However, it is the poor performance of beef animals on pasture and forages, rather than their efficiency in the use of grain, that now limits their use. If we are to have a significant diversion of grain land to pasture and leguminous forages, the expense will have to be met by a much greater value for the nitrogen obtained, or a much greater efficiency of roughage conversion by the animals. Otherwise, we can expect dairy products, poultry, and swine to gain in relative importance.

SOIL CONSERVATION

Soil erosion is among the most serious and long-standing issues facing American agriculture, and it is also one of the more nebulous (52). The principal losses are a result of runoff, but wind can also be a significant factor (69a). Concerns over erosion losses were evident among colonial farmers (18), but little was done except to move to new sites and cultivate hillsides less frequently. Settlement of the West reminded people of the finiteness of the land and helped erosion and land conservation become activist political issues between 1890 and 1920 (44). That activism was part of a larger movement concerned with a broad array of social ills, a movement that drew its main strength from sources other than the farms. The subject emerged again during the 1930s (because of drought and the Depression) and became institutionalized in the federal Soil Conservation Service (SCS) and the antecedents of the Agricultural Stabilization and Conservation Service (ASCS) (37). SCS is basically an advisory and planning agency, while ASCS administers action programs through local committees, making payments to farmers for approved practices in conservation and production. The payment programs, while assuming the attributes of pork-barrel politics, appear to recognize that land stewardship is in the public interest, but this principle is far from clear (37). At issue are farmer's responsibilities for externalities, the public's interest in future productivity, and the stream of income that land ownership deserves (11).

Dramatic examples of erosive gullying in the Piedmont and the loessal soils of western Tennessee, western Iowa, and the Palouse are numerous. The Piedmont represents perhaps the most extreme case of regional erosion—an average of 15 cm of soil was lost during 100–200 yr of farming by traditional methods (ca. $10–20 \text{ t ha}^{-1}\text{yr}^{-1}$), with much of the land destroyed by gullying (95). Erosion rates elsewhere are much lower, but precise quantification has proved difficult. The usual method involves a predictive calculation of sheet and rill erosion with the Universal Soil Loss Equation (USLE) (52, 108), which was developed from early studies by SCS (never heavily endowed for research) with strip watersheds at 49 locations. The USLE is a statistical regression, derived mainly from ca. 10,000 plot-years at those sites, involving a slope and its length, soil erodibility, rainfall intensity, cropping practice, and preventive measures such as terracing. The *Transactions of the ASAE* provide numerous examples of the continuing updating and validation of the USLE. The cropping factor is complex since erosion varies with the type and direction of cultivation, the surface residues, and the vegetative cover during each season of the year (97). The predicted "loss" is perhaps better termed an index of predicted movement, since the major portion of the moved soil remains on the land. The bed and suspended flows of streams and rivers are derived more from down-cutting of banks and channels and from gullying. The flows delivered to the oceans from the continental United States (including dissolved minerals) are an order of magnitude smaller than the average USLE erosion rate (49a, 52). Moreover, while predicted movement may provide reasonable estimates for a region (i.e. for a watershed or for one of SCS's Major Land Resource Areas), close agreement with a particular site is rare (52, 108), and the equation has not been fitted to Mediterranean climates (59).

In apposition to the loss estimate is the *tolerable rate of loss*. This term unfortunately has little foundation in research. It hinges mainly on depth of profile and nutrient content. Deep loess is assigned a large tolerance; shallow soils with infertile substrata have only small tolerances. Rates of soil formation are not considered. While recovery of the organic content of surface horizons can be rapid (0.02–0.05 yr^{-1}; see 6, 92), the weathering of parent materials and the formation and movement of secondary clay minerals have smaller and more uncertain time constants (8, 39, 46). Older (traditional) views of soil quality placed heavy emphasis on the top soil and its content of organic nitrogen content. With the possibility of replenishing nutrients from external sources and restoring humus quickly, the criteria of excellence are shifting from nutrient content to water-holding capacity. Pierce et al (70), for example, used the latter in developing a new method for defining tolerance. Regardless of the method by which it is estimated, tolerance depends upon the planning horizon since the loss leading to an acceptable reduction (say 4%) in future productivity is finite. For long horizons (200 yr), the permitted annual loss becomes so small as to preclude the acceptability of farming on many sites with more than a 6% slope, while most lesser slopes can be farmed indefinitely.

Uncertainties about the loss equation, tolerance specification, and future values leave open many questions. The reassuring fact is that only 30% of our cropland is on slopes greater than 6% (52). Much of that is loess with enormous tolerance and/or is under some type of conservation practice. But it is also clear that the estimated erosion rates there are too great according to present criteria. "No-tillage" methods, despite problems in nutrient cycling and increased use of herbicides, may represent an alternative for hill lands (69, 97), but the best alternative is to return them to grass.

Similar uncertainties exist for the landowner. The present worth of possible future benefits invariably proves to be much less than the costs of conservation (11). Some widely used conservation techniques cost little or at least less than traditional methods (e.g. contour farming with chisel plows that leave heavy surface residues). Others, such as drainage, terracing, and more frequent use of sod crops on sloping lands, involve either large investments or significant reductions in income. The use of legume forages (a sod crop) in traditional rotations is one of the most effective means of reducing erosion on sloping lands (108), but changes in the animal industry and our recent emphasis on grains for export plus conservation tillage methods have expanded grain production on hill lands. (In Iowa, over 1×10^6 ha of permanent pasture was converted to cropland between 1970 and 1980.) It is difficult to be enthusiastic about further expansion of grain production (e.g. for gasohol or fuels from biomass). From a farmer's perspective, however, high grain prices represent a temporary opportunity—hill lands can always be returned to productive grass in the future. In fact, the chisel plow, which permits a few years of safe cropping on slopes as great as 15%, has become a powerful tool for renovating degraded hill pastures.

There is general agreement that, without government payments, economic incentives for conservation are small (37), yet conservation practices are used on a large percentage of the susceptible land. The land ethic is still alive. While the type of ownership (e.g. landlord, owner-operator, corporation) has little influence on erosion rates (54), farm size does have an effect. In one study, large farms with a mean size of 280 ha were found to have adopted minimum tillage methods more quickly than smaller farms (mean of 56 ha) (55). Nationally, larger farms have less erosion (54), while small farms receive the bulk of conservation payments (37). We can speculate that the larger farms have greater access to information and capital; they may also occupy the safer lands. It is also much easier to specify and implement a conservation plan for a large farm. Field sizes and shapes can be altered to isolate problems, the entire area needing treatment is more apt to be within the same ownership, and the sacrifice of grassed waterways and unfarmed areas has a smaller relative effect on income.

SCALE AND OWNERSHIP

Land distribution in colonial times followed several patterns, and there was a large range in the size of holdings (18). This variation was a product partly of the great diversity among immigrants and partly of the distribution of land by the proprietors of royal grants. William Penn obtained personal title to all of Pennsylvania in payment for a royal debt, while Georgia, the last of the royal colonies, was placed in the hands of James Oglethorpe and other trustees for the settlement of persecuted Protestants and worthy indigents. A different pattern emerged as the western lands were opened by fee and grant, first in units of 16–130 ha (40–320 acres) and later of 65 ha (160 acres) (76). A 65 ha farm required substantial capital and considerable time for development (18), so farm sizes remained static for a time. However, family farms also included large plantations in the South and 20,000 ha Mexican grants in the West. Contrasts in the scale of arable farming continue today. Some of the differences in scale are related to site productivity and cropping patterns—50 ha of grapes in California, 1400 ha of fallow wheat in western Kansas, and 180 ha of maize in Illinois will generate similar sales values.

Americans have always felt ambivalent about farm size. We seem to admire the industry required to amass and manage a larger holding, but we are also concerned about the concentration of land wealth. That concern may have some of its roots in Adam Smith's view that nature's efforts gave added return to human labor only in agriculture (88), leading David Ricardo to formulate the concept of land rent, whose value would be proportional to the scarcity of food. Coupled with Malthusian concepts, one can paint a picture of wealthy farmers as the final bastions of affluence among the starving masses. Marxists employ the concept of Ricardian rent in justifying collective farms, but they have yet to reach a consensus about small farms. Some argue that such petty bourgeois (i.e. small farmers) really only work for wages and thus are a special proletariat (21). The American attitude was influenced more directly by Thomas Jefferson's belief that wide ownership of land would provide the best protection of other liberties (76).

In 1978, the 126,500 farms in Iowa averaged 108 ha in size, but they included 33,700 farms (27% of the total number) with an average size of only 15.5 ha, or less than 4% of the state's farmland (98). In contrast, 86% of the farmland was held by 68,900 farms (54% of the total number) with an average size of 170 ha. Even these larger farms were small by business standards, since few would employ more than one hired laborer. There were 42,600 farms (34%) with sales (i.e. production) of less than $20,000, contributing only 4.2% of the total agricultural sales. Income can only be estimated from the Census data, but the moderate-sized farms (170 ha) appear to have averaged $25,000–$30,000 net income in 1978 (a good year) while the quartile with the smallest sales had incomes below $1000. The return on equity for all farms, with no allowance for owner labor or management was above the national average of 3.8% but far below the return in the industrial and service sectors. It appears that most farmers discount the rental value of their land and view it as a long-term trust for children or as an eventual capital gain. These data for Iowa are representative of the national situation. In 1982, 60% of our 2.4×10^6 farms had sales between $1,000 and $20,000. These small farms lost 1.1×10^9 in farming but gained 27×10^9 in nonfarm income. This classification of farms is expanding. It includes hobby farms, retirement acreage, part-time farmers (berry and fruit gardens, etc), and estate remanents, and thus represents an urban extension; but its real nature is unknown. The very small income from small farms reflects the low income potential of the land, a greater amount of depreciable improvements (barns, wells, fences, drains) per ha, and the fact that a greater proportion of the landscape is given to farmsteads, access roads, and fences, or is marginal for farming. At the other end of the scale, 1% of the US farms had 30% of all sales in 1982 (a poor year) and 60% of net income. These included the large, vertically integrated (including distribution and marketing) fruit and vegetable operations and large farming and ranching corporations. Also included are the large feedlot and poultry operations, which are major purchasers of farm and ranch commodities. Their sales thus represent a double accounting of farm sales. Again, the true picture is unclear.

Similar diversity exists in ownership and has remained unchanged in recent decades. Family and partnership units (82% of land) coexist with "corporate" and communal approaches (98). Corporations held 12% of the land in 1978. Most such corporations were actually family farms with only a few stockholders; other corporations accounted for only 1.5% of the land. The marginal value of land to family farmers (i.e. to neighbors and young farmers) is greater than its current income potential. Farmland, a nondepreciable asset, continues to be a poor investment for outsiders, particularly when it is leveraged (i.e. when returns are well below the discount rate).

Regardless of ownership, interesting scale relations seem to exist related to the information problem outlined earlier. Large economic units allow for the presence of specialized management skills at the center—from bookkeepers and other financial experts to professional agronomists. While the large unit may excel in central planning and financial and labor management, actual field operations and the associated detailed field information may become remote. Moderate-sized economic units suffer some from dependence on ad hoc advisory services (annual visits to the tax accountant and farm advisor) but gain an advantage in field management because of the greater skills of owner-operators.

At their career peaks, owner-operators seem more than a match for the larger units. Problems arise in reaching that peak and maintaining it, however, and it is not surprising that most successful farmers served long apprenticeships and had financial backing from parental farmers. The increasing need for business skills is being felt even here, though, and the type of university education now sought by farm youth emphasizes agricultural economics rather than plant or animal science. One can predict that relative technological competence will not be as important in the future as in the past. The influence and complexities of both micro- and macroeconomics have become dominant. Increasing depreciable capital and operating expenses are one aspect, but tax complexities, government programs, and regulations have become a major concern. Environmental and social regulations add a significant management burden, and tomorrow's farmers are abandoning courses in entomology in favor of those in business law.

CONCLUSION

American agriculture has undergone continuous change since its inception. The demographic and technological changes of the past 40 years are particularly striking. Widespread use of tractors and nitrogen fertilizer and sharply improved efficiencies in animal performance greatly increased the productivities of both labor and land. Concepts of carrying capacity have changed, and it is now possible to sustain a large population from a small land base. In theory, that means that agriculture can be restricted to the least sensitive lands and, given energy for nutrient cycling or supply, can be sustained indefinitely. It also means that agriculture and our food supply are now dependent upon social stability in the larger society for a continued flow of goods and services.

That we continue to farm more land than we need, including hill lands, indicates that we are still in transition.

I have tried to focus on mainstream issues, avoiding those I consider temporary aberrations. Irrigation from the Ogallala aquifer is a temporary phenomenon, and that land will eventually return to grazing or dryland methods. Similarly, salinity problems are small in extent and easily solved by abandonment or proper drainage. In general, American farming is relatively free of major problems and irreversibilities. The breadth of this essay limited it to a sampling of the literature, and thus I could only hint at the depth of scientific and technological knowledge in each area. I have deliberately avoided pesticide issues, which are now institutionalized by regulatory systems. Their use will increase or decline with changing knowledge. In addition, I have skirted carefully through the maze of traditional social, economic, and political issues (beginning at least with the Romans), despite temptations to deal with the Goldschmidt hypothesis and other assorted topics. While my focus is on the health of the land and the biology it sustains, my review and personal experiences suggest that the social health of rural society is reasonably good, and improving.

The basic issues of sustainability and what type of agriculture will be superior in the future rests on the availability of energy. Given the genetic superiority of present plants and animals and our increased scientific knowledge, we will not return, with energy limitations, to the old ways, but will evolve towards better, new-farming systems. With large energy reserves, numerous alternatives to fossil sources, and enormous discretionary use in other sectors of our society, farmers are not likely to retreat soon from the use of tractors and fertilizers. With shortages, agriculture will either obtain its small energy requirements by allocation or competitively through increases in food prices. Changes in the area under production will come first, with further intensification on the best lands as one strategy. The role of legumes is uncertain since their importance could change dramatically with genetic changes in symbiont or animal digestion efficiency. Finally, as in the past, change will come mainly through farmer innovation and initiative. With more than 2 million independent farmers, nonconventional ways for better farming will be thoroughly explored.

ACKNOWLEDGMENTS

I greatly appreciate the literature review suggestions made by I. Buddenhagen, W. J. Chancellor, A. McCalla, P. Martin, J. Randall, and M. J. Singer. This manuscript was prepared by the traditional longhand method; thereafter, I am afraid, the process was highly automated.

Literature Cited

1. Allard, R. W., Jain, S. K. 1962. Population studies in predominantly self-pollinated species. II. Analysis of quantitative genetic changes in bulk-hybrid populations of barley. *Evolution* 16:90–101

2. Allos, H. F., Bartholomew, W. V. 1959. Replacement of symbiotic fixation by available nitrogen. *Soil Sci.* 87:61–66

3. Austin, R. B., Bingham, J. Blackwell, R. D., Evans, L. T., Ford, M. A. 1980. Genetic improvements in winter wheat yields since 1900 and associated physiological changes. *J. Agric. Sci.* 94: 675–89

4. Avlani, P. K., Chancellor, W. J. 1977. Energy requirements for wheat production and use in California. *Trans. ASAE* 20:429–37

5. Azevado, J., Stout, P. R. 1974. *Farm Animal Manures: An Overview of their Role in the Agricultural Environment, Calif. Agric. Exp. Stn. Man. 44.* Berkeley: Univ. Calif. 109 pp.

6. Baldock, J. O., Musgrave, R. B. 1980. Manure and mineral fertilizer effects in continuous and rotational sequences in Central New York. *Agron. J.* 72:511–18

7. Banks, V. J., Beale, C. L. 1973. *Farm Population Estimates, 1910–70, US Dep. Agric. Stat. Bull. No. 573.* Washington DC: USGPO. 47 pp.

7a. Bezdicek, D. F., Power, J. F., Keeney, D. R., Wright, M. J., eds. 1984. *Organic Farming: Current Technology and its Role in a Sustainable Agriculture.* Madison, Wis.: Am. Soc. Agron. 192 pp.

8. Bockheim, J. G. 1980. Solution and use of chronofunctions in studying soil development. *Geoderma* 24:71–85

9. Boeringa, R., ed. 1980. *Alternative Methods of Agriculture.* Amsterdam: Elsevier. 199 pp.

10. Brewster, D. 1979. The family farm: A changing concept. In *Structure Issues of American Agriculture,* pp. 74–79. Washington DC: Econ. Stat. Coop. Serv., US Dep. Agric.

11. Bromley, D. W. 1982. The rights of society versus the rights of landowners and operators. In *Soil Conservation Policies, Institutions, and Incentives,* ed. H. G. Halcrow, E. O. Heady, M. L. Cotner, pp. 219–32. Ankeny, Iowa: Soil Conserv. Soc. Am. 329 pp.

12. Buringh, P., van Heemst, H. D. J. 1977. An estimation of world food production based on labour-oriented agriculture. Wageningen, The Netherlands: Cent. World Food Mark. Res. 46 pp.

13. Bywater, A. C., Baldwin, R. L. 1980. Alternative strategies in food-animal production. In *Animals, Feed, Food and People,* ed. R. L. Baldwin, pp. 1–29. Boulder, Colo.: Westview. 149 pp.

14. Cardwell, V. G. 1982. Fifty years of Minnesota corn production: source of yield increase. *Agron. J.* 74:984–90

15. Cervincka, V., Chancellor, W. J., Coffelt, R. J., Curley, R. G., Dobie, J. B. 1974. *Energy Requirements for Agriculture in California, Jt. Study Rep. Calif. Dep. Food Agric. & Univ. Calif.* Davis: Univ. Calif. 151 pp.

16. Chancellor, W. J. 1981. Substituting information for energy in agriculture. *Trans. ASAE* 24:802–7, 813

17. Chiarappa, L., ed. 1971. *Crop Loss Assessment Methods, FAO Manual.* Farnham, England: Commonw. Agric. Bur.

18. Cochrane, W. W. 1979. *The Development of Agriculture. A Historical Analysis.* Minneapolis: Univ. Minn. Press. 464 pp.

19. Costanza, R. 1980. Embodied energy and economic evaluation. *Science* 210: 1219–24

20. Dean, G. W., Carter, H. O., Wagstaff, H. R., Olayide, S. O., Ronning, M., Bath, D. L. 1972. *Production Functions and Linear Programming Models for Dairy Cattle Feeding, Giannini Found. Monogr. No. 31, Calif. Agric. Exp. Stn.* Berkeley: Univ. Calif. 52 pp.

21. de Janvry, A. 1981. *The Agrarian Question and Reformism in Latin America.* Baltimore: Johns Hopkins Univ. Press. 311 pp.

22. de Wit, C. T., van Heemst, H. D. J. 1976. Aspects of agricultural resources. In *Chemical Engineering in a Changing World,* ed. W. T. Koetsier, pp. 125–45. Amsterdam: Elsevier. 550 pp.

23. Doering, O. C. III. 1977. *An Energy Based Analysis of Alternative Production Methods and Cropping Systems in the Corn Belt, Natl. Sci. Found. Rep. (NSF-AER 75-1826).* West Lafayette, Ind.: Purdue Univ. 43 pp.

24. Donald, C. M., Hamblin, J. 1976. The biological yield and harvest index of cereals as agronomic and plant breeding criteria. *Adv. Agron.* 28:361–405

25. Donald, C. M., Hamblin, J. 1983. The convergent evolution of annual seed crops in agriculture. *Adv. Agron.* 36:97–143

26. Duffus, J. E. 1978. The impact of yellows control on California sugarbeets. *J. Am. Soc. Sugar Beet Technol.* 20:1–5

27. Duncan, W. G. 1971. Leaf angles, leaf

area, and canopy photosynthesis. *Crop Sci.* 11:482–85

28. Eberhardt, S. A., Russell, W. A. 1966. Stability parameters for comparing varieties. *Crop Sci.* 6:36–40

29. Economic Research Service, US Department of Agriculture. 1974. *The U.S. Food and Fiber Sector: Energy Use and Outlook. (Prep. for Senate Comm. Agric. For.)* Washington DC: USGPO. 111 pp.

30. Evans, H. J., Barber, L. E. 1977. Biological nitrogen fixation for food and fiber production. *Science* 197:332–39

31. Finlay, K. W., Wilkinson, G. N. 1963. The analysis of adaptation in a plant breeding program. *Aust. J. Agric. Res.* 14:742–54

32. Frey, K. J., Browning, J. A., Simons, M. D. 1977. Management systems for host genes to control disease. *Ann. NY Acad. Sci.* 287:255–74

33. Frissel, M., ed. 1978. *Cycling of Mineral Nutrients in Agricultural Ecosystems. Agro-Ecosystems* 4:1–354 (Spec. issue)

34. Garner, H. V., Dyke, G. V. 1969. *The Broadbalk Yields, Rothamsted Exp. Stn. Rep. 1968.* Part 2:26–49

35. Gause, G. F. 1934. *The Struggle for Existence.* Baltimore: Williams & Wilkins. 163 pp.

36. Goodman, D. 1975. The theory of diversity-stability relationships in ecology. *Q. Rev. Biol.* 50:237–66

37. Halcrow, H. G., Heady, E. O., Cotner, M. L., eds. 1982. *Soil Conservation Policies, Institutions, and Incentives.* Ankeny, Iowa: Soil Conserv. Soc. Am. 330 pp.

38. Hamilton, W. H. III, Watt, K. E. F. 1970. Refuging. *Ann. Rev. Ecol. Syst.* 1:263–86

39. Hardin, J. 1982. A quantitative index of soil development from field descriptions: Examples from a chronosequence in central California. *Geoderma* 28:1–28

40. Harlan, H. V., Martini, M. L. 1938. The effect of natural selection in a mixture of barley varieties. *J. Agric. Res.* 57:189–99

41. Hathaway, D. E., Beegle, J. A., Bryant, W. K. 1968. *The People of Rural America.* Bur. Census, US Dep. Comm. Washington DC: USGPO. 289 pp.

42. Heady, E. O. 1982. The adequacy of agricultural land: A demand-supply perspective. In *The Cropland Crisis. Myth or Reality?,* ed. P. R. Crosson, pp. 23–56. Baltimore: Johns Hopkins Univ. Press. 250 pp.

43. Heichel, G. H., Barnes, D. K. 1984. Opportunities for meeting crop nitrogen needs from symbiotic nitrogen fixation. See Ref. 7a, pp. 49–59

44. Held, R. B., Clawson, M. 1965. *Soil Conservation in Perspective.* Baltimore: Johns Hopkins Univ. Press. 344 pp.

45. Herendeen, R. A. 1973. *The Energy Cost of Goods and Services, ORNL-NSF-EP-58.* Oak Ridge, Tenn.: Oak Ridge Natl. Lab.

46. Jenny, H. 1980. *The Soil Resource: Origin and Behavior; Ecol. Studies.* Vol. 37. New York: Springer-Verlag. 377 pp.

47. Jewell, C. A. 1976. The impact of America on English agriculture. *Agric. Hist.* 50:125–36

48. Johnson, J. W., Welch, L. F., Kurtz, L. T. 1975. Environmental implications of N fixation. *J. Environ. Qual.* 4:303–6

49. Johnson, W. A., Stoltzfus, V., Craumer, P. 1977. Energy conservation in Amish agriculture. *Science* 198:373–78

49a. Judson, S., Ritter, D. F. 1964. Rates of denudation in the United States. *J. Geophys. Res.* 69:3395–401

50. Kampmeijer, P., Zadoks, J. C. 1977. *EPIMUL, a Simulator of Foci and Epidemics in Mixtures of Resistant and Susceptible Plants, Mosaics and Multilines, Simulation Monogr.* Wageningen, The Netherlands: Pudoc. 50 pp.

51. King, D., Roughgarden, J. 1982. Multiple switches between vegetative and reproductive growth. *J. Theor. Biol.* 44:23–34

52. Larson, W. E., Pierce, F. J., Dowdy, R. H. 1983. The threat of soil erosion to long-term crop production. *Science* 219:458–65

53. LaRue, T. A., Patterson, T. G. 1981. How much nitrogen do legumes fix. *Adv. Agron.* 34:15–38

54. Lee, L. K. 1980. The impact of landownership factors on soil conservation. *J. Agric. Econ.* 62:1070–76

55. Lee, L. K. 1983. Land tenure and adoption of conservation tillage. *J. Soil Water Conserv.* 38:166–68

56. Loomis, R. S. 1978. Ecological dimensions of medieval agrarian systems: An ecologist responds. *Agric. Hist.* 52:478–83

57. Loomis, R. S. 1983. Crop manipulations for efficient use of water: An overview. In *Limitations to Efficient Use of Water in Crop Production,* ed. H. M. Taylor, W. R. Jordan, T. R. Sinclair, pp. 345–74. Madison, Wis.: Am. Soc. Agron. 538 pp.

58. Loomis, R. S. 1983. Productivity of agricultural systems. In *Physiological Plant Ecology IV, Ecosystem Processes, Encycl. Plant Physiol., N.S.,* ed. O. L. Lange, P. S. Nobel, C. B. Osmond, H. Ziegler, pp. 151–72. Berlin/Heidelberg: Springer-Verlag. 644 pp.

59. Loomis, R. S. 1983. An ecological overview. In *A Guidebook to California Agriculture*, ed. A. F. Scheuring, pp. 389–404. Berkeley: Univ. Calif. Press. 413 pp.

60. Marshall, D. R., Allard, R. W. 1974. Performance and stability of mixtures of grain sorghum. I. Relationship between level of genetic diversity and performance. *Theor. Appl. Genet.* 44:145–52

61. Martin, P. L. 1983. Labor-intensive agriculture. *Sci. Am.* 249(4):54–59

62. Moe, P. W. 1981. Energy metabolism of dairy cattle. *J. Dairy Sci.* 64:1120–39

63. Munson, R. D., Doll, J. P. 1959. The economics of fertilizer use in crop production. *Adv. Agron.* 11:133–69

64. National Research Council, Committee on the Genetic Vulnerability of Major Crops. 1972. *Genetic Vulnerability of Major Crops*. Washington DC: Natl. Acad. Sci. 304 pp.

65. National Research Council, Committee on Animal Nutrition. 1969. *United States-Canadian Tables of Feed Composition, Publ. No. 1684*. Washington DC: Natl. Acad. Sci. 92 pp.

66. National Research Council, Committee on Agricultural Production Efficiency. 1975. *Agricultural Production Efficiency*. Washington DC: Natl. Acad. Sci. 199 pp.

67. Olshan, M. A. 1981. Modernity, the folk society, and the Old Order Amish: An alternative interpretation. *Rural Sociol.* 46:297–309

68. Olson, K. D., Langley, J., Heady, E. O. 1982. Widespread adoption of organic farming practices: Estimated impacts on U.S. agriculture. *J. Soil Water Conserv.* 37:41–45

69. Phillips, R. E., Blevins, R. L., Thomas, G. W., Frye, W. W., Phillips, S. H. 1980. No-tillage agriculture. *Science* 208:1108–13

69a. Péwé, T. L., ed. 1981. *Desert Dust: Origin, Characteristics, and Effect on Man*. Spec. Pap. 186. Boulder, Colo.: Geol. Soc. Am. 303 pp.

70. Pierce, F. J., Larson, W. E., Dowdy, R. H., Graham, W. A. P. 1983. Productivity of soils: Assessing long-term changes due to erosion. *J. Soil Water Conserv.* 38:39–44

70a. Pimentel, D., Berardi, G., Fast, S. 1984. Energy efficiencies of farming wheat, corn, and potatoes organically. See Ref. 7a, pp. 151–61

71. Pimentel, D., Hurd, L. E., Belloti, A. C., Forster, M. J., Oha, I. N., Sholes, O. D., Whitman, R. J. 1973. Food production and the energy crisis. *Science* 182:443–49

72. Pimentel, D., Oltenacu, P. A., Nesheim, M. C., Krummel, J., Allen, M. S., Chick, S. 1980. The potential for grass-fed livestock: Resource constraints. *Science* 207:843–48

73. Poleman, T. T. 1975. World food: A perspective. *Science* 188:510–18

74. Quisenberry, K. S., Reitz, L. P. 1974. Turkey wheat: The cornerstone of an empire. *Agric. Hist.* 48:98–114

75. Rappaport, R. A. 1971. The flow of energy in an agricultural society. *Sci. Am.* 224(3):117–32

76. Raup, P. M. 1972. Societal goals in farm size. In *Size, Structure and Future of Farms*, ed. A. G. Ball, E. O. Heady, pp. 3–18. Ames, Iowa: Iowa State Univ. Press. 404 pp.

77. Reed, C. O. 1928. Agricultural machinery and implements, United States. In *Encyclopaedia Britannica*, 1:373–79. 14th ed.

78. Reiger, J. H., Beegle, J. A., Fulton, P. N. 1978. Diaspora and adaptation: A case study of youth from a low-income rural area. In *Patterns of Migration and Population Change in America's Heartland, Mich. Agric. Exp. Stn. Res. Bull. No. 344*, pp. 52–63. 104 pp.

79. Reitz, L. P., Salmon, S. C. 1968. Origin, history, and use of Norin 10 wheat. *Crop Sci.* 7:686–89

80. Riggs, T. J., Hanson, P. R., Start, N. D., Miles, D. M., Morgan, C. L., Ford, M. A. 1981. Comparison of spring barley varieties grown in England and Wales between 1800 and 1980. *J. Agric. Sci.* 97:599–610

81. Robinson, R. A. 1976. *Plant Pathosystems, Adv. Ser. Agric. Sci. 3*. Berlin/Heidelberg: Springer-Verlag. 184 pp.

81a. Ruttan, V. W. 1982. *Agricultural Research Policy*. Minneapolis: Univ. Minn. Press. 369 pp.

82. Saloutos, T. 1976. The immigrant contribution to American Agriculture. *Agric. Hist.* 50:45–67

83. Schultz, T. W. 1964. *Transforming Traditional Agriculture*. New Haven, Conn.: Yale Univ. Press. 212 pp.

84. Simmonds, N. W. 1979. *Principles of Crop Improvement*. London: Longmans. 408 pp.

85. Sims, N. L. 1940. *Elements of Rural Sociology*. New York: Crowell. 690 pp.

86. Sinclair, R. 1967. Von Thünen and urban sprawl. *Ann. Assoc. Am. Geogr.* 57:72–87

87. Slicher van Bath, B. H. 1963. *The Agrarian History of Western Europe, A.D. 500–1850*. London: Arnold. 364 pp.

88. Smith, A. 1950. [1776]. *Inquiry into the Nature and Causes of the Wealth of Na*

tions, ed. E. Cannan. Vol. 1. London: Methuen. 524 pp. 6th ed.

89. Speth, J. D., Spielmann, K. A. 1983. Energy source, protein metabolism, and hunter-gatherer subsistence strategies. *J. Anthropol. Archaeol.* 2:1–31

90. Stanford, E. H., Laude, H. M., Enloe, J. A. 1960. Effect of harvest dates and location on the genetic composition on the Syn1 generation of Pilgrim ladino clover. *Agron. J.* 52:149–52

91. Steinhart, J. S., Steinhart, C. W. 1974. Energy use in the U.S. food system. *Science* 184:307–16

92. Stevenson, F. J. 1982. Origin and distribution of nitrogen in soil. In *Nitrogen in Agricultural Soils, Agron. No. 22,* ed. F. J. Stevenson, pp. 1–42. Madison, Wis.: Am. Soc. Agron. 940 pp.

93. Touchton, J. T., Hoeft, R. G., Welch, L. F. 1979. Effect of nitrapyrin on nitrification of broadcast-applied urea, plant nutrient concentrations, and corn yield. *Agron. J.* 71:787–91

94. Trenbath, B. R. 1974. Biomass productivity of mixtures. *Adv. Agron.* 26:177–210

95. Trimble, S. W. 1974. *Man-Induced Soil Erosion on the Southern Piedmont, 1700–1970.* Ankeny, Iowa: Soil Conserv. Soc. Am. 180 pp.

96. Tweeten, L. 1983. The economics of small farms. *Science* 219:1037–41

97. Unger, P. W., McCalla, T. M. 1980. Conservation tillage methods. *Adv. Agron.* 33:1–58

98. US Department of Commerce, Bureau of the Census. 1981. *Census of Agriculture, 1978.* Vol. 1, part 51. Washington DC:

USGPO. 965 pp.

99. US Department of Commerce, Bureau of the Census. 1981. *Population Profile of the United States: 1980.* Current Population Reports, Ser. P-20, No. 363. Washington, DC: USGPO. 56 pp.

100. US Department of Commerce, Bureau of the Census. 1983. *Fertility of American Women: June 1981.* Current Population Reports, Ser. P-20, No. 378. Washington DC: USGPO. 63 pp.

101. Vanderplank, J. E. 1982. *Host-Pathogen Interactions in Plant Disease.* New York: Academic. 207 pp.

102. van Heemst, H. D. J., Merkelijn, J. J., van Keulen, H. 1981. Labour requirements in various agricultural systems. *Q. J. Int. Agric.* 20:178–201

103. Van Soest, P. J. 1982. *Nutritional Ecology of the Ruminant.* Corvallis, Ore.: O & B. 373 pp.

104. Ward, G. M. 1980. Energy, land and feed constraints on beef production in the 80's. *J. Anim. Sci.* 51:1051–64

105. Welch, L. F. 1976. The Morrow plots—hundred years of research. *Ann. Agron.* 27:881–90

106. Whitson, R. E., Kay, R. D., Le Pori, W. A., Rister, E. M. 1981. Machinery and crop selection with weather risk. *Trans. ASAE* 24:288–91, 295

107. Williams, W. A., Morse, M. D., Ruckman, J. E. 1972. Burning vs. incorporation of rice crop residues. *Agron. J.* 64:467–68

108. Wischmeier, W. H., Smith, D. D. 1978. *Predicting Rainfall Erosion Losses—A Guide to Conservation Planning. US Dep. Agric. Handb. No. 537.* 58 pp.

Ann. Rev. Ecol. Syst. 1984. 15:479–99

ASSOCIATIONS AMONG PROTEIN HETEROZYGOSITY, GROWTH RATE, AND DEVELOPMENTAL HOMEOSTASIS

J. B. Mitton and M. C. Grant

Department of Environmental, Population, and Organismic Biology, University of Colorado, Boulder, Colorado 80309

INTRODUCTION

In 1954 I. Michael Lerner published *Genetic Homeostasis*, a compendium of observations on the relationships among heterozygosity, growth rate and other measures of performance, and developmental homeostasis. These observations were primarily made on cultivated plants and domesticated animals, and the experiments usually contrasted inbred, predominantly homozygous lines with the highly heterozygous progeny of crosses between inbred lines. Highly heterozygous individuals and lines generally exhibited the traits that breeders strive to fix in their strains. In comparison with inbred, homozygous lines, they usually had superior growth rates; often attained greater size; and generally had more buffered developmental processes, resulting in lower morphological variation (e.g. 8). Contrary results, however, are known (e.g. 10, 89). These results—and similar studies (20, 21, 88)—have influenced the thinking of plant and animal breeders, but the impact of these studies on population biology and population genetics has, as yet, been slight.

The mechanism underlying the phenomenon of heterosis has received a great deal of attention from empiricists and theoreticians (see 111 for a review). Summarizing briefly, the antithesis of heterosis—inbreeding depression—has been viewed as the consequence of either increased homozygosity at a large

479

0066-4162/84/1120-0479$02.00

number of overdominant loci or the increased expression of a small number of deleterious, recessive alleles.

The causal mechanisms that underly the general pattern of increased vigor, increased developmental homeostasis, and superior yields in highly heterozygous individuals continues to be the subject of a complex controversy. The term *heterosis* was coined by G. H. Shull in 1918 specifically for the purpose of reducing confusion about the distinction between the observed phenomenon of hybrid vigor and its underlying mechanisms (90). He offered this terminological change in order to move away from the commonly used term *heterozygosis*, which explicitly attributed the effect to the level of heterozygosity. He did not intend to downgrade the significance of heterozygosity as an explanatory mechanism, but he did want the phenomenon to be identified and discussed independently of its underlying cause.

A number of early workers specifically studied heterosis by assessing the effect of heterozygosity per se (see reviews in 10, 16, 43, 100, 109; experimental studies in 2, 39). These scholars approached the phenomenon in conceptually distinct, complementary ways by (*a*) examining the effects on individual genotypes (developmental homeostasis, for example) and (*b*) studying the effects on populations constructed of genotypic mixtures (stability and genetic homeostasis). The simplest interpretation of the results is that individual organisms with comparatively higher levels of heterozygosity also have comparatively higher levels of individual homeostasis (e.g. stability and consistency), although there are a number of examples in which this general pattern does not hold (1, 10, 91, 78). At an individual level, "yield," in its many facets, is strongly correlated with the level of heterozygosity (8, 77, 89). Indeed, this phenomenon is the foundation upon which the strategy of developing inbred lines and the subsequent hybrid synthesis so common in commercially developed crop plants is primarily based. At the population level, mixtures of genotypes commonly have yields and productivities at approximately the same level as the weighted averages of their individual components, but they are generally superior in stability and consistency of performance (2, 78, 91).

There are three primary hypotheses about the causal genetic properties that account for the observed increases in performance and individual homeostasis: (*a*) the effect is due to heterozygosity through the mechanism of dominance, (*b*) the effect is due to heterozygosity through the mechanism of overdominance, and (*c*) the effect is due to fortuitous combinations of particular genes. Taking applied breeding studies as a whole, we judge that heterozygosity per se is the single most important explanatory factor (91), even though numerous studies have clearly shown it is not a complete explanation; some unexplained variability in performance among lines of equal heterozygosities always remains (e.g. 1, 2). The extent to which dominance or overdominance is the principal mechanism of action of heterozygosity remains a matter of disagreement.

There is additional evidence that contributes to our understanding of the significance of heterozygosity. Polyploidy, a major feature of plant evolution, is thought to be an important factor in many groups largely because of its effects on individual levels of heterozygosity (e.g. 38, 46, 61). The presence of additional chromosome sets permits the establishment, for example, of fixed heterozygosity. Some theoretical work (97, 98) indicates that heterosis due to overdominance could be intimately tied to the evolution of gene duplications. Most students of polyploidy attribute a major role to the increased level of individual genetic diversity produced by polyploidy, although there are studies showing that some of the effects are due simply to the presence of extra chromosomes, not to an increased number of different alleles (e.g. 79). Furthermore, a major group of plants—the conifers—that show little or no polyploidy have the highest known levels of protein heterozygosity (32). There are obviously several evolutionary routes to the creation and maintenance of high levels of heterozygosity.

The positive association between heterozygosity and fitness has also been clearly observed in species that alternate sexual and parthenogenetic reproduction. Minor fitness advantages enjoyed by heterozygous genotypes may go undetected in obligately sexual species, but these advantages can be amplified by an extended period of parthenogenetic reproduction. Heterosis has been observed in crosses between inbred strains of *Daphnia magna* (35), and excesses of heterozygous genotypes are common in natural populations of this species.

Our interpretation of the literature leads us to believe that roughly 70–80% of the effects on growth and developmental stability can be attributed to heterozygosity per se, about 15–20% to the effects of specific gene combinations, and the remainder to as yet unidentified causes. These proportions, of course, vary somewhat from organism to organism.

PATTERNS IN NATURAL POPULATIONS

The observation that heterozygosity strongly influences both vigor and stability has been generally accepted in the applied literature for many years. How widespread the phenomenon is in natural populations was still an open question in the early 1960s, largely because the tools needed to assess the degree of heterozygosity were extremely time consuming, cumbersome, and limited to a few, especially tractable organisms.

Within the last two decades, population biologists have made a series of empirical observations that link the heterozygosity of protein polymorphisms with viability, growth rate, developmental stability, and physiological variables such as oxygen consumption. These data are consistent with Lerner's (58) general observations, and they have spurred a critical examination of how the theory of population genetics accounts for the determination of fitness. Furthermore, these relationships offer a new opportunity to explore the mechanisms underlying heterosis and inbreeding depression. These observations are of particular interest to ecologists and population biologists because they place heterosis and inbreeding depression in the arena of the natural population.

EMPIRICAL OBSERVATIONS

The relationships between heterozygosity and developmental stability were first identified in crosses between inbred strains, but they are also evident within populations when heterozygosity is measured with single chromosomes or single protein polymorphisms. For example, *Drosophila* heterozygous for their second chromosomes develop lower levels of morphological variability in wings than do homozygous individuals (88).

Protein Heterozygosity and Developmental Stability

Protein heterozygosity was first related to morphological variability in the marine fish, *Fundulus heteroclitus* (70). A single protein polymorphism was used to divide a population sample into two groups, one heterozygous and the other homozygous for that locus. Meristic variation at 7 scale and fin ray characters was used to quantify the morphological variability of the two groups, which were then compared. This process was repeated for each of 5 loci in 2 populations. In most of these comparisons, the heterozygotes exhibited lower phenotypic variation than the homozygotes.

Protein heterozygosity and developmental stability have been shown to be related in the monarch butterfly, *Danaus plexippus* (22). Six polymorphisms and two morphological characters—length of the forewing and size of the forewing spot—were used in this study. Each of the polymorphisms was used one at a time to separate individuals into classes of homozygotes and heterozygotes, and the classes were then compared. The mean values of the morphological characters did not vary among genotypes, but the variances of the characters were associated with genotypic classes. Overall, heterozygotes had lower variances in 19 of 24 tests, and 5 of these tests were statistically significant. Thus, heterozygous individuals clustered more closely about the means of the character distributions, but the protein genotypes did not influence the direction of deviation from the mean.

Fluctuating asymmetry, the nondirectional imbalance between bilaterally paired characters, is a measure of developmental stability that is correlated with protein heterozygosity. More heterozygous individuals and populations exhibit greater symmetry than more homozygous ones.

The relationship between protein heterozygosity and fluctuating asymmetry has been demonstrated most rigorously in the rainbow trout, *Salmo gairdneri*

(56). The researchers measured bilateral symmetry with 5 characters taken from gill rakers, mandibular pores, and fin rays and estimated heterozygosity with 13 protein polymorphisms. The data consisted of genotypes at each of the 13 polymorphisms and the 5 bilateral characters for each individual. The number of heterozygous loci was negatively correlated with both the number of asymmetric characters and the magnitude of the asymmetry. When each locus was tested individually, 2 of the 13 loci were significantly related to the asymmetry. Even when these 2 loci were removed from the analysis, however, the remaining polymorphisms were still correlated with the number of asymmetric characters. Thus, developmental stability was found to be associated with both the heterozygosity of specific loci and protein heterozygosity in general.

Handford, the author who first tested for relationships between protein heterozygosity and developmental stability in homeotherms, questioned the generality of the preceding observations. He compared morphological variability and protein heterozygosity in the rufous-collared sparrow, *Zonotrichia capensis*, but these sets of characters were independent in this data set (33). Handford noted that all previous positive results had been discovered in poikilotherms, and he postulated that homeotherms would not exhibit relationships between heterozygosity and developmental stability. This study was later criticized for being based on a small sample (23), but a second study on birds (24) directly addressed Handford's hypothesis. The authors found that enzyme heterozygosity at 4 loci was related to variability in limb size in the house sparrow, *Passer domesticus*, with more heterozygous individuals exhibiting less variability.

Natural selection does not influence all morphological characters with the same intensity, for selection on some characters is extreme but undetectable at others. Strong stabilizing selection modifies the number of caudal fin rays in the guppy *Poecilia reticulata* (5), for example, and this selection is also related to protein heterozygosity (6). More heterozygous individuals are most abundant at the central phenotype (27 caudal fin rays), and more homozygous individuals are more common in fin ray classes above and below 27. Selection on caudal fin rays continues throughout the life cycle of this fish, and differential mortality strengthens this relationship between heterozygosity and the caudal fin ray number. McAndrew, Ward & Beardmore (63) compared heterozygosity and the variability of caudal, anal, and dorsal fin rays in a large study of the plaice, *Pleuronectes platessa*. Despite careful and thorough analyses of the data, no hint of any relationship between heterozygosity and morphological variability was found. The authors compared this negative result with the strong relationship found in the guppy (6). A study is needed to determine whether the relationships between heterozygosity and developmental stability are stronger for characters under stabilizing selection than for those unaffected by selection.

Although the most direct tests for relationships between heterozygosity and developmental stability are conducted within populations, the same relationships can be revealed in comparisons among populations. This fact was first demonstrated in an analysis of geographic variation of 15 populations of the side-blotched lizard on islands in the Gulf of California (95). Four independent scale characters were used to measure fluctuating asymmetry, and genotypes at 18 protein loci were used to measure heterozygosity. Populations with higher levels of heterozygosity had lower levels of asymmetry. A similar result was found in two species of fresh water bivalves (44). Geographically peripheral populations had lower levels of heterozygosity and higher levels of asymmetry.

An unusual phenomenon was used opportunistically to study the relationship between heterozygosity and morphological variability in a sexual diploid species, *Poeciliopsis monacha,* and a parthenogenetic triploid species of live-bearing fishes, *P. 2 monacha-lucida* (103). Both species live in the Arroyo de los Platano of Mexico. The triploid species produces unreduced triploid eggs without recombination and uses the sperm of *P. monacha* to initiate development, but it does not incorporate any of the genetic material from the sperm into the egg. Thus, this species is comprised of a series of clones. Founder events had produced a cline in heterozygosity that was detected with 25 protein loci, with low levels of heterozygosity in small upstream populations and much higher levels in the larger downstream populations. Heterozygosity at these 25 loci ranged from 0.1% to 8.1% in the sexual species, and it was 52% in the triploid clone. Variation in the physical environment was associated with the cline in heterozygosity among the sexual populations. Upstream populations lived in small pools that occasionally dried up and were later recolonized, and downstream populations lived in permanent environments that always had running water. The internal standard in this experiment was the triploid clone, which was a genetic constant replicated in heterogeneous environments. Bilateral symmetry, measured with 8 paired characters (scale counts, fin rays, and premaxillary and dentary tooth counts) did not vary among the populations of the triploid clone, suggesting that the differences in the physical environment did not substantially influence the development of these characters. In contrast to this pattern of homogeneity, the populations of the sexual species exhibited different levels of symmetry, with the more homozygous populations having more asymmetry.

Protein Heterozygosity and Growth Rate

The first observations linking protein heterozygosity directly to growth rate were reported in two studies of the American oyster, *Crassostrea virginica* (93, 112). The design and results of these experiments are quite similar. Oyster spat or larvae were collected and transferred to trays anchored in a bay so that all of the animals were approximately the same age and in a similar environment with the same opportunity to feed. One year later, the animals were removed from the tray, and the largest and the smallest—or the fastest and slowest growing individuals—were weighed. Genotypes were determined for 5 loci in the first experiment and 7 loci in the second. In both experiments, a single polymorphism, *GOT-1*, exhibited a pattern of variation distinct from all the rest. Genotypes at this locus had different weights, but the distribution of genotypes fit Hardy-Weinberg expectations. At the other loci, the smallest weight classes had few heterozygotes. Heterozygotes at each of the loci tended to be heavier than homozygotes, and an individual's weight was positively correlated with its number of heterozygous loci. The growth advantage associated with enzyme heterozygosity occurred in older age classes as well (92).

Similar results were reported for the Pacific oyster, *Crassostrea gigas* (26), but in this study, animals were collected as adults from natural populations. Most loci had deficiencies of heterozygotes, and there was a negative correlation—among populations—between the oysters' mean weight and the inbreeding coefficient. Within populations, heterozygotes clearly tended to weigh more than homozygotes at each locus, and weight was positively correlated with individual heterozygosity.

The growth rate of another marine pelecypod, the blue mussel, *Mytilus edulis,* is also associated with protein heterozygosity (52). In this study a seasonal raft was moored in spring and removed in fall to obtain mussels of similar age that shared the same environment. Each individual was measured, and then its genotype was determined for 5 polymorphic enzymes. The shell lengths ranged from less than 2 mm to more than 20 mm. The mean growth rate increased with individual heterozygosity, while the variance in growth rate decreased. A laboratory study of *Mytilus edulis* revealed no association between protein heterozygosity and growth within progeny of a pair cross (7).

An association between protein heterozygosity and growth was reported for the tiger salamander, *Ambystoma tigrinum* (80). In 5 of 7 natural populations sampled for young larvae, there was a positive correlation between heterozygosity at 8 protein loci and snout-vent length, a classical measure of size. An experiment was conducted in the laboratory to determine whether heterozygous individuals' larger size could be attributed to greater growth rates. A single male and a single female were chosen by genotype, and 462 offspring were produced by a pair cross. The offspring were arbitrarily assigned to 4 replicate populations kept in separate aquaria and allowed to grow until a substantial variance in size among individuals developed. In 2 of the 4 replicates, there were significant, positive correlations between size and heterozygosity. Thus, the correlation between heterozygosity and size in natural populations may be attributable to differential growth rates. The correlation disappears later in the larval period, however, perhaps owing to differential metamorphosis of genotypes.

A nexus of relationships involving individual heterozygosity, weight, fecundity, and fetal growth have been reported for the white-tailed deer, *Odocoileus virginianus* (15). Adult females with high levels of enzyme heterozygosity weigh more than more homozygous individuals; they also enjoy an advantage in fecundity (42), with highly heterozygous individuals having a higher rate of twinning. The weight of females is positively correlated with the weight of their fetuses. In addition, fetal growth rates are positively correlated with individual fetal heterozygosity.

There is some evidence of an association between protein heterozygosity and growth rate in domesticated animals. For example, sheep heterozygous for isocitrate dehydrogenase have been reported to grow 10% faster than homozygotes (4). Growth, food consumption, and protein heterozygosity were examined in pigs, and the more heterozygous individuals gained weight more quickly while consuming less food (62). The enzyme glucose-phosphate isomerase is also associated with meat quality and weight gain in pigs (87). This correlation is difficult to interpret, however, because this enzyme is tightly linked to blood group loci that are also consistently associated with meat quality and production.

A relationship has also been reported between individual heterozygosity and fetal growth in man (9). This study was replicated with 93 babies sampled from a hospital in Rome and 98 from a hospital in New Haven. Heterozygosity was recorded for 5 loci in Rome and 4 loci in New Haven, and the babies were placed in three categories: preterm; full term, normal weight; full term, low weight. Preterm babies did not differ from full term normal weight, babies, but the light babies had a significantly lower level of heterozygosity than the other groups.

A relationship between protein heterozygosity and growth rate in plants was first reported in quaking aspen, *Populus tremuloides* (72). The elevation, age of largest ramet, sex, and genotype at three enzyme loci was recorded for 104 clones of aspen in the Front Range of the Colorado Rocky Mountains. The growth rate of each of these clones of aspen was estimated as the average width of the annual rings, which were measured from cores taken from 5 ramets of each clone; the effect of the age of the ramet was removed statistically. There was a significant, positive relationship between the number of heterozygous proteins and the estimated growth rate.

Protein heterozygosity is related to the variability of growth rate in ponderosa pine *(Pinus ponderosa)* and lodgepole pine *(Pinus contorta)* as well, but in neither of these trees is heterozygosity related to the mean growth rates of mature trees (31, 47, 48, 71, 73). In each of these studies, genotypes and growth rates were compared within a single population. Cores were extracted from more than 100 trees, and protein genotypes were obtained from needle tissue for 6 protein polymorphisms in ponderosa pine and 4 in lodgepole pine.

Analysis of covariance was used to control for the effects age and slope aspect. In ponderosa pine, highly heterozygous individuals exhibit higher growth variability than predominantly homozygous individuals, and in lodgepole pine, the opposite occurs. These contrary results do not necessarily indicate that no generalizations can be made from these sorts of studies, for the associations between heterozygosity and growth rate may reflect yet another variable—cone production (73). Further studies on ponderosa pine do reveal a negative relationship between the relative growth rate and relative levels of female cone production. In addition, more heterozygous individuals, as a group, have lower variance in the relative levels of female cone production than more homozygous individuals (Y. B. Linhart & J. B. Mitton, unpublished manuscript).

Protein heterozygosity is also associated with characteristics of seedling growth in ponderosa pine (M. A. Farris & J. B. Mitton, unpublished manuscript). In this greenhouse study of seedling growth, approximately 20 seeds were collected from each of 8 different trees near Boulder, Colorado, and the seeds were weighed and allowed to germinate and grow in perlite-filled, clear plastic tubes. After germination the lengths of the shoot and root were measured every 2 days. At the termination of the experiment, the genotype of each seedling for each of the 11 polymorphisms was determined from needle tissue. Seedling heterozygosity was not related to shoot growth, but a positive association appeared between seedling heterozygosity and root length after the megagametophyte, the nutrient storage tissue, separated from the seedling.

Protein heterozygosity is associated with diameter growth in the pitch pine, *Pinus rigida* (57). Trees from 8 natural populations were cored in order to estimate the diameter growth, and the genotypes of these trees were inferred for 21 proteins from haploid megagametophytic tissue. The relationship between heterozygosity and diameter growth was highly dependent upon the age of the stand, with the correlation between individual heterozygosity and growth rate becoming increasingly positive as the mean age of the stand increased. The authors speculated that it was in the most mature stands that light, space, nutrients, and water became limiting, and under these conditions of stiff competition, the variation in potential for growth among genotypes was expressed.

Relationships among protein heterozygosity and growth rate have now been reported in marine invertebrates, the tiger salamander, white-tailed deer, pigs, sheep, man, quaking aspen, and several species of conifers. The reports summarized above all compare growth characteristics among individuals within the same population, and in virtually every case, the most highly heterozygous individuals enjoy some advantage. The phenomenon appears to be general.

Now that the generality of the phenomenon has been established, we can focus upon the mechanism underlying the phenomenon. Do the enzymes employed in these studies influence metabolic rates, or are they simply convenient chromosome markers, providing information on heterozygosity at many other loci or perhaps provide only a ranking of individuals by their degree of inbreeding? To address this problem, we must get closer to the level of enzyme action.

Protein Heterozygosity and Oxygen Consumption

If enzyme polymorphisms directly influence such gross measures of fitness as developmental stability and growth rate, then the associations between these measures and heterozygosity must reflect the underlying influences of the enzyme polymorphisms upon metabolism. This line of reasoning led Koehn & Shumway (54) to follow up the studies of the relationships between heterozygosity and growth rate in the American oyster (93, 112) with an examination of the relationship between heterozygosity and oxygen consumption. The animals used in this study were collected in a different year and from a different population than those in the earlier studies, and the five protein polymorphisms in this study only partially overlap those in the earlier ones. Oxygen consumption was measured in the laboratory under both normal temperature and salinity conditions and under a combination of high temperature and salinity conditions that constituted a stress to the oysters. After controlling for the effects of size, oxygen consumption was measured for each individual under normal and stress conditions, and the genotypes at each of the 5 enzyme polymorphisms were determined. Oxygen consumption under both sets of conditions was highly correlated with the number of heterozygous loci. Both regression lines have a negative slope, but the regression line for the stress measurement has a much steeper slope. Under stress, the energy demand of the 5 locus homozygote is 2.5 times as great as that of the 5 locus heterozygote. When the data are analyzed one locus at a time, the results are consistent across loci. In each case, the oxygen demand of the heterozygotes is significantly smaller than that of homozygotes. These data are consistent with the information on growth rate. If all animals consume the same amount of energy, highly heterozygous individuals' lower oxygen consumption should leave more energy to invest in growth.

The positive relationship between enzyme heterozygosity and growth rate in the tiger salamander prompted laboratory studies of the relationship between heterozygosity and oxygen consumption (J. B. Mitton, C. Carey, and T. D. Kocher, unpublished manuscript). Resting oxygen consumption and active oxygen consumption were measured, and the genotype of each animal was obtained for each of 11 polymorphic loci. The number of heterozygous loci was negatively correlated with resting oxygen consumption, an association analogous to the relationships reported for oysters (54). The number of heterozygous loci was also related to oxygen consumption under forced activity, but this relationship was positive, i.e. highly heterozygous individuals consumed more oxygen. Thus, highly heterozygous individuals consume less oxygen at rest and more oxygen in vigorous exercise than more homozygous ones.

INTERPRETATIONS AND IMPLICATIONS

Crop Plants

The causes of heterosis have been investigated most thoroughly in the crop plant literature (25). In the opening chapter of the most recent comprehensive review of heterosis, Jinks took perhaps the strongest position and said: "most of the critical evidence from biometrical genetical analyses points to dispersion as the major cause of heterosis [in crop plants]" (40, p. 44). The term *dispersion* may not be familiar to biologists working with natural populations. Dispersion and *association* are quantitative genetic terms describing the genome-wide tendency for all alleles that have, say, a positive effect on growth to be associated in one pure bred line (complete association) or to be distributed equally between two pure bred lines (complete dispersion). Jinks asserted that most pure bred lines fall somewhere between these two extremes and that heterosis in crosses between strains is due to the assembly of a "correct gene content" rather than to heterozygosity per se. Therefore, a breeder would do equally well in assembling either a highly homozygous line with the "correct gene content" or a highly heterozygous line containing the same genes. Jinks (40) presents an impressive case, both theoretically and empirically, to support his view that dispersion is the primary cause of heterosis.

How do these results apply to natural populations? We believe that their relevance is limited. First and most importantly, natural populations differ dramatically from pure breeding stocks (32). Second, it seems unlikely that levels of dispersion would be correlated with levels of heterozygosity in natural populations—a necessary condition if dispersion is to explain variation in growth and development. Each generation of sexual reproduction reshuffles the gene pool, randomizing the distribution of alleles at different loci.

Natural Populations

Why is variation in the number of heterozygous proteins related to developmental stability, growth rate, and oxygen consumption in natural populations? Three possibilities are commonly presented in the literature: (*a*) the enzymes mark blocks of chromosomes and are fortuitously linked to genes directly affecting growth and development; (*b*) protein polymorphisms constitute a sample of genes whose heterozygosity reflects a continuum between highly inbred (low heterozygosity) and randomly outbred (high heterozygosity) individuals; and (*c*) the genotypes of enzyme polymorphisms typically exhibit different kinetic characteristics; these differences affect the flow of energy through metabolic pathways and thereby influence growth, development, and oxygen consumption. Although this issue will not be resolved here, theoretical considerations and other empirical data make some of these possibilities more likely, others less so.

Most of the empirical observations indicate the existence of associations between other variables and the level of individual heterozygosity or the number of heterozygous enzyme polymorphisms. What is heterozygosity really measuring? When gel electrophoresis was introduced to population geneticists (60), proteins were presented as a random sample of the genome. This supposition is clearly incorrect. Different groups of proteins (structural, storage, metabolic, nonmetabolic, and regulatory) have different levels of genetic variation 41, 64, 83), and proteins with different quaternary structures and subunit sizes vary regularly in their levels of genetic variation (34, 51, 104, 105). The genetic variability of proteins is no more representative of the genetic variability of the entire genome than antigenic loci are representative of genes influencing morphological variation. Genetic variation at all loci is subject to mutation and genetic drift—there may be associations among genetic variabilities in these sets of loci, but there do not have to be. Furthermore, individual heterozygosity at a few (or a few dozen) loci may be correlated with individual heterozygosity at a larger number of loci (perhaps 20 to 100), but a small number of loci cannot accurately rank individuals within a population for the individual heterozygosity of the entire genome (13, 75). Heterozygosity at a dozen enzyme polymorphisms is much more likely to estimate enzyme heterozygosity within one or two metabolic pathways than at several thousand variable loci. This line of reasoning suggests that the metabolic pathways sampled in electrophoretic analyses are likely to influence variability in growth rates.

Genes are linked on chromosomes, and natural selection acting upon one gene will influence the frequencies of tightly linked polymorphic genes (59, 101). But linkage is not sufficient justification for attributing the genotypic associations between a marker gene and physiological characteristics to other unseen, undetected loci. Associations will only be evident if the genotypes at the linked genes are not independently distributed, that is, if there is linkage disequilibrium between loci. Linkage is a common phenomenon; linkage disequilibrium is not. Surveys of natural populations generally reveal little or no linkage disequilibrium (76), and studies of linked gene systems reveal that the linkage disequilibrium generated by stochastic processes decays more quickly than expected due to the superior fitness of highly heterozygous individuals (3, 14). The distribution of genotypes at one locus will occasionally reflect selection at other loci, but this is not likely to be the primary explanation of so general a phenomenon (3).

Ledig et al (57) have recently presented a useful portrayal of a major school of thought on the significance of the empirically observed relationship between heterozygosity and various morphological parameters. As a consequence of their work with pitch pine—in which they observed positive associations between heterozygosity and growth rate in mature stands—they proposed that the primary explanatory factor was the degree of outcrossing. It is well recognized that inbreeding by plants with predominantly outcrossing mating systems results in severe inbreeding depression. Ledig et al made two assertions: (a) the enzyme loci observed in electrophoretic studies are basically unimportant in determining growth rate or stability and serve only as useful markers for heterozygosity at loci that really do matter, and (b) it is the presence of homozygosity at particular loci that results in the comparative reduction in growth or stability, rather than the general heterozygosity that is the key attribute of the individuals studied. Their argument that protein polymorphisms serve only as markers is significantly weakened by the results described above. The view that the polymorphic enzymes do matter is also supported by the kinetic work described below. Is it simply a matter of semantics to argue over whether it is the degree of heterozygosity or of homozygosity that is most important? The empirical data indicate that the major factor is the level of heterozygosity (or 1 − level of homozygosity). Inbreeding and selfing decrease the level of individual heterozygosity, but individuals with all levels of individual heterozygosity can be produced by complete outcrossing. In addition, the levels of heterozygosity can be varied by mechanisms other than variation in the mating system. For these reasons, we argue that the level of heterozygosity is the more general concept and that the level of inbreeding is an important special case.

Kinetic studies of enzyme polymorphisms generally reveal that proteins produced by different genotypes at a locus perform differently; the kinetic differences are often consistent with patterns of geographic variation or gross physiological measures. The first example of this sort of study was a kinetic analysis of a serum esterase polymorphism in the sucker *Catostomus clarkii*, and a survey of the geographic variation in the allelic frequencies of this polymorphism (49). The esterase genotypes differed in the temperature at which they expressed their maximum velocity, and the temperature of maximum velocity corresponded to the average climatic temperature in the field where that genotype was most common. This study provided the model for a genre of studies that has become increasingly sophisticated, both in the examination of enzyme kinetics and in the field and laboratory tests of predictions taken from kinetic data (11, 12, 17–19, 37, 50, 53, 66, 94, 106–108; reviewed in 55, 65).

Perhaps the most complete story of the influence of a polymorphic enzyme locus upon geographic distribution, physiology, and demography is that of lactate dehydrogenase (LDH) in *Fundulus heteroclitus*. The *LDH-B* locus, the LDH most common in the heart, has two common alleles in *F. heteroclitus*. The genotypes have distinct kinetic properties, and the different temperature optima for kinetic variables can be used to describe the latitudinal cline in gene frequencies along the East Coast between Maine and North Carolina (86). LDH is associated with viability differences within the life cycle (74), and heterozygotes at this locus exhibit lower morphological variance than homozygotes, suggesting that LDH exerts an influence upon developmental stability (70). The three *LDH* genotypes differ in the levels of ATP associated with red blood cells (81, 85), which has led to several testable hypotheses concerning the physiology of whole animals. Varying the level of ATP associated with red blood cells should alter the amount of oxygen delivered to tissues. This hypothesis was tested for both the timing of egg hatching (which is dependent upon oxygen tension) and adult swimming endurance. Homozygotes differed in their swimming endurances, and as predicted, the differences were temperature-dependent (19). Furthermore, the time until hatching of eggs was dependent upon the *LDH* genotype, and there was a regular progression of hatching times from one homozygote, to the heterozygote, to the other homozygote (18). This study is the most complete picture available of the ways in which a single enzyme locus can influence development, time of hatching, swimming endurance, and viability differentials, but there is no reason to suspect that this enzyme in this species is unique. The differential effect of *LAP* genotypes on the osmoregulation of the blue mussel, *Mytilus edulis* (50) and the influence of PGI upon the differential survival and flight activity of *Colias eurytheme* (106–108) are also well documented.

The Paradox of Kinetic Intermediacy and Overdominance

A general finding of enzyme kinetic studies is the intermediacy of heterozygotes (28). The common result in the empirical studies summarized here is overdominance—heterozygotes enjoy higher growth rates and greater regularity in their development. Is it possible to translate biochemical intermediacy into fitness superiority?

Evolutionary biologists recognize the major significance of environmental variation, in both space and time (36). Let us imagine a life cycle as a series of chained events (for example, germination, seedling establishment, growth, reproduction) in which the fitnesses of genotypes vary but the heterozygote is always intermediate between homozygotes. Let *fAA1* and *faa1* represent the fitnesses of genotypes *AA* and *aa* during event 1. In a series of 2 events, if *fAA1* = *faa2*, *faa1* = *fAA2*, and *fAa* is always intermediate, then *fAa* > *fAA* = *faa*, for all *f*. In the example in Table 1, the fitnesses of homozygotes reverse in successive chained events within a life cycle, leaving heterozygotes with the highest fitness. For a greater number of chained events, the conditions leading to heterozygote superiority are somewhat less restrictive. This theme is presented in detail by Gillespie (27, 28), who concludes that this fitness configuration and variation can result in overdominance and developmental homeostasis.

The theory of fitness determination from a large pool of polymorphic loci has been undergoing constant revision ever since population geneticists were introduced to electrophoretic data. Initially, attention focused upon the genetic load produced if selection maintained all or most of the polymorphisms. The error in this way of thinking was quickly discovered (45, 68, 99), and the development of rank-order or threshold selection models reduced concern over genetic load and made these discussions more biologically relevant (69, 110). The most recent developments in models of fitness determination are exciting, for they predict associations between components of fitness and individual heterozygosity. For loci that are maintained polymorphic by balancing selection (a most important modifying clause)—in the great majority of cases examined with computer simulation—the fitness of an individual increases with the number of its heterozygous loci (29, 30, 102). This is the general pattern in the empirical data being reviewed here: Among individuals within a population, the more heterozygous individuals exhibit superior growth rates and enhanced developmental stability. These studies do not prove that the theory is correct or that enzymes are maintained by balancing selection, but the congruence between theoretical expectation and empirical observation certainly leaves room for optimism.

Heterosis and overdominance seem to be enhanced in fluctuating environments and most clearly seen when stocks or genotypes are examined across a range of environments. This generalization appears to be consistent with empirical observations on domesticated species (89, 91), theoretical discussions (27, 36), and experimental results (67, 82, 84). Furthermore, it is consistent with the advantages accruing to kinetically intermediate heterozygotes in fluctuating environments. Kinetic and physiological analyses of enzyme polymorphisms have revealed that these polymorphisms can have major physiological effects (55).

TABLE 1   A chain of intermediate fitnesses can result in superior fitness

Genotype	Fitness in event I	Fitness in event II	Product
AA	1.0	0.4	0.40
Aa	0.7	0.7	0.49
aa	0.4	1.0	0.40

Negative Results

The relationships between protein heterozygosity and both growth and development appear to be general, but they are certainly not universal. They have been detected with statistical methods, and they are discovered more easily under some circumstances than others. For example, heterozygosity and mean growth rates are associated in young oysters and pine seedlings, but not in adult ponderosa or lodgepole pines. Virtually all of the surplus energy of larval oysters and pine seedlings is put into growth, while the surplus energy of adult trees in partitioned between reproduction and growth. Such differential apportionment of energy may weaken the relationship between growth and heterozygosity.

The association between heterozygosity and developmental stability is not expected in all characters (96), for some are physically constrained in their development (consider bones in the skull), while others are not, and characters differ in their degree of canalization. Soulé (96) predicts that the relationship between heterozygosity and developmental stability will be strongest in unconstrained characters with low canalization and small coefficients of variation. The negative results reported for plaice (63) may fit this prediction.

We predict, as others have before us, that the advantages of heterozygosity will increase with environmental heterogeneity. There is a positive association between heterozygosity and growth in blue mussels in a natural environment (52), but none was detected in the laboratory (7). The different degrees of environmental variation in these studies may explain these disparate observations.

RECOMMENDATIONS FOR FURTHER RESEARCH

Attempting to chart the course of future basic research is probably foolhardy and maybe even undesireable. Nevertheless, we feel that there are several directions of research that will be particularly fruitful.

First, we believe that analyses within populations that contrast individuals from different sections of the heterozygosity axis will be most informative under stressful conditions. For example, water use efficiency in plants may vary with heterozygosity under conditions of low soil moisture but not otherwise. Thus, it will be easier to distinguish the characteristics of highly heterozygous from those of highly homozygous individuals during certain seasons and life cycle stages.

Second, we believe that the degree of heterozygosity will have the greatest impact on characters that are significant components of fitness. Growth rate, emphasized above, certainly contributes to fitness in most organisms. We recommend that other variables, such as respiration rate, photosynthetic rate, water use efficiency, fecundity, age and size at first reproduction, and scope for growth, should be examined for their response to different levels of enzyme heterozygosity.

Third, we believe that a comparative search to determine the limits of generality of the phenomenon would be worthwhile. One approach would be to compare the strength of associations with heterozygosity among populations (or species) that differ dramatically in their degree of environmental heterogeneity. We predict that there is a positive correlation between the strength of such correlations and environmental heterogeneity.

Finally, we believe that the generality of associations between heterozygosity and growth justifies placing the phenomenon in an applied and predictive context. For example, we anticipate that the costs of an electrophoretic screening of forest tree seedlings prior to restocking would be more than repaid by the enhanced growth rate of highly heterozygous individuals chosen for planting.

CONCLUDING REMARKS

Enzymes are protein catalysts that control the flow through metabolic pathways. Many enzymes in natural populations are polymorphic, and for those enzymes whose kinetics have been investigated, there are typically differences in performance among genotypes. Such differences can also be detected at the level of the whole animal or plant. Empirical studies have revealed variation among heterozygosity classes in growth rates, developmental stability, and oxygen consumption. These observations cover a remarkably broad taxonomic range: gymnosperms, angiosperms, invertebrates, and vertebrates, including humans. We believe that there is sufficient evidence on hand to state that individual organisms' level of heterozygosity is a major organizing principle in natural populations of both plants and animals.

Many of the empirical studies summarized here report associations among individual heterozygosity (i.e. the number of heterozygous loci) and measures of performance such as growth rate and developmental stability. While these observations may have been unexpected a decade ago, they are now predicted by fitness determination models in population genetics for those loci whose variation is maintained by balancing selection. Instead of being bewildered by a vast number of multilocus genotypes, we should examine a single axis of genetic variation composed of polymorphic enzymes—the continuum of individual heterozygosity. We are optimistic that these studies will provide fresh insights in population biology and valuable tools for plant and animal breeders.

ACKNOWLEDGMENTS

We wish to acknowledge the support provided by a fellowship from the John Simon Guggenheim Foundation to JBM.

Literature Cited

1. Adams, M. W., Shank, D. B. 1959. The relationship of heterozygosity to homeostasis in maize hybrids. *Genetics* 44:777–86

2. Allard, R. W. 1961. Relationship between genetic diversity and consistency of performance in different environments. *Crop Sci.* 1:127–33

3. Asmussen, M. A., Clegg, M. C. 1981. Dynamics of the linkage disequilibrium function under models of gene-frequency hitchhiking. *Genetics* 99:337–56

4. Baker, C. M. A., Manwell, C. 1977. Heterozygosity of the sheep: Polymorphism of 'malic enzyme', isocitrate dehydrogenase (Nadp), catalase and esterase. *Aust. J. Biol. Sci.* 30:127–40

5. Beardmore, J. A., Shami, S. A. 1976. Parental age, genetic variation and selection. In *Population Genetics and Ecology*, ed. S. Karlin, E. Nevo, pp. 3–22. New York: Academic

6. Beardmore, J. A., Shami, S. A. 1979. Heterozygosity and the optimum phenotype under stabilizing selection. *Aquilo Ser Zool.* 20:100–10

7. Beaumont, A. R., Beveridge, C. M., Budd, M. D. 1983. Selection and heterozygosity within single families of the mussel *Mytilus edulis* (L.). *Mar. Biol. Lett.* 4:151–61

8. Becher, H. C., Geiger, H. H., Morgenstern, K. 1982. Performance and phenotypic stability of different hybrid types in winter rye. *Crop Sci.* 22:340–44

9. Bottini, E., Gloria-Bottini, F., Lucarelli, P., Polzonetti, A., Santoro, F., Varveri, A. 1979. Genetic polymorphisms and intrauterine development: Evidence of decreased heterozygosity in light for dates human newborn babies. *Experientia* 35:1565–67

10. Bradshaw, A. D. 1965. Evolutionary significance of phenotypic plasticity in plants. *Adv. Genet.* 13:115–55

11. Burton, R. S., Feldman, M. W. 1983. Physiological effects of an allozyme polymorphism: Glutamate-pyruvate transaminase and response to hyperosmotic stress in the copepod *Tigriopus californicus*. *Biochem. Genet.* 21:239–51

12. Cavener, D. R., Clegg, M. T. 1981. Evidence for biochemical and physiological differences between enzyme genotypes in *Drosophila melanogaster*. *Proc. Natl. Acad. Sci. USA* 78:4444–47

13. Chakraborty, R. 1981. The distribution of the number of heterozygous loci in an individual in natural populations. *Genetics* 98:461–66

14. Clegg, M. T., Kidwell, J. F., Horch, C. R. 1980. Dynamics of correlated genetic systems. V. Rates of decay of linkage disequilibria in experimental populations of *Drosophila melanogaster*. *Genetics* 94:217–34

15. Cothran, E. G., Chesser, R. K., Smith, M. H., Johns, P. E. 1983. Influences of genetic variability and maternal factors on fetal growth in white-tailed deer. *Evolution* 37:282–91

16. Crow, J. F. 1948. Alternative hypotheses of hybrid vigor. *Genetics* 33:477–87

17. Day, T. H., Hillier, P. C., Clarke, B. 1974. The relative quantities and catalytic activities of enzymes produced by alleles at the alcohol dehydrogenase locus in *Drosophila melanogaster*. *Biochem. Genet.* 11:155–65

18. DiMichele, L., Powers, D. A. 1982. LDH-B genotype-specific hatching times of *Fundulus heteroclitus* embryos. *Nature* 296:563–65

19. DiMichele, L., Powers, D. A. 1982. Physiological basis for swimming endurance differences between LDH-B genotypes of *Fundulus heteroclitus*. *Science* 216:1014–16

20. Dobzhansky, T., Spassky, B. 1963. Genetics of natural populations. XXXIV. Adaptive norm, genetic load, and genetic elite in *D. pseudoobscura*. *Genetics* 48:1467–85

21. Dobzhansky, T., Wallace, B. 1953. The genetics of homeostasis in *Drosophila*. *Proc. Natl. Acad. Sci. USA* 39:162–71

22. Eanes, W. F. 1978. Morphological variance and enzyme heterozygosity in the monarch butterfly. *Nature* 276:263–64

23. Eanes, W. F. 1981. Enzyme heterozygosity and morphological variance. *Nature* 290:609–10

24. Fleischer, R. C., Johnston, R. F., Klitz, W. J. 1983. Allozymic heterozygosity and morphological variation in house sparrows. *Nature* 304:628–30

25. Frankel, R., ed. 1983. *Heterosis: Reappraisal of Theory and Practice*. Berlin: Springer-Verlag. 290 pp.

26. Fujio, Y. 1982. A correlation of heterozygosity with growth rate in the Pacific oyster, *Crassostrea gigas*. *Tohoku J. Agric. Res.* 33:66–75

27. Gillespie, J. 1977. A general model to account for enzyme variation in natural populations. IV. The quantitative genetics of viability mutants. In *Measuring Selection in Natural Populations*, ed. F. B. Christiansen, T. M. Fenchel, pp. 301–14. Berlin: Springer-Verlag. 564 pp.

28. Gillespie, J. H., Langley, C. H. 1974. A general model to account for enzyme variation in natural populations. *Genetics* 76:837–48

29. Ginzburg, L. R. 1979. Why are heterozygotes often superior in fitness? *Theor. Popul. Biol.* 15:264–67

30. Ginzburg, L. R. 1983. *Theory of Natural Selection and Population Growth.* Menlo Park, Calif: Benjamin/Cummings. 155 pp.

31. Grant, M. C., Mitton, J. B., Linhart, Y. B. 1982. Ecological and evolutionary studies of forest trees in Colorado. In *Ecological Studies in the Colorado Alpine: A Festschrift for John W. Marr, Univ. Colo. Occas. Pap. No. 37,* ed. J. Halfpenny, pp. 96–100

32. Hamrick, J., Linhart, Y. B., Mitton, J. B. 1979. Relationships between life history characteristics and electrophoretically detectable genetic variation in plants. *Ann. Rev. Ecol. Syst.* 10:173–200

33. Handford, P. 1980. Heterozygosity at enzyme loci and morphological variation. *Nature* 286:261–62

34. Harris, H., Hopkinson, D. A., Edwards, Y. H. 1977. Polymorphism and the subunit structure of enzymes: A contribution to the neutralist-selectionist controversy. *Proc. Natl. Acad. Sci. USA* 74:698–701

35. Hebert, P. D. N., Ferrari, D. C., Crease, T. J. 1982. Heterosis in *Daphnia*: A reassessment. *Am. Nat.* 119:427–34

36. Hedrick, P. W., Ginevan, M. E., Ewing, E. P. 1976. Genetic polymorphism in heterogeneous environments. *Ann. Rev. Ecol. Syst.* 7:1–32

37. Hoffmann, R. J. 1983. Temperature modulation of the kinetics of phosphoglucose isomerase genetic variants from the sea anemone *Metridium senile.* *J. Exp. Zool.* 227:361–70

38. Jackson, R. C. 1976. Evolution and systematic significance of polyploidy. *Ann. Rev. Ecol. Syst.* 7:209–34

39. Jain, S. K., Allard, R. W. 1960. Population studies in predominantly self-pollinated species. I. Evidence for heterozygote advantage in a closed population of barley. *Proc. Natl. Acad. Sci. USA* 46:1371–77

40. Jinks, J. L. 1983. Biometrical genetics of heterosis. In *Heterosis: Reappraisal of Theory and Practice,* ed. R. Frankel, pp. 1–46. Berlin: Springer-Verlag. 290 pp.

41. Johnson, G. B. 1974. Enzyme polymorphism and metabolism. *Science* 184:28–37

42. Johns, P. E., Baccus, R., Manlove, M. N., Pinder, J. E. III, Smith, J. H. 1977. Reproductive patterns, productivity, and genetic variability in adjacent white-tailed deer populations. *Proc. Ann. Conf. Southeast. Assoc. Game Fish Comm.* 31: 167–72

43. Jones, D. F. 1958. Heterosis and homeostasis in evolution and applied genetics. *Am. Nat.* 92:321–28

44. Kat, P. W. 1982. The relationship between heterozygosity for enzyme loci and developmental homeostasis in peripheral populations of aquatic bivalves (Unionidae). *Am. Nat.* 119:824–32

45. King, J. L. 1967. Continuously distributed factors affecting fitness. *Genetics* 55:483–92

46. Klekowski, E. J. 1966. Evolutionary significance of polyploidy in Pteridophyta. *Science* 153:305–7

47. Knowles, P., Mitton, J. B. 1980. Genetic heterozygosity and radial growth variability in *Pinus contorta. Silvae Genet.* 29:114–17

48. Knowles, P., Grant, M. C. 1981. Genetic patterns associated with growth variability in ponderosa pine. *Am. J. Bot.* 68:942–46

49. Koehn, R. K. 1969. Esterase heterogeneity: Dynamics of a polymorphism. *Science* 163:943–44

50. Koehn, R. K., Bayne, B. L., Moore, M. N., Siebenaller, J. F. 1980. Salinity related physiological and genetic differences between populations of *Mytilus edulis. Biol. J. Linn. Soc.* 14:319–34

51. Koehn, R. K., Eanes, W. F. 1978. Molecular structure and protein variation within and among populations. *Evol. Biol.* 11:39–100

52. Koehn, R. K., Gaffney, P. M. 1984. Genetic heterozygosity and growth rate in *Mytilus edulis. Mar. Biol.* In press

53. Koehn, R. K., Newell, R. J. E., Immerman, F. 1980. Maintenance of an aminopeptidase allele frequency cline by natural selection. *Proc. Natl. Acad. Sci. USA* 77:5385–89

54. Koehn, R. K., Shumway, S. R. 1982. A genetic/physiological explanation for differential growth rate among individuals of the American oyster *Crassostrea virginica* (Gmelin). *Mar. Biol. Lett.* 3: 35–42

55. Koehn, R. K., Zera, A. J., Hall, J. G. 1983. Enzyme polymorphism and natural selection. In *Evolution of Genes and Proteins,* ed. M. Nei, R. K. Koehn, pp. 115–36. Sunderland, Mass: Sinauer. 331 pp.

56. Leary, R. F., Allendorf, F. W., Knudsen, K. L. 1983. Developmental stability and enzyme heterozygosity in rainbow trout. *Nature* 301:71–72

57. Ledig, F. T., Guries, R. P., Bonefield, B. A. 1983. The relation of growth to

heterozygosity in pitch pine. *Evolution* 37:1227–38

58. Lerner, I. M. 1954. *Genetic Homeostasis*. Edinburgh: Oliver & Boyd. 134 pp.

59. Lewontin, R. C. 1964. The interaction of selection and linkage. I. General considerations: Heterotic models. *Genetics* 49: 49–67

60. Lewontin, R. C., Hubby, J. L. 1966. A molecular approach to the study of genetic heterozygosity in natural populations. II. Amount of variation and degree of heterozygosity in natural populations of *Drosophila pseudoobscura*. *Genetics* 54: 595–609

61. MacIntyre, R. J. 1976. Evolution and ecological value of duplicate genes. *Ann. Rev. Ecol. Syst.* 7:421–68

62. Makaveev, T., Venev, I., Baulov, M. 1978. Investigations on activity level and polymorphisms of some blood enzymes in farm animals with different growth energy. II. Correlations between homo- and heterozygosity of some protein and enzyme phenotypes and fattening ability and slaughter indices in various breeds of fattened pigs. *Genet. Sel.* 10:229–36

63. McAndrew, B. J., Ward, R. D., Beardmore, J. A. 1982. Lack of relationship between morphological variance and enzyme heterozygosity in the plaice, *Pleuronectes platessa*. *Heredity* 48:117–25

64. McConkey, E. H., Taylor, B. J., Phan, D. 1979. Human heterozygosity: A new estimate. *Proc. Natl. Acad. Sci. USA* 76:6500–4

65. McDonald, J. F. 1983. The molecular basis of adaptation: A critical review of relevant ideas and observations. *Ann. Rev. Ecol. Syst.* 14:77–102

66. McDonald, J. F., Anderson, S. M., Santos, M. 1980. Biochemical differences between products of the *Adh* locus in Drosophila. *Genetics* 95:1013–22

67. McDonald, J. F., Ayala, F. J. 1974. Genetic response to environmental heterogeneity. *Nature* 250:572–74

68. Milkman, R. D. 1967. Heterosis as a major cause of heterozygosity in nature. *Genetics* 74:727–34

69. Milkman, R. D. 1978. Selection differentials and selection coefficients. *Genetics* 88:391–403

70. Mitton, J. B. 1978. Relationship between heterozygosity for enzyme loci and variation of morphological characters in natural populations. *Nature* 273:661–62

71. Mitton, J. B. 1983. Conifers. In *Isozymes in Plant Genetics and Breeding, Part B*, ed. S. Tanksley, T. Orton, pp. 443–72. Amsterdam: Elsevier. 472 pp.

72. Mitton, J. B., Grant, M. C. 1980. Observations on the ecology and evolution of quaking aspen, *Populus tremuloides*, in the Colorado Front Range. *Am. J. Bot.* 67:202–9

73. Mitton, J. B., Knowles, P., Sturgeon, K. B., Linhart, Y. B., Davis, M. 1981. Associations between heterozygosity and growth rate variables in three western forest trees. In *Proc. Symp. Isozymes North Am. For. Trees & For. Insects, US Dep. Agric. Gen. Tech. Rep. PSW–48*, ed. M. T. Conkle, pp. 27–34. Washington DC: USGPO

74. Mitton, J. B., Koehn, R. K. 1975. Genetic organization and adaptive response of allozymes to ecological variables in *Fundulus heteroclitus*. *Genetics* 79:97–111

75. Mitton, J. B., Pierce, B. A. 1980. The distribution of individual heterozygosity in natural populations. *Genetics* 95: 1043–54

76. Mukai, T., Mettler, L. E., Chigusa, S. L. 1971. Linkage disequilibrium in a local population of *Drosophila melanogaster*. *Proc. Natl. Acad. Sci. USA* 68:1065–69

77. Pfahler, P. L. 1966. Heterosis and homeostasis in rye (*Secale cereale* L.). I. Individual plant production of varieties and intervarietal crosses. *Crop Sci.* 6: 397–401

78. Pfahler, P. L., Linshens, H. F. 1979. Yield stability and population diversity in oats (*Avena* sp.). *Theor. Appl. Genet.* 54:1–5

79. Pfeiffer, T. L., Schrader, E., Bingham, E. T. 1980. Physiological comparisons of isogenic diploid-tetraploid, tetraploid-octoploid alfalfa populations. *Crop Sci.* 20:299–302

80. Pierce, B. A., Mitton, J. B. 1982. Allozyme heterozygosity and growth in the tiger salamander, *Ambystoma tigrinum*. *J. Hered.* 73:250–53

81. Place, A. R., Powers, D. A. 1979. Genetic variation and relative catalytic efficiencies: Lactate dehydrogenase B allozymes of *Fundulus heteroclitus*. *Proc. Natl. Acad. Sci. USA* 76:2354–58

82. Powell, J. R. 1971. Genetic polymorphisms in varied environments. *Science* 174:1035–36

83. Powell, J. R. 1975. Protein variation in natural populations. *Evol. Biol.* 8:79–119

84. Powell, J. R., Wistrand, H. 1978. The effect of heterogeneous environments and a competitor on genetic variation in *Drosophila*. *Am. Nat.* 112:935–47

85. Powers, D. A., Greaney, G. S., Place, A. R. 1979. Physiological correlation between lactate dehydrogenase genotype and haemoglobin function in killifish. *Nature* 277:240–41

86. Powers, D. A., Place, A. R. 1978. Biochemical genetics of *Fundulus heteroclitus* (L.). I. Temporal and spatial variation in gene frequencies of LDH-B, MDH-A, GPI-B and PGM-A. *Biochem. Genet.* 16:593–607

87. Rasmusen, B. A. 1981. Blood groups and pork production. *BioScience* 31:512–15

88. Robertson, F. W., Reeve, E. C. R. 1952. Heterozygosity, environmental variation and heterosis. *Nature* 17:286–87

89. Rowe, P. R., Andrew, R. H. 1964. Phenotypic stability for a systematic series of corn genotypes. *Crop Sci.* 6:563–66

90. Shull, G. H. 1948. What is "heterosis"? *Genetics* 33:439–46

91. Simmonds, N. W. 1962. Variability in crop plants, its use and conservation. *Biol. Rev.* 37:422–65

92. Singh, S. M. 1982. Enzyme heterozygosity associated with growth at different developmental stages in oysters. *Can. J. Genet. Cytol.* 24:451–58

93. Singh, S. M., Zouros, E. 1978. Genetic variation associated with growth rate in the American oyster *(Crassostrea virginica)*. *Evolution* 32:342–53

94. Somero, G. N. 1978. Temperature adaptation of enzymes: Biological optimization through structure-function compromises. *Ann. Rev. Ecol. Syst.* 9:1–19

95. Soulé, M. 1979. Heterozygosity and developmental stability: Another look. *Evolution* 33:396–401

96. Soulé, M. E. 1982. Allomeric variation. I. The theory and some consequences. *Am. Nat.* 120:751–64

97. Spofford, J. B. 1969. Heterosis and the evolution of duplications. *Am. Nat.* 103:407–32

98. Spofford, J. B. 1972. A heterotic model for the evolution of duplications. *Brookhaven Symp. Biol.* 23:121–43

99. Sved, J. A., Reed, T. E., Bodmer, W. F. 1967. The number of balanced polymorphisms that can be maintained in a natural population. *Genetics* 55:469–81

100. Thoday, J. M. 1955. Balance, heterozygosity and developmental stability. *Cold Spring Harbor Symp. Quant. Biol.* 20:318–26

101. Thomson, G. 1977. The effect of a selected locus on linked neutral loci. *Genetics* 85:753–88

102. Turelli, M., Ginzburg, L. 1983. Should individual fitness increase with heterozygosity? *Genetics* 104:191–209

103. Vrijenhoek, R. C., Lerman, S. 1982. Heterozygosity and developmental stability under sexual and asexual breeding systems. *Evolution* 36:768–76

104. Ward, R. D. 1977. Relationship between enzyme heterozygosity and quaternary structure. *Biochem. Genet.* 15:123–35

105. Ward, R. D. 1978. Subunit size of enzymes and genetic heterozygosity in vertebrates. *Biochem. Genet.* 16:799–810

106. Watt, W. B. 1979. Adaptation at specific loci. I. Natural selection in phosphoglucose isomerase of *Colias* butterflies: Biochemical and population aspects. *Genetics* 87:177–94

107. Watt, W. B. 1983. Adaptation at specific loci. II. Demographic and biochemical elements in the maintenance of the *Colias* PGI polymorphism. *Genetics* 103:691–724

108. Watt, W. B., Cassin, R. C., Swan, M. S. 1983. Adaptation at specific loci. III. Field behavior and survivorship differences among *Colias* PGI genotypes are predictable from in vitro biochemistry. *Genetics* 103:725–39

109. Whaley, W. G. 1944. Heterosis. *Bot. Rev.* 10:461–98

110. Wills, C. 1981. *Genetic Variability.* Oxford: Clarendon. 312 pp.

111. Wright, S. 1977. *Evolution and the Genetics of Populations,* Vol. 3, *Experimental Results and Evolutionary Deductions.* Chicago: Univ. Chicago Press. 613 pp.

112. Zouros, E., Singh, S. M., Miles, H. E. 1980. Growth rate in oysters: An overdominant phenotype and its possible explanations. *Evolution* 34:856–67

Ann. Rev. Ecol. Syst. 1984. 15:501–22

THE APPLICATION OF ELECTROPHORETIC DATA IN SYSTEMATIC STUDIES

Donald G. Buth

Department of Biology, University of California, Los Angeles, California 90024

INTRODUCTION

The data base generally known as *electrophoretic data* is widely acknowledged to be of value to systematics (1, 2, 8, 17, 24, 79, 86). Although starch-gel electrophoresis of enzymes has become the established method of generating this data base, the analysis of electrophoretic data has remained varied and at times openly contested (28, 30, 55, 65, 67, 69, 82, 83). The treatment of these data has produced interesting results that have been perceived to demonstrate severe limitations on the nature and application of this data base at various taxonomic levels. Differing opinions on data treatment often obscure the multistep nature of these treatments. Many studies purporting to compare systematic treatments of electrophoretic data actually confuse the issue by simultaneously varying procedures at several levels, e.g. data transformation, coding, *and* method of analysis. Several options exist for each step, and these potential differences affect the results of comparative studies in various ways.

The historical perspective employed here is essential for understanding the development of issues and the refinement of procedures. This review also assesses the unnecessary limitations that have been imposed on the application of electrophoretic data in systematic studies and identifies the issues and problems that fuel the controversy over analytical procedures. Finally, recommendations for the phylogenetic treatment of electrophoretic data are presented.

0066-4162/84/1120-0501$02.00

Isozymes and Allozymes

Markert & Moller (50, p. 753) introduced the concept of *isozymes,* which they defined as "the different molecular forms in which proteins may exist with the same enzymatic specificity." The combination of zone electrophoresis and histochemical staining provided the technical means of resolving molecular variation, which initially was widely applied in the fields of developmental biology and genetics. In his recent review, Markert (49, p. 1) noted the utility of isozymes in the investigation of many kinds of biological problems, clearly recognizing that "sometimes isozymes are the subject of investigations, and sometimes they are used simply as tools to investigate other problems." The latter is certainly true in biochemical systematics, yet the application of these tools to systematic problems on a broad scale has been limited. Avise (1) attributed this early paucity of applications in systematics to emphases in other areas, e.g. the neutralist-selectionist controversy. Certainly some comparative work involving isozyme "patterns" did appear, but most of these studies were concerned with the biochemical identification of taxa rather than the elucidation of their systematic relationships. A problem has clearly existed regarding the large-scale quantification of isozyme data, and this problem has not been resolved in a satisfactory manner.

The field of population genetics developed rapidly as a primary consumer of isozyme technology. These applications, however, necessitated a partitioning of the isozyme concept, since they only used the relevant *allozyme* subset—defined by Prakash et al (70, p. 843) as "the variant proteins produced by allelic forms of the same locus, to avoid the now common confusion with 'isozymes' which are the various polymers produced from monomers specified by different loci." The development of genetic distance/similarity coefficients allowed the summarization of allozyme data for intersample comparisons. Although this quantification may have initially been designed with intraspecific studies in mind—that is, the gamma level of systematic study sensu Mayr (51)—quantified allozyme data were soon applied to comparative studies of taxa [or the beta level of study sensu Mayr (51)]. In a review of the systematic value of electrophoretic data, Avise (1) recognized the difference between isozyme and allozyme data sets but limited the discussion to allozymes only. However, isozyme data can also be used in systematic studies (17, 90). Some applications are discussed in this review.

Levels of Application

Bush & Kitto (10) evaluated several molecular methods for estimating the levels of genetic divergence between taxa (based on the degree of sensitivity and ease of analysis) and concluded that gel electrophoresis is the best method for comparing races, species, and closely related genera. Avise (1, p. 477) also noted that "it is doubtful that overall genic similarities determined by electrophoresis will be of great systematic value much beyond the level of the genus." Regardless of the systematic approach, inferences of relationships must be based on *some* characteristics shared among taxa. Many investigators will only use electrophoretic data up to the point at which the allelic divergence is sufficient at all loci to render the taxa totally different; after this point, there is no longer any means to assess relationships. Some have considered this limitation on electrophoretic data to "lower taxonomic levels" as a severe restriction on its utility in systematics. Indeed, for many researchers this limitation has been used to justify the development of new molecular methods to study higher level systematic problems. However, electrophoretic data are taxonomically limited only in those cases where comparisons of allozymes are made. Limiting systematic applications to the allozyme subset of all potential characters imposes an unnecessary restriction because many isozyme-level characteristics—e.g. the number of loci in a multilocus system, regulation of expression, and heteropolymer assembly—can be applied to higher taxonomic problems (17, 90).

Approach to Systematics

Avise's (1) perspective on the systematic value of electrophoretic data was based on the observation that the levels of genetic similarity among conspecific populations are high, whereas comparative values among species are, in general, much lower. He therefore deduced that closely related species may be arranged according to the percentages of shared alleles or genotypes. Avise stated that "many readers will recognize the electrophoretic technique as a phenetic, as opposed to a phyletic, approach to systematics" (1, p. 478). Mickevich & Johnson (55) asserted that Avise had "thoroughly confused" the important distinction between types of analysis (i.e. approaches to systematics) and sources of data (i.e. techniques of data collection). Electrophoretic methods yield data; they are not an approach to systematics. Equating electrophoretic data with the phenetic approach in systematics unnecessarily limits their application. It should be noted that in subsequent allozyme studies, Avise and his colleagues employed phyletic (i.e. phylogenetic) methods (4, 5, 67). These and numerous other studies have used electrophoretic data to infer the phylogenies of a broad array of organisms. Even so, there has been no concensus on the appropriate treatment of electrophoretic data in phylogenetic studies, and problems have persisted at several analytical stages. Recognizing such problems will facilitate their ultimate solution and improve phylogenetic treatment of electrophoretic data.

THE TREATMENT OF ALLOZYME DATA

According to Barrowclough (7, p. 226), "the analysis of electrophoretic data is straightforward." His statement must be viewed, however, within the context of what may be called a stereotyped procedure, which consists of (*a*) interpreting electromorphs on gels in terms of modern molecular genetics (allozymes), (*b*) computing allele frequencies at various loci, (*c*) converting allelic frequency data into a measure of genetic distance among populations, and (*d*) clustering these coefficients. I refer to this procedure as a stereotype (or "standard procedure" to some) because of its current widespread use and the fact that these basic steps were outlined a decade ago by Avise (1). Many current texts—whether from the vantage point of systematics (e.g. 31) or population genetics (e.g. 44)—promote this orientation toward treatment of electrophoretic data. But this procedure is just one of several that may be pursued. Each has its strengths and weaknesses, as will be discussed below.

Transformation to Distance Data

In discussing the conversion of particulate allozyme data to distance coefficients, Hedrick (44, p. 70) argued that "these measures help to consolidate the data into manageable proportions and aid one in visualizing general relationships among the groups. Although some information is lost by reducing arrays of frequency data to a single value, patterns among populations obscured by the mass of numbers may become apparent by utilizing genetic distance or similarity values." Those choosing to use distance coefficients were initially concerned about the accuracy of measurement. One way of increasing the accuracy of measurement is to enhance the sample size by increasing the number of individuals examined, the number of loci scored, or both. It has been demonstrated, however, that genetic distance estimates are far more severely affected by the number of loci sampled than by the number of individuals sampled (40, 64). The resolution of this issue allowed a more cost-effective planning of "distance-oriented" electrophoretic studies. Other considerations in designing studies, which are discussed below, include the choice of the distance/similarity coefficient and of the clustering algorithm, a subject of some controversy.

CHOICE OF DISTANCE/SIMILARITY COEFFICIENT   Several measures of genetic distance and similarity have been proposed. Many are highly correlated with one another, although they are often based on different biological or mathematical assumptions (1, 43, 44). For various reasons, two coefficients have come to predominate in the systematic literature: those proposed by Nei (63) and by Rogers (72). Each of these coefficients has mathematical properties that may influence their selection, use, and interpretation. Nei's genetic distance coefficient *(D)* is said to measure a biological property—the mean number of electrophoretically detectable substitutions per locus that have accumulated since the two populations diverged from their common ancestor. This coefficient can be corrected for error due to small sample sizes (64), but it is nonmetric (28); it does not satisfy the triangle inequality [when comparing the distances among three taxa A, B, and C, the distance (A,C) must be less than or equal to the distance (A,B) plus the distance (B,C)]. As a result of nonmetricity, the branch length may be negative, an undesirable and biologically uninterpretable result for a coefficient used in reconstructing phylogenies. Rogers's (72) similarity coefficient *(S)* has no biological premise; rather, it estimates the mean geometric distance between allele frequencies, summarizing this information across all loci. Although it has not yet been modified to correct for small sample sizes, this coefficient satisfies the triangle inequality and thus has fewer theoretical restrictions on its use in certain clustering algorithms.

CLUSTERING OF COEFFICIENTS   Several clustering algorithms are available for treating the distance matrices of allozyme data, including the Fitch and Margoliash (37) method, the unweighted pair-group method with arithmetic averages (UPGMA) (81), the Distance Wagner procedure (27), and the modified Farris method (83). Felsenstein (29) discussed the maximum likelihood estimation of evolutionary trees. He also considered other methods that deal with quantitative characters including allele frequencies which have not been used as extensively in systematic studies.

Several studies have attempted to evaluate some of the dendrograms produced by these clustering procedures by looking at the "goodness of fit" between the input distances in the data matrix and the output distances from the tree. Prager & Wilson (69) compared the Fitch-Margoliash method, UPGMA, and the Distance Wagner procedure and concluded that the Fitch-Margoliash method is superior. Swofford (82) reevaluated and corrected this study and reached the opposite conclusion: The Distance Wagner procedure provided the best fit of the data. Swofford also asserted that UPGMA can yield a phylogeny only when the rates of evolutionary divergence are relatively homogeneous, "a condition all too often unsatisfied in real data sets" (82, p. 26).

Tateno et al (83) used a different approach in comparing the Fitch-Margoliash method, UPGMA, the Distance Wagner procedure, and the modified Farris method in applications to a simulated phylogeny. These investigators found the Distance Wagner and modified Farris methods superior under the condition that the coefficients of variation of branch lengths are large (i.e. for "distantly related species"), concluding that "any tree-making method is likely to have errors in obtaining the correct topology with a high probability, unless all lengths of the true tree are sufficiently long" (83, p. 387).

Nei et al (65) simultaneously compared five genetic distance measures (including Nei's and Rogers's coefficients) and three clustering methods (UPGMA and the Distance Wagner and modified Farris methods), using a simulated phylogeny. They concluded that the best combination is to use Nei's standard distance coefficients clustered using UPGMA. They also found no improvement in accuracy from employing the metric Rogers's coefficient versus the nonmetric Nei's coefficient. Comparing this study with Swofford's (82) is interesting, yet it does not solve the problem. Nei et al (65) maintained that Swofford's goodness-of-fit criterion is a poor choice for evaluating trees, preferring instead to compare generated trees to a "known" tree. However, when the "known" tree is generally constructed with descendants of equal distance from branch points (65), an algorithm that calculates the branch lengths based on such a presupposition might be expected to perform well. When the analysis is biased in favor of equal rates of evolution, as in Nei et al's (65) study, the outcome will clearly be affected. Thus, Swofford's (82) and Nei et al's (65) studies are not necessarily comparable.

PROBLEMS   Some propose that it may be desirable to match a phylogenetic method such as the rate-independent Distance Wagner procedure with the use of a metric coefficient, e.g. Rogers's (72) coefficient. Others prefer to use UPGMA clustering of Nei's (63) coefficient, thereby assuming the constant evolutionary rates that are necessary for this combination to yield a phylogeny. Given Farris's (28) discussion of distance data in general, however, these arguments may not be relevant. Farris noted that (a) information is lost during the reduction of particulate allozyme data to distance measures, (b) many distance measures infer distances that cannot be realized in phylogeny reconstructions (e.g. negative branch lengths or unlikely ancestral conditions), and (c) neither metric nor nonmetric distances yield physically interpretable branch lengths. His (28) critique employed a "path-length interpretation" (30) of the observed distance between the hypothesized ancestors and the descendants along the branch of the tree. Branch lengths must correspond to the set of hypothetical ancestors and fit the observed distances, placing some constraints on possible distance values. These constraints, as well as other aforementioned problems, led Farris (28) to conclude that measures of allele frequency distance cannot be appropriately analyzed by fitting branch lengths to a distance matrix.

Felsenstein (30) took a different view regarding the use of distance methods for inferring phylogenies. He agreed with Farris's interpretation if the distances are viewed with the meaning that Farris ascribed to them. Felsenstein (30) went on to say, however, that the "path-length interpretation" is not necessarily the only one, and if the distances are viewed from a statistical standpoint, the resulting "expected distance interpretation" need not conform to the restrictions Farris (28) outlined. Felsenstein then discussed the potentially more serious problems with the assumptions of additivity and independence used under the expected distance interpretation, especially as "these assumptions are dubious for many kinds of data often analyzed by distance methods" (30, p. 23). Although it is difficult to say where this new interpretation will lead, viewing numerical analytical procedures from more aspects usually contributes to our understanding of both data and procedures, and this case should be no exception. Considerable work remains to be done in this area, and other views on the use of distance data will be forthcoming (J. S. Rogers, personal communication; D. L. Swofford, personal communication).

Transformation to Character/State Data

The use of character/state data has several advantages over the use of distance data, e.g. homoplasious steps may be identified, and actual branch lengths have physical meaning. But the use of character/state data is not without its problems. How to designate allozyme characters has not been obvious, and coding and ordering states within such characters remain problematic (57).

CODING   Fundamental problems with the coding of electrophoretic data into characters and states involve recognizing what comprises a character and, secondarily, determining how states may be partitioned within a particular character scheme. Several methods of achieving these goals have been proposed.

The allele as character   In the first treatment of character/state allozyme data in a phylogenetic study, Mickevich & Johnson (55) employed a coding scheme in which each allelic entity was recognized as a separate character, each of which had two possible states: "present," if its frequency was 0.05 or greater, or "absent," if its frequency was less than 0.05. This method of coding was later called the independent allele model by Mickevich & Mitter (56), who outlined its rationale:

> There is no a priori reason to rule out the occurrence of any combination of the observed alleles. Since the number of alleles at a locus can and does vary widely among taxa, one may regard the appearance or disappearance of individual alleles as independent events. Thus, each allele becomes a separate character whose state in each taxon is to be scored in some way (56, p. 47).

Mickevich & Johnson (55) recognized that an alternative method of coding allozyme data under an independent alleles model might employ the actual allele frequencies rather than the presence or absence of alleles. They noted, however, that "although it might seem that frequency coding would yield greater precision, presence-absence coding measures an evolutionarily more significant variable. Since selection can alter the frequencies of only those alleles that are present, acquisition of an allele, for cladistics, may be more important than subsequent modification of allelic frequency" (55, p. 261). Farris (28) also doubted that

> details on frequencies are useful at all. At any systematically worthwhile level, say subspecies or higher, it is nearly always true that a locus that differs appreciably between two taxa differs just about completely; entirely separate sets of alleles will be present [3, 6, 91]. This being the case, there is not much comparative information in the frequencies beyond simple presences (28, p. 22).

The distributions of loci with respect to genetic identity, which Ayala et al (6) examined across species, illustrate Farris's point. At most loci, complete allelic identity or complete difference is the rule; only a small proportion or loci share partial identity across species. This issue, however, is one of *grain* rather than of precision or evolutionary importance. To choose not to use allelic frequency information is simply to advocate a coarse-grained approach to coding; that is, to code the allelic characters with the barest minimum of information gathered or only distinguishing between presence and absence. With most loci, this procedure will work quite well, given the complete difference between species that often exists. Though an investigator may doubt the accuracy of the frequency estimates and choose not to take them seriously, an argument can be made that allelic frequency differences (if measured accurately) are real biological phenomena; information about them is lost under a coarse-grained coding scheme. Mickevich & Mitter (56) decided that either frequency coding or presence/absence coding may be used, depending upon whether or not the frequencies are regarded as cladistically informative.

Applications of frequency coding of allozyme data have been rare, however. Simon (80) used frequency coding in her study of some cicadas in which there were no taxon-specific alleles. Buth (11, 12, 14) applied frequency coding to entire allozyme data sets for several fish groups that included complete allelic differences at some loci. Mickevich & Mitter (56) conducted an empirical test of these coding alternatives; in most cases, they yielded the same cladogram. Further modifications of the frequency-based trees in incongruent cases (e.g. setting branch lengths less than 1.0 equal to zero) eliminated the branching sequence differences between even more cases. Mickevich & Mitter (56) therefore concluded that frequency coding provides no comparative information beyond the presence of alleles. This conclusion is surprising if one considers the method used to suppress the expression of additional information retrieved from the frequency-coded data set. Mickevich & Mitter (56) properly identified the independent alleles model as biologically unrealistic, however, because this model can yield a hypothetical intermediate without alleles at a locus. An alternative is to use the locus, rather than the allele, as a character.

The locus as character   Recognizing the locus as the biologically appropriate unit for character designation presented a new problem: how to recognize character states from data bases consisting of sets of electromorph frequencies (57). Mickevich & Mitter (57, p. 170) "discovered that the only logical recourse seems to be to treat entire electrophoretic profiles (e.g. allele frequency distributions) for each locus, as observed in individual taxa, as character states," i.e. "each locus becomes a single character whose states are combinations of alleles" (56, pp. 48–49). Although this result is desirable, states recognized in this fashion may be vulnerable to sampling error. In many instances, limitations on sampling may prevent the acquisition of an ideal data base; sampling error becomes a problem particularly for researchers who deal with rare or difficult to obtain organisms. As a result, some investigators choose to control for equivalent sample sizes among taxa, thus making it possible to use all data in a comparable fashion. Alternatively, one may employ some frequency boundary—e.g. a minimum frequency of 0.05 for an allele to be recognized as a significant contributor to an array at a locus—to minimize frequency differences due to sample sizes.

A hypothetical data set (Table 1) is encoded, using the locus as the character and entire allelic arrays (no minimum frequency requirement) as the states (Table 2). This encoding procedure, which is referred to as *qualitative coding*, has some shortcomings, however. Substantial differences in allele frequency of potential statistical significance can be obscured by this coding practice. An example of this (Table 2) shows that at Ldh-A, taxa A and C would be considered as having the same state "a+b" and taxa B and D, the same state "a+b+c," even though the allele frequencies are quite different in all cases. This particular example suggests that qualitative coding may be coarse-grained, may miss potential information available via frequencies, and may impart misinformation on the nature of "states" that are shared. A related problem is illustrated at the Ldh-C locus (Tables 1 and 2). The appearance of numerous minor alleles at this locus yields the designation of a different state for every taxon, so the coding might be considered to be too "fine-grained."

An alternative coding procedure illustrated in Table 3 may be referred to as *quantitative coding*. This procedure addresses the question of whether or not the states expressed by two taxa are statistically identical. Such states would be recognized as different following statistical evaluation of the individuals in the genotypic arrays of the sample, rather than via frequencies alone. Such statistical tests may require the assumption of Hardy-Weinberg equilibrium and may identify as significantly different two samples of identical allele frequencies differing only in the distribution of individuals in the genotypic arrays (D. L. Swofford, personal communication). If this situation confounds the application of this coding procedure, tests involving the frequency data alone must be employed. Alternatively, differences in genotypic arrays (i.e. possible equilibrium differences) among samples of equivalent allele frequencies may be viewed as informative and worthy of recognition as separate states. Quantitative coding can also be viewed as a course-grained procedure, as shown by the treatment of the data for the Ldh-B and Ldh-C loci of Table 1 in Table 3. No state differences are recognized among the four taxa, although shared or unique minor alleles might suggest otherwise.

Thus, qualitative and quantitative coding may exhibit quite different arrays of character states. Both coding procedures are vulnerable to problems of minimal and differential sample sizes and may be criticized as being too

Table 1 Sample allele frequency data set with alleles lettered at each locus

Taxon	Locus		
	Ldh-A	Ldh-B	Ldh-C
Taxon A			
F (a)	0.50	0.60	0.50
F (b)	0.50	0.40	0.50
Taxon B			
F (a)	0.10	0.50	0.50
F (b)	0.80	0.40	0.45
F (c)	0.10	0.05	0.05
F (d)		0.05	
Taxon C			
F (a)	0.95	0.45	0.50
F (b)	0.05	0.46	0.45
F (c)		0.05	
F (d)		0.04	0.05
Taxon D			
F (a)	0.95	0.40	0.55
F (b)	0.04	0.55	0.40
F (c)	0.01	0.03	
F (d)		0.02	
F (e)			0.05

coarse-grained or too fine-grained in certain situations. Empirical tests using real data sets are necessary to determine the value of these procedures.

ORDERING   Among the difficult aspects of dealing with allozyme data, Mickevich & Mitter (57) recognized that the ordering of character states—after they have been identified—is critically important. If only two states are recognized at a particular character, their association is obvious. When more than two states exist at a character, the pattern of linkage in states is less obvious because several combinations become possible. Mickevich & Mitter (56, 57) have developed several ordering methods, all employing the locus as the character, for these more complex situations. The first method, employing the "shared allele model," assumes that electromorphs with the same electrophoretic mobility are identical. Sharing identical allelic products forms the basis of the relationship between states (i.e. of the combination of alleles). The evolutionary ordering of these states is determined by applying a parsimony criterion that joins states so as to minimize the total amount of allelic evolution, i.e. the transformation series is developed by joining each state with the one that shares the most alleles. According to this model, the shared presence of alleles is more informative than their shared absence.

Table 2 Qualitative coding of allelic presence in sample data set from Table 1

Taxon	Locus		
	Ldh-A	Ldh-B	Ldh-C
Taxon A	$(a + b)$	$(a + b)$	$(a + b)$
Taxon B	$(a + b + c)$	$(a + b + c + d)$	$(a + b + c)$
Taxon C	$(a + b)$	$(a + b + c + d)$	$(a + b + d)$
Taxon D	$(a + b + c)$	$(a + b + c + d)$	$(a + b + e)$

The second method, the "minimum allele turnover model," is similar to the shared allele model in its determination of relatedness in allelic combinations and its utilization of a transformation series that minimizes the total number of allelic changes. This model, however, minimizes *losses* of alleles as well as gains, a possible shortcoming if one considers a shared presence to be more informative than a shared loss. In the third method, the "relative mobility model," the user determines the relatedness of allelic combinations based on the relative mobilities of allelic products, assuming that a mobility difference reflects a sequence difference and that multiple differences have accumulated in stepwise fashion (57). In the fourth, "systematic" method, Mickevich (54) applied transformation series analysis to order states that consisted of observed allelic combinations, again based on the locus as character. This method "can be regarded as fitting, in a recursive fashion, both transformation series and cladogram simultaneously to the observations" (57, p. 170). This method differs from the three previously discussed models that fix the transformation series independently of the cladogram. Further, the identity of electromorphs with equivalent electrophoretic mobilities is not assumed. Computer algorithms for this method are not generally available (none have been published), and much remains to be learned about the actual process used in the few published applications.

Empirical tests by Mickevich & Mitter (57) demonstrated the inferiority of the shared alleles model; of the three remaining ordering models, the minimum

Table 3 Quantitative coding of statistical differences in sample data set from Table 1

Taxon	Locus		
	Ldh-A	Ldh-B	Ldh-C
Taxon A	○[a]	■	●
Taxon B	☆	■	●
Taxon C	□	■	●
Taxon D	□	■	●

[a] Character states are indicated by arbitrarily chosen symbols; no ordering is to be inferred. (Adequate sample size to conduct statistical tests is assumed.)

allele turnover model performed best as determined by the combined criteria of "boldness" and "fit" (57). Although a transformation between the observed conditions that implies the least amount of evolutionary change may hold the greatest promise for applications in systematics, Mickevich & Mitter (57, p. 176) concluded that the minimum allele turnover model "is not the final statement on the matter, but will be a step toward a theory governing the evolution of these perplexing features."

CLUSTERING   Tests of clustering algorithms have not been applied to character/state allozyme data with the same rigor as they have been to distance-oriented studies (65, 69, 82, 83). Mickevich & Johnson's (55) comparison of minimum-length Wagner trees and UPGMA phenograms in an empirical test of the expectation of congruence using multiple data sets (including allozymes) was among the first to address this problem. Riska (71) and Colless (23), however, have pointed out a number of procedural errors in Mickevich & Johnson's (55) study that suggest that additional studies of this kind are necessary. Nevertheless, the construction of minimum-length Wagner trees (26) continues to be the clustering method of choice for those using allozyme data in particulate form (41, 58–60).

ROOTING   Assigning evolutionary direction, or polarity, is an important procedural step that may be employed when character-state trees are being developed or, alternatively, to root a Wagner network (48) yielding a Wagner tree. Farris (27) discussed the estimation of roots for undirected trees (i.e. networks), calculated by the Wagner algorithm. He outlined three methods for rooting networks; (a) using outgroups, (b) assuming that evolutionary rates are homogeneous, and (c) minimizing the variance of evolutionary rates. The use of outgroups assumes monophyly for the group under study, a group that must be well-enough defined so that taxa external to this group (i.e. outgroups) can be clearly identified. The utility of rooting with outgroups may be limited by the amount of information supplied by such groups. Primitive traits must be shared between the outgroup and the group under study for this rooting procedure to be useful. Often a single outgroup is used to identify the rooting point on a network. The accuracy of the placement of this root may be tested, however, by the use of additional outgroups (e.g. 14). If the group under study is indeed monophyletic and all of the outgroups are informative, the same rooting point should be recognized by all outgroups, increasing confidence in the placement of the root (for a more detailed discussion of outgroup analysis, see 87).

If the second method—assuming the homogeneity of evolutionary rates—is adopted, then the two most divergent OTUs [Operational Taxonomic Units sensu Sneath & Sokal (81)] in the network must have had the same rate of evolutionary change, and the root may be placed at the midpoint of the path between these OTUs. In the third method, rooting the network so as to minimize the variance of evolutionary rates also requires the assumption that there is some limitation on the possible variation of such rates.

The Wagner procedure does not contain any assumptions about evolutionary rates, so it can be used to test for their heterogeneity. Therefore, rooting procedures involving assumptions about these rates or placing limitations on their variances generally are inferior to the outgroup rooting procedure. The use of outgroups is the method of choice (J. S. Farris, personal communication) for phylogenetic studies. Although Farris (27) did not specifically discuss allozymes, it has also been shown that outgroup analysis is an appropriate method to root trees using allozyme data (8). The other rooting methods Farris (27) discussed may have to be employed to root a network, however, if the outgroups cannot be clearly identified or are otherwise uninformative (J. S. Farris, personal communication).

Brush (9, p. 163) noted that "it is often difficult to establish primitive-derived states in regard to biochemical characters," especially if one is attempting to establish polarity by making inferences about the functional attributes of the characters themselves. After character states have been coded and ordered, the evolutionary direction of the order can be identified via outgroup analysis. Patton & Avise (67) discussed a "qualitative Hennigian approach" to the analysis of allozyme data, emphasizing its utility as a nonnumerical alternative to quantitative analyses. Their method, however, is basically an outline of the application of Hennigian principles (45) to the choice of evolutionary polarity among characters states. Buth et al (20) had previously employed such a rooting system for allozyme data, but they did not specifically identify it as "Hennigian." But their specific application, as well as Patton & Avise's (67) general exposition of the method, may be quite limited due to the nature of the coding of the characters being analyzed. Both of these studies recognized the allele rather than the locus as the character; the shortcomings of this practice were discussed earlier. Murphy et al (62) applied these Hennigian principles in their study of the relationships among lizards of the genus *Eumeces*, recognizing the locus as the character for their allozyme data set.

PROBLEMS   Several problems arise at a number of the analytical steps involved in the character/state treatment of allozyme data. Coding procedures have been improved with the recognition that the locus is the character, yet considerable improvement is necessary for the recognition of states. Both qualitative and quantitative coding should be developed further and empirically tested. The ordering process is currently the most problematic, but it is also receiving considerable attention (54, 56, 57; D. L. Swofford, personal communication). The development of ordering procedures is necessary to accommodate complex branching characters in phylogenetic studies. Although Mickevich & Mitter (56, 57) did not discuss this subject specifically, their treatment of branching characters involved partitioning each branch of a complex character state tree into a series of separate characters, each originating at a particular node of the branching character-state tree (M. F. Mickevich, personal communication). Other branches of the original character-state tree would not be relevant to the order and direction of a particular branch (the new character), and elements on the other branches would simply be coded as sharing a primitive condition with the origin (root) of the new character. Some investigators may find this partitioning of characters unsatisfying, as it does impose new conditions to the concept of the locus as character. Perhaps new methods will be developed that can incorporate the entire branched character as a unit in the construction of Wagner trees.

THE TREATMENT OF ISOZYME DATA

The use of allozymes involves only part of the genetic information available in electrophoretic data. For example, insights into phylogenetic relationships can be obtained using characters that reflect additional genetic information available from the number and developmental expression of isozymes encoded in multilocus systems (88). Characters such as these have been applied in systematic studies for many years, but they have never achieved the visibility of allozyme data. One possible reason for this restricted application is the fortuitous nature of the discovery of isolated, systematically informative isozyme characters. These characters were used virtually one at a time until quite recently when entire data sets of isozyme characters began to be employed (e.g. see 34).

Kinds of Data

Three major catagories of isozyme characters have been recognized (15): (a) the number of genes controlling a multilocus enzyme system, (b) the regulation of enzyme expression, including tissue specificity and intensity of expression (i.e. qualitative and quantitative differentiation at the regulatory level), and (c) the ability, or lack thereof, of heteropolymer assembly in multimeric enzymes (due to interaction at the intralocus or interlocus level). These characters have already been used in systematic contexts in various groups of organisms. These characters and their uses are reviewed below. (For background information and further details see 17, 49, 88–90).

THE NUMBER OF GENES   The origin of new genes through gene duplication is a major force in evolution (66). The degree to which duplicated genes are related to one another depends upon the nature and period of their origin. New isozyme loci may be formed through tandem duplication (of a single gene) or polyploidization (of the entire genome). The appearance of a new gene, or genes, may serve as a synapomorphy in a particular lineage and thus be of considerable systematic value. Likewise, the loss of a gene in an already duplicated multilocus system can also constitute an evolutionary step and serve as a synapomorphy.

Genomic enhancement via tandem duplication is presumably uncommon, yet it has been reported at a number of taxonomic levels. In the diploid characid fish, *Cheirodon axelrodi,* a presumed tandem duplication of a glucosephosphate isomerase locus has been reported in 1 of 132 specimens examined (46). In the diploid cobitid fish, *Acanthopsis choirorhynchus,* one of the glucosephosphate isomerase loci exhibits a tandem duplication shared by all members of this species (32). Several genera of the tetraploid fish family Catostomidae share a duplication in the mitochondrial aspartate aminotransferase system that is presumably of tandem origin (13, 25). These examples indicate a directed sequence of events for a tandemly duplicated gene from its origin as a polymorphism within a single species, to its establishment as an apomorphic character of that species, to its retention as a synapomorphy in daughter species (16).

Polyploidization, either as allopolyploidy or autopolyploidy, is an important sympatric speciation process (38, 47, 66), which has played a greater role in certain groups of organisms, e.g. plants (42) and fishes (16, 75), than in others. A polyploidization event marks an important evolutionary step—an autapomorphy for newly formed species and eventual synapomorphy for any resulting daughter species. Duplicated loci, whether of tandem or polyploid origin, may persist in lineages for indefinite periods, although the allozymes of these loci may diverge in form and function. Thus, the locus number character may be useful in those lineages in which duplications have persisted for some time. An example of the utility of this character is the allotetraploidization event marking the origin of the catostomid fishes (85). Duplicate loci retained for at least 50 million years are interpreted as synapomorphies linking the extant species of the family Catostomidae (33). Fisher et al (36) discussed the evolution of several multilocus isozyme systems in the chordates, with gene duplications appearing as synapomorphies at a number of taxonomic levels among the lower vertebrates.

Thus far, changes in gene number have been discussed only in the context of gains of duplicated genes. The loss of expression by one member of a duplicated gene pair (functional diploidization) is also a possibility as long as the remaining gene continues to perform its original function (66). Therefore, among polyploid lineages, functional diploidization may yield characters of systematic value; the retained duplicated condition is clearly primitive in nature, whereas the reduction in the functional gene number is the derived condition. Ferris & Whitt (34) and Buth (13, 14) have applied gene reduction, or silencing, as a criterion in phylogenetic studies of tetraploid catostomid fishes.

GENE REGULATION   Each cell of a developing organism has the genetic potential to express all structural gene products. Yet spatial and temporal variation in isozyme expression appears to be the rule rather than the exception (68, 88–90). Systematists can take advantage of the tissue-specific nature of gene regulation in single-locus or multilocus enzyme systems by recognizing the expression of a given enzyme locus in a given tissue is scorable as a two-state (presence/absence) character. One of the best examples of the application of this kind of character is Shaklee & Whitt's (77) study of Ldh-C expression in the gadiform fishes. The Ldh-C locus is present in nearly all bony fishes; it is expressed in many tissues of morphologically primitive fishes, but it has quite restricted tissue expression in many teleosts. Expression of the Ldh-C locus is limited to either the eye *or* liver in many of these morphologically advanced fishes (78). Shaklee & Whitt (77, p. 564) recognized that "the widespread existence of the two derived LDH-C states (eye vs. liver) throughout the advanced teleosts suggests that, at a relatively early stage of evolution, canalization of the C gene was fixed into either of these restricted modes." Different families of gadiform fishes express Ldh-C products in either eye or liver, suggesting that the Gadiformes are not monophyletic (77). In addition, the differential patterns of expression of creatine kinase loci among specific tissues are of systematic value in several groups of fishes (35) and in amphibians and reptiles (22).

HETEROPOLYMER ASSEMBLY   Many enzymes are multimeric in nature; the enzyme is comprised of subunits that must be assembled prior to the initiation of enzyme function. If more than one kind of subunit is produced—as would occur in the heterozygous condition for a single locus or if multiple products of the same multilocus system are proximate—the different subunits may interact to yield heteropolymers. The heteropolymers exhibit intermediate electrophoretic mobilities relative to the respective homopolymers, with the exact mobility depending upon the number of each kind of subunit that was used to form the heteropolymer. The ability to form heteropolymers is, in general, presumed to be the primitive condition. It is assumed that newly mutated alleles and duplicated loci at early stages of allelic differentiation yield products that are structurally and metabolically similar, allowing the formation of heteropolymers. The derived condition—the loss of the ability to form heteropolymers—can be of systematic value. Such a character might be coded in a simple two-state (presence/absence) fashion.

    Gorman (39) was among the first to apply the heteropolymer assembly character in a systematic context. He examined the interactive expression of products of the tetrameric lactate dehydrogenases Ldh-A ("muscle form M") and Ldh-B ("heart form H") in lacertid lizards, noting the absence of the asymmetric heterotetramers ABBB and AAAB. Many other lizard families express all five expected LDH isozymes (AAAA, AAAB, AABB, ABBB, and BBBB) leading Gorman to suggest that this partial restriction of heterotetramer assembly may be a synapomorphy for the Lacertidae. Murphy (61) reported that the restriction of formation of just one (ABBB) asymmetrical heterotetramer of lactate dehydrogenase defined the lizard family Scincidae. Both partial and total restriction of LDH heterotetramer assembly has been shown to be of systematic utility among closely related darters of the genus *Etheostoma* (20) and for identification of families of cypriniform fishes, especially the Gyrinocheilidae (21). Toledo & Ribeiro (84) noted a correlation of increasing restriction in LDH heterotetramer assembly and phyletic advancement in advanced teleost fishes.

Applications and Problems

Isozyme characters may be treated like any other characters in terms of coding, ordering, etc. In many of the examples above, researchers have utilized two-state character coding, usually presence or absence of a particular condition. Other coding options are available and perhaps, in some cases, more desirable. For example, the regulation character of tissue-specific expression might possibly be coded with the locus as the character and the array of tissues in which it is expressed as the state. Such coding may not necessarily yield two-state characters, so that problems in ordering the states may result. An important limitation on the use of many of these isozyme characters should also be noted. With the coded examples provided, many of the derived conditions involve the loss or absence of the trait, e.g. gene silencing, restriction (or absence) of expression in a particular tissue, and restriction (or absence) of heteropolymer assembly. Such loss characters are especially vulnerable to homoplasy and should be used with appropriate caution.

CONCLUSIONS AND RECOMMENDATIONS

This review has considered the various types of information inherent in allozyme and isozyme data, as well as the options for phylogenetic treatment of these data in systematic studies. Although several analytical stages await development and refinement, researchers can pursue phylogenetic studies incorporating several alternative treatments of their data. Buth (18) outlined a number of recommendations for the phylogenetic application of allozyme data in systematic studies. These recommendations, placed in the context of the points raised in this review, are listed below.

1.   Researchers should incorporate a "geographic sampling strategy" (18) when making collection plans for a particular study. Although small samples

(in terms of the number of individuals) are adequate for measures of heterozygosity and genetic distance (40, 64), the extension of this concept to "exemplar" studies of taxa (i.e. based on extremely limited geographic sampling) by assuming that there is no geographic variation within taxa is premature. Many studies have demonstrated limited intraspecific geographic differentiation using genetic distance measures (1). Using distance measures can obscure statistically significant differences, however, if the nature of the allelic differentiation is quantitative rather than qualitative (19). As Selander & Whittam (76, p. 112) noted, populations of "a great variety of organisms, ranging from bacteria to humans . . ., are strongly structured genetically and . . . their evolution cannot be understood without reference to this structure."

2.   Raw data—in the form of sums of individuals exhibiting each scored genotype (e.g. 19)—should be published in phylogenetic studies so that they can be utilized to develop new or improved analytical procedures. Currently, expected documentation only includes simple allozyme frequencies. These frequencies cannot be reevaluated for quantitative coding or for relevant frequency limits under a qualitative coding scheme. Worse yet, some studies just present distance matrices, precluding any possibility of alternative character/state treatments.

3.   The coding of allozyme data should involve the recognition of the locus as the character (56). My preference is to use allozyme as well as isozyme data in particulate fashion, encoding these data as characters and states. This method is preferable because it yields the maximal information content and because of the other considerations Farris (28) discussed. The interpretation of distance treatments Felsenstein (30) advanced must also be considered, however, and it may be the forerunner of an enhanced debate on how to use genetic distance data in phylogenetic studies. Until these methods have been refined or new developments supersede them, I recommend the use of *both* qualitative and quantitative coding of particulate allozyme data.

4.   The problem of ordering is currently the most critical issue in this research area. I recommend that researchers explicitly state their ordering procedures and document their data, as I recommended above. These data can help us develop improved analytical procedures.

5.   Analyses of electrophoretic data should not include limiting assumptions regarding the rate of evolution. Analyses free of rate assumptions will play an important role in the study of evolutionary rates and patterns of differentiation among taxa. The construction of minimum length Wagner trees (26) is recommended, and a strong effort should be made to improve computing programs.

6.   Outgroups should be used to ascertain the evolutionary direction in phylogenetic studies. Only if outgroups are demonstrated to be uninformative should one use the rooting alternatives Farris (27) discussed.

7. Homoplasious steps revealed in phylogenetic treatment of allozyme data should be examined for possible introgressive origins (18). Although some groups of organisms are more prone to introgressive hybridization than others, this phenomenon has received limited attention among phylogenetic systematists. For some groups, the introgression may come from sympatric but not closely related organisms requiring the investigator to look further than the group under study to determine the origin of the homoplasy.

8. The pursuit of phylogenetic relationships can optimally be achieved if two independent data sets are treated comparably and examined for congruence. Congruence studies are currently the subject of considerable controversy (23, 52, 53, 55, 73, 74), which is beyond the scope of this review. Both allozyme and isozyme data continue to be valuable alternatives for use in congruence studies.

ACKNOWLEDGMENTS

I have benefited greatly from discussing many of these issues with J. S. Farris, S. D. Ferris, M. F. Mickevich, C. Mitter, M. M. Miyamoto, R. W. Murphy, W. J. Rainboth, D. L. Swofford, and G. S. Whitt, although some may not agree with the interpretations I have presented here. C. B. Crabtree, R. H. Matson, R. W. Murphy, and especially W. J. Rainboth critically evaluated the manuscript and made many helpful suggestions. This study was supported by the UCLA Department of Biology Fisheries Program.

Literature Cited

1. Avise, J. C. 1974. Systematic value of electrophoretic data. *Syst. Zool.* 23:465–81

2. Avise, J. C. 1976. Genetic differentiation during speciation. In *Molecular Evolution*, ed. F. J. Ayala, 7:106–22. Sunderland, Mass: Sinauer. 277 pp.

3. Avise, J. C., Ayala, F. J. 1976. Genetic differentiation in speciose versus depauperate phylads: Evidence from the California minnows. *Evolution* 30:46–52

4. Avise, J. C., Patton, J. C., Aquadro, C. F. 1980. Evolutionary genetics of birds. I. Relationships among North American thrushes and allies. *Auk* 97:135–47

5. Avise, J. C., Patton, J. C., Aquadro, C. F. 1980. Evolutionary genetics of birds. II. Conservative protein evolution in North American sparrows and relatives. *Syst. Zool.* 29:323–34

6. Ayala, F. J., Tracey, M. L., Barr, L. G., McDonald, J. F., Pérez-Salas, S. 1974. Genetic variation in natural populations of five *Drosophila* species and the hypothesis of the selective neutrality of protein polymorphisms. *Genetics* 77:343–84

7. Barrowclough, G. F. 1983. Biochemical studies of microevolutionary processes. In *Perspectives in Ornithology*, ed. A. H. Brush, G. A. Clark Jr., 7:223–61. Cambridge: Cambridge Univ. Press. 560 pp.

8. Baverstock, P. R., Cole, S. R., Richardson, B. J., Watts, C. H. S. 1979. Electrophoresis and cladistics. *Syst. Zool.* 28:214–19

9. Brush, A. H. 1979. Comparison of egg-white proteins: Effect of electrophoretic conditions. *Biochem. Syst. Ecol.* 7:155–65

10. Bush, G. L., Kitto, G. B. 1978. Application of genetics to insect systematics and analysis of species differences. In *Biosystematics in Agriculture*, ed. J. A. Romberger, 6:89–118. Montclair, NJ: Allanheld, Osmun. 340 pp.

11. Buth, D. G. 1979. Biochemical systematics of the cyprinid genus *Notropis*. I. The subgenus *Luxilus*. *Biochem. Syst. Ecol.* 7:69–79

12. Buth, D. G. 1979. Genetic relationships among the torrent suckers, genus *Thoburnia*. *Biochem. Syst. Ecol.* 7:311–16
13. Buth, D. G. 1979. Duplicate gene expression in tetraploid fishes of the tribe Moxostomatini (Cypriniformes, Catostomidae). *Comp. Biochem. Physiol.* 63B: 7–12
14. Buth, D. G. 1980. Evolutionary genetics and systematic relationships in the catostomid genus *Hypentelium*. *Copeia* 1980: 280–90
15. Buth, D. G. 1981. Cladistic treatment of isozyme (rather than allozyme) data. Presented at 2nd Ann. Meet. Willi Hennig Soc., Ann Arbor, Mich.
16. Buth, D. G. 1983. Duplicate isozyme loci in fishes: Origins, distribution, phyletic consequences and locus nomenclature. In *Isozymes: Current Topics in Biological and Medical Research*, ed. M. C. Rattazzi, J. G. Scandalios, G. S. Whitt, 10:381–400. New York: Liss. 436 pp.
17. Buth, D. G. 1984. Use of isozyme characters in systematic studies. *Syst. Zool.* Submitted
18. Buth, D. G. 1984. Allozymes of the cyprinid fishes: Variation and application. In *Evolutionary Genetics of Fishes*, ed. B. J. Turner, 11:561–90. New York: Plenum
19. Buth, D. G., Crabtree, C. B. 1982. Genetic variability and population structure of *Catostomus santaanae* in the Santa Clara drainage. *Copeia* 1982:439–44
20. Buth, D. G., Burr, B. M., Schenck, J. R. 1980. Electrophoretic evidence for relationships and differentiation among members of the percid subgenus *Microperca*. *Biochem. Syst. Ecol.* 8:297–304
21. Buth, D. G., Rainboth, W. J., Joswiak, G. R. 1983. Restriction of interlocus heteropolymer assembly in *Gyrinocheilus aymonieri* (Cypriniformes: Gyrinocheilidae). *Isozyme Bull.* 16:55
22. Buth, D. G., Murphy, R. W., Miyamoto, M. M., Lieb, C. S. 1985. Creatine kinases of amphibians and reptiles: Evolutionary and systematic aspects of gene expression. *Copeia.* In press
23. Colless, D. H. 1980. Congruence between morphometric and allozyme data for *Menidia* species: A reappraisal. *Syst. Zool.* 29:288–99
24. Corbin, K. W. 1983. Genetic structure and avian systematics. In *Current Ornithology*, Vol. 1, ed. R. F. Johnston, Ch. 8, pp. 211–44. New York: Plenum
25. Crabtree, C. B., Buth, D. G. 1981. Gene duplication and diploidization in tetraploid catostomid fishes *Catostomus*

fumeiventris and *C. santaanae*. *Copeia* 1981:705–8
26. Farris, J. S. 1970. Methods of computing Wagner trees. *Syst. Zool.* 19:83–92
27. Farris, J. S. 1972. Estimating phylogenetic trees from distance matrices. *Am. Nat.* 106:645–68
28. Farris, J. S. 1981. Distance data in phylogenetic analysis. In *Advances in Cladistics*, Vol. 1, *Proc. 1st Meet. Willi Hennig Soc.*, ed. V. A. Funk, D. R. Brooks, pp. 3–23. New York: NY Bot. Gard. 250 pp.
29. Felsenstein, J. 1981. Evolutionary trees from gene frequencies and quantitative characters: Finding maximum likelihood estimates. *Evolution* 35:1229–42
30. Felsenstein, J. 1984. Distance methods for inferring phylogenies: A justification. *Evolution* 38:16–24
31. Ferguson, A. 1980. *Biochemical Systematics and Evolution*. New York: Wiley. 194 pp.
32. Ferris, S. D., Whitt, G. S. 1977. Duplicate gene expression in diploid and tetraploid loaches (Cypriniformes, Cobitidae). *Biochem. Genet.* 15:1097–112
33. Ferris, S. D., Whitt, G. S. 1977. Loss of duplicate gene expression after polyploidisation. *Nature* 265:258–60
34. Ferris, S. D., Whitt, G. S. 1978. Phylogeny of tetraploid catostomid fishes based on the loss of duplicate gene expression. *Syst. Zool.* 27:189–203
35. Fisher, S. E., Whitt, G. S. 1979. Evolution of the creatine kinase isozyme system in the primitive vertebrates. In *The Biology and Physiology of the Living Coelacanth, Occas. Pap. Calif. Acad. Sci. No. 134*, ed. J. E. McCosker, M. D. Lagios, 12:142–59
36. Fisher, S. E., Shaklee, J. B., Ferris, S. D., Whitt, G. S. 1980. Evolution of five multilocus isozyme systems in the chordates. *Genetica* 52/53:73–85
37. Fitch, W. M., Margoliash, E. 1967. Construction of phylogenetic trees. *Science* 155:279–84
38. Futuyma, D. J. 1979. *Evolutionary Biology*. Sunderland, Mass: Sinauer. 565 pp.
39. Gorman, G. C. 1971. Evolutionary genetics of island lizard populations. In *Yearb. Am. Philos. Soc.*, pp. 318–19
40. Gorman, G. C., Renzi, J. R. 1979. Genetic distance and heterozygosity estimates in electrophoretic studies: Effects of sample size. *Copeia* 1979:242–49
41. Gorman, G. C., Buth, D. G., Soulé, M., Yang, S. Y. 1983. The relationships of the Puerto Rican *Anolis*: Electrophoretic and karyotypic studies. In *Advances in Herpetology and Evolutionary Biology*, ed. A. G. J. Rhodin, K. Miyata, pp.

626–42. Cambridge, Mass: Mus. Comp. Zool., Harvard Univ. 725 pp.

42. Hart, G. E. 1983. Genetics and evolution of multilocus isozymes in hexaploid wheat. In *Isozymes: Current Topics in Biological and Medical Research*, ed. M. C. Rattazzi, J. G. Scandalios, G. S. Whitt, 10:365–80. New York: Liss. 436 pp.

43. Hedrick, P. W. 1975. Genetic similarity and distance: Comments and comparisons. *Evolution* 29:362–66

44. Hedrick, P. W. 1983. *Genetics of Populations*. Boston: Sci. Books Int. 629 pp.

45. Hennig, W. 1966. *Phylogenetic Systematics*. Urbana: Univ. Ill. Press. 263 pp.

46. Kuhl, P., Schmidtke, J., Weiler, C., Engel, W. 1976. Phosphoglucose isomerase isozymes in the characid fish *Cheirodon axelrodi*: Evidence for a spontaneous gene duplication. *Comp. Biochem. Physiol.* 55B:279–81

47. Lewis, W. H., ed. 1980. *Polyploidy: Biological Relevance*. New York: Plenum. 583 pp.

48. Lundberg, J. C. 1972. Wagner networks and ancestors. *Syst. Zool.* 21:398–413

49. Markert, C. L. 1983. Isozymes: Conceptual history and biological significance. See Ref. 42, 7:1–17

50. Markert, C. L., Moller, F. 1959. Multiple forms of enzymes: Tissue, ontogenetic, and species specific patterns. *Proc. Natl. Acad. Sci. USA* 45:753–63

51. Mayr, E. 1969. *Principles of Systematic Zoology*. New York: McGraw-Hill. 428 pp.

52. Mickevich, M. F. 1978. Taxonomic congruence. *Syst. Zool.* 27:143–58

53. Mickevich, M. F. 1980. Taxonomic congruence: Rohlf and Sokal's misunderstanding. *Syst. Zool.* 29:162–76

54. Mickevich, M. F. 1982. Transformation series analysis. *Syst. Zool.* 31:461–78

55. Mickevich, M. F., Johnson, M. S. 1976. Congruence between morphological and allozyme data in evolutionary inference and character evolution. *Syst. Zool.* 25: 260–70

56. Mickevich, M. F., Mitter, C. 1981. Treating polymorphic characters in systematics: A phylogenetic treatment of electrophoretic data. See Ref. 28, pp. 45–58

57. Mickevich, M. F., Mitter, C. 1983. Evolutionary patterns in allozyme data: A systematic approach. In *Advances in Cladistics*, Vol. 2, *Proc. 2nd Meet. Willi Hennig Soc.*, ed. N. I. Platnick, V. A. Funk, pp. 169–76. New York: Columbia Univ. Press. 218 pp.

58. Miyamoto, M. M. 1981. Congruence among character sets in phylogenetic studies of the frog genus *Leptodactylus*. *Syst. Zool.* 30:281–90

59. Miyamoto, M. M. 1983. Biochemical variation in the frog *Eleutherodactylus bransfordii*: Geographic patterns and cryptic species. *Syst. Zool.* 32:43–51

60. Miyamoto, M. M. 1983. Frogs of the *Eleutherodactylus rugulosus* group: A cladistic study of allozyme, morphological, and karyological data. *Syst. Zool.* 32:109–24

61. Murphy, R. W. 1984. Phylogenetic inferences of lactate dehydrogenase tetramer patterns in skinks (Reptilia: Scincidae). *Isozyme Bull.* 17:63

62. Murphy, R. W., Cooper, W. E., Richardson, W. S. 1983. Phylogenetic relationships of the North American five-lined skinks, genus *Eumeces* (Sauria: Scincidae). *Herpetologica* 39:200–11

63. Nei, M. 1972. Genetic distance between populations. *Am. Nat.* 106:283–92

64. Nei, M. 1978. Estimation of average heterozygosity and genetic distance from a small number of individuals. *Genetics* 89:583–90

65. Nei, M., Tajima, F., Tateno, Y. 1983. Accuracy of estimated phylogenetic trees from molecular data. II. Gene frequency data. *J. Mol. Evol.* 19:153–70

66. Ohno, S. 1970. *Evolution by Gene Duplication*. New York: Springer-Verlag. 160 pp.

67. Patton, J. C., Avise, J. C. 1983. An empirical evaluation of qualitative Hennigian analyses of protein electrophoretic data. *J. Mol. Evol.* 19:244–54

68. Philipp, D. P., Parker, H. R., Whitt, G. S. 1983. Evolution of gene regulation: Isozymic analysis of patterns of gene expression during hybrid fish development. See Ref. 42, 10:193–237

69. Prager, E. M., Wilson, A. C. 1978. Construction of phylogenetic trees for proteins and nucleic acids: Empirical evaluation of alternative matrix methods. *J. Mol. Evol.* 11:129–42

70. Prakash, S., Lewontin, R. C., Hubby, J. L. 1969. A molecular approach to the study of genic heterozygosity in natural populations. IV. Patterns of genic variation in central, marginal and isolated populations of *Drosophila pseudoobscura*. *Genetics* 61:841–58

71. Riska, B. 1979. Character variability and evolutionary rate in *Menidia*. *Evolution* 33:1001–4

72. Rogers, J. S. 1972. Measures of genetic similarity and genetic distance. *Univ. Tex. Publ.* 7213:145–53

73. Rohlf, F. J., Sokal, R. R. 1980. Comments on taxonomic congruence. *Syst. Zool.* 29:97–101

74. Rohlf, F. J., Colless, D. H., Hart, G. 1983. Taxonomic congruence re-examined. *Syst. Zool.* 32:144–58

75. Schultz, R. J. 1980. Role of polyploidy in the evolution of fishes. See Ref. 47, pp. 313–40

76. Selander, R. K., Whittam, T. S. 1983. Protein polymorphism and the genetic structure of populations. In *Evolution of Genes and Proteins*, ed. M. Nei, R. K. Koehn, 5:89–114. Sunderland, Mass: Sinauer. 331 pp.

77. Shaklee, J. B., Whitt, G. S. 1981. Lactate dehydrogenase isozymes of gadiform fishes: Divergent patterns of gene expression indicate a heterogeneous taxon. *Copeia* 1981:563–78

78. Shaklee, J. B., Kepes, K. L., Whitt, G. S. 1973. Specialized lactate dehydrogenase isozymes: The molecular and genetic basis for the unique eye and liver LDHs of teleost fishes. *J. Exp. Zool.* 185:217–40

79. Shaklee, J. B., Tamaru, C. S., Waples, R. S. 1982. Speciation and evolution of marine fishes studied by the electrophoretic analysis of proteins. *Pac. Sci.* 36:141–57

80. Simon, C. M. 1979. Evolution of periodical cicadas: Phylogenetic inferences based on allozymic data. *Syst. Zool.* 28:22–39

81. Sneath, P. H. A., Sokal, R. R. 1973. *Numerical Taxonomy*. San Francisco: Freeman. 573 pp.

82. Swofford, D. L. 1981. On the utility of the distance Wagner procedure. See Ref. 28, pp. 25–43

83. Tateno, Y., Nei, M., Tajima, F. 1982. Accuracy of estimated phylogenetic trees from molecular data. I. Distantly related species. *J. Mol. Evol.* 18:387–404

84. Toledo, S. A., Ribeiro, A. F. 1978. Directional reduction of lactate dehydrogenase isozymes in teleosts. *Evolution* 32:212–16

85. Uyeno, T., Smith, G. R. 1972. Tetraploid origin of the karyotype of catostomid fishes. *Science* 175:644–46

86. Wake, D. B. 1981. The application of allozyme evidence to problems in the evolution of morphology. In *Evolution Today, Proc. 2nd Int. Congr. Syst. Evol. Biol.*, ed. G. G. E. Scudder, J. L. Reveal, pp. 257–70. Pittsburgh: Hunt Inst. Bot. Doc., Carnegie-Mellon Univ. 486 pp.

87. Watrous, L. E., Wheeler, Q. D. 1981. The out-group comparison method of character analysis. *Syst. Zool.* 30:1–11

88. Whitt, G. S. 1981. Evolution of isozyme loci and their differential regulation. See Ref. 86, pp. 271–89

89. Whitt, G. S. 1981. Developmental genetics of fishes: Isozymic analyses of differential gene expression. *Am. Zool.* 21: 549–72

90. Whitt, G. S. 1983. Isozymes as probes and participants in developmental and evolutionary genetics. See Ref. 42, 10:1–40

91. Zimmerman, E., Kilpatrick, C., Hecht, B. 1978. The genetics of speciation in the rodent genus *Peromyscus*. *Evolution* 32: 565–76

Ann. Rev. Ecol. Syst. 1984. 15:523–75

OPTIMAL FORAGING THEORY: A CRITICAL REVIEW

Graham H. Pyke

Department of Vertebrate Ecology, The Australian Museum, 6–8 College Street, New South Wales 2000, Australia

INTRODUCTION

Proponents of optimal foraging theory attempt to predict the behavior of animals *while they are foraging;* this theory is based on a number of assumptions (133, 155, 210, 231). First, an individual's contribution to the next generation (i.e. its "fitness") depends on its behavior while foraging. This contribution may be measured genetically or culturally as the proportion of an individual's genes or "ideas", respectively, in the next generation. In the former case, the theory is simply an extension of Darwin's theory of evolution.

Second, it is assumed that there should be a heritable component of foraging behavior, i.e. an animal that forages in a particular manner should be likely to have offspring that tend to forage in the same manner. This heritable component can be either the actual foraging responses made by an animal or the rules by which an animal learns to make such responses. In other words, optimal foraging theory may apply regardless of whether the foraging behavior is learned or innate. Given these first two assumptions, it follows that the proportion of individuals in a population foraging in ways that enhance their fitness will tend to increase over time. Unless countervailed by sufficiently strong group selection (see 287, 242), foraging behavior will therefore evolve, and the average foraging behavior will increasingly come to be characterized by those characteristics that enhance individual fitness.

The third assumption is that the relationship between foraging behavior and fitness is known. This relationship is usually referred to as the *currency* of fitness (231). In general, any such currency will include a time scale, although in some cases it may be assumed that fitness is a function of some *rate*.

0066–4162/84/1120–0523$02.00

The fourth assumption is that the evolution of foraging behavior is not prevented by genetic constraints such as the physical linkage of genes that affect a number of traits or the effects of single genes on multiple traits (155). Such genetic linkage of traits could impede the rate of evolution, but it is assumed that mutations will eventually arise that circumvent these linkages.

The fifth assumption is that the evolution of foraging behavior is subject to "functional" constraints that have been realistically determined. When focusing on foraging behavior (rather than on foraging traits in general), it is assumed that the morphologies and physical properties of the animals are known and evolutionarily fixed. Assumptions may also have to be made about the level of information available to an animal and about the animal's ability to store and process this information (e.g. 125, 189, 268). Animals might obtain information through either direct experience or observation of others (e.g. 274). Such functional constraints can also be considered as evolutionary variables within a broader framework.

The final assumption is that foraging behavior evolves more *rapidly* than the rate at which the relevant conditions change (210). Thus, the evolution and adaptation of foraging behavior should approximately reach completion with individuals foraging in ways close to (i.e. statistically indistinguishable from) those that *maximize* their expected fitness, subject to any functional constraints. In this sense, it is hypothesized that animals forage "optimally."

This approach to animal foraging behavior began to develop about 18 years ago with papers by MacArthur & Pianka (150) and Emlen (64). The number of papers published annually that either include the *optimal foraging* in the title or clearly develop or test predictions based on optimal foraging theory appears to have increased rapidly between 1973 and 1981 and to have decreased since then (Figure 1). It will be interesting to see what happens in the future.

During the last 18 years, the literature on optimal foraging theory has been reviewed a number of times (46, 76, 132–135, 137, 199, 209, 210, 229, 231). Nevertheless, at this stage another review seems timely. Past reviews have tended to take a relatively positive view of the value of optimal foraging theory and to take apparent tests of predictions derived from this theory at face value. Optimal foraging theory has engendered considerable controversy, however (e.g, 118, 178). Furthermore, in many optimal foraging studies, the theory and the observations may not be appropriately matched because, for example, the assumptions are unrealistic or unsupported or the mathematical calculations are incorrect (see 209 and discussion below). It is therefore time, I believe, to review the various points of view concerning optimal foraging theory, to derive criteria for evaluating *studies* that attempt to test this theory, and to begin to judge its usefulness on the basis of studies that meet these criteria. In order to achieve these goals, it is also necessary to review both the theoretical and empirical developments that have occurred. I shall attempt below to carry out

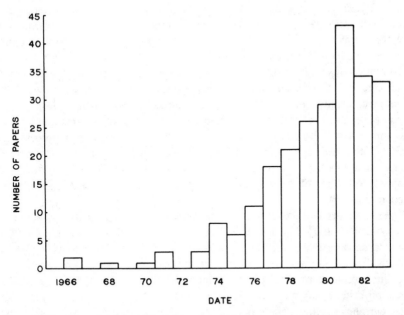

Figure 1 Yearly numbers of published papers from author's reprint collection that either contain the words *optimal foraging* in the title or include the development or tests of optimal foraging predictions.

such a review. I will deal with the behavior of animals *while they are foraging* but not with the *amount* of time that animals allocate to foraging nor with *when* animals choose to forage (see 38, 56, 92, 93, 102, 198). I shall also omit consideration of the recent attempts to relate optimal foraging theory to experiments on "matching" (see 101, 106) and to apply it in a variety of contexts (e.g. 1, 39, 62, 94, 95, 191, 192, 207, 235, 286).

ATTITUDES TOWARD OPTIMAL FORAGING THEORY

The most critical view of optimal foraging theory is that it is "tautological" (178) or "*not* scientific" (118). This view arises from the following properties of this theory: First, when predictions and observations do not agree, it is not clear which assumptions are at fault (155, see also 50). Authors have tended to rationalize such discrepancies between observed and predicted results by attributing them to faulty assumptions regarding constraints or the currency of fitness rather than those assumptions about the heritability of behavior, the lack of genetic constraints, or the rate of evolution. They thereby seem to invite the criticism (e.g. 85, 149) that they are unwilling to abandon certain assumptions. Second, it is difficult to obtain independent tests of most, if not all, of the assumptions that underlie optimal foraging theory. Probably few would doubt the assumption about the heritability of foraging behavior, but confirmatory experiments (e.g. 54) are mostly lacking. Foraging behavior almost certainly affects animal fitness, but unless the effects of this behavior on survival and reproduction can be demonstrated for animals that differ only in terms of selected aspects of their foraging, the currency of fitness cannot be verified. Under carefully controlled circumstances, an animal will do as it pleases, and consequently, it is difficult to obtain more than a *minimum* estimate of an animal's capabilities for information acquisition, storage and processing. Elaborate breeding experiments would be necessary to verify the existence of any linkage between foraging behavior and other traits. There also does not appear to be any information available on rates of evolution and of changes in conditions.

To view optimal foraging theory as tautological or unscientific seems unreasonable. Like other scientific theories, it is based on assumptions. The existence of a number of assumptions in the case of this theory makes it difficult, but not impossible, to evaluate it fully. Some of the assumptions are easier to relate to observations than others. A sound knowledge of an animal's natural history may well determine how realistic the assumptions concerning the currency of fitness and functional constraints are (see 210), but it is of little help with regard to the other assumptions. It is also relatively easy to alter the theoretical treatments of foraging and the data collection schemes to accommodate alterations in the assumptions concerning fitness currency and functional constraints. For example, the assumption that an animal is simply maximizing the net rate of energy gain can easily be modified to allow for a minimum requirement of some nutrient (e.g. see 188). At present, however, there is no obvious way to incorporate deviations in other assumptions into the theory. Such deviations should lead to "suboptimal" foraging behavior, but unless some way can be found to predict the magnitude and direction of any departures from optimal behavior, there seems little point in pursuing this (but see 90, 213, 238, 239). The most logical approach to differences between predictions and observations is therefore to explore the consequences of variation in the currency and constraints assumptions and to devise experiments and observations from which the most realistic assumptions about these variables can be deduced. If the most realistic currency and constraints assumptions and the most careful development of theory do not lead to reasonably close agreement between observed and predicted results a reasonable fraction of the time, then optimal foraging theory should properly be judged as not very useful (see below). Therefore, its usefulness cannot be determined on the basis of only one or a few studies (see also 280).

A second view of optimal foraging theory, which is also unflattering, is that it is doomed to failure because of the complexity of the natural world. In other words, according to this view, the effort to develop this theory is hopeless because one or more of the basic assumptions is almost certain to be false. This view would seem to be overly pessimistic, however, and if it were applied in analogous fashion to the motion of falling objects, it would surely produce a table of falling coefficients, rather than the law of gravity. The success of many optimal foraging studies also argues against this view.

A third view of this theory, highly flattering but clearly inaccurate, is that there are already so many studies supporting its predictions that it can be regarded as well-established and verified; hence there is no need for further development and tests of its predictions. However, the large and growing number of studies in which some discrepancy has been found between the predictions of optimal foraging theory and observations—especially in terms of precise quantitative predictions—mitigates strongly against this view.

A final view of this theory, to which I subscribe, is that it is still too early to pronounce judgment and that further development and testing are warranted. Though the number of "optimal foraging studies" is already quite large (Figure 1), none of them fulfills all of the criteria for evaluating this theory outlined below. When those studies that come closest to satisfying all the criteria are examined, some patterns emerge in terms of the degree of success of the theory, and these patterns suggest avenues for future investigation.

THE DEVELOPMENT OF THEORY

In 1977 it was reasonable to consider development (and tests) of optimal foraging theory in the following four roughly independent categories: (a) diet, (b) patch choice, (c) when to leave a patch; and (d) movement (e.g. 210). Some 7 years later, these categories are still useful and convenient, but several new developments have occurred. The first of these is the addition of a fifth category, namely central place foraging (e.g. 179), which deals with situations in which an animal has a central foraging base, such as a nest, from which it goes on foraging bouts and to which it returns with at least some of what it has gathered while foraging. In such situations, the choices of where to forage (patch choice), when to leave a patch, and what to eat (diet) are interrelated, and hence a separate category is required.

A second general development has been an increased realization of the potential importance of the stochastic or variable nature of the world (e.g. 23, 159, 174, 195, 253). From the outset of optimal foraging theory, it was realized that most, if not all, parameters—such as the amount of energy obtained from an individual of a particular prey type—are random variables (i.e. they take different values with certain probabilities). In the initial models, this kind of randomness was adequately dealt with by the use of mean values for the various parameters (e.g. 32). In some situations, however, the fitness of a foraging animal is a function not only of the mean values of the various parameters but also of their variances. In these situations, an animal should make its foraging decisions on the basis of these variances as well as the means. Animals that prefer lower variance, all else being equal, are termed "risk averse," while animals that prefer higher variance are "risk prone" (23).

The variable nature of the world also means that animals cannot be omniscient and that they will have only an imperfect knowledge of what they may encounter at a future time and place (199). In such situations, an animal should be a statistician—collecting and storing information as it forages and using this information to make foraging decisions. For example, there may be a positive correlation among the nectar volumes in flowers on the same plant, in which case an animal feeding on this nectar might base its decisions about when to move to another plant on the amount of nectar obtained so far from the present plant. Therefore, it should be less likely to change plants the more nectar it has obtained from the flowers of the present plant.

It has also been recognized that the world is always changing and that animals should accordingly devote time to "sampling" their environment in order to obtain necessary information for subsequent foraging decisions (245). For example, if the relative quality of two food patches switches from time to time, then an animal that can feed in either patch should always spend some time in what at the time might be the worse patch so that it could make the appropriate adjustments when conditions change. At present, however, there appears to have been no mathematical development of such sampling regimes.

A final general development has been the alteration of a number of the assumptions in the original foraging models. For example, in the original model of optimal diet (see 210), it was assumed that animals recognize distinct food types almost instantaneously. More recently, recognition time has been included, which has altered the original predictions of the theory (63, 111).

1. Risk Aversion and Risk Proneness

In general, the fitness of a foraging animal will be some function of a number of foraging parameters such as the amounts of energy obtained from the various food types, the handling times involved, the rates of encounter, and so on. If these foraging parameters are random variables, the average or expected fitness will usually depend not only on the mean values of the parameters but also on their variances. Optimal foraging theory began with two kinds of exception to this rule.

The first kind of exception resulted from the initial assumption that the fitness of a foraging animal is a linear function of the net rate of food gain while foraging. In algebraic terms, it was implicitly assumed (e.g. 210) that fitness $F = a + bR$, where R is the net rate of food gain and a and b are constants. In this case, since $E(F) = a + bE(R)$, any variance associated with R does not affect the expected fitness, and the maximization of the expected fitness is equivalent to the maximization of the expected net rate of food gain. As the following example illustrates, however, the situation becomes quite different when fitness is assumed to be a nonlinear function of R (or its equivalent) (see also 23, 214, 252). Suppose that fitness $F = 0$ if $R < T$ (i.e. some threshold) and $F = 1$ (i.e. an arbitrary constant) if $R > T$. Suppose also that R is a random variable with probability distribution $P(R = r) = P(r)$. Then the expected value of F is simply the *probability* that the animal meets it minimum food requirements (i.e. $\int_T^\infty P(r)dr$), which depends on the general shape of the probability distribution for R. Suppose further that the variance of the probability distribution for R (but not the mean) depends on which of two (or more) foraging possibilities the animal chooses and that the probability distributions of R are always symmetric. Then the choice that maximizes the expected fitness will depend on whether the mean R (i.e. $E(R)$) is greater or less than the threshold T. If $E(R) > T$, the animal should opt for the smaller variance (i.e. be risk averse); but if $E(R) < T$, the animal should be risk prone and preferentially select the higher variance. In other words, if an animal is likely to starve, its best chance for survival will come from the most variable situation (assuming no differences in means). On the other hand, if it can expect to exceed its food requirements, it should prefer the lowest variance.

In general terms, the optimal response to variance in R (assuming a constant mean) depends on the shape of the relationship between fitness and R at the point where R equals its average or expected value. If the relationship is convex (i.e. bowed downwards), an animal should be risk prone; if it is concave (i.e. bowed upwards), an animal should be risk averse. If fitness is a sigmoidal (i.e. S-shaped) function of R and the frequency distributions of R are symmetric with the same mean, then an animal should be risk averse or prone to the right or left of the inflection point, respectively (from 23). If both the mean and variance of R depend on an animal's foraging decisions, the situation is more complicated (253).

The second kind of exception to the general rule that the expected fitness will depend on the means and variances of the various foraging parameters arises from the assumption that foraging events are independent of one another and of previous foraging decisions. Suppose, for example, that encounter rates with different kinds of food types are not affected by any aspect of past history, such as the food types previously encountered or consumed. Then the foraging process can be modeled as a renewal process, and the expected rate of food gain depends only on the averages of the food gain from each food type, the handling time for each food type, and the time between encounters with food items (e.g. 32, 33; see below). In this case, if fitness increases linearly with food gain, the optimal diet will depend on these averages and not on any associated variances. On the other hand, if foraging events are influenced by past history, the variances and covariances of any foraging parameters should affect the expected rate of food gain and hence the optimal foraging strategy (e.g. 87, 174, 202).

2. Optimal Diet

The original and simplest model of optimal diets was based on the following assumptions: (a) The fitness of a forager increases linearly with the expected rate of food intake, where food value is measured in calories or weight. The costs of handling and searching for food are assumed to be equal. (b) Each food type has an associated average food value and average handling time, both of which are known to the forager. (c) The forager requires negligible time to recognize food types and does not make any mistakes in doing so. (d) Handling and searching for food are mutually exclusive. In other words, the forager decides whether or not to eat a particular food item at the instant it encounters that item. This assumption was implicit during the early developments of optimal diet theory. (e) The rates of encounter with the different food types are constant and independent of each other and of past history. (f) Handling times and food yields are independent of past history. (g) Food items are encountered sequentially rather than simultaneously. (h) Food items, if eaten, are totally consumed. (i) Dietary choices are constant. (j) The foraging time is long compared with the time spent handling and searching for food. (k) There are no constraints on total food volume or the rate of food intake.

Under these assumptions, foraging can be thought of as a renewal process where the renewal event is the recommencement of searching for a food item and where the expected food gains and times taken between renewal events are independent of past experience. Therefore, the optimal diet depends only on the average values of the food gains and handling times for each food type and on the encounter rates with the different food types (32, 253). It also follows (see 117 and references in 210) that in the optimal diet, a food type is either always eaten or always ignored; the optimal diet is found by starting with the food type having the highest average food gain to average handling time ratio (i.e. the highest rank) and adding food types with successively lower ratios until R reaches a maximum. Consequently, the optimal diet in the *present* case has the following testable properties (210):

1. Whether or not a food type should be eaten is independent of its abundance and depends only on the *absolute* abundances of food types of higher rank. An animal should never specialize on a relatively low-ranked food type regardless of its abundance (see also 68, 236).
2. As the abundance of a relatively high-ranked food type increases, lower-ranked types should eventually be dropped from the diet, starting with the lowest-ranked type included. So, increasing overall food abundance should

lead to greater specialization. Increasing abundance of relatively low-ranked food types may, however, have no effect on the optimal diet.
3. As mentioned above, a food type is either completely included in the optimal diet or completely excluded from it—animals should never exhibit "partial preferences."

Many studies have attempted to test one or more of these three predictions (e.g. 51, 52, 59, 67, 78, 81, 84, 141, 241, 246, 248, 257, 266). A much smaller number have tried to determine the exact optimal diet using the above model and to compare this diet with the observed one (65, 66, 75, 82, 83, 136, 163, 190, 263, 278). (These studies will be evaluated below.) There are also many recent studies that examine the dietary preferences of animals and the factors correlated with these preferences (e.g. 10, 18, 57, 60, 61, 74, 112, 114, 121, 122, 148, 151, 164, 220, 222, 223, 225, 255, 260, 265, 276). Further developments in optimal diet theory have occurred through alterations to the assumptions in the above model.

DIFFERENTIAL COSTS OF HANDLING AND SEARCHING FOR FOOD   The simplest possible variation of the above model is to assume that fitness is an increasing linear function of the *net* rate of food gain rather than the gross rate. The distinction is real so long as the costs of handling and searching for food are different. This alteration produces potential differences in the exact optimal diet, but it does not change the three more qualitative predictions outlined above (31).

NUTRIENT CONSTRAINTS   Significant departures from the above optimal diet predictions result from the simplest of models that include nutrient requirements. Suppose, for example, that fitness is maximized when the rate of food gain is maximized, subject to the constraint that the rate of gain of some nutrient must at least equal some threshold value. In this case (see 188), "partial preferences" should result (i.e. some food types, when encountered, should be eaten with probabilities between 0 and 1), and the preference for a particular food type should depend not only on the abundances of more preferred food types but also on its own abundance (and probably the abundances of less preferred food types). Similar predictions arise from other models that include nutrient requirements (153, 211, 212). After allowing for nutrient requirements, increases in abundance of preferred food types should still lead to greater dietary specialization (from 188).

RECOGNITION TIME, MISIDENTIFICATION, AND CRYPTICITY   The optimal diet model can easily be modified to allow for the time required for recognition of food types by adding recognition times to the handling times (107, 111).

When an animal encounters a food item, it *must* spend some time (which may vary with the food type) to recognize the food type, and it may also incur handling time if it chooses to eat the food item. This version of the optimal diet model does not predict partial preferences, but it does predict that whether or not a particular food type is included in the diet should depend on its own abundance (absolute and/or relative), as well as the abundances of more preferred food types (63, 107, 111). In fact, a suboptimal food type may be included in the optimal diet if it becomes abundant enough (111). Furthermore, food types may be ranked in terms of energy/handling time ratios, increasing the abundance of preferred food types should lead to greater dietary specialization, and food types should be added or deleted from the optimal diet in rank order. Allowing for misidentifiction of food types leads to the same set of predictions. The same predictions also arise if recognition time is only required when an animal specializes (see 66).

Crypticity of food types can lead to yet another variation in the predictions of the optimal diet model (66, 111). Hughes (111) allowed for the possibility that an animal might mistake inedible objects for an otherwise valuable food type, spending some time before discovering the error and rejecting the object. If the density of the more valuable food type were sufficiently low, the optimal diet might consist of specialization on less valuable food types with total exclusion of the more valuable but cryptic one. Hence, with decreases in the abundance of the more valuable food type, the forager might first switch from specializing on the more valuable food type to generalizing and then to specializing on the less valuable food type. Otherwise, this version of the optimal diet model produces the same predictions as the model including recognition times and misidentification.

THE DEPENDENCE OF DIET ON THE DEGREE OF SATIATION   An animal's diet may not be constant but may depend on its degree of satiation. Suppose, for example, that an animal requires only a small amount of food to reach "satiation" (i.e. the food level above which further increases do not enhance fitness) and that it has just encountered a low-value food item that would normally be excluded from the diet. If fitness is maximized when the time required to reach satiation is minimized, then the optimal strategy might be to consume the food item rather than to continue searching for a more preferred food type (221). In other words, an animal might specialize until it has almost reached satiation and then expand its diet. Richards (221) showed that for two prey types A and B, such diet expansion should only occur if the animal requires less than the amount of food in the higher ranked item to reach satiation. This variation in the optimal diet model is therefore most applicable to animals that forage for food items that are large relative to total intake.

THE EFFECTS OF FORAGER EXPERIENCE ON HANDLING TIMES AND FOOD
YIELDS   As the rate of encounter with a food type increases, the handling time
for that food type will probably decrease (43, 111, 157, 281); handling time
may also decrease with increasing "hunger" (e.g. 129, 277).

Hughes (111) modified the original diet model by setting each handling time
equal to a function of the encounter rate. He then showed that partial preferences should not develop; that whether or not a particular food type is included
in the optimal diet depends on its own abundance, as well as on the abundances
of more preferred food types: and that as changes in the abundance of food
types occur, the optimal diet could switch from specialization on one food type
to specialization on another, with or without generalization in between (111).
McNair (157) obtained similar results.

Food yields per food item may also be influenced by an animal's past diet
(22, 183). Though this possibility apparently has not been modeled, it is clear
that allowing for it considerably complicates the determination of optimal diets
and probably requires knowledge of an animal's past diet. Rates of encounter
with detected food items may also depend on past experience. Animals might,
for example, develop "search images" whereby they are more (or less) likely to
detect food types they have encountered before (e.g. 157, 166, 184). The
formation of such search images may explain frequency-dependent dietary
selection (e.g. 73, 105, 167).

NONRANDOM ENCOUNTERS WITH FOOD ITEMS   In the original optimal diet
model, it was assumed that encounter rates with food types are constant and
independent of past history and of each other. This type of food encounter
pattern is termed random, and it produces a probability distribution (negative
exponential) of the time between an animal's beginning to look for food and the
next food encounter independent of past history (e.g. 187, 219). This probability distribution still could depend on past history, however. Suppose, for
example, that there is a single food type, items of which "arrive" when an
animal is both handling and searching for food. Food items that arrive during
search time have been "encountered." Suppose further that the handling time
for a food item and the time interval between successive food arrivals are
independent random variables. The expected interval between the time when
the animal finishes handling a food item and simultaneously recommences
searching and when it encounters the next food item will then depend on the
length of the handling time. This expected time interval would be very short if,
for example, handling times and interarrival times were constant, with the
former slightly shorter than the latter.

In general, the relationship between past history and the time interval
between the point when an animal recommences searching and its next food

encounter will be complex. Continuing the above example, suppose that successive handling times and successive interarrival times are all independent of one another. Even then, it is possible that any number N of food items will arrive during the handling time for the last item. The probability density function for the required time interval T will be a function of N and the handling time H. Hence, the expected time interval between the end of handling one food item and the end of handling the next (i.e. $T + H$) will depend on the probability density function of H and the conditional probability function of N given $H = h$ and $N = n$. The situation becomes even more complicated if there are two or more food types, each with its own associated probability distributions of handling times and interarrival times.

Krebs et al (136) considered the following special case. They supposed that there are two food types with constant values $2E$ and E, that the interarrival time t between successive food items is constant, that the handling times associated with each food type are random variables $(H1, H2)$ such that $2E/E(H1) > E/E(H2)$, and that food types occurred in one of the following three sequences, where L = large and S = small: $L, S, L, S,—L,L,L,S,L,L,L,S,—L,S,S,L,S,S,—$. They defined p as the probability that $H2 < t$ and argued that the optimal strategy of the animal is to specialize on the larger food type if $p < 1/2$, independent of the sequence of food types. As Rechten et al (219) pointed out, however, this formulation is incorrect; the correct optimal strategy apparently has not been determined. They argue that if there is a single food type and if an animal recommences searching *at random* during any interarrival interval, then the expected time until the next food encounter is $0.5 [\mu + (\sigma^2 \div \mu)]$, where μ is the rate of food arrival and σ^2 is the variance of the interarrival time. But recommencement times are only likely to be distributed randomly over interarrival times if the average handling times are much larger than the interarrival times. Consequently, Rechten et al's deductions for the case of two food types are unlikely to be correct in most cases.

McNair (156) considered a more general situation by assuming that the search time (after handling time has ended) required to encounter a food item has a probability distribution that depends on both the prey types last encountered and the one that will be found next. In this case, prey types are not necessarily added to the optimal diet in decreasing order of the energy to handling time ratio; whether or not a food type is included in the optimal diet will depend on its own abundance instead (156).

Pulliam (187) also examined nonrandom encounters with food items by assuming that an animal encounters clumps of food; that every item in these clumps can be consumed without further search time; and that during searching time, encounters with clumps occur at random. The qualitative predictions of the simple optimal diet model remain unchanged, but if overall food densities are constant, increased clumping of food items tends to shift the optimal diet towards greater specialization (187).

DECLINING RATES OF FOOD ENCOUNTER   If an animal forages amongst patches and if the rate of encounter with a food type declines with the removal of that food type from a patch, then the optimal diet within a patch and the optimal time of departure from it are interrelated. Under these circumstances, the optimal diet may depend on the time spent in a patch so far and on the abundances of nonpreferred food types, and it may consist of partial preferences (99). The optimal diet may also depend on the average time spent in a patch (99), just as foraging strategies may, in general, depend on the time scale involved (see 48, 115, 210, 224).

OTHER POSSIBLE VARIATIONS IN THE OPTIMAL DIET MODEL   The optimal diet model has not been modified to allow for simultaneous encounters with food items. In such situations, however, partial preferences may be optimal because a food item will probably be eaten if it is encountered along with less preferred items or ignored if a more preferred item is present.

The optimal diet model might also be modified to allow for nonlinear relationships between fitness and the expected rate of food intake (i.e. to include risk) or to allow fitness to be affected by factors other than food (e.g. 177). Some theoretical treatments of diets have not been included in the above discussion (e.g. 17, 48, 89, 181, 249–251, 258).

SUMMARY OF PREDICTIONS   The qualitative predictions that arise from the simple optimal diet model break down under relatively straightforward and realistic variations to the model. Partial rather than absolute preferences are expected if fitness depends on more than one food value (e.g. energy and nutrients or the energy intake mean and variance). Preferences for food types that depend on the abundance of each food type rather than on the abundances of better food types should develop if fitness depends on more than one food value, if recognition of food types requires time or is imprecise, if food types are cryptic, or if handling times or food yields depend on experience. In the last two situations, the optimal diet may also involve specialization on an inherently inferior food type or switching between specialized diets. This breakdown of the simple qualitative predictions means that much care should be exercised in testing them.

3. Optimal Patch Choice

Patch choice is analogous to dietary choice and can be modeled in exactly the same manner if two conditions are met: first, food patches must always be encountered before being accepted or rejected and second, the within-patch food yield and time spent must be independent of which patch types are accepted and of the overall rate of food gain. There are no obvious examples that satisfy all of these assumptions, however, so it is not surprising that apparently no one has focused on this kind of patch choice.

In general, one of the above assumptions will not be met. The various possible departures from these assumptions determine an array of different patch-choice situations that I shall consider below.

The simplest kind of patch-choice situation involves the following assumptions: (a) the locations and "qualities" of all patches are known to the foraging animal; (b) these locations and qualities are constant; (c) there is no resource depletion during the time the animal spends in a patch; and (d) fitness is an increasing linear function of the animal's net rate of food gain. Under these assumptions, the animal's optimal strategy is to spend all its time in the most conveniently located patch with the greatest net rate of food gain. A variety of patch-choice situations can be obtained by altering one or another of the above assumptions as follows:

THE CURRENCY OF FITNESS   Fitness may be a nonlinear function of the net rate of food gain or may depend on the rates of both nutrient and food gains. If it is not a linear function of the net rate of food gain, then the expected fitness for a particular strategy will depend on the mean, variance, and possibly the general shape of the frequency distribution of that net rate. Suppose, for example, that the expected fitness is given by $E(F) = aE(R) - b\,V(R)$, where $V(R)$ is the variance of R and a and b are positive constants (i.e. an animal is risk averse) (e.g. 214, 215). Suppose also that there are two patch types, that the amount of food obtained during time t_i in patch type i has mean $R_i t_i$ and variance $V_i t_i$ (i.e. each time period is independent of other time periods), that an animal spends a proportion p of total time T in patch 1, and that the costs of travel between patches are negligible. Then the animal's overall rate of food gain will have mean $E(R) = pR_1 + (1 - p)R_2$ and variance $V(R) = (pV_1 + (1-p)V_2)/T$, and the expected fitness $E(F)$ will be maximized with respect to p when $p = 1$ or 0, depending on whether $aT(R_1 - R_2)$ is greater than or less than $b(V_1-V_2)$. In other words, the animal should allocate all of its time to one or another patch type depending on the difference in means relative to the difference in variances between the two patch types [i.e. $(R_1-R_2)/(V_1-V_2)$] and the relative importance in determining fitness of the mean and variance of the total amount of food obtained during the foraging time T (i.e. aT/b). If one of the patch types has both the highest R and the lowest V, then the animal should spend all its time in a patch of that type. If the expected fitness is not a linear function of the mean and variance of the rate of food gain, however, the optimal strategy may consist of allocating time to both patches.

An identical situation prevails if fitness depends on the rate of intake of two or more food values, such as food weight and the amount of some nutrient. If fitness is a linear function of the rate of intake of food and a nutrient, then the optimal strategy will be to allocate all available time to one patch type, the identity of which will depend on the quantitative details of the fitness function and the available patch types. If the fitness function is nonlinear, then allocating time to more than one patch may be optimal. Patch choice may also affect other aspects of fitness such as predation risk to a forager (e.g. 91, 100, 161, 163, 238).

IMPERFECT KNOWLEDGE ABOUT THE QUALITY OF PATCHES   If an animal does not know the quality of the available patch types, then its optimal allocation of time should depend on how much foraging time remains, on its experience so far in each patch type, and on any a priori knowledge about the kinds of patches available (138). If a lot of foraging time remains, an animal should devote some time to sampling the available patch types before deciding how to allocate the remaining time. If its experience indicates that one patch type is much worse than the others, the animal should stop foraging in that patch type. If the animal knows the array of available patch qualities but not which ones are which, it should spend less time sampling that if it knows only that patch quality varies.

There is no comprehensive theoretical treatment of this foraging problem. Krebs et al (138), however, have considered the following special case for two patches: (a) In each patch, an animal's foraging consists of trials for which the probability of obtaining a food reward is a constant but unknown P_i. In their examples, P_1 and P_2 are (0.50, 0.00), (0.40, 0.10), (0.35, 0.15), and (0.30, 0.20). (b) The P_i each have an a priori beta probability distribution with parameters $(\alpha_i, \beta_i) = (0, 0), (0, 2)$, or $(2, 0)$. This distribution has the desirable property that the a posteriori probability distribution of P_i after n_i trials with r_i successes in patch i also has a beta distribution with new parameters $(\alpha_i + r_i, \beta_i + n_i - r_i)$ (138). (c) The time and effort required by an animal to move between the two patches is negligible. (d) An animal samples each of the two patches equally and then exploits just one of them. (e) The total number of foraging trials is $2N$ of which M are spent sampling the two patches. (f) An animal adopts the value of M that maximizes the expected number of food rewards during the $2N$ foraging trials. Krebs et al (138) then derived the optimum M. For each combination a_1, a_2, n, where n is the number of foraging trials carried out so far in each patch and a_i is the number of food rewards obtained so far in each patch, they determined whether the expected future gain was greater if the animal allocated the next two trials to sampling each patch and then specialized on the apparent better patch (Es) or if it specialized immediately (Ed). They worked backwards in n from $n=N$ and found that, if only two trials remained, specialization was better than continued sampling for all combinations of a_1 and a_2. They also found that the higher the value of n, the more "consistently higher" (138) Ed was than Es (in, presumably, the proportion of a_1, a_2 combinations).

It is not clear from their paper, however, how they subsequently derived the optimal M. One possible way to determine it would be to estimate the probability associated with each combination of a_1, a_2, and n, use these probabilities to calculate the expected difference between Ed and Es for each n, and set M equal to the n at which this expected difference first becomes negative as n decreases from N. Krebs et al reported that the optimum M decreases as the difference between P_1 and P_2 increases. They also found that for a given a_1, a_2 combination, the likelihood that Ed would exceed Es decreases as the total number of trials ($2N$) increases. That is, sampling should last longer for larger N's.

TEMPORAL VARIATION IN PATCH QUALITY   If the qualities of the available patches vary over time, the optimal strategy may not be to allocate all of the available time to one patch, but it will sometimes include time spent sampling the various patches (245). However, no theoretical model of this situation has apparently been developed.

4. Optimal Patch Departure Rules

As an animal spends time in a food patch, it may obtain information about the quality of the patch *while* depleting the food available in it. An animal may therefore leave a patch because of information gained or resource depletion or both. The failure to appreciate fully these dual reasons for patch departure has resulted in some confusion concerning the development and testing of predictions concerning rules of patch departure.

Charnov (33) developed the first model of patch departure, and he incorporated seven assumptions: (*a*) fitness increases linearly with the expected net rate of energy intake; (*b*) the expected net energy gain from a patch $g_i(t)$ depends on the patch type i and is a continuous function of amount of time t allocated to the patch, (*c*) the slope of $g_i(t)$ decreases with increasing t because of resource depletion; (*d*) the foraging animal "knows" the quality of each patch (i.e. the function $g_i(t)$ for each i); (*e*) the average travel time between patches (t_b) is known to the animal; (*f*) the proportion of visited patches of type i (P_i), and the energy costs per unit time in traveling between patches (e_b) and while searching within a patch (e_w) are constant and known to the forager; (*g*) an animal bases its decision to depart from a patch only on the patch type and the amount of time spent in the patch. From these assumptions, it follows that the foraging process can once again be modeled as a renewal process (33, 34) and that the forager should allocate time T_i to each patch of type i such that the instantaneous or "marginal" rate of net energy gain (i.e. $\partial g_i(t)/\partial t$, when $t = T_i$) has dropped by time T_i to the overall rate for the habitat (33). Hence, the instantaneous rates of net energy gain at the times of departure should be equal for each patch type and should be higher in habitats whose overall net rate of energy intake is higher (33). The times allocated to patches should therefore be greater for consistently better patches but should decrease as habitat quality increases (33). Essentially the same theory and predictions have been derived by Cook & Hubbard (42) and Parker & Stuart (182). This theory has been modified in the following ways:

DIFFERENT POSSIBLE DEPARTURE RULES   An animal's departure may depend on the amount of food obtained in a patch rather than on the length of time spent there. Assuming that food occurs in discrete food items, the amount of food obtained in a patch is a discrete rather than a continuous variable. From Charnov's theoretical work (33), it follows that in such a situation an animal should leave a patch as soon as it obtains an amount of food such that the marginal rate of energy gain from the amount of food consumed equals the overall rate in the habitat. So, the interval between the time the last food item was obtained in a patch and departure from that patch [i.e. the "giving-up times" (139)] should be zero.

It might also be assumed in the case of discrete food items that an animal's departure from a patch depends only on a giving-up time for each patch type (139). The expected marginal rate of energy gain in a patch should then depend on *both* the patch type and the time since the last food item. Krebs et al (139) predicted that giving-up times should be the same in different patch types within a habitat. This prediction should only be correct in the above model, however, if the departure decisions and the expected marginal rates of energy gain in the patches depend solely on the giving-up times. In general, giving-up times should be longer in patches that are consistently better than others (158).

Iwasa et al (113) have compared the above three departure rules, involving time per patch, food per patch, and giving-up time, respectively, for the case when identical food items are encountered at random within patches that all have the same initial number of food items. They also assume that the food handling times are negligible. When the maximum rates of food gain are determined for each departure rule, the optimized fixed-number (i.e. the fixed amount of food) strategy is best, followed by the fixed-time strategy and then by the fixed giving-up-time strategy (113). This conclusion seems reasonable, since the instantaneous rate of food gain should be most *directly* affected by the number of food items removed (or remaining), less directly by the elapsed time, and least directly by the time since the last food item was consumed. If patch quality varies but the animal knows the quality of each patch before entering it and is able to adopt a different number, time, or giving-up-time threshold for each patch type, then the three optimal strategies should yield the same result. McNair (158) found, however, that it is possible to construct situations in which the fixed giving-up-time strategy is better than the fixed-time strategy.

THE DEPENDENCE OF FITNESS ON THE MEAN AND VARIANCE OF ENERGY INTAKE   If fitness is a nonlinear function of an animal's net rate of energy intake and if some of the parameters that determine this rate are random variables, then the net rate of energy intake will also be a random variable; and the expected fitness will be a function of its mean, variance, and possibly the general shape of its frequency distribution (see above). Stephens & Charnov (253) considered this possibility for the above model of patch departure. They assumed that the expected fitness is maximized when the probability that an animal obtains less than some threshold amount of food is minimized; that all of the parameters in the model are constants, except travel time between patches; that there is one patch type; and that encounters with patches occur at random so that between-patch travel time has a mean of $1/q$ and a variance of $1/q^2$. Then they used renewal theory (see 47) to derive the mean (μ_τ) and variance (σ_τ^2) for the energy gain (e_τ) during a foraging period of length τ. They had previously shown that if τ is large relative to the time between foraging decisions, then e_τ will be approximately normally distributed; also minimizing the probability of starvation will be equivalent to maximizing the ratio $(R - \mu_\tau)/\sigma_\tau$, where R is the net energy requirement. Finally, they contrasted the optimal time spent in each patch if the probability of starvation is minimized (t_1) with that if the mean rate of energy gain is maximized (t_2). They found that there is a critical level of R (R^*) such that t_1 is greater or less than t_2 if R is less or greater than R^*, respectively (253).

UNKNOWN PATCH QUALITY   The situation is rather different if the foraging animal does not know the qualities of patches before it enters and samples them. The animal may leave a patch long before there has been any resource depletion if the initial sampling of the patch suggests that it is inferior to others. To develop models of optimal patch departure, it is therefore necessary to consider the sampling procedures that animals might adopt. Sampling by a foraging animal may be defined as acquisition and storage of information that is correlated with future foraging success. It should be widespread, since food densities should be correlated both temporally and spatially.

The array of possible kinds of information that an animal may usefully acquire while foraging is large. For example, an animal feeding on discrete food items may store the magnitudes of all time intervals between successive food items, the time intervals between arrival in a patch and the first encounter with a food item, and the time since the last food item was encountered, and it may also store the food values of all food items encountered. This information may be correlated with the value of and time to the next food item. Assumptions about the level of information foraging animals possess are critical in the development of models of optimal patch departure. Since an animal's ability to store information may decay over time, assumptions concerning an animal's memory are also important.

The range of ways in which animals might use their information is also large. It is usually assumed that foraging animals are constantly making decisions based on the values of a number of parameters (e.g. 159, 199). These relationships between the parameters and the decisions are termed decision rules (e.g. 199). The general aim of optimal foraging models is to determine the optimal decision rules.

Several distinct optimal patch departure models have been developed and these are considered separately below.

ANIMALS FORAGING FOR NECTAR IN FLOWER CLUSTERS   For an animal that is foraging for nectar, any cluster of flowers such as an inflorescence or a plant may be thought of as a patch. Since these animals do not remain indefinitely in a single patch, they must be making decisions about whether to stay in or leave the patch. Such decisions could be made on a continuous basis, but it seems more likely that they are made just after the animal has consumed the nectar in a flower. As an animal visits more flowers relative to the number available in the patch, the probability of revisiting flowers should increase. There should also be a positive correlation between the nectar volume in any two flowers within the same cluster because these flowers are likely to have been visited previously at about the same time, and they may have relatively similar rates of nectar production. Consequently, the amount of nectar that the animal can expect at the next flower on the present cluster should rise with increases in either the amount of nectar obtained at the present flower or the number of flowers available in the cluster, and it should decrease with increases in the number of flowers within the cluster visited so far.

With this view of foraging by a nectarivore in mind, I considered the following model for departure from a flower cluster (194):

1. Fitness is assumed to increase linearly with the overall net rate of energy gain.
2. The probability p that the next flower visited by the animal within the present flower cluster is a revisit increases as the number of flowers already visited within the cluster (n_v) increases and as the number of available flowers (n_a) decreases.
3. The nectar volume obtained by the animal at the present flower is positively correlated with the nectar in the next flower that the animal would visit if it continued to forage within the present flower cluster.

4. After consuming the nectar in a flower, the animal decides whether to visit another flower in the same flower cluster or to leave and visit a flower on another cluster.
5. The animal leaves a cluster if the amount of nectar obtained from the present flower (or the average nectar obtained from the last m flowers) is less than a threshold (T), which depends on n_v and n_a. If the animal is foraging optimally, it should be more likely to visit another flower within the present cluster the greater its expected nectar at the next flower within that cluster is. Consequently, the optimal threshold T should increase with either an increase in n_v or a decrease in n_a.
6. The time and energetic costs incurred by an animal in moving between flowers within a cluster, moving between clusters, or removing nectar from a flower are all constant.
7. The animal obtains little or no nectar at a revisited flower.
8. The animal is assumed to know the above relationships and parameters and the overall rate of energy gain in the habitat.

It follows from points 2, 3, and 7 above that the probability density function for the amount of nectar (μ_n) in the next flower that the animal would visit if it continued to forage within the present cluster is given by $f(\mu_n|\mu_p,n_v,n_a) = f(\mu_n,\mu_p,n_v,n_a) = f$, where μ_p is the amount of nectar obtained at the present flower (or the mean amount obtained at the last m flowers). If $E_n = E(\mu_n|\mu_p,n_v,n_a) = \int \mu_n f \, d\mu_n$ is the expected value of μ_n, then it would also be the case that $\partial E_n/\partial \mu_p > 0$, $\partial E_n/\partial n_v < 0$, and $\partial E_n/\partial n_a > 0$. In order to find the set of optimal thresholds for different numbers of visited flowers and available flowers, it is apparently necessary to carry out computer simulations of the above foraging model (e.g. 202). Three solutions have been deduced without proof, however, from the "marginal value theorem" that Charnov (33) developed, which was outlined above for situations in which patch qualities are known (103, 195; J. M. Pleasants, personal communication).

In all cases, the optimal threshold T for a given n_v and n_a is assumed to depend on the animal's overall rate of energy gain in the habitat (R) and the times and costs required to move between flowers within a cluster (t_f and $c_f t_f$) and to handle a flower and remove its nectar contents (t_h and $c_h t_h$). The three proposed solutions for the optimal T are given by the following equations:

$$E(\mu_n|\mu_p \geq T) - (c_f t_f + c_h t_h) = R(t_f + t_h), \qquad 1.$$

i.e. the expected rate of energy gain obtained by moving to the next flower on the present cluster, *given that the animal chooses to do so*, is equal to the overall rate of energy gain in the habitat (195). This statement is *not* equivalent, as

Hodges (103) implied, to the prediction that an animal should leave a plant whenever the expected rate of energy gain obtained by probing the next flower within the present cluster is less than the overall rate in the habitat.

$$E(\mu_n | \mu_p = T) - (c_f t_f + c_h t_h) = R(t_f + t_h),$$ 2.

i.e. the threshold T is such that if the animal obtains *exactly* that amount of nectar at the present flower and chooses to visit another flower on the present flower cluster, its expected rate of energy gain for so doing is equal to the overall rate of energy gain in the habitat (103; J. M. Pleasants, personal communication).

$$\frac{E(\mu_n | \mu_p = T) - (c_f t_f + c_h t_h)}{t_f + t_h} = \frac{\mu - (c_c t_c + c_h t_h)}{t_c + t_h}$$ 3.

where t_c and $t_c c_c$ are the time and cost required to move between flowers on different flower clusters and μ is the expected nectar at the *first* flower on another cluster (103). Hodges (103) allowed t_h to depend on the amount of nectar energy obtained. He justified only considering the first flower on another flower cluster on the basis of his observations that the animals he studied (bumblebees) visited one flower per flower cluster most of the time.

Though each of these proposed foraging strategies will produce one desired result—i.e. the likelihood of departure from a flower cluster will decrease with increases in the amount of nectar obtained at the present flower—none is the correct optimal strategy (202). Nor do we know the extent to which these proposed strategies tend to be good approximations of the correct one.

Computer simulations indicate that the optimal thresholds will usually rise with increases in the numbers of visited flowers and fall with increases in the number of available flowers, even if there is no immediate risk of revisiting a flower (202). In other words, if a flower cluster still has many unvisited flowers, the animal should tend to continue sampling the cluster before rejecting it. This finding is analogous to the sampling situation Krebs et al (138) consider.

FORAGING IN PATCHES FOR RANDOMLY ENCOUNTERED PREY ITEMS   Oaten (174) considered the optimal patch departure rule in the following situation: (a) Fitness increases linearly with the expected rate of food intake. (b) Food consists of prey items that are all identical and that occur in patches that differ in terms of the initial number of prey present. Food items are not replaced as they are eaten. (c) The proportion p_k of patches encountered by the animal that initially contain k prey is known to the animal. (d) The animal also knows the probability density function $f(t_1, \ldots, t_j, k)$ for the time between arrival in a patch and the first food encounter (t_1) and the time t_i between the $(i-1)$th and ith food encounter, given that the patch initially contains k food items and that the animal obtained j food items from the patch. (e) The time required for the animal to travel between patches has an expected value that is known to the animal. (f) The animal's decision concerning whether or not to stay in its present patch depends on the time intervals $t_1, \ldots, t_j$ up to the last food item obtained and the time v_{j+1} since the last food item was encountered. More precisely, it is assumed that after j food items the animal employs a threshold V_j (i.e. the giving-up time) that is a function of $t_1, \ldots, t_j$ such that if and when $v_{j+1} = V_j$, the animal leaves its present patch. (g) The handling times of food items are negligible.

Based on these assumptions, Oaten (174) derived implicit equations for V_j and showed that an optimally foraging animal will tend to stay longer in a patch than an animal that leaves when its instantaneous probability of capture per unit time has fallen to the overall rate in the habitat. Continued foraging in the patch provides not only food but valuable information as well (159, 174).

Green (87) modified Oaten's model by adopting the following assumptions: (a) Each patch consists of n bits (i.e. places where food items may be found). (b) The animal takes unit time to search each bit. (c) Within each patch, the number of bits containing one food item follows a binomial distribution with parameters n and p. Remaining bits contain no food. The value of p then determines the patch type. (d) In terms of patch quality, the animal knows only that p is distributed over patches with a beta distribution whose parameters are α and β. (e) As the animal searches a patch, the probability that it will find food in the next bit does not change. (f) The animal will leave its present patch if it has searched t_k bits and found only k food items.

Green (87) then compared the rate of food gain for the optimal t_k's for three alternative cases: (a) The naive strategy—The animal is assumed to learn nothing about the quality of its present patch from its experience in it; the optimal strategy is to search all n bits in each patch visited. (b) The omniscient strategy—The animal is assumed to know the qualities of each patch before visiting them. Therefore, the animal should thoroughly search the best patches and ignore the others (as in the optimal diet model above). (c) The instantaneous rate strategy—The animal is assumed to leave a patch when the probability of finding a prey in the next bit falls below some critical value. In this case, the optimal critical value is typically lower than the overall rate of food gain in the habitat. An optimal forager in Green's model should therefore tend to stay longer in each patch than an animal that adopts the best instantaneous rate strategy. Green (87) found that the omniscient strategy is the best if it can be used, the naive strategy is the worst, and for the parameter values he considered, the instantaneous rate strategy is almost as good as the strateagy of employing the optimal t_k's.

Iwasa et al (113) also developed a modified version of Oaten's (174) model. They made the following additional assumptions: (a) Within each patch encounters with food items occur at random and (b) The instantaneous rate of food encounter in a patch is proportional to the number of food items remaining in the patch. They then showed that the expected number of food items remaining in a patch, given that n have been eaten in time t (including any time since last food item), is a function of the p_i's (i.e. the distribution of patch quality) and t and n, but it does not *also* depend on the $t_1, \ldots, t_j$ and $v_{(j+1)}$ of Oaten's model. The optimal strategy is therefore to leave a patch whenever t and n are such that this expected number of remaining food items is less than some threshold (n_t) (113). Iwasa et al also showed that this optimal departure rule becomes a fixed-number (per patch) strategy when all patches have the same known number of food items and a fixed-time strategy when the number of food items per patch has a Poisson distribution. It is not clear whether a fixed-number, a fixed-time, or a fixed giving-up-time strategy generally would, when optimized, yield the highest rate of food gain in the present model. They suggested, however, that the fixed giving-up-time strategy is the best of the three only when the distribution of food is highly contagious.

Stewart-Oaten (254) modified his earlier model (174) in the same manner as Iwasa et al (113) and showed that if the number of food items per patch has a Poisson distribution, then the optimal strategy is to "leave after constant time T" (i.e. a fixed-time strategy). He also demonstrated that in this situation, the following three strategies are all equivalent to the optimal strategy: (a) employ a department rule such that the average instantaneous capture rate at the time of departure from a patch is equal to the overall capture rate; (b) leave a patch when the instantaneous probability of capture per unit time is equal to the overall capture rate; and (c) leave when this instantaneous capture probability is equal to an optimum threshold. McNamara (159) also developed two special cases of Oaten's foraging situation model (174).

FORAGING IN PATCHES THAT VARY OVER TIME   If patch quality varies over time, then an animal may sometimes do better if it moves to a new patch than if it remains where it is. The optimal strategy will be determined by the frequency and magnitude of *potential* changes in patch quality and the extent to which patch quality tends to remain constant over time. For example, if patch quality may change at time t_i and if patch qualities at time t_{n+1} are likely to be the same as at time t_n, then an animal should remain in its present patch at time t_n if it is in a high quality patch and move to a new patch if it is in a low quality patch. This strategy would be called "win-stay" (see 288). In contrast, if patch qualities tend to reverse themselves every time they change, an animal should adopt a "win-leave" strategy (see 288) and leave its patch if it was of high quality during the last time interval.

Janetos (119) developed a model of this foraging situation based on the following assumptions: (*a*) Fitness increases linearly with the expected rate of food gain. (*b*) There are two kinds of patches, good and bad, which may change in quality after each day. Good and bad patches change quality with probabilities *a* and *b* respectively. (*c*) The daily food gains in good and bad patches are *G* and *B* respectively. (*d*) The cost of changing patches is *C* (measured in food units). Janetos considered the two possible strategies: (*a*) sit-and-wait, where the animal never moves from its present patch, and (*b*) active forager (win-stay), where the animal remains in its present patch for another day if its has just had a good day and changes patches otherwise. Two other possible strategies that Janetos does not consider are: (*c*) active forager (win-shift), where the animal changes patches if it just had a good day and otherwise remains in its present patch for another day and (*d*) active forager (always shift), where the animal changes patches at the end of every day. The average daily food gain from strategy (*d*) (i.e. $(bG+aB-C)/(a+b)$) is always less than that from strategy (*a*) (i.e. $(bG+aB)/(a+b)$ (see 119). Consequently, we need not consider strategy (*d*) further. Strategy (*b*) is better than strategy (*a*) if $(G-B)/C > (a+b)^2/b(a+b-1)$. Similarly, strategy (*c*) is better than strategy (*a*) if $(G-B)/C > (a+b)^2/a(a+b-1)$. Since $1-a-b$ must be either positive or negative and since $(G-B)/C$ is positive, it follows that the optimal strategy is either (*b*) or (*c*). Using Janetos's methods, (119) it can easily be shown that if $(a+b)<1$, strategy (*b*) is better than strategy (*c*) as long as $[(G-B)/C > (a-b)(a+b)^2/ab[1-(a+b)^2]]$; and if $(a+b)>1$, (*b*) is better if the reverse inequality holds. Consequently, if $a+b<1$ and $a<b$, then strategy (*b*) is always optimal; whereas if $a+b>1$ and $a>b$, then strategy (*c*) is always optimal. In the other cases, the optimal strategy depends on the difference between good and bad patches relative to the cost of moving between patches and on the magnitudes of *a* and *b*. Janetos & Cole (120) consider two other possible strategies.

5. Optimal Movements

Many animals undergo movement relative to the medium in which they are foraging. This movement may result from activities of the foraging animal, as in the case of a mobile animal that walks, flies, swims, etc., while in search of food or of a stationary feeder that exerts some control over the rate at which its foraging medium passes by (e.g. a suspension feeder that controls the rate of movement of water through its filtering apparatus). Movement may also occur independently of an animal's behavior as, for example, in the case of stream-living animals that rely on the movement of the water to bring them food. This section will deal with the active movement of foraging animals.

THE PATTERN OF MOVEMENT   As animals move from one place to another while foraging they exhibit various movement patterns (see references in 209,

210). If, for example, the movements of an animal are divided (naturally or artificially) into linear segments, then there may be some tendency for the segment lengths and angular changes in direction to assume particular values. There may also be some relationship between the present rate of food intake and the size of the linear segments and direction changes. Many animals, for example, exhibit "area-restricted searching" whereby they alter their movements in response to food so that they tend to remain in the local area.

Two models dealing with such movement patterns have been developed (see 36, 193, 194). One assumes that the animal does not know where food may occur (36, 193), while the other assumes that the animal knows the exact locations of a number of "resource points" that may contain food (194). Both models assume that fitness increases linearly with the rate of food gain.

In the first model, the foraging animal is visualized as moving among points on a uniform bounded grid in the following manner (193): (a) Movements can only occur between a point in the grid and one of its four nearest neighbours (Hence movements are of constant length). (b) The direction of a movement depends only on the direction of the previous movement. (c) The animal turns right, continues straight ahead, turns left, or goes backwards with probabilities P_r, P_s, P_l, and P_b respectively. These probabilities are obtained by discrete approximation to a normal distribution that has a mean angle 0° and is truncated at $+180°$ and $-180°$. The probabilities are then uniquely determined by $P_r - P_b$, which in this case is the directionality of movement (see 146). (d) The animal obtains no food at revisited grid points. The optimal directionality minimizes path recrossing. (36, 193). Using computer simulation, I showed that this optimal directionality increases with increases in the size of the grid and with decreases in the length of the foraging bout and that it depends on the behavior of the animal at the grid boundary (193). For realistic boundary behavior and a wide range of grid sizes and bout lengths, the model predicts directionalities between 0.8 and 1.0 (i.e. movements with relatively little turning) (1930).

In the second model, the movements are visualized as follows (194): (i) Food occurs at "resource points," which are randomly distributed but the locations of which are known to (e.g. can be seen by) the foraging animal. (b) The animal cannot tell how much food is available at a resource point before it gets there. (c) The animal chooses the next resource point just before leaving the present one. (d) The animal moves linearly between resource points. (e) The animal chooses the next resource point by aiming its departure in some direction relative to the direction of the last movement, scanning a sector of angular width $2w$ about this aimed direction and then choosing the closest resource point. Heinrich (98) pointed out, however, that animals may not always choose the closest detected resource point. (f) The difference between the arrival and aimed departure directions and the width of the scanning sector may depend on the direction of previous movement or on the amount of food obtained at the present resource point. In other words, the animal is able to "remember" certain information. These relationships are adjusted in order to maximize fitness. (g) Food is renewed at a constant, relatively low rate. (h) There is a positive correlation between the amounts of food in neighboring resource points. The closer the points are, the higher this correlation will be. I have argued that this correlation pattern arises from the tendency of the animal (or other animals) to move between closely neighboring resource points, thereby creating similarities among neighboring points in terms of the time since the last visit (194). (i) The area containing the resource points is large, so that encounters with its boundaries can be ignored.

I went on to deduce that the optimal rule of movement will have the following properties (194): (a) The mean angular change in direction should be 0°. (b) The animal should alternate right and left hand turns. (c) The width of the scanning sector should increase as the amount of food obtained at a resource point increases. Consequently, the variance of the angular change in direction should increase simultaneously (i.e. the directionality should decrease). (d) The frequency distribution of the distances moved between resource points should resemble a negative geometric distribution. If the animal cannot obtain or store some of the information it is assumed to know in the model, then these predictions will have to be modified accordingly (194). In addition, as Zimmermann (296) pointed out, the optimal directionality should decrease if the rate of food renewal is rapid or if the animal removes only a fraction of the available food during a visit to a resource point. In some circumstances, random movement (i.e. no directionality) could be expected. The exact optimal rule of movement has not been determined for the present model (209). Such an exercise would almost certainly require a large-scale computer simulation (209).

THE SPEED OF MOVEMENT OF MOBILE ANIMALS   As foraging animals travel faster, the rates of energy expenditure and of encounters with the locations of food items should increase (e.g. 77, 171, 204, 273, 275). If food items are cryptic, however, the probability that encountered food items will actually be detected should decrease at the same time. Gendron & Staddon (77) developed a model incorporating all of these potential consequences of increasing speed, which is based on the following assumptions: (a) Fitness increases linearly with the net rate of energy gain. (b) Searching for and handling food items are mutually exclusive activities. (c) There is only one food type. (d) The rate of encounter with food items is $S \times D$, where D is the density of food items and S is the search rate (measured by the area searched per unit time). (e) The probability P_d that the animal detects an encountered food item is: $P_d = [\, 1 - (S/M)^k\,]^{(1/k)}$, where M is the search rate at which no food items are detected and k is a conspicuousness index. (f) The rate of energetic expenditure while searching increases linearly with the search rate. (g) The rate of energetic expenditure while handling food items is negligible. (h) The animal adopts the search rate that maximizes the net rate of energy gain.

Gendron & Staddon discovered that the optimal search rate could only be found using computer simulations based on particular parameter values but noted that it is lower than the rate that maximizes the gross rate of energy gain. They also modified the above model to allow for two food types that are eaten whenever encountered and detected and found that the optimal search rate in this case depends on the density and crypticity of each food type, as well as on their energetic values and handling times.

I developed a simpler model by omitting any consideration of crypticity and implicitly assuming that all encountered food items are detected (204). The optimal search rate is then the maximum speed that the animal can sustain over the search period. If the energetic cost of searching at this search rate increases more rapidly than a linear function would, however, the optimal search rate in my model may be some intermediate speed.

FILTERING RATE BY STATIONARY SUSPENSION FEEDERS   For animals that obtain their food by filtering water (or air), increases in the filtering rate should be accompanied by increases in the rate of ingestion of food particles and in the energetic costs of filtering and by a decrease in the energy absorption efficiency for each particle ingested. Three similar models of this foraging situation have been developed (142, 144, 256). All of them incorporate the following assumptions: (a) Fitness increases linearly with the net rate of energy gain. (b) The rate of food (energy) ingestion is proportional to the filtering rate F (measured as volume per unit time). (c) The energetic cost of filtering, E_F, is: $E_F = bF^x$, where b and x are constants ($x=3$ in 142; $x=2$ in 144). (d) The animal adopts the filtering rate that maximizes the net rate of energy gain. The three models basically differ only in their assumptions concerning absorption efficiency. Lam & Frost (142) assumed simply that energy gain is equal to the amount of energy ingested (i.e. the absorption efficiency is 100%) and that energetic costs increase with increasing body length. They derived the optimal filtering rate and showed that it increases with increasing particle density or decreasing body length. Lehman (144), in contrast, assumed that there is a constant number N of food particles in the animal's gut. Consequently, the time t that each particle spends in the gut is given by $N = FDt$, where D is the particle density. He also assumed that the gut has a maximum volume and that the amount of energy obtained per ingested particle increases with increasing passage time (t) in the gut. Therefore, the animal should maintain a full gut, and the optimal filtering rate is found by plotting the net rate of energy gain against the filtering rate after assigning particular values to the various parameters. Finally, Taghon (256) assumed that the fraction of energy obtained from ingested particles decreases with increases in the filtering or ingestion rates. He calculated the optimal filtering rate for constant absorption efficiency and for absorption efficiencies that decline linearly or exponentially with rising filtering rates. He concluded that in all cases, the optimal filtering rate increases with increasing energy value per food particle.

FORAGING MODE   Alternate foraging modes have received little attention to date (6, 170, 172). Norberg (170) considered two alternate modes, one of which resulted in a higher rate of energy gain at a given food density and in a higher rate of energy expenditure than the other. He showed that the higher the food density, the more likely it is that the more efficient, but more expensive, strategy will lead to a higher net rate of energy gain (170). He also considered the special case of birds that climb or hop vertically in trees while searching for food and then fly between trees (172). He assumed that the birds could hop up each tree and fly to the next tree using gliding as much as possible (Strategy A), hop down each tree and fly up to the top of the next (Strategy B), or alternate hopping up and down each tree and fly horizontally between trees (Strategy C). He showed that Strategy B is never optimal; Strategy C produces a better net rate of energy gain than Strategy A if and only if the distance between successively visited trees is less than about half the distance coverable in gliding flight, with height loss equal to the foraging height zone within each tree (172).

6. Optimal Central Place Foraging

For a central place forager, one cannot consider diet, patch choice, departure from patches, and movement rules independently of one another. The transportation distance for food gathered will depend on the animal's location at the time of return, and this location may depend on previous movements or on where the animal has chosen to forage. At the same time, the frequency of trips back to the central place will depend on the animal's diet and on the amount of food obtained at each location.

Present models of central place foraging focus on just one aspect of foraging: (a) movements while searching for food, (b) the relationships between diet and distance from the central place, and (c) the association between distance and both patch choice and the rule of departure from a patch. Consequently, I shall deal separately with each of these foraging categories. In all three cases, it is assumed that the cost of transporting food items to the central place depends only on the time required, and not on the size of the load (cf 54).

SEARCHING MOVEMENTS   According to Morrison's (165) model, a foraging animal leaves its central place, searches for a patch of food using the movement rule described below, returns to the central place with a food item, and then makes $(n-1)$ more trips to and from the patch, obtaining the same amount of food each time; then it resumes the search for another patch. His model is based on the following assumptions (165): (*a*) Fitness increases linearly with the expected rate of food gain. (*b*) The locations of patches are initially unknown. (*c*) The animal searches for patches by traveling a constant distance (D) in between turns, which results in new movement directions that are *independent* of previous events. During the search, the animal detects all patches that occur within a detection radius (r). Every time the animal changes direction, it re-searches areas searched during the previous movement interval. The overlap between areas searched during nonsuccessive movement intervals is assumed to be negligible. (*d*) The animal adopts the distance (D) that produces maximum fitness. Morrison then showed that as D increases, the rate of searching new areas increases, the expected search time to find a patch consequently decreases, the expected final distance from the central place increases, and the costs of each visit to the patch therefore increase. He derived an expression for the average time per feeding visit to a patch and showed that this is minimized (i.e. the expected rate of food gain is maximized) for values of D that decrease as n increases. In other words, if an animal revisits a patch many times, it should adopt a movement rule that keeps it near its central place. Because the locations of patches are initially unknown, in the present model the animal will tend to visit patches more distant than the nearest patch (165).

DIET VS DISTANCE FROM CENTRAL PLACE   Schoener (232) considered the situation where an animal encounters food items that differ in terms of their net energy yield and handling time and where any chosen food item is transported to a central place. He assumed that fitness increases linearly with the average net rate of energy gain. The optimal strategy is obtained in the same manner as the optimal diet in the above diet models; encountered food items should be ranked according to the ratio: (net energy gain − transportation cost) ÷ (handling time + transportation time). He also assumed that both net energy gain and handling time are functions of prey length. For several choices for these functions, if transportation time is independent of prey length, then as transportation time (i.e. distance) increases the length of the best prey should increase and the range of prey lengths taken should shift upwards. These effects can be reversed, however, if transportation time increases with prey length (232).

Lessells & Stephens (145) considered an animal that forages in patches and adopted the following assumptions: (*a*) Fitness increases linearly with the expected rate of food gain. (*b*) Patches occur at different distances (i.e. travel times t_T) from a central place. (*c*) Each chosen food item is transported to the central place. (*d*) Within a patch, the animal accepts the first food item of "value" (i.e. energy) C or greater, where C may vary with transportation time or distance. (*e*) Handling times other than transportation time are negligible. (*f*)

The animal "chooses" the relationship between C and t_T that maximizes fitness. For each C, there will be an expected energy gain e and a search time t_s per patch; the optimal C is obtained graphically from a plot of e against t_s in essentially the same way as Charnov's marginal value theorem is solved (see 33, 145). Lessells & Stephens also showed that the optimal threshold C increases with increases in the distance of a patch from the central place. In other words, the animal should tend to take food items of higher energy value from more distant patches.

PATCH CHOICE AND PATCH DEPARTURE VS DISTANCE FROM CENTRAL PLACE   If there is no resource depletion during foraging in available patches, then an animal should forage in the closest patch exclusively. If resource depletion does occur, however, then the animal should forage in a number of patches, and the overall allocation of *time* to the various patches will depend not only on patch choices but also on the rules for departure from each patch. Consequently, in the context of central place foraging, patch choice and the patch departure rule must be considered together. In all of the models discussed below, the animal's diet is assumed to be independent of patch location.

Andersson (4) formulated a continuous model for this foraging situation based on eight assumptions: (*a*) Fitness increases linearly with the amount of food obtained during a fixed time S or decreases with the time required to obtain a fixed amount of food. (The same results should apply if the rate of food gain is the currency of fitness.) (*b*) Stationary, identical food items are distributed randomly and with density λ throughout the total foraging area, which is circular with radius R. (*c*) When searching, the animal is surrounded by a detection area of radius r_d, which is much smaller than R. (*d*) $t(r)$ is the total time that each point of distance r from the central place remains within the animal's detection area. (*e*) Given that a food item occurs at a certain point at distance r from the central place, the item will be discovered and removed with probability $P[t(r)]$ where P increases with increasing t_r but with decreasing slope (i.e. diminishing returns set in as the animal spends more and more time searching at each point). (*f*) Removed food items are not replaced. (*g*) Each food item obtained at distance r from the central place entails a transportation time $C_t(r)$. (*h*) The animal adopts the functional relationship $t(r)$ that maximizes fitness. Andersson then derived an expression for the average time required per food item for food obtained from the circular belt $(r, r+dr)$ given $t(r)$; defined *marginal cost* as the cost per food item from this belt during an additional time $dt(r)$ as $dt(r)$ tends to 0; and showed that for general forms of $P[t(r)]$, the total search time has been optimally allocated if and only if this marginal food cost is equal throughout the foraging area. In addition, if $P(t) = 1 - \exp[-at(r)]$, where a is the instantaneous rate of detection, and $C_t(r) = r/vz$, where v is the transport velocity and z is the number of food items per load, then the optimal $t(r)$ is given by:

$$t(r)=[\ln (M - r/vz)a \ \lambda\pi r_d^2]/a, \qquad\qquad 4.$$

where M is the marginal food cost under optimal time allocation. Finally, Andersson found that over a variety of parameter values, the optimal $t(r)$ (i.e. the time per unit area) decreases approximately linearly with r (4).

Several authors (80, 128, 179) have developed a discrete model that deals with the same foraging situation. They assume that after an animal has foraged for time t in a patch, its expected food gain is $f(t)$, where $f'(t)>0$ and $f''(t)<0$; that after foraging in each patch, the animal transports the food obtained to a central place at a time cost T (i.e. the time for a return trip), and that fitness increases linearly with the expected rate of food gain. These assumptions form the basis for a mathematical model identical to Charnov's (33) model of the allocation of time to patches, and the solution is found in the same manner as for Charnov's model (80, 128). If there is only *one* patch, since T increases with the increasing distance d of a patch from the central place, the optimal time spent in the visited patch and amount of food obtained there will both increase with increasing d (80, 128). If, however, there are two or more patches at different distances, then the animal should spend decreasing amounts of time per patch as the distance from the central place increases (see 33). Whether the time per patch (i.e. time per unit area) should decrease linearly with distance (as predicted in 4) will depend on the relationships between time and energy yield per patch and between patch abundance and distance.

The above model has not been modified to allow for some (but not instant) food renewal. However, food renewal underlies a simple model Evans (70) developed. He made the following assumptions: (*a*) An animal forages in food patches that last one day. (*b*) Patches of food arise anew at the beginning of each day. (*c*) A unit area (e.g. 1 m^2) contains a food patch on any given day with probability p. (*d*) The animal, after leaving its central place, searches each consecutive circular belt (perhaps by moving in a spiral) until it finds a patch, and then it forages in that patch for the rest of the day. The probability P_k that the animal will forage in the circular band between $(k-1)$ and k distance units (corresponding to area units) is given then by $P_k= [1-q^{(2k-1)}]q^{(2k-2)}$, where q $= 1-p$. In this case, the search time per unit area, which is proportional to $P_k/(2k-1)$, might follow the same general pattern, but it would reach a peak earlier than P_k. In contrast to Andersson's (4) predictions, any decrease would not be linear.

EVALUATING OPTIMAL FORAGING THEORY

1. Criteria for Accepting Optimal Foraging Studies

Because of both the many assumptions that form the basis for any optimal foraging predictions and the difficulty of devising independent, direct tests of some of these assumptions, the usefulness of the optimal foraging approach will only become clear after a large number of optimal foraging studies have been conducted. Not all studies, however, provide equally valid contributions to the overall evaluation of optimal foraging theory, and I propose that only those studies that satisfy the following criteria be considered acceptable.

1. When the assumptions about the foraging animal can be assessed independently, they should be as realistic as possible, and they should be justified. Such assumptions normally include those made about the currency of fitness, the information possessed by the animal, and any behavioral constraints. For many herbivores, for example, it would not be realistic to assume that fitness is measured solely in terms of the net rate of energy gain (see 72). Very few studies explicitly justify the assumptions made, however (e.g. 194, 197).

2. The foraging model should correspond as closely as possible to the actual foraging situation. It is inappropriate, for example, to apply a model based on an assumption of random food encounters to a situation in which food encounters occur after constant intervals (219) or to use Charnov's (33) marginal value theorem in situations where an animal is likely to be sampling rather than depleting each patch.

3. The predictions should follow logically from the underlying model. In some cases, the mathematics may be controversial (e.g. 34, 259, 264).

4. Parameters that determine the optimal diet should be estimated in an unbiased manner. For example, encounter rates with food types should be determined "from the animal's point of view."

5. Because many foraging parameters are random variables, there will usually be some error associated with any optimal foraging predictions. Consequently, such predictions should usually be determined and expressed in terms of means and standard errors or of confidence intervals. I know of no published studies in which this has been done.

6. Confidence intervals should also be determined for observed foraging behavior, followed by appropriate statistical comparisons of the observed and predicted confidence intervals.

7. In laboratory studies, the experimental foraging situation should mimic natural foraging situations as closely as possible (206), even though unnatural foraging situations may, of course, tell us much about the capabilities of animals.

8. In laboratory studies, animals should be given sufficient experience with the experimental foraging situation to allow their behavior to reach an equilibrium. Like Bayesian statisticians, animals are likely to have a priori assessments of food distributions, among other factors, and their foraging decisions will therefore be influenced by past experience (e.g. 116). Only after an animal's behavioral response to a foraging situation has reached an equilibrium is it reasonable to assume that this response has not been affected by the animal's experience before the experiment began. Even then, this assumption may not be valid. Studies of animals with limited experience may, however, tell us much about how they learn to forage.

Unfortunately, there appear to be no studies (including my own) that satisfy all of these criteria. However, studies that fail to include confidence intervals for predictions but satisfy the other criteria can provide some idea of the usefulness of optimal foraging theory. If there is no significant difference between the confidence interval for the observed behavior and a *point* prediction (e.g. the mean) for the optimal behavior, then the same result would hold if a confidence interval had been used for the optimal behavior. If the first difference is significant, however, using a predicted confidence interval for the optimal behavior may fail to reveal a significant difference between the observed and predicted values. Failure to develop confidence intervals for predictions should therefore lead to a bias towards rejecting optimal foraging hypotheses.

In the following sections, I shall review tests of optimal foraging theory, concentrating on those studies that come closest to satisfying the above criteria.

2. Diet

THE CURRENCY OF FITNESS   Most optimal diet models assume that fitness is positively correlated with the gross or net rate of food or energy gain. There is a growing literature that indicates, however, that nutrients affect growth and/or the maintenance and food preferences of herbivores (11, 13, 27, 28, 72, 127, 162, 168, 169, 175, 180, 186, 224, 226, 227, 233, 234, 265, 282, 283), granivores (40, 86, 148, 190), and predators (88, 185). Some nectarivorous animals may maximize their net rates of energy gain while foraging (e.g. 194), but the small amounts of amino acids and other nutrients in nectars may have nutritional significance (8, 9). For animals foraging on food items that appear to differ only in terms of their size, it is probably realistic to express fitness as food or energy gain (e.g. 63, 66, 82, 107, 136, 138, 139, 199, 216, 217, 269, 271, 278, 284, 296).

Most optimal diet models are also based on the assumption (usually implicit) that the relation between fitness and food or energy gain is linear and consequently that fitness is maximized when the expected net rate of food or energy gain is maximized. It is also always assumed that dietary choice does not influence an animal's ability to perform other tasks, such as avoiding predators. For the foraging situations considered so far, this seems a reasonable assumption.

THE DETECTION OF FOOD ITEMS   In most optimal diet models, it is assumed that animals cannot determine the values of food items at a distance and that encounters occur sequentially rather than simultaneously. This assumption is likely to be unrealistic for many animals, however. In particular, among those animals for which optimal diet predictions have been tested, detection and evaluation of food items at a distance occurs in nectarivores (e.g. 152, 195,

200, 203, 271), fish (e.g. 75, 78, 266, 278), back swimmers (81), and swallows (263). The time an animal needs to move to a detected food item should be included in the handling time for that item, so the value of a food item may depend on the distance at which it is detected. If a number of food items are encountered simultaneously, an animal may neglect a food item that it often eats if a better item is present. The only diet choice situations considered so far for which food evaluation at a distance should not occur and for which the net rate of food or energy gain should be a reasonable currency of fitness are predation by crabs on mussels (63), predation by great tits on pieces of mealworm presented through a small window over a conveyor belt (66, 107, 136), pigeons "searching" for food in a Skinner box by pecking at a key (143), and redshank searching for worms of different sizes (82). In all these cases, encounters with food items should be sequential rather than simultaneous. I shall discuss the results of these studies below and then briefly consider the results of other studies.

STUDIES IN WHICH FOOD ENCOUNTERS ARE SEQUENTIAL, FOOD EVALUA-TION AT A DISTANCE DOES NOT OCCUR, AND FOR WHICH THE NET RATE OF FOOD OR ENERGY GAIN IS A REASONABLE CURRENCY OF FITNESS   It is usually assumed in optimal diet theories that the expected time to the next food encounter is independent of the time spent handling the last food item. This assumption will be correct if food encounters occur at random but incorrect if they are nonrandom and if the encounter process continues while the animal is handling food. In the conveyor belt studies by Krebs et al (136), Houston et al (107), Erichsen et al (66), and Rechten et al (218), the interval between food encounters was constant and the encounter process (i.e. the movement of the conveyor belt) continued while the bird handled the food. These studies therefore require modifications in the optimal diet theory, as Krebs et al (136) and Rechten et al (219) have attempted to do. Neither modification appears to be correct (see 219 and above), but I shall assume below that the correct theory would produce negligible changes in the predictions generated during the above conveyor belt studies.

In these studies and in Lea's (143) Skinner box study, animals (birds in all cases) were sequentially presented with two types of food items, which could be accepted or rejected. Depending on the values of the various parameters, the optimal diet was therefore either always to accept the better food type and never accept the other (i.e. specialize) or to accept both food types (i.e. generalize) (see the above discussion of optimal diet theory).

Krebs et al (136) varied the encounter rates with the 2 food types and tested their predictions on 5 individual birds in each of 5 experimental tests. They found that in all 25 instances, the optimal diet predictions were upheld in the sense that the birds preferentially selected the better food type when specialization was predicted (assuming that food types were instantly and correctly recognizable) and did not select preferentially when generalization was predicted. They also discovered, however, that when specialization was predicted, the birds continued to include some of the worse food type in their diets. In other words, the birds exhibited partial preferences.

Houston et al (107) varied the encounter rates and in some cases added a recognition time to each food type. They tested 4 individual birds in 3 experimental treatments; the observed and predicted dietary selections were consistent with each other in 11 of the 12 instances. In one instance, a bird was predicted to specialize but did not. They also found partial preferences.

Erichsen et al (66) included food items that resembled the better food type but yielded no return and varied the encounter rates with the 3 food types. As the birds always rejected the fake food items, this was essentially a two-food-type situation. They tested 5 birds in 2 treatments; the observed and predicted dietary selections were consistent 9 out of 10 instances. In one case, a bird was selective when it should not have been. They too reported partial preferences.

Rechten et al (218) used essentially the same experimental design as Krebs et al (136), alternating two food types that were regularly spaced on the conveyor belt. They found that when specialization is predicted, the birds make two kinds of "errors," namely, rejecting a profitable item (RP error) and taking an unprofitable item (TU error). They also discovered that at higher food presentation rates (i.e. shorter distances between food items), there were more RP errors and fewer TU errors, as one would expect if the birds sometimes misidentify the food types. They showed, however, that the birds could have adopted error probabilities that would have yielded higher rates of food gain, so misidentification does not completely account for the observed partial preferences. Rechten et al also observed that on the average, the birds rejected more items per item taken than would be expected if the birds were including sampling in their strategy and that hungry birds came closer to the optimal foraging predictions than partially satiated birds.

Lea's (143) results contrast with the relatively favorable ones obtained in the conveyor-belt studies. He varied the search time between the handling of one food item and the encounter with the next, the handling times associated with each food type, and the magnitude of the food reward associated with each food type. As expected, the 6 pigeons he studied became less and less selective as the search time lengthened; but they did not change from specializing to generalizing in a predicted stepwise manner, and they did not demonstrate a rapid decrease in selectivity at the predicted threshold search time. Lea also found, contrary to expectation, that the rate of encounter with the worse prey type affected its level of inclusion in the diet.

The different levels of success of optimal foraging theory in the above conveyor-belt and Skinner-box studies could be due to a greater similarity between the conveyor-belt foraging situation and the natural foraging situations of the birds in the studies. In the conveyor-belt situation, a bird sees potential food items through a small open window above the conveyor belt and either allows an item to pass or else picks up the item for eating and/or inspection (e.g. 136). Food value is varied by changing the size of a food item; handling time is varied by enclosing the food items in containers from which they must be extracted; recognition time is varied by making the containers clear or opaque. In Lea's (143) Skinner-box study, the pigeon made dietary decisions (i.e. "searches") by pecking a certain number of times on a key, chose to accept a particular food type by pecking on a key and then waited for a period of time before gaining access to a food bin for another period of time. The conveyor-belt studies seem to mimic natural foraging situations more closely than the Skinner-box study.

Elner & Hughes (63) studied different-sized crabs feeding on mussels of different sizes. When mussel availability was unlimited, the crabs chose mussel sizes close to the optimal size and that the crabs included bigger and smaller mussels as the supply of optimally sized mussels was depleted. Since crabs must spend some time recognizing mussel size, the abundance of a particular mussel size should determine whether it is included in the optimal diet (see the above discussion of optimal diet theory). Elner & Hughes discovered that the foraging of the crabs was consistent with this expectation. Contrary to expectation, however, they also found some partial preferences.

OTHER STUDIES IN WHICH FOOD EVALUATION AT VARIOUS DISTANCES MAY NOT OCCUR   Granivorous birds that essentially search areas of ground beneath them may not evaluate food at different distances. If the density of seeds is sufficiently low, seed encounter should also be sequential rather than simultaneous. However, seeds of different species are apparently not equivalent in terms of their nutrients (e.g. 40, 86), so the theory of optimal diets for granivores should take nutrient constraints into account (188, 190).

Pulliam (190) studied chipping-sparrows that were feeding on seeds in oak woodland. Some seed species were never eaten by caged or wild birds, and Pulliam assumed that these seeds were nutritionally unsuitable. When attention was restricted to the remaining seed types, he discovered that as predicted, seed types with ratios of energy yield to handling time above a threshold value tended to be included in the birds diet and vice versa. He also found partial preferences, which he interpreted in terms of the nutritional qualities of the seeds.

EXPERIMENTAL STUDIES WITH NECTARIVORES   A nectarivore will often be able to see many flowers or plants from its present location. Thus, the optimal diet theory must be modified to allow for detection at a distance and for simultaneous encounters. Waddington & Holden (271) assumed that in such situations, a nectarivore that can only plan one step at a time should always choose the next flower (or plant, etc.) so as to achieve the maximum expected ratio of net energy gain divided by the time required to move to and exploit the next flower. As pointed out above, this will not be precisely the optimal strategy, but I shall assume that it is a good approximation.

Waddington & Holden (271) observed honeybees foraging for nectar among two types of randomly distributed artificial flowers that differed in color and shape. They varied the average nectar yield per flower of each type by varying the proportions of flowers containing 2µl of unscented 25% sucrose solution, varied the densities of the 2 flower types, and for each of 7 cases, determined the proportions of visits to each flower type if the bees foraged optimally. The observed and predicted proportions were all similar, although a few significant differences were found (271).

Real and his colleagues (216, 217) also observed nectarivores (bumblebees and wasps) foraging for nectar among two kinds of differently colored, randomly distributed artificial flowers. In both studies, the densities of each flower type were equal, and the value of each type was varied by changing the probability distribution of the nectar volume per flower. The insects exhibited flower preferences that increased with increases in the average nectar yield per flower and decreased with increases in the variance (216, 217).

Waddington et al (269) also considered the importance of reward variance in determining flower preferences. They observed bumblebees foraging among four flowers of two color types arranged in a square with diagonally opposite flowers of different colors. Each time a bee left a flower, the nectar in that flower was replenished according to a probability distribution of nectar rewards. The bumblebees preferred flower types with constant rewards to ones with the same mean but variable rewards (269). In addition, the bumblebees exhibited partial rather than absolute flower preferences (269).

In all of the above studies of nectarivore foraging, the animals were risk averse and did not always maximize their expected net rate of energy gain. This result seems surprising in view of the large numbers of flowers (or plants) that many nectarivores (especially bees) are likely to visit during a day.

Marden & Waddington (152) observed honeybees choosing between two artificial flowers that differed in color and sometimes distance. When the flowers were equidistant from a bee, 14 of the 15 bees always chose one color or the other, while the last one showed no color preference. When the flowers were at different distances from a bee (but not in different directions), 8 of 10 bees tended to choose the closest flower independently of its color; one bee was initially constant to a particular color and later tended to choose the closest flower; another bee remained color constant. Thus, almost all the bees eventually tended to forage optimally. The preferences were partial rather than absolute, however.

OTHER STUDIES   In other diet studies, fewer of the criteria outlined above are satisfied. Nevertheless, these studies yield results that tend to be similar to those just discussed. Many studies have found that animals exhibit partial rather than absolute food preferences (e.g. 3, 51, 59, 76, 86, 104, 115, 141, 263, 266, 278, 294). Some studies have indicated that an animal's preference for a particular food type depends only on the abundances of better food types (e.g. 52, 53), while others have found that the abundance of the particular food type is also important (e.g. 78, 115, 263, 266).

3. Patch Choice

THE CURRENCY OF FITNESS   The same comments made about optimal diets as to whether or not it is realistic to assume that fitness increases linearly with the net rate of food or energy gain also apply to optimal patch choice. In addition, fitness associated with patch choice should often depend on predation, etc.

PATCH CHOICE IN THE ABSENCE OF RESOURCE DEPLETION WITH KNOWN LOCATION AND QUALITY OF PATCHES   Most studies of patch choice focus on patches where there is no resource depletion, the locations are known, and factors other than food mass or energy are unimportant. In the earliest of these studies, the patches differed in the average rate of food or energy gain, and the variance was ignored. The animals preferred the patch with the highest rate of food or energy gain, but preferences were partial rather than absolute (147, 243, 245, 279, 291). Recently, several studies have considered how the variance in the rate of food intake affects patch preferences. Two kinds of birds—juncos and white-crowned sparrows—were found to prefer a patch with a relatively low variance in the rate of food intake (i.e. they were risk averse) when they could expect to meet their food (i.e. energy) requirements and to be risk prone when unable to do so. Preferences in these studies were also partial rather than absolute (24–26).

In another set of recent studies, the patches differed not only in the food supplied but also in predation risk (91, 100, 161, 163, 238). Sticklebacks in the simulated presence of an avian predator tended to feed in low rather than in high density swarms where the fish are less able to pay attention to predator approaches. In the absence of the predator, the fish preferred the high density swarms. (161). Sparrows sometimes prefer a distant patch providing shelter from predators to a closer, exposed patch (91). Small bluegills did not switch habitats when expected, and this may have been due to differential predation risk (163). Patch choice by back swimmers was related to a balance between feeding rate and predation risk (238).

If there is no resource depletion within patches and if fitness increases linearly with the rate of food or energy gain, then the optimal strategy is to forage all the time in the best patch. If fitness also depends on some other aspect of food intake such as the variance in the rate of intake or the intake rate of some nutrient, however, then allocating time to more than one patch may be optimal (see the discussion of theories of optimal patch choice). The presence of other factors affecting fitness may explain the partial preferences observed in the above studies.

Two studies indicate the importance of factors other than energy. Belovsky (12) studied the foraging behavior of moose, for which diet and patch choice are equivalent because the various food types occur in patches. He assumed that moose are subject to constraints in the maximum feeding time available each day, daily rumen processing capacity, sodium requirements, and energy metabolism. He found that a model based on the assumption that fitness increases linearly with daily energy gain accurately predicts the amounts of aquatic vegetation, deciduous leaves, and forbs consumed by a moose each day. In turn, nectar-gathering workers of social bee species appeared to choose sunflower cultivars on the basis of energy production per plant, whereas solitary bees, which collect pollen as well as nectar from these plants, did not (260).

PATCH CHOICE WHEN RESOURCES ARE DEPLETED AND RENEWED   In most natural situations, an animal will deplete the available food resources as it forages in a patch. At the same time or in between foraging bouts, these resources may be renewed. The optimal patch choice (assuming the locations of patches are known) will therefore depend on the patterns of resource depletion and renewal. For example, if resource renewal is rapid relative to depletion, then an animal should always forage in the best patch. If renewal is slow, the optimal strategy would be to avoid recently depleted patches.

Cole et al (37) explored the extent to which hummingbirds can learn the appropriate patch choice strategy if both resource depletion and renewal occur. Their patches were two artificial flowers, one of which supplied nectar while the other did not. In between foraging bouts, the positions of the two flower types either remained constant or were switched. The birds had an initial bias towards shifting between flower locations on successive bouts, and they learned this strategy more rapidly than they learned to return to the same flower location. The authors attributed this difference to the relatively slow renewal of flower nectar, so that the "win-shift" strategy is more appropriate under natural conditions. Two other studies concluded that animals distribute themselves between two patches so that the rates of food or energy intake in the patches are equal (160, 200).

4. Rules of Departure from a Patch

The amount of time an animal spends in a patch may affect its rate of food intake, its predation risk, and the time it spends elsewhere performing other important tasks, such as monitoring potential intruders. Consequently, these factors may all enter into determining the currency of fitness for an animal's rule of departure from a patch. So far, the currency for all models of patch departure has been the net rates of food or energy gain; the tests of these models have involved situations for which this is a reasonable choice. Other factors— e.g. predation (97), intrusions (123), and thermoregulation (49)—will clearly have to be included in some situations, however. If one only considers food, an animal may leave a patch because its rate of food or energy intake has decreased through depletion of the available food or through movement into poorer regions of the patch, or because sampling experience indicates that it is a relatively poor patch. The theory of patch departure has concentrated, however, on resource depletion rather than on sampling (see the above discussion of the optimal rules of departure). It is assumed that the locations and qualities of patches are known. The theory should therefore be tested only in situations where this assumption is justified. In the discussion below, I shall first consider studies that satisfy this requirement, then a study that deals mostly with sampling, and finally research that involves both depletion and sampling (see also 209).

RESOURCE DEPLETION IN PATCHES OF KNOWN QUALITY AND LOCATION   When food items are discrete, the optimal rule of patch departure will depend on the informaton used by the animal. An animal may base its departure decisions on the amount of food obtained so far, the amount of time spent in the patch, the time since the last food encounter, or other intercapture intervals, among other factors. In the absence of independent tests, assumptions about information use must be taken at face value.

Krebs et al (139) were the first to attempt to evaluate optimal patch departure. In one of their experiments, the locations and qualities of the patches should have been known to the foraging animals (chickadees). All of the patches were artificial pine cones containing a single food item (mealworm pieces) randomly assigned to one of six covered holes. The authors found that the birds obtained 0.26 food items per patch on average. This number is much smaller than the optimum of 1.0 that would prevail if the birds base their departure on the number of food items obtained so far. It could be close to the optimum, however, if the birds base their departure from a patch on the total time spent in the patch or on the time since consuming the last (and first in this case) food item. Krebs et al carried out some other experiments indicating that the chickadees based their departures on the time since the last food item was consumed (i.e. on the giving-up time) rather than on the time spent in a patch or the amount of food obtained so far. Zach & Falls (292) reported similar results for ovenbirds. Bond's (20) analysis of the probability distributions of giving-up times suggested that departure decisions may be affected by other factors as well. Giving-up times for caddis larvae were unaffected by the previous level of feeding (262). Giving-up times of parasitoid wasps were approximately constant in patches of different qualities, and the wasps reduced the prey to about the same level in each patch (110).

Several studies have implicitly assumed that foraging animals base their departure from patches on the time spent in a patch so far (42, 45, 123, 135). Cowie (45) varied the predicted times per patch by changing interpatch travel times and found that the observed average times and the predicted times were similar. Cook & Hubbard (42) considered six patches that differed in food density and predicted the percentages of the foraging animal's time that should be spent in each patch type. They found that the animals spent some time in patches they should not have visited at all, but that otherwise the difference between the observed and predicted values was minimal. Krebs & Cowie (135) applied this same approach to Smith & Sweatman's (245) and Zach & Falls's (291) data (see 209). Resource depletion was minimal in the latter two studies, however, so I included them in the previous discussion of patch choice. Kacelnik et al (123) studied great tits that could obtain food in a patch by hopping on a perch. They simulated resource depletion by requiring increasing numbers of hops to obtain the next food item. The most common numbers of hops were equal to the predicted number for 3 out of 4 birds, one bird having considerably longer hopping bouts than predicted. In all of these studies, the average or most common time spent per patch agrees reasonably well with the predicted time per patch, but the time spent per patch is not a constant as was predicted. A number of studies have obtained results that support more qualitative predictions of optimal patch departure (e.g. 16, 41, 58, 79, 124, 140, 237).

PATCHES OF UNKNOWN QUALITY   To nectivores that feed at flowers in inflorescences or on plants, these clusters of flowers represent patches. As such an animal moves among the flowers within a patch, it may be increasingly likely to revisit a flower (e.g. 195, 202, 208), it may visit flowers that tend to contain less nectar than others (e.g. 103, 196, 205), or it may obtain information about the otherwise unknown quality of its present patch (e.g. 103, 195, 202). Such foraging situations are therefore ideal for developing and testing models of optimal patch departure. In general, however, these models must include sampling as well as depletion.

The only study in which sampling and depletion are included in both the model and the data collection is one of mine (202). I observed honey eaters feeding at inflorescences that almost always had seven flowers in a regular arrangement. As these birds visited flowers within an inflorescence, the probability of a revisit increased slightly until seven flowers had been visited, and then it became close to one. There was a positive correlation among the nectar volumes of flowers from the same inflorescence, so the birds could have estimated patch quality on the basis of their sampling experience in a patch. I assumed that fitness increases linearly with the expected net rate of energy gain; that the birds base their departure decisions on nectar thresholds with respect either to the nectar volume obtained from the last flower visited or to the average nectar volume obtained from the flowers visited within a patch so far; and that these thresholds will vary with the number of flowers that have been visited. I used a computer simulation based on the observed probabilities of flower revisitation, the patterns of nectar distribution, and the times associated with various activities of the birds to derive the optimal nectar thresholds and from them, the optimal frequency distributions of the numbers of flowers probed per inflorescence. The observed and predicted frequency distributions were qualitatively similar but significantly different. The direction of this discrepancy is what would be expected if the birds tend to be transient (202), a characteristic verified by subsequent bird banding (G. H. Pyke & H. F. Recher, unpublished data).

Other attempts have been made to test predictions of optimal patch departure using foraging by nectarivores (15, 103, 108, 109, 195, 284). In all these cases, however, the development of the predictions is inappropriate to the foraging situation (see the above discussion of optimal patch departure theory; also 209, 230). The same is true for Davies' study (52) of spotted flycatchers foraging among nondepletable patches of flying insects (see 209).

5. Rules of Movement

MOVEMENT PATTERNS AND SPEED OF MOVEMENT   Relatively little attention has been given to developing and testing models of optimal patterns of movement (see 209, 210, 270). Many studies have shown that animals become less directional in movements or tend to engage in area-restricted searching after their encounter with large amounts of food or food odor (19, 21, 55, 69, 194, 210, 247, 267, 285, 289, 293, 295). Such behavior is expected whenever food is patchily distributed (e.g. 210). However, there are as yet no predictions as to exactly what the directionality, or more generally, the rule-governing patterns, of movement should be in different situations (see the above discussion of optimal movement patterns). Consequently, quantitative tests of the predictions are not feasible. The degree of directionality exhibited by animals varies widely (2, 130, 196, 203, 210, 228, 240, 296–298).

In addition, where animals can see food items at a distance, they tend to move to the closest and the best items (best in terms of the rate of food gain) (e.g. 152, 176, 203, 297, 298). Animals may also tend to alternate left and right turns (e.g. 194) or to turn in one directon rather than the other (e.g. 35).

The speed of movement of foraging animals has received even less attention than movement patterns (see 204). In one of the few studies, Ware (272) found that the average swimming speed of bleak, a planktivorous fish, is close to the speed that maximizes the net rate of energy gain.

FORAGING MODE   Animals sometimes exhibit a variety of foraging modes while feeding in the same area and on the same food types. Kestrels, for example, may hunt while hovering or while sitting on a perch (225). Hummingbirds and honey eaters may perch and hop or hover while feeding among flowers within an inflorescence or plant (201). I deduced that in two observed foraging situations, hovering hummingbirds and perching honey eaters would have obtained lower net rates of energy gain if they had adopted the alternative mode of foraging (201). Hovering seems to enable a bird to move more rapidly between flowers within an inflorescence than perching and hopping, but at a higher energetic cost (201). Zach (290) found that crows, which break whelks by dropping them on rocks, minimized the total amount of ascending flight required when choosing the height of the drop. No other tests of optimal foraging modes have apparently been carried out.

6. Central Place Foraging

All quantitative tests of optimal central place foraging have assumed that fitness increases linearly with the rate of food or energy intake. Other factors may also be important (e.g. 154), however.

For central place foragers, patches may differ in their quality and distance from the central place, and animals should vary their time and load size per patch accordingly (see the above discussion of optimal central place foraging). Several researchers found that the rate of food gain in a patch decreases as more time is spent in the patch, and they have used this relationship and the theory of optimal central place foraging (see 80, 128, 179) to derive predicted times and load sizes for different patches. In these cases—as required by the theory—there is effectively only one patch, and the animal is not simultaneously choosing among patches. Close agreement between the observed and predicted load sizes was reported (30, 126, 128, 261). However, chipmunks consistently spent less time and collected smaller loads than predicted (80). In all cases, the load sizes and patch times increased with increasing distance or time from the central place (80, 126, 128, 261). This qualitative result is consistent with Carlson & Moreno's (29) and Nunez's (173) findings.

A central place forager must also choose which patches to visit, and it would be expected to allocate different amounts of time to patches at different distances from the central place. But the predicted pattern of time allocation is sensitive to assumptions about the knowledge of patch locations and the rate of food renewal (see above), so quantitative tests of such predictions seem premature. Central place foraging has also been the subject of a number of other studies (e.g. 5, 7, 14, 44, 71, 96, 131, 224, 244).

CONCLUSIONS

Optimal foraging theory can only be expected to be useful when its assumptions, mathematical development, and testing are appropriate for the studies to which it is applied. Future optimal foraging studies should therefore deal explicitly with these problems and satisfy the kind of criteria I have presented above. As in many other areas of research, the achievement of these goals should be enhanced by better communication between theoreticians and empiricists.

Up to this point, the predictions of optimal foraging theory have been supported to some extent but not completely. In many cases, the (correct) prediction is some kind of all-or-nothing behavioral response, yet animals invariably exhibit more gradual and incomplete responses (e.g. partial diet preferences, variable patch times). Many explanations for these discrepancies have been proposed (e.g. sampling, mistakes), but further development and testing of these rationalizations remain to be carried out.

If attention is restricted to those studies that provide genuine tests of optimal foraging theory, the level of agreement between predicted and observed foraging behavior has been reasonably good, except for the findings of gradual rather than all-or-nothing behavioral responses. Consequently, the optimal foraging approach seems worth pursuing. Nevertheless, there are still extremely few studies that come close to meeting my proposed critiera, and it is therefore premature to form a firm opinion regarding the usefulness of the approach.

ACKNOWLEDGMENTS

Harry Recher provided helpful comments on an earlier version of this paper. Melinda Brouwer, Debbie Bushell, Judy Recher, Martin Shivas, and Grace Sevkowski assisted with proofreading.

Literature Cited

1. Abrams, P. A. 1982. Functional responses of optimal foragers. *Am. Nat.* 120:382–90
2. Adams, G. 1981. Search paths of fireflies in two dimensions. *Fla. Entomol.* 64:66–73
3. Allen, P. L. 1983. Feeding behaviour of *Asterias rubens* (L) on soft bottom bivalves: A study in selective predation. *J. Exp. Mar. Biol. Ecol.* 70:79–90
4. Andersson, M. 1978. Optimal foraging area: Size and allocation of search effort. *Theor. Popul. Biol.* 13:397–409
5. Andersson, M. 1981. Central place foraging in the whinchat, *Saxicola rubetra. Ecology* 62:538–44
6. Andersson, M. 1981. On optimal predator search. *Theor. Popul. Biol.* 19:58–86
7. Aronson, R. B., Givnish, T. J. 1983. Optimal central-place foragers: A comparison with null hypotheses. *Ecology* 64:395–99
8. Baker, H. G. 1977. Chemical aspects of pollination of woody plants in the tropics. In *Tropical Trees as Living Systems,* ed. P. B. Tomlinson, M. Zimmerman. London/New York: Cambridge Univ. Press
9. Baker, H. G., Baker, I. 1975. Studies of nectar-constitution and pollinator-plant coevolution. In *Coevolution of Animals and Plants,* ed. I. E. Gilbert, P. H. Raven. Austin: Univ. Tex. Press

10. Barnard, C. J., Stephens, H. 1981. Prey size selection by lapwings in lapwing-gull associations. *Behaviour* 77:1–22
11. Batzli, G. O., Cole, F. R. 1979. Nutritional ecology of microtine rodents—digestibility of forage. *J. Mammal.* 60: 740–50
12. Belovsky, G. E. 1978. Diet optimization in a generalist herbivore, the moose. *Theor. Popul. Biol.* 14:105–34
13. Belovsky, G. E. 1981. Food plant selection by a generalist herbivore: The moose. *Ecology* 62:1020–30
14. Bernstein, R. A. 1982. Foraging-area size and food density: Some predictive models. *Theor. Popul. Biol.* 22:309–23
15. Best, L. S., Bierzychudek, P. 1982. Pollinator foraging on foxglove *(Digitalis purpurea)*—a test of a new model. *Evolution* 36:70–79
16. Bibby, C. J., Green, R. E. 1980. Foraging behaviour of migrant pied flycatchers, *Ficedula hypoleuca*, on temporary territories. *J. Anim. Ecol.* 49:507–21
17. Bobisud, L. E., Potratz, C. J. 1976. One-trial versus multi-trial learning for a predator encountering a model-mimic system. *Am. Nat.* 110:121–28
18. Bohl, E. 1982. Food supply and prey selection in planktivorous Cyprinidae. *Oecologia* 53:134–38
19. Bond, A. B. 1980. Optimal foraging in a uniform habitat: The search mechanism of the green lacewing. *Anim. Behav.* 28: 10–19
20. Bond, A. B. 1981. Giving-up as a Poisson process—the departure decision of the green lacewing. *Anim. Behav.* 29: 629–30
21. Brunner, J. F., Burts, E. C. 1975. Searching behaviour and growth rates of *Anthocoris nemoralis* (Hemiptera: Anthocoridae), a predator of the pear Psylla, *Psylla pyricola*. *Ann. Entomol. Soc. Am.* 68:311–15
22. Calow, P. 1975. The feeding strategies of two freshwater gastropods, *Ancylus fluviatilis* Mull. and *Planorbis contortus* Linn. Pulmonata) in terms of ingestion rates and absorption efficiencies. *Oecologia* 20:33–49
23. Caraco, T. 1980. On foraging time allocation in a stochastic environment. *Ecology* 61:119–28
24. Caraco, T. 1981. Energy budgets, risk, and foraging preferences in dark-eyed juncos. *Behav. Ecol. Sociobiol.* 8:213–17
25. Caraco, T. 1982. Aspects of risk-aversion in foraging white-crowned sparrows. *Anim. Behav.* 30:719–27
26. Caraco, T., Martindale, S., Whittam, T. S. 1980. An empirical demonstration of risk-sensitive foraging preferences. *Anim. Behav.* 28:820–30
27. Carefoot, T. H. 1967. Growth and nutrition of *Aplysia punctata* feeding on a variety of marine algae. *J. Mar. Biol. Assoc. UK* 47:565–89
28. Carefoot, T. H. 1973. Feeding, food preference, and the uptake of food energy by the supralittoral isopod, *Ligia pallasii*. *Mar. Biol.* 18:228–36
29. Carlson, A., Moreno, J. 1981. Central place foraging in the wheatear *Oenanthe oenanthe*—an experimental test. *J. Anim. Ecol.* 50:917–24
30. Carlson, A., Moreno, J. 1982. The loading effect in central place foraging. *Behav. Ecol. Sociobiol.* 11:173–83
31. Charnov, E. L. 1973. *Optimal foraging: Some theoretical explorations.* PhD thesis. Univ. Wash., Seattle. 95 pp.
32. Charnov, E. L. 1976. Optimal foraging: Attack strategy of a mantid. *Am. Nat.* 110:141–51
33. Charnov, E. L. 1976. Optimal foraging: The marginal value theorem. *Theor. Popul. Biol.* 9:129–36
34. Charnov, E. L. 1981. Marginal value: An answer to Templeton & Lawlor. *Am. Nat.* 117:394
35. Cheverton, J. 1982. Bumblebees may use a suboptimal arbitrary handedness to solve difficult foraging decisions. *Anim. Behav.* 30:934–35
36. Cody, M. L. 1971. Finch flocks in the Mohave Desert. *Theor. Popul. Biol.* 2: 142–58
37. Cole, S., Hainsworth, F. R., Kamil, A. C., Mercier T., Wolf, L. L. 1982. Spatial learning as an adaptation in hummingbirds. *Science* 217:655–57
38. Collier, G., Hirsch, E., Kanarek, R. 1977. The operant revisited. In *Handbook of Operant Behaviour*, ed. W. K. Honig, J. E. R. Staddon, pp. 28–52. New York: Prentice-Hall
39. Comins, H. N., Hassell, M. P. 1979. The dynamics of optimally foraging predators and parasitoids. *J. Anim. Ecol.* 48:335–51
40. Conley, J. B., Blem, C. R. 1978. Seed selection by Japanese Quail, *Coturnix coturnix japonica*. *Am. Midl. Nat.* 100: 135–40
41. Cook, R. M., Cockrell, B. J. 1978. Predator ingestion rate and its bearing on feeding time and the theory of optimal diets. *J. Anim. Ecol.* 47:529–47
42. Cook, R. M., Hubbard, S. F. 1977. Adaptive searching strategies in insect parasites. *J. Anim. Ecol.* 46:115–26
43. Cornell, H. 1976. Search strategies and the adaptive significance of switching in

some general predators. *Am. Nat.* 110: 317–20

44. Covich, A. P. 1976. Analyzing shapes of foraging areas: Some ecological and economic theories. *Ann. Rev. Ecol. Syst.* 7:235–57

45. Cowie, R. J. 1977. Optimal foraging in the great tits *(Parus major). Nature* 268:137–39

46. Cowie, R. J., Krebs, J. R. 1979. Optimal foraging in patchy environments. In *Population Dynamics, 20th Symp. Br. Ecol. Soc., London, 5–7 April, 1978,* ed. R. M. Anderson, B. D. Turner, L. R. Turner.

47. Cox, D. R. 1962. *Renewal Theory.* London: Methuen

48. Craig, R. B., DeAngelis, D. L., Dixon, K. R. 1979. Long- and short-term dynamic optimization models with application to the feeding strategy of the Loggerhead Shrike. *Am. Nat.* 113:31–51

49. Crowder, L. B., Magnuson, J. J. 1981. Cost-benefit analysis of temperature and food resource use: A synthesis with examples from the fishes. In *Behavioral Energetics: Vertebrate Costs of Survival, Ohio State Univ. Biosci. Colloq. No. 7,* ed. W. P. Aspey, S. I. Lustick. Columbus: Ohio State Univ. Press

50. Curio, E. 1983. Time-energy budgets and optimization. *Experientia* 39:25–34

51. Davidson, D. W. 1978. Experimental tests of optimal diet in two social insects. *Behav. Ecol. Sociobiol.* 4:35–41

52. Davies, N. B. 1977. Prey selection and the search strategy of the spotted flycatcher. *(Muscicapa striata):* A field study on optimal foraging. *Anim. Behav.* 25:1016–33

53. Davies, N. B. 1977. Prey selection and social behaviour in wagtails (Aves: Motacillidae). *J. Anim. Ecol.* 46:37–57

54. DeBenedictis, P. A., Gill, F. B., Hainsworth, F. R., Pyke, G. H., Wolf, L. L. 1978. Optimal meal size in hummingbirds. *Am. Nat.* 112:301–16

55. Downing, J. A. 1981. In situ foraging responses of 3 species of littoral cladocerans. *Ecol. Monogr.* 51:85–103

56. Drummond, H., Burghardt, G. M. 1983. Geographic variation in the foraging behaviour of the garter snake, *Thamnophis elegans. Behav. Ecol. Sociobiol.* 12:43–48

57. Dunstone, N. 1978. Fishing strategy of the mink *(Mustela vison)*—Time-budgeting of hunting effort. *Behaviour* 67:157–77

58. Dunstone, N., O'Connor, R. J. 1979. Optimal foraging in an amphibious mammal. 1. The aqualung effect. *Anim. Behav.* 27:1182–94

59. Ebersole, J. P., Wilson, J. C. 1980. Optimal foraging—the responses of *Peromyscus leucopus* to experimental changes in processing time and hunger. *Oecologia* 46:80–85

60. Eggers, D. M. 1977. The nature of prey selection by planktivorous fish. *Ecology* 58:46–59

61. Eggers, D. M. 1982. Planktivore preference by prey size. *Ecology* 63:381–90

62. Ellis, J. E., Wiens, J. A., Rodell, C. F., Anway, J. C. 1976. A conceptual model of diet selection as an ecosystem process. *J. Theor. Biol.* 60:93–108

63. Elner, R. W., Hughes, R. N. 1978. Energy maximization in the diet of the shore crab *Carcinus maenas. J. Anim. Ecol.* 47:103–16

64. Emlen, J. M. 1966. The role of time and energy in food preference. *Am. Nat.* 100:611–17

65. Emlen, J. M., Emlen, M. G. R. 1975. Optimal choice in diet: Test of a hypothesis. *Am. Nat.* 109:427–35

66. Erichsen, J. T., Krebs, J. R., Houston, A. I. 1980. Optimal foraging and cryptic prey. *J. Anim. Ecol.* 49:271–76

67. Erlinge, S. 1981. Food preference, optimal diet and reproductive output in stoats *Mustela erminea* in Sweden. *Oikos* 36:303–15

68. Estabrook, G. F., Dunham, A. E. 1976. Optimal diet as a function of absolute abundance, relative abundance, and relative value of available prey. *Am. Nat.* 110:401–13

69. Evans, H. F. 1976. The searching behaviour of *Anthocoris confusus* (Reuter) in relation to prey density and plant surface topography. *Ecol. Entomol.* 1:163–69

70. Evans, R. M. 1982. Efficient use of food patches at different distances from a breeding colony in black-billed gulls. *Behaviour* 79:28–38

71. Fagerstrom, T., Moreno, J., Carlson, A. 1982. Load size & energy delivery in birds feeding nestlings—constraints on and alternative strategies to energy-maximization. *Oecologia* 56:93–98

72. Freeland, W. J., Janzen, D. H. 1974. Strategies in herbivory by mammals: The role of plant secondary compounds. *Am. Nat.* 108:269–89

73. Fullick, T. G., Greenwood, J. D. 1979. Frequency dependent food selection in relation to two models. *Am. Nat.* 113: 762–65

74. Galen, C., Kevan, P. G. 1983. Bumblebee foraging and floral scent dimorphism: *Bombus kirbyellus* Curtis (Hymenoptera: Apidae) and *Polemonium visco-*

sum Nutt. (Polemoniaceae). *Can. J. Zool.* 61:1207–13

75. Gardner, M. B. 1981. Mechanisms of size selectivity by planktivorous fish: A test of hypothesis. *Ecology* 62:571–78

76. Garton, E. V. 1979. Implications of optimal foraging theory for insectivorous forest birds. In *Role of Insectivorous Birds in Forest Ecosystems,* ed. J. G. Dickson, R. N. Conner, R. R. Fleet, J. C. Kroll, J. G. Jackson. New York: Academic

77. Gendron, R. P., Staddon, J. E. R. 1983. Searching for cryptic prey: The effect of search rate. *Am. Nat.* 121:172–86

78. Gibson, R. M. 1980. Optimal prey-size selection by 3-spined sticklebacks *(Gasterosteus aculeatus)*—a test of the apparent-size hypothesis. *Z. Tierpsychol.* 52:291–307

79. Giller, P. S. 1980. The control of handling time and its effects on the foraging strategy of a Heteropteran predator, *Notonecta. J. Anim. Ecol.* 49:699–712

80. Giraldeau, L. A., Kramer, D. L. 1982. The marginal value theorem—a quantitative test using load size variation in a central place forager, the eastern chipmunk, *Tamias striatus. Anim. Behav.* 30:1036–42

81. Gittelman, S. H. 1978. Optimum diet and body size in backswimmers (Heteroptera: Notonectidae, Pleidae). *Ann. Entomol. Soc. Am.* 71:737–47

82. Goss-Custard, J. D. 1977. Optimal foraging and the size selection of worms by redshank, *(Tringa totanus),* in the field. *Anim. Behav.* 25:10–29

83. Goss-Custard, J. D. 1977. The energetics of prey selection by redshank, *Tringa totanus* (L.), in relation to prey density. *J. Anim. Ecol.* 46:1–19

84. Goss-Custard, J. D. 1977. Responses of redshank, *Tringa totanus,* to absolute and relative densities of two prey species. *J. Anim. Ecol.* 46:867–74

85. Gould, S. J., Lewontin, R. C. 1979. The spandrels of San Marco and the Panglossian paradigm: A critique of the adaptationist programme. *Proc. R. Soc. London Ser. B* 205:581–98

86. Green, R. 1978. Factors affecting the diet of farmland skylarks, *Alauda arvensis. J. Anim. Ecol.* 47:913–28

87. Green, R. F. 1980. Bayesian birds: A simple example of Oaten's stochastic model of optimal foraging. *Theor. Popul. Biol.* 18:244–56

88. Greenstone, M. H. 1979. Feeding behaviour of free-living wolf spiders optimises dietary proportions of the essential amino acids. *Nature* 282:501–3

89. Griffiths, D. 1975. Prey availability and

the food of predators. *Ecology* 56:1209–14

90. Griffiths, D. 1981. Sub-optimal foraging in the ant-lion *Macroleon quinquemaculatus. J. Anim. Ecol.* 50:697–702

91. Grubb, T. C., Greenwald, L. 1982. Sparrows and a brushpile—foraging responses to different combinations of predation risk and energy cost. *Anim. Behav.* 30:637–40

92. Hainsworth, F. R. 1978. Feeding: Models of costs and benefits in energy regulation. *Am. Zool.* 18:701–14

93. Hainsworth, F. R., Tardiff, M. F. Wolf, L. L. 1981. Proportional control for daily energy regulation in hummingbirds. *Physiol. Zool.* 54:452–62

94. Hassell, M. P. 1980. Foraging strategies, population models and biological control—a case study. *J. Anim. Ecol.* 49:603–28

95. Hassell, M. P., Southwood, T. R. E. 1978. Foraging strategies of insects. *Ann. Rev. Ecol. Syst.* 9:75–98

96. Hegner, R. E. 1982. Central place foraging in the white-fronted bee-eater. *Anim. Behav.* 30:953–63

97. Heinrich, B. 1979. Foraging strategies of caterpillars: Leaf damage and possible predator avoidance strategies. *Oecologia* 42:325–37

98. Heinrich, B. 1983. Do bumblebees forage optimally, and does it matter? *Am. Zool.* 23:273–81

99. Heller, R. 1980. On optimal diet in a patchy environment. *Theor. Popul. Biol.* 17:201–14

100. Heller, R., Milinski, M. 1979. Optimal foraging of sticklebacks on swarming prey. *Anim. Behav.* 27:1127–41

101. Heyman, G. M. 1983. Optimization theory: Close but no cigar. *Behav. Anal. Lett.* 3:17–26

102. Hixon, M. A. 1982. Energy maximizers and time minimizers: Theory and reality. *Am. Nat.* 119:596–99

103. Hodges, C. M. 1981. Optimal foraging in bumblebees—hunting by expectation. *Anim. Behav.* 29:1166–71

104. Horn, M. H. 1983. Optimal diets in complex environments—feeding strategies of 2 herbivorous fishes from a temperate rocky intertidal zone. *Oecologia* 58:345–50

105. Horsley, D. T., Lynch, B. M., Greenwood, J. J., Hardman, B., Mosely, S. 1979. Frequency-dependent selection by birds when the density of prey is high. *J. Anim. Ecol.* 48:483–90

106. Houston, A. I. 1983. Optimality theory and matching. *Behav. Anal. Lett.* 3:1–15

107. Houston, A. I., Krebs, J. R., Erichsen, J. T. 1980. Optimal prey choice and dis-

crimination time in the great tit. (*Parus major* L.). *Behav. Ecol. Sociobiol.* 6:169–75

108. Howell, D. J., Hartl, D. L. 1980. Optimal foraging in glossophagine bats: When to give up. *Am. Nat.* 115:696–704

109. Howell, D. J., Hartl, D. L. 1982. In defense of optimal foraging by bats: A reply to Schluter. *Am. Nat.* 119:438–39

110. Hubbard, S. F., Cook, R. M. 1978. Optimal foraging by parasitoid wasps. *J. Anim. Ecol.* 47:593–604

111. Hughes, R. N. 1979. Optimal diets under the energy maximization premise: The effects of recognition time and learning. *Am. Nat.* 113:209–21

112. Hughes, R. N., Seed, R. 1981. Size selection of mussels by the blue crab *Callinectes sapidus:* Energy maximizer or time minimizer? *Mar. Ecol.* 6:83–89

113. Iwasa, Y., Higashi, M., Yanamura, N. 1981. Prey distribution as a factor determining the choice of optimal foraging strategy. *Am. Nat.* 117:710–23

114. Jaeger, R. G., Barnard, D. E. 1981. Foraging tactics of a terrestrial salamander: Choice of diet in structurally simple environments. *Am. Nat.* 117:639–64

115. Jaeger, R. G., Joseph, R. G., Barnard, D. E. 1981. Foraging tactics of a terrestrial salamander—sustained yield in territories. *Anim. Behav.* 29:1100–5

116. Jaeger, R. G., Rubin, A. M. 1982. Foraging tactics of a terrestrial salamander—judging prey profitability. *J. Anim. Ecol.* 51:167–76

117. Jaenike, J. 1978. On optimal oviposition behavior in phytophagous insects. *Theor. Popul. Biol.* 14:350–56

118. Jander, R. 1982. Random and systematic search in foraging insects. In *The Biology of Social Insects*, ed. M. D. Breed, C. D. Michener, H. E. Evans. Boulder, Colo: Westview

119. Janetos, A. C. 1982. Active foragers vs. sit-and-wait predators: A simple model. *J. Theor. Biol.* 95:381–85

120. Janetos, A. C., Cole, B. J. 1981. Imperfectly optimal animals. *Behav. Ecol. Sociobiol.* 9:203–10

121. Jenkins, S. H. 1980. A size-distance relation in food selection by beavers. *Ecology* 61:740–46

122. Johnson, D. R., Campbell, W. V., Wynne, J. C. 1980. Fecundity and feeding preference of the two-spotted spider mite on domestic and wild species of peanuts. *J. Econ. Entomol.* 73:575–76

123. Kacelnik, A., Houston, A. I., Krebs, J. R. 1981. Optimal foraging and territorial defence in the great tit *(Parus major). Behav. Ecol. Sociobiol.* 8:35–40

124. Kamil, A. C., Peters, J., Lindstrom, F. J.

1981. An ecological perspective on the study of the allocation of behavior. In *2nd Ann. Harvard Symp. Quant. Anal. Behav.*

125. Kamil, A. C., Yoerg, S. J. 1982. Learning and foraging behaviour. In *Perspectives in Ethology,* ed. P. P. G. Bateson, P. H. Klopfer, 5:325–64. New York: Plenum Press

126. Kasuya, E. 1982. Central place water collection in a Japanese paper wasp, *Polistes chinensis antennalis. Anim. Behav.* 30:1010–14

127. Kenward, R. E., Sibly, R. M. 1977. A woodpigeon *(Columba palumbas)* feeding preference explained by a digestive bottleneck. *J. Appl. Ecol.* 14:815–26

128. Killeen, P. R., Smith, J. P., Hanson, S. J. 1981. Central place foraging in *Rattus norvegicus. Anim. Behav.* 29:64–70

129. Kislalioglu, M., Gibson, R. N. 1976. Prey "handling time" and its importance in food selection by the 15-spined stickleback, *Spinachia spinachia* (L.). *J. Exp. Mar. Biol. Ecol.* 25:151–58

130. Kitching, R. L., Zalucki, M. P. 1982. Component analysis and modelling of the movement process: Analysis of simple tracks. *Res. Popul. Ecol.* 24:224–38

131. Kramer, D. L., Nowell, W. 1980. Central place foraging in the eastern chipmunk, *Tamias striatus. Anim. Behav.* 28: 772–78

132. Krebs, J. R. 1973. Behavioral aspects of predation. In *Perspectives in Ethology,* ed. P. P. Bateson, P. H. Klopfer. New York: Plenum

133. Krebs, J. R. 1978. Optimal foraging: Decision rules for predators. In *Behavioural Ecology: An Evolutionary Approach,* ed. J. R. Krebs, N. B. Davies. Oxford: Blackwell

134. Krebs, J. R. 1980. Optimal foraging, predation risk, and territory defence. *Ardea* 68:83–90

135. Krebs, J. R., Cowie, R. J. 1976. Foraging strategies in birds. *Ardea* 64:98–116

136. Krebs, J. R., Erichsen, J. T., Webber, J. I., Charnov, E. L. 1977. Optimal prey selection in the great tit *(Parus major). Anim. Behav.* 25:30–38

137. Krebs, J. R., Houston, A. I., Charnov, E. L. 1980. Some recent developments in optimal foraging. In *Foraging Behavior; Ecological, Ethological and Psychological Approaches,* ed. A. C. Kamil, T. Sargent, pp. 3–18. New York: Garland STPM

138. Krebs, J. R., Kacelnik, A., Taylor, P. 1978. Test of optimal sampling by foraging great tits. *Nature* 275:27–31

139. Krebs, J. R., Ryan, J. C., Charnov, E. L. 1974. Hunting by expectation or optimal

foraging? A study of patch use by chickadees. *Anim. Behav.* 22:953–64

140. Kruse, K. C. 1983. Optimal foraging by predaceous diving beetle larvae on toad tadpoles. *Oecologia* 58:383–88

141. Lacher, T. E. Jr., Willig, M. R., Mares, M. A. 1982. Food preference as a function of resource abundance with multiple prey types: An experimental analysis of optimal foraging theory. *Am. Nat.* 120:297–316

142. Lam, R. K., Frost, B. W. 1976. Model of copepod filtering response to changes in size and concentration of food. *Limnol. Oceanogr.* 21:490–500

143. Lea, S. E. G. 1979. Foraging and reinforcement schedules in the pigeon—optimal and non-optimal aspects of choice. *Anim. Behav.* 27:875–86

144. Lehman, J. T. 1976. The filter-feeder as an optimal forager, and the predicted shapes of feeding curves. *Limnol. Oceanogr.* 21:501–16

145. Lessells, C. M., Stephens, D. W. 1983. Central place foraging—Single prey loaders again. *Anim. Behav.* 31:238–43

146. Levin, D. A., Kerster, N. W., Niedzlek, M. 1971. Pollinator flight directionality and its effect on pollen flow. *Evolution* 35:113–18

147. Lewis, A. R. 1980. Patch use by gray squirrels and optimal foraging. *Ecology* 61:1371–79

148. Lewis, A. R. 1982. Selection of nuts by gray squirrels and optimal foraging theory. *Am. Midl. Nat.* 107:250–57

149. Lewontin, R. C. 1979. Fitness, survival and optimality. In *Analysis of Ecological Systems*, ed. D. H. Horn, R. Mitchell, G. R. Stairs. Columbus: Ohio State Univ. Press

150. MacArthur, R. H., Pianka, E. R. 1966. On optimal use of a patchy environment. *Am. Nat.* 100:603–9

151. Magnhagen, C., Wiederholm, A. M. 1982. Food selectivity versus prey availability—a study using the marine fish *Pomatoschistus microps*. *Oecologia* 55:311–15

152. Marden, J. H., Waddington, K. D. 1981. Floral choices by honeybees in relation to the relative distances to flowers. *Physiol. Entomol.* 6:431–35

153. Marten, G. C. 1973. An optimization equation for predation. *Ecology* 54:92–101

154. Martindale, S. 1982. Nest defense and central place foraging: A model and experiment. *Behav. Ecol. Sociobiol.* 10:85–89

155. Maynard Smith, J. 1978. Optimization theory in evolution. *Ann. Rev. Ecol. Syst.* 9:31–56

156. McNair, J. N. 1979. A generalized model of optimal diets. *Theor. Popul. Biol.* 15:159–70

157. McNair, J. N. 1981. A stochastic foraging model with predator training effects. II. Optimal diets. *Theor. Popul. Biol.* 19:147–62

158. McNair, J. N. 1982. Optimal giving-up times and the marginal value theorem. *Am. Nat.* 119:511–29

159. McNamara, J. 1982. Optimal patch use in a stochastic environment. *Theor. Popul. Biol.* 21:269–88

160. Milinski, M. 1979. Evolutionarily stable feeding strategy in sticklebacks. *Z. Tierpsychol.* 51:36–40

161. Milinski, M., Heller, R. 1978. Influence of a predator on the optimal foraging behaviour of sticklebacks (*Gasterostrus aculeatus* L.). *Nature* 275:642–44

162. Mitchell, R. 1981. Insect behavior, resource exploitation and fitness. *Ann. Rev. Entomol.* 26:373–96

163. Mittelbach, G. G. 1981. Foraging efficiency and body size: A study of optimal diet and habitat use by bluegills. *Ecology* 62:1370–86

164. Molles, M. C., Pietruszka, R. D. 1983. Mechanisms of prey selection by predaceous stoneflies—roles of prey morphology, behaviour and predator hunger. *Oecologia* 57:25–31

165. Morrison, D. W. 1978. On the optimal searching strategy for refuging predators. *Am. Nat.* 112:925–34

166. Mueller, H. 1971. Oddity and specific searching image more important than conspicuousness in prey selection. *Nature* 233:345–46

167. Murdoch, W. W. 1969. Switching in general predators: Experiments on predator specificity and stability of prey populations. *Ecol. Monogr.* 39:335–54

168. Myers, J. H. 1979. The effects of food quantity and quality on emergence time in the cinnabar moth. *Can. J. Zool.* 57:1150–56

169. Nicotri, M. E. 1980. Factors involved in herbivore food preference. *J. Exp. Mar. Biol. Ecol.* 42:13–26

170. Norberg, R. A. 1977. An ecological theory on foraging time and energetics and choice of optimal food-searching method. *J. Anim. Ecol.* 46:511–29

171. Norberg, R. A. 1981. Optimal flight speed in birds when feeding young. *J. Anim. Ecol.* 50:473–77

172. Norberg, R. A. 1983. Optimal locomotion modes of foraging birds in trees. *Ibis* 125:172–80

173. Nunez, J. A. 1982. Honeybee foraging strategies at a food source in relation to its

distance from the hive and the rate of sugar flow. *J. Apic. Res.* 21:139–50

174. Oaten, A. 1977. Optimal foraging in patches: A case for stochasticity. *Theor. Popul. Biol.* 12:263–85

175. Oates, J. T., Waterman, P. G., Choo, G. M. 1980. Food selection by the South Indian leaf-monkey, *Presbytis johnii*, in relation to leaf chemistry. *Oecologia* 45: 45–56

176. O'Brien, W. J., Slade, N. A., Vinyard, G. L. 1976. Apparent size as the determinant of prey selection by bluegill sunfish *(Lepomis macrochirus)*. *Ecology* 57: 1304–10

177. Ohguchi, O., Aoki, K. 1983. Effects of colony need for water on optimal food choice in honey-bees. *Behav. Ecol. Sociobiol.* 12:77–84

178. Ollason, J. G. 1980. Learning to forage—optimally? *Theor. Popul. Biol.* 18: 44–56

179. Orians, G. H., Pearson, N. E. 1979. On the theory of central place foraging. In *Analysis of Ecological Systems,* ed. D. J. Horn, G. R. Stairs, R. D. Mitchell, pp. 155–77. Columbus: Ohio State Univ. Press

180. Owen-Smith, N., Novellie, P. 1982. What should a clever ungulate eat? *Am. Nat.* 119:151–78

181. Palmer, A. R. 1981. Predator errors, foraging in unpredictable environments, and risk: The consequences of prey variation in handling time versus net energy. *Am. Nat.* 118:908–15

182. Parker, G. A., Stuart, R. A. 1976. Animal behavior as a strategy optimizer: Evolution of resource assessment strategies and optimal emigration thresholds. *Am. Nat.* 110:1055–76

183. Partridge, L. 1981. Increased preferences for familiar foods in small mammals. *Anim. Behav.* 29:211–16

184. Pietrewicz, A. T., Kamil, A. C. 1979. Search image formation in the Blue Jay *(Cyanocitta cristata). Science* 204:1332–33

185. Poston, H. A. 1976. Optimum level of dietary biotin for growth, feed utilization, and swimming stamina of fingerling lake trout *(Salvelinus namaycush). J. Fish. Res. Board Can.* 33:1803–6

186. Price, M. R. S. 1978. The nutritional ecology of Coke's hartebeest *(Alceplaphus buselaphus cokei)* in Kenya. *J. Appl. Ecol.* 15:33–49

187. Pulliam, H. R. 1974. On the theory of optimal diets. *Am. Nat.* 108:59–75

188. Pulliam, H. R. 1975. Diet optimization with nutrient constraints. *Am. Nat.* 109: 765–68

189. Pulliam, H. R. 1980. Learning to forage optimally. See Ref. 137

190. Pulliam, H. R. 1980. Do chipping sparrows forage optimally? *Ardea* 68:75–82

191. Pulliam, H. R. 1980. On digesting a theory. *Auk* 97:418–20

192. Pulliam, H. R. 1981. Optimal management of optimal foragers. In *Renewable Resource Management,* ed. T. J. Vincent, J. M. Skowronski. Berlin: Springer-Verlag

193. Pyke, G. H. 1978. Are animals efficient harvesters? *Anim. Behav.* 26:241–50

194. Pyke, G. H. 1978. Optimal foraging: Movement patterns of bumblebees between inflorescences. *Theor. Popul. Biol.* 13:72–97

195. Pyke, G. H. 1978. Optimal foraging in hummingbirds: Testing the marginal value theorem. *Am. Zool.* 18:739–52

196. Pyke, G. H. 1978. Optimal foraging in bumblebees and coevolution with their plants. *Oecologia* 36:281–93

197. Pyke, G. H. 1979. Optimal foraging in bumblebees: Rule of movement between flowers within inflorescences. *Anim. Behav.* 27:1167–81

198. Pyke, G. H. 1979. The economics of territory size and time budget in the golden-winged sunbird. *Am. Nat.* 114:131–45

199. Pyke, G. H. 1980. Optimal foraging in nectar-feeding animals and coevolution with their plants. See Ref. 137

200. Pyke, G. H. 1980. Optimal foraging in bumblebees: Calculation of net rate of energy intake and optimal patch choice. *Theor. Popul. Biol.* 17:232–46

201. Pyke, G. H. 1981. Why hummingbirds hover and honeyeaters perch. *Anim. Behav.* 29:861–67

202. Pyke, G. H. 1981. Honeyeater foraging: A test of optimal foraging theory. *Anim. Behav.* 29:878–88

203. Pyke, G. H. 1981. Optimal foraging in hummingbirds: Rule of movement between inflorescences. *Anim. Behav.* 29: 889–96

204. Pyke, G. H. 1981. Optimal travel speeds of animals. *Am. Nat.* 118:475–87

205. Pyke, G. H. 1981. Optimal foraging in bumblebees: Rule of departure from an inflorescence. *Can. J. Zool.* 60:417–28

206. Pyke, G. H. 1981. Hummingbird foraging on artificial inflorescences. *Behav. Anal. Lett.* 1:11–15

207. Pyke, G. H. 1981. Optimal nectar production in a hummingbird-pollinated plant. *Theor. Popul. Biol.* 20:326–43

208. Pyke, G. H. 1982. Foraging in bumblebees: Rule of departure from an inflorescence. *Can. J. Zool.* 60:417–28

209. Pyke, G. H. 1983. Animal movements: An optimal foraging approach. In *The Ecology of Animal Movement,* ed. I. R.

Swingland, P. J. Greenwood. Oxford: Clarendon

210. Pyke, G. H., Pulliam, H. R., Charnov, E. L. 1977. Optimal foraging: A selective review of theory and tests. *Q. Rev. Biol.* 52:137–54

211. Rapport, D. J. 1971. An optimization model of food selection. *Am. Nat.* 105:575–88

212. Rapport, D. J. 1980. Optimal foraging for complementary resources. *Am. Nat.* 116:324–46

213. Rapport, D. J., Turner, J. E. 1977. Economic models in ecology. *Science* 195:367–73

214. Real, L. 1980. On uncertainty and the law of diminishing returns in evolution and behaviour. In *Limits to action: The Allocation of Individual Behavior*, ed. J. E. R. Staddon. New York: Academic

215. Real, L. 1980. Fitness, uncertainty, and the role of diversification in evolution and behavior. *Am. Nat.* 115:623–38

216. Real, L. 1981. Uncertainty and pollinator—plant interactions: The foraging behavior of bees and wasps on artificial flowers. *Ecology* 62:20–26

217. Real, L., Ott, J., Silverfine, E. 1982. On the tradeoff between the mean and the variance in foraging: Effect of spatial distribution and color preference. *Ecology* 63:1617–23

218. Rechten, C., Avery, M., Stevens, A. 1983. Optimal prey selection—why do great tits show partial preferences. *Anim. Behav.* 31:576–84

219. Rechten, C., Krebs, J. R., Houston, A. I. 1981. Great tits and conveyor belts—a correction for non-random prey distribution. *Anim. Behav.* 29:1276–77

220. Reichman, O. J. 1977. Optimization of diets through food preferences by heteromyid rodents. *Ecology* 58:454–57

221. Richards, L. J. 1983. Hunger and the optimal diet. *Am. Nat.* 122:326–34

222. Rissing, S. W. 1981. Prey preferences in the desert horned lizard: Influence of prey foraging method and aggressive behavior. *Ecology* 62:1031–40

223. Robinson, S. K., Holmes, R. T. 1982. Foraging behavior of forest birds: The relationships among search tactics, diet, and habitat structure. *Ecology* 63:1918–31

224. Rockwood, L. L. 1977. Foraging patterns and plant selection in Costa Rican leaf cutting ants. *J. NY Entomol. Soc.* 85:222–33

225. Rudolph, S. G. 1982. Foraging strategies of American Kestrels during breeding. *Ecology* 63:1268–76

226. Rusterholz, M., Turner, D. C. 1978. Experiments on nutritional wisdom of roe deer. *Rev. Suisse Zool.* 85:718–30

227. Savory, C. J. 1977. The food of red grouse chicks *Lagopus l. scoticus. Ibis* 119:1–9

228. Scheibling, R. E. 1981. Optimal foraging of *Oreaster reticulatus* (L.) (Echinodermata: Asteroidea). *J. Exp. Mar. Biol. Ecol.* 51:173–85

229. Schluter, D. 1981. Does the theory of optimal diets apply in complex environments? *Am. Nat.* 118:139–47

230. Schluter, D. 1982. Optimal foraging in bats: Some comments. *Am. Nat.* 119:121–25

231. Schoener, T. W. 1971. Theory of feeding strategies. *Ann. Rev. Ecol. Syst.* 11:369–404

232. Schoener, T. W. 1979. Generality of the size-distance relation in models of optimal feeding. *Am. Nat.* 114:902–14

233. Scriber, J. M. 1981. Sequential diets, metabolic costs, and growth of *Spodoptera eridania* (Lepidoptera, Noctuidae) feeding upon dill, lima bean and cabbage. *Oecologia* 51:175–80

234. Scriber, J. M., Slansky, F. Jr. 1981. The nutritional ecology of immature insects. *Ann. Rev. Entomol.* 26:183–211

235. Sih, A. 1977. Optimal foraging theory used to deduce the energy available in the environment. *Biotropica* 9:216

236. Sih, A. 1979. Optimal diet: The relative importance of the parameters. *Am. Nat.* 113:460–63

237. Sih, A. 1980. Optimal foraging: Partial consumption of prey. *Am. Nat.* 116:281–90

238. Sih, A. 1980. Optimal behavior: Can foragers balance two conflicting demands? *Science* 210:1041–43

239. Sih, A. 1982. Optimal patch use: Variation in selective pressure for efficient foraging. *Am. Nat.* 120:666–85

240. Sirota, Y. 1978. A preliminary simulation model of movement of larvae of *Culex pipiens molestus* (Diptera: Culicidae). II. Experimental studies on the dispersal of insects. *Res. Popul. Ecol.* 19:170–80

241. Sites, J. W. Jr. 1978. The foraging strategy of the dusky salamander, *Desmognathus fuscus* (Amphibia, Urodela, Plethodontidae): An empirical approach to predation theory. *J. Herpetol.* 12:373–83

242. Slobodkin, L. B. 1974. Prudent predation does not require group selection. *Am. Nat.* 108:665–78

243. Smith, J. N. M., Dawkins, R. 1971. The hunting behaviour of individual great tits in relation to spatial variations in their food density. *Anim. Behav.* 19:695–706

244. Smith, J. P., Maybee, J. S., Maybee, F. M. 1979. Effects of increasing distance to food and deprivation level on food

hoarding in *Rattus norvegicus. Behav. Neural Biol.* 27:302–18

245. Smith, J. N. M., Sweatman, H. P. A. 1974. Food searching behavior of tit mice in patchy environments. *Ecology* 55: 1216–32

246. Stamps, J., Tanaka, S., Krishman, V. V. 1981. The relationship between selectivity and food abundance in a juvenile lizard. *Ecology* 62:1079–92

247. Stanton, M. L. 1982. Searching in a patchy environment: Foodplant selection by *Colias periphyle* butterflies. *Ecology* 63:839–53

248. Stein, R. A. 1977. Selective predation, optimal foraging, and the predator-prey interaction between fish and crayfish. *Ecology* 58:1237–53

249. Stenseth, N. C. 1981. Optimal food selection: Some further considerations with special reference to the grazer-hunter distinction. *Am. Nat.* 117:457–75

250. Stenseth, N. C., Hansson, L. 1979. Optimal food selection: A graphic model. *Am. Nat.* 113:373–89

251. Stenseth, N. C., Hansson, L., Myllymäki, A. 1977. Food selection of the field vole *Microtus agrestis. Oikos* 29:511–24

252. Stephens, D. W. 1981. The logic of risk-sensitive foraging preferences. *Anim. Behav.* 29:628–29

253. Stephens, D. W., Charnov, E. L. 1982. Optimal foraging: Some simple stochastic models. *Behav. Ecol. Sociobiol.* 10: 251–63

254. Stewart-Oaten, A. 1982. Minimax strategies for a predator-prey game. *Theor. Popul. Biol.* 22:410–24

255. Sutherland, W. J. 1982. Do oystercatchers select the most profitable cockles. *Anim. Behav.* 30:857–61

256. Taghon, G. L. 1981. Beyond selection: Optimal ingestion rate as a function of food value. *Am. Nat.* 118:202–14

257. Taghon, G. L. 1982. Optimal foraging by deposit-feeding invertebrates—roles of particle size and organic coating. *Oecologia* 52:295–304

258. Taghon, G. L., Self, R. F. L., Jumars, P. A. 1978. Predicting particle selection by deposit feeders: A model and its implications. *Limnol. Oceanogr.* 23:752–59

259. Templeton, A. R., Lawlor, L. R. 1981. The fallacy of the averages in ecological optimization theory. *Am. Nat.* 117:390–93

260. Tepedino, V. J., Parker, F. D. 1982. Interspecific differences in the relative importance of pollen and nectar to bee species foraging on sunflowers. *Environ. Entomol.* 11:246–50

261. Tinbergen, J. M. 1981. Foraging decisions in starlings (*Sturnus vulgaris* L.). *Ardea* 69:1–67

262. Townsend, C. R., Hildrew, A. G. 1980. Foraging in a patchy environment by a predatory net-spinning caddis larva—a test of optimal foraging theory. *Oecologia* 47:219–21

263. Turner, A. K. 1982. Optimal foraging by the swallow (*Hirundo rustica* L.)—prey size selection. *Anim. Behav.* 30:862–72

264. Turrelli, M., Gillespie, J. H., Schoener, T. W. 1982. The fallacy of the fallacy of the averages in ecological optimization. *Am. Nat.* 119:879–84

265. Vadas, R. L. 1977. Preferential feeding—optimization strategy in sea urchins. *Ecol. Monogr.* 47:337–71

266. Visser, M. 1982. Prey selection by the 3-spined stickleback (*Gasterosteus aculeatus* L.) *Oecologia* 55:395–402

267. Waage, J. K. 1979. Foraging for patchily-distributed hosts by the parasitoid, *Nemeritis canescens. J. Anim. Ecol.* 48: 353–71

268. Waddington, K. D. 1982. Information used in foraging. See Ref. 118

269. Waddington, K. D., Allen, T., Heinrich, B. 1981. Floral preferences of bumblebees (*Bombus edwardsii*) in relation to intermittent versus continuous rewards. *Anim. Behav.* 29:779–84

270. Waddington, K. D., Heinrich, B. 1980. Patterns of movement and floral choice by foraging bees. See Ref. 137

271. Waddington, K. D., Holden, L. R. 1979. Optimal foraging: On flower selection by bees. *Am. Nat.* 114:179–96

272. Ware, D. M. 1975. Growth, metabolism, and optimal swimming speed of a pelagic fish. *J. Fish Res. Board Can.* 32:33–41

273. Ware, D. M. 1978. Bioenergetics of pelagic fish: Theoretical change in swimming speed and ration with body size. *J. Fish. Res. Board Can.* 35:220–28

274. Weigl, P. D., Hanson, E. V. 1980. Observational learning and the feeding behaviour of the red squirrel *Tamiasciurus hudsonicus:* The ontogeny of optimization. *Ecology* 61:213–18

275. Weihs, D. 1975. An optimum swimming speed of fish based on feeding efficiency. *Isr. J. Technol.* 13:163–69

276. Wells, H., Wells, P. H., Smith, D. M. 1981. Honeybee responses to reward size and colour in an artificial flower patch. *J. Apic. Res.* 20:172–79

277. Werner, E. E. 1974. The fish size, prey size, handling time relation in several sunfishes and some implications. *J. Fish. Res. Board Can.* 31:1531–36

278. Werner, E. E., Hall, D. J. 1974. Optimal foraging and the size selection of prey by the bluegill sunfish (*Lepomis macrochirus*). *Ecology* 55:1042–52

279. Werner, E. E., Hall, D. J. 1979. Forag-

ing efficiency and habitat switching in competing sunfishes. *Ecology* 60:256–64

280. Werner, E. E., Mittelbach, G. G. 1981. Optimal foraging: Field tests of diet choice and habitat switching. *Am. Zool.* 21:813–29

281. Werner, E. E., Mittelbach, G. G., Hall, D. J. 1981. The role of foraging profitability and experience in habitat use by the bluegill sunfish. *Ecology* 62:116–25

282. Westoby, M. 1974. An analysis of diet selection by large generalist herbivores. *Am. Nat.* 108:290–304

283. Westoby, M. 1978. What are the biological bases of varied diets? *Am. Nat.* 112:627–31

284. Whitham, T. G. 1977. Coevolution of foraging in *Bombus* and nectar dispensing in *Chilopsis:* A last dreg theory. *Science* 197:593–96

285. Williamson, C. E. 1981. Foraging behaviour of a freshwater copepod—frequency changes in looping behavior at high and low prey densities. *Oecologia* 50:332–36

286. Wilson, D. S. 1976. Deducing the energy available in the environment: An application of optimal foraging theory. *Biotropica* 8(2):96–103

287. Wilson, D. S. 1978. Prudent predation: A field study involving three species of tiger beetles. *Oikos* 31:128–36

288. Wolf, L. L., Hainsworth, F. R. 1983. Economics of foraging strategies in sunbirds and hummingbirds. In *Behavioral*

Energetics: The Cost of Survival in Vertebrates, ed. W. P. Aspey, S. I. Lustick. Columbus: Ohio State Univ. Press

289. Yano, E. 1978. A simulation model of searching behaviour of a parasite. *Res. Popul. Ecol.* 22:105–22

290. Zach, R. 1979. Shell dropping—decision making and optimal foraging in Northwestern crows. *Behaviour* 68:106–17

291. Zach, R., Falls, J. B. 1976. Ovenbird (Aves: Parulidae) hunting behavior in a patchy environment: An experimental study. *Can. J. Zool.* 54:1863–79

292. Zach, R., Falls, J. B. 1976. Do ovenbirds (Aves: Parulidae) hunt by expectation? *Can. J. Zool.* 54:1894–1903

293. Zach, R., Falls, J. B. 1977. Influence of capturing a prey on subsequent search in the ovenbird (Aves: Parulidae). *Can. J. Zool.* 55:1958–69

294. Zach, R., Falls, J. B. 1978. Prey selection by captive ovenbirds (Aves: Parulidae). *J. Anim. Ecol.* 47:929–43

295. Zach, R., Falls, J. B. 1979. Foraging and territoriality of male ovenbirds (Aves: Parulidae) in a heterogeneous habitat. *J. Anim. Ecol.* 48:33–52

296. Zimmerman, M. 1979. Optimal foraging: A case for random movement. *Oecologia* 43:261–67

297. Zimmerman, M. 1981. Optimal foraging, plant density and the marginal value theorem. *Oecologia* 49:148–53

298. Zimmerman, M. 1982. Optimal foraging: Random movement of pollen collecting bumblebees. *Oecologia* 53:394–98

SUBJECT INDEX

CUMULATIVE INDEXES

CONTRIBUTING AUTHORS, VOLUMES 11–15

597

CHAPTER TITLES, VOLUMES 11–15

599